Waves and Optics

AF662214

Waves and Optics

N. Singh
Retired Professor of Physics
Department of Physics, M.D. University,
Rohtak, (Haryana)

TECHSAR™
NEW DELHI

Published and Printed by
Techsar Pvt. Ltd.
4435-36/7, Ansari Road, Daryaganj
New Delhi-110 002 (India)
E-mail: tech_info@techsarworld.com
Website: www.techsarworld.com

ISBN: 978-93-90620-18-0

© 2024 I.K. International Pvt. Ltd.

All rights reserved. No part of this publication may be reproduced, stored in a retrieval system, or transmitted in any form or any means: electronic, mechanical, photocopying, recording, or otherwise, without the prior written permission from the publisher.

This Book is Dedicated to

all my friends and well-wishers

Preface

Waves and optics is a branch of physics in which we study the properties of sound and visible light. Both sound and light propagate in wave form. Sound needs medium to travel. While light needs no medium to travel, i.e., light can travel in vacuum also. **Sound waves** are longitudinal mechanical waves generated by density or pressure fluctuations. In other words, a **sound wave is a pressure disturbancethat** moves through air or water, or whatever the host medium happens to be. In brief, sound waves are a ***pressure disturbance moving through a medium. Without the medium, there can be no sound.*** **Sound intensity** is measured in decibels (dB).

Light waves are electromagnetic waves. These waves are associated with varying electric and magnetic fields. The two fields are perpendicular to each other and also perpendicular to the direction of the wave travel. Each field vibrates at the same frequency, which is equal to the frequency of the wave. The ratio, E/H = c [$c = 2.998 \times 10^{-8}$ m/s], is equal to the velocity of light. The light waves are transverse waves. Electromagnetic waves remain unaffected by electric and magnetic fields.

Light and sound are two most familiar natural phenomena experienced in our life. Our visual and sound contact with world around us is affected by the direction of light and sound through two of our sense organs. We see different colours spread in nature because of light. The red colourrising sun and setting sun, the green colour of the plants, the blue colour of the sky or blue colour of the deep ocean, all are light phenomena. Similarly, we hear different sounds spread in nature because of sound. The chirping of birds specially in the morning and evening, the sound of whistle coming from the train, communication among human being and among animals, melodious music in motion pictures, all are because of sound. Quite often, vision and sound simulate our thinking and enrich our lives in many ways. Due to our ability to see and hear, we become aware of the existence of an infinite variety of objects around us and their behavior. It would be no exaggeration to say that light and sound serve as a catalyst in the formation of life on the earth and plays a vital role in sustaining it in the planet.

For understanding the nature of light and sound, the best brains have been engaged ever since the humans moved out of the cave. In "Waves and Optics", we study the wave nature of light and sound, and several phenomena that occur when light and sound interact with matter.Theknowledge of light and sound has opened several new applications and many channels of research and development. For instance, the development of lasers facilitated revolutionary applications in healthcare, industry, communication, computing, and many more. New imaging technology helped to create holographic and night vision devices. Optical fibre has completely changed the way of communication with the world around us. Similarly, communication among people in a house, office, teaching class, at far distances and in motion pictures is because of sound. Because of these reasons, both optics and sound are important subjects of research. The basic knowledge of the subject prepares students to study advanced courses and leads to specialized training in the fields of research and development.

A course on "waves and optics" is the core of the B.Sc. (honours) curriculum in physics and general undergraduate degree of science of every university in India. The classroom lectures on "waves and optics" invariably become abstract, dull and uninteresting due to lack of demonstrations. Through this book, we have made a sincere effort to bridge the gap between the theory and its practical applications. This book inspires students to apply the acquired theoretical knowledge to real-life applications systematically starting from simple and familiar concepts/topics before proceeding to general cases and less familiar situations. However, no compromise has been made on rigorous mathematical equations. The fundamental concept of each article is presented in a simple language for easy understanding. All important formulae have been derived from first principles so that the book can also be used for self-study. Solved examples are given in each chapter of the book to clarify difficult concepts. To test the students' understanding of the subject, unsolved questions and objective type questions are given at the end of each chapter.

Every care has been taken to produce an error-free book. Yet some omissions might have escaped my attention. I shall be grateful to the readers for bringing such errors and omissions to my notice. To improve the content and presentation of the book author will welcome suggestions from readers.

N. Singh

Acknowledgements

I wish to express my thanks to Dr. H.S. Hans, Ex Head Department of Physics, Panjab University, Chandigarh for providing facilities during my research work leading to my Ph.D. degree. I am grateful to Dr. K. N. Pathak, Ex Prof. of Physics and Vice chancellor of P.U. Chandigarh for his encouragement for writing the book. My sincere thanks are due to Dr. Satya Prakash, Ex Prof. of Physics, P.U. Chandigarh and Ex Vice chancellor of Jiwaji University, Gwalior for helping me in learning basics of theoretical physics during my Ph.D. thesis work. My sincere thanks are due to my senior colleges Dr. Jahan Singh and Dr. S.P. Singh, both Ex Head of Department of Physics, M.D. University, Rohtak (Haryana) for helping me in various ways. I would like to thank my colleagues Dr. Joginder Singh, Ex Prof. of Physics, P.A.U. Ludhiana, Dr. V.K. Jindal, Ex Prof. of Physics, P.U. Chandigarh, and Dr. D.K. Chaturvedi, Ex Prof. of Physics, Kurukhetra University Kurukhetra for helping me in learning physics.I wish to express my thanks to Dr. K.C. Chaturvedi, Warden Chaube Hostel and hostel mates Dr. M.K. Sharma and Mr. A.K. Chaturvedi to keep my moral always highand helping me in various ways to reach at this stage.

I appreciate my wife: Sushma, daughter Dr. Monika, son, Er. Nishant, and daughter-in- law, Er. Kritika, for their patience and invaluable cooperation while writing the book. I am grateful to my parent-in-laws, Er. Ram Krishna, Mrs. Asha Krishna, Son-in-law, Anurag Krishna, and grandsons, Akshaj Krishna, Aarab Krishna, Aaditya Krishna, Kayansh, granddaughter, Kanisha, who always has been source of encouragement and happiness for me while writing the manuscript.

N. Singh

Contents

CHAPTER 1

Superposition of Collinear Harmonic Oscillators

1.1 INTRODUCTION

Both light and sound waves are travelling waves. The standing waves are formed by the superposition of waves. The phenomenon of beats and interference patterns are formed by superposition of waves. Both the phenomena play important role in daily life of human beings. In this chapter, the definition and theory of travelling waves, standing waves and phenomenon of beats is given.

When two or more travelling waves overlap, the resultant wave is algebraic sum of displacement of these waves. This phenomenon is called the *principle of superposition.* This phenomenon can occur for all types of waves, like light waves, sound waves, radio waves, water waves, mechanical waves, etc.

A *standing (stationary) wave* is a resultant wave due to superposition of two travelling waves moving in opposite direction having same frequency, speed and amplitude.

Two waves are said to be *coherent waves* if the phase difference between two waves at a point in space remains always constant.

The intensity at the point of superposition depends on the phase difference ϕ between two waves.

When $\cos \phi = 1$, I will be maximum, i.e.,

$$I_{\max} = I_1 + I_2 + 2\sqrt{I_1 I_2} = \left(\sqrt{I_1} + \sqrt{I_2}\right)^2$$

When $\cos \phi = -1$, I will be minimum, i.e.,

$$I_{\min} = I_1 + I_2 - 2\sqrt{I_1 I_2} = \left(\sqrt{I_1} - \sqrt{I_2}\right)^2$$

Here I_1 and I_2 are intensities of two waves. Since intensity varies with phase difference (ϕ) in space, so the interference pattern is formed by the superposition of coherent waves.

Two waves are said to be *incoherent waves* if the phase difference between the two waves at a point in space varies randomly throughout the observation time. The resultant amplitude does not remain constant with time. In the incoherent superposition of waves, the resultant intensity is directly proportional to the sum of the squares of the amplitudes of individual waves, i.e., $I = k\,(a_1^2 + a_2^2)$. Here a_1 and a_2 are amplitudes of two waves. That means in case of incoherent superposition, the resultant intensity is equal to the sum of the intensities of individual waves and does not change with time. So, interference pattern is not visible with superposition of incoherent waves.

When two waves of slightly different frequency overlap, the algebraic sum of displacements gives rise to beats. The phenomenon of beats is a kind of distinct sensation, which can be heard sometimes when two musicians are playing their music instruments. There are lots of applications of beats. For example:

(i) Tuning the musical instruments

(ii) Police radar to trap speeding vehicle

(iii) Determination of frequency of tuning fork in physics laboratory

1.2 LINEARITY AND SUPERPOSITION PRINCIPLE

When two or more travelling waves overlap, the resultant wave is algebraic sum of displacement of these waves. This phenomenon is called the *principle of superposition.* This phenomenon can occur for all type of waves, like light waves, sound waves, radio waves, water waves, mechanical waves, etc.

Let us consider two simple harmonic travelling waves of same frequency moving in the same direction. Let a_1 and a_2 be amplitudes of the waves and is the phase difference between them at any point x in the medium. The displacements, and of the two waves are written as

$$Y_1 = a_1 \sin(\omega t - kx) \quad \text{...(1.1a)}$$

and

$$Y_2 = a_2 \sin(\omega t - kx + \phi) \quad \text{...(1.1b)}$$

where, ω, and x are angular frequency and position of the waves. k is the momentum of the waves. According to the principle of superposition of waves, the resultant displacement of a wave is given by

$$\begin{aligned} Y &= Y_1 + Y_2 \\ &= a_1 \sin(\omega t - kx) + a_2 \sin(\omega t - kx + \phi) \\ &= a_1 \sin(\omega t - kx) + a_2 \sin(\omega t - kx)\cos(\phi) + a_2 \cos(\omega t - kx)\sin(\phi) \\ &= \sin(\omega t - kx)\,[a_1 + a_2 \cos(\phi)] + \cos(\omega t - kx)[a_2 \sin(\phi)] \end{aligned} \quad \text{...(1.2)}$$

Taking

$$a_1 + a_2 \cos(\phi) = R \cos\theta \qquad ...(1.3a)$$

and

$$a_2 \sin(\theta) = R \sin\theta \qquad ...(1.3b)$$

Equation (1.2) becomes

$$Y = R \sin(\omega t - kx) \cos\theta + R \cos(\omega t - kx) \sin\theta$$

$$= R \sin(\omega t - kx + \theta) \qquad ...(1.3)$$

The resultant harmonic wave has amplitude R.

Dividing $R \sin\varphi$ by $R \cos\varphi$ given in Eqs. (1.3a) by (1.3b), we get

$$\tan\theta = \frac{a_2}{a_1 + a_2} \tan\phi \qquad ...(1.4)$$

When

$$a_1 = a_2 = a$$

$$\tan\theta = \frac{1}{2} \tan\theta \qquad ...(1.5)$$

Resultant amplitude

Squaring and adding Eqs. (2.3a) and (2.3b), we get

$$R^2 [\sin^2\theta + \cos^2\theta] = [a_1 + a_2 \cos(\phi)]^2 + a_2^2 \sin^2\phi$$

or

$$R^2 = a_1^2 + 2a_1 a_2 \cos(\phi) + a_2^2 \cos^2\phi + a_2^2 \sin^2\phi$$

$$= a_1^2 + 2a_1 a_2 \cos(\phi) + a_2^2$$

or

$$R = \sqrt{a_1^2 + 2\, a_1\, a_2 \cos(\phi) + a_2^2} \qquad ...(1.6)$$

It is to be noted that the resultant amplitude of two waves, given by Eq. (1.6), is the function of phase difference between the two waves.

Intensity

The intensity at the point of superposition is proportional to square of the amplitude of the resultant wave given by Eq. (1.6), i.e.

$$I \propto R^2$$

or

$$I = CR^2$$

or

$$I = C\, [a_1^2 + 2\, a_1\, a_2 \cos(\phi) + a_2^2] \qquad ...(1.7)$$

where, C is the constant of proportionality.

If $I_1 = a_1^2$ and $I_2 = a_2^2$, then

$$I = I_1 + 2\sqrt{I_1\, I_2} \cos(\phi) + I_2 \qquad ...(1.8)$$

Here I_1 and I_2 are intensities of two waves. The intensity at the point of superposition depends on the phase difference (ϕ) between the two waves.

Case 1. When $\cos \phi = 1$, I will be maximum, i.e.,

$$I_{max} = I_1 + I_2 + 2\sqrt{I_1 I_2} = \left(\sqrt{I_2} + \sqrt{I_2}\right)^2 \qquad ...(1.9)$$

Case 2. When $\cos \phi = -1$, I will be minimum, i.e.,

$$I_{min} = I_1 + I_2 - 2\sqrt{I_1 I_2} = \left(\sqrt{I_1} - \sqrt{I_2}\right)^2 \qquad ...(1.10)$$

Maximum intensity of n coherent waves will be

$$(I_{max})_{co} = \left(\sqrt{I_0} + \sqrt{I_0} + \sqrt{I_0} + ...\right)^2 = n^2 I_0 \qquad ...(1.11)$$

Average intensity due to two incoherent waves is given by

$$(I_R)_{inco} = I_1 + I_2 \qquad ...(1.12)$$

Intensity of n identical incoherent n waves will be

$$(I_R)_{inco} = I_0 + I_0 + ... = n\, I_0 \qquad ...(1.13)$$

1.3 STANDING WAVE

A standing (stationary) wave is a resultant wave due to superposition of two travelling waves moving in opposite direction having same frequency, speed and amplitude.

Formation of standing waves obtained due to superposition of wave A and B are shown in Fig. 1.1. Initially (i.e., $t = 0$), A and B are in phase to each other as shown by dashed lines overlapping each other. The resultant wave due to superposition of A and B waves is shown by solid line in Fig. 1.1(a). One-eighth of period later (i.e., $t = T/8$, $x = \lambda/8$ or $\pi/4$) wave A has moved to the left and wave B has moved to the right. Their positions and resultant is shown in Fig. 1.1(b). At $t = T/4$ (i.e., $x = \lambda/4$ or $\pi/2$) Waves A and B are completely out of phase from each other and therefore, the resultant becomes zero at every point as shown in Fig. 1.1(c). The resultant waves A and B at time $t = 3T/8$ (i.e., $x = 3\lambda/8$ or $3\pi/4$) and at time $t = T/2$ (i.e., $x = \lambda/2$ or π) are shown in Figs. 1.1(d) and 1.1(e).

In this standing wave, the displacement is always zero points, N_1, N_2 and N_3, as shown in Fig. 1.1(e). The positions of zero amplitude in the stationary wave are called nodes. There are points of maximum amplitude half way between two nodes in this standing wave. The positions of maximum amplitudes in the standing wave are called *antinodes*.

Figure 1.2 shows the resultant displacement changes between $t = 0$ and $t = T/2$. The displacement is always zero at nodes. Half way between two nodes in this standing wave there are positions of maximum amplitude. These positions of maximum amplitudes in the

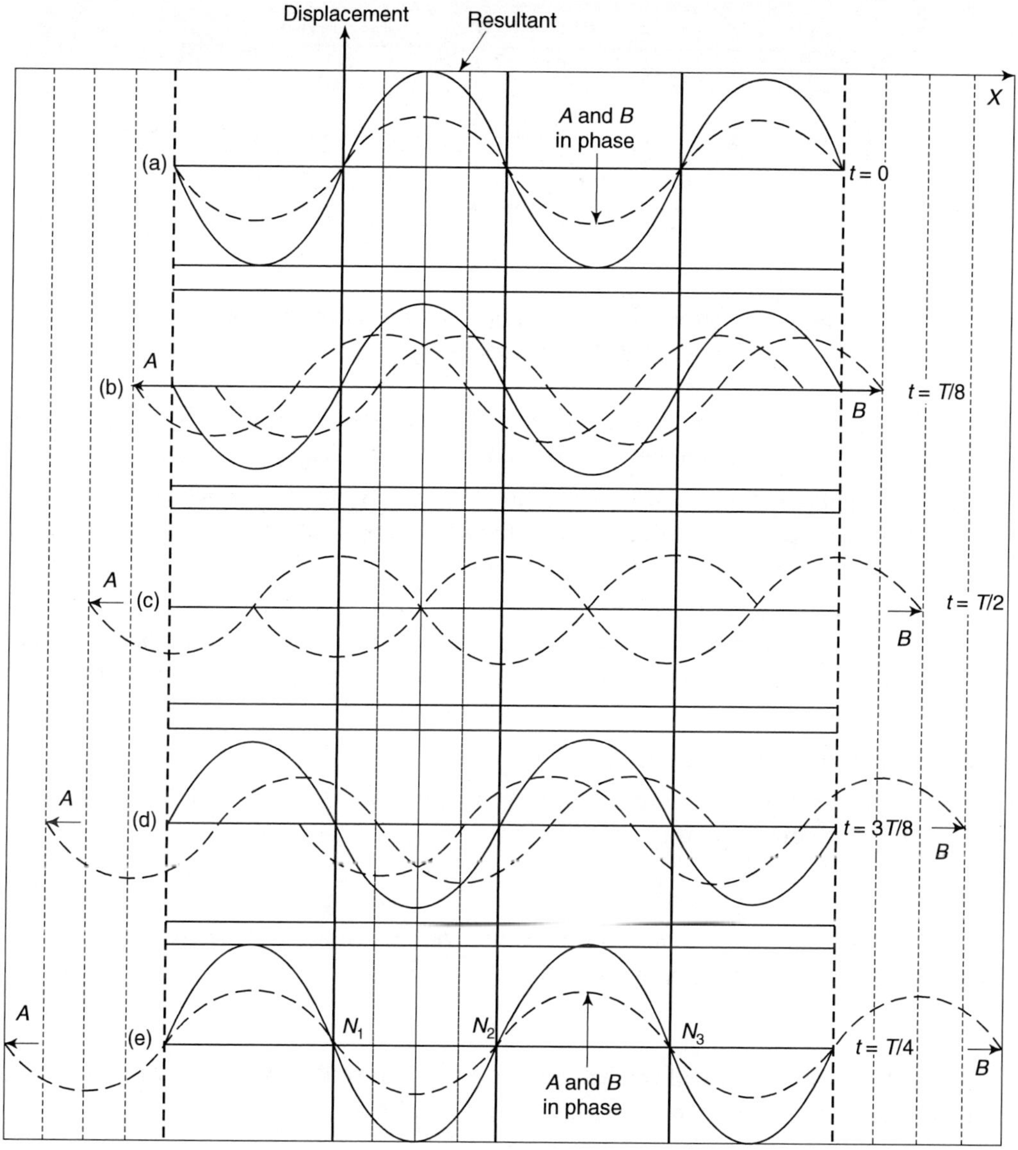

Fig. 1.1 Shows stationary wave when two waves moving in opposite direction superpose to each other. N_1, N_2 and N_3 are nodes. Distance $N_1 N_2 = N_2 N_3 = \lambda/2$, i.e., separation between two adjacent nodes is $\lambda/2$. Maximum position of resultant wave are called antinodes. The separation between two adjacent antinodes is $\lambda/2$.

standing wave are called antinodes (A). The positions of zero amplitude of standing wave are called nodes and positions of maximum amplitude are called antinodes. The following points are to be noted in case of stationary wave:

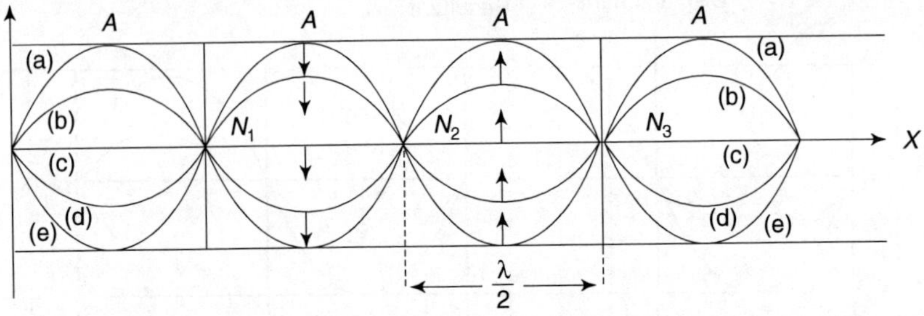

Fig. 1.2 Shows successive displacement of stationary waves. *A* and *N* are antinodes and nodes

(i) The separation of two adjacent nodes is $\lambda/2$, i.e., the distance $N_1 N_2 = N_2 N_3 = \lambda/2$. Similarly, the separation between two adjacent antinodes is $\lambda/2$.

(ii) The amplitude of a stationary wave, unlike progressive wave, depends on position. It ranges from zero at the nodes to $2a$ at the antinodes, where a is the amplitude of each progressive wave producing the stationary wave.

(iii) At any one time, all the particles between two adjacent nodes have the same phase, while each particle between the next pair of adjacent nodes is out of phase (i.e., have phase difference of π radians) with respect to particles between previous pair of nodes.

Mathematical treatment of stationary waves

Let the displacement (Y_1) of a travelling wave of frequency (ω) and amplitude (a) moving from right to left be written as

$$Y_1 = a \sin (\omega t - kx) \qquad ...(1.14)$$

Let the displacement $(Y)_2$ of another travelling wave of same frequency (ω) and amplitude (a) moving from left to right be written as

$$Y_2 = \sin (\omega t + kx) \qquad ...(1.15)$$

A standing wave is formed by the addition of two travelling waves moving in opposite direction, i.e., the displacement of stationary wave will be

$$\begin{aligned} Y &= Y_2 + Y_2 \\ &= a \sin (\omega t - kx) + a \sin (\omega t + kx) \\ &= a [\sin (\omega t) \cos (kx) - \cos (\omega t) \sin (kx)] + a [\sin (\omega t) \cos (kx) + \cos (\omega t) \sin (kx)] \\ &= 2\, a \sin (\omega t) \cos (kx) \\ &= A \sin (\omega t) \qquad ...(1.16) \end{aligned}$$

where, $2a$ (cos (kx) = A and $k = \dfrac{2\pi}{\lambda}$ is the wave vector of the travelling wave.

Note that the resultant wave given by Eq. (1.16) is composed of two components

(i) $\sin(\omega t)$ shows that the amplitude varies with time t at a particular position.

(ii) $\cos(kx)$ shows that the amplitude varies with position x at a particular time.

Case 1. When $A = 2a$, i.e., when amplitude is maximum

$$\cos(kx) = \pm 1$$

This implies that $kx = m\pi$, where, $m = 0, 1, 2, ...$

or
$$\frac{2\pi}{\lambda}x = m\pi \quad \text{or} \quad x = m\frac{\lambda}{2}$$

i.e., when $x = 0, \frac{\lambda}{2}, \frac{2\lambda}{2}, ...$, the amplitude ($A$) is maximum or $A = 2a$. The positions, at which the amplitude is maximum, are called antinodes.

That is at positions $x = 0, \frac{\lambda}{2}, \frac{2\lambda}{2}, ...$, there are antinodes. The separation between two adjacent antinodes is $\frac{\lambda}{2}$.

Case 2. When $A = 0$, i.e., when amplitude is minimum

$$\cos(kx) = 0$$

This implies that $kx = (2m - 1)\frac{\pi}{2}$, where, $m = 1, 2, 3, ...$

or
$$\frac{2\pi}{\lambda}x = (2m - 1)\frac{\pi}{2} \quad \text{or} \quad x = (2m - 1)\frac{\lambda}{4}$$

i.e., when $x = \frac{\lambda}{4}, \frac{3\lambda}{4}, \frac{5\lambda}{4}, ...$, the amplitude ($A$) is minimum or $A = 0$. The positions, at which the amplitude is zero, are called nodes.

That is, at positions $x = \frac{\lambda}{4}, \frac{3\lambda}{4}, \frac{5\lambda}{4}, ...$, there are nodes. The distance between two adjacent nodes is $= \frac{\lambda}{2}$.

Summary

- A stationary wave is the resultant wave of two travelling waves of same frequency, speed and amplitude moving in opposite directions.
- A node is a point of zero displacement.

- An antinode is point of maximum displacement.
- Separation of two adjacent nodes = separation of two adjacent antinodes = $\frac{\lambda}{2}$.
- All particles between two adjacent notes are in the same phase, but completely out of phase with particles between next pairs of adjacent nodes.

1.4 SUPERPOSITION OF TWO WAVES OF SAME FREQUENCY AND CONSTANT PHASE DIFFERENCE, BUT TRAVELLING IN THE SAME DIRECTION

Let us consider two simple harmonic travelling waves of same frequency moving in same direction. Let a_1 and a_2 be amplitudes of the waves and ϕ is the phase difference between them at any point x in the medium. The displacements, Y_1 and Y_2 of the two waves are written as

$$Y_1 = a_1 \sin(\omega t - kx) \quad \text{...(1.17a)}$$

and

$$Y_2 = a_2 \sin(\omega t - kx + \phi) \quad \text{...(1.17b)}$$

where, ω, and x are angular frequency and position of the waves and k is the momentum of the waves. According to the principle of superposition of waves, the resultant displacement of a wave is given by

$$\begin{aligned} Y &= Y_1 + Y_2 \\ &= a_1 \sin(\omega t - kx) + a_2 \sin(\omega t - kx + \phi) \\ &= a_1 \sin(\omega t - kx) + a_2 \sin(\omega t - kx)\cos(\phi) + a_2 \cos(\omega t - kx)\sin(\phi) \\ &= \sin(\omega t - kx)[a_1 + a_2\cos(\phi)] + \cos(\omega t - kx)[a_2 \sin(\phi)] \end{aligned} \quad \text{...(1.18)}$$

Taking

$$a_1 + a_2 \cos(\phi) = R\cos\theta \quad \text{...(1.19a)}$$

and

$$a_2 \sin(\phi) = R\sin\theta \quad \text{...(1.19b)}$$

Equation (1.2) becomes

$$\begin{aligned} Y &= R\sin(\omega t - kx)\cos\theta + R\cos(\omega t - kx)\sin\theta \\ &= R\sin(\omega t - kx + \theta) \end{aligned} \quad \text{...(1.20)}$$

The resultant harmonic wave has amplitude R.

Dividing $R \sin \varphi$ by $R \cos \varphi$ given in Eqs. (1.19a) by (1.19b), we get

$$\tan\theta = \frac{a_2 \sin(\phi)}{a_1 + a_2\cos(\phi)} \quad \text{...(1.21)}$$

When $a_1 = a_2 = a$

$$\tan\theta = \tan\left(\frac{\phi}{2}\right) \quad \text{...(1.22)}$$

Resultant amplitude

Squaring and adding Eqs. (1.19a) and (1.19b), we get

$$R^2 [\sin^2 \theta + \cos^2 \theta] = [a_1 + a_2 \cos (\phi)]^2 + a_2^2 \sin^2 \phi$$

or
$$R^2 = a_1^2 + 2\, a_1\, a_2 \cos (\phi) + a_2^2 \cos^2 \phi + a_2^2 \sin^2 \phi$$
$$= a_1^2 + 2\, a_1\, a_2 \cos (\phi) + a_2^2$$

or
$$R = \sqrt{a_1^2 + 2\, a_1\, a_2 \cos (\phi) + a_2^2} \quad ...(1.23)$$

It is to be noted that the resultant amplitude of two waves, given by Eq. (1.23), is the function of phase difference between two waves.

Intensity

The intensity at the point of superposition is proportional to the square of the amplitude of the resultant wave is given by Eq. (1.23), i.e.,

$$I \propto R^2$$

or
$$I = CR^2$$

or
$$I = C\,[a_2^2 + 2\, a_1\, a_2 \cos (\phi) + a_2^2] \quad ...(1.24)$$

where, C is the constant of proportionality.

If
$$I_1 = a_1^2 \text{ and } I_2 = a_2^2, \text{ then}$$
$$I = I_1 + 2\sqrt{I_1\, I_2} \cos (\phi) + I_2 \quad ...(1.25)$$

Here I_1 and I_2 are intensities of two waves. The intensity at the point of superposition depends on the phase difference (ϕ) between two waves.

Case 1. When $\cos \phi = 1$, I will be maximum, i.e.,

$$I_{max} = I_1 + I_2 + 2\sqrt{I_1\, I_2} = \left(\sqrt{I_1} + \sqrt{I_2}\right)^2 \quad ...(1.26)$$

Case 2. When $\cos \phi = -1$, I will be minimum, i.e.,

$$I_{min} = I_1 + I_2 - 2\sqrt{I_1\, I_2} = \left(\sqrt{I_1} - \sqrt{I_2}\right)^2 \quad ...(1.27)$$

1.5 SUPERPOSITION OF TWO WAVES OF SAME FREQUENCY AND CONSTANT PHASE DIFFERENCE, BUT TRAVELLING IN THE OPPOSITE DIRECTION

Two travelling waves of equal frequency (ω) and amplitude (a) moving in opposite direction are

$$Y_1 = a \sin (\omega t - kx) \quad ...(1.28a)$$
$$Y_2 = \sin (\omega t + kx) \quad ...(1.28b)$$

where, $k = \dfrac{2\pi}{\lambda}$ is the wave vector of the travelling waves. When these two waves moving in opposite direction overlap, due to superposition they form the standing waves.

A standing wave can be formed by the addition of two travelling waves, i.e.,

$$Y = Y_1 + Y_2$$

$$= a \sin(\omega t - kx) + a \sin(\omega t + kx)$$

$$= a [\sin(\omega t) \cos(kx) - \cos(\omega t) \sin(kx)] + a [\sin(\omega t) \cos(kx) + \cos(\omega t) \sin(kx)]$$

$$= 2a \sin(\omega t) \cos(kx)$$

$$= A \sin(\omega t) \qquad ...(1.29)$$

The resultant wave (Y) given by Eq. (1.29) represents standing wave with amplitude $A = 2a \cos(kx)$. The amplitude (A) is a function of time and position, x.

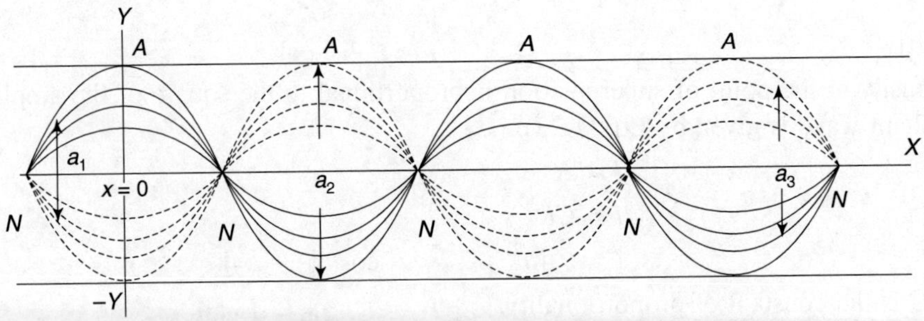

Fig. 1.3 Appearance of stationary waves produced in a string. A and N points are antinodes and nodes

Case 1. When amplitude is maximum, i.e., $A = 2a$

$$|\cos(kx)| = 1$$

This implies that

$$kx = n\pi, \quad \text{for } n = 0, 1, 2, 3, ...$$

$$x = \frac{n\pi}{k} = \frac{n\pi}{\frac{2\pi}{\lambda}} = \frac{n\lambda}{2}$$

When $$x = 0, \frac{\lambda}{2}, \frac{2\lambda}{2}, \frac{3\lambda}{2}, ...$$

Case 2. When amplitude is minimum, i.e., $A = 0$

$$|\cos(kx)| = 0$$

This implies that

$$kx = (2n-1)\frac{\pi}{2}, \quad \text{for } n = 1, 2, 3, ...$$

$$x = \frac{(2n-1)}{k}\frac{\pi}{2} = \frac{(2n-1)}{\frac{2\pi}{\lambda}}\frac{\pi}{2} = \frac{(2n-1)\lambda}{4}$$

When $x = \frac{\lambda}{4}, \frac{3\lambda}{4}, \frac{5\lambda}{4}, ...$

The amplitude is maximum or displacement of the particles of the medium is maximum at position $x = 0, \frac{\lambda}{2}, \frac{2\lambda}{2}, \frac{3\lambda}{2}, ...$, and amplitude or displacement of the particles of the medium is minimum at position $x = \frac{\lambda}{4}, \frac{3\lambda}{4}, \frac{5\lambda}{4}, ...$ The points at which amplitude is maximum are called antinodes and the position at which amplitude is minimum are called nodes. Thus, nodes occur at $x = \frac{\lambda}{4}, \frac{3\lambda}{4}, \frac{5\lambda}{4}, ...$ and antinodes occur at $x = 0, \frac{\lambda}{2}, \frac{2\lambda}{2}, \frac{3\lambda}{2}, ...$

The stationary wave is shown in Fig. 1.1. We notice from Eq. (1.29) that time for which sin (ωt) = 1, and |cos (kx) = 1, the displacement is maximum for all particles at positions $\frac{\lambda}{2}, \frac{2\lambda}{2}, \frac{3\lambda}{2}, ...,$. At the time when sin $(\omega t) = 0$, all the particles of the medium passes through their mean position as shown in Fig. 1.1.

1.6 SUPERPOSITION OF *N* COHERENT WAVES OF SAME FREQUENCY AND DIFFERENT PHASE DIFFERENCE

Two waves are said to be coherent waves if the phase difference between the two waves at a point in space remains always constant. Similarly, n waves are said to be coherent waves if the phase difference among n waves at a point in space remains always constant.

Let us consider the superposition of n number of coherent waves in the same direction having amplitudes $a_1, a_2, a_3, ..., a_n$. The displacement of n waves having angular frequency (ω) and phase angle $\phi_1, \phi_2, \phi_3,..., \phi_n$ can be written as

$$Y_1 = a_1 \sin(\omega t - \phi_1)$$
$$Y_2 = a_2 \sin(\omega t - \phi_2)$$
$$Y_3 = a_3 \sin(\omega t - \phi_3)$$
$$Y_4 = a_4 \sin(\omega t - \phi_4)$$
$$\cdots\cdots\cdots$$
$$Y_n = a_n \sin(\omega t - \phi_n)$$

The resultant displacement $(\overline{Y_1})$ of two waves (Y_1 and Y_2) is

$$\begin{aligned} \overline{Y_1} &= Y_1 + Y_2 \\ &= a_1 \sin(\omega t - \phi_1) + a_2 \sin(\omega t - \phi_2) \\ &= a_1 [\sin \omega t \cos \phi_1 - \cos \omega t \sin \phi_1] + a_2 [\sin \omega t \cos \phi_2 - \cos \omega t \sin \phi_2] \\ &= [a_1 \cos \phi_1 + a_2 \cos \phi_2] \sin \omega t - [a_1 \sin \phi_1 + a_2 \sin \phi_2] \cos \omega t \end{aligned} \quad ...(1.30)$$

Let $\quad R_1 \cos \theta_1 = a_1 \cos \phi_1 + a_2 \cos \phi_2$

and $\quad R_1 \sin \theta_1 = a_1 \sin \phi_1 + a_2 \sin \phi_2$

Equation (1.24) can be written as

$$\overline{Y}_1 = R_1\,[\cos\theta \sin\omega t - \sin\theta \cos\omega t]$$

or

$$\overline{Y}_1 = R_1 \sin(\omega t - \theta_1) \qquad ...(1.31)$$

Equation (1.31) represents the resultant of two waves, Y_1 and Y_2. Sum of squares of $R_1 \cos\theta_1$ and $R_1 \sin\theta_1$ will be

$$R_1^2 = (a_1 \sin\phi_1 + a_2 \sin\phi_2)^2 + (a_1 \sin\phi_1 + a_2 \sin\phi_2)^2 \qquad ...(1.32)$$

Equation (1.32) represents the square of the amplitude of resultant of two waves.

Dividing $R_1 \sin\theta_1$ by $R_1 \cos\theta_1$, the phase angle of resultant n waves is

$$\tan\theta_1 = \frac{a_1 \sin\phi_1 + a_2 \sin\phi_2}{a_1 \sin\phi_1 + a_2 \sin\phi_2} \qquad ...(1.33)$$

The resultant displacement $(\overline{Y}_2)$ of two waves, $\overline{Y}_1$ and Y_3 is

$$\begin{aligned}\overline{Y}_2 &= \overline{Y}_1 + Y_3\\ &= R_1 \sin(\omega t - \theta_1) + a_3 \sin(\omega t - \phi_3)\\ &= R_1\,[\sin\omega t \cos\theta_1 - \cos\omega t \sin\theta_1] + a_3\,[\sin\omega t \cos\phi_3 - \cos\omega t \sin\phi_3]\\ &= [R_1 \cos\theta_1 + a_3 \cos\phi_3] \sin\omega t - [R_1 \sin\theta_1 + a_3 \sin\phi_3] \cos\omega t \qquad ...(1.34)\end{aligned}$$

Let

$$R_2 \cos\theta_2 = R_1 \cos\theta_1 + a_3 \cos\phi_3 = a_1 \cos\phi_1 + a_2 \cos\phi_2 + a_3 \cos\phi_3$$

and

$$R_2 \sin\theta_2 = R_1 \sin\theta_1 + a_3 \sin\phi_3 = a_1 \sin\phi_1 + a_2 \sin\phi_2 + a_2 \sin\phi_2$$

Equation (1.34) can be written as

$$\overline{Y}_2 = R_2\,[\cos\theta_2 \sin\omega t - \sin\theta_2 \cos\omega t]$$

or

$$\overline{Y}_2 = R_2 \sin(\omega t - \theta_2) \qquad ...(1.35)$$

Similarly, the resultant of supposition of $\overline{Y}_2$ and Y_4 will be

$$\overline{Y}_3 = R_3 \sin(\omega t - \theta_3) \qquad ...(1.36)$$

and the resultant of supposition of $\overline{Y}_{n-1}$ and Y_n will be

$$\overline{Y} = R \sin(\omega t - \theta) \qquad ...(1.37)$$

where, $R_3 \cos\theta_3 = R_2 \cos\theta_2 + a_4 \cos\phi_4 = a_1 \cos\phi_1 + a_2 \cos\phi_2 + a_3 \cos\phi_3 + a_4 \cos\phi_4$

and $R_3 \sin\theta_3 = R_2 \sin\theta_2 + a_4 \sin\phi_4 = a_1 \sin\phi_1 + a_2 \sin\phi_2 + a_2 \sin\phi_2 + a_4 \sin\phi_4$

$$\begin{aligned}R\cos\theta &= R_{n-2} \cos\theta_{n-2} + a_n \cos\phi_n\\ &= a_1 \cos\phi_1 + a_2 \cos\phi_2 + a_3 \cos\phi_3 + a_4 \cos\phi_4 + ... + a_n \cos\phi_n \qquad ...(1.38)\end{aligned}$$

and

$$\begin{aligned}R\sin\theta &= R_{n-2} \sin\theta_{n-2} + a_n \sin\phi_n\\ &= a_1 \sin\phi_1 + a_2 \sin\phi_2 + a_3 \sin\phi_3 + a_4 \sin\phi_4 + ... + a_n \sin\phi_n \qquad ...(1.39)\end{aligned}$$

Sum of squares of $R\cos\theta$ and $R\sin\theta$ will be

$$\begin{aligned}R^2 = {} & (a_1 \cos\phi_1 + a_2 \cos\phi_2 + a_3 \cos\phi_3 + a_4 \cos\phi_4 + ... + a_n \cos\phi_n)^2\\ & + (a_1 \sin\phi_1 + a_2 \sin\phi_2 + a_3 \sin\phi_3 + a_4 \sin\phi_4 + ... + a_n \sin\phi_n)^2 \qquad ...(1.40)\end{aligned}$$

Equation (1.40) represents the square of the amplitude of resultant of n waves.

Dividing $R \sin \theta$ by $R \cos \theta$, the phase angle of resultant n waves is

$$\tan \theta = \frac{a_1 \sin \phi_1 + a_2 \sin \phi_2 + a_3 \sin \phi_3 + a_4 \sin \phi_4 + ... + a_n \cos \phi_n}{a_1 \cos \phi_1 + a_2 \cos \phi_2 + a_3 \cos \phi_3 + a_4 \cos \phi_4 + ... + a_n \cos \phi_n} \quad ...(1.41)$$

Here, $a_1, a_2, a_3, ..., a_n$ amplitudes of n number of coherent waves travelling in the same direction with same frequency, $\omega . \phi_1, \phi_2, \phi_3, ..., \phi_n$ are the phase angles of n waves.

1.7 SUPERPOSITION OF *N* COHERENT WAVES OF EQUAL FREQUENCY AND SAME PHASE DIFFERENCE

N waves are said to be coherent waves if the phase difference between the two waves at a point in space remains constant throughout the observation time. Coherent superposition is defined as the superposition of coherent waves.

Let us consider the superposition of n number of coherent waves in the same direction having amplitudes $a_1, a_2, a_3, ..., a_n$. The displacement of N waves having angular frequency (ω) and phase angle (ϕ) can be written as

$$Y_1 = a_1 \sin (\omega t - \phi)$$
$$Y_2 = a_2 \sin (\omega t - \phi)$$
$$Y_3 = a_3 \sin (\omega t - \phi)$$
$$Y_4 = a_4 \sin (\omega t - \phi)$$
$$..........................$$
$$Y_n = a_N \sin (\omega t - \phi)$$

The resultant displacement $(\overline{Y_1})$ of two waves, Y_1 and Y_2, is

$$\overline{Y_1} = Y_1 + Y_2$$
$$= a_1 \sin (\omega t - \phi) + a_2 \sin (\omega t - \phi)$$
$$= a_1 [\sin \omega t \cos \phi - \cos \omega t \sin \phi] + a_2 [\sin \omega t \cos \phi - \cos \omega t \sin \phi]$$
$$= [a_1 \cos \phi + a_2 \cos \phi] \sin \omega t - [a_1 \sin \phi + a_2 \sin \phi] \cos \omega t \quad ...(1.42)$$

Let $R_1 \cos \theta_1 = a_1 \cos \phi + a_2 \cos \phi$

and $R_1 \sin \theta_1 = a_1 \sin \phi + a_2 \sin \phi$

Equation (1.42) can be written as

$$\overline{Y_1} = R_1 [\cos \theta \sin \omega t - \sin \theta \cos \omega t]$$

or

$$\overline{Y_1} = R_1 \sin (\omega t - \theta_1) \quad ...(1.43)$$

Equation (1.43) represents the resultant of two waves, Y_1 and Y_2. Sum of squares of $R_1 \cos \theta_1$ and $R_1 \sin \theta_1$ will be

$$R_1^2 = (a_1 \sin \phi + a_2 \sin \phi)^2 + (a_1 \cos \phi + a_2 \cos \phi)^2 \quad ...(1.44)$$

Equation (1.44) represents the square of the amplitude of resultant of two waves.

Dividing $R_1 \sin\theta_1$ by $R_1 \cos\theta_1$, the phase angle of resultant n waves is

$$\tan\theta_1 = \frac{a_1 \sin\phi + a_2 \sin\phi}{a_1 \cos\phi + a_2 \cos\phi} \quad ..(1.45)$$

The resultant displacement ($\bar{Y}_2$) of two waves, $\bar{Y}_1$ and Y_3, is

$$\begin{aligned} \bar{Y}_2 &= \bar{Y}_1 + Y_3 \\ &= R_1 \sin(\omega t - \theta_1) + a_3 \sin(\omega t - \phi) \\ &= R_1 [\sin\omega t \cos\theta_1 - \cos\omega t \sin\theta_1] + a_3 [\sin\omega t \cos\theta - \cos\omega t \sin\theta] \\ &= [R_1 \cos\theta_1 + a_3 \cos\phi] \sin\omega t - [R_1 \sin\theta_1 + a_3 \sin\phi] \cos\omega t \quad ...(1.46) \end{aligned}$$

Let $R_2 \cos\theta_2 = R_1 \cos\theta_1 + a_3 \cos\phi = a_1 \cos\phi + a_2 \cos\phi + a_3 \cos\phi$

and $R_2 \sin\theta_2 = R_1 \sin\theta_1 + a_3 \sin\phi = a_1 \sin\phi + a_2 \sin\phi + a_2 \sin\phi$

Equation (1.46) can be written as

$$\bar{Y}_2 = R_2 [\cos\theta_2 \sin\omega t - \sin\theta_2 \cos\omega t]$$

or

$$\bar{Y}_2 = R_2 \sin(\omega t - \theta_2) \quad ...(1.47)$$

Similarly, the resultant of supposition of $\bar{Y}_2$ and Y_4 will be

$$\bar{Y}_3 = R_3 \sin(\omega t - \theta_3) \quad ...(1.48)$$

and the resultant of supposition of $\bar{Y}_{n-1}$ and Y_n will be

$$\bar{Y} = R \sin(\omega t - \theta) \quad ...(1.49)$$

where, $R_3 \cos\theta_3 = R_2 \cos\theta_2 + a_4 \cos\phi_4 = a_1 \cos\phi + a_2 \cos\phi + a_3 \cos\phi + a_4 \cos\phi$

and $R_3 \sin\theta_3 = R_2 \sin\theta_2 + a_4 \sin\phi = a_1 \sin\phi + a_2 \sin\phi + a_3 \sin\phi + a_4 \sin\phi$

$$\begin{aligned} R\cos\theta &= R_{n-2} \cos\theta_{n-2} + a_n \cos\phi \\ &= a_1 \cos\phi + a_2 \cos\phi + a_3 \cos\phi + a_4 \cos\phi + ... + a_N \cos\phi \quad ...(1.50) \end{aligned}$$

and

$$\begin{aligned} R\sin\theta &= R_{n-2} \sin\theta_{n-2} + a_n \sin\phi \\ &= a_1 \sin\phi + a_2 \sin\phi + a_2 \sin\phi + a_4 \sin\phi + ... + a_N \sin\phi \quad ...(1.51) \end{aligned}$$

Sum of squares of $R\cos\theta$ and $R\sin\theta$ will be

$$R^2 = (a_1 + a_2 + a_3 + a_4 + ... + a_N)^2 \quad ...(1.52)$$

Equation (1.52) represents the square of the amplitude of resultant of n waves.

Dividing $R\sin\theta$ by $R\cos\theta$, the phase angle of resultant n waves is

$$\tan\theta = \frac{a_1 \sin\phi + a_2 \sin\phi + a_3 \sin\phi + a_4 \sin\phi + ... + a_N \sin\phi}{a_1 \cos\phi + a_2 \cos\phi + a_3 \cos\phi + a_4 \cos\phi + ... + a_N \cos\phi} \quad ...(1.53)$$

Here, $a_1, a_2, a_3, ..., a_N$ are amplitudes of n number of coherent waves travelling in the same direction with same frequency, $\omega \cdot \phi$ is the phase angle of N waves.

The intensity at a point due to superposition on N coherent waves is

$$I \propto R^2 \text{ or } I = kR^2 = k\,(a_1 + a_2 + a_3 + a_4 + ... + a_N)^2$$

where, k is the constant of proportionality. Thus, the intensity at a point due to superposition of N coherent waves is square of the sum of amplitudes all waves.

1.8 SUPERPOSITION OF *N* HARMONIC WAVES OF EQUAL FREQUENCY AND SAME PHASE DIFFERENCE: ANOTHER METHOD

Superposition of *N* harmonic waves of equal frequency

N harmonic waves are said to be coherent waves if the phase difference between the two waves at a point in space remains constant throughout the motion. The resultant wave of superposition of N coherent harmonic waves is also a harmonic wave. Analytical theory of superposition of N coherent harmonic waves is given below.

Let us consider the superposition of n number of coherent harmonic waves in the same direction having amplitudes $a_1, a_2, a_3, ..., a_n$. The displacement of n waves having angular frequency (ω) and phase angle (ϕ) can be written as

$$\begin{aligned}
Y_1 &= a_1 \sin(\omega t - \phi) \\
Y_2 &= a_2 \sin(\omega t - \phi) \\
Y_3 &= a_3 \sin(\omega t - \phi) \\
Y_4 &= a_4 \sin(\omega t - \phi) \\
&\cdots\cdots \\
Y_n &= a_n \sin(\omega t - \phi)
\end{aligned}$$

The resultant displacement ($\bar{Y}_1$) of two waves, Y_1 and Y_2, is

$$\begin{aligned}
\bar{Y}_1 &= Y_1 + Y_2 \\
&= a_1 \sin(\omega t - \phi) + a_2 \sin(\omega t - \phi) \\
&= a_1 [\sin \omega t \cos \phi - \cos \omega t \sin \phi] + a_2 [\sin \omega t \cos \phi - \cos \omega t \sin \phi] \\
&= [a_1 \cos \phi + a_2 \cos \phi] \sin \omega t - [a_1 \sin \phi + a_2 \sin \phi] \cos \omega t \\
&= [a_1 + a_2] \cos \phi \sin \omega t - [a_1 + a_2] \sin \phi \cos \omega t \qquad ...(1.54)
\end{aligned}$$

Let $R_1 = [a_1 + a_2]$

Equation (1.54) can be written as

$$\bar{Y}_1 = R_1 [\cos \phi \sin \omega t - \sin \phi \cos \omega t]$$

or

$$\bar{Y}_1 = R_1 \sin(\omega t - \phi) \qquad ...(1.55)$$

Equation (1.55) represents the resultant of two waves, Y_1 and Y_2. Square of R_1

$$R_1^2 = [a_1 + a_2]^2 \qquad (1.56)$$

represents the square of the amplitude of resultant of two waves.

The resultant displacement ($\bar{Y}_2$) of two waves, $\bar{Y}_1$ and Y_3, is

$$\bar{Y}_2 = \bar{Y}_1 + Y_3$$

$$= R_1 \sin(\omega t - \phi) + a_3 \sin(\omega t - \phi)$$

$$= R_1 [\sin \omega t \cos \phi - \cos \omega t \sin \phi] + a_3 [\sin \omega t \cos \phi - \cos \omega t \sin \phi]$$

$$= [R_1 + a_3] \cos \phi \sin \omega t - [R_1 + a_3] \sin \phi \cos \omega t \qquad ...(1.57)$$

Let $R_2 = R_1 + a_3 = a_1 + a_2 + a_3$

Equation (1.57) can be written as

$$\bar{Y}_2 = R_2 [\cos \phi \sin \omega t - \sin \phi \cos \omega t]$$

or

$$\bar{Y}_2 = R_2 \sin(\omega t - \phi) \qquad ...(1.58)$$

Similarly, the resultant of supposition of $\bar{Y}_2$ and Y_4 will be

$$\bar{Y}_3 = R_3 \sin(\omega t - \phi)$$

and the resultant of supposition of $\bar{Y}_{n-1}$ and Y_n will be

$$\bar{Y} = R \sin(\omega t - \phi) \qquad ...(1.59)$$

where, $R_3 = R_2 + a_4 = a_1 + a_2 + a_3 + a_4$

$$R = R_{n-2} + a_n = a_1 + a_2 + a_3 + a_4 + ... + a_n$$

and

Square of R will be

$$R^2 = (a_1 + a_2 + a_3 + a_4 + ... + a_n)^2 \qquad ...(1.60)$$

Equation (1.60) represents the square of the amplitude of resultant of n waves.

Here, a_1, a_2, a_3, ..., a_n are amplitudes of n number of coherent waves travelling in the same direction with same frequency, $\omega \cdot \phi$ is the phase angles of n waves.

The intensity at a point due to superposition on N coherent waves is

$$I \propto R^2 \quad \text{or} \quad I = kR^2 = k\,(a_1 + a_2 + a_3 + a_4 + ... + a_n)^2$$

where, k is the constant of proportionality. Thus, the intensity at a point due to superposition of N coherent waves is square of the sum of amplitudes all waves.

1.9 SUPERPOSITION OF COHERENT WAVES OF EQUAL FREQUENCY

Two waves are said to be coherent waves if the phase difference between the two waves at a point in space remains constant throughout the observation time. In the superposition of coherent waves, the resultant intensity (I) is directly proportional to the square of the vector sum of amplitudes of individual waves, i.e., $I = k\,(a_1 + a_2 + a_3 + a_4 + ...)^2$.

In coherent superposition, the resultant amplitude does not change at a point with time or space. The resultant amplitudes defined in Eq. (1.6) or Eq. (1.26) or Eq. (1.34) remain constant with time. The superposition of waves always has constant phase difference, ϕ. Therefore,

$$\frac{d\phi}{dt} = 0$$

The amplitude defined by Eq. (1.6) is

$$R^2 = a_1^2 + a_2^2 + 2\, a_1\, a_2 \cos(\phi) \quad ...(1.61)$$

where, R is the amplitude of resultant wave of superposition of two waves.

a_1 is the amplitude of the first wave.

a_2 is the amplitude of the second wave.

Φ is the phase of two superposing waves.

In coherent superposition, the phase of two waves (Φ) does not change with time. Therefore, the time average of cos Φ, at a point will be cos Φ, i.e.,

$$\langle \cos \Phi \rangle = \cos \Phi \quad ...(1.62)$$

Hence, the amplitude of resultant wave of superposition of two waves of the coherent superposition, R does not change with time. If $I_1 = a_1^2$ and $I_2 = a_2^2$, therefore the intensity (I) at the point of superposition will be

$$I = I_1 + I_2 + 2\sqrt{I_1}\sqrt{I_2} \cos \phi \quad ...(1.63)$$

If the amplitude is $a_1 = a_2$, then the intensity $I_1 = I_2 = I_0$, therefore,

$$I = 2I_0 (1 + \cos \phi) = 4I_0 \cos^2 \frac{\phi}{2} \quad ...(1.64)$$

The interference pattern due to superposition of two coherent waves ofthe same amplitude is visible and is governed by Eq. (1.64).

1.10 SUPERPOSITION OF INCOHERENT WAVES OF EQUAL FREQUENCY

Two waves are said to be incoherent waves if the phase difference between the two waves at a point in space varies randomly throughout the observation time. The resultant amplitude does not remain constant with time. In the incoherent superposition of waves, the resultant intensity is directly proportional to the sum of the squares of the amplitudes of individual waves, i.e., $I = k\,(a_1^2 + a_1^2 + a^2{}_1 + a^2{}_1 + ...)$. That means in case of incoherent superposition, the resultant intensity is equal to the sum of the intensities of individual waves and changes with time.

The resultant amplitude of a wave due to superposition of two waves of amplitudes a_1 and a_2, and having angular frequency ω_1 and ω_2 is given by Eq. (1.21).

$$R^2 = a_1^2 + a_2^2 + 2\, a_1\, a_2 \cos(\omega_1 - \omega_2)\, t$$

or

$$R^2 = a_1^2 + a_2^2 + 2\, a_1\, a_2 \cos \phi \quad ...(1.65)$$

The phase difference, $\phi = (\omega_1 - \omega_2)t$ between two waves varies with time. Hence,

$$\frac{d\phi}{dt} \neq 0$$

The value of cos ϕ varies between –1 to +1 with respect to time. Therefore, the time average of cos ϕ at a point will be 0, i.e.,

$$<\cos \phi> = 0$$

Hence, for incoherent superposition, Eq. (1.40) becomes

$$R^2 = a_1^2 + a_2^2 + 2\, a_1\, a_2 \times 0$$

or

$$R^2 = a_1^2 + a_2^2$$

If $I_1 = a_1^2$ and $I_2 = a_2^2$, then intensity (I) at the point of superposition will be

$$I = I_1 + I_2 \qquad \text{...(1.66)}$$

If the amplitudes of two incoherent waves are equal, i.e., $a_1 = a_2$, then the intensity $I_1 = I_2 = I_0$, therefore,

$$I = I_0 + I_0 = 2\, I_0 \qquad \text{...(1.67)}$$

Thus, the resultant intensity is equal to the sum of two individual intensities of the superposing waves. Equation (1.66) shows that intensity will not vary from point to point on the screen and the interference pattern will not be visible if two superposing waves are incoherent.

In the analogy of Eq. (1.64), the interference pattern produced due to superposition of two incoherent waves of equal amplitudes is given by

$$I = 4\, I_0 <\cos^2 \frac{\phi}{2}> \qquad \text{...(1.68)}$$

In Eq. (1.43), $<\cos^2 \frac{\phi}{2}>$ is the time average of $\cos^2 \frac{\phi}{2}$. The value of $\cos^2 \left(\frac{\phi}{2}\right)$ varies between minimum value 0 and maximum value, 1. Therefore, the value of $<\cos^2 \frac{\phi}{2}> = 0.5$. Now Eq. (1.68) becomes

$$I = 4\, I_0 \times 0.5 = 2\, I_0 = I_0 + I_0$$

1.11 BEATS KEITH AND GIBBS

Superposition of two waves of nearly same frequency

When two waves of slightly different frequency overlap, the algebraic some of displacements gives rise to beats. The phenomenon of beats is a kind of distinct sensation, which can be heard sometimes when two musicians are playing their musical instruments.

Let f_1 and f_2 be frequencies of two instruments.The beat frequency depends on the difference of the frequencies of two waves. The closer are the two frequencies the lower will be beat frequency. The beat frequency will be zero when they are perfectly in tune.The human ear is normally very sensitive to pitch and the two notes have to be well within one semitone for true beats to be heard.

Formula for beat frequency

Let us consider two travelling waves of angular frequencies (ω_1 and ω_2) having amplitudes a_1 and a_2.

$$Y_1 = a_1 \sin(\omega_1 t) \quad ...(1.69a)$$

$$Y_2 = a_2 \sin(\omega_2 t) \quad ...(1.69a)$$

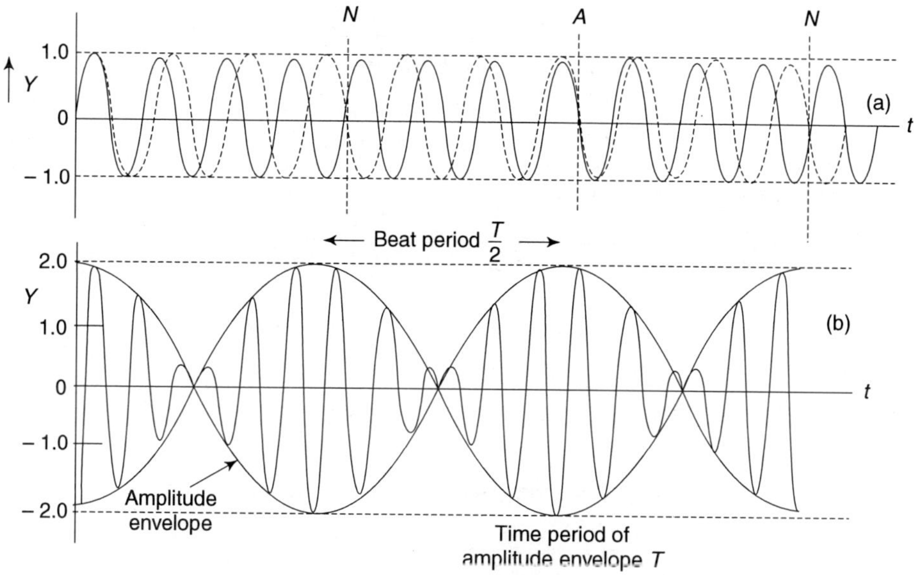

Fig. 1.4 Superposition of two harmonic waves is shown in (a). One of frequency (ω_1) and another of frequency (ω_2) give rise to beats as shown in (b)

The resultant sum of Y_1 and Y_2 is

$$\begin{aligned} Y &= Y_1 + Y_2 \\ &= a_1 \sin \omega_1 t) + a_2 \sin(\omega_2 t) \\ &= a_1 \sin(\omega_1 t) + a_2 [\sin\{\omega_1 t - (\omega_1 t - \omega_2 t)\}] \\ &= a_1 \sin(\omega_1 t) + a_2 [\sin(\omega_1 t)\cos(\omega_1 t - \omega_2 t) - \cos(\omega_1 t)\sin(\omega_1 t - \omega_2 t) \\ &= \sin(\omega_1 t)[a_1 + a_2 \cos(\omega_1 t - \omega_2 t)] - a_2 \cos(\omega_1 t)\sin(\omega_1 t - \omega_2 t) \quad ...(1.70) \end{aligned}$$

Let $$R\cos(\theta) = a_1 + a_2 \cos(\omega_1 - \omega_2)t \quad ...(1.71a)$$

and $$R\sin(\theta) = a_2 \sin(\omega_1 - \omega_2)t \quad ...(1.71b)$$

Substituting from Eq. (1.71a) and (1.71b), Eq. (1.70) becomes

$$\begin{aligned} Y &= R\cos(\theta)\sin(\omega_1 t) - R\sin(\theta)\cos(\omega_1 t) \\ &= R\sin(\omega_1 t - \theta) \quad ...(1.72) \end{aligned}$$

Dividing R sin (θ) by R cos (θ), we get

$$\tan\theta = \frac{a_2 \sin(\omega_1 - \omega_2)t}{a_1 + a_2 \cos(\omega_1 - \omega_2)t} \quad ...(173)$$

From Eq. (1.73) it is clear that phase angle (θ) is function of time, therefore, phase angle of resultant wave given by Eq. (1.72) varies with time. Adding squares of R cos (θ) and R sin (θ), we get

$$R^2 = [a_1 + a_2 \cos(\omega_1 - \omega_2)t]^2 + [a_2 \sin(\omega_1 - \omega_2)t]^2$$

or

$$R = \sqrt{a_1^2 + a_2^2 + 2\, a_1 a_2 \cos(\omega_1 - \omega_2)t}$$

or

$$R = \sqrt{a_1^2 + a_2^2 + 2\, a_1 a_2 \cos 2\pi(f_1 - f_2)t} \quad ...(1.74)$$

When $a_1 = a_2$ the amplitude becomes

$$R = a^2 \sqrt{2[1 + \cos 2\pi(f_1 - f_2)t]} = 2\, a^2 \cos 2\pi\left(\frac{f_1 - f_2}{2}\right)t \quad ...(1.75)$$

From this equation it is clear that amplitude (R) of resultant wave varies with time.

Case 1. Amplitude (R) is maximum

when $\cos 2\pi(f_1 - f_2)t = 1$

This implies that

$$2\pi(f_1 - f_2)t = 2n\pi, \text{ where } n = 0, 1, 2, 3, ...$$

or

$$t = \frac{n}{f_1 - f_2} \quad ...(1.76)$$

Equation (1.76) shows that the amplitude of the resultant wave (R) is maximum at time

$$t = 0, \frac{1}{f_1 - f_2}, \frac{2}{f_1 - f_2}, \frac{3}{f_1 - f_2}, ...$$

Maximum value of the amplitude

$$R = \sqrt{a_1^2 + a_2^2 + 2\, a_1 a_2} = a_1 + a_2$$

Case 2. Amplitude (R) is minimum

when $\cos 2\pi(f_1 - f_2)t = -1$

This implies that

$$2\pi(f_1 - f_2)t = (2n - 1)\pi, \text{ where } n = 1, 2, 3, ...$$

or

$$t = \frac{(2n - 1)}{2(f_1 - f_2)} \quad ...(1.77)$$

Equation (1.77) shows that the amplitude of the resultant wave (R) is minimum at time

$$t = \frac{1}{2(f_1 - f_2)}, \frac{3}{2(f_1 - f_2)}, \frac{5}{2(f_1 - f_2)}, \ldots$$

Minimum value of the amplitude

$$R = \sqrt{a_1^2 + a_2^2 - 2\, a_1\, a_2} = a_1 - a_2$$

Thus, maxima and minima occur alternately after equal intervals of time. The time interval between the two consecutive minima and minima is equal to $\dfrac{1}{(f_1 - f_2)}$.

The time interval (or time period T) between any two consecutive maxima or two consecutive minima is $T = \dfrac{1}{(f_1 - f_2)}$. So the frequency of beats, $f = (f_1 - f_2)$. Therefore, the number of maxima (beats) produced per second is $(f_1 - f_2)$. That is, the number of maxima or beats produced per second is equal to the difference between the frequencies of superposing waves.

The knowledge of beat frequency has lot of practical applications.

Beat Intensity

The intensity of wave motion is proportional to the square of the amplitude, and therefore

Intensity, $\propto$ (amplitude)^2

Putting the value from Eq. (1.75)

$$\text{Intensity} \propto 4\, a^2 \cos^2 2\pi\, \frac{f_1 - f_2}{2}\, t$$

Using the identity $2\cos^2\theta = 1 + \cos 2\theta$

$$\text{Intensity} \propto 2\, a^2\, [1 + \cos (2\pi\, (f_1 - f_2))]$$

This equation indicates that the intensity varies with beat frequency $(f_1 - f_2)$. From Eq. (1.75), one can note the amplitude variation as frequency $(f_1 - f_2)/2$ varies, i.e., the amplitude varies at half the rate of the intensity. This is because the intensity is maximum when the amplitude has its maximum value, $2a$, and when it has its minimum value, $-2a$.

1.12 APPLICATIONS OF BEATS

(i) Tuning the musical instruments

Musicians use the phenomenon of beats for tuning their instruments. First they play two instruments, one by one, and adjust the frequency of one instrument in such a way that its sound appears to them similar to the sound of the other instrument. When the sounds of the two instruments become nearly equal, then they play two instruments simultaneously and hear beats. Now they adjust the frequency of one in such a way that the number of beats per second goes on decreasing slowly until the beats disappears. Now the frequencies of the two instruments are exactly equal.

(ii) Police radar to trap speeding vehicle

The outgoing signal from police patrol vehicle is reflected from the speeding vehicle. The outgoing and reflected signals are received by the detector at the patrol vehicle get overlapped. Then the speed of the vehicle is determined from the beat frequency using the Doppler shift as follows:

In the radar speed trap a beam is transmitted from a stationary source, reflected from a moving vehicle and then detected by a stationary receiver.

To determine the velocity of the vehicle the reflected signal is made to overlap with the outgoing signal. By knowing the beats, the beat frequency is found. The velocity (v) of a vehicle moving towards or away from a radar speed trap is given by the formula

$$v = \frac{c\Delta f}{2f}$$

where, f is the frequency of the transmitter and Δf is the observed frequency change or Doppler shift. To know more about Doppler shift see Section 4.10 of Chapter 4.

(iii) Determination of frequency of tuning fork in physics laboratory

One can find the exact frequency of a forkof nearly equal frequency using the phenomenon of beats, if the frequency (f_1) of one tuning fork is known. For finding the unknown frequency of a fork, both the tuning forks are sounded together and the beats are noted. Suppose there are x beats in 1 second. Then the frequency of the tuning fork will be either $(f_1 + x)$ or $(f_1 - x)$. Now one prong of this fork,whose frequency is not known, is loaded with small wax so that the frequency is slightly lowered. Again, the two forks are sounded together and beats are noted. If the number of beats per second decreases, then it means that the frequency of fork is higher than the frequency of the known fork, i.e., the frequency of the fork is $(f_1 + x)$. On the other hand, if on loading with wax, the number of beats per second increases, then the frequency of the fork is $(f_1 - x)$.

Formulas at a Glance

S.No.	Subject	Formula	Interpretation
1.	The resultant amplitude of two waves of equal frequency	$R = \sqrt{a_1^2 + a_2^2 + z\, a_1\, a_2 \cos(\phi)}$	a_1 and a_2 are amplitudes of the waves and ϕ is the phase difference between them.
2.	Maximum intensity of superposition of two waves of equal frequency	$I_{max} = I_1 + I_2 + 2\sqrt{I_1 I_2}$ $= \left(\sqrt{I_1} + \sqrt{I_2}\right)^2$	$I_1 = a_1^2$ and $I_2 = a_2^2$

Contd.

3.	Maximum intensity of superposition of two waves of equal frequency	$I_{min} = I_1 + I_2 - 2\sqrt{I_1 I_2}$ $= \left(\sqrt{I_1} - \sqrt{I_2}\right)^2$	
4.	Maximum intensity n coherent waves of same amplitudes and same frequency	$(I_{max})_{co} = \left(\sqrt{I_0} + \sqrt{I_0} + \sqrt{I_0} + ...\right)^2$ $= n^2 I_0$	$I_0 = a_0^2$
5.	Average intensity due to two incoherent waves is	$(I_R)_{inco} = I_1 + I_2$	
6.	Intensity of n identical incoherent n waves of equal amplitudes and equal frequencies will be	$(I_R)_{inco} = I_0 + I_0 + ... = nI_0$	
7.	The resultant of two wave of two waves of frequencies f_1 and f_2	$Y = A \sin 2\pi \dfrac{f_1 + f_2}{2} t$	First factor shows a slow variation of amplitude $A = 2a \cos 2\pi \dfrac{f_1 - f_2}{2} t$, while the second factor shows rapid variation of displacement. The amplitude frequency is $\dfrac{f_1 - f_2}{2}$. Beat frequency f = Number of beats per second $= f_1 - f_2$

SOLVED EXAMPLES

Example 1.1 Two sitar strings, A and B, playing a note slightly out of tune produce beats of frequency 7 Hz. The tension of the string B is slightly increased and beat frequency is found to decrease to 4 Hz. What is the original frequency of B, if the frequency of A is 429 Hz.

***Solution*:** Let frequency of strings A and B be f_A and f_B.

Frequency f_A 429 Hz

Strings A and B produce 7 beats, so frequency of thestring B will be

$$f_B = 429 \pm 7$$

That means, the frequency (f_B) is either 436 Hz or 422 Hz

When, we increase the tension of a string, its frequency increases. If frequency (f_B) is greater than the frequency, f_A, then further increase in f_B will result in an increase in the beat frequency. But the beat frequency is found to decrease by 4 Hz. This shows that $f_B < f_A$. Hence, the original frequency of string *B* is 422 Hz.

Let f^T_B be the frequency of string *B* after its tension is increased

So $$f_A - f^T_B = 4$$

Since $f_A = 429$ Hz, so we get $f^T_B = 425$ Hz.

Ans. The original frequency of string *B* is 422 Hz.

Example 1.2 *A*, *B* and *C* are three tuning forks. The frequency of *A* is 351 Hz. *A* and *B* produce 5 beats per second, while *B* and *C* produce 4 beats per second. When *A* is loaded with some wax, it produces 2 beats per second with *B*, and 6 beats per second with *C*. What are the frequencies of *B* and *C*.

***Solution*:** Let the frequencies of fork *A*, *B*, *C* be f_A, f_B and f_C, respectively.

The frequency of fork *A* is $f_A = 351$ Hz.

Forks *A* and *B* produce 5 beats, so frequency of $f_B = 351 \pm 5$.

That is, frequency f_B is either 346 Hz or 356 Hz.

Let the frequency of fork A loaded with some wax be f^L_A.

The fork *A* loaded with some wax and fork *B* produce 2 beats.

This is possible only when $f_A > f_B$.

So the frequency, $f_B = 346$ Hz.

Now $f^L_A - f_B = 2$, so $f^L_A = 348$ Hz.

Now forks *B* and *C* produce 4 beats, so $f_C = 346 \pm 4$ Hz. That means the frequency f_C will be either 350 Hz or 342 Hz. When $f_C = 350$, $f_C - f^L_A = 2$, when $f_C = 342$, $f_C - f^L_A = 6$.

The second condition is found to be true, so $f_C = 342$ Hz.

Ans. The frequency of fork *B* is 346 Hz and frequency of fork *C* is 342 Hz.

Example 1.3 Calculate the speed of sound of wavelengths 1.01 m and 1.02 m produce 7 beats in 2 seconds when two instruments are played.

***Solution*:** Formula used

$$v = f\lambda,$$

where, *v*, *f* and λ are velocity, frequency and wavelength of wave.

Given: $\lambda_1 = 1.01$ m, $\lambda_2 = 1.02$ m

Let *v* be the speed of the sound and f_1 and f_2 are the frequencies of two waves, then

$$f_1 = \frac{v}{\lambda_1} = \frac{v}{1.01} \text{ and } f_2 = \frac{v}{\lambda_2} \frac{v}{1.02}$$

Number of beats per second, $f_1 - f_2 = \frac{7}{2}$

Putting the values of f_1 and f_2

$$\frac{v}{1.01} - \frac{v}{1.02} = \frac{7}{2} \quad \text{or} \quad v = \frac{1.01 \times 1.021}{0.01} \times \frac{7}{2} = 360.57 \text{ m/s}$$

Ans. Speed of sound is 360.57 m/s

Example 1.4 Calculate the velocity of wave, when two waves of lengths 90 and 91 cm are moving with same velocity and beat frequency 5.

***Solution*:** Formula used

$$v = f\lambda,$$

where, v, f and λ are velocity, frequency and wavelength of wave.

Given: $\lambda_1 = 0.90$ m, $\lambda_2 = 0.91$ m

Let v be the speed of the sound and f_1 and f_2 be the frequencies of two waves, then

$$f_1 = \frac{v}{\lambda_1} = \frac{v}{0.90} \text{ and } f_2 = \frac{v}{\lambda_2} = \frac{v}{0.91}$$

Number of beats per second, $f_1 - f_2 = 5$

Putting the values of f_1 and f_2

$$\frac{v}{0.90} - \frac{v}{0.91} = 5 \qquad \text{or} \quad v = \frac{0.90 \times 0.91}{0.01} \times 5 = 409.5 \text{ m/s}$$

Ans. Speed of sound is 409.5 m/s

Example 1.5 Two travelling waves of equal frequency superimpose at a point. If the ratios of intensity of individual waves are 25 and 16, what is the ratio of their amplitudes? Find the ratio of maximum and minimum intensities.

***Solution*:** Let I_1 and I_2 be the intensities of two waves and their amplitudes are a_1 and a_2.

Given: $I_1 = 25, \; I_2 = 16$

The relation between amplitude and intensity is

$$I_1 = a_1^2 \text{ and } I_2 = a_2^2$$

So the values of $a_1 = 5$ and $a_2 = 4$.

Therefore, ratio of amplitudes is $\frac{a_1}{a_2} = \frac{5}{4}$

The ratio of maximum and minimum intensities is

$$\frac{I_{max}}{I_{min}} = \frac{\left(\sqrt{I_1}+\sqrt{I_2}\right)^2}{\left(\sqrt{I_1}-\sqrt{I_2}\right)^2} = \frac{(a_1+a_2)^2}{(a_1-a_2)^2} = \frac{81}{3} = \frac{9}{1}$$

Ans. The ratio of amplitudes is $\frac{5}{4}$ and ratio of maximum and minimum intensities is $\frac{9}{1}$.

QUESTIONS FOR EXAMINATION

1.1 Differentiate between progressive wave and stationary wave.

1.2 What are nodes and antinodes? Explain how the standing waves are formed?

1.3 State the principle of superposition.

1.4 Find the resultant wave due to superposition of two waves of the same frequency travelling in the same direction.

1.5 Explain the formation of standing wave in a string clamped at both ends and discuss the various modes of vibrations.

1.6 Show that the resultant of two travelling waves of the same amplitude and frequency that travel in opposite direction is a stationary wave.

1.7 Find the resultant wave due to superposition of two waves of the same frequency travelling in opposite direction.

1.8 Prove that in incoherent superposition of waves, the resultant intensity is equal to the summation of the intensities of individual waves.

1.9 Find the resultant wave due to superposition of N waves of the same frequency travelling in the same direction.

1.10 What do you mean by coherent superposition? What are its characteristics?

1.11 What do you mean by incoherent superposition? What are its characteristics?

1.12 Differentiate between coherent and incoherent superposition.

1.13 What are beats? Explain how beats are produced? What is the necessary condition to produce beats?

[**Hint:** Necessary condition to produce beats is that the frequency of two waves should nearly same.]

1.14 What are beats? Prove that the number of beats per second produced by two sources of sound is equal to the difference in the frequency $(f_1 - f_2)$ of the two sources.

1.15 What are beats? How are they produced? Find the expression for beat frequency.

1.16 Distinguish between beats and interference produced by two sound source.

[***Ans.*** In beats at a given position, intensity varies periodically with time with periodicity $T = [1/(f_1 - f_2)]$, while in interference at a given time, intensity varies periodically with position with periodicity (λ).]

1.17 Find the superosition of N harmonic waves and show that intensity due to N superposition of N harmonic waves of same frequency and constant phase is the sum of square of the amplitudes of all simple harmonic waves.

OBJECTIVE QUESTIONS

1.1 Standing waves are produced by superposition of

(a) Two waves having the same frequency travelling in the same direction.
(b) Two waves having the same frequency travelling in the opposite direction.
(c) Two waves having different frequencies travelling in the same direction.
(d) Two waves having different frequencies travelling in the same direction.

1.2 Interference patterns are produced by superposition of

(a) Two waves having the same frequency travelling in the same direction.
(b) Two waves having the same frequency travelling in the opposite direction.
(c) Two waves having different frequencies travelling in the opposite direction.
(d) Two waves having different amplitudes travelling in the same direction.

1.3 Beats are produced by superposition of

(a) Two waves having the same frequency travelling in the same direction.
(b) Two waves having the same frequency travelling in the opposite direction.
(c) Two waves having different frequencies travelling in same direction.
(d) Two waves having different amplitudes travelling in the same direction.

1.4 Two coherent harmonic waves of equal frequencyare travelling in a medium in the same direction. If I_1 and I_2 are the intensities of individual waves, then the sum of maximum and minimum intensities is:

(a) $I_1 + I_2$ (b) $\left(\sqrt{I_1}+\sqrt{I_2}\right)^2$ (c) $\left(\sqrt{I_1}-\sqrt{I_2}\right)^2$ (d) $4\sqrt{I_1}\sqrt{I_2}$

1.5 Two coherent harmonic waves of equal frequencies are travelling in a medium in the same direction. If a_1 and a_2 are the amplitudes of these waves, then the sum of maximum and minimum intensities at a point is:

(a) $a_1^2 + a_2^2$ (b) $(a_1 + a_2)^2$ (c) $(a_1 - a_2)^2$ (d) $4\, a_1\, a_2$

1.6 Two coherent harmonic waves of equal frequency are travelling in a medium in the same direction. If I_1 and I_2 are the intensities of individual waves, then the ratio of maximum and minimum intensities at a point is:

(a) $\dfrac{I_1 + I_2}{I_1 - I_2}$ (b) $\dfrac{\left(\sqrt{I_1}+\sqrt{I_2}\right)^2}{\left(\sqrt{I_1}-\sqrt{I_2}\right)^2}$ (c) $\dfrac{\left(\sqrt{I_1}-\sqrt{I_2}\right)^2}{\left(\sqrt{I_1}+\sqrt{I_2}\right)^2}$ (d) $\dfrac{\sqrt{I_1}+\sqrt{I_2}}{\sqrt{I_1}-\sqrt{I_2}}$

1.7 Two coherent harmonic waves of equal frequency are travelling in a medium in the same direction. If a_1 and a_2 are the amplitudes of these waves, then the ratio of maximum and minimum intensities at a point is:

(a) $\dfrac{a_1^2 + a_2^2}{a_1^2 - a_2^2}$ (b) $\dfrac{(a_1 + a_2)^2}{(a_1 - a_2)^2}$ (c) $\dfrac{(a_1 - a_2)^2}{(a_1 + a_2)^2}$ (d) $\dfrac{a_1 + a_2}{a_1 + a_2}$

1.8 Two incoherent harmonic waves of equal frequency are travelling in a medium in the same direction. If I_1 and I_2 are the intensities of individual waves, then the sum of maximum and minimum intensities at a point is:

(a) $I_1 + I_2$ (b) $\sqrt{I_1} + \sqrt{I_2}$ (c) $\sqrt{I_1} - \sqrt{I_2}$ (d) $I_1 - I_2$

1.9 Two travelling waves of same frequency superimpose at a point If ratio of the intensities of two waves are 16 and 9. then ratio of maximum and minimum intensities will be

(a) $\frac{4}{3}$ (b) $\frac{49}{1}$ (c) $\frac{7}{1}$ (d) $\frac{16}{9}$

1.10 Two sound progressive waves given by $Y_1 = 3 \sin (402 \pi t)$ and $Y_2 = 2 \sin (408 \pi t)$ are situated very near to the ears of a person. He will hear

(a) 3 beats per second with intensity ratio of maxima and minima 3/2

(b) 3 beats per second with intensity ratio of maxima and minima 25/1

(c) 6 beats per second with intensity ratio of maxima and minima 5/1

(d) 6 beats per second with intensity ratio of maxima and minima 3/2

1.11 The number of beats produced by oscillations $Y_1 = a \sin (322 \pi t)$ and $Y_2 = a \sin (328 \pi t)$ is

(a) 4 (b) 6 (c) 2 (d) 3

1.12 A source of unknown frequency produces 6 beats per second with a source of 252 Hz and 8 beats per second with a source of 266 Hz. The frequency of unknown source is

(a) 272 Hz (b) 254 Hz (c) 262 (d) 258

1.13 Two waves of lengths 80 and 81 cm are moving with the same velocity produce 10 beats in 2 seconds. The velocity of the wave is

(a) 318 m/s (b) 321 m/s (c) 324 m/s (d) 350 m/s

1.14 The intensity due to superposition of N (even number) simple harmonic waves of the same frequency and constant phase having amplitudes, a_1 and a_2, is

(a) $[a_1 - a_2 + a_3 - \cdots - a_N]^2$ (b) $[a_1 + a_2 + a_3 + \cdots + a_N]^2$

(c) $[a_1^2 + a_2^2 + a_3^2 + \cdots + a_N^2]$ (d) $[a_1^2 - a_2^2 + a_3^2 - \cdots - a_N^2]$

Answers

1.1 (b) 1.2 (a) 1.3 (c) 1.4 (d) 1.5 (d) 1.6 (b) 1.7 (b) 1.8 (a)

1.9 (b) 1.10 (b) 1.11 (d) 1.12 (d) 1.13 (c) 1.14 (b)

CHAPTER 2

Superposition of Two Perpendicular Harmonic Oscillations

2.1 INTRODUCTION

When two sinusoidal or cosine waves superimpose perpendicular to each other, the resultant of these will be figures of various shapes. The shapes of these figures depend upon the ratio of frequency $\left(\frac{f_x}{f_y}\right)$ and phase difference of two waves. These figures are called the *Lissajous' figures*.

For example, the Lissajous' figure for two signals which have the same frequency, $\frac{f_x}{f_y} = \frac{1}{1}$, and have zero phase difference in a staight line at 45° to the horizontal as shown in Fig. 2.1(i). The Lissajous' figure with $\frac{f_x}{f_y} = \frac{1}{1}$ and phase difference of $\frac{\pi}{4}$ is an ellipse as shown in Fig. 2.1 (iv). The Lissajous' figure with $\frac{f_x}{f_y} = \frac{1}{1}$ and phase difference of $\frac{\pi}{2}$ is a circle as shown in Fig. 2.2 (i).

The analytical expression of these Lissajous' figures is given by Eq. (2.6). When two sinusoidal signals are applied to the horizontal plate X and vertical plate Y of cathode ray oscilloscope (CRO), for same frequency and phase difference 0°, 45° and 90°, one can see a straight line at 45° with horizontal axis, an ellipse and a circle, respectively. The Lissajous' figures are used to find the phase difference of two waves superimposed perpendicular to each other.

2.2 GRAPHICAL AND ANALYTICAL METHODS

Consider two monochromatic light waves

$$Y_1 = a_1 \sin(\omega t - kx) \quad \text{...(2.1a)}$$

and

$$Y_2 = a_2 \sin(\omega t - kx + \delta) \quad \text{...(2.1b)}$$

where, a_1 and a_2 are the amplitudes of the waves, ω, and x are angular frequency and position of the waves, k is the momentum of the photon and δ is the phase difference between two waves. According to the principle of superposition of waves, the resultant is given by

$$\begin{aligned} Y &= Y_1 + Y_2 \\ &= a_1 \sin(\omega t - kx) + a_2 \sin(\omega t - kx + \delta) \\ &= a_1 \sin(\omega t - kx) + a_2 \sin(\omega t - kx)\cos(\delta) + a_2 \cos(\omega t - kx)\sin(\delta) \\ &= \sin(\omega t - kx)[a_1 + a_2\cos(\delta)] + \cos(\omega t - kx)[a_2 \sin(\delta)] \quad \text{...(2.2)} \end{aligned}$$

Taking

$$a_1 + a_2\cos(\delta) = R\cos\varphi \quad \text{...(2.3a)}$$

and

$$a_2 \sin(\delta) = R\sin\varphi \quad \text{...(2.3b)}$$

equation (2.2) becomes

$$\begin{aligned} Y &= R\sin(\omega t - kx)\cos\varphi + R\cos(\omega t - kx)\sin\varphi \\ &= R\sin(\omega t - kx + \varphi) \quad \text{...(2.3)} \end{aligned}$$

The resultant harmonic wave has amplitude R.

Dividing $R \sin \varphi$ by $R \cos \varphi$ given in Eq. (2.3a) by (2.3b), we get

$$\tan\varphi = \frac{a_2}{a_1 + a_2}\tan\delta \quad \text{...(2.4)}$$

When $a_1 = a_2 = a$

$$\tan\varphi = \frac{1}{2}\tan\delta \quad \text{...(2.5)}$$

Analytical expression of Lissajous' figures

Squaring Y_1 and Y_2 and then adding, we get

$$\frac{Y_1^2}{a_1^2} + \frac{Y_2^2}{a_2^2} = \sin^2(\omega t - kx) + \sin^2(\omega t - kx + \delta)$$

$$\begin{aligned} &\sin^2(\omega t - kx) + [\sin(\omega t - kx)\cos\delta + \cos(\omega t - kx)\sin\delta]^2 \\ &= \sin^2(\omega t - kx) + \sin^2(\omega t - kx)\cos^2\delta + \cos^2(\omega t - kx)\sin^2\delta \\ &\quad + 2\sin(\omega t - kx)\cos(\omega t - kx)\sin\delta\cos\delta \\ &= \sin^2(\omega t - kx)[1 + \cos^2\delta] + \cos^2(\omega t - kx)\sin^2\delta + \frac{1}{2}\sin 2(\omega t - kx)\sin 2\delta \end{aligned}$$

$$= \sin^2(\omega t - kx)[1 + \cos^2\delta] + \cos^2(\omega t - kx)[1 - \cos^2\delta] + \frac{1}{2}\sin 2(\omega t - kx)\sin 2\delta$$

$$= 1 + [\sin 2 (\omega t - kx) - \cos^2 (\omega t - kx) \cos^2 \delta + \frac{1}{2} \sin 2 (\omega t - kx) \sin 2 \delta$$

$$= 1 - \cos 2 (\omega t - kx) \cos^2 \delta + \frac{1}{2} \sin 2 (\omega t - kx) \sin 2 \delta$$

$$= 1 + [2 \sin^2 (\omega t - kx) - 1] \cos^2 \delta + \frac{1}{2} \sin 2 (\omega t - kx) \sin 2 \delta$$

or

$$\frac{Y_1^2}{a_1^2} + \frac{Y_2^2}{a_2^2} = 1 + \left[2\frac{Y_1^2}{a_1^2} - 1\right] \cos^2 \delta + \frac{Y_1}{a_1}\sqrt{1 - \frac{Y_1^2}{a_1^2}} \sin 2 \delta \quad \text{...(2.6)}$$

Equation (2.6) is the analytical expression. Other methods for getting Eq. (2.6) are given in **Appendix A**.

Case 1. When $\delta = 0$, this Eq. (2.6) reduces to the equation of a straight line

$$Y_2 = \frac{a_2}{a_1} Y_1 \text{ as } \sin 0 = 0 \text{ and } \cos 0 = 1 \quad \text{...(2.7)}$$

Case 2. When $a_1 = a_2 = a$, Eq. (2.7) reduces to the equation of a straight line inclined at an angle 45° from the horizontal axis.

$$Y_2 = \frac{a_2}{a_1} Y_1 \quad \text{...(2.8)}$$

Case 3. When $\delta = \frac{\pi}{2}$, Eq. (2.6) reduces to the equation of an ellipse

$$\frac{Y_1^2}{a_1^2} + \frac{Y_2^2}{a_2^2} = 1 \text{ as } \sin \pi = 0 \text{ and } \cos \frac{\pi}{2} = 0 \quad \text{...(2.9)}$$

Case 4. When $a_1 = a_2 = a$, Eq. (2.9) reduces to the equation of a circle

$$Y_1^2 + Y_2^2 = a^2 \quad \text{...(2.10)}$$

Case 5. When $0 < \delta < \frac{\pi}{2}$, Eq. (2.6) reduces to the equation of an ellipse

$$\frac{Y_1^2}{a_1^2} + \frac{Y_2^2}{a_2^2} = 1 \quad \text{...(2.11)}$$

where major axis of ellipse is

$$a_1' = a_1 \sqrt{1 + \left[2\frac{Y_1^2}{a_1^2} - 1\right] \cos^2 \delta + \frac{Y_1}{a_1}\sqrt{1 - \frac{Y_1^2}{a_2^2}} \sin 2\delta} \quad \text{...(2.12)}$$

and minor axis of the ellipse is

$$a_2' = a_2 \sqrt{1 + \left[2\frac{Y_1^2}{a_1^2} - 1\right]\cos^2\delta + \frac{Y_1}{a_1}\sqrt{1 - \frac{Y_1^2}{a_2^2}}\sin 2\delta} \quad ...(2.13)$$

Resultant amplitude

Squaring and adding Eq. (2.3a) and (2.3b), we get

$$R^2 [\sin^2 \varphi + \cos^2 \varphi] = [a_1 + a_2 \cos(\delta)]]^2 + a_2^2 \sin^2 \delta$$

or

$$R^2 = a_1^2 + 2\, a_1\, a_2 \cos(\delta) + a_2^2 \cos^2 \delta + a_2^2 \sin^2 \delta$$

$$= a_1^2 + 2\, a_1\, a_2 \cos(\delta) + a_2^2$$

or

$$R = \sqrt{a_1^2 + 2\, a_1\, a_2 \cos(\delta) + a_2^2} \quad ...(2.14)$$

The resultant amplitude of two waves is given by Eq. (2.14).

Intensity

The resultant intensity at the point of superposition is proportional to the square of the amplitude of the resultant wave is given by Eq. (2.14), i.e.,

$$I \propto R^2$$

or

$$I = CR^2$$

or

$$I = C\,[a_1^2 + 2\, a_1\, a_2 \cos(\delta) + a_2^2] \quad ...(2.15)$$

where, C is the constant of proportionality.

If $I_1 = a_1^2$ and $I_2 = a_2^2$, then

$$I = I_1 + 2\sqrt{I_1\, I_2}\cos(\delta) + I_2 \quad ...(2.16)$$

Here I_1 and I_2 are intensities of two waves and δ is the phase difference between the two waves.

Case 1. When $\cos\delta = 1$, I will be maximum, i.e.,

$$I_{max} = I_1 + I_2 + 2\sqrt{I_1\, I_2} = \left(\sqrt{I_1} + \sqrt{I_2}\right)^2 \quad ...(2.17)$$

Case 2. When $\cos\delta = -1$, I will be minimum, i.e.,

$$I_{min} = I_1 + I_2 - 2\sqrt{I_1\, I_2} = \left(\sqrt{I_1} - \sqrt{I_2}\right)^2 \quad ...(2.18)$$

Maximum intensity of n coherent waves will be

$$(I_{max})_{co} = \left(\sqrt{I_0} + \sqrt{I_0} + \sqrt{I_0} + ...\right)^2 = n^2\, I_0 \quad ...(2.19)$$

Average intensity due to two incoherent waves is given by

$$(I_R)_{inco} = I_1 + I_2 \quad ...(2.20)$$

Intensity of identical incoherent n waves will be

$$(I_R)_{inco} = I_0 + I_0 + ... = n\, I_0 \quad ...(2.21)$$

2.3 LISSAJOUS' FIGURES WITH EQUAL AND UNEQUAL FREQUENCY

When two sinusoidal signals of same amplitude are given to the horizontal plate X and vertical plate Y of the cathode ray oscilloscope (CRO), which are perpendicular to each other, then a series of patterns are visible on the screen. These figures are known as Lissajous' figures. The shapes of these figures are determined by the ratio of frequency and phase difference of the two signals. For each frequency ratio the phase difference of the signals affects the shape of the figure.

For example, the Lissajous' figure for two signals which have the same frequency, $\frac{f_X}{f_Y} = \frac{1}{1}$ and have zero phase difference show a straight line at 45° to the horizontal axis as shown in Fig. 2.1(i). On the other hand, the Lissajous' figure with frequency ratio $\frac{f_X}{f_Y} = \frac{2}{1}$ and phase difference of $\frac{\pi}{4}$ crosses the vertical axis 4 times and the horizontal axis twice as shown in Fig. 2.1(v), where the ratio of frequency $\frac{f_X}{f_Y} = \frac{4}{2} = \frac{2}{1}$. The Lissajous' figure with frequency ratio $\frac{f_X}{f_Y} = \frac{3}{2}$ and zero phase difference crosses the vertical axis 3 times and the horizontal axis twice as shown in Fig. 2.1(iii).

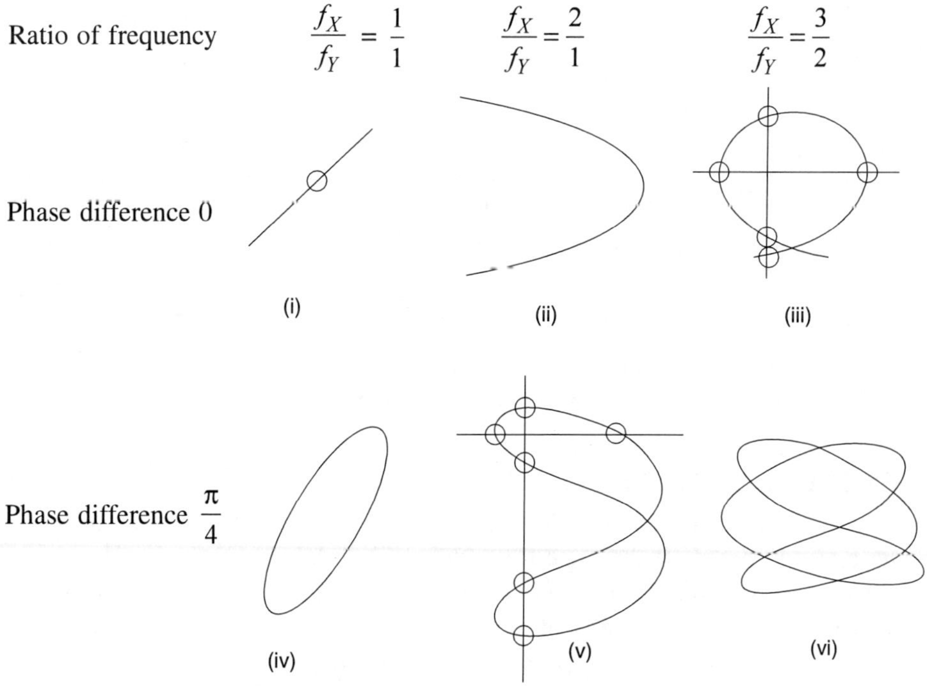

Fig. 2.1 Lissajous' figures on a CRO produced by two signals applied on its horizontal and vertical plates

The frequency ratio of a Lissajous' figure can be determined by ratio of the number of loops in x and y direction, i.e., by

$$\text{Frequency ratio} \qquad \frac{f_x}{f_y} = \frac{\text{Number of loops in } y\text{-direction}}{\text{Numner of loops } x\text{-direction}}$$

$$\text{or} \qquad \frac{f_X}{f_Y} = \frac{\text{Number of times the loop crosses the vertical axis}}{\text{Number of times the loop crosses the horizontal axis}}$$

For example, two signals which have the same frequency, i.e., $\frac{f_X}{f_Y} = \frac{2}{2} = \frac{1}{1}$ and have $\pi/2$ phase difference show a circle on the CRO screen as shown in Fig. 2.2(i). In case of circle, the vertical axis crosses it 2 times and the horizontal axis crosses it twice. Similarly, for frequency ratio, $\frac{f_X}{f_Y} = \frac{6}{4} = \frac{3}{1}$ and phase difference of $\pi/2$ produce a Lissajous' figure as shown in Fig. 2.2(iii). This Lissajous' figure crosses the vertical axis six times and the horizontal axis four times.

Ratio of frequency $\quad \frac{f_X}{f_Y} = \frac{1}{1} \quad \frac{f_X}{f_Y} = \frac{2}{1} \quad \frac{f_X}{f_Y} = \frac{3}{2}$

Phase difference $\frac{\pi}{2}$

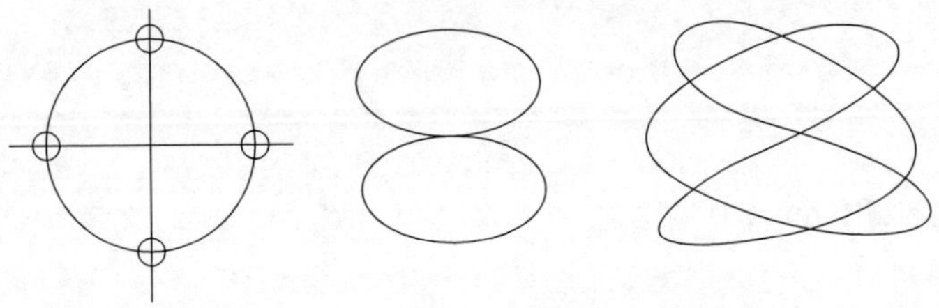

(i) (ii) (iii)

Fig. 2.2 Lissajous' figures on a CRO produced by two signals applied on its horizontal and vertical plates

2.4 USES OF LISSAJOUS' FIGURES

Lissajous' figures displayed on an oscilloscope can be used to give a quick estimate of the relative phase of two signals at the same frequency. Figure 2.3 shows the Lissajous' figures due to superposition of two perpendicular cosine signals. Let the displacement of one wave X along horizontal axis and displacement of second wave Y along vertical axis are

$$X = \cos(\omega t), \qquad Y = \cos(\omega t + \delta)$$

Where, the phase angle (δ) is indicated in degrees in the top right of each Lissajous' figures. In this case, the oscilloscope is adjusted so that the X and Y signals have exactly the same amplitude (8 divisions peak-to-peak) and the pattern is accurately centred on the screen. The Lissajous' figures for phase difference $\delta = 0$, 30°, 45°, 60° and 90° are shown in Fig. 2.3.

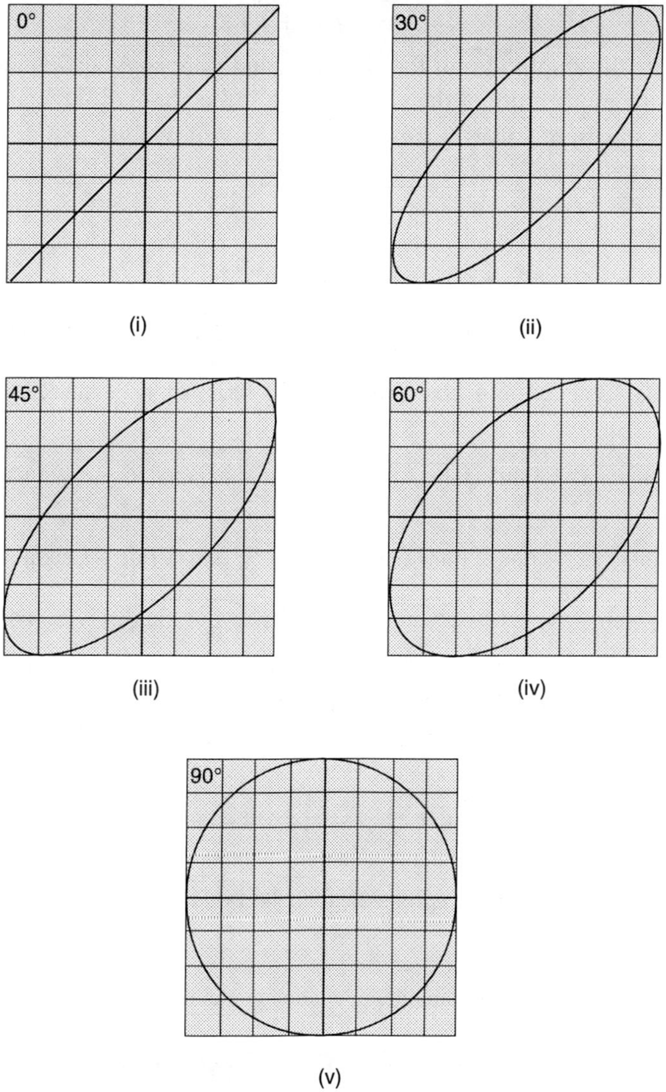

Fig. 2.3 Lissajous' figures on a CRO produced by two signals applied on its horizontal and vertical plates

APPENDIX A

Evaluation of Analytical Expression for Lissajous' Figures

Method 1

Consider two monochromatic light waves

$$Y_1 = a_1 \sin(\omega t - kx) \qquad \text{...(A1)}$$

and $$Y_2 = a_2 \sin(\omega t - kx + \delta) \quad ...(A2)$$

where, a_1 and a_2 are the amplitudes of the waves, ω, and x are angular frequency and position of the waves, k is the momentum of the photon and δ is the phase difference between the two waves. Taking the square of wave $Y_2 = a_2 \sin(\omega t - kx + \delta)$, we get

$$\frac{Y_2^2}{a_2^2} = \sin^2(\omega t - kx + \delta)$$

$$= [\sin(\omega t - kx)\cos\delta + \cos(\omega t - kx)\sin\delta]^2$$

$$= \sin^2(\omega t - kx)\cos^2\delta + \cos^2(\omega t - kx)\sin^2\delta + 2\sin(\omega t - kx)\cos(\omega t - kx)\sin\delta\cos\delta$$

$$= \sin^2(\omega t - kx)\cos^2\delta + \cos^2(\omega t - kx)[1 - \cos^2\delta] + \frac{1}{2}\sin 2(\omega t - kx)\sin 2\delta$$

$$= \cos^2(\omega t - kx) + [\sin^2(\omega t - kx) - \cos^2(\omega t - kx]\cos^2\delta + \frac{1}{2}\sin 2(\omega t - kx)\sin 2\delta$$

$$= \cos^2(\omega t - kx) - \cos 2(\omega t - kx)\cos^2\delta + \frac{1}{2}\sin 2(\omega t - kx)\sin 2\delta$$

$$= 1 - \sin^2(\omega t - kx) + [2\sin^2(\omega t - kx) - 1]\cos^2\delta + \sin(\omega t - kx)\cos(\omega t - kx)\sin 2\delta$$

$$= 1 - \frac{Y_1^2}{a_1^2} + \left[2\frac{Y_1^2}{a_1^2} - 1\right]\cos^2\delta + \frac{Y_1}{a_1}\sqrt{1 - \frac{Y_1^2}{a_1^2}}\sin 2\delta$$

$$\frac{Y_1^2}{a_1^2} + \frac{Y_2^2}{a_2^2} = 1 + \left[2\frac{Y_1^2}{a_1^2} - 1\right]\cos^2\delta + \frac{Y_1}{a_1}\sqrt{1 - \frac{Y_1^2}{a_1^2}}\sin 2\delta \quad ...(A3)$$

Equation (A3) is the analytical expression for Lissajous' figures.

Method 2

Consider two monochromatic light waves

$$X = a_1 \cos(\omega t), \qquad Y = a_2 \cos(\omega t + \delta)$$

where, a_1 and a_2 are the amplitudes of the waves, ω, and x are angular frequency and position of the waves. k is the momentum of the photon and δ is the phase difference between two waves.

Squaring equation $Y = a_2 \sin(\omega t + \delta)$

$$Y^2 = a_2^2 [\cos(\omega t + \delta)]^2$$

$$\frac{Y^2}{a_2^2} = [\cos(\omega t)\cos\delta - \sin(\omega t)\sin\delta]^2$$

$$= \left[\frac{x}{a_1}\cos\delta - \sqrt{1 - \frac{x^2}{a_1^2}}\sin\delta\right]^2$$

$$= \frac{X^2}{a_1^2}\cos^2\delta + \left(1 - \frac{X^2}{a_1^2}\right)\sin^2\delta - \frac{X}{a_1}\sqrt{1 - \frac{X^2}{a_1^2}}\sin 2\delta$$

$$= \frac{X^2}{a_1^2}\cos^2\delta + \left(1 - \frac{X^2}{a_1^2}\right)(1 - \cos^2\delta) - \frac{X}{a_1}\sqrt{1 - \frac{X^2}{a_1^2}}\sin 2\delta$$

$$= \frac{X^2}{a_1^2}\cos^2\delta + \left(1 - \frac{X^2}{a_1^2}\right) - \left(1 - \frac{X^2}{a_1^2}\right)\cos^2\delta - \frac{X}{a_1}\sqrt{1 - \frac{X^2}{a_1^2}}\sin 2\delta$$

$$\frac{X^2}{a_1^2} + \frac{Y^2}{a_2^2} = 1 - \cos^2\delta + 2\frac{X^2}{a_1^2}\cos^2\delta - \frac{X}{a_1}\sqrt{1 - \frac{X^2}{a_1^2}}\sin 2\delta$$

$$= 1 + \left[2\frac{X^2}{a_1^2} - 1\right]\cos^2\delta - \frac{X}{a_1}\sqrt{1 - \frac{X^2}{a_1^2}}\sin 2\delta \qquad \text{...(A4)}$$

Equation (A4) is the analytical expression for Lissajous' figures.

Method 3

Consider two monochromatic light waves

$$Y_1 = a_1 \cos(\omega t - kx) \qquad \text{...(A5)}$$

and

$$Y_2 = a_2 \cos(\omega t - kx + \delta) \qquad \text{...(A6)}$$

where, a_1 and a_2 are the amplitudes of the waves, ω, and x are angular frequency and position of the waves. k is the momentum of the photon and δ is the phase difference between the two waves. Adding the square of these waves, we get

$$\frac{X^2}{a_1^2} + \frac{Y^2}{a_2^2} = \cos^2(\omega t - kx) + \cos^2(\omega t - kx + \delta)$$

$$= \cos^2(\omega t - kx) + [\cos(\omega t - kx)\cos\delta - \sin(\omega t - kx)\sin\delta]^2$$

$$= \cos^2(\omega t - kx) + \cos^2(\omega t - kx)\cos^2\delta + \sin^2(\omega t - kx)\sin^2\delta$$
$$- 2\sin(\omega t - kx)\cos(\omega t - kx)\sin\delta\cos\delta$$

$$= \cos^2(\omega t - kx)[1 + \cos^2\delta] + \sin^2(\omega t - kx)[1 - \cos^2\delta]$$
$$- \frac{1}{2}\sin 2(\omega t - kx)\sin 2\delta$$

$$= 1 + [\cos^2(\omega t - kx) - \sin^2(\omega t - kx)\cos^2\delta - \frac{1}{2}\sin 2(\omega t - kx)\sin 2\delta$$

$$= 1 + \cos 2(\omega t - kx)\cos^2\delta - \frac{1}{2}\sin 2(\omega t - kx)\sin 2\delta$$

$$= 1 + [2\cos^2(\omega t - kx) - 1]\cos^2\delta - \sin(\omega t - kx)\cos(\omega t - kx)\sin 2\delta$$

$$= 1+\left[2\frac{X^2}{a_1^2}-1\right]\cos^2\delta-\frac{X}{a_1}\sqrt{1-\frac{X^2}{a_1^2}}\sin 2\delta \qquad ...(A7)$$

Equation (A7) is the analytical expression for Lissajous' figures.

Method 4

Taking the square of wave $Y_2 = a_2 \cos(\omega t - kx + \delta)$, we get

$$\frac{Y_2^2}{a_2^2} = \cos^2(\omega t - kx + \delta)$$

$$= [\cos(\omega t - kx)\cos\delta - \sin(\omega t - kx)\sin\delta]^2$$

$$= \cos^2(\omega t - kx)\cos^2\delta + \sin^2(\omega t - kx)\sin^2\delta - 2\sin(\omega t - kx)\cos(\omega t - kx)\sin\delta\cos\delta$$

$$= \cos^2(\omega t - kx)\cos^2\delta + \sin^2(\omega t - kx)[1 - \cos^2\delta) - \frac{1}{2}\sin 2(\omega t - kx)\sin 2\delta$$

$$= \sin^2(\omega t - kx) + [\cos^2(\omega t - kx) - \sin^2(\omega t - kx)]\cos^2\delta - \frac{1}{2}\sin 2(\omega t - kx)\sin 2\delta$$

$$= 1 - \cos^2(\omega t - kx) + [2\cos^2(\omega t - kx) - 1]\cos^2\delta - \sin(\omega t - kx)\cos(\omega t - kx)\sin 2\delta$$

$$= 1 - \frac{Y_1^2}{a_1^2} + \left[2\frac{Y_1^2}{a_1^2} - 1\right]\cos^2\delta - \sqrt{1-\cos^2(\omega t - kx)}\cos(\omega t - kx)\sin 2\delta$$

$$\frac{Y_1^2}{a_1^2} + \frac{Y_2^2}{a_2^2} = 1+\left[2\frac{Y_1^2}{a_1^2}-1\right]\cos^2\delta-\frac{Y_1}{a_1}\sqrt{1-\frac{Y_1^2}{a_1^2}}\sin 2\delta \qquad ...(A8)$$

Equation (A8) is the analytical expression for Lissajous' figures.

Since there is phase difference of $\pi/2$ between waves $Y_1 = a_1 \sin(\omega t - kx)$ and $Y_1 = a_1 \cos(\omega t - kx)$, so

$$\sin 2\left(\delta + \frac{\pi}{2}\right) = \sin(2\delta + \pi) = -\sin 2\delta \text{ and } \cos 2\left(\delta + \frac{\pi}{2}\right) = \cos(2\delta + \pi) = -\cos 2\delta$$

Formulas at a Glance

S.No.	Subject	Formula	Explanation
1.	Analytical expression for Lissajous' figures	$\frac{Y_1^2}{a_1^2} + \frac{Y_2^2}{a_2^2} = 1+\left[2\frac{Y_1^2}{a_1^2}-1\right]\cos^2\delta +$ $\frac{Y_1}{a_1}\sqrt{1-\frac{Y_1^2}{a_1^2}}\sin 2\delta$ where, $Y_1 = a_1\sin(\omega t - kx)$ and $Y_2 = a_2\sin(\omega t - kx + \delta)$	a_1 and a_2 are the amplitudes of the waves. ω, and x are angular frequency and position of the waves. k is the momentum of the photon and δ is the phase difference between two waves.

Contd.

2.	When $\delta = 0$	$Y_2 = \frac{a_2}{a_1} Y_1$	This formula reduces to the equation of a straight line.
3.	When $\delta = 45°$ from the horizontal axis	$Y_2 = Y_1$	$a_1 = a_2$
4.	When $\delta = \frac{\pi}{2}$	$\frac{Y_1^2}{a_1^2} + \frac{Y_2^2}{a_2^2} = 1$	Equation reduces to the equation of an ellipse
5.	When $a_1 = a_2 = a$	$Y_1^2 + Y_2^2 = a^2$	Equation to reduces to the equation of a circle
6.	When $0 < \delta < \frac{\pi}{2}$	$\frac{Y_1^2}{a_1^2} + \frac{Y_2^2}{a_2^2} = 1$ Where major axis of ellipse is $a_2' = a_1 \sqrt{1 + \left[2\frac{Y_1^2}{a_1^2} - 1\right] \cos^2 \delta + \frac{Y_1}{a_1}\sqrt{1 - \frac{Y_1^2}{a_1^2}} \sin 2\delta}$ and minor axis of the ellipse is $a_2' = a_2 \sqrt{1 + \left[2\frac{Y_1^2}{a_1^2} - 1\right] \cos^2 \delta + \frac{Y_1}{a_1}\sqrt{1 - \frac{Y_1^2}{a_1^2}} \sin 2\delta}$	Equation reduces to the equation of an ellipse
7.	Resultant amplitude	$R = \sqrt{a_1^2 + 2 a_1 a_2 \cos(\delta) + a_2^2}$	$Y = Y_1 + Y_2$
8.	Intensity	$I = I_1 + 2\sqrt{I_1 I_2} \cos(\delta) + I_2$	$I_1 = a_1^2$ and $I_2 = a_2^2$
9.	When $\cos\delta = 1$, I will be maximum	$I_{max} = I_1 + I_2 + 2\sqrt{I_1 I_2} = \left(\sqrt{I_1} + \sqrt{I_2}\right)^2$	
10.	When $\cos\delta = -1$, I will be minimum	$I_{min} = I_1 + I_2 - 2\sqrt{I_1 I_2} = \left(\sqrt{I_1} - \sqrt{I_2}\right)^2$	
11.	Maximum intensity of n coherent waves	$(I_{max})_{co} = \left(\sqrt{I_0} + \sqrt{I_0} + \sqrt{I_0} + ...\right)^2 = n^2 I_0$	
12	Average intensity due to two incoherent waves	$(I_R)_{inco} = I_1 + I_2$	

13.	Intensity of n identical incoherent waves	$(I_R)_{inco} = I_0 + I_0 + ... = n\, I_0$	

QUESTIONS FOR EXAMINATION

2.1 Derive the formula for Lissajous' figures.

2.2 Show that the equation for Lissajous' figures reduces to a straight line, a parabola and a circle.

2.3 Derive the formula for resultant amplitude of superposition of travelling waves having phase difference.

2.4 Derive the formula for intensity in case of Lissajous' figures.

2.5 Show that maximum intensity is obtained by superposition of two travelling waves having zero phase difference.

2.6 Show that minimum intensity is obtained by superposition of two travelling waves having phase difference π.

2.7 Drive formula for maximum intensity of n coherent waves.

2.8 Derive formula for intensity of n identical incoherent waves.

2.9 Derive the formula for average intensity of two incoherent waves.

2.10 Draw the Lissajous' figures by superposition of two waves of equal frequencyand phase difference of zero and $\pi/4$.

2.11 Draw the Lissajous' figures by superposition of two waves of having frequency ratio $\frac{f_x}{f_y} = \frac{2}{1}$ and for phase difference of zero and $\pi/4$.

2.12 Draw the Lissajous' figures by superposition of two waves of having frequency ratio $\frac{f_x}{f_y} = \frac{3}{2}$ and for phase difference of zero and $\pi/4$.

2.13 Draw the Lissajous' figures by superposition of two waves of having frequency ratio $f_x : f_y$ is 1:1, 2:1 and 3:2 and for phase difference of $\pi/2$.

2.14 Draw the Lissajous' figures due to superposition of two cosine signals having phase difference 0°, 30°, 45°, 60° and 90°.

2.15 What are the uses of Lissajous' figures?

OBJECTIVE QUESTIONS

2.1 The Lissajous' figure due to superposition of cosine waves of same frequency at 90° and phase difference 30° will be

(a) circle (b) ellipse (c) straight line (d) parabola

2.2 The Lissajous' figure due to superposition of cosine waves of same frequency at 90° and phase difference 45° will be

(a) circle (b) parabola (c) straight line (d) ellipse

2.3 The Lissajous' figure due to superposition of cosine waves of same frequency at 90° and phase difference 60° will be

(a) circle (b) parabola (c) ellipse (d) straight line

2.4 The Lissajous' figure due to superposition of cosine waves of same frequency at 90° and phase difference 90° will be

(a) circle (b) parabola (c) ellipse (d) straight line

2.5 The Lissajous' figure due to superposition of perpendicular travelling waves of same frequency and phase difference zero will be

(a) circle (b) parabola (c) ellipse (d) straight line

2.6 The Lissajous' figure due to superposition of perpendicular travelling waves of frequency ratio $\frac{f_x}{f_y} = \frac{2}{1}$ and phase difference zero will be

(a) circle (b) parabola (c) ellipse (d) straight line

2.7 The Lissajous' figure due to superposition of travelling waves of frequency ratio 1 : 1 at 90° and phase difference $\pi/4$ will be

(a) circle (b) parabola (c) ellipse (d) straight line

2.8 When phase difference is $\pi/4$ and frequency ratio $\frac{f_x}{f_y} = \frac{3}{2}$, then how many times Lissajous' figure crosses the vertical axis and the horizontal axis?

(a) 4 and 6 (b) 3 and 2 (c) 3 and 8 (d) 6 and 4, respectively

2.9 When phase difference is zero and frequency ratio is $\frac{f_x}{f_y} = \frac{3}{2}$, then how many times Lissajous' figure crosses the vertical axis and the horizontal axis

(a) 4 and 6 (b) 3 and 2 (c) 3 and 8 (d) 6 and 4, respectively

2.10 When phase difference is $\pi/4$ and frequency ratio is $\frac{f_x}{f_y} = \frac{2}{1}$, then how many times Lissajous' figure crosses the vertical axis and the horizontal axis

(a) 4 and 6 (b) 3 and 2 (c) 4 and 2 (d) 2 and 3, respectively

2.11 When phase difference is zero and frequency ratio is $\frac{f_x}{f_y} = \frac{2}{1}$, then how many times Lissajous' figure crosses the vertical axis and the horizontal axis?

(a) 1 and 2 (b) 2 and 1 (c) 2 and 4 (d) 3 and 2, respectively

2.12 Average intensity due to n incoherent waves is given by

(a) $n^2 I_0$ (b) $n I_0$ (c) $n \sqrt{I_0}$ (d) I_0

2.13 Average intensity due to n coherent waves is given by

(a) $n^2 I_0$ (b) $n I_0$ (c) $n \sqrt{I_0}$ (d) I_0

Answers

2.1 (b)	2.2 (d)	2.3 (d)	2.4 (a)	2.5 (d)	2.6 (b)	2.7 (c)	2.8 (d)
2.9 (b)	2.10 (c)	2.11 (b)	2.12 (b)	2.13 (a)			

CHAPTER 3

Wave Motion: General

3.1 INTRODUCTION

A wave which propagates in a medium continuously in the same direction without any change in its amplitude is called a *plane progressive wave*. A plane progressive wave is also called *progressive wave* or *travelling wave*. The propagation of the wavein the medium space is periodic, that is why the wave is called a *simple harmonic travelling wave* or just *simple harmonic wave*. The travelling waves can be both longitudinal waves and transverse waves. Mathematically, a travelling wave is described by sinusoidal function.

The travelling transverse waves can be produced by throwing a stone in a water pond as water waves called ripples and in a string by moving it (say) up and down with the help of the hand. The longitudinal waves or sound waves can be produced in an organ pipe by blowing air in it.

The speed of the travelling wave and the speed of particle oscillation is different. The speed of the wave decides the motion of energy in the form of wave in the medium. The speed of particle of the medium decides the speed of oscillation about its mean position. The particles of the medium do not move with the wave. They just vibrate periodically about their mean position. However, the oscillation of particles is responsible for transferring the energy (disturbance) in the form of waves in the medium.

When a longitudinal travelling wave is reflected from a rigid surface of a closed pipe, the reflected wave suffers a phase change of π, i.e., the incident and reflected waves are out of phase. When a longitudinal travelling wave is reflected from anopen end of the pipe, the reflected wave suffers no phase change, i.e., the incident and reflected waves are in the same phase. The incident and reflected sound waves on superposition producestanding waves. The special patterns of sound formed by the standing waves are called *normal modes* or *musical notes.*

The musical organ pipes like flute, whistle, clarinet, etc., are based on different notes produced by the overlapping of sound waves. Due to compression and rarefaction, the longitudinal sound waves are also called pressure waves. Analytical description various sound notes produced by organ pipes is given in this chapter.

3.2 PLANE PROGRESSIVE WAVE

The constituents of the medium vibrate periodically about their mean position. These are called *harmonic oscillators*. These harmonic oscillators transfer energy from one oscillator to another oscillator. The periodic vibrations of constituents of the medium and energy transfer from one oscillator to another oscillator can be represented by a wave. Since, the gap between oscillators is very small so distance between them can be considered almost continuous. Thus the process by which energy transfers from one point to another point in the medium is called a wave. The wave decides the displacement of the constituent of medium at every location (x) and at a given time (t) during its motion. The constituents of the medium vibrate periodically about their mean position and do not move in the direction of the wave. It is the energy (disturbance) that moves in the form of a wave.

A wave which propagates in a medium continuously in the same direction without any change in its amplitude is called a *plane progressive wave*. A plane progressive wave is also called *progressive wave* or *travelling wave*. The propagation of the wavein the medium space is periodic, that is why the wave is called a *simple harmonic plane progressive wave* or *simple harmonic progressive wave* or *simple harmonic travelling wave* or just *simple harmonic wave*. Mathematically, a travelling wave is described by sinusoidal function.

3.3 LONGITUDINAL AND TRANSVERSE WAVES

Energy (disturbance) transfer in the medium takes place in the form of waves. These waves can be divided into two categories: longitudinal waves and transverse waves.

3.3.1 Longitudinal Waves

In a longitudinal wave *i* particle of the medium oscillates in the direction of the motion of the propagation of the wave,for example, a sound wave is a longitudinal wave. Sound waves can propagate in air, gas and solids. During the propagation of sound waves, particles of the medium vibrate periodically (but do not move) about their mean position in the direction in which sound wave travels. Sound wavesmoves in the form of compression and rarefaction periodically. The particles get compressed in the region of compression. They get decompressed (rarefied) in the adjacent region of rarefaction. Motion of sound wave in the form of compression and rarefaction is shown in Fig. 3.1.

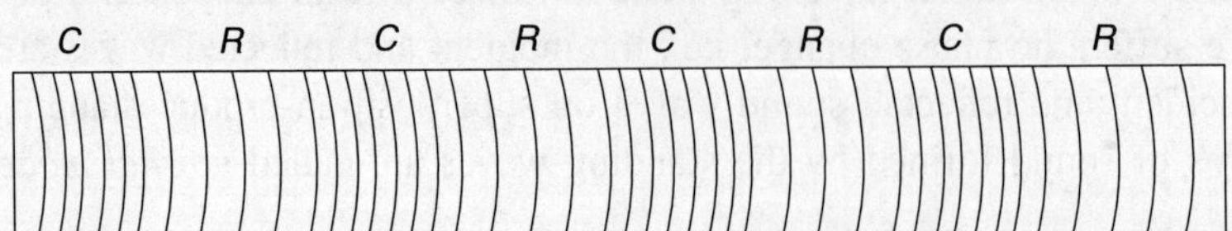

Fig. 3.1 Propagation of longitudinal waves. *C* represents a compression region and *R* represents a rarefaction region

3.3.2 Transverse Waves

In a transverse wave particle of the medium oscillates in perpendicular (i.e., transverse) direction to the direction of propagation of the wave. Figure 3.2 shows a transverse wave moving in x-direction. A transverse wavemoves in the form of crests and troughs, while the particles vibrates periodically in Y direction. One should note that the particles do not move in X direction, they just oscillate about their mean position. In fact, the disturbance or energy is transferred in the form of waves. For example, light, X-rays, γ-rays, radio waves, microwaves, string waves, etc., are transverse waves.

All electromagnetic waves are transverse waves. In this case the electromagnetic field oscillates and no material is necessary as they can move in vacuum.

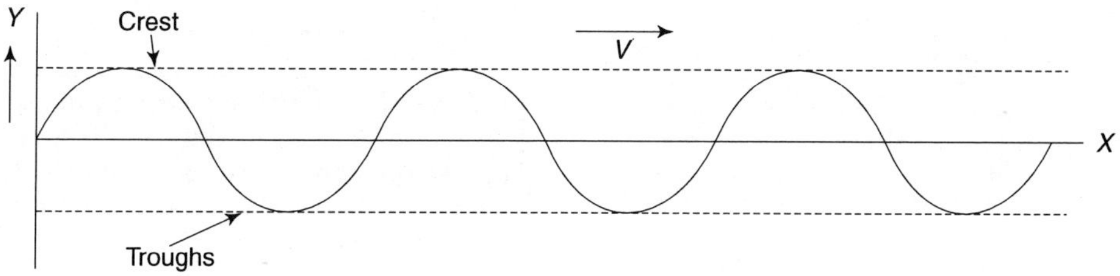

Fig. 3.2

3.4 WATER WAVE OR RIPPLES

Water waves are transverse waves. In the propagation of water waves, water molecules oscillate about their mean position. One should note that the particles do not move in the direction of wave motion, they just oscillate about their mean position in the direction perpendicular to the direction in which the water wave travels. Water waves move in the form of crests and troughs perodically. At the position of crest, the water molecules get raised above the normal level of the water surface, while the water molecules get depressed below the normal level of the water surface at the position of troughs.Alternate crests and troughs in the wave motion are shown in Fig. 3.3.

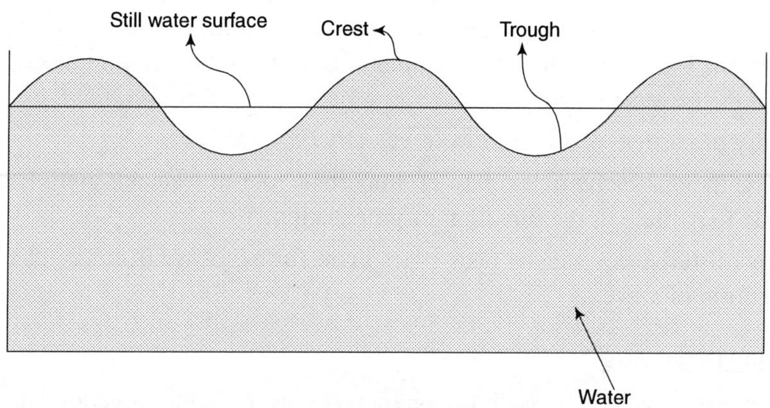

Fig. 3.3 Water waves or ripples

3.5 COMPARISON BETWEEN LONGITUDINAL AND TRANSVERSE WAVES

	Longitudinal waves	Transverse waves
1.	In longitudinal waves the constituents of the medium oscillate about their mean position in the direction of propagation of the wave.	In transverse waves the constituents of the medium oscillate about their mean position in the perpendicular direction of propagation of the wave.
2.	In longitudinal wave, the waves travel in the form of compression and rarefaction alternately.	In transverse wave, the wave travel in the form of crest and trough alternately.
3.	They need a medium to travel.	Some transverse waves need medium, while some transverse do not need medium.
4.	They cannot be polarized.	They can be polarized.
5.	For example, sound waves, waves in organ pipes, seismic waves.	For example, transverse strings and water waves need medium. All types of electromagnetic waves do not need medium. They can travel in vacuum such as light, radio waves, X-rays, γ-rays.

3.6 CHARACTERISTICS OF PROGRESSIVE WAVES

The following are the characteristics of progressive waves.

1. A progressive wave is the disturbance produced in the medium by the repeated periodic motion of the constituents of the medium.
2. It is the progressive wave which travels. The constituents of the medium do not move. They oscillate about their mean position.
3. There is continuous phase difference between the constituents of the medium when progressive wave moves.
4. The velocity of the progressive wave remains constant throughout the medium.
5. When a progressive wave propagates, each constituent of the medium oscillates about its mean position with the same amplitude and time period. The velocity of the constituent is different at different points. It is maximum at the mean position and zero at the extreme positions.
6. Velocity of particle is different from velocity of wave.
7. The oscillations of particles play an important role in transmission of energy from particle to particle in the direction of propagation.
8. Energy (disturbance) transfer takes place in the form of wave from one place to another in the form of wave.

3.7 TRAVELLING WAVE

A wave which propagates in a medium continuously in the same direction without change in its amplitude is called a *plane progressive wave*. A plane progressive wave is also called *progressive wave* or *travelling wave*. The propagation of the wavein the medium space is periodic, that is why the wave is called a *simple harmonic travelling wave* or just *simple harmonic wave*. Mathematically, a travelling wave is described by sinusoidal function.

Wave equation

Suppose, a travelling simple harmonic wave is moving in a medium along positive direction of x-axis. The variation of displacement as a function of position at any particular time is shown in Fig. 3.4.

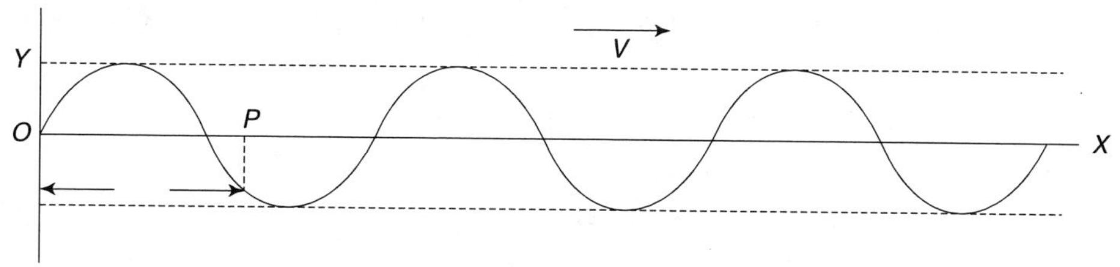

Fig. 3.4

When a wave, moving in the positive direction of x-axis, passes through the position of the particle at origin O (where, $x = 0$), then simple harmonic motion of the displacement (Y) of the particle about mean, position at O along y-direction, at any time t is given by equation $Y = A \sin \omega t$,

where, maximum displacement of the particle along y-direction A is called the amplitude of the wave and ω is the angular velocity of periodic vibration of the particle about its mean position.

Alternative forms of the wave equation

During motion of the wave (say) along x-axis, successive particles start oscillating about their mean positions at a fixed time later than the time of preceding particle. Therefore, as wave move away from O, the phase of the oscillation of particles with respect to the particle at O goes on increasing. For example, if phase change of the wave at O is zero, then phase of the wave at distance $\lambda/4$, $\lambda/2$, $3\lambda/4$ and λ is $\pi/2$, π, $3\pi/4$ and 2π, respectively. If ϕ is the phase change of the particle P at distance x from the origin O, then the displacement of the particle P at any instant (t) is given by

$$Y = A \sin (\omega t - \phi)$$

At distance λ, there is phase change of 2π

Therefore, at distance x, there will be phase change, $\frac{2\pi}{\lambda} x = kx$

where, $k = \frac{2\pi}{\lambda}$ is called the propagation constant. Now wave equation can be written as

$$Y = A \sin (\omega t - kx) \quad \text{...(3.1)}$$

Since $\omega = 2\pi f = \frac{2\pi}{T}$, where, T is the time period of the oscillation of the particle, Eq. (3.1) can be written as

$$Y = A \sin\left(\frac{2\pi}{T}t - \frac{2\pi}{\lambda}x\right)$$

or
$$Y = A \sin 2\pi\left(\frac{t}{T} - \frac{x}{\lambda}\right) \quad ...(3.2)$$

The frequency of the particle oscillation $f = \frac{v}{\lambda} = \frac{1}{T}$. Putting for $\frac{1}{T}$, Eq. (3.2) becomes

$$Y = A \sin 2\pi\left(\frac{v}{\lambda}t - \frac{x}{\lambda}\right)$$

or
$$Y = A \sin \frac{2\pi}{\lambda}(vt - x) \quad ...(3.3)$$

Equations (3.1–3.3), are the equations of a plane wave progressive simple harmonic wave moving along the positive x-direction.

If the wave travelling along the negative direction of x-axis, the Eqs. (3.1–3.3), are written as

$$Y = A \sin(\omega t + kx)$$

$$Y = A \sin 2\pi\left(\frac{t}{T} + \frac{x}{\lambda}\right)$$

$$Y = A \sin \frac{2\pi}{\lambda}(vt + x)$$

In these equations, v is the speed of propagation of the wave, not the speed of particle.

For convenience let us consider that Eq. (3.1) is a transverse travelling wave moving in +ve x-direction. If ϕ is the phase difference between the travelling wave given by Eq. (3.1) and another travelling wave along +ve x-direction, then the equation of the other wave will be

$$Y = A \sin(\omega t - kx + \phi) \quad ...(3.4)$$

Equation (3.4) has all the properties of the travelling wave.

Evaluation of phase difference, ϕ

The constant ϕ, the argument of the sine function, can be obtained by using the identity

$$\sin(C + D) = \sin C \cos D + \cos C \sin D$$

to get

$$Y(x, t) = a[\sin(\omega t - kx) + \cos(\omega t - kx) \sin\phi]$$
$$= a \cos\phi \sin(\omega t - kx) + a \sin\phi \cos(\omega t - kx)$$
$$= A \sin(\omega t - kx) + B \cos(\omega t - kx) \quad ...(3.5)$$

That is, travelling wave is the combination of sine and cosine functions.

In Eq. (5), $A = a \cos \phi$ and $B = a \sin \phi$, so that $\tan \phi = \frac{B}{A}$ or $\phi = \tan^{-1}\left(\frac{B}{A}\right)$ and

$a = \sqrt{A^2 + B^2}$

Displacement of the constituents of medium at time (*t*) and position, *x*

To understand that Eq. (3.4) represents a sinusoidal travelling wave, let us consider a fixed time, $t = t_0$. Then the argument sine function in Eq. (4) is simply kx + constant. Thus, the shape of the travelling wave is a sine wave of function x at any fixed time, t_0. Now, let us consider a fixed location, $x = x_0$. Then the argument of sine function in Eq. (3.4) is simply constant, ωt. Thus, displacement Y varies sinusoidally with t at any fixed location, x_0. That is, the constituents of the medium at different positions execute simple harmonic motion. Finally, as t increases x must increase in the positive direction to keep $(kx - \omega t + \phi)$ constant. Thus, Eq. (3.4) represents a sinusoidal (harmonic or periodic) wave travelling along the positive direction of x-axis. On the other hand, the function

$$Y(x, t) = a \sin(\omega t + kx + \phi) \quad \text{...(3.6)}$$

represents a travelling wave in the negative direction of x-axis. Here, $Y(x, t)$ is the displacement of constituents of medium at position x and t a is amplitude of the wave. ω is the angular velocity. k is the angular wave number. Φ is initial phase angle at $x = 0$ and $t = 0$.

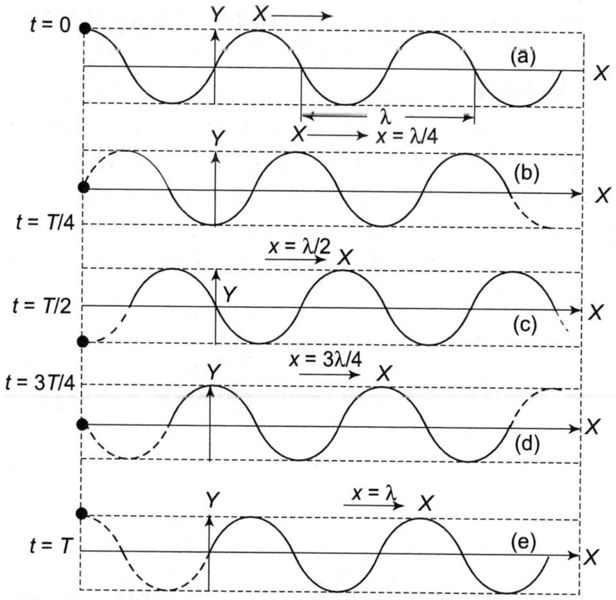

Fig. 3.5 A harmonic wave progressing along the positive of x-axis at different times

Figure 3.5 shows the plot of Eq. (3.4) for different values at time intervals $T/4$. Here, T is time period. In a wave, crest is the point of maximum positive displacement, the trough is the point of maximum negative displacement. To see how a wave travels, we can fix attention on a crest and see how it moves with time. In the figure, this is shown by a cross (×) on the crestat $t = 0$. We can see the motion of a particular constituent of the medium at a fixed location, say at the origin of the x-axis. This is shown by a solid dot (•) at maximum displacement ($Y = a$). The plots of Fig. 3.5 show that with time, the solid dot (•) at the origin moves periodically, i.e., the particle at the origin oscillates about its mean position as the wave progresses. At time $t = T/4$, (×) moves to the right by distance ($\lambda/4$) and dot (•) moves vertically down at $Y = 0$. At time $t = T/2$, (×) moves to the right by distance ($\lambda/2$) and dot (•) moves further downwards to maximum displacement, i.e., $Y = -a$. At time $t = T/2$, (×) moves to right by distance ($\lambda/2$) and dot (•) moves upwards at $Y = a$. At time $t = 3T/4$, (×) moves to right by distance ($3\lambda/4$) and dot (•) moves upwards to $Y = 0$. At time $t = T$, (×) moves to the right by distance (λ) and dot (•) moves upwards to maximum displacement, $Y = a$. During the time the solid dot (•) has completed one full oscillation, the crest has moved further by a distance. This is true for constituents at any other location also.

3.8 SPEED OF THE WAVE AND SPEED OF CONSTITUENT OF MEDIUM

The speed of the wave, $v = \dfrac{dx}{dt}$, is the speed of the energy disturbance movingalong x-direction (in one dimension), while speed of the constituent of the medium, $u = \dfrac{dY}{dt}$, is the rate of displacement of the constituent in y-direction or perpendicular to the direction of wave motion.

Speed of the travelling wave

To obtain the speed of propagation of a travelling wave, let usconsider a travelling wave at time t as shown in Fig. 3.6. In time interval Δt, the entire wave pattern shifts to the right (positive direction of x-axis) by a distance Δx. It is convenient to look at the motion of the crest of the wave. In particular, the crest shown by A moves a distance Δx in time Δt. The speed of the wave v is then $\Delta x/\Delta t$. The wave will move with the same speed v (otherwise the wave pattern will not be fixed).

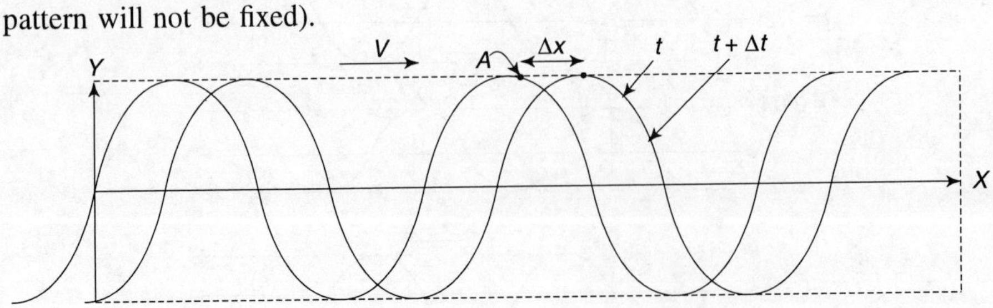

Fig. 3.6 Progression of a harmonic wave from time t to $t + \Delta t$, where, t is a small interval. The wave pattern as a whole shifts to the right. The crest of the wave for a point with any fixed phase moves right by distance Δx in time Δt

As the wave moves, each point on the wave such as A retains its displacement Y. The points on the string do not retain their displacement, but points on the waves do so. For each wave point, the argument of sine function must be constant, i.e.

$$kx - \omega t = \text{constant} \quad ...(3.7)$$

As time increases from t to $t + \Delta t$, the position of point x also increases to $x + \Delta x$ in positive x-direction for keeping the argument of the wave constant. So, we can write

$$\omega t - kx = k\,(x + \Delta x) - \omega\,(t + \Delta t)$$

or

$$k\Delta x - \omega \Delta t = 0$$

Taking Δx and Δt vanishingly small, this equation gives

$$\frac{dx}{dt} = \frac{\omega}{k}$$

$$\frac{dx}{dt} = v = \frac{\omega}{k} \quad ...(3.8)$$

Putting $\omega = 2\,\pi f = \frac{2\pi}{T}$ and $k = \frac{2\pi}{\lambda}$, Eq. (3.8) becomes

or

$$v = \frac{\omega}{k} = \frac{\frac{2\pi}{T}}{\frac{2\pi}{T}} = \frac{\lambda}{T} = \lambda f \quad ...(3.9)$$

Equation (3.9), a general relation for all progressive wave. T is time period of the wave or the time for one oscillation of wave.

When time period of wave is T, the distance moved by the wave $vT = \lambda$ (wavelength of the wave). That is, Eq. (3.9) indicates that the wave pattern moves a distance equal to one wavelength in the time T of full oscillation. The velocity of travelling wave is same in one medium. But it is different for different medium. That is, the speed of a mechanical wave depends on mass density in general (linear mass density for strings or wires) and elastic properties (Young's modulus for linear medium like strings or wires, bulk modulus in general) of the medium. Equation (3.9) relates wavelength λ to frequency f for the fixed speed of the wave. In the medium both transverse and longitudinal waves can travel. The transverse and longitudinal waves will have different speeds of in a medium.

Speed of constituent of medium of travelling wave

The travelling wave is written as

$$Y = Y_m \sin\,(\omega t - kx) \quad ...(3.10)$$

As time increases from t to $t + \Delta t$, the vertical displacement at position x also increases to $Y + \Delta Y$. In $t \to 0$ limit, we can write

$$\lim_{t \to \infty} \frac{\Delta Y}{\Delta t} = \frac{dY}{dt}$$

Differentiating Eq. (3.10) w.r.t. t, we get

$$\frac{dY}{dt} = \omega Y_m \cos(\omega t - kx)$$

The transverse speed, $u = \frac{dY}{dt}$ of the string element will be

or $$u = \omega Y_m \sqrt{1 - \sin^2(kx - \omega t)}$$

or $$u = \omega Y_m \sqrt{1 - \frac{Y^2}{Y_m^2}}$$

or $$u = \omega \sqrt{Y_m^2 - Y^2}$$

or $$u = \omega r$$

Here ω is the angular velocity of the particle $u = +\omega r$ is the linear velocity of the particle. Plus sign shows that the particle is moving in clockwise direction. $u = -\omega r$ is the linear velocity of the particle, when it is moving in anticlockwise direction. One should note that while taking differentiation of Y w.r.t. time, the position of the constituent is taken constant.

When particle is at position $x = 0$, the equation of travelling wave is

$$Y = Y_m \sin(\omega t)$$

Differentiating this equation w.r.t. t, we will get velocity of particle, $u = -\omega r$. This conclusion is true for the particle at any position x.

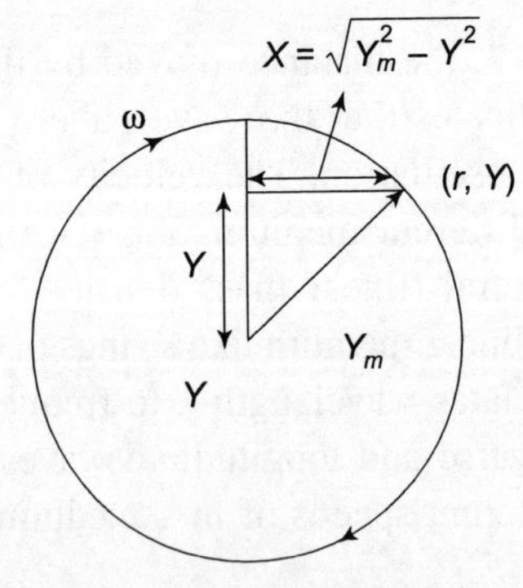

Fig. 3.7 Shows a periodic motion of particle about its mean position

Case 1. When $Y = 0$, i.e., when, vertical displacement of particle is zero

$$u_{max} = \omega Y_m$$

i.e., speed of the particle u is maximum when $Y = 0$, i.e., when displacement of particle is zero.

Case 2. When $Y = Y_m$, i.e., when, vertical displacement of particle is maximum

$$u_{min} = 0$$

i.e., speed of the particle u is minimum when displacement of particle is equal to amplitude of the wave.

From these two cases we can infer that the particle vibrates periodically about its mean position along the perpendicular direction of wave motion. The particle exhibits simple harmonic motion. The travelling wave is a sinusoidal wave and represents the simple periodic harmonic motion. The vertical displacement Y of the particle and position of travelling wave at $t = 0$, $T/2$, $3T/4$, T is shown in Fig. 3.5.

3.9 WAVE PACKET

Wave particle duality principle was put forward by de Broglie. That is, electron has both particle and wave nature. Therefore, sometimes electron behaves like a particle and at other occasions it behaves like a wave. First of all Schrödinger explained the dual nature by representing the motion of particle by a wave packet. A *wave packet* is the resultant of superposition of a group of waves with nearly the same wavelength and same amplitude that interfere constructively over localized region of space where the particle is located. Everywhere out of this localized region of space they interfere destructively making the amplitude zero. The velocity of motion of wave packet is called *group velocity*.

Let us consider two waves the motion of particle y_1 and y_2 having same amplitude A but having different angular frequencies (ω or ω') and phase velocities (v or v'):

$$y_1 = A \sin (kx - \omega t) \quad ...(3.11)$$

$$y_2 = A \sin (k'x - \omega' t) \quad ...(3.12)$$

where, wave vector, $k = \dfrac{2\pi}{\lambda}$ and $k' = \dfrac{2\pi}{\lambda'}$. The resultant wave is given by

$$y = y_1 + y_2 = A \sin (kx - \omega t) + \sin (k'x - \omega' t) \quad ...(3.13)$$

Using the trigonometric formula

$$\sin C + \sin D = 2 \sin \frac{C + D}{2} \cos \frac{C - D}{2}$$ Eq. (3.13) becomes

$$y = 2A \sin \frac{(k + k')x - (\omega + \omega')t}{2} \cos \frac{(k - k')x - (\omega - \omega')t}{2}$$

Taking $\dfrac{\omega + \omega'}{2} \cong \omega$, $\dfrac{k + k'}{2} \cong k$, $\omega - \omega' = d\omega$, $k - k' = dk$

$$y = 2A \sin (kx - \omega t) \cos \left(\frac{dx}{2} x - \frac{d\omega}{2} t\right)$$

or

$$y = 2A \cos\left(\frac{dx}{2} x - \frac{d\omega}{2} t\right) \sin (kx - \omega t) \quad ...(3.14)$$

Equation (3.14) is the analytical expression for the group of waves, i.e., the *wave packet* formed by a *group of waves*.

Equation (3.14) shows that the resultant wave is an amplitude modulated wave of angular frequency (ω) and propagation vector, *k*. The *phase velocity* of the wave is

$$v_p = \frac{\omega}{k}$$

The amplitude of the wave,

$$R = 2\,A \cos\left(\frac{dx}{2}x - \frac{d\omega}{2}t\right) \qquad ...(3.15)$$

which varies sinusoidally as shown in Fig. 3.8.

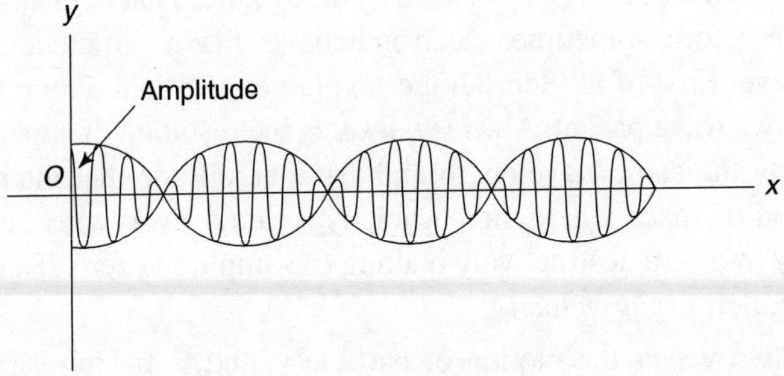

Fig. 3.8

This variable of amplitude produces successive wave groups.

3.10 PHASE VELOCITY AND GROUP VELOCITY

Phase velocity

The velocity of an individual wave is called the phase velocity. It is also the velocity with which the each component wave moves inside the wave packet. Let x be the position of a point on the wavefront of the wave. The phase velocity (v_p) is defined as

$$v_p = \frac{dx}{dt} \qquad ...(3.16)$$

The argument of travelling wave defined by Eq. (3.11) is

$$kx - \omega t = \text{constant}$$

Differentiating it w.r.t. t

$$\frac{dx}{dt} = \frac{\omega}{k}$$

or

$$v_p = \frac{\omega}{k} \qquad ...(3.17)$$

If f is the frequency of the wave and wave vector, $k = \dfrac{2\pi}{\lambda}$, where λ is the wavelength of the wave, we can also get Eq. (3.17) by using the relation $\omega = 2\pi f$ and $v = f\lambda$ as follows:

$$\omega = 2\pi \frac{v}{\lambda} = k\,v \quad \text{or } v = \frac{\omega}{k}$$

Here v is the linear velocity of the wave. That is, phase velocity of wave is same as its linear velocity.

Group Velocity

The velocity with which the resultant of group of wave or packet moves is called the *group velocity*, v_g. At $x = 0$ and $t = 0$, the maximum amplitude as given by Eq. (3.14), is

$$R_{max} = 2\,A$$

In general the condition from Eq. (3.14) for movement of wave of amplitude ($2A$) by one cycle is

$$\frac{dk}{2}x - \frac{d\omega}{2} = 0$$

or

$$\frac{dx}{dt} = \frac{d\omega}{dk}$$

or

$$v_g = \frac{d\omega}{dk} \qquad ...(3.18)$$

Movement of wave packet is depicted in Fig. 1.10. The velocity of wave packet propagated in medium is called group velocity. A wave packet in quantum mechanics represents the classical particle.

If the wave packet moves like a classical particle, which follows Newtonian mechanics, the group velocity (v_g) must begiven by classical relation

$$\frac{d\omega}{dk} = \frac{p}{m} \qquad ...(3.19)$$

where, $p = \hbar k$ is the momentum of particle and m is its mass. Now Eq. (3.19) can be written as

$$\frac{d\omega}{dk} = \frac{\hbar k}{m} \qquad ...(3.20)$$

Integrating Eq. (3.20)

$$\int d\omega = \frac{\hbar}{m}\int k\,dk + \text{constant}$$

or

$$\omega = \frac{\hbar k^2}{2m} + \text{constant}$$

or
$$\hbar\omega = \frac{\hbar^2 k^2}{2m} + C \qquad ...(3.21)$$

The first term in r.h.s. of Eq. (3.21) is kinetic energy of the particle. The second term C is integration constant and has dimensions of energy. Constant C may be interpreted as the *potential energy* of particle. Equation (3.21) may also be written as

$$\hbar\omega = \frac{\hbar^2 k^2}{2m} + V(r) = E \qquad ...(3.22)$$

Total energy (E) of the particle will be, if the angular frequency (ω) of the wave remains constant during the motion of the particle. Similarly, E will remain constant, if potential energy ($V(r)$) and velocity $\left(v = \frac{\hbar k}{m}\right)$ of the particle are constant. All these conditions are satisfied when an electron moves in circular orbits around the nucleus.

Problem 3.1 The harmonic wave given by $y_1 = 12 \sin(5\pi - 4x)$ and $y_2 = 16 \sin(7\pi - 6x)$ are superposed to produce a single wave. Find the phase speed of each wave and group velocity of the resultant wave.

Solution: Formula required

Phase velocity

$$v_p = \frac{\omega}{k}$$

Group velocity

$$v_g = \frac{d\omega}{dk}$$

Given: The frequency of wave y_1 is $\omega_1 = 5\pi$

The frequency of wave y_y is $\omega_2 = 7\pi$

The wave vector of wave y_1 is $k_1 = 4$

The wave vector of wave y_2 is $k_2 = 6$

Calculation:

Phase velocity of y_1 wave is

$$v_p^1 = \frac{\omega_1}{k_1} = \frac{5\pi}{4} = 3.925 \text{ cm/s, if } x \text{ is taken in centimetres.}$$

Phase velocity of y_2 wave is

$$v_p^2 = \frac{\omega_2}{k_2} = \frac{7\pi}{6} = 3.663 \text{ cm/s, if } x \text{ is taken in centimetres}$$

Group velocity of resultant wave

$$v_g = \frac{d\omega}{dk} = \frac{\omega_2 - \omega_1}{k_2 - k_1} = \frac{7\,\pi - 5\,\pi}{6-4} = \pi = 3.14 \text{ cm/s if } x \text{ is taken in centimetres.}$$

Ans. The phase velocity of the y_1 wave is 3.925 cm/s, if x is taken in centimetres.

The phase velocity of the y_2 wave is 3.663 cm/s, if x is taken in centimetres.

The group velocity of the resultant wave is 3.14 cm/s, if x is taken in centimetres.

Relation between group velocity and phase velocity

Group velocity, $v_g = \dfrac{d\omega}{dk}$ ...(3.23)

Phase velocity, $v_p = \dfrac{\omega}{k}$ ...(3.24)

or $$\omega = v_p^k \quad \text{or} \quad d\omega = dv_p\, k + v_p\, dk$$

or $$\frac{d\omega}{dk} = \frac{dv_p}{dk} k + v_p \qquad \text{...(3.25)}$$

The group velocity

$$v_g = v_p + k\frac{dv_p}{d\lambda} \times \frac{d\lambda}{dk} = v_p + k\frac{dv_p}{d\lambda} \times -\frac{2\pi}{k^2} \text{ as } \lambda = \frac{2\pi}{k}$$

or $$v_g = v_p - \frac{2\pi}{k}\frac{dv_p}{d\lambda}$$

This equation shows that v_g is less than v_p when the medium is dispersive, i.e., when the phase velocity (v_p) of each wave in the group is different. Whereas, in a non-dispersive medium, each wave travels with the constant speed. Hence, in such medium

$$\frac{dv_p}{d\lambda} = 0$$

Therefore, in non-dispersive medium, $v_g = v_p$. This result is true for electromagnetic waves in vacuum and elastic waves in a homogeneous medium.

Case 1. $\dfrac{dv_p}{d\lambda} = 0 \Rightarrow v_g = v_p$

If the phase velocity of the component waves is independent of the wavelength, i.e., $\dfrac{dv_p}{d\lambda} > 0$, then group velocity and phase velocity are equal as in the case of electromagnetic waves propagating in vacuum. In this case, wave packet moves with the velocity of the component waves making up the wave packet, $\dfrac{dv_p}{d\lambda} = 0$ in non-dispersive media.

Case 2. $\frac{dv_p}{d\lambda} > 0 \Rightarrow v_g < v_p$

If the phase velocity of the component waves increases with increase of the wavelength, i.e., $\frac{dv_p}{d\lambda} > 0$, the group velocity is less than the phase velocity as the case of an electromagnetic wave propagating in a transparent medium like water or glass. In this case, the wave packet moves with a velocity less than that of the component waves making up the wave packet. When white light passes through a glass prism, constituent wavelengths are separated out and travel through the prism with different phase velocities. This phenomenon is called *normal dispersion* and the medium is called a *normal dispersive medium.*

Case 3. $\frac{dv_p}{d\lambda} < 0 \Rightarrow v_g > v_p$

If the phase velocity of the component waves increases with decrease of the wavelength, i.e., $\frac{dv_p}{d\lambda} < 0$, then the group velocity is more than the phase velocity as in the case of an electromagnetic wave propagating in an ionized media. In this case, the wave packet moves with a speed more than that of component waves making up the wave packet. This phenomenon is called an *anomalous dispersion* and the medium is called *anomalous dispersive medium.*

3.11 DIFFERENTIAL FORM OF WAVE EQUATION

The wave equation is given by Eq. (3.1):

$$Y(x, t) = a \sin(\omega t - kx) \qquad ...(3.26)$$

Differentiating partially Eq. (3.26) w.r.t. position x, we get

$$\frac{\partial Y}{\partial x} = -ak \cos(\omega t - kx)$$

Further differentiating this equation partially w.r.t. position x, we get

$$\frac{\partial^2 Y}{\partial x^2} = -ak^2 \sin(\omega t - kx) = -k^2 Y \qquad ...(3.27)$$

Differentiating partially Eq. (3.26) w.r.t. time variable t, we get

$$\frac{\partial Y}{\partial t} = 2\, a\, \omega \cos(\omega t - kx)$$

Further differentiating this equation partially w.r.t. time variable t reduces to

$$\frac{\partial^2 Y}{\partial t^2} = -a\omega^2 \sin(\omega t - kx) = -\omega^2 Y \qquad ...(3.28)$$

Combining Eqs. (3.27) and (3.28), we get

$$\frac{\omega^2}{k^2}\frac{\partial^2 Y}{\partial x^2} = \frac{\partial^2 Y}{\partial t^2}$$

$$\frac{\partial^2 Y}{\partial x^2} = \frac{1}{v^2}\frac{\partial^2 Y}{\partial t^2} \qquad ...(3.29)$$

where, $v = \frac{\omega}{k}$ is the velocity of the wave.

In three dimensions, wave equation is written as

$$\nabla^2 Y = v^2 \frac{\partial^2 Y}{\partial t^2} \qquad ...(3.30)$$

where, $\nabla^2 = \frac{\partial^2}{\partial x^2} + \frac{\partial^2}{\partial y^2} + \frac{\partial^2}{\partial z^2}$ is a Laplacian operator.

The general form of differential wave equation in one dimension is written as

$$\frac{\partial^2 Y}{\partial x^2} = \frac{1}{v^2}\frac{\partial^2 Y}{\partial t^2} + \frac{1}{\xi}\frac{\partial Y}{\partial t} \qquad ...(3.31)$$

and in three dimensions is given by

$$\nabla^2 Y = v^2 \frac{\partial^2 Y}{\partial t^2} + \frac{1}{\xi}\frac{\partial Y}{\partial t} \qquad ...(3.32)$$

In these equations ξ is called diffusivity. It takes of the energy loss of the wave in the medium.

3.12 DISPLACEMENT AND PRESSURE OF LONGITUDINAL WAVES

A sound wave is a longitudinal mechanical wave. It can be described in terms of the longitudinal displacement of the particles of the medium (called displacement wave). It can also be described in terms of the excess pressure generated due to compression and rarefaction (called pressure wave).

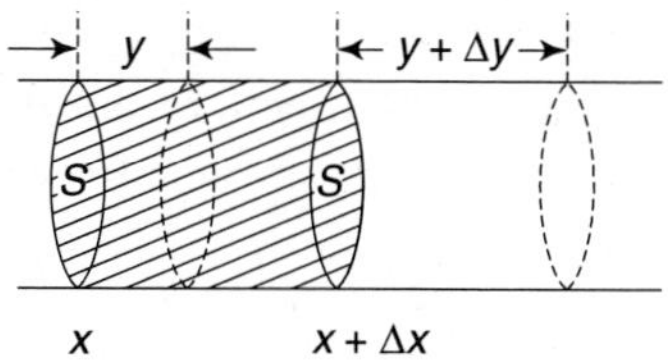

Fig. 3.9

Let us consider a sound wave travelling in the +ve x-direction in a medium. At time t and position x, the particle is displaced by amount Y in the x-direction. Then the displacement wave will be described by

$$Y = A \sin(\omega t - kx) \qquad ...(3.33)$$

Let us assume that the element of medium is confined within x and $x + \Delta x$. If S is the area of cross section, the volume element within Δx will be $V = S\,\Delta x$. As the wave passes, the ends at x and $x + \Delta x$ are displaced by amounts Y and $Y + \Delta Y$, so that increase in volume of the element will be $\Delta V = S\,\Delta Y$. Therefore, volume strain for the element under consideration will be

$$\frac{\Delta V}{v} = \frac{S\,\Delta Y}{S\,\Delta x} = \frac{\Delta Y}{\Delta x} \qquad ...(3.34)$$

The bulk modulus of medium is defined as

$$B = \frac{p}{\dfrac{-\Delta V}{v}} \qquad ...(3.35)$$

Then stress or excess pressure will be

$$p = B\left[\frac{-\Delta V}{v}\right] \qquad ...(3.36)$$

Putting the value of $\dfrac{\Delta V}{v} = \dfrac{\Delta Y}{\Delta x}$ in Eq. (3.36), we get

$$p = -B\frac{\Delta Y}{\Delta x} \qquad ...(3.37)$$

From Eq. (3.33), we can write

$$\frac{\Delta Y}{\Delta x} = -Ak\cos(\omega t - kx)$$

or

$$p = AkB\cos(\omega t - kx)$$

or

$$p = p_0\cos(\omega t - kx) \qquad ...(3.38)$$

where, initial pressure is $p_0 = AkB$

From Eqs. (1) and (6) it is clear that:

1. A sound wave may be considered either a displacement wave, $Y = A$ sun $(\omega t - kx)$ or pressure wave, $p = p_0 \cos(\omega t - kx)$.
2. The pressure wave and displacement wave are out of phase by $\pi/2$ with each other. The displacement wave will be maximum when pressure wave is minimum or vice versa. This is clear from Fig. 3.10.
3. The amplitude of pressure wave is

$$p_0 = AkB = Ak\rho v^2 \qquad \text{as} \quad v = \sqrt{\frac{B}{\rho}}$$

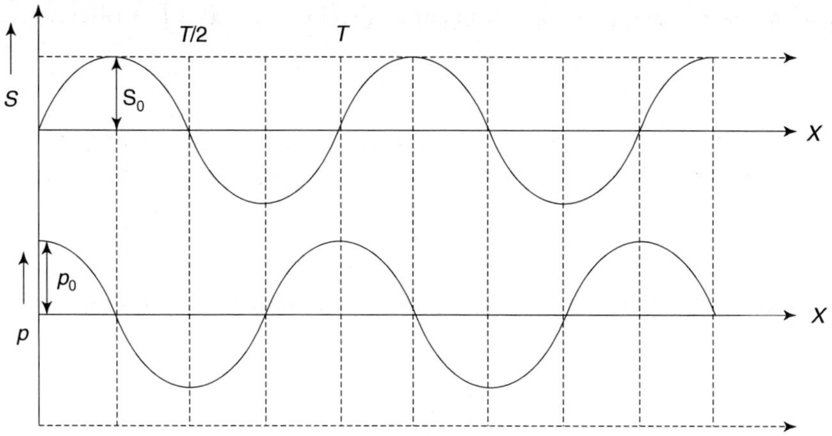

Fig. 3.10

or $$p_0 = AkB = \rho v A\omega \quad \text{as} \quad v = \frac{\omega}{k} \qquad ...(3.39)$$

4. Sound sensors like ear or mike detect pressure changes. Pressure wave is preferred over displacement wave to describe the sound.

3.13 MODES OF VIBRATIONS IN ORGAN PIPES

The simplest organ pipe is a cylindrical tube of uniform diameter. One end of an organ pipe is always open while the other may be closed or open. The pipe which is closed at one end is called 'closed end organ pipe'. The pipe which is open at both ends is called 'open end organ pip'. If both ends of the organ pipe are closed, then it becomes a cavity.

Normally, organ pipe is filled with air. Closed and open pipes are used to produce different sounds by vibration of air column. Such sound producing tubes are called organ pipes. For example, flute, whistle, clarinet, etc., are the instruments (organ pipes) which are used to produce different sounds.

3.13.1 Longitudinal Standing Waves and Normal Modes in a Closed Organ Pipe

A closed organ pipe of length L lying along the x-axisis shown in Fig. 5.30. The closed end of the pipe is $x = 0$ and the open end of the pipe is at $x = L$. When it is blown from its open end (at $x = L$), a longitudinal (sound) wave travels in the air of the pipe towards the closed end (at $x = 0$). The closed end reflects longitudinal waves back towards the open end, when the incident and the reflected waves moving in opposite directions in the air column superpose and form stationary longitudinal waves. At the closed end of the pipe, the air particles have no freedom to vibrate. Hence, a node is formed at the closed end. On the other hand, the air particles have the highest freedom to vibrate at the open end of the pipe. Therefore, an antinode is formed at the open end. Because of the boundary conditions the air in an organ pipe closed at one end oscillates only in some special patterns which are called *normal modes.*

The incident wave moving along the negative direction of *x*-axis is written as

$$Y_1 = a \sin \frac{2\pi}{\lambda} (vt + x)$$

The wave reflected at the closed end of the pipe suffers a phase change of π and moves along the positive direction of *x*-axis. Therefore, the reflected wave is written as

$$Y_2 = a \sin \left[\frac{2\pi}{\lambda} (vt - x) + \pi \right] = - a \sin \frac{2\pi}{\lambda} (vt - x)$$

After superposition, the resultant displacement (*Y*) at time *t* and position *x* will be linear sum of two, i.e.,

$$\begin{aligned} Y &= Y_1 + Y_2 \\ &= a \left[\sin \frac{2\pi}{\lambda} (vt + x) - \sin \frac{2\pi}{\lambda} (vt - x) \right] \\ &= 2a \cos \left(\frac{2\pi\, vt}{\lambda} \right) \sin \left(\frac{2\pi\, x}{\lambda} \right) \end{aligned} \qquad ...(3.40)$$

At the closed end of the pipe, i.e., at $x = 0$, Eq. (3.40) can be written as

$$Y = 2a \cos \left(\frac{2\pi\, vt}{\lambda} \right) \sin (0) = 0 \qquad \text{as } \sin (0) = 0 \qquad ...(3.41)$$

for all values of *t*. Since displacement $Y = 0$, so a node is formed at the closed end.

At the open end of the pipe, i.e., at $x = L$, Eq. (3.40) is written as

$$Y = 2a \cos \left(\frac{2\pi\, vt}{\lambda} \right) \sin \left(\frac{2\pi\, L}{\lambda} \right) \qquad ...(3.42)$$

The displacement (*Y*) will be maximum, when

$$\sin \left(\frac{2\pi\, L}{\lambda} \right) = \pm 1$$

This equation implies that

$$\frac{2\pi\, L}{\lambda} = (2m - 1) \frac{\pi}{2}, \text{ where } m = 1, 2, 3, ...$$

or

$$\lambda = \frac{4\, L}{2\, m - 1} \qquad ...(3.43)$$

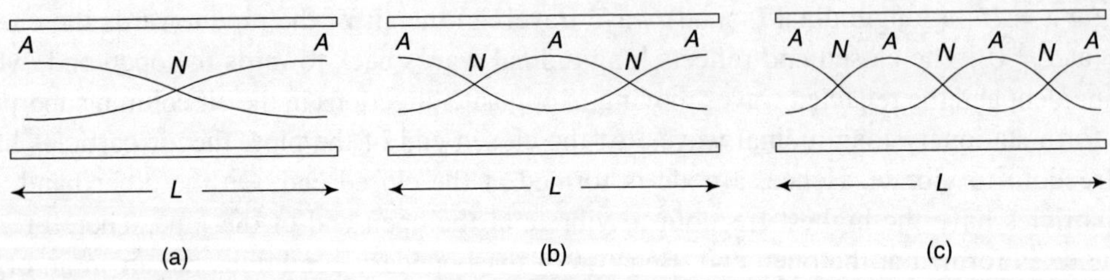

Fig. 3.11

First mode of vibration. For $m = 1$ and $\lambda = \lambda_1$ from Eq. (3.43), we get

$$\lambda_1 = 4L \qquad \text{or } L = \frac{\lambda_1}{4}$$

where, λ_1 is the wavelength of the standing waves set up in the air column for $m = 1$. Mode vibration for $m = 1$ is shown in Fig. 3.11 (a). An antinode (*A*) is formed at the open end and a node (*N*) is formed at the closed end. The frequency of first mode of vibration is

$$f_1 = \frac{v}{\lambda_1} = \frac{v}{4L} \qquad \text{...(3.44)}$$

where, v is velocity of sound. This is the lowest frequency of mode of vibrations of the air column in the closed pipe and is called the *fundamental, frequency*. The sound note produced in the closed pipe is called the *fundamental tone* or the *first harmonic* of the closed pipe. It is produced when air is blown 'lightly' at the open end of the pipe.

Second mode of vibration. For $m = 2$ and $\lambda = \lambda_2$ from Eq. (3.43), we get

$$\lambda_2 = \frac{4L}{3} \qquad \text{or} \quad L = \frac{3\lambda_2}{4}$$

where, λ_2 is the wavelength of the standing waves set up in the air column for $m = 2$.

The second mode of vibration is shown in Fig. 3.11(b). In this mode, two additional modes [one antinode (*A*) and one node (*N*)] are formed between closed and open ends of the pipe. The frequency of vibration of second mode is

$$f_2 = \frac{v}{\lambda_2} = \frac{3v}{4L} = 3 f_1 \qquad \text{...(3.45)}$$

Equation (3.45) shows that the frequency of the second mode is three times that of the fundamental frequency. The sound note produced in the pipe is called the *first overtone* or the *third harmonic* of the closed pipe. The second note and higher overtones are produced when air is blown 'forcefully' at the open end of closed pipe.

Third mode of vibration. For $m = 3$ and $\lambda = \lambda_2$ from Eq. (3.43), we get

$$\lambda_3 = \frac{4L}{5} \qquad \text{or} \quad L = \frac{5\lambda_3}{4}$$

where, λ_3 is the wavelength of the standing waves set up in the air column for $m = 3$.

The third mode of vibration is shown in Fig. 3.11(c). In the third mode, four additional nodes [two antinodes (*A*) and two node (*N*)] are produced between two ends of the closed pipe. The frequency of vibration in the third mode is

$$f = \frac{v}{\lambda_3} = \frac{5v}{4L} = 5 f_1 \qquad \text{...(3.46)}$$

Equation (3.46) shows that the frequency of the third mode is five times that of fundamental frequency. The sound note produced in the pipe is called the *second overtone* or the *fifth harmonic* of the closed pipe.

Similarly, the frequency of the mth mode of vibration is

$$f_m = \frac{v}{\lambda_m} = \frac{(2m-1)v}{4L} = (2m-1) f_1 \qquad ...(3.47)$$

From the calculation of the first, third and fifth overtones and (2 m – 1) overtones it is clear that relationship among the frequencies of the tones and the overtones of the closed pipe is

$$f_1 : f_2 : f_3 : .... : f_m :: 1:3:5: .... (2m-1)$$

The ratio of frequencies of all modesof vibration shows that a closed pipe produces only odd harmonics.

3.13.2 Longitudinal Standing Waves and Normal Modes in an Open Organ Pipe

Anopen organ pipe of length L lying along the x-axis withone open end at $x = 0$ and another at $x = L$. When it is blown from its open end at $x = L$, a longitudinal (sound) wave travels in the air of the pipe towards the end at $x = 0$. From this end it is reflected back. The stationary longitudinal waves are formed. The incident and the reflected waves travelling in the air column in opposite directions superpose each other. Since both the ends of the pipe are open, the particles of air at both the ends have freedom to vibrate. Therefore, antinodes are formed at both open ends. Because of the boundary conditions the air in an open organ pipe oscillates only in some special patterns which are called *normal modes.*

The incident wave moving from end at $x = 1$ to end $x = 0$, i.e., moving along the negative direction of x-axis is written as

$$Y_1 = a \sin \frac{2\pi}{\lambda}(vt + x)$$

The wave reflected at the open end of the pipe (i.e., at $x = 0$) suffers no phase change and moves along the positive direction of x-axis, i.e., from open end at $x = 0$ to open end $x = L$. Therefore, the reflected wave is written as

$$Y_2 = a \sin \frac{2\pi}{\lambda}(vt - x)$$

The resultant displacement (Y) at time t and position x is given by the sum of Y_1 and Y_2, i.e.,

$$\begin{aligned} Y &= Y_1 + Y_2 \\ &= a\left[\sin\frac{2\pi}{\lambda}(vt + x) + \sin\frac{2\pi}{\lambda}(vt - x)\right] \\ &= 2a \sin\left(\frac{2\pi vt}{\lambda}\right)\cos\left(\frac{2\pi x}{\lambda}\right) \qquad ...(3.48) \end{aligned}$$

At the open end of the pipe, i.e., at $x = 0$, Eq. (3.49) becomes

$$Y = 2a\sin\left(\frac{2\pi vt}{\lambda}\right)\cos(0) = 2a\sin\left(\frac{2\pi vt}{\lambda}\right) \text{ as } \cos(0) = 1 \qquad ...(3.49)$$

This value of displacement (Y) always has maximum value for all values of t. Clearly, an antinode is formed at $x = 0$ open end.

At the open end of the pipe, i.e., at $x = L$, Eq. (3.48) is written as

$$Y = 2\,a \sin\left(\frac{2\pi\, vt}{\lambda}\right) \cos\left(\frac{2\pi\, L}{\lambda}\right) \qquad \text{...(3.50)}$$

The displacement (Y) in Eq. (3.50) will always be maximum, if

$$\cos\left(\frac{2\pi\, L}{\lambda}\right) = \pm 1$$

This implies that

$$\frac{2\pi\, L}{\lambda} = m\,\pi, \qquad \text{where } m = 1, 2, 3, ...$$

or

$$\lambda = \frac{2\,L}{m} \qquad \text{...(3.51)}$$

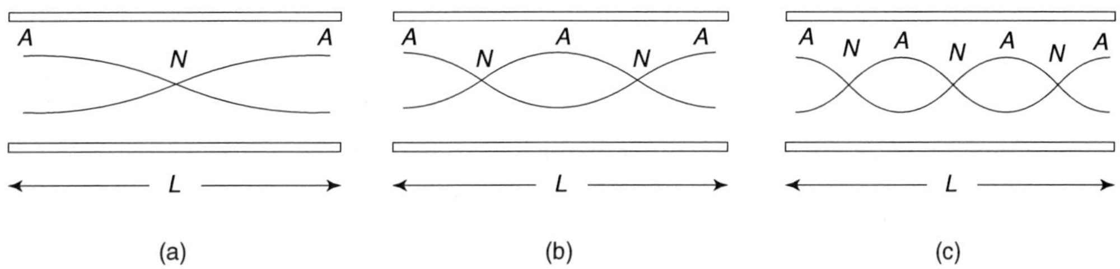

Fig. 3.12

First mode of vibration. For $m = 1$ and $\lambda = \lambda_1$, from Eq. (3.51) we get

$$\lambda_1 = 2\,L \qquad \text{or} \qquad L = \frac{\lambda_1}{2}$$

where, λ_1 is the wavelength of the standing waves set up in the air column for $m = 1$.

First mode of vibration is shown in Fig. 3.12(a). It has two antinodes (A) at two open ends and one node (N) in the middle of the pipe. The frequency of vibration in first mode of vibration is

$$f_1 = \frac{v}{\lambda_1} = \frac{v}{2\,L} \qquad \text{...(3.52)}$$

where, v is velocity of sound. The frequency of first mode of vibration lowest. The frequency of this mode is called the *fundamental frequency* of an open organ pipe. The sound note produced in the pipe is called the *fundamental tone*or the *first harmonic* of the open pipe. Fundamental frequency is produced when air is blown 'lightly' at the open end of the pipe.

Second mode of vibration. For $m = 2$ and $\lambda = \lambda_2$, from Eq. (3.52), we get

$$\lambda_2 = \frac{2L}{2} \quad \text{or} \quad L = \frac{2\lambda_2}{2}$$

where, λ_2 is the wavelength of the standing waves set up in the air column for $m = 2$. The second mode of vibration is shown in Fig. 3.12(b). In second mode, three additional nodes [one antinodes (A) in the middle of pipe and two nodes (N)] are formed between two antinodes at the open ends of the pipe. The frequency of vibration of second mode is

$$f_2 = \frac{v}{\lambda_2} = \frac{2v}{2L} = 2 f_1 \qquad ...(3.53)$$

Equation (3.53) shows that the frequency of the second mode is two times the fundamental frequency. The sound note produced in the open pipe is called the *first overtone* or the *second harmonic* of the open organ pipe. The irst overtone and higher overtones are produced when air is blown 'forcefully' at the open end of the pipe.

Third mode of vibration. For $m = 3$ and $\lambda = \lambda_3$, from Eq. (3.51), we get

$$\lambda_3 = \frac{2L}{3} \quad \text{or} \quad L = \frac{3\lambda_3}{2}$$

where, λ_3 is the wavelength of the standing waves set up in the air column for $m = 3$. Third mode of vibration is shown in Fig. 3.12(c). In this mode, five additional nodes [four antinodes (A): two antinodes and three nodes (N)] are formed between two open ends of the pipe. The frequency of the third mode vibration is

$$f_3 = \frac{v}{\lambda_3} = \frac{3v}{2L} = 3 f_1 \qquad ...(3.54)$$

Equation (3.54) shows that the frequency of the third mode is three times that of the fundamental frequency. The sound note produced in the pipe is called the *second overtone*'or the*third harmonic* of the open organ pipe.

Similarly, the frequency of the mth mode of vibration is

$$f_m = \frac{v}{\lambda_m} = \frac{mv}{2L} = m f_1 \qquad ...(3.55)$$

From this calculation of the first, second, third and m^{th} harmoinic it is clear that the frequencies of the tones and overtones of the open pipe has the relationship:

$$f_1 : f_2 : f_3 : .... : f_m ... :: 1 : 2 : 3 : .... m$$

The ratio of frequencies of modes of vibrations of an open pipe produces both even and odd harmonics. In conclusion we can say that in a closed organ pipe only odd harmonics are present, even harmonics are missing while in an open organ pipe all harmonics are present. Therefore, the musical sounds produced by an open organ pipe arealmost double than the musical sounds produced by a closed organ pipe.

3.14 LONGITUDINAL STANDING WAVES AND NORMAL MODES IN AIR COLUMNS PARTIALLY FILLED WITH WATER

The longitudinal standing waves can be produced in air columns by resonance effects, just like vibrations of a string tied at one end andfree at other end. Standing waves of this type are easily produced in air columns such as glass tubes partially filled with water. The length of the air column can be adjusted by changing the water level in the tube. Because of the boundary conditions, an air column partially filled with water, can oscillate only in some special patterns which are called *normal modes.*

There is a node at therigid end (say $x = 0$) of the air column in touch with water, so displacement of particles is zero. This can be explained as follows.

Let the incident wave moving in +ve x-direction be

$$Y_1 = a \sin (\omega t - kx)$$

The reflected wave from rigid end is out of phase so can be written as

$$Y_2 = a \sin (\omega t - kx + \pi) = - a \sin (\omega t - kx)$$

The resultant displacement of two waves is

$$Y = Y_1 + Y_2 = 0$$

However, the pressure at node at rigid point touching water is maximum, because the pressure due to compression of incident waves and pressure due to compression of reflected waves are added up. The pressure at antinode is minimum, because the pressure due to rarefaction of incident waves and pressure due to rarefaction of reflected waves are subtracted. The two waves travelling in opposite directions are in phase here, so there are no pressure fluctuations.

From the expression of the wave function given by Eq. (3.56) we note that at the end, $x = L$, there will be an antinode or displacement will be maximum when

$$\left|\sin \frac{2\pi L}{\lambda}\right| = 1 \qquad \text{...(3.56)}$$

This condition defines the wavelengths of normal modes in an air column when one end of the tube is open. The condition given by Eq. (3.56) implies that

$$\frac{2\,\pi L}{\lambda} = (2\,m - 1)\,\frac{\pi}{2}, \; m = 1, 2, 3, \ldots$$

or

$$\lambda = \frac{4L}{(2\,m-1)}$$

or

$$\lambda = 4\,L, \frac{4L}{3}, \frac{4L}{5}, \ldots, \frac{4\,L}{(2\,m-1)}$$

In Fig. 3.13(a) to (c) we have shown some special patterns of the sound waves called normal modes in an air column with the open end.

Modes of vibration of standing waves shown in Fig. 3.13(a) are called the first harmonic or the fundamental with wavelength

$$\lambda_1 = 4\,L \text{ and frequency } f_1 = v/\lambda_1 = v/4\,L$$

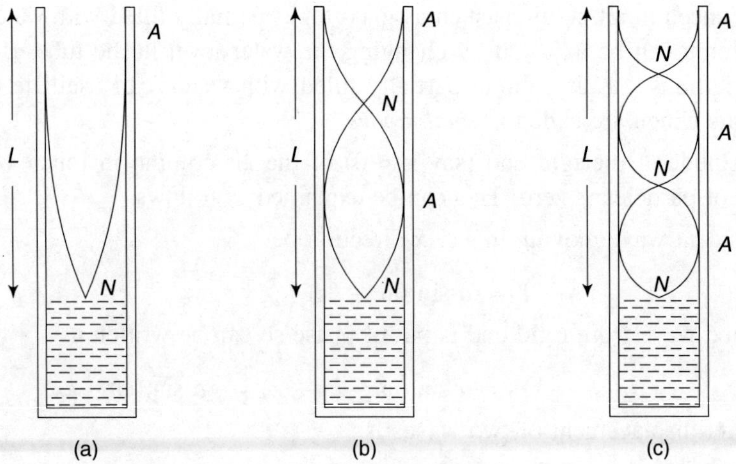

Fig. 3.13

Modes of vibration of standing waves shown in Fig. 3.13(b) are called the second harmonic or the first overtone with wavelength

$\lambda_2 = 4\,L/3$ and frequency $f_2 = v/\lambda_2 = 3v/4\,L$ or $f_2 = 3\,f_1$, i.e., frequency of second harmonic is three times of the fundamental frequency.

Modes of vibration of standing waves shown in Fig. 3.13(c) are called the third harmonic or the second overtone with wavelength.

$\lambda_3 = 4\,L/5$ and frequency $f_3 = v/\lambda_3 = 5\,v/4\,L$ or $f_3 = 5\,f_1$, i.e., frequency of the third harmonic is five times of the fundamental frequency.

In general, modes of vibration of standing waves of mth harmonic or the $(m - 1)^{\text{th}}$ overtone with wavelength

$\lambda_3 = 4\,L/(2\,m - 1)$ and frequency $f_m = v/\lambda_m = (2\,m - 1)\,v/4\,L$ or $f_m = (2\,m - 1)\,f_1$, i.e., frequency of mth harmonic is $(2\,m - 1)$ times of the fundamental frequency.

Another interesting property of superposition of waves is the phenomenon of beats. We will discuss it in Section 4.14.

3.15 MODES OF VIBRATIONS OF ORGAN PIPES IN BRIEF

The simplest organ pipe is a cylindrical tube of uniform diameter. One end of the organ pipe is always open while the other may be closed or open. The pipe which is closed at one end is called *closed endorgan pipe*. The pipe which is open at both the ends is called *open end organ pipe*. If both the ends of the organ pipe are closed, then it becomes a cavity.

Normally, organ pipe contains air. The incident wave travelling along the axis of the pipe will be reflected back from the other end because the medium is bounded. The incident and reflected waves on superposition produces stationary wave.The mechanical waves in an organ pipe are always longitudinal–stationary.

If reflection is from closed end of an organ pipe, the closed end will be a node as particles at node are permanently at rest. If reflection is from the open end where particles of air are free to move, the antinode will be produced in an organ pipe. In short, closed end is always a node while free end, an antinode.

Normally, organ pipe is filled with air. Closed and open pipes are used to produce different sounds by vibration of air column. Such sound producing tubes are called organ pipes. For example, flute, whistle, clarinet, etc., are the instruments (organ pipes) which are used to produce different sounds.

(A) Closed end Organ Pipe

In case of closed end organ pipe, as closed end will always be a node, while free end an antinode. In this case, the minimum distance between a node and antinode is $\lambda/4$. Only those longitudinal stationary waves are formed for which length of the pipe is equal to odd integral multiple of $\frac{\lambda}{4}$, i.e.,

$$L = \frac{\lambda}{4} \quad \text{or} \quad L = \frac{3\lambda}{4} \quad \text{or} \quad L = \frac{5\lambda}{4}$$

i.e.,

$$\lambda = \frac{4L}{1} \quad \text{or} \quad \lambda = \frac{4L}{3} \quad \text{or} \quad \lambda = \frac{4L}{5}$$

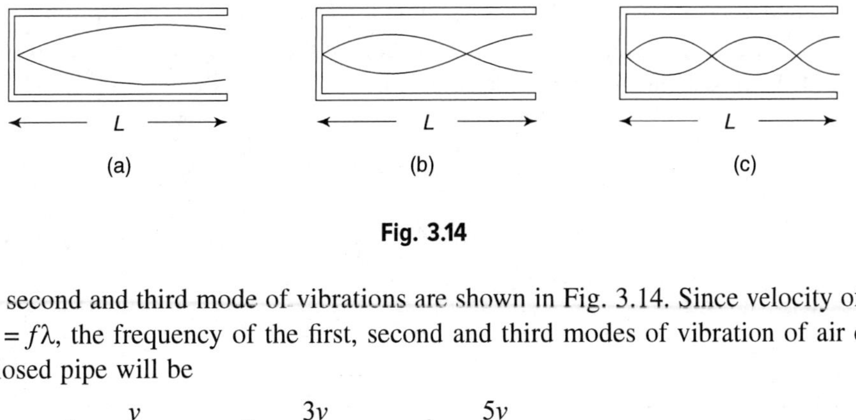

Fig. 3.14

First, second and third mode of vibrations are shown in Fig. 3.14. Since velocity of sound wave, $v = f\lambda$, the frequency of the first, second and third modes of vibration of air column in the closed pipe will be

$$f_1 = \frac{v}{4L} \quad \text{or} \quad f_2 = \frac{3v}{4L} \quad \text{or} \quad f_3 = \frac{5v}{4L}$$

So if $\quad f_1 = f_c \left(= \frac{v}{4L} \right) \quad \text{or} \quad f_2 = 3f_c \quad \text{or} \quad f_3 = 5f_c$

In case of closed end organ pipe, the following observations can be made.

1. Maximum possible wavelength is 4 L.
2. Fundamental frequency is $f_C = v/4\ L$. The frequency of higher harmonics are only odd multiples of fundamental frequency, i.e., even harmonics are absent.
3. First mode of vibration is called fundamental or first harmonic, second mode vibration is called third harmonic or first overtone, third mode of vibration is called fifth harmonic or second overtone, and so on.
4. The modes of vibration of closed end organ pipe are identical to that of a rod (string) is clamped at one end except that the waves in rod can be longitudinal or transverse, whilewaves are always longitudinal in an organ pipefilled with air and the waves in string are transverse only.

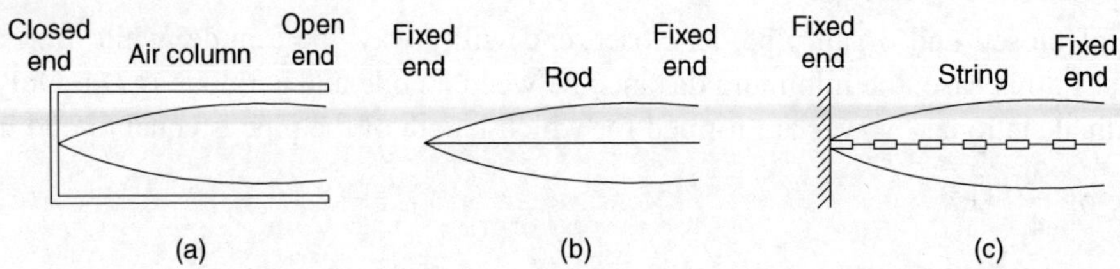

Fig. 3.15

(B) Open end Organ Pipe

In case of open end organ pipe, there are antinodes at both ends. The minimum distance between two consecutive antinodes is λ/2. Only those stationary waves are formed for which the length of the pipe is integral multiple of λ/2, i.e.,

$$L = \frac{\lambda}{2} \quad \text{or} \quad L = \frac{2\lambda}{2} \quad \text{or} \quad L = \frac{3\lambda}{2}$$

i.e.,

$$\lambda = \frac{2L}{1} \quad \text{or}\ \lambda = \frac{2L}{2} \quad \text{or} \quad \lambda = \frac{2L}{3}$$

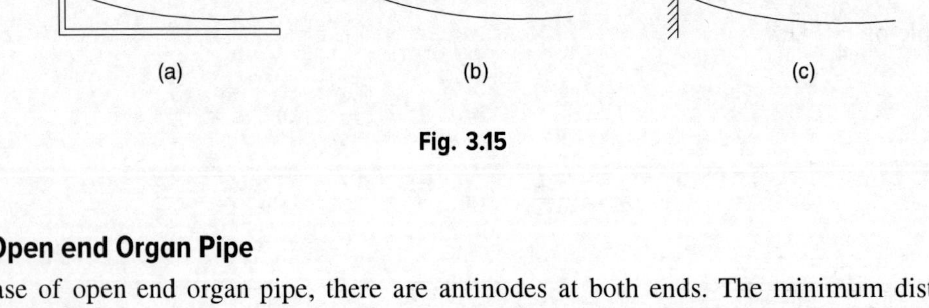

Fig. 3.16

First, second and third modes of vibration are shown in Fig. 3.16. Since $v = f\lambda$, the frequency of the first, second and third modes of vibration of air column in the pipe will be

$$f_1 = \frac{v}{2L} \quad \text{or} \quad f_2 = \frac{2v}{2L} \quad \text{or} \quad f_3 = \frac{3v}{2L}$$

So if $$f_1 = f_0 \left(= \frac{v}{2L}\right) \quad \text{or} \quad f_2 = 2f_0 \quad \text{or} \quad f_3 = 3f_0$$

In case of open end organ pipe, the following observations can be made:

1. Maximum possible wavelength is $2L$.
2. Fundamental frequency is $f_0 = v/2L$. The higher harmonics are integral multiple of fundamental frequency, that is, all harmonics are present in an open organ pipe.
3. First mode of vibration is called fundamental or first harmonic, second mode of vibration is called second harmonic or first overtone, third mode of vibration is called fifth harmonic or second overtone, and so on.
4. The mode of vibration of open end organ pipes are identical to that of a rod clamped in the middle as shown in Fig. 3.17 except that the waves in a rod can be longitudinal or transverse, while waves are always longitudinal in an organ pipe filled with air.

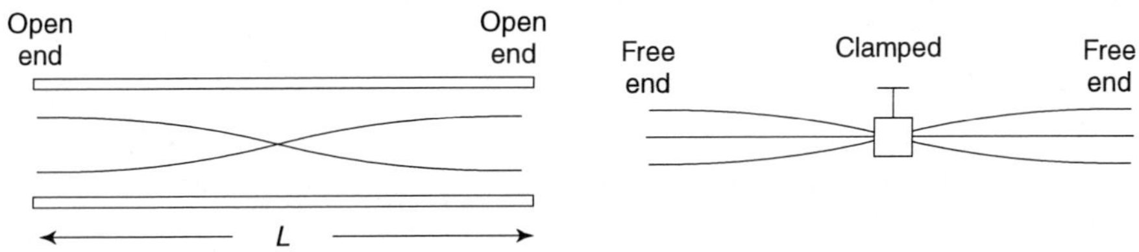

(a) Open end organ pipe

(b) Rod clamed at the middle

Fig. 3.17

Formulas at a Glance

	Subject	Formula	Explanation
1.	Travelling wave	$Y(x, t) = A \sin(kx - \omega t + \phi)$ x is the position of the constituents of the medium (Y) is the displacement of the constituents about its mean position.	ϕ in the argument is the phase of the wave. ω and k are angular frequency and angular wave vector, respectively. A is the amplitude of the wave.
2.	Speed of the travelling wave	$v = \frac{\omega}{k} = \frac{\lambda}{T} = \lambda f$	T and are time period and frequency of the travelling wave.

Contd.

3.	Equation of wave packet	$y = R \sin (kx - \omega t)$, where $R = 2A \cos\left(\frac{dk}{2}x - \frac{d\omega}{2}t\right)$ is amplitude of the equation of the wave packet.	$dk = k - k'$ and $d\omega = \omega - \omega'$, ω and ω', k and k'. ω and k are angular frequency and angular wave vector, respectively, of one wave. ω' and k' are angular frequency and angular wave vector, respectively, of the second wave.
4.	Phase velocity	$v_p = \frac{\omega}{k}$	
5.	Group velocity	$v_g = \frac{d\omega}{dk}$	$dk = k - k'$ and $d\omega = \omega - \omega'$
6.	Equation relating the phase and group velocity	$v_g = v_p - \frac{2\pi}{k}\frac{dv_p}{d\lambda}$	
7.	Pressure wave	$p = p_0 \cos (\omega t - kx)$	Where initial pressure is $p_0 = AkB = \rho v A\omega$, B is bulk modulus
8.	Velocity of wave in medium	$v = \sqrt{\frac{B}{\rho}}$	B is bulk modulus
9.	Wavelength of the normal modes of vibrations in an open organ pipe filled with air	$\lambda = \frac{2L}{m}$, $m = 1, 2, 3, ...$	Where, L is length of air column
10.	Wavelengths of various harmonics in case of open organ pipe	$\lambda_1 = 2L, \lambda_2 = \frac{2L}{2L_1}, \lambda_3 = \frac{2L}{3}$, etc. $L = \frac{\lambda_1}{2}, L = \frac{2\lambda_2}{2}, L = \frac{3\lambda_3}{2}$	$\lambda_1 \rightarrow$ wavelength of first harmonic $\lambda_2 \rightarrow$ wavelength of second harmonic $\lambda_3 \rightarrow$ wavelength of third harmonic
11.	Wavelength of the normal modes of vibrations in a closed organ pipe partially filled with water	$\lambda = \frac{4L}{(2m-1)}$, $m = 1,2,3, ...$	Where, L is length of air column
12.	Wavelengths of various harmonics in case of open organ pipe	$\lambda_1 = 4L, \lambda_2 = \frac{4L}{3}, \lambda_3 = \frac{4L}{5}$ $L = \frac{\lambda_1}{4}, L = \frac{3\lambda_2}{4}, L = \frac{5\lambda_3}{4}$	$\lambda_1 \rightarrow$ wavelength of the first harmonic $\lambda_2 \rightarrow$ wavelength of second harmonic $\lambda_3 \rightarrow$ wavelength of third harmonic

Contd.

13.	Wavelength of the normal modes of vibrations in a closed organ pipe partially filled with water	$\lambda = \frac{4L}{(2m-1)}$, $m = 1, 2, 3, ...$	Where L is length of air column
14.	Various harmonic frequencies	$f_1 = \frac{v}{4L}$, $f_2 = \frac{3v}{4L}$, $f_3 = \frac{5v}{4L}$ or $f_1 = f_C \left(= \frac{v}{4L}\right)$, $f_2 = 3f_C$, $f_3 = 5f_C$	$v = f\lambda$, $f_1 \rightarrow$ wavelength of first harmonic $f_2 \rightarrow$ wavelength of second harmonic $f_3 \rightarrow$ wavelength of third harmonic
15.	Frequencies of various harmonics in case of open organ pipe	$f_1 = \frac{v}{2L}$, $f_2 = \frac{2v}{2L}$, $f_3 = \frac{3v}{2L}$ or $f_1 = f_0 \left(= \frac{v}{2L}\right)$, $f_2 = 2f_0$, $f_3 = 3f_0$	v is velocity of wave $v = f\lambda$
16.	Frequencies of various harmonics in case of closed organ pipe	$f_1 = \frac{v}{4L}$, $f_2 = \frac{3v}{4L}$, $f_3 = \frac{5v}{4L}$ or $f_1 = f_C \left(= \frac{v}{4L}\right)$, $f_2 = 3f_C$, $f_3 = 5f_C$	$f_1 \rightarrow$ wavelength of first harmonic $f_2 \rightarrow$ wavelength of second harmonic $f_3 \rightarrow$ wavelength of third harmonic
17.	After end correction the frequency of closed organ pipe	$f_C = \frac{v}{4(L+0.6\,r)}$	r is radius of the pipe, velocity of wave $v = f\lambda$
18.	After end correction the frequency of open organ pipe	$f_0 = \frac{v}{2(L+1.2\,r)}$	r is radius of the pipe, velocity of wave, $v = f\lambda$

SOLVED EXAMPLES

Example 3.1 A pipe 27 cm long is open at both the ends. Which harmonic mode of the pipe is resonantly excited by a 1.2 kHz source? Will resonance with the same source be observed if one end of the pipe is closed? Take the speed of sound in air to be 324 m/s.

Solution: Formula required

$$\lambda = \frac{2L}{m},\ v = f\lambda,\ m = \frac{2L}{v}$$

where, L is the length of the pipe. m stands for m^{th} the number of harmonic. If v and f are velocity and frequency of sound in air, then wavelength of sound is

$$\lambda = v/f$$

Putting this value of λ, formula for m^{th} harmonic will be

$$m = \frac{2\,Lf}{v}$$

Given: Length of the pipe, $L = 27$ cm
Frequency of source is 1.2 kHz
Speed of sound is 324 m/s
Calulation:

$$m = \frac{2\,L\,f}{v} = m = \frac{2 \times 0.27 \times 1200}{324} = 2$$

So a source of frequency 1.1 kHz will resonantly excite the second harmonic of the open pipe.

When one end of pipe is closed, the formula for m^{th} harmonic is

$$(2\,m - 1) = \frac{4\,L}{\lambda},\ v = f\lambda,$$

Putting the value of λ, the formula for m^{th} harmonic with closed pipe is

$$(2\,m - 1) = \frac{4\,Lf}{v}$$

$$m = 1\ 1 = \frac{4\,Lf}{v}\quad m = 2\quad 3 = \frac{4\,Lf}{v},\ m = 3\quad 5 = \frac{4\,Lf}{v},$$

That is, in closed pipe, only odd number of harmonics are present. Since $\frac{Lf}{v} = \frac{0.27 \times 1200}{324}$ $= 1$, so no harmonic will be present in closed pipe with a source of frequency 1.2 kHz.

Ans. Source of frequency 1.2 kHz will resonantly excite thesecond harmonic of the open pipe. No harmonic will be present in closed pipe with a source of frequency 1.2 kHz.

Example 3.2 A tuning fork having frequency of 330 Hz is vibrated first above a cylindrical tube. The height of the tube is 110 cm. Water is slowly poured in. What is the minimum height of water required for resonance? Speed of sound wave is 330 m/s.

Solution: Let v be the speed of sound and L is the length of the column of pipe.

Formula for the nth order fundamental

$$f = n\,\frac{v}{4L},\ \text{where } n = 1, 3, 5 \ldots.$$

Given: Height of tube $L = 110$ cm

Speed of the sound wave, $v = 340$ m/s $= 340 \times 100$ cm/s

Length of air column in the pipe is

$$L = n\frac{v}{4f} = n\frac{330 \times 1000}{4 \times 330} = 25\ n \text{ cm, where } n = 1, 3, 5...$$

i.e. $L = 25, 75, 125, ...$

Length of the tube is 110 cm. The length of air column must be less than 110 cm. The length of the air column can be 25 cm or 75 cm.

When water is filled then let h be the height of water and L be the length of air column, then

$$L + h = 110 \quad h = 110 - L$$

Height of water (h) will be minimum when L is maximum. Maximum value of $L_{max} = 75$ cm

Therefore, minimum height of water, $h_{min} = 110 - L_{max} = 35$ cm.

Ans. Minimum height of water column for resonance is 35 cm.

Example 3.3 The air column in a pipe closed at one end is made to vibrate in its second overtone by a tuning fork of frequency 440 Hz. The speed of sound in air is 330 m/s. End corrections may be neglected. Let P_0 denote the mean pressure at any point in the pipe, and ΔP_0 the maximum amplitude of the pressure variation.

(a) Find the length (L) of the air column.

(b) What is the amplitude of pressure variation in the middle of the column?

(c) What are the maximum and minimum pressures at the open end of the pipe?

(d) What are the maximum and minimum pressures at the closed end of the pipe?

Solution: (a) In case of closed organ pipe as fundamental frequency is ($v/4L$) and only odd harmonics are present, second overtone will mean fifth harmonics and so

$$f = \frac{5v}{4L} = 440 \text{ Hz (given)}$$

and hence $$L = \frac{5 \times 330}{4 \times 440} = \frac{15}{16} \text{ m} \qquad \textbf{\textit{Ans.}}$$

(b) Variation of pressure amplitude of standing pressure waves along the length of the column with $x = 0$ at its open end is written as

$$P = \Delta P_0 \sin kx = \Delta P_0 \sin\left(\frac{2\pi}{\lambda}x\right) \text{ as } \left(k = \frac{2\pi}{\lambda}\right)$$

The second overtone is at $L = \frac{5}{4}\lambda$,

At the middle of the pipe, $x = \frac{L}{2} = \frac{5}{8}\lambda$ and hence at the middle the pressure is

$$P = \Delta P_0 \sin\left(\frac{2\pi}{\lambda} \times \frac{5}{7}\lambda\right) \Delta P_0 \sin\left(\frac{5\pi}{4}\right)$$

or $$P = \Delta P_0 \times \frac{1}{\sqrt{2}} = \frac{\Delta P_0}{\sqrt{2}} \quad \textbf{\textit{Ans.}}$$

At free end $x = 0$, $P = 0$, i.e., the amplitude of pressure wave is zero (as it is a node), so

$$P_{max} = P_{min} = P_0 \pm 0 = P_0 \quad \textbf{\textit{Ans.}}$$

(d) At closed end $x = \frac{5}{4}\lambda$, so the amplitude of pressure at closed end is

$$\lceil P \rceil = \left| \Delta P_0 \sin\left(\frac{2\pi}{\lambda} \times \frac{5}{4}\lambda\right) \right| = \Delta P_0$$

So $$P_{max} = P_0 + \Delta P_0 \quad \text{and} \quad P_{min} = P_0 - \Delta P_0 \quad \textbf{\textit{Ans.}}$$

Example 3.4 A tube of certain diameter and length 49 cm is open at both ends. Its fundamental frequency of resonance is 340 Hz. The velocity of sound in air is 340 m/s. Estimate the diameter of the tube. One end of the tube is now closed. Calculate the frequency of resonance of the tube.

Solution: Formula for fundamental frequency of resonance for an organ pipe open at both ends is

$$f_1 = \frac{v}{2L}$$

Taking open end correction, formula for fundamental frequency of resonance for an organ pipe open at both ends is

$$f_1 = \frac{v}{2(L + 2 \times 0.6\, r)}$$

where, r is the radius of pipe.

Given: Length of the tube, $L = 49$ cm

Fundamental resonance frequency, $f_1 = 340$ Hz

Speed of sound, $v = 340$ m/s $= 340 \times 100$ cm

Calculation:

Putting the values of various quantities in the formula, we get

$$340 = \frac{340 \times 100}{2(49 + 2 \times 0.6\ \ )}$$

or $$49 + 2 \times 0.6\, r = 50$$

or $$r = \frac{10}{12} \text{ cm}$$

The diameter of pipe, $d = 2\,r = 2 \times \frac{10}{12} = 1.67$ cm

Taking open end correction, formula for the fundamental frequency of resonance for an organ pipe open at one end is

$$f_1 = \frac{v}{4\,(L + 0.6\,r)}$$

where, r is the radius of pipe.

$$f_1 = \frac{340 \times 100}{4\left(49 + 0.6 \times \dfrac{10}{12}\right)} = \frac{340 \times 100}{198} = 171.7 \text{ Hz}$$

Ans. Diameter of the pipe is 1.67 cm

For one end closed pipe the fundamental frequency is 171.7 Hz.

Example 3.5 For a certain organ pipe, three successive resonance frequencies are observed at 405, 567 and 729 Hz, respectively. Taking the speed of the sound in air to be 324 m/s (a) explain whether the pipe is closed at one end or open at both ends, and (b) determine the fundamental frequency and length of the pipe.

Solution: Formula for fundamental frequency of resonance for an organ pipe open at one end is

$$f_1 = \frac{v}{4L}$$

Given: Three successive resonance frequencies: 405, 567 and 729 Hz

Speed of sound, v = 324 m/s

(a) The given resonance frequencies are in the ratio

$$405 : 567 : 729 \quad \text{or} \quad 5 ; 7 ; 9$$

Clearly these are odd integers so the given pipe is a one end closed pipe.

(b) From the ratio of frequencies it is clear that the frequency of 5th harmonic (2nd overtone) is 425 Hz, i.e.,

$$405 = 5\,f_1 \quad \text{or} \quad f_1 = 81 \text{ Hz}$$

Putting this value in the formula

$$f_1 = \frac{v}{4L}$$

The length of the pipe

$$L = \frac{v}{4f} = \frac{324}{4 \times 81} = 1 \text{ m}$$

Ans. (a) Pipe is closed at one end

(b) Length of pipe is 1 m

Example 3.6 The first overtone of an open organ pipe beats with the first overtone of a closed organ pipe with a beat frequency of 2.4 Hz. The fundamental frequency of the closed organ pipe is 115 Hz. Find the lengths of the pipes. The speed of sound is 345 m/s.

Solution: Let L be the length of organ pipe and v be the speed of sound.

Formula for frequency of resonance for organ pipe closed at one end is

$$f_C = (2\,n - 1)\,\frac{v}{4L},\; n = 1, 2, 3\; ....$$

Formula for frequency of resonance for organ pipe open at both ends is

$$f_0 = n\,\frac{v}{2L},\; n = 1\; 2, 3, ...$$

Given: Beat frequency, $f = 2.2$ Hz

Fundamental frequency of closed organ pipe, $f_C = 115$ Hz.

Speed of sound, $v = 345$ m/s

Calculation:

For $n = 1$, fundamental resonance frequency of closed organ pipe

$$f_C = \frac{v}{4L_C} = 115 \quad \text{or} \quad L_C = \frac{345}{4 \times 115} = 0.75 \text{ m}$$

Frequency of the first overtone or the second fundamental frequency of a closed organ pipe is

$$3\,f_C = \frac{3v}{4L_C} = 3 \times 115 = 345$$

Frequency of the first overtone or the second fundamental frequency of an open organ pipe is

$$2f_0 = \frac{2v}{2L_0}$$

But frequency

$$3\,f_C - 2\,f_0 = \frac{3v}{4L_C} - \frac{2v}{2L_0} = \pm\, 2.4 \quad \text{or} \quad \frac{345}{L_0} = 345 \mp 2.4$$

or
$$L_0 = \frac{345}{345 \mp 2.4} \Rightarrow L_0 = 1.0747 \text{ m} \quad \text{or} \quad 0.9349$$

Ans. Length of the pipe is 1.0747 m or 0.9349 m

Example 3.7 A pipe of length 1.5 m closed at one end is filled with a gas and it resonates in its fundamental with a tuning fork. Another pipe of the same length but open at both ends is filled with air and it also resonates in its fundamental with the same tuning fork. Calculate the velocity of the sound at 0°C in the gas. The velocity of the sound in air is 360 m/s at 30°C, where the experiment is performed.

Solution: Formula for frequency of resonance for organ pipe closed at one end is

$$f_C = (2n - 1)\frac{v}{4L}, \; n = 1, 2, 3 \ldots.$$

Formula for frequency of resonance for organ pipe open at both ends is

$$f_A = n\frac{v}{2L}, \; n = 1\ 2, 3, \ldots.$$

Given: Length of the pipe, $L = 1.5$ m

Speed of sound in air at 30°C, $v_A = 360$ m/s

Calculation:

If v_G is velocity of sound in gas, then fundamental resonance frequency of organ pipe closed at one end,

$$f_C = \frac{v_G}{4L}$$

If v_A is velocity of sound in air, then fundamental resonance frequency for organ pipe open at both ends,

$$f_A = \frac{v_A}{2L}$$

If f is the frequency of tuning fork, then according to the problem at 30°C,

$$f = f_C = f_A \text{ or } \frac{v_G}{4L} = \frac{v_A}{2L} \text{ or } v_G = 2\,v_A = 2 \times 360 = 720 \text{ m/s}$$

Let velocity of sound $T = 30$°C and 0°C are $(v_g)_T$ and $(v_g)_0$.

So
$$\frac{(v_g)_T}{(v_g)_0} = \sqrt{\frac{273+30}{273}} = \sqrt{\frac{303}{273}}$$

or
$$(v_g)_0 = (v_g)_T\sqrt{\frac{303}{273}} = 720\sqrt{\frac{303}{273}} = 684 \text{ m/s}$$

Ans. Speed of sound in gas at 0°C is 684 m/s

SHORT ANSWER QUESTIONS

3.1 What are longitudinal and transverse waves?

3.2 Write the difference between longitudinal and transverse waves.

3.3 What are water waves or ripples?

3.4 What are travelling waves?

3.5 Write the characteristics of travelling waves.

3.6 Define the ratio of mode of vibrations in open end and closed end organ pipes?

Ans. In case of closed end organ pipe higher harmonics are odd integral multiple of fundamental frequency, f_C, while in case of open end organ pipe all harmonics are integral multiples of fundamental frequency, f_0.

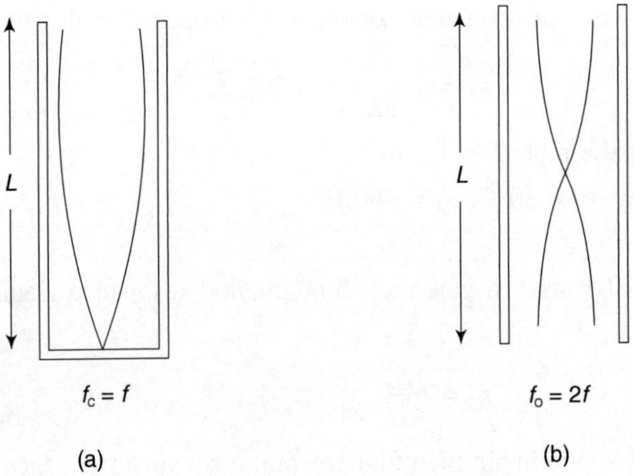

Fig. 3.18

3.7 Find the ratio of fundamental frequency of identical closed end and open end organ pipes.

Ans. $f_C = v/4L$ and $f_0 = v/2L$,

Therefore, $f_0 = \dfrac{v}{2L} = 2\left(\dfrac{v}{4L}\right) = 2f_C$

This relation shows that if a closed organ pipe is converted into an open organ pipe its frequency will become double and if open organ pipe is converted into closed organ pipe its frequency will become half.

3.8 Find the ratio of fundamental frequency of an open end organ pipe and the same organ pipe when half submerged in water as shown in Fig. 3.19(b).

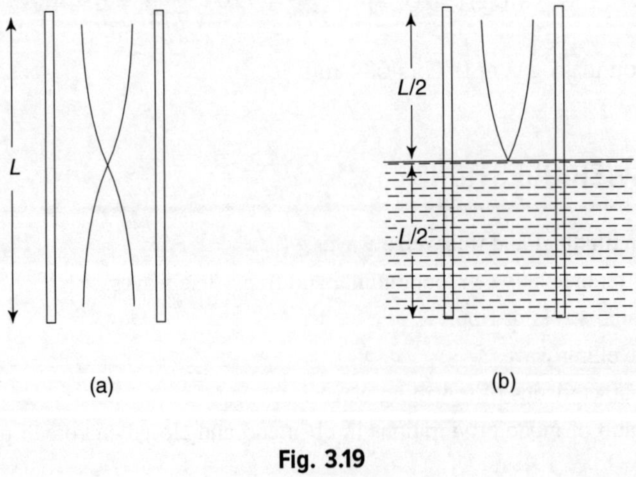

Fig. 3.19

Ans. Its fundamental frequency will be

$$f_C = \frac{v}{4\left(\frac{1}{2}\right)} = \frac{v}{2L} = f_0$$

i.e., if half part of an open end organ pipe is submerged in water, its fundamental frequency becomes equal to the fundamental frequency of open pipe, i.e., frequency will remain unchanged.

3.9 What is end correction? Find analytical relation in case of open organ closed pipes.

Ans. Due to vibrations of molecules they have finite momentum. Because of finite momentum of air molecules in organ pipes reflection takes place not exactly at open end of closed organ pipe but somewhat above it; so in a closed open pipe, antinode is not formed exactly at free end but slightly above it. This distance above the end of open end is called end correction or Reyleigh correction e. If r is the radius of the pipe, then end correction, $e = 0.6\,r$. If L is the length of closed organ pipe, by including the end correction its effective length must be $L + 0.6\,r$. In case of open organ pipe both ends are open, so effective length should be considered $L + 2 \times 0.6\,r$ instead of L. By considering the end correction,

$$f_C = \frac{v}{4\,(L + 0.6\,r)} \quad \text{and } f_0 = \frac{v}{2\,(L + 1.2\,r)}$$

The effective length of the pipe will be less for narrower pipes. That is why for a given v and L narrower the pipe higher will the frequency or pitch and sound will be shriller.

3.10 Prove that $f \propto \frac{1}{\lambda}$ or $f \propto \frac{1}{L}$ when v = constant for an organ pipe (closed or open)

Ans. $f_C = v/4\,L$ for closed organ pipe
and $f_0 = v/2\,L$ for open organ pipe
If v = constant for an organ pipe (closed or open), then

$f \propto \frac{1}{\lambda}$ or $f \propto \frac{1}{L}$ [as $\lambda \propto L$]

Therefore, with decrease in length of vibrating air column its frequency or pitch will increase. This is why the pitch increases gradually as an empty vessel is filled slowly.

3.11 Show that $f \propto v$ for organ pipe, when λ = constant

Ans. $f_C = v/4\,L$ and $\lambda = 4\,L$ for closed organ pipe
and $f_0 = v/2\,L$ and $\lambda = 2\,L$ for open organ pipe.
If λ = constant, then for a given length of organ pipe
$f \propto v$
Therefore,

(a) With rise in temperature as velocity increases (as $v \propto \sqrt{T}$), so the frequency or pitch also increases.

(b) With change in gas in the pipe as v changes ($v \propto \sqrt{\gamma/M}$ for constant pressure), so f also changes.

(c) With increase in moisture as v increases, so the frequency or pitch also increases.

3.12 Show that $v \propto \lambda$ for organ pipe, when f = constant

Ans. $f_C = v/4\,L$ and $\lambda = 4\,L$ for closed organ pipe

and $f_0 = v/2\,L$ and $\lambda = 2\,L$ for open organ pipe.

If f = constant for an organ pipe

$v \propto \lambda$ because $L \propto \lambda$ and $v \propto L$

$$v = f\lambda$$

$$f = \frac{v}{\lambda} = \text{constant},$$

Thus the frequency of an organ pipe will remain unchanged if the ratio of speed of sound in it to its length remains constant.

3.13 Discuss the reflection of longitudinal pressure wave from closed and open end in an organ pipe with emphasis on change in phase if any.

Ans. When an incident wave is reflected from the rigid closed end, a phase change of π is introduced in the reflected wave. Therefore, resultant displacement of the particles is zero. The point of zero displacement is called node. Explanation of this argument in detail is as follows. In case of reflection from a rigid boundary when a longitudinal compression reaches the boundary, it will press the boundary with maximum pressure and will be reflected back as compression, as the boundary is not free to move. Due to reflection at rigid boundary a phase of π radians will be introduced in the reflected wave, i.e., incident and reflected waves will be out of phase at rigid boundary. As a result displacement of the particles will be zero. The point of zero is called a node.

When incident wave is reflected from the open end, because there is atmospheric pressure at open end, no phase change is introduced in the reflected wave, i.e., incident and reflected waves are in the same phase. Therefore, resultant displacement of the particles is maximum. The point of maximum displacement is called an antinode. Explanation of this argument in detail is as follows.

In case of reflection from a open end when a longitudinal rarefaction wave reaches the open end, it will press the open end with minimum pressure and will be reflected back as rarefaction, as there is atmospheric pressure at the open end. Due to reflection at open end no phase change will be introduced in the reflected wave, i.e., incident and reflected waves will be in the same phase at open end. As a result displacement of the particles will be maximum. The point of maximum is called an antinode.

3.14 Where will a person hear maximum sound at (displacement) node or antinode?

Ans. Perception of sound is due to pressure variations and as at node displacement is minimum, pressure will be maximum while at antinode as displacement is maximum, pressure will be minimum. So sound will be maximum at nodes (which is actually pressure antinode).

3.15 Two organ pipes of same length open on both ends produce sound of different pitch if their radii are different. Why?

Ans. Due to end correction, the frequency of open organ pipe depends on its radius according to the relation

$$f = \frac{v}{(2\,(L + 1.2\,r)}$$

So the pitch will be different for different radii and narrower the pipe higher will be the pitch.

3.16 Explain why the quality of sound from an open pipe is different from from a closed pipe of same fundamental frequency.

Ans. Quality of sound basically depends on the number of overtones present. In case of closed organ pipe overtones are only odd multiples of fundamental frequency while in case of open pipe they are all integral multiples of fundamental frequency. So the number of overtones in open pipe are more than in closed pipe and so the quality of sound in two pipes will be different.

3.17 Can we ever construct an organ pipe whose frequency does not change with temperature? If no why? If yes, under what conditions?

Ans. In case of an organ pipe $f \propto v/L$. So frequency will not change with temperature if v/L is independent of temperature.

Now as v and L both depend on temperature so f will be dependent on temperature if

$$\frac{v_1}{L_1} = \frac{v_2}{L_2} \quad \text{or} \quad \frac{v_1}{v_2} = \frac{L_1}{L_2}$$

Now as $v = \sqrt{T}$ and $L = L_0\,(1 + \alpha\,\theta)$

$$\sqrt{\frac{273+\theta_2}{273+\theta_1}} = \frac{1+\alpha\,\theta_2}{1+\alpha\,\theta_1}$$

or $$\left(1+\frac{\theta_2}{273}\right)^{1/2} \times \left(1+\frac{\theta_1}{273}\right)^{-1/2} = (1 + \alpha\,\theta_2) \times (1 + \alpha\,\theta_1)^{-1}$$

or $$\left(1+\frac{1}{2}\frac{\theta_2}{273}\right) \times \left(1-\frac{1}{2}\frac{\theta_1}{273}\right) = (1 + \alpha\,\theta_2) \times (1 - \alpha\,\theta_1)$$

or $$1+\frac{1}{2}\frac{(\theta_2-\theta_1)}{273} = 1 + \alpha\,(0_2 - 0_1)$$

or $$\alpha = \frac{1}{546} \text{ per °C}$$

So the frequency of organ pipe will be independent of temperature vibrations if the coefficient of linear expansion of its material is 1/ 546 per °C.

3.18 What is the phase difference between the constituents of medium on the two sides of a node?

Ans. π

3.19 What is the phase difference between a node and its nearest antinode in a stationary wave?

Ans. $\pi/2$

3.20 The fundamental frequency of a sound source is 150 Hz and the source produces all the harmonics. State, with reasons, with which of the following frequencies this source will be in resonance of 150 Hz, 200 Hz, 300, 400 Hz, 500 Hz, 600 Hz.

Ans. 150 Hz, 300 Hz, 600 Hz

3.21 An organ pipe closed at one end resonates with a tuning fork of frequency 125 Hz. With what other frequencies can it resonate? If the other end is also opened, then?

Ans. 125 × (3, 5, 7, ...); 250 × (1, 2, 3, 4,....)

3.22 How will the frequency of a note emitted by an organ pipe is affected by decrease in temperature?

Ans. As frequency $n \propto v/L$, since velocity of sound (v) decreases with decrease in temperature, so the frequency of an organ pipe also decreases with decrease in temperature and vice versa.

3.23 A tuning fork produces resonance in a closed pipe, but it does not produce resonance with an open pipe of same length. Why?

Ans. Due to end correction, the fundamental frequency of an open pipe is not exactly double the fundamental frequency of the closed pipe of same length.

3.24 What is the ratio between their lengths, if the frequency of the fundamental note of a closed organ pipe and that of an open organ pipe are the same?

Ans. Let f be the fundamental frequency of the organ pipes, L_1 is length of closed organ pipe and L_2 is length of open organ pipe, λ_1 and λ_2 are corresponding wavelengths. If v is velocity of sound, then.

$$v = f\lambda_1 \text{ and } v = f\lambda_2$$

The length of closed organ pipe, $L_1 = \frac{\lambda_1}{4} = \frac{v}{4f}$

The length of open organ pipe is $L_2 = \frac{\lambda_2}{2} = \frac{2}{2f}$

So ratio $L_1 : L_2 :: 1 : 2$

3.25 If the frequency of the first overtone of a closed organ pipe and that of an open organ pipe are the same, what is the ratio between their lengths?

Ans. Let f be the frequency of the first overtone of the organ pipes, L_1 is the length of closed organ pipe and L_2 be length of open organ pipe, λ_1 and λ_2 are corresponding wavelengths. If v is velocity of sound, then.

$$v = f\lambda_1 \text{ and } v = f\lambda_2$$

The length of closed organ pipe, $L_1 = \frac{3\lambda_1}{4} = \frac{3v}{4f}$

The length of open organ pipe is, $L_2 = \frac{2\lambda_2}{2} = \frac{2v}{2f}$

So ratio $L_1 : L_2 :: 3 : 4$

3.26 The equation of a plane progressive wave is $Y = 0.03 \sin 2\pi (332t - x)$.

(i) If this wave is reflected from a rigid end, then what will be the equation of the reflected wave? The amplitude remains 80% after reflection.

(ii) If this wave is reflected from a free end, then what will be the equation of the reflected wave? The amplitude remains 90% after reflection.

Ans. (i) $Y = -0.024 \sin 2\pi (332t + x)$, (ii) $Y = 0.027 \sin 2\pi (332t + x)$

3.27 The equation of longitudinal stationary wave is $Y = 5 \cos \left(\frac{2\pi x}{8}\right) \sin (60\pi t)$, where x and y are in cm and t in second. Calculate the amplitude, frequency and speed of the progressive wave. What is the distance between two consecutive nodes?

Ans. 2.5 cm, 30 /s, 240 cm/s, 4 cm.

3.28 The equation of longitudinal stationary wave produced in a closed organ pipe is $Y = 4\sin\left(\frac{2\pi x}{8}\right)\cos(120\,\pi t)$, where, x and y are in cm and t in second. Find

(i) frequency, amplitude, wavelength and speed of the original progressive wave;

(ii) separation between two consecutive antinodes; and

(ii) equation of original progressive wave.

Ans. (i) $n = 60$ Hz, $a = 2$ cm, $\lambda = 8$ cm, 2 cm, 480 cm/s, (ii) 4 cm, and (iii) $Y = 2\sin\left(\frac{2\pi x}{8} - 120\,\pi t\right)$

QUESTIONS FOR EXAMINATION

3.1 Find the speed of the travelling waves.

3.2 What is wave packet? Define the phase velocity and group velocity.

3.3 What are phase velocity and group velocity? Establish a relation between phase velocity and group velocity.

3.4 Write the conditions for equal, less than and more than group velocity of a wave.

3.5 What is wave equation? Find the differential form of wave equation.

3.6 Find the pressure due to longitudinal waves.

3.7 Discuss the longitudinal standing waves in an air column partially filled with water.

3.8 Find the wavelengths of various harmonics of the longitudinal standing waves in an open organ pipe.

3.9 Find the wavelengths of various harmonics of the longitudinal standing waves in a closed organ pipe.

3.10 Sketch standing wave patterns for fundamental and second overtone for the vibration of air column in a tube (a) closed at one end, and (b) open at both ends.

3.11 Show that ratio of frequency for harmonics of closed organ pipe is 1 : 3 : 5

3.12 Show that ratio of frequency for harmonics of open organ pipe is 1 : 2 : 3

3.13 Show the positions of nodes and antinodes in an open organ pipe with the help of a diagram when the air column inside it produces third overtone.

3.14 Show with the help of a diagram the positions of nodes and antinodes in an organ pipe closed at one end when the second harmonic of its fundamental frequency is produced.

3.15 Explain the formation of standing waves in an organ pipe closed at one end and discus the various modes of vibration.

3.16 Explain the formation of standing waves in an open organ pipe and discus the various modes of vibration.

3.17 What do you mean by harmonics? With the help of a diagram show that only odd harmonics are produced in a closed organ pipe.

Note: The tones are called 'harmonics', if the frequencies of the fundamental tone and the overtones produced by a sound source are in harmonic progression.

3.18 Derive the formula $\frac{(2m-1)v}{4L}$ of the frequency of the vibrating column in closed organ pipe of length L, where, m is an integer and v is the velocity of sound in air.

3.19 Drive the formula $\frac{mv}{4L}$ of the frequency of the vibrating columnin an open organ pipe of length L, where, m is an integer and v is the velocity of sound in air.

OBJECTIVE QUESTIONS

3.1 Differential of wave equation is

(a) $\nabla^2 Y = \frac{1}{v^2}\frac{\partial^2 Y}{\partial x^2}$ (b) $\nabla^2 Y = v^2 \frac{\partial^2 Y}{\partial t^2}$ (c) $\nabla^2 Y = v^2 \frac{\partial^2 Y}{\partial x^2}$ (d) $\nabla^2 Y = \frac{1}{v^2}\frac{\partial^2 Y}{\partial t^2}$

3.2 $\frac{dv_p}{d\lambda} = 0$ implies that

(a) $v_g \neq v_p$ (b) $v_g > v_p$ (c) $v_g = v_p$ (d) $v_g < v_p$

3.3 $\frac{dv_p}{d\lambda} > 0$ implies that

(a) $v_g \neq v_p$ (b) $v_g > v_p$ (c) $v_g = v_p$ (d) $v_g < v_p$

3.4 $\frac{dv_p}{d\lambda} < 0$ implies that

(a) $v_g \neq v_p$ (b) $v_g > v_p$ (c) $v_g = v_p$ (d) $v_g < v_p$

3.5 A stationary wave is set up by holding a tuning fork near the open endin an air column of a glass tube partially filled with water. The open end of the tube will be

(a) Always a node (b) Always an anti-node
(c) Sometimes a node sometimes an anti-node (d) Neither a node nor an anti-node

3.6 In a long cylindrical tube the water level is adjusted and the air column above it is made to resonate with a vibrating tuning fork kept at the open end. Maximum sound is heard when the air column lengths are equal to

(a) $\frac{\lambda}{4}, \frac{\lambda}{2}, \frac{3\lambda}{4}$ (b) $\frac{\lambda}{4}, \lambda, \frac{3\lambda}{4}$ (c) $\frac{\lambda}{2}, \frac{3\lambda}{2}, \frac{5\lambda}{2}$ (d) $\frac{\lambda}{4}, \frac{3\lambda}{4}, \frac{5\lambda}{4}$

3.7 If the speed of sound in air is v, then fundamental frequency of the air column in a pipe of length L closed at one end will be

(a) $f = v/2L$ (b) $f = v/4L$ (c) $f = 3v/4L$ (d) $f = v/L$

3.8 If the velocity of sound in air is 340 m/s, the frequency of the fundamental note emitted by a tube of length 1 m closed at one end is

(a) 85 Hz (b) 240 Hz (c) 340 Hz (d) 170 Hz

Hint: When closed at one end, length of air column, $L = \frac{\lambda}{4} = 1$ m

The fundamental frequency, $f_C = \frac{v}{4L} = \frac{340}{4 \times 1} = 85$ Hz, as $v = f$

3.9 If oil of density higher than water is filled in place of water in a resonance tube, its frequency will

(a) decrease (b) increase

(c) depend on the density of material of tube (d) Remains unchanged

Hint: In resonance tube the lower part of it is filled with water for reflecting the sound wave. In the upper part the air column resonates with tuning fork frequency. So, if water is replaced by oil having higher density than water, resonant frequency of air column will not change in resonance tube.

3.10 In an open organ pipe the fundamental note is produced when its length is

(a) $\frac{\lambda}{4}$ (b) λ (c) $\frac{3\lambda}{4}$ (d) $\frac{\lambda}{2}$

3.11 The velocity of sound in air is 360 m/s. The fundamental frequency of an open organ pipe and of length 0.4 m will be

(a) 900 Hz (b) 450 Hz (c) 360 Hz (d) 225 Hz

Hint: When pipe open at bothlength air column, $L = \frac{\lambda}{2} = 0.3$ m

The fundamental frequency, $f_C = \frac{v}{2L} = \frac{360}{2 \times 0.4} = 450$ Hz, as $v = f$

3.12 Frequencies of open pipe open are

(a) Even multiples fundamental frequency

(b) Odd multiples fundamental frequency

(c) Fractional multiples fundamental frequency

(d) Integral multiples fundamental frequency

3.13 An open pipe produces a fundamental note. When one of its ends is closed, even then a fundamental note is emitted. The frequency of the note will be

(a) double (b) half (c) same (d) one-fourth

3.14 A cylindrical tube, open at both the ends, has fundamental frequency f in air. The tube is dipped vertically in water so that half of it is filled with water. The fundamental frequency of the air column will be

(a) $f/2$ (b) $3f/4$ (c) f (d) $2f$

Hint: Fundamental frequency of pipe open at both ends of length L: $f_0 = \frac{v}{2L} = f$

The half of the tube is dipped in water.

Now, fundamental frequency closed at one end of length $\frac{L}{2}$: $f_{01} = \frac{v}{4\frac{L}{2}} = 2\frac{v}{L} = f_0 = f$

3.15 If the fundamental frequency of a pipe closed at one end is 524 Hz, the fundamental frequency of a pipe of the same dimensions but open at both ends will be

(a) 1048 Hz (b) 524 Hz (c) 262 Hz (d) 131 Hz

Hint: When closed at one end resonates with fundamental frequency, $L = \frac{\lambda}{4}$

The fundamental frequency, $f_C = \frac{v}{4L} = 524$ Hz, as $v = f\lambda$

Fundamental frequency of pipe open at both ends $f_0 = \frac{v}{2L} = 2f_C = 1048$ Hz

3.16 An organ pipe vibrates in fundamental resonance with air, hydrogen and methane. The velocities of sound in the three cases are different, because of in three cases

(a) Frequency changes

(b) Wavelength changes

(c) Wavelength and frequency both change

(d) Wavelength and frequency both remain unchanged

Hint: Velocity of sound, $v = \sqrt{\frac{\gamma p}{\rho}}$ depends on density. Frequency of resonance tube, $f = \frac{v}{4L}$

3.17 The displacement of a particle in a string stretched in the x-direction is represented by Y. Among the following expressions for Y, those describing wave motion are:

(a) cos (**kx**) sin (ωt) (b) $k^2 x^2 - w^2 t^2$

(c) $\cos^2 (kx + \omega t)$ (d) $\cos (k^2 x^2 - w^2 t^2)$

3.18 A wave represented by the equation $y = A \cos (kx + \omega t)$ is superposed with another wave to form a standing wave such that the point $x = 0$ is node. The equation for other wave is

(a) $A \sin (kx + \omega t)$ (b) $-A \cos (kx - \omega t)$ (c) $-A \cos (kx + \omega t)$ (d) $A \sin (kx + \omega t)$

3.19 In an open organ pipe at both ends and length L, wavelength of the first overtone produced is

(a) $L/4$ (b) $3\,L/4$ (c) $L/2$ (d) L

3.20 A cylindrical resonance tube, open at both ends, has a fundamental frequency (f) in air. Half of the length of the tube is dipped vertically in water. The fundamental frequency of the air column is now

(a) $f/2$ (b) $3\,f/4$ (c) f (d) $2f$

3.21 A closed organ pipe produced a fundamental frequency of 386 Hz. This organ pipe is cut into two equal parts. The fundamental frequencies produced of these parts will be

(a) 193 Hz, 386 Hz (b) 386 Hz, 772 Hz (c) 193 Hz, 772 Hz (d) 772 Hz, 1544 Hz

3.22 The ratio of frequency of harmonics of closed organ pipe is

(a) 1 : 3: 4 (b) 1: 3 : 6 (c) 1; 3 : 5 (c) 1 :3 : 7

3.23 The ratio of frequency of harmonics of closed organ pipe is

(a) 1 : 3: 2 (b) 1: 3 : 6 (c) 1; 3 : 5 (d) 1 :2 : 3

3.24 In a closed organ pipe, the fundamental frequency is f. The ratio of the frequencies of the next three overtones is

(a) 3 : 4 : 5 (b) 2 : 3 : 4 (c) 3 : 5 : 7 (d) 3 : 7 : 9

3.25 An open pipe is suddenly closed at one end with the result that the frequency of the third harmonic of the closed pipe is found to be higher by 150 Hz than the fundamental frequency of the open pipe. The fundamental frequency of the open pipe is

(a) 300 Hz (b) 400 Hz (c) 150 Hz (d) 600 Hz

Hint: The fundamental frequency of open organ pipe, $f = \dfrac{v}{2L}$

The frequency of the third harmonic of closed organ pipe is

$$f + 150 = \frac{3v}{4L} = \frac{3}{2}f \text{ or } f = 300 \text{ Hz}$$

3.26 The frequencies of the first overtone of a closed and an open organ pipe are equal. The length of the pipes are in the ratio

(a) 1 : 2 (b) 2:3 (c) 3 : 4 (d) 4 : 5

Hint: The frequency of second harmonic of open organ pipe, $f_0 = \dfrac{2v}{2\,L_1}$

The frequency of the second harmonic of closed organ pipe, $f_c = \dfrac{3v}{4L_2}$

$$f_c = f_0 \text{ or } f_c = \frac{3v}{4L_2} = \frac{2v}{2L_1} \text{ or } \frac{L_1}{L_2} = \frac{4}{3}$$

3.27 The fundamental frequency of an open organ pipe is 300 Hz. The first overtone of this pipe is the same as the first overtone of a closed organ pipe. If the speed of sound is 330 m/s, then the length of the closed organ pipe is

(a) 51.5 cm (b) 41.25 cm (c) 61.5 cm (d) 71.25 cm

Hint: Fundamental frequency of open organ pipe, $f_0 = \dfrac{v}{2L_1} = 300$ Hz

The frequency of second harmonic of open organ pipe, $f_{10} = \dfrac{2v}{2L_1} = \dfrac{v}{L_1} = 600$ Hz

The frequency of second harmonic of closed organ pipe, $f_c = \dfrac{3v}{4L_2}$

$$f_{1c} = f_{10} = \frac{3v}{4L_2} = \frac{v}{L_1} = 600$$

or $$L_2 = \frac{3v}{4\times 600} = \frac{3\times 330}{4\times 600} = 0.4125 \text{ m} = 41.25 \text{ cm}$$

3.28 An open organ pipe of length 33 cm resonates with the frequency of 1000 Hz. If the speed of sound is 330 m/s, then this frequency is

(a) fundamental frequency of the organ pipe

(b) first overtone of the pipe

(c) second overtone of the pipe

(d) fourth overtone of the pipe

Hind: Fundamental frequency of open organ pipe, f_0 = 1000 Hz

$$f_0 = \frac{v}{2L_1} \text{ or } L_1 = \frac{v}{2f_0} = \frac{330}{2 \times 1000} = 0.165 \text{ m}$$

Frequency first overtone of open organ pipe

$$f_{10} = \frac{2v}{2L_1} \text{ or } L_1 = \frac{v}{f_{10}} = \frac{330}{1000} = 0.33 \text{ m}$$

That is, frequency is the first overtone of the pipe.

3.29 A closed organ pipe (P_1) vibrating in its first overtone and other open organ pipe (P_2) vibrating in its third overtone resonate with a given tuning fork. The ratio of the length of the organ pipes (P_1 and P_2) is

(a) 8/3 (b) 3/8 (c) 16/3 (d) 3/16

Hint: Frequency of third overtone (fourth harmonic) of an open organ pipe $P_2\, f_{30} = \frac{4v}{2L_2}$

The frequency of first overtone (second harmonic) of closed organ pipe $P_1\, f_{1c} = \frac{3v}{4L_1}$

$$f_{1c} : f_{30} \text{ or } \frac{3v}{4L_1} = \frac{4v}{2L_3} \text{ or } \frac{L_1}{L_2} = \frac{3}{8}$$

3.30 A closed organ pipe of length L and an open organ pipe contain gases of densities, ρ_1 and ρ_2, respectively. The compressibility of gases is same in both the pipes which are vibrating in their first overtone with the same frequency. The length of the open organ pipe is

(a) $\frac{3L}{4}$ (b) $\frac{4L}{3}$ (c) $\frac{4L}{3}\sqrt{\frac{\rho_1}{\rho_2}}$ (d) $\frac{4L}{3}\sqrt{\frac{\rho_2}{\rho_1}}$

Hint: Velocity of sound in a gas, $v = \sqrt{\frac{\gamma P}{\rho}}$

Let the velocity of sound in the gas of density, $\rho_1 : v_1 = \sqrt{\frac{\gamma P}{\rho_1}}$

and the velocity of sound in the gas of density, $\rho_2 : v_2 = \sqrt{\frac{\gamma P}{\rho_2}}$

The ratio, $\frac{v_1}{v_2} = \sqrt{\frac{\rho_2}{\rho_1}}$

Fundamental frequency of closed organ pipe of length L, $f_c = \frac{v}{4L}$

Fundamental frequency of open organ pipe of length L_1, $f_0 = \frac{v}{2L}$

Frequency of the first overtone (second harmonic) of closed organ pipe of length L in gas of density, ρ_1:

$$f_{1c} = \frac{3v_1}{4L}$$

Frequency of first over tone (second harmonic) of open organ pipe of length L_1 in gas of density, ρ_2

$$: f_{i0} = \frac{2v_2}{2L_1}$$

$$f_{1c} = f_{1c} \text{ or } \frac{3v_1}{4L} = \frac{2v_2}{2L_1} \text{ or } L_1 = \frac{4L}{3}\frac{v_2}{v_1} = \frac{4L}{3}\sqrt{\frac{\rho_1}{\rho_2}}$$

3.31 How many nodes and antinodes will be there, when the closed organ pipe produces the third overtone?

(a) three nodes, three antinodes
(b) three nodes, four antinodes
(c) four nodes, three antinodes
(d) four nodes, four antinodes

Hint: In closed organ pipe, the fundamental frequency, $f = \frac{v}{4L}$ and $L = \frac{\lambda}{4}$

In closed organ pipe, the frequency of the third overtone (fourth harmonic), $f_3 = \frac{7v}{4L}$ and

$L = \frac{7\lambda}{4}$

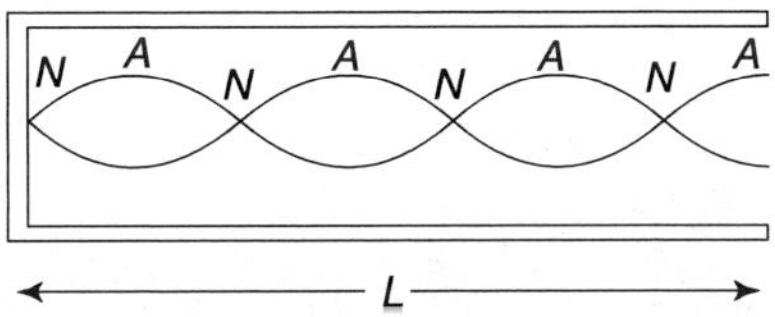

Fig. 3.20

3.32 Velocity of sound in air is 320 m/s. A closed organ pipe of length 1 m does not resonate with frequency

(a) 80 Hz (b) 240 Hz (c) 320 Hz (d) 400 Hz

Hint: Fundamental frequency of closed organ pipe, $f_c = \frac{v}{4L} = \frac{320}{4 \times 1} = 80$

The frequency of the second harmonic of closed organ pipe, $f_{c1} = \frac{3v}{4L_2} = 3\,f_c = 240$ Hz

The frequency of the third harmonic of closed organ pipe, $f_{c1} = \frac{5v}{4L_2} = 5\,f_c = 400$ Hz

3.33 The length of the air column 0.15 m resonates in the fundamental mode with a tuning fork. When this length is changed to 0.50 m, the same tuning fork resonates with the first overtone. The end correction is

(a) 0.020 m (b) 0.025 m (c) 0.05 m (d) 0.012 m

Hint: End correction: In the resonance tube, the antinode is not formed exactly at the open end but slightly outside at a distance, *e*. Hence, the length of the air column in the first and second states of resonance are $(L_1 + e)$ and $(L_2 + e)$. Then

$$L_1 + e = \frac{\lambda}{4} \text{ and } L_2 + e = \frac{3\lambda}{4}$$

$$\frac{L_2 + e}{L_1 + e} = 3 \text{ or } (L_2 + e)$$

$$= 3\,(L_1 + e) \text{ or } 2\,e = L_2 - 3\,L_1 \text{ or } e = \frac{L_2 \quad L_1}{}$$

$$e = \frac{L_2 - 3L_1}{2} = e = \frac{0.50 - 3 \times 0.15}{2} = 0.025$$

3.34 The diameter of resonance column tube is 4 cm. The frequency of tuning fork is 525 Hz. The speed of sound at temperature 38°C is 336 m/s. When the first resonance occurs, the length of air column from water level is

(a) 14.6 cm (b) 15.8 cm (c) 16.2. cm (d) 17.5 cm

Hint: Fundamental frequency of resonance column, $f = \frac{v}{4L}$ or $L = \frac{v}{4f_c}$

The end correction: $e = 0.3\,d = 0.3 \times 4 = 1.2$ cm

Effective length of the column, $L + e = \frac{\lambda}{4} = \frac{v}{4f} = \frac{336}{4 \times 525} = 0.16 \text{ m} = 16 \text{ cm}$

or $$L = 16 \text{ cm} - 1.2 \text{ cm} = 15.8 \text{ cm}$$

Answers

3.1 (d)	3.2 (c)	3.3 (d)	3.4 (b)	3.5 (b)	3.6 (d)	3.7 (b)	3.8 (a)
3.9 (d)	3.10 (d)	3.11 (b)	3.12 (d)	3.13 (b)	3.14 (c)	3.15 (a)	3.16 (a)
3.17 (a)	3.18 (b)	3.19 (d)	3.20 (c)	3.21 (d)	3.22 (b)	3.23 (c)	3.24 (c)
3.25 (a)	3.26 (c)	3.27 (b)	3.28 (b)	3.29 (b)	3.30 (c)	3.31 (d)	3.32 (c)
3.33 (b)	3.34 (b)						

CHAPTER 4

Wave Motion in a String

4.1 INTRODUCTION

In a stretched string when a transverse wave travels, it is reflected back on reaching the end. The reflected wave has π phase difference and same frequency, speed and amplitude. The incident and reflected waves on superposition result in a standing wave or stationary wave.

If stationary wave is plucked or struck, a number of different stationary waves are produced simultaneously. Only specific nodes are possible. These are discussed in Section 4.8.

The patterns of energy which move from one place to another, without physical transfer or flow of matter, are called *waves*. Most familiar waves such as waves in a string, water waves, sound waves, seismic waves, etc., are called *mechanical waves*. These waves require a medium for propagation, they cannot travel through vacuum. They propagate through oscillations of constituent particles and depend on the elastic properties of the medium. The electromagnetic waves such as visible light waves, X-rays, γ-rays, radio waves do not need any medium to travel. They can travel in vacuum.

A third kind of wave is so called the *matter waves*. They propagate through the constituents of matter such as electrons, protons, neutrons, atoms and molecules. They rise in quantum mechanical description of matter.

The periodic motion of the wave is equivalent to the periodic motion of a particle in a circle. Let r be the radius of the circle and ω, the angular frequency. The velocity (v) of the wave particle is written as

$$v = \omega r = 2\pi f r = \frac{2\pi v r}{\lambda} \text{ or } \lambda = 2\pi r$$

Here, f and λ are the frequency and wavelength of the wave, respectively.

In case of string vibration, if T is the tension on the string and m is mass per unit length of the string, then the velocity of wave propagation in the string is given by

$$v = \sqrt{\frac{T}{m}}$$

The proof of this equation is given in Appendix A.

4.2 TRANSVERSE WAVES IN A STRING

The vibrations of a string are transverse. The transmission of the transverse wave of a rope through a slit is shown in Fig. 5.1 of Section 5.2 of Chapter 5. It is observed that transverse wave formed by the rope does not pass through the slit, if the slit is kept perpendicular to the plane of vibrations as shown in Fig. 5.1(a). When the slit is kept parallel to plane of vibrations then the transverse wave formed by the rope passes through the slit as shown in Fig. 5.1(b). The transverse wave formed by the rope is called plane polarized wave. The meaning of the plane of polarization is that the vibrations of the transverse wave are restricted to one plane only. Plane of polarization is perpendicular to the direction of the motion of a string.

4.3 STANDING WAVES

A standing (stationary) wave is a resultant wave due to superposition of two travelling waves moving in opposite direction having the same frequency, speed and amplitude. A standing wave is also called stationary wave, because the particles at some points of the resultant wave always remain stationary.

Formation of standing waves obtained due to superposition of waves A and B are shown in Fig. 4.1. Initially (i.e., $t = 0$), A and B are in phase with each other (shown by dashed lines) overlap each other. The resultant wave due to superposition of A and B waves is shown by solid line in Fig. 4.1.(a). One-eighth of period later (i.e., $t = T/8$, $x = \lambda/8$ or $\pi/4$) wave A has moved to the left and wave B has moved to the right. Their positions and resultant is shown in Fig. 4.1(b). At $t = T/4$ (i.e., $x = x = \lambda/4$ or $\pi/2$), Waves A and B are completely out of phase from each other and therefore, the resultant becomes zero at every point as shown in Fig. 4.1(c). The resultant waves, A and B, at time $t = 3\ T/8$ (i.e., $x = 3\lambda/8$ or $3\pi/4$) and at time $t = T/2$ (i.e., $x = \lambda/2$ or π) are shown in Fig. 4.1(d) and 4.1(e).

In this standing wave, the displacement is always zero points N_1, N_2 and N_3 as shown in Fig. 4.1 (e). The positions of zero amplitude in the stationary wave are called nodes. There arepointsof maximum amplitude half-way between two nodes in this standing wave. The positions of maximum amplitudes in the standing wave are called antinodes.

Figure 4.2 shows the resultant displacement changes between $t = 0$ and $t = T/2$. The displacement is always zero at nodes. Half-way between the two nodes in this standing wave there are positions of maximum amplitude. These positions of maximum amplitudes in

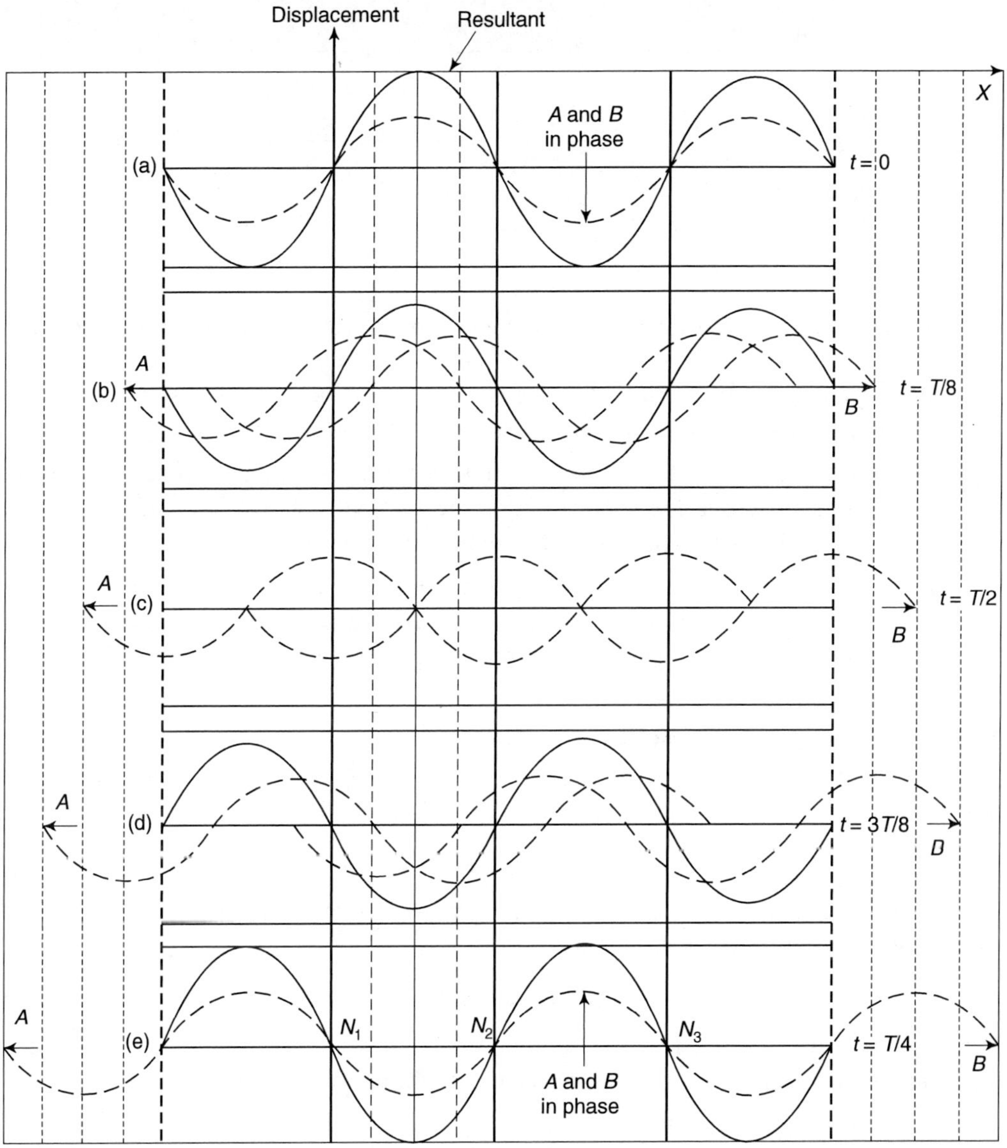

Fig. 4.1 Stationary wave when two waves moving in opposite direction superpose to each other. N_1, N_2 and N_3 are nodes. Distance $N_1 N_2 = N_1 N_2 = \lambda/2$, i.e., separation between two adjacent nodes is $\lambda/2$. Maximum position of resultant wave is called antinode. The separation between two adjacent antinodes is $\lambda/2$

the standing wave are called antinodes (A). The positions of zero amplitude of standing wave are called nodes and positions of maximum amplitude are called antinodes. The following points are to be noted in case of a stationary wave:

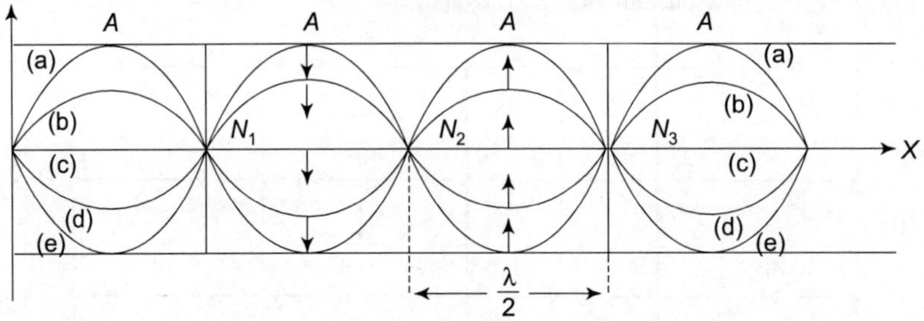

Fig. 4.2 Successive displacement of stationary waves. *A* and *N* are antinodes and nodes

(i) The separation of two adjacent nodes is $\lambda/2$, i.e., the distance, $N_1 N_2 = N_1 N_2 = \lambda/2$. Similarly, the separation between two adjacent antinodes is $\lambda/2$.

(ii) The amplitude of a stationary wave, unlike progressive wave, depends on position. It ranges from zero at the nodes to $2a$ at the antinodes, where, a is the amplitude of each progressive waves producing the stationary wave.

(iii) At any one time, all the particles between two adjacent nodes have the same phase, while each particle between the next pair of adjacent nodes is out of phase (i.e., have phase difference of π radians) with respect to particles between previous pair of nodes.

Mathematical treatment of stationary waves

Let the displacement (Y_1) of a travelling wave of frequency (ω) and amplitude (a) moving from right to left is written as

$$Y_1 = a \sin (\omega t - kx) \qquad ...(4.1a)$$

Let the displacement (Y_2) of another travelling wave of same frequency (ω) and amplitude (a) moving from left to left is written as

$$Y_2 = \sin (\omega t + kx) \qquad ...(4.1b)$$

A standing wave is formed by the addition of two travelling waves moving in opposite direction, i.e., the displacement of stationary wave will be

$$Y = Y_1 + Y_2$$

$$= a \sin (\omega t - kx) + a \sin (\omega t + kx)$$

$$= a [\sin (\omega t) \cos kx) - \cos (\omega t) \sin (kx)] + a [\sin (\omega t) \cos (kx) + \cos (\omega t) \sin (kx)]$$

$$= 2 a \sin (\omega t) \cos (kx)$$

$$= A \sin (\omega t) \qquad ...(4.2)$$

where, $2 a \cos (kx) = A$ and $k = \dfrac{2\pi}{\lambda}$ is the wave vector of the travelling wave.

Note that the resultant wave given by Eq. (4.2) is composed of two components:

(i) sin (ωt) – this shows that the amplitude varies with time t at a particular position.

(ii) cos (kx) – this shows that the amplitude varies with position x at a particular time.

Case 1. When $A = 2\,a$, i.e., when amplitude is maximum

$$\cos(kx) = \pm 1,$$

This implies that $kx = m\,\pi$, where, $m = 0, 1, 2, ...$

or
$$\frac{2\pi}{\lambda}\,x = m\pi \text{ or } x = m\,\frac{\lambda}{2}$$

i.e., when $x = 0, \frac{\lambda}{2}, \frac{2\lambda}{2}$,...., the amplitude ($A$) is maximum or $A = 2\,a$. The positions, at which the amplitude is maximum, are called antinodes.

That is at positions $x = 0, \frac{\lambda}{2}, \frac{2\lambda}{2}$,...., there are antinodes. The separation between two adjacent antinodes is $\frac{\lambda}{2}$.

Case 2. When $A = 0$, i.e., when amplitude is minimum

$$\cos(kx) = 0$$

This implies that $kx = (2\,m - 1)\,\frac{\pi}{2}$, where, $m = 1, 2, 3,....$

or
$$\frac{2\pi}{\lambda}\,x = (2\,m - 1)\,\frac{\pi}{2} \text{ or } x = (2\,m - 1)\,\frac{\lambda}{4}$$

i.e., when $x = \frac{\lambda}{4}, \frac{3\lambda}{4}, \frac{5\lambda}{4}$,....., the amplitude ($A$) is minimum or $A = 0$. The positions, at which the amplitude is zero, are called nodes.

That is, at positions $x = \frac{\lambda}{4}, \frac{3\lambda}{4}, \frac{5\lambda}{4}$,...., there are nodes. The distance between two adjacent nodes is $= \frac{\lambda}{2}$.

Summary

A stationary wave is the resultant wave of two travelling waves of same frequency, speed and amplitude moving in opposite directions.

A node is a point of zero displacement.

An antinode is point of maximum displacement.

Separation of two adjacent nodes = separation of two adjacent antinodes = $\frac{\lambda}{2}$.

All particles between two adjacent notes are in the same phase, but completely out of phase with particles between next pairs of adjacent nodes.

4.4 CHARACTERISTICS OF STATIONARY WAVES

The following are the characteristics of the stationary waves:

1. A stationary wave is produced when two wave of same wavelength, same time period, same frequency and same speed moving in opposite directions are superposed.
2. The points at which particles do not oscillate are called nodes.
3. The points, at which the displacement of the particle of the medium during oscillation is maximum, are called antinodes.
4. The distance between any two adjacent nodes is half of the wavelength.
5. The distance between any two adjacent antinodes is half of the wavelength.
6. The distance between any two adjacent nodes and antinodes is one-fourth of the wavelength.
7. Except at nodes, each constituent of the medium executes simple harmonic oscillation about its mean position with time period equal to the time period of the wave motion.
8. Amplitudes of all constituents of the medium are not the same. It is zero at the nodes and maximum at the antinodes.

4.5 COMPARISON BETWEEN PROGRESSIVE AND STATIONARY WAVES

	Progressive waves	Stationary waves
1.	A progressive wave is produced due to the oscillation of the constituents of the medium.	A stationary wave is formed when two identical progressive waves travelling in opposite direction are superimposed.
2.	The velocity of waves depends on the properties of the medium.	The waves remain stationary, i.e., they do not move.
3.	Each constituent of the medium executes periodic motion about their mean position with the same amplitude.	Except the node, all constituents of the medium execute simple harmonic motion with varying amplitude.
4.	There is a continuous change of the phase from constituent to constituent.	All constituents between two adjacent nodes are at the same phase, but they differ in phase by from those in the preceding as well as succeeding segments.
5.	At any instant all the constituents do not come together to the mean position, they pass their mean position in succession but with the same velocity.	All constituents pass their mean position at a time, but with different velocities.
6.	No constituents of the medium is at rest permanently.	At nodes the particles are at rest.

Contd.

7.	Throughout the medium the compression and rarefaction move from point to point.	The compression and rarefaction do not move from point to point. At certain fixed equidistance points, they simply appear and disappear.
8.	In case of longitudinal waves wavelength consistsof a compression and a rarefaction, in the case of transverse waves it consists of a crest and a trough.	The wavelength is the distance between two nodes or two antinodes.
9.	The transmission of energy occurs across every plane in the direction of propagation of the waves.	Flow of energy does not occur across any plane.
10.	The equation of a stationary wave is written as $Y_1 = a \sin (\omega t \mp kx)$, where, $k = \frac{2\pi}{\lambda}$, a is the amplitude of the travelling wave, ω is the angular frequency, and λ is the wavelength of the wave.	The equation of a stationary wave is written as $Y = A \sin (\omega t)$, where, $A = 2a \cos (kx)$ is the amplitude of the stationary wave.

4.6 WAVE EQUATION OF A STRING

Let the displacement (Y_1) of a travelling wave of frequency (ω) and amplitude (a) at position x and time (t) moving from right to left is written as

$$Y_1 = a \sin \frac{2\pi}{\lambda} (vt - x) \qquad ...(4.3)$$

Let the displacement (Y_2) of reflected travelling wave of same frequency (ω) and amplitude (a) at position x and time (t) moving from left to right is written as

$$Y_2 = \sin \frac{2\pi}{\lambda} (vt + x) \qquad ...(4.4)$$

The linear sum of two travelling waves moving in opposite direction will be

$$Y = Y_1 + Y_2$$

$$= a \sin \frac{2\pi}{\lambda} (vt - x) + a \sin \frac{2\pi}{\lambda} (vt + x)$$

$$= a \left[\sin \frac{2\pi}{\lambda} (vt) \cos \left(\frac{2\pi}{\lambda} x \right) - \cos \frac{2\pi}{\lambda} (vt) \sin \left(\frac{2\pi}{\lambda} \right) x \right]$$

$$- a \left[\sin \frac{2\pi}{\lambda} (vt) \cos \left(\frac{2\pi}{\lambda} x \right) + \cos \frac{2\pi}{\lambda} (vt) \sin \left(\frac{2\pi}{\lambda} \right) x \right]$$

$$= -2a \cos \frac{2\pi}{\lambda} (vt) \sin \left(\frac{2\pi}{\lambda} x \right)$$

$$= A \cos \frac{2\pi}{\lambda} (vt) \qquad ...(4.5)$$

where, amplitude of the wave, $A = -2\,a\,\sin\left(\frac{2\pi}{\lambda}x\right)$. Equation (4.5) is an equation of standing wave a standing wave, because the positions of zero amplitude and maximum amplitude remains fixed. The positions of zero and maximum amplitudes are called nodes.

4.7 DIFFERENTIAL FORM OF WAVE EQUATION OF STRING

The wave equation of a string is given by Eq. (4.5):

$$Y(x, t) = -2\,a\cos\frac{2\pi}{\lambda}(vt)\sin\left(\frac{2\pi}{\lambda}x\right) \qquad ...(4.6)$$

Differentiating partially Eq. (4.6) w.r.t. position x, we get

$$\frac{\partial Y}{\partial x} = -\frac{2\pi}{\lambda}\times 2\,a\cos\frac{2\pi}{\lambda}(vt)\cos\left(\frac{2\pi}{\lambda}x\right)$$

Further differentiating partially Eq. (4.6) w.r.t. position x, one gets

$$\frac{\partial^2 Y}{\partial x^2} = \left(\frac{2\pi}{\lambda}\right)^2 \times 2\,a\cos\frac{2\pi}{\lambda}(vt)\sin\left(\frac{2\pi}{\lambda}x\right) = \left(\frac{2\pi}{\lambda}\right)^2 Y \qquad ...(4.7)$$

Differentiating partially Eq. (4.6) w.r.t. position t, we get

$$\frac{\partial Y}{\partial t} = \frac{2\pi v}{\lambda}\times 2\,a\sin\frac{2\pi}{\lambda}(vt)\sin\left(\frac{2\pi}{\lambda}x\right)$$

Further differentiating partially Eq. (4.6) w.r.t. position t reduces to

$$\frac{\partial^2 Y}{\partial t^2} = \left(\frac{2\pi\, v}{\lambda}\right)^2 \times 2\,a\cos\frac{2\pi}{\lambda}(vt)\sin\left(\frac{2\pi}{\lambda}x\right) = \left(\frac{2\pi\, v}{\lambda}\right)^2 Y \qquad ...(4.8)$$

Putting the value of Y from Eq. (4.8), Eq. (4.7) becomes

$$\frac{\partial^2 Y}{\partial x^2} = \frac{1}{v^2}\frac{\partial^2 Y}{\partial t^2} \qquad ...(4.9)$$

Equation (4.9) is the differential form of wave equation of string.

Putting the value of velocity of wave, $v = \sqrt{\frac{T}{m}}$

$$T^2\frac{\partial^2 Y}{\partial x^2} = m^2\frac{\partial^2 Y}{\partial t^2} \qquad ...(4.10)$$

where, m is the mass of string per unit length and T is the tension of the string.

In three dimensions, wave equation of a string is written as

$$\nabla^2 Y = \frac{1}{v^2}\frac{\partial^2 Y}{\partial t^2} \qquad ...(4.11)$$

and
$$T^2\nabla^2 Y = m^2 \frac{\partial^2 Y}{\partial t^2} \qquad ...(4.12)$$

where, $\nabla^2 = \frac{\partial^2}{\partial x^2} + \frac{\partial^2}{\partial y^2} + \frac{\partial^2}{\partial z^2}$ is a Laplacian operator.

4.8 NORMAL MODES OF STRETCHED STRING

Standing waves can be produced in stretched strings or in air columns by resonance effects. *Standing waves* are also known as *stationary waves*, because particles at some points of standing wave always remain stationary. Because of the boundary conditions a vibrating system (a string fixed at two ends or a string fixed at one end and free at the other end or an air column partially filled with water) can oscillate only in some special patterns which are called *normal modes.*

Let us consider a stretched string whose both ends are clamped as shown in Figs. 4.3(a) to (c), so they cannot oscillate, i.e., they remain at rest permanently. The vibrations of stretched string whose only one end is fixed will be considered in Section 4.9. When a string under tension is set into vibration, transverse harmonic waves propagate along it from one end to another end. If the length of string is finite, harmonic transverse waves propagate in one direction, then a reflected transverse harmonic wave propagates in opposite direction. The harmonic waves of same frequency, speed and amplitude moving in opposite direction on superposition produce standing waves of large amplitude.

The string vibrates in such a way that fixed points are called *nodes* as fixed points are not free to move, while the point between two adjacent nodes is called an *antinode* as here the displacement is maximum as shown in Fig. 4.3.

Mathematical treatment of stationary waves

Let the displacement (Y_1) of a travelling wave of frequency (ω) and amplitude (a) at position x and time t moving from right to left is written as

$$Y_1 = a \sin \frac{2\pi}{\lambda}(vt - x) \qquad ...(4.13)$$

Let the displacement (Y_2) of reflected travelling wave of same frequency (ω) and amplitude (a) at position x and time t moving from left to left is written as

$$Y_2 = - \sin \frac{2\pi}{\lambda}(vt + x) \qquad ...(4.14)$$

The linear sum of two travelling waves Y_1 and Y_2 moving in opposite direction, will be [see Section 4.6]

$$Y = A \cos \frac{2\pi}{\lambda}(vt) \qquad ...(4.15)$$

where, amplitude of the wave $A = - 2\, a \sin\left(\frac{2\pi}{\lambda} x\right)$. The function of Eq. (4.15) does not describe a moving wave, because x and t dependence of the function is not the combination of $(vt \pm x)$. That is, the function of Eq. (4.15) describes a standing wave, because the positions of zero amplitude and maximum amplitude remain fixed. The positions of zero and maximum amplitudes are called nodes.

In Fig. 4.3, we have shown the standing waves in a string clamped at both the ends.

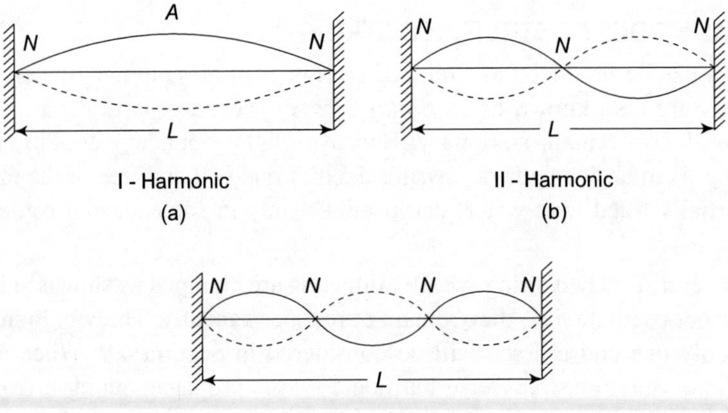

III - Harmonic

(c)

Fig. 4.3

The positions of zero amplitude (node) can be found as follows.

The displacement at ends $x = 0$, because $\sin(0) = 0$. If L is the length of string, the displacement at end $x = L$ will be

$$A = 2\,a\,\sin\left(\frac{2\pi}{\lambda}L\right)$$

The amplitude at $x = L$ will also be zero, i.e., $A = 0$ at $x = L$, only when

$$\sin\left(\frac{2\pi}{\lambda}L\right) = 0$$

This implies that $\frac{2\pi}{\lambda}L = m\,\pi$, where, $m = 1, 2, 3, ...$

or $$\lambda = \frac{2L}{m}$$

i.e., when $\lambda = 2\,L, \frac{2L}{2}, \frac{2\lambda}{3}, ...$

The mode with wavelength, $\lambda_1 = 2\,L$ and frequency, $f_1 = v/2\,L$ is called the fundamental frequency, or the first harmonic or pitch of the tone.

The next mode with wavelength $\lambda_2 = L$ and frequency $f_2 = v/L = 2\,f_1$ is called the second harmonic or first overtone.

The next mode with wavelength $\lambda_3 = \frac{2L}{3}$ and frequency $f_3 = 3\ v/2L = 3f_1$ is called the third harmonic or second overtone.

In general, mode with wavelength $\lambda_m = \frac{2L}{m}$ and frequency $f_m = m\ v/2L = m\ f_1$ is called the mth harmonic or $(m - 1)^{th}$ overtone

If we examine the right-hand side of Eq. (4.15), it shows that each point on the string is executing a simple harmonic motion (SMH). The amplitude of the SMH at point x is $2a$ sin $(2\pi\ x/\lambda)$. The phase at time $t = 0$ for all these SMH's is zero. Therefore, when standing waves have been set up in a string clamped at both ends, the points in the string vibrate in phase and execute SMH. There are points in the string which do not oscillate at all. These points are called the nodes located at:

In case of the first harmonic of wavelength

$$\lambda_1 = 2L \text{ and frequency, } f_1 = v/2L$$

There are two nodes: One at $x = 0$ and the second at $x = L$ as shown in Fig. 4.3(a).

In case of the second harmonic of wavelength

$\lambda_2 = L$ and frequency, $f_2 = v/L = 2f_1$, there are three nodes: $x = 0$, $x = \frac{L}{2}$ and $x = L$ as shown in Fig. 4.3(b).

In case of the third harmonic of wavelength

$\lambda_3 = 2L/3$ and frequency $f_3 = 3v/2L = 3f_1$, there are four nodes: at $x = 0$, $x = L/3$, $x = \frac{2L}{3}$ and $x = L$ as shown in Fig. 4.3(c)

In case of mth harmonic of wavelength

$\lambda_m = \frac{2L}{m}$ and frequency, $f_m = m\ v/2L = m\ f_1$, there are $(m + 1)$ nodes: at $x = 0$, L/m, $\frac{2L}{m}, \frac{3L}{m}, \frac{4L}{m}, \frac{(m-1)\,L}{m}, L.$

Separation of two nodes is equal to L/m.

There are points in the string which execute SMH with maximum amplitude, $2a$. These points, called antinodes, are in midway between successive nodes.

The positions of maximum amplitude (antinode) can be found as follows:

The displacement at end $x = 0$ is zero, because sin (0) = 0. If L is the length of string, the displacement at the end $x = L$ will also be zero. The displacement at any position x is given as

$$A = 2\,a\,\sin\left(\frac{2\pi}{\lambda}x\right)$$

The amplitude of the standing wave will be maximum, i.e., $A = 2a$, when

$$\sin\left(\frac{2\pi}{\lambda}x\right) = \pm 1$$

This implies that $\frac{2\pi}{\lambda}x = (2m-1)\frac{\pi}{2}$, where, $m = 1, 2, 3,....$

or $$\lambda = \frac{4x}{(2m-1)} \text{ or } x = \frac{(2m-1)\lambda}{4}.$$

or $$x = \frac{\lambda}{4}, \frac{3\lambda}{4}, \frac{5\lambda}{4}, ...\frac{(2m-1)\lambda}{4}$$

In case of the first harmonic of wavelength

$\lambda_1 = 2L$ and frequency, $f_1 = v/2L$, there is one maximum called antinode at $x = \frac{\lambda_1}{4} = \frac{L}{2}$ as shown in Fig. 4.3(a).

In case of the second harmonic of wavelength

$$\lambda_2 = L \text{ and frequency, } f_1 = v/L = 2f_1$$

There are two antinodes: at $x = \frac{\lambda_2}{4} = \frac{L}{4}, x = \frac{3\lambda_2}{4} = \frac{3L}{4}$

In case of the third harmonic of wavelength

$$\lambda_3 = \frac{2L}{3} \text{ and frequency, } f_3 = 3v/2L = 3f_1$$

There are three nodes: at $x = \frac{\lambda_3}{4} = \frac{1}{4}\frac{2L}{3}, x = \frac{3\lambda_3}{4} = \frac{3}{4}\frac{2L}{3}, x = \frac{5\lambda_3}{4} = \frac{5}{4}\frac{2L}{3}$

In general, for m^{th} harmonic of wavelength

$$\lambda_m = \frac{2L}{m} \text{ and frequency, } f_m = m\,v/2L = m f_1$$

There are m antinodes: at $x = \frac{\lambda_m}{4} = \frac{1}{4}\frac{2L}{m}, x = \frac{3\lambda_m}{4} = \frac{3}{4}\frac{2L}{m}, x = \frac{5\lambda_m}{4} = \frac{5}{4}\frac{2L}{m}, ...$

$$x = \frac{(2m-1)\lambda_m}{4} = \frac{(2m-1)}{4}\frac{2L}{m}$$

In case of mth harmonic, the separation of two antinodes is equal to $\frac{\lambda_m}{2} = \frac{L}{m}$

According to the principle of superposition a stretched string tied at both ends can vibrate simultaneously in more than one mode. Which mode is strongly excited depends on where the string is plucked or bowed. Musical instruments like the sitar and violin are designed on this principle.

Summary

- A stationary wave is the resultant wave of two travelling waves of same frequency, speed and amplitude moving in opposite directions.
- A node is a point of zero displacement.
- An antinode is point of maximum displacement.
- Separation of two adjacent nodes = separation of two adjacent antinodes = $\frac{\lambda}{2}$.
- All particles between two adjacent nodes are in the same phase, but completely out of phase with particles between the next pairs of adjacent nodes.

4.9 STANDING WAVES IN A STRING: FIXED AT ONE END AND FREE AT THE OTHER END

We next study the modes of vibrations of a string tied at one end and with the other end free. The string vibrates in such a way that fixed points are nodesas fixed points are not free to move, while the free end is an antinode as here the displacement is maximum as shown in Fig. 4.4.

From the expression of the wave function given by Eq. (4.15) we note that the end $x = L$ will be an antinode if and only if

$$\left|\sin \frac{2\pi L}{\lambda}\right| = 1 \qquad ...(4.16)$$

This condition defines the wavelengths of normal modes in a string with one end free. The condition given by Eq. (4.16) implies that

$$\frac{2\pi L}{\lambda} = (2\,m - 1)\,\frac{\pi}{2}, \qquad m = 1, 2, 3,....$$

or

$$\lambda = \frac{4L}{(2\,m - 1)}$$

or

$$\lambda = 4\,L, \frac{4L}{3}, \frac{4L}{5}, ..., \frac{4L}{(2\,m-1)}$$

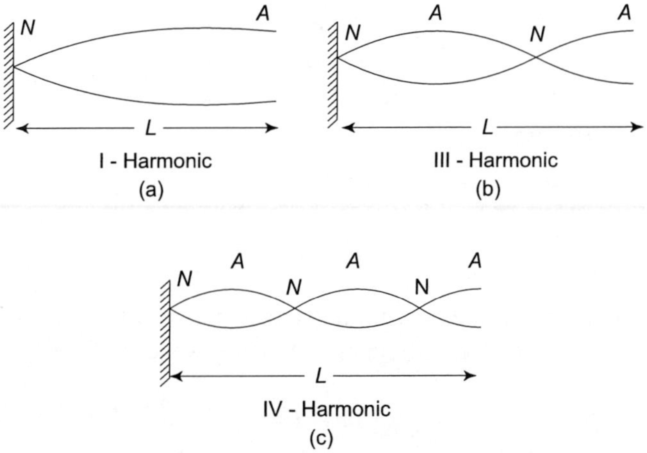

I - Harmonic
(a)

III - Harmonic
(b)

IV - Harmonic
(c)

Fig. 4.4

In Fig. 4.4(a) to (c) we have shown some normal modes in a string with one end free.

Standing waves shown in Fig. 4.4(a) are called the first harmonic or the fundamental with wavelength

$$\lambda_1 = 4\,L \text{ and frequency, } f_1 = v/\lambda_1 = v/4\,L.$$

Standing waves shown in Fig. 4.4(b) are called the second harmonic or the first overtone with wavelength

$\lambda_2 = 4\,L/3$ and frequency, $f_2 = v/\lambda_2 = 3\,v/4\,L$ or $f_2 = 3\,f_1$, i.e., frequency of the second harmonic is three times of the fundamental frequency.

Standing waves shown in Fig. 4.4(c) are called the third harmonic or the second overtone with wavelength,

$\lambda_3 = 4\,L/5$ and frequency, $f_3 = v/\lambda_3 = 5\,v/4\,L$ or $f_3 = 5\,f_1$, i.e., frequency of the third harmonic is five times of the fundamental frequency.

In general, standing waves of mth harmonic or the $(m-1)^{th}$ overtone with wavelength,

$\lambda_3 = 4\,L/(2\,m - 1)$ and frequency, $f_m = v/\lambda_m = (2\,m - 1)\,v/4\,L$ or $f_m = (2\,m - 1)\,f_1$, i.e., frequency of mth harmonic is $(2\,m - 1)$ times of the fundamental frequency.

4.10 VIBRATIONS OF STRINGS

When a string under tension is set into vibration, transverse harmonic waves propagate along it. If the length of string is finite, reflected wave will also exists and the standing waves of large amplitude are formed due to multiple reflections.These standing waves are called *resonant standing waves* and are the waves of finite length, whose either one or both ends are fixed.

The fixed points of the string at which the string is not free to move are called nodes, while the point of plucking or free end, at which the displacement of the string is maximum, is called an antinode. The formation of standing waves is shown in Fig. 4.5.

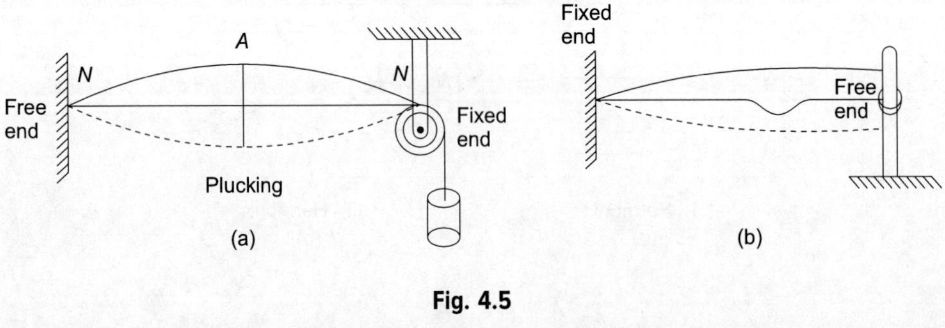

Fig. 4.5

From Fig. 4.5 one note that the minimum distance between two nodes is $\lambda/2$, so when a string is fixed at both the ends [see article 4.8] standing waves will be formed.

$$L = \frac{\lambda}{2} \quad \text{or} \quad L = \frac{2\lambda}{2} \quad \text{or} \quad L = \frac{3\lambda}{2}$$

i.e., $2\,L = n\lambda$ or $\lambda_n = \dfrac{2L}{n}$ with $n = 1, 2, 3$ ...(4.17)

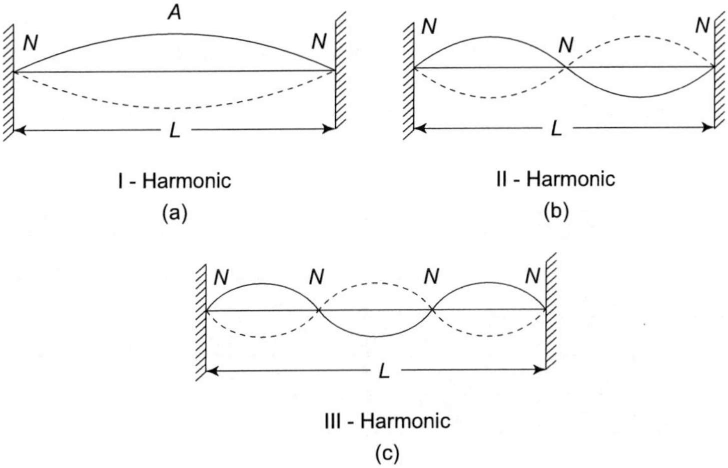

Fig. 4.6

In a string of length L only those standing waves can exist for which $\lambda = 2\,L/1, 2\,L/2, 2\,L/3,....,$ and maximum possible wavelength of standing wave is $2\,L$.

In case of waves, $v = f\lambda$ or $f = v/\lambda$. For waves along a string $v = \sqrt{T/m}$, so $f = \sqrt{T/m}\,/\lambda$. Substituting the value of $\lambda = 2\,L$ from Eq. (4.17), the frequency of string vibrations will be

$$f_n = \frac{n}{2L}\sqrt{T/m} \qquad \text{with } n = 1, 2, 3, ... \qquad ...(4.18)$$

Thus, in case of vibrations of strings, a number of natural frequencies are possible and if we take

$$f = \frac{1}{2L}\sqrt{\frac{T}{m}}, \text{ where } f_n = nf \qquad ...(4.19)$$

i.e., $f_1 = f, \quad f_2 = 2\,f, \quad f_3 = 3\,f$, etc.

So in case of vibrations of strings, higher frequencies are integral multiple of f, i.e., vibrations of strings form a harmonic series.

The frequency will be minimum when $n = 1$, i.e.,

$$f_{\min} = f_1 = f = \frac{1}{2L}\sqrt{\frac{T}{m}}$$

In this case the string will vibrate with one loop with nodes at the ends and antinode in the middle of the string as shown in Fig. 4.6(a). The minimum possible frequency, with which a string vibrates, is called its *fundamental frequency.* The corresponding mode of vibration is called *fundamental mode* or the *first harmonic*. The higher possible frequencies are called *overtones*. The first overtone with $f_2 = \frac{2}{2L}\sqrt{\frac{T}{m}}$ is called the second harmonic. The second overtone with $f_3 = \frac{3}{2L}\sqrt{\frac{T}{m}}$ is called the third harmonic. All these harmonics are summarized in Table 4.1 for ready reference.

Table 4.1

Harmonic	Mode	Antinodes or loops	Nodes	Frequency	Wavelength
First	Fundamental	1	2	f	$2L/1$
Second	I overtone	2	3	$2f$	$2L/2$
Third	II overtone	3	4	$3f$	$2L/3$
....	....	....	....	....	....
n^{th}	$(n-1)^{th}$ overtone	n	$(n+1)^{th}$	nf	$2L/n$

In case of frequency of vibrating string, the following points are worth noting.

1. Since a string has many natural frequencies (all integral multiples of fundamental frequency), so when it is excited with a tuning fork (or a vibrating body), the string will be in resonance with the given body if any one of its natural frequencies coincides with that of the body.
2. The frequency of vibrating string is given by the formula $f = (1/2\,L)\ \sqrt{T/m}$, so from this equation we can infer that
   (a) $f \propto (1/L)$ if T and m are constant
   (b) $f \propto \sqrt{T}$ if L and m are constant
   (c) $f \propto (1/(1/\sqrt{m})$ if T and L are constant
   These laws of vibrations of string are known as Mersenne's laws of vibration of string. According these laws, the frequency of a string can be changed by changing its length, tension or mass per unit length.
3. If M is the mass of a string of length L, then mass per unit length, $m = M/L$

So
$$f = \frac{1}{2L}\sqrt{\frac{T}{m}} = \frac{1}{2L}\sqrt{\frac{T}{\left(\frac{M}{L}\right)}} = \frac{1}{2}\sqrt{\frac{T}{ML}} \qquad \text{...(4.20)}$$

If the radius of string is r and the density of string is ρ or its mass per unit area is $m = \pi r^2 \rho$.

So
$$f = \frac{1}{2L}\sqrt{\frac{T}{\pi r^2 \rho}}, \quad \text{i.e.,} \quad f = \frac{1}{2L}\sqrt{\frac{T}{\pi \rho}} \qquad \text{...(4.21)}$$

4. If the string vibration is n^{th} harmonic, its frequency will be nf. The number of loops in the string or antinodes will be n. Total number of nodes (including two at the ends) will be $(n + 1)$. For example, in case of 3rd harmonic of a vibrating string, frequency of vibration will be $3f$, number of antinodes will be 3, then total number of nodes $= 3 + 1 = 4$ and $\lambda = 2L/3$.
5. Let us consider the vibrations of composite string. It is a string made up by joining two strings of different lengths, cross sections and densities. The composite string will have same tension throughout, the joint will be a node. The lowest frequency called fundamental frequency of the string will be

$$f_C = n_1 f_1 = n_2 f_2 \quad ..(4.22)$$

The higher harmonics will be integral multiples of the lowest frequency, f_C.

SOLVED EXAMPLES

Example 4.1 Explain, is it possible to monitor the temperature of a wire by measuring its frequency of vibration?

Solution: Yes, it is possible to monitor the temperature of a wire by measuring its frequency of vibration. Let us consider a wire suspended vertically from a rigid support and loaded at the free end. The frequency of the wire will be given by

$$f = \frac{n}{2L}\sqrt{\frac{T}{m}} = \frac{p}{2}\sqrt{\frac{T}{ML}} \text{ as } \frac{M}{L} = m$$

For a given wire, mass of a wire and tension in it remains constant so

$$f = \frac{K}{\sqrt{L}}, \text{ where } K = \frac{n}{2}\sqrt{\frac{T}{M}}$$

Differentiating it with respect to L

$$\Delta f = K\left[-\frac{1}{2}L^{-\frac{3}{2}}\Delta L\right]$$

or
$$\Delta f = \frac{K}{\sqrt{L}} \times -\frac{1}{2}\frac{\Delta L}{L}$$

or
$$\frac{\Delta f}{f} = -\frac{1}{2}\frac{\Delta L}{L}$$

Let the change in temperature be $\Delta\theta$. Dividing the above equation by $\Delta\theta$, the change in frequency is given by

$$\frac{\Delta f}{f} = \frac{1}{2}\frac{\Delta L \Delta\theta}{L \Delta\theta}$$

or
$$\frac{\Delta f}{f} = -\frac{1}{2}\alpha\Delta\theta, \ \alpha = \frac{\Delta L}{L\Delta\theta}$$

or
$$\Delta\theta = -\frac{2}{\alpha}\frac{\Delta f}{f}$$

From this equation, we can say that the change in temperature depends on fractional change of frequency. So, by measuring fractional change in vibrational frequency, we can monitor the temperature of the wire. If there is fractional increase in frequency, the temperature will decrease and vice versa,

Example 4.2 A sonometer wire has a length of 88 cm between two fixed ends. Where should the two bridges be placed so as to divide the wire into three segments whose fundamental frequencies are in the ratio 1 : 2 : 3?

Solution: Formula frequency of vibration of sonometer wire,

$$f = \frac{1}{2L}\sqrt{\frac{T}{m}}$$

where, L is the length of the wire between two fixed ends, T is tension of the wire, and m is mass of the wire per unit length.

Given: Length of the wire, L = 88 cm

Ratio of wire segments, 1 : 2 : 3

Calculation: Since $f \propto \frac{1}{L}$, so for fundamental frequencies in the ratio of 1 : 2 : 3, the length of the wire should be in the ratio $1 : \frac{1}{2} : \frac{1}{3}$, i.e., $L_1 : L_2 : L_3 :: 6 : 3 : 2$

If the common factor is x, then length of the string

$$6x + 3x + 2x = 88 \text{ or } 11x = 88 \text{ or } x = 8$$

So
$$L_1 = 6 \times 8 = 48 \text{ cm}, L_2 = 3 \times 8 = 24 \text{ cm}, L_3 = 2 \times 8 = 16 \text{ cm}$$

Ans. The length of three segments is 48 cm, 24 cm, and 16 cm.

Example 4.3 A sonometer wire fixed at one end has a solid mass (M) hanging from its other end to produce tension in it. It is found that 75 cm length of the wire produces a certain fundamental frequency when plucked. When the same mass (M) is hanging in water completely submerged in it, it is found that the length of the wire has to be changed by 5 cm in order to will produce the same fundamental frequency. Calculate the density of the material of mass M. Density of water is 1.0001×10^3 kg/m^3.

Solution: Formula frequency of vibration of sonometer wire

$$f = \frac{1}{2L}\sqrt{\frac{T}{m}}$$

where, L is the length of the wire between two fixed ends, T is tension of the wire and m is mass of the wire per unit length.

Given: Length of sonometer wire, L = 75 cm = 0.75 m and resonance fundamental frequency is f

Mass of the wire is M

Tension of wire, $T = Mg$

Density of water is 1.00001×10^3 kg/m^3.

Calculation: Resonance frequency of vibration will be

$$f = \frac{1}{2 \times 0.70}\sqrt{\frac{Mg}{m}} \qquad ...(4,23)$$

When suspended mass is submerged in water

$$T' = M'g = Mg\left[1 - \frac{\sigma}{\rho}\right] \qquad ...(4,24)$$

As tension is decreased, to produce the same fundamental frequency, length of the wire must also be decreased, i.e., $L' = 0.75 - 0.05 = 0.70$ m

$$f = \frac{1}{2 \times 0.70}\sqrt{\frac{Mg\left[1 - \frac{\sigma}{\rho}\right]}{m}}$$

From Eqs. (4.23) and (4.24), we get

$$1 - \frac{\sigma}{\rho} = \left(\frac{0.70}{0.75}\right)^2 \text{ or } \frac{\sigma}{\rho} = 1 - \frac{196}{225} = \frac{29}{225}$$

or
$$\rho = \frac{225}{29} \times 1.0001 \times 10^3 = 7.76 \text{ kg/m}^3$$

Ans. Density of mass M is density of water is 7.76×10^3 kg/m^3.

Example 4.4 (a) Find the fundamental frequency of the sonometer wire if fundamental frequency of sonometer wire increases by 5 Hz if its tension is increased by 21% keeping the length constant. (b) Find the change in the fundamental frequency of the sonometer wire, when the length of the wire is increased by 20% keeping the original tension in the wire.

Solution: Formula frequency of vibration of sonometer wire

$$f = \frac{1}{2L}\sqrt{\frac{T}{m}} \qquad ...(4.25)$$

where, L is the length of the wire between two fixed ends, T is tension of the wire and m is the mass of wire per unit length.

Calculation: (a) In this case length of the wire is kept constant.

The fundamental frequency increases when tension is increased by 21%.

If the fundamental frequency and tension of the sonometer wire are f and T, respectively, then the new fundamental frequency and tension of sonometer wire will be

$$f' = f + 5 \text{ and } T' = T + \frac{21}{100}T = 1.21\,T, \text{ respectively,}$$

Formula for new frequency when tension is increased

$$f' = \frac{1}{2L}\sqrt{\frac{T'}{m}} \qquad ...(4.26)$$

From Eqs. (4.25) and (4.26), we have

$$\frac{f'}{f} = \sqrt{\frac{T'}{T}} \text{ or } \frac{f+5}{f} = \sqrt{\frac{1.21\,T}{T}} \text{ or } 1 + \frac{5}{f} = \sqrt{1.21} = 1.1$$

or

$$\frac{5}{f} = 0.1 \text{ or } f = 50 \text{ Hz}$$

(b) In this case tension of the wire is kept constant.

The length of the wire increased by 25%.

If length and tension of the wire are L and T, respectively, then the new fundamental frequency and length of sonometer wire will be

$$f' \text{ and } L' = L + \frac{25}{100}L = 1.25\,L, \text{ respectively,}$$

Formula for new frequency when length is increased,

$$f' = \frac{1}{2L'}\sqrt{\frac{T}{m}} \qquad ...(4.27)$$

From Eqs. (4.25) and (4.27), we have

$$\frac{f'}{f} = \frac{L}{L'} \text{ or } \frac{f'}{f} = \frac{1}{1.25} \text{ or } f' = \frac{f}{1.25} = \frac{50}{1.25} \text{ or } f' = 40 \text{ Hz}$$

Changed fundamental frequency, $\Delta f = 40 - 50 = -10$ Hz,
i.e., fundamental frequency decreases by 10 Hz.

Ans. (a) Fundamental frequency of the wire is 50 Hz.

(b) Fundamental frequency decreases by 10 Hz.

Example 4.5 A wire of density 3.5×10^3 kg/m^3 is stretched between two clamps 2 m apart and is subjected to an extension of 6.4×10^{-4} m. What will be the lowest frequency of transverse vibrations in the wire. Young's modulus of the wire is 7×10^{10} N/m^2.

Solution: Formula frequency of vibration of sonometer wire,

$$f = \frac{1}{2L}\sqrt{\frac{T}{m}}$$

where, L is length of the wire between two fixed ends, T is tension of the wire and m is mass of the wire per unit length.

Given: Density of wire, $\rho = 3.5 \times 10^3$ kg/m^3

Length of wire between two clamps, $L = 2$ m

Young's modulus of the wire, $Y = 7 \times 10^{10}$ N/m^2.

Extension of the wire, $\Delta L = 6.4 \times 10^{-4}$ m

Calculation: Frequency fundamental vibration is

$$f = \frac{1}{2L}\sqrt{\frac{T}{m}} = \frac{1}{2L}\sqrt{\frac{T}{\rho A}}$$

where, $m = \rho A$, A is cross-sectional area of the wire.

Here tension due to elasticity of wire,

$$Y = \frac{\text{Stress}}{\text{Strain}} \text{ or Stress} = Y \times \text{Strain or } \frac{T}{A} = Y \times \frac{\Delta L}{L}$$

or $$T = YA \times \frac{\Delta L}{L}$$

$$f = \frac{1}{2L}\sqrt{\frac{YA}{\rho A} \times \frac{\Delta L}{L}} = \frac{1}{2L}\sqrt{\frac{Y}{\rho} \times \frac{\Delta L}{L}}$$

$$= \frac{1}{2 \times 2}\sqrt{\frac{7 \times 10^{10}}{3.5 \times 10^3} \times \frac{6.4 \times 10^{-4}}{2}} = 20 \text{ Hz}$$

Ans. Fundamental frequency is 20 Hz.

Example 4.6 A steel wire of length 2 m, mass 0.5 kg and uniform cross-sectional area 2×10^{-6} m^2 is rigidly fixed at both the ends. The temperature of the wire is lowered by 25 °C. If the transverse waves are set up by plucking the string in the middle, calculate the frequency of the fundamental mode of the vibration. The Young's modulus and expansion coefficient of the wire are 2×10^{11} N/m^2 and 1.44×10^{-5}/°C.

Solution: Formula frequency of vibration of the sonometer wire

$$f = \frac{1}{2L}\sqrt{\frac{T}{m}}$$

where, L is the length of the wire between two fixed ends, T is tension of the wire and m is mass of the wire per unit length.

Given: Length of steel wire = 2 m

Mass of the wire, $M = 0.5$ kg

Cross-sectional area of wire $A = 2 \times 10^{-6}$ m^2

Young's modulus of the wire $Y = 2 \times 10^{11}$ N/m^2

Coefficient of expansion, $\alpha = 1.44 \times 10^{-5}$/°C

Change in temperature, $\Delta\theta = 25$ °C

Calculation: Frequency fundamental vibration is

$$f = \frac{1}{2L}\sqrt{\frac{T}{m}} = \frac{1}{2L}\sqrt{\frac{TL}{M}} = \frac{1}{2}\sqrt{\frac{T}{ML}}$$

where, $m = \frac{M}{L}$, A is cross-sectional area of wire.

Here tension due to elasticity of wire,

$$Y = \frac{\text{Stress}}{\text{Strain}} \text{ or Stress} = Y \times \text{Strain or } \frac{T}{A} = Y \times \frac{\Delta L}{L}$$

or

$$T = YA \times \frac{\Delta L}{L} = YA \times \alpha\Delta\theta \text{ as thermal strain, } \alpha D\theta = \frac{\Delta L}{L}$$

$$f = \frac{1}{2}\sqrt{\frac{YA}{ML} \times \alpha\Delta\theta}$$

$$= \frac{1}{2}\sqrt{\frac{2 \times 10^{11} \times 2 \times 10^{-6}}{0.5 \times 2} \times 25 \times 1.44 \times 10^{-5}} = 6 \text{ Hz}$$

Ans. Fundamental frequency is 6 Hz.

Example 4.7 An aluminium wire of cross-sectional area 1.2×10^{-6} m^2 is joined to a copper wire of the same cross section. This compound wire is stretched on a sonometer, pulled by a weight of 11 kg. The total length of the compound wire between the two bridges is 2.0 m of which the aluminium wire is 0.8 m and rest is the copper wire. Transverse vibrations are set up in the wire by using an external force of variable frequency. Find the lowest frequency of excitation for which standing waves are formed such that the joint in the wire is a node. What is the total number of nodes observed at this frequency excluding the two at the ends of the wire? The density of aluminium is 2.9×10^3 kg/m^3 and that of copper 1.1601×10^3 kg/m^3.

Solution: Formula frequency of vibration of sonometer wire,

$$f = \frac{1}{2L}\sqrt{\frac{T}{m}} = \frac{1}{2L}\sqrt{\frac{T}{\rho A}}, \text{ as } m = \rho A$$

where, L is the length of the wire between two fixed ends, T is tension of the wire and m is mass of the wire per unit length.

Given: Cross-sectional area of wire $A = 1.2 \times 10^{-6}$ m^2

Mass of attached weight attached, $M = 11$ kg

Total length of the wire, $L = 1.5$ m

Length of aluminium wire, $X = 0.6$ m

Density of aluminium wire, $\rho_{Al} = 2.9 \times 10^3$ kg/m^3

Density of copper wire, $\rho_{Cu} = 1.1601 \times 10^4$ kg/m^3

Calculation: Total length of the wire is 2.0 m. Out of this length of aluminium wire is $L_{Al} = 0.8$ m. So the length of the copper wire, $L_{Cu} = 2.0 - 0.8 = 1.2$ m. The tension in the whole wire is same, $T = Mg = 10$ g N.

Let f_{Al} and f_{Cu} *abe* the fundamental frequencies of aluminium and copper wires. So

$$f_{Al} = \frac{1}{2L_{Al}}\sqrt{\frac{T}{\rho_{Al}A}} \text{ and } f_{Cu} = \frac{1}{2L_{Cu}}\sqrt{\frac{T}{\rho_{Cu}A}}$$

Since the total wire is a composite wire, so the whole wire will vibrate with fundamental frequency,

$$f = n_{Al}\, f_{Al} = n_{Cu}\, f_{Cu}$$

Putting the values of f_{Al} and f_{Cu},

$$\frac{n_{Al}}{2L_{Al}}\sqrt{\frac{T}{\rho_{Al}\, A}} = \frac{n_{Cu}}{2L_{Cu}}\sqrt{\frac{T}{\rho_{Cu}A}}$$

or $$\frac{n_{Al}}{2\times 0.8}\sqrt{\frac{T}{A\times 2.9\times 10^2}} = \frac{n_{Cu}}{2\times 1.2}\sqrt{\frac{T}{A\times 1.1601\times 10^4}}$$

or $$\frac{n_{Al}}{n_{Cu}} = \frac{2}{3}\sqrt{\frac{2.9\times 10^2}{11.6\times 10^2}} = \frac{2}{3}\times\frac{1}{2} = \frac{1}{3}$$

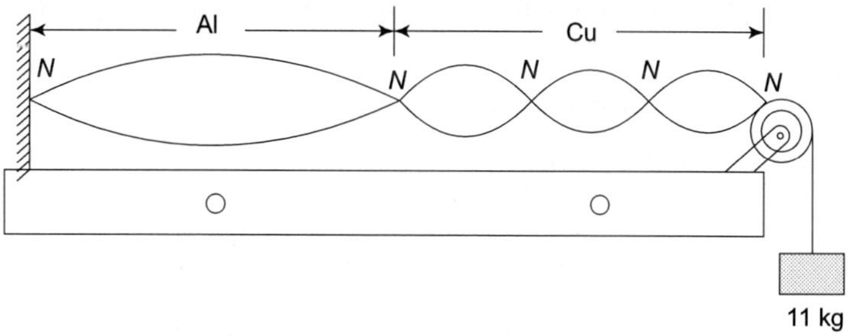

Fig. 4.7

From the ratio of n_{Al} and n_{Cu}, we infer that for fundamental frequency of composite string, $n_{Al} = 1$ and $n_{Cu} = 3$, i.e., aluminium string will vibrate in the first harmonic loop (one loop), i.e., fundamental mode, while copper string in third harmonic (3 loops), i.e., 2nd overtone as shown in Fig. 4.7. Therefore,

$$f = f_{Al} = 3\, f_{Cu} = 3.$$

The frequency of fundamental mode,

$$F = f_{\text{Al}} = \frac{1}{2 \times 0.8}\sqrt{\frac{11 \times 9.8}{1.2 \times 10^{-6} \times 2.9 \times 10^{3}}} \cong 110 \text{ Hz } (3f_{\text{Cu}})$$

This implies that that total number of modes in the string will be 5 and so the number of nodes excluding the nodes at the ends = 5 – 2

Ans. There are 3 nodes other than two nodes at the ends.

Fundamental frequency is 110 Hz

Example 4.8 Fundamental frequency of two identical sonometer wires is 300 Hz, when the same tension is applied. What fractional increase in the medium of one wire would produce 6 beats per second, when both wires vibrate together?

Solution: Formula frequency of vibration of sonometer wire,

$$f = \frac{1}{2L}\sqrt{\frac{T}{m}} = K\sqrt{T}, \text{ where, } K = \frac{1}{2L}\sqrt{\frac{1}{m}} \qquad \text{...(4.28)}$$

where, L is the length of the wire between two fixed ends, T is tension of the wire and m is mass of the wire per unit length.

Given: Fundamental frequency of sonometer wire, $f = 600$ Hz

Number of beats occurred, $\Delta f = 6$ per second

Calculation: Differentiating Eq. (4.28)

$$\Delta f = K\left[\frac{1}{2}T^{-1/2}\right]\Delta T \qquad \text{...(4.29)}$$

Dividing Δf by f, we get the fractional change in fundamental frequency,

$$\frac{\Delta f}{f} = \frac{1}{2}\frac{\Delta T}{T} \qquad \text{...(4.30)}$$

is due to fractional change in tension

$$\frac{\Delta T}{T} = 2 \times \frac{\Delta f}{f} = 2 \times \frac{6}{300} = 0.04$$

Percentage fraction change in tension, $\frac{\Delta T}{T} = 0.04 \times 100 = 4\%$

Ans. For 4% fractional change in tension, 5 beats are produced.

4.11 MELDE'S EXPERIMENT

If a string is forced to vibrate by plucking, it vibrates freely at all of its natural frequencies given by Eq. (4.27),

$$f_n = \frac{n}{2L}\sqrt{\frac{T}{m}}, \qquad \text{...(4.31)}$$

where, n = 1,2, 3, ..., T is tension of the string, m is the mass per unit length of the string. On the other hand if the string is forced to vibrate at one of the natural frequencies, it will vibrate with large amplitude. This can be demonstrated by the Melde's experiment shown in Fig. 4.8.

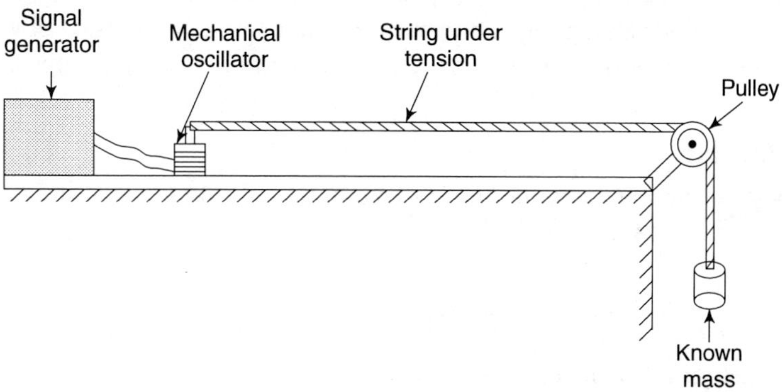

Fig. 4.8 Apparatus for Melede's experiment

The frequency of the signal generator is slowly increased. At first, it starts vibrating with small amplitude. After some time, at frequency f_1 the string vibrates with large amplitude in the form of a single loop as shown in the Fig. 4.9(a). If the frequency is increased beyond this value, the amplitude of the vibrations dies. When the forcing frequency reaches $2f_1$, the string again vibrates with large amplitude, but this time it vibrates with two loops as shown in Fig. 4.9(b). At frequency $3f_1$ it vibrates with three loops, etc. Substituting the relevant values of L, T and m in Eq. (4.30b) it is found that the frequencies, f_1, $2f_1$ and $3f_1$, are respectively equal to the frequencies of the first, second and third harmonics of the string. This is an example of resonance. In the string resonance occurs only tothose forced frequencies which are equal to its natural frequencies of the vibrations.

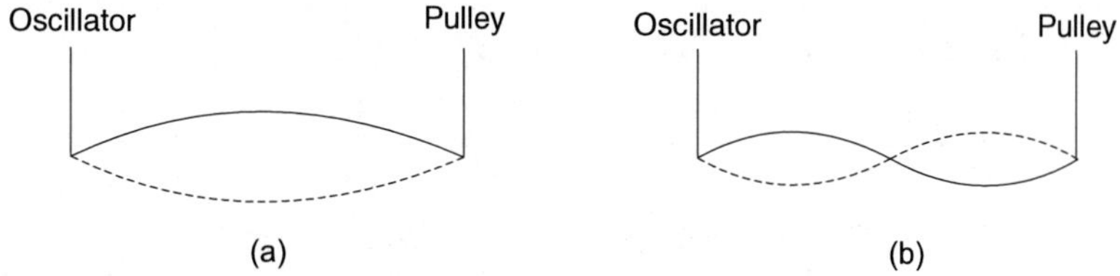

Fig. 4.9 Modes of vibration in Melede's experiment

Notes:

1. If the amplitude of vibration of the oscillator is small in comparison to that of string, then the string behaves (almost) as if it is fixed at the attachment point of the oscillator.
2. If the reflected waves are almost as strong as the incident waves, then the displacement at the nodes is not exactly zero.
3. By stroboscopic illumination,vibrations of the string can be seen in slow motion. So by stroboscopic illumination one can easily seenodes and antinodes.

Note: Stroboscope is an instrument for studying periodic motion or determining speeds of rotation by shining a bright light at intervals so that a moving or rotating object appears stationary.

A *stroboscope*, also known as a *strobe*, is an instrument used to make a cyclically moving object appear to be slow-moving, or stationary. When a rotating or vibrating object is observed with a stroboscope at its vibration frequency (or a submultiple of it), it appears stationary.

4.12 SONOMETER

The frequency, f_1, of the fundamental mode of vibration of a stretched string can be obtained by putting $n = 1$ in Eq. (4.31), as

$$f_1 = \frac{1}{2L}\sqrt{\frac{T}{m}} \qquad ...(4.32)$$

From this equation one can see that

(i) $f_1 \propto 1/L$, if T and m are constant

(ii) $f_1 \propto \sqrt{T}$, if L and m are constant

(iii) $f_1 \propto 1/\sqrt{m}$, if T and L are constant

These relationships are sometimes called the *laws of vibration of stretched strings*. They may be verified experimentally by using a sonometer as shown below:

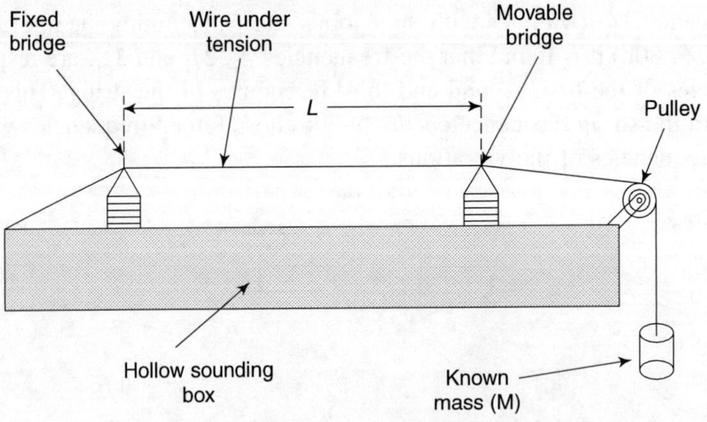

Fig. 4.10 Sonometer

To verify $f_1 \propto 1/L$

Let us selectsuitable values of T and m. Now the position of the movable bridge is altered so that the length, L, vibrates at the same note as the tuning fork of the known frequency. If the experimenter is not sufficiently 'musical' to detect whether the two notes have the same pitch, he can make use of a resonance technique. A small piece of paper in the form of an inverted *vee* (V) is placed on the centre of the wire, and the stem of the vibrating tuning fork is held against one of the bridges. This forces the wire to vibrate, and if its length is such that its fundamental frequency of vibration is equal to the frequency of the tuning fork, the wire vibrates with large amplitude and throws the paper off the wire. The procedure is repeated using tuning forks of other known frequencies, and without altering T and m. A graph of f_1 against $1/L$, is a straight line, which passes through the origin. This shows that $f_1 \propto 1/L$.

To verify $f_1 \propto \sqrt{T}$

Let L be kept constant at some suitable value, and the mass, M, and therefore the tension, T ($= Mg$) is altered, so that when the wire is forced to vibrate by a tuning fork of known frequency, it produces the same vibrations as that of tuning fork. The procedure is repeated using tuning forks of other known frequencies by changing M. A graph of f_1 versus $\sqrt{T}$ is a straight line and passes through the origin. Straight line shows that $f_1 \propto \sqrt{T}$.

To verify $f_1 \propto 1/\sqrt{m}$

When T and f_1 are constant, then from Eq. (4.32), we have

$$L \propto 1/\sqrt{m}$$

The mass per unit length, m, of a wire can be found by weighing it. The length, L, of the wire is adjusted so that it is forced to vibrate with the same note as that of tuning fork. The procedure is repeated using wires of different masses per unit length with the same tuning fork. Tension and frequency of vibration are kept constant each time, i.e., each wire is applied with the same tension, and the length of the wire is adjusted until it vibrates at the same frequency when forced by the same tuning fork. A graph of L versus $1/\sqrt{m}$ will be a straight line and passes through the origin. This straight line graph verifies that $f_1 \propto 1/\sqrt{m}$.

4.13 PLUCKED AND STRUCK STRINGS

The ends of the string are fixed, so there will be nodes at both the ends of the string. When a transverse wave is forced to travel along a stretched string, the wave is reflected on reaching the ends of the string. The incident and reflected waves have the same speed, frequency and amplitude, and therefore on superposition the resultant wave will be a stationary wave.

If a stretched string is forced to vibrate by plucking or striking, a number of different stationary waves are formed simultaneously. Only specific modes of vibration are possible. These modes of vibration are described in the following section.

4.13.1 Modes of Vibration

Let length of the string be L. The ends of a stretched string are fixed, and therefore the both ends of the string must have displacement nodes. The three simplest modes of vibration of a string are shown in Fig. 4.11.

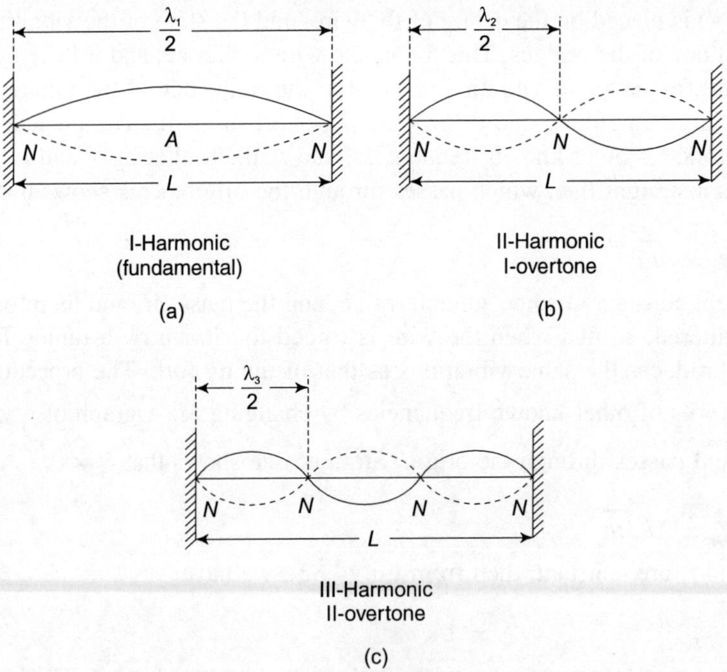

Fig. 4.11 Modes of vibration of stretched string

The first mode of vibration is called the fundamental mode. The frequency, at which it vibrates, is called the fundamental frequency. The fundamental tone is called the first harmonic, while the first overtone and the second overtone are called the second harmonic and the third harmonic, respectively. Let λ_1, λ_2, and λ_3 be the wavelengths of the first, second and third harmonics, respectively. It is to be noted that the separation of two adjacent nodes is equal to half the wavelength, then from Fig. 4.11 we have

$$\frac{\lambda_1}{2} = L, \quad \frac{\lambda_2}{2} = \frac{L}{2}, \quad \frac{\lambda_3}{2} = \frac{L}{3}$$

In general, the wavelength of nth harmonic is

$$\frac{\lambda_n}{2} = \frac{L}{n} \qquad \text{...(4.33)}$$

If v is the speed of either tone of the travelling waves that have produced the stationary wave, then $v = f_n \lambda_n$. So, the frequency of nth harmonic will be

$$f_n = \frac{nv}{2L}, \qquad \text{...(4.34)}$$

Here one should note that velocity v is same for all tones.

The frequency, f_1, first harmonic can be obtained by putting $n = 1$ in Eq. (4.34), as

$$f_1 = \frac{v}{2L}$$

Therefore, Eq. (4.34) can be written as

$$f_n = n f_1, \qquad n = 1, 2, 3,....$$

i.e., the frequency of the each overtone isan integer multiple of fundamental frequency.

If T is the tension of the string and m is the mass per unit length, the velocity of sound,

$$v = \sqrt{\frac{T}{m}} \qquad ...(4.35)$$

Putting this value of v, the frequency of nth harmonics will be

$$f_n = \frac{n}{2L}\sqrt{\frac{T}{m}} \qquad ...(4.36)$$

where, $n = 1, 2, 3, 4,....$

4.13.2 Musical Instruments

When a guitar string is plucked or piano string is struck, transverse waves travel along the strings and are reflected on reaching the ends. The energy of any wave is not sufficient so the wavelength associated with it does not give rise to one of the allowed stationary waves is very quickly dissipated. Frequencies of the stationary waves of a string are given by Eq. (4.36). The strings vibrate with all these frequencies simultaneously.

The fundamental modc has thc largcst amplitude of vibration, and therefore, it isthe dominant frequency. The relative amplitudes of the various overtones depend on the particular instrument being played. Various overtones give an instrument its characteristic sound. Characteristics of sound are discussed in Chapter 7.

4.14. ENERGY OF VIBRATING STRING, TRANSFER OF ENERGY

A wave on a stretched stringis set up by providing energy for the motion of a string. As the wave moves away from us, it transports that energy in the form of both kinetic energy and elastic potential energy. Let us discuss each form in detail.

Kinetic energy

An element of the string of mass (m) oscillates in simple harmonic motion in transverse direction with velocity (u) to the wave motion as the wave passes through it. The kinetic energy is associated with this element. When the element is passing through its $Y = 0$ position as shown in Fig. 4.12, its transverse velocity and hence its kinetic energy is a maximum. When the element is at its extreme position, $Y = Y_m$, its transverse velocity and thus its kinetic energy is zero.

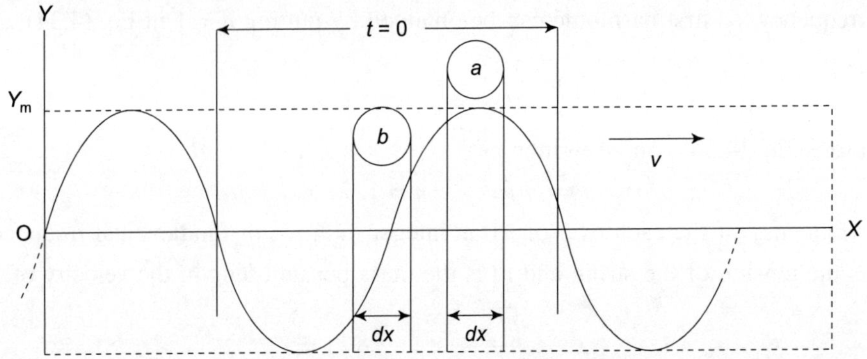

Fig. 4.12 A diagram of a travelling wave on a string at a time $t = 0$. String element 'a' is at displacement $y = Y_m$, and string element 'b' is at displacement $Y = 0$. The kinetic energy of the string element at each position depends on the transverse velocity of the element. The potential energy depends on the amount by which the string element is stretched as the wave passes through it

Elastic potential energy

To send a sinusoidal wave along a straight string, the wave is stretched by the string element of length dx in transverse direction. The string element of length dx oscillates in simple harmonic motion in transverse direction with velocity (u) to the wave motion as the wave passes through it. This element has the kinetic energy. When the element is passing through its mean position in transverse direction, its length must increase and decrease in a periodic way, if the string element is to fit the sinusoidal waveform. Elastic potential energy associated with element of length dx changes, just like it changes for a spring.

When the string element is at its $Y = Y_m$ position as shown in Fig. 4.12, its length has its normal undisturbed value, dx, so its elastic potential energy is zero. However, when the element passes through its $Y = 0$ position, it is stretched to its maximum extent, and its elastic potential energy then is maximum.

Both kinetic energy and elastic potential energy of the oscillating string element are maximum at $Y = 0$. The energy of spring element is zero at maximum displacement regions, and is maximum at zero displacement regions. As the wave travels along the string, the tension in the string continuously does work to transfer energy from regions with energy to regions with no energy.

The rate of energy transfer

The travelling wave is written as

$$Y = Y_m \sin (kx - \omega t)$$

The transverse speed (u) of the string element can be obtained by differentiating equation w.r.t. t, i.e.,

$$u = \frac{dY}{dt} = -\omega Y_m \cos (kx - \omega t) \qquad ...(4.37)$$

Let ρ be the mass demsity of the string, then kinetic energy (dE) associated with a string element of length (dx) is given by

$$dE = \frac{1}{2}\rho\, S dx\, u^2 \qquad ...(4.38)$$

where, S is cross-sectional area of the string.

The rate of change of kinetic energy of a string element,

$$\frac{dE}{dt} = -\frac{1}{2}\rho\, S\, u^2 \frac{dx}{dt}$$

Putting the value of transverse speed (u) of the element of string and speed of the wave, $\frac{dx}{dt} = v$, Eq. (4.38) becomes

$$\frac{dE}{dt} = \frac{1}{2}\rho\, S\, v\, \omega^2\, Y_m^2 \cos^2 (kx - \omega t) \qquad ...(4.39)$$

The average rate at which kinetic energy is transported is

$$\overline{\left(\frac{dE}{dt}\right)} = \frac{1}{2}\rho\, S\, v\, \omega^2\, Y_m^2\, \overline{\cos^2 (kx - \omega t)}$$

The average value of the square of a cosine function over a wavelength is $\frac{1}{2}$, so the average rate at which kinetic energy transported will be

$$\overline{\left(\frac{dE}{dt}\right)} = \frac{1}{4}\rho S\, v\, \omega^2\, Y_m^2 \qquad ...(4.40)$$

In an oscillatıng system such as a pendulum, the average kinetic energy and the average potential energy are equal.

The average power, which is the sum of the average rate of kinetic energy and the potential will be

$$\bar{p} = \overline{\left(\frac{dE}{dt}\right)} = \frac{1}{2}\rho\, S\, v\, \omega^2\, Y_m^2 \qquad ...(4.41)$$

The factor ρS and v in this equation depend on the material and tension of the string. The factors ω and Y_m depend on the process that generates the wave. The dependence of the average power of a wave on the square of its amplitude and also on the square of its angular frequency is a general result, true for waves of all types.

If m is the mass per unit length of the string, so $m = \rho S$, then Eq. (4.41) can be written as

$$\bar{p} = 2\overline{\left(\frac{dE}{dt}\right)} = \frac{1}{2} m v \omega^2\, Y_m^2$$

Intensity

The intensity is defined as the average energy transmitted per unit to normal area per second, i.e., power per unit area. Since S is the cross-sectional area of string, so

$$I = \frac{2}{S}\overline{\left(\frac{dE}{dt}\right)} = \frac{1}{2}\rho v\omega^2 Y_m^2 \qquad ...(4.42)$$

In case of pressure waves, initial pressure amplitude,

$$p_0 = Y_m kB = Y_m k\rho v^2 \text{ as } v = \sqrt{B/\rho}$$

or $$p_0 = Y_m \rho v kv = Y_m \rho v \text{ w} \quad \text{as} \quad kv = \frac{2\pi}{\lambda} v = 2\pi f = \omega$$

Therefore, $$I = \frac{1}{2}\rho v \omega^2 \left[\frac{p_0}{\rho v\omega}\right]^2 = \frac{1}{2}\frac{p_0^2}{\rho v} \qquad ...(4.43)$$

The formula for intensity of sound in terms of displacement and pressure amplitude is given by Eqs. (4.42) and (4.43), respectively.According to these equations for a given source and medium

$$I \propto y_m^2 \text{ or } I \propto p_0^2 \qquad ...(4.44)$$

That is, intensity of sound is proportional square of amplitude, Y_m, of the wave and proportional to the square of initial pressure, p_0.

Decibel and sound level

The SI unit of intensity of sound is W/m^2. The human ear responds to sound intensities over a wide range, i.e., from 10^{-12} W/m^2 to 1 W/m^2. So instead of specifying intensity of sound in W/m^2, we use a logarithmic scale of intensity called the *sound level* defined as

$$SL = 10 \log\left[\frac{I}{I_0}\right] \qquad ...(4.45)$$

where, I_0 is the threshold of the human year, i.e., 10^{-12} W/m^2. The sound level defined in this way is expressed in decibel (dB). A sound of intensity I_0 has an $SL = 10 \log [I_0/I_0] = 0$ dB, while sound at the upper range of human hearing called, threshold of pain, has an intensity of 1 W/m^2 or $SL = 10 \log [1/10^{-12}] = 120$ dB.

We also use dB as a relative measure to compare different sounds with one another, rather than with reference intensity. Let I_1 and I_2 be intensities of two sounds, so

$$SL_1 - SL_2 = 10 \log\left[\frac{I_1}{I_0}\right] - 10 \log\left[\frac{I_2}{I_0}\right]$$

or $$SL_1 - SL_2 = 10 \log\left[\frac{I_1}{I_2}\right] \qquad ...(4.46)$$

If the intensity ratio of two sounds is 2, they differ in *SL* by 10 log 2 = 3 dB. Here, it must be kept in mind that ratio of two intensities corresponds to the difference in their sound levels (and not ratio of sound levels).

Note: While solving problems related to the intensity of sound along with the above, also remember that:

1. As intensity (energy flux), $I = \dfrac{\Delta \bar{E}}{S\Delta t}$, while energy density, $U = \dfrac{\Delta \bar{E}}{\Delta V}$,

   So the ratio of intensity and energy density,

$$\frac{1}{U} = \frac{\Delta \bar{E}}{S\Delta t} \times \frac{S\Delta L}{\Delta \bar{E}} = v \qquad \left[\text{as } \Delta V = S\Delta L \text{ and } \frac{\Delta L}{\Delta t} = v\right]$$

is equal to the transverse velocity of element, ΔL.

The intensity I (energy flux) = $U \times v$ = Energy density × velocity ...(4.47)

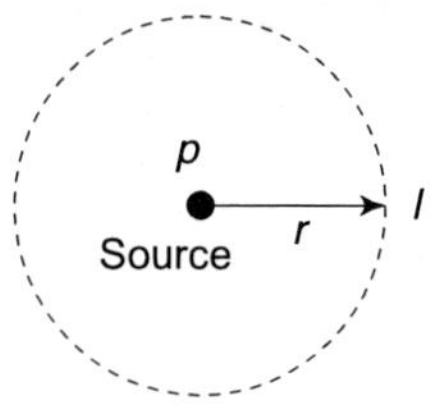

Fig. 4.13

2. With the increase in distance from the source the total energy or power transmitted remains the same but intensity decreases. For an isotropic point source of power, P, intensity, I, at a distance, r, covering surface area, $S = 4\pi r^2$, from point source will be

$$I = \frac{P}{S} = \frac{P}{4\,\pi r^2} \qquad ...(4.48)$$

   Equating Eqs. (4.42) and (4.48), we get

$$\frac{1}{2}\rho\, v\, \omega^2\, Y_m^2 = \frac{P}{4\,\pi\, r^2}$$

Which shows that $Y_m^2 \propto (1/r^2)$ is a particular sound source. Thus, for spherical waves,

$$Y_m \propto (1/r) \text{ and } I \propto (1/r^2)$$

That is, amplitude of the wave decreases with distance, but the intensity of sound decreases with the square of radial distance.

In case of electromagnetic waves (e.g., for light or radio waves),

$$I = \frac{1}{\mu_0}(EB) \quad \text{with} \quad \frac{E}{B} = c \quad \text{and} \quad c = \frac{1}{\sqrt{\mu_0\, \epsilon_0}}$$

Formulas at a Glance

	Subject	Formula	Interpretation
1.	Travelling wave	$Y(x, t) = A \sin(kx - \omega t)$, x is the position of the constituents of the medium, Y is the displacement of the constituents about its mean position.	ω and k are angular frequency and angular wave vector, respectively. A is the amplitude of the wave.
2.	Speed of the travelling wave	$v = \frac{\omega}{k} = \frac{\lambda}{T} = \lambda f$	T and ω are time period and frequency of the wave.
3.	Equation of the standing wave	wave $y = R \sin(\omega t)$	$R = 2A \cos(kx)$ is amplitude of the equation of the wave packet.
4.	Condition for maxima (anitnode) in case of standing wave	$x = m\frac{\lambda}{2}$	$m = 0, 1, 2, \ldots.$
5.	Condition for minima (nodes) in case of standing wave	$x = (2m - 1)\frac{\lambda}{4}$	$m = 1, 2, 3, \ldots..$
6.	Differential form of standing wave	$\frac{\partial^2 Y}{\partial x^2} = v^2 \frac{\partial^2 Y}{\partial t^2}$ or $m^2 \frac{\partial^2 Y}{\partial x^2} = T^2 \frac{\partial^2 Y}{\partial t^2}$	$v = \sqrt{\frac{T}{m}}$, m is the mass of string per unit length and T is the tension of the string.
7.	Equation for normal modes of stretched string	$Y = R \cos \frac{2\pi}{\lambda}(vt)$	$R = 2A \sin\left(\frac{2\pi}{\lambda}x\right)$ is amplitude of the normal modes.
8.	Wavelength of normal; mode of stretched string	$\lambda = \frac{2L}{m}$	$m = 1, 2, 3, \ldots.$
9.	Condition for position of antinodes	$x = \frac{(2m-1)\lambda}{4}$	$m = 1, 2, 3, \ldots.$
10.	Formular for frequency of nth normal mode	$f_n = \frac{n}{2L}\sqrt{\frac{T}{m}}$	$n = 1, 2, 3, \ldots..$ L is length of string. T is tension of the string, m is mass per unit length.
11.	Wavelength of normal mode of string free at one end	$\lambda = \frac{4L}{(2m-1)}$	$m = 1, 2, 3, \ldots.$

Contd.

12.	Average kinetic energy = Potential energy of constituent of the string	$\overline{\left(\frac{dE}{dt}\right)} = \frac{1}{4}\rho S v \omega^2 Y_m^2$ $\rho \rightarrow$ mass density of string, $S \rightarrow$ cross-sectional area of the string	$v = \frac{dx}{dt} \rightarrow$ velocity of string $\omega \rightarrow$ angular velocity of string, $Y_m \rightarrow$ amplitude of the wave.
13.	Average power transferred	$\overline{p} = 2\overline{\left(\frac{dE}{dt}\right)} = \frac{1}{2}\rho S v \omega^2 Y_m^2$	
14.	Intensity of transferred energy	$I = \frac{1}{2}\rho v \omega^2 \left[\frac{p_0}{\rho v \omega}\right]^2 = \frac{1}{2}\frac{p_0^2}{\rho v}$	where, $p_0 = Y_m \rho v kx$ minimum intensity that can be heard by ear.
15.	Formular for sound level	$SL = 10 \log\left[\frac{I}{I_0}\right]$ Intensity, $I = \frac{\text{Power}}{\text{Areas}} = \frac{P}{S}$	$I_0 = 10^{-12}$ W/m^2 the minimum intensity that can be heard by ear.
16.	Formula to compare the sound level of two sounds	$SL_1 - SL_2 = 10 \log\left[\frac{I_1}{I_2}\right]$	I_1 and I_2 are intensities of two sounds.

Example 4.9 A string has a linear density (μ) of 520 g/m and is stretched with a tension (T) of 40 N. A wave whose frequency (f) and amplitude (Y_m) are 125 Hz and 9.5 mm, respectively, is travelling along the string. At what average rate the wave transporting energy along the string?

Solution: Formula required linear density

Average power

$$\overline{p} = \frac{1}{2}\mu v \omega^2 Y_m^2$$

Given: Linear density, μ = 520 g/m = 0.520 kg/m

Tension of the string, T = 40 N

Frequency of wave. f = 125 Hz

Amplitude of the wave, Y_m = 9.5 mm = 0.0095 m

Calculation: Angular velocity,

$$\omega = 2\pi f = 250\ \pi \text{ radian/s} = 785 \text{ radian/s}$$

Velocity of wave, $v = \sqrt{\frac{T}{\mu}} = \sqrt{\frac{40}{0.520}} = 8.77$ m/s

Average power,

$$\overline{p} = \frac{1}{2} \times 0.520 \times 8.77 \times (785)^2 \times (0.0095)^2 = 126.8 \text{ W}$$

Ans. Average power of wave is 1126.8 W.

Example 4.10 The faintest sound the human can detect at a frequency of 1.2 kHz (for which the ear is most sensitive) corresponds to an intensity of about 10^{-12} W/m^2 (the so called threshold of hearing). Determine the pressure amplitude and maximum displacement associated with this sound assuming the density of air 1.4 kg/m^3 and velocity of sound in air 330 m/s.

Solution: If initial power is p_0, then intensity,

$$I = \frac{1}{2}\frac{p_0^2}{\rho v}$$

Pressure amplitude,

$$p_0 = Y_m \rho v\, kv = Y_m \rho v\, \omega$$

Given: Frequency, $f = 1.2$ kHz $= 1.2 \times 10^3$ Hz

Intensity, $I = 10^{-12}$ W/m^2

Density of air, $\rho = 1.4$ kg/m^3

Speed of sound, $v = 330$ m/s

Calculation: Initial pressure amplitude

$$p_0 = \sqrt{I \times 2\rho v} = \sqrt{10^{-12} \times 2 \times 1.4 \times 330} = 3.04 \times 10^{-5} \text{ N/m}^2$$

The maximum displacement

$$Y_m = \frac{p_0}{\rho v \omega} = \frac{3.04 \times 10^{-5}}{1.4 \times 330 \times 2\pi \times 1.2 \times 10^3} = 1.71 \times 10^{-11} \text{ m}$$

Ans. Initial pressure amplitude is 3.04×10^{-5} N/m^2

The maximum displacement is 1.71×10^{-11} m

Example 4.11 What is the maximum possible sound level in dB of sound waves in air? Given that density of air is 1.3 kg/m^3, $v = 332$ m/s and atmospheric pressure, $P = 1.01 \times 10^5$ N/m^2.

Solution: If initial power is p_0, then intensity,

$$I = \frac{1}{2}\frac{p_0^2}{\rho v}$$

Sound levelis defined as

$$SL = 10 \log\left[\frac{I}{I_0}\right] \quad ..(4.49)$$

where, I_0 is the threshold of the human year, i.e., 1.12×10^{-12} W/m^2.

Given: For maximum possible sound intensity, pressure amplitude of the wave will be equal to the atmospheric pressure, i.e.,

$$p_0 = P = 1.02 \times 10^5 \text{ N/m}^2$$

Density of air, $\rho = 1.4$ kg/m^3

Speed of sound, $v = 330$ m/s

Calculation:

$$I = \frac{1}{2}\frac{p_0^2}{\rho v} = \frac{(1.02 \times 10^5)^2}{2 \times 1.4 \times 330} = 1.126 \times 10^7 \text{ W/m}^2$$

Sound level,

$$SL = 10 \log\left[\frac{I}{I_0}\right] = 10 \log\left[\frac{10^7}{10^{-12}}\right] = 190 \text{ dB}$$

Ans. The maximum possible sound level of sound waves in air is 190 dB

Example 4.12 (a) The power of sound from the speaker of a radio is 25 mW. By turning the knob of volume control the power of sound is increased to 500 mW. What is the power increase in dB as compared to the original power?

(b) How much more intense is an 85 dB sound than a 15 dB whisper?

Solution: If I_1 and I_2 are initial and final intensities, then sound level

Increase, $\Delta SL = SL_2 - SL_1 = 10 \log\left[\frac{I_2}{I_1}\right]$

Given: (a) Power of sound level of radio speaker, $I_1 = 25$ mW

By turning knob power of sound level of radio speaker, $I_2 = 500$ mW

(b) $SL_2 = 85$ dB, $SL_1 = 15$ dB

Calculation: (a) Sound level increase, $\Delta SL = SL_2 - SL_1 = 10 \log\left[\frac{I_2}{I_1}\right]$

$$= 10 \log\left[\frac{500}{25}\right] = 13 \text{ dB}$$

(b) $SL_2 - SL_1 = 10 \log\left[\frac{I_2}{I_1}\right]$; $85 - 15 = 10 \log\left[\frac{I_2}{I_1}\right]$ or $\log\left[\frac{I_2}{I_1}\right] = 7$

or $$\frac{I_2}{I_1} = 10^7$$

Ans. (a) Sound level increase is 13 dB

(b) Ratio of intensities is 10^7

Example 4.13 A window whose area is 1 m^2 opens on a street where the street noises result in an intensity level at the window of 50 dB. How much 'acoustic power' enters the window via the sound waves. Now if an acoustic absorber is fitted at the window, how much energy from the street will it collect in ten hours?

Solution: Sound level, $SL = 10 \log \left[\frac{I}{I_0}\right]$

Power, $P = \frac{I}{A}$

Given: Area of window, $A = 1\ m^2$

Intensity level of noise, $SL = 50$ dB

Minimum intensity can be heard by ear, $I_0 = 10^{-12}$ W

Time, $t = 4$ hours $= 10 \times 60 \times 60$ seconds

Calculation:

$$50 = 10 \log \left[\frac{I}{I_0}\right] \text{ or } \log \left[\frac{I}{I_0}\right] = 5$$

or
$$I = 10^{-1} \times 10^{5} = 10^{-7}\ W/m^2 = 0.1\ \mu W/m^2$$

$$P = IA = 0.1 \times 1 = 0.1\ \mu W$$

$$E = P\,t = 0.1 \times 10^{-6} \times 10 \times 60 \times 60 = 36 \times 10^{-4}\ J = 3.6\ mJ$$

Ans. Power entering the window is 0.1 μW.

Energy collected in 5 hours is 3.6 mJ

Example 4.14 A dog while barking delivers about 1.2 mW of power. If this power is uniformly distributed over a hemispherical area, what is the sound level at a distance of 6 m? What would the sound level if instead of 1 dog, 7 dogs start barking at the same time, each delivering 1.2 mW of power?

Solution: Sound level, $SL = 10 \log \left[\frac{I}{I_0}\right]$

Power, $I = \frac{1}{S}$

Given: Power delivered by dog, $P = 1.2$ mW

Radius of hemisphere, $r = 6$ m

Minimum intensity can be heard by ear, $I_0 = 10^{-12}$ W

Calculation: Surface area of semi-sphere, $S = 2\pi r^2 = 2 \times 3.14 \times 6 \times 6 = 226\ m^2$

$$I = \frac{P}{S} = \frac{10^{-3}}{226} = 6.37 \times 10^{-6}\ W/m^2 = 4.42\ \mu W/m^2$$

$$SL = 10 \log \left[\frac{I}{I_0}\right] = 10 \log \left[\frac{4.42 \times 10^{-6}}{10^{-12}}\right]$$

$$= 10 \times [\log 4.42 + 6] = 10 \times [0.65 + 6] = 66.5\ dB$$

When 7 dogs are barking at the same level, $I_2 = 7\,I$, so

$$SL_2 - SL_1 = 10 \log \left[\frac{I_2}{I}\right] = 10 \log \left[\frac{7I}{I}\right]$$

or $$SL_2 = SL_1 + 10 \log [7] = 68 + 10 \times 0.85 = 76.5 \text{ dB}$$

Ans. Sound level due to one dog is 66.5 dB.

Sound level due to 7 dogs is 76.5 dB.

Appendix A

Motion in a vertical circle

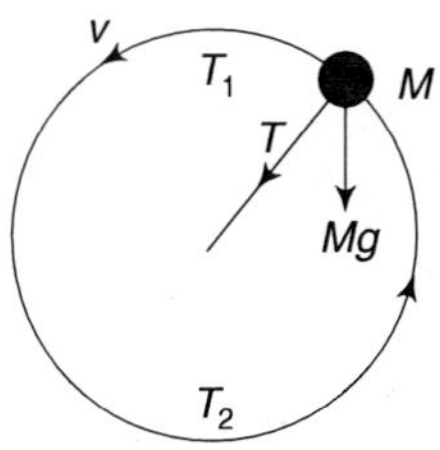

Fig. 4.14

Let us consider a motion of a particle of mass M in vertical circle of radius r as shown in Fig. 4.12. One should know that for given linear velocity (v) of rotation the centripetal force is constant. Let ω be the angular velocity of the object. If the object makes 1 round in τ seconds, the frequency of the object is

$$f = \frac{1}{\tau}$$

The angular velocity of the object,

$$\omega = 2\pi f = 2\pi \frac{1}{\tau} = \frac{2\pi r}{r\tau} = \frac{v}{r} \text{ or } v = \omega r$$

The centripetal force,

$$M\omega^2 r = \frac{M\omega^2 r^2}{r} = \frac{Mv^2}{r}$$

Since the contribution due to gravitational attraction varies as the particle makes one revolution so does the tension in the string. At the top of the circle the centripetal force is provided by both tension in the string, T_1, and the gravitational attraction, Mg, so at top of circle the tension and weight of the particle is balanced by the centripetal force, i.e.,

$$\frac{Mv^2}{r} = T_1 + Mg \quad \text{or} \quad T_1 = \frac{Mv^2}{r} - Mg$$

At the bottom of the circle, the tension, T_1, and the gravitational attraction act in opposite direction and therefore, at the bottom of circle, the tension minus the weight of the particle is balanced by the centripetal force, i.e.,

$$\frac{Mv^2}{r} = T_2 - Mg \quad \text{or} \quad T_2 = \frac{Mv^2}{r} + Mg$$

From these two equations it is clear that tension (T_2) at the bottom of the circleis maximum. In case of micro particles like atom or molecule, we can neglect the gravitational attraction, so the above equations can be written as

$$\frac{Mv^2}{r} = T \quad \text{or } mv^2 = T \quad \text{or } v = \sqrt{\frac{T}{m}}$$

where, $m = M/r$.

SHORT ANSWER QUESTIONS

4.1 What is the characteristic form of a wave function which represents (i) a travelling wave, (ii) a standing wave?

Ans. (i) Travelling (progressive) wave is characterised by wave functions of the form $Y = f(vt \pm x)$. For example, $Y = A \sin(vt \pm x)$, $Y = A \cos(vt \pm x)$ are travelling waves.

(ii) A stationary (standing) wave is characterized by wave functions of the form $Y = f(x) f'(t)$. For example, the superposition of two travelling waves,

$$\begin{aligned} Y &= Y_1 + Y_2 \\ &= a \sin(vt - x) + a \sin(vt + x) \\ &= a [\sin(vt) \cos(x) - \cos(vt) \sin(x)] + a [\sin(vt) \cos(x) + \cos(vt) \sin(x)] \\ &= 2a \cos(x) \sin(vt) \end{aligned}$$

is the standing wave.

4.2 Write the distance between (i) a node and the nearest antinode, and (ii) two successive antinodes in a stationary wave, in terms of wavelength.

Ans. (i) $\frac{\lambda}{4}$ (ii) $\frac{\lambda}{2}$

4.3 What is the phase difference between a node and its nearest antinode in a stationary wave?

Ans. $\frac{\pi}{2}$ radian.

4.4 Why the bats can judge distances, directions, sizes, etc. of the obstacles far beyond in the night?

Ans. Bats emit high frequency ultrasonic waves and can also detect these waves reflected from objects that come in their way. Hence, they can fly around without colliding with the objects in the night.

4.5 On what factors does the speed of a transverse wave in a stretched string depend?

4.6 Why do the musicians playing sitar, violin, etc., press the wires at different places by their left hand?

Ans. By doing so, the length of the vibrating wire is changed and notes of different frequencies are produced.

4.7 The weight suspending from a sonometer wire is increased four times. Will the frequency of the wire become exactly twice?

Ans. Yes, provided the length of the sonometer wire remains same. By increasing the weight four times, it is expected that the length of the wire slightly increases, so the frequency decreases slightly.

4.8 Two strings of equal lengths are stretched by equal forces. The strings are made of the same material. If the diameters are in the ratio 3 : 2, what will be the ratio of their fundamental frequencies?

Ans. 2 : 3.

4.9 Draw a diagram showing the formation of the second overtone in a stretched string.

[**Hint:** The string producing is the second overtone vibrate in three segments.

QUESTIONS FOR EXAMINATION

4.1 What are standing waves? Derive the equation for standing waves. Find the condition for getting nodes and antinodes.

4.2 What are characteristics of standing waves? Write the difference between the travelling waves and standing waves.

4.3 What are the velocity of constituents of medium and velocity of wave? Find differential form of the standing wave equation.

4.4 What are normal modes of a stretched string? Derive the equation of motion of standing waves. Find the wavelengths of various modes of standing waves in a stretched string.

4.5 What are normal modes of a stretched string? Find the wavelengths of various modes of standing waves in a stretched string free at one end.

4.6 What are normal modes of a stretched string? Derive the equation of motion of standing waves. Find the wavelengths and frequency of various modes of standing waves in a stretched string.

4.7 Discuss Mede's experiment. Describe its utility.

4.8 What is a sonometer? How it is used to verify the formula, $f_1 = \frac{1}{2L}\sqrt{\frac{T}{m}}$?

4.9 What is the difference in velocity of particle and velocity of wave? Derive the formula, $v = \sqrt{\frac{T}{m}}$.

4.10 Derive the formula of power of energy transfer through wave motion in the medium. Also find the relation between energy intensity and power.

4.11 What is decibel? Write the formula for sound level. Also derive the formula to compare sound level of two sounds.

4.12 Write the formula for the fundamental frequency of a transverse wave in a stretched string and explain with its help how the frequency of the string depends on the length and radius of the string. Write the frequency of the first overtone produced along with fundamental frequency (n) of the string.

4.13 Explain the formation of a standing wave in a string clamped at both ends and describe the various modes of vibrations.

4.14 How does the frequency of the transverse oscillations of a stretched string depend on the tension and density of the material of the string? Give the necessary formula. Show in a diagram the position of nodes and antinodes on the string when it vibrates to produce the third harmonic of its fundamental frequency.

4.15 What do you understand by harmonics? Explain, giving necessary sketch, that both type of harmonics, even and odd, are produced in a stretched string.

4.16 Describe a sonometer. How is the frequency of a running tuning fork is determined?

OBJECTIVE QUESTIONS

4.1 Which of the following properties of a wave is independent of the

(a) speed (b) wavelength (c) frequency (d) amplitude

4.2 In an standing wave in air, the variation of pressure at the node is

(a) maximum (b) minimum

(c) zero (d) average of maximum and minum

4.3 In an standing wave the strain is

(a) maximum at nodes (b) minimum at nodes

(c) maximum at antinodes (c) minimum at antinodes

4.4 A uniform string of length L and M is fixed at both ends under tension, T. Then it can vibrate with frequency given by the formula

(a) $f = \frac{1}{2}\sqrt{\frac{T}{ML}}$ (b) $f = \frac{1}{2L}\sqrt{\frac{T}{M}}$ (c) $f = \frac{1}{2}\sqrt{\frac{T}{M}}$ (d) $f = \frac{1}{2}\sqrt{\frac{M}{LT}}$

4.5 A sonometer wire of density d and radius r is held between two bridges at a distance L apart. The wire has a tension T. The fundamental frequency of the wire will be

(a) $f = \frac{1}{2Lr}\sqrt{\frac{T}{\pi d}}$ (b) $f = \frac{r}{2L}\sqrt{\frac{\pi d}{T}}$ (c) $f = \frac{1}{2Lr}\sqrt{\frac{d}{\pi T}}$ (c) $f = \frac{1}{2L}\sqrt{\frac{d}{T}}$

4.6 A sitar wire is replaced by another wire of same length and material but of triple radius. If the tension in the wire remains the same, the frequency will become

(a) Nine times (b) Three times (c) One-third (d) One-ninth

4.7 If the tension in a sonometer wire is increased by a factor of four, the fundamental frequency of vibration changes by a factor of

(a) 4 (b) $\frac{1}{4}$ (c) 2 (d) $\frac{1}{2}$

4.8 The frequency of the note produced by plucking a given string increases because

(a) the length of the string increases.

(b) the tension of the string increases.

(c) the tension in the string decreases.

(d) the tension and length of the string decreases proportionately.

4.9 A stretched wire of length 60 cm vibrates with a fundamental note of frequency 256 Hz. If the length of the wire is reduced to 15 cm, while the tension is kept constant, the fundamental frequency of the wire will be

(a) 64 Hz (b) 256 Hz (c) 512 Hz (d) 1024 Hz

4.10 A stretched wire of length 88 cm is divided into three segments whose frequencies are in the ratio 1 : 3 : 4; the length of the segments must be in the ratio

(a) 16 : 24 : 48 (b) 24 : 48 : 18 (c) 24 : 16 : 48 (d) 48 : 24 : 16

4.11 The speed of transverse waves in a stretched string is 700 cm/s. If the string is 2 m long, the frequency with which it resonates in fundamental mode is

(a) $\frac{7}{2}$ Hz (b) $\frac{7}{4}$ Hz (c) 7 Hz (d) $\frac{2}{7}$ Hz

Answers

4.1 (d) 4.2 (a) 4.3 (a) 4.4 (a) 4.5 (a) 4.6 (c) 4.7 (c) 4.8 (b)
4.9 (d) 4.10 (d) 4.11 (b)

CHAPTER 5

Velocity of Waves

5.1 INTRODUCTION

A wave is formed by collective oscillations of molecules or atoms, constituents of the medium, about their mean position by gaining energy from external source. A wave propagates energy from one place to another without transferring molecules or atoms. For example, if we drop a little pebble in the pond of water, the water surface gets disturbed. The disturbance propagates in the form of circles. If we continue to drop the pebbles in the pond, we see that circles rapidly moving outward from the point where the water surface is disturbed. It gives a sensation as if the water is moving outwards from the point of disturbance. If we put the cork pieces on the disturbed surface, it is found that cork pieces move up and down but do not move from the centre of the disturbance. This shows that the water mass does not flow outward with the circles. Thus, by dropping a pebble in the pond *the disturbance moves out in the form of crest of circles or water waves.*

Amplitude, wavelength and phase are used to characterize the waves. Waves can be one, two or three dimensional according to the number of dimensions in which they propagate energy. For example, waves moving along strings are one dimensional. Surface or ripples on water are two dimensional, while sound or light waves travelling radially from a point source are three dimensional.

The waves are of two types: (i) mechanical, and (ii) non-mechanical. The waves which require medium for their propagation are called *mechanical waves*. Sometimes, they are also called *elastic waves*, because elastic density and density of the medium play an important role. For example, waves of a string, sound waves are mechanical waves. The waves which do not require the medium for their propagation are called *non-mechanical waves*. For example, light waves, heat (infrared) waves and radio waves are non-mechanical waves as they do not require any medium for their propagation. In fact, all electromagnetic waves like X-rays, γ-rays, microwaves, light waves, infrared and ultraviolet waves are non-mechanical waves.

A third kind of wave is called *matter waves*. The energy of electrons, protons, neutrons, atoms and molecules, called the constituents of matter, transfer through matter waves. The matter waves are useful in several devices of modern technology. For example, matter waves associated with it are employed in the operation of electron microscope.

In this chapter, we will study mechanical waves which require a material medium for their propagation.

Light is a electromagnetic wave and exhibits interference and diffraction phenomena. But these phenomena do not indicate that light wave is a transverse wave. Phenomenon of polarization shows that light wave is a transverse wave. Polarization is a technique of restricting the vibrations of light wave to one direction or one plane perpendicular to the direction of its propagation.

Transverse waves: According to quantum mechanics, transverse wave is that wave in which the direction of the oscillation of particle is perpendicular to the direction of propagation of the wave. For example, light and sound waves. Transverse waves can also be produced in a string. The transverse waves can be polarized.

Longitudinal waves: In a longitudinal wave in the direction of oscillation of particles is parallel to the direction of propagation of the wave. For example, sound wave. A sound wave can be both transverse and longitudinal. Longitudinal waves cannot be polarized.

5.2 TRANSVERSE VIBRATIONS IN A STRING

The transmission of transverse wave of a rope through a slit is shown in Fig. 5.1. To show the plane of polarization, let us consider that one end of the rope is fixed at point B after

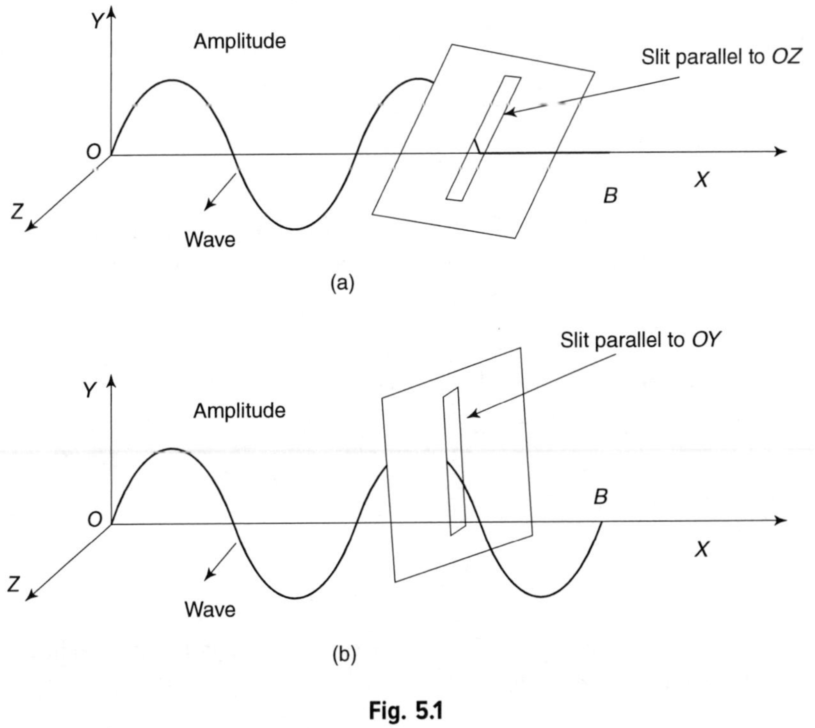

Fig. 5.1

passing the rope through a slit. The width of the slit is more than the diameter of the rope. The other end point (*O*) of the rope is forced to vibrate along *YZ* plane as shown in Fig. 5.1(a). It is observed that transverse wave formed by the rope does not pass through the slit, if the slit is kept perpendicular to the plane of vibrations, i.e., in *ZX* plane. When the slit is kept parallel to the plane of vibrations, i.e., *YZ* plane, then the transverse wave formed by the rope passes through the slit as shown in Fig. 5.1(b). The transverse wave formed by the rope is called plane polarized. Plane of polarized means that the vibrations of the transverse wave are restricted to one direction.

5.3 VELOCITY OF TRANSVERSE VIBRATIONS OF STRETCHED STRINGS

Let *m* be the mass per unit length of a string. When tension (*T*) is applied, the velocity of a transverse wave propagating along a string is given by

$$v = \sqrt{\frac{T}{m}}$$

To prove this, let us consider a small segment of the pulse of a string of length (ΔL) forming an arc of the circle of radius ® as shown in Fig. 5.2.

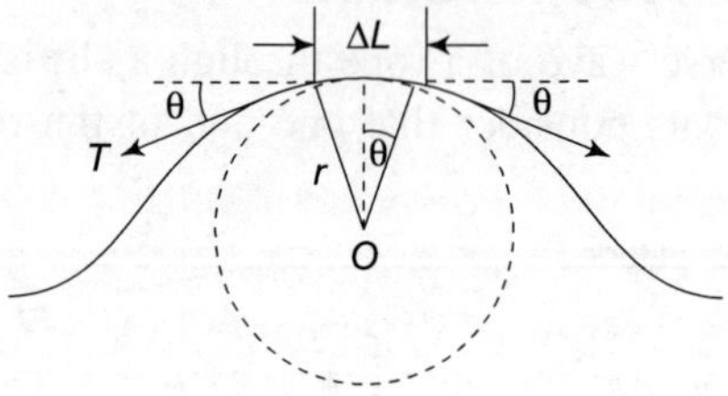

Fig. 5.2

A force equal to tension (*T*) in the string pulls this segment tangentially at each end. The horizontal components ($T \cos \theta$) of these forces cancel each other. While the vertical components ($T \sin \theta$) add to form a restoring force,

$$F = 2\, T \sin \theta$$

For small θ, sin θ = θ, so restoring force becomes

$$F = T\, 2\theta$$

From the geometry of Fig. 5.2, $2\theta = \dfrac{\Delta L}{r}$, so

$$F = T\, \frac{\Delta L}{r} \qquad ...(5.1)$$

The string element (ΔL) is moving with velocity (v) in an arc of the circle, it has a centripetal force towards the centre of the circle,

$$F = \Delta m\, a = \frac{\Delta m v^2}{r} \quad ...(5.2)$$

where, mass of the segment ΔL is $\Delta m = m \times \Delta L$. The acceleration of the wave is given by

$$a = \frac{v^2}{r}$$

At equilibrium, the centripetal force balances the force due to tension, therefore,

$$\frac{m\Delta L\, v^2}{r} = T\,\frac{\Delta L}{r}$$

The velocity of transverse vibrations of stretched string,

$$v = \sqrt{\frac{T}{m}} \quad ...(5.3)$$

From Eq. (5.2), we can infer that the velocity of a transverse wave propagating along a string depends only on the tension and mass per unit length, m. It is independent on the frequency of the wave.

If A is the cross section of the wire, L is the length of the wire, and ρ is mass per unit volume, then mass of the wire,

$$M = AL\rho$$

So, mass per unit length

$$m = A\rho$$

Now, the velocity of transverse vibrations of stretched string will be

$$v = \sqrt{\frac{T}{A\rho}}$$

or

$$v = \sqrt{\frac{S}{\rho}} \quad ...(5.4)$$

where, stress on the string, $S = \frac{T}{A}$.

If transverse mechanical wave is propagating through a solid, its velocity is given by

$$v = \sqrt{\frac{\eta}{\rho}} \quad ...(5.5)$$

where, η is the bulk modulus (modulus of rigidity) of the solid and ρ is the mass per unit volume of the solid.

SOLVED EXAMPLES

Example 5.1 A wire of uniform cross section is stretched between two points 1.5 m apart. The wire is fixed at one end and a weight of 4 kg is hung over a pulley at the other end produces fundamental frequency of 700 Hz. (a) What is the velocity of transverse waves propagating in the wire? (b) If now the suspended weight is submerged in a liquid of density 3/4 that of the weight, what will be the velocity and frequency of the waves propagating along the wire?

Solution: Formula required

Velocity of transverse wave in wire, when tension (T) is applied, is

$$v = \sqrt{\frac{T}{m}}$$

where, m is mass of wire per unit length. Let v_A and v_B *be* the velocities of wire, when tensions (T_A and T_B) are applied, then

$$v_A = \sqrt{\frac{T_A}{m}} \text{ and } v_B = \sqrt{\frac{T_B}{m}}\text{, the ratio of velocities, } \frac{v_B}{v_A} = \sqrt{\frac{T_B}{T_A}} \text{ or } v_B = v_A = \sqrt{\frac{T_B}{T_A}}$$

Given: Length of wire, $L = 1.5$ m, fundamental frequency of wire, $f = 700$ Hz, tensions, T_A = mg, $m = 4$ kg

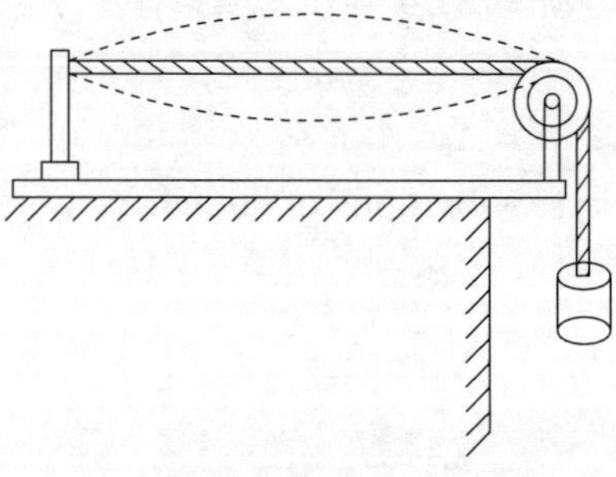

Fig. 5.3

Ratio of densities of liquid and weight, $\frac{\sigma}{\rho} = \frac{3}{4}$. So tension of the wire weight is submerged in liquid,

$$T_B = m'g = m\left(1 - \frac{\sigma}{\rho}\right)g = m\left(1 - \frac{3}{4}\right)g$$

(a) For fundamental vibration of wire, $L = \frac{\lambda}{2}$ or $\lambda = 2\,L = 2 \times 1.5 = 3$ m

Velocity of fundamental wave, $v_A = f\lambda$ or $v_A = 700 \times 3 = 2100$ m/s

Velocity of wave when weight is submerged in liquid

$$v_B = v_A\sqrt{\frac{T_B}{T_A}} = v_A\sqrt{\frac{m'g}{mg}} = v_A\sqrt{\frac{m'}{m}} = 2100\sqrt{1-\frac{3}{4}} = 2100\times\frac{1}{2} = 1050 \text{ m/s}$$

Since wavelength, λ = cost = 3 m for fundamental wave, the frequency of wave (f_B) of wave when the wire is submerged in liquid is given by equation

$$v_B = f_B\,\lambda \quad \text{or} \quad f_B = \frac{v_B}{\lambda} = \frac{1050}{3} = 350 \text{ Hz}$$

Ans. (a) Velocity of fundamental wave when tension is due to weight 49 kg = 2100 m/s

(b) Velocity and frequency of fundamental wave when 4 kg weight is submerged in liquid are 1050 m/s and 350 Hz.

Example 5.2 A wire of mass 9.8 gm per metre passes over a frictionless pulley fixed on the top of an inclined frictionless plane which makes an angle of 30° with the horizontal. Masses, M_1 and $M_{2,}$ are tied at the two ends of the wire. The mass M_1 rests on the plane and the mass M_2 hangs vertically downwards. The whole system is in equilibrium. Now a transverse wave propagates along the wire with a velocity of 100 m/s. Find the value of masses, M_1 and M_2. [g = 9.8 m/s^2]

Solution: Formula required

Velocity of transverse wave in wire, when tension (T) is applied,

$$v = \sqrt{\frac{T}{m}}$$

where, m is mass of wire per unit length.

At equilibrium, from Fig. 5.13, we have

$$M_1\,g\sin\theta = T \quad \text{and} \quad M_1\,g\cos\theta = \text{R}$$

and $$M_2\,g = \text{T}$$

Given: Mass per metre, m = 4.9 gm = 9.8 × 10^{-3} kg, g = 9.8 m/s^2, v = 200 m/s, θ = 30°

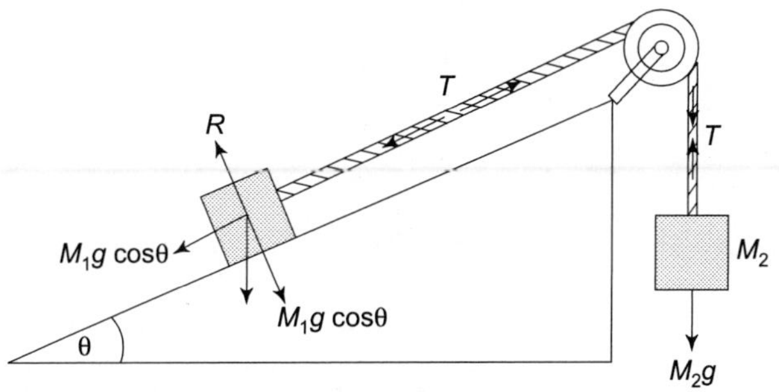

Fig. 5.4

From the formula of velocity of wave, $T = mv^2 = 4.9 \times 10^{-3} \times (200)^2 = 196\ N$. Putting the value of T,

$$M_{2g} = T \quad \text{or} \quad M_2 = \frac{196}{9.8} = 20 \text{ kg}$$

$$M_1 g \sin\theta = T \quad \text{or} \quad M_1 = \frac{196}{9.8 \times \sin 30} = \frac{196}{9.8 \times \frac{1}{2}} = 40 \text{ kg}$$

Ans. Mass $M_1 = 40$ kg, $M_2 = 20$ kg

Example 5.3 A uniform rope of length 9 m and mass 5 kg hangs vertically from a rigid support. A block of mass 4 kg is attached to the free end of the rope. A transverse pulse of wavelength 0.07 m is produced at the lower end of the rope. What is the wavelength of the pulse when it reaches the top of the rope.

Solution: Formula required

Velocity of transverse wave in rope, when tension (T) is applied,

$$v = \sqrt{\frac{T}{m}}$$

where, m is mass of rope per unit length.

Given: Length of rope, $L = 9$ m, mass of rope, $M = 5$ kg, mass of suspended block, $M_B = 4$ kg, $\lambda_A = 0.06$ m.

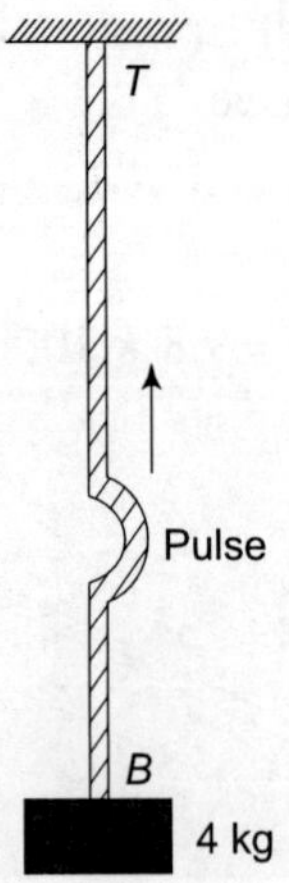

Fig. 5.5

Let the tension of the rope with mass $M_B = (5 + 4)$ kg and with mass $M_A = 4$ kg be T_B and T_A. Let the corresponding velocities of wave in the rope be v_T and v_A. So the ratio of velocities, v_B and v_A, is written as

$$v_B = v_A\sqrt{\frac{T_B}{T_A}} = v_A\sqrt{\frac{M_B g}{M_A g}} = v_A\sqrt{\frac{9}{4}} = \frac{3}{2} = 1.5$$

If f_B, λ_B and f_A, λ_a are the corresponding frequencies and wavelengths, then

$$\frac{v_B}{v_A} = \frac{f_B\,\lambda_B}{f_A\,\lambda_A} = 1.5 \quad \text{or} \quad \frac{\lambda_B}{\lambda_A} = 1.5$$

In this problem, $f_B = f_A$ as frequency is the characteristic of the source, so

$$\lambda_B = 1.5\,\lambda_A = 1.5 \times 0.07 \text{ m} = 0.105 \text{ m}$$

Ans. The wavelength of the pulse is 0.09 m.

Example 5.4 A uniform rope of mass 0.1 kg and length 2.45 m hangs from a ceiling. (a) Find the speed of the transverse wave in the rope at a point 0.5 m distant from the lower end. (b) Calculate the time taken by the wave to travel the full length of the rope. [$g = 9.8$ m/s^2]

Solution: Formula required

Velocity of transverse wave in rope, when tension (T) is applied,

$$v = \sqrt{\frac{T}{m}}$$

where, m is mass of the rope per unit length.

Given: Mass of the rope, $M = 0.15$ kg, length of the rope, $L = 2.25$ m, the distance from lower end $x = 0.6$ m, $g = 9.8$ m/s^2.

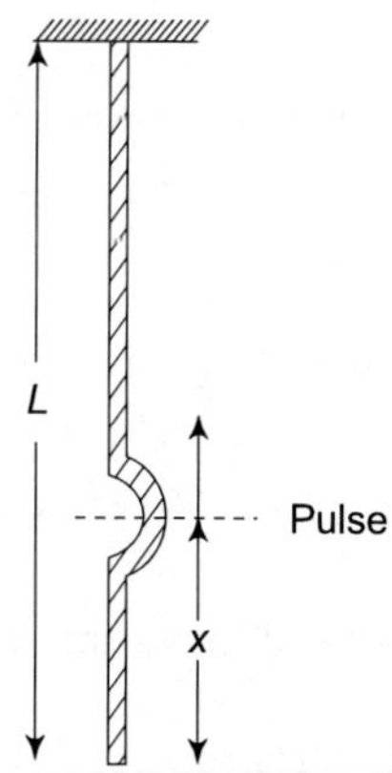

Fig. 5.6

Let the mass per unit length be, $m = \frac{M}{L}$.

(a) As the string has mass and it is suspended vertically, tension in it will be different at different points. For a point at distance x from the free end, the mass of length x of the string will be $\frac{M}{L}x$ or mx, so the tension at point x will be

$$T = \left(\frac{M}{L}x\right)g = (mx)g$$

$$v = \sqrt{\frac{T}{m}} = \sqrt{\frac{mxg}{m}} = \sqrt{xg} = \sqrt{0.6 \times 9.8} = 2.02 \text{ m/s}$$

(b) From part (a) it is clear that the tension is different at different points of the string, so velocity of wave is different at different points. At point x, let the wave travel a distance dx in time dt. So

$$v = \frac{dx}{dt} \quad \text{or} \quad \sqrt{xg} = \frac{dx}{dt}$$

$$\int dt = \int \frac{dx}{\sqrt{xg}} \quad \text{or} \quad t = \frac{1}{\sqrt{g}}\int_0^L x^{-1/2}dx$$

or

$$t = 2\sqrt{\frac{L}{g}} = 2\sqrt{\frac{2.25}{9.8}} = 0.96 \text{ sec.}$$

Ans. (a) Velocity at point 0.5 from the free end is 2.02 m/s.

(b) Time taken by wave to travel full length of string is 0.96 sec.

Example 5.5 A copper wire is held at the two ends by rigid supportsat 30°C, the wire is taut with negligible tension. Find the speed of the transverse waves in this wire at 10°C if $Y = 1.3 \times 10^{11}$ N/m², $\alpha = 1.7 \times 10^{-5}$ per °C, and $\rho = 9 \times 10^3$ kg/m³.

Solution: Formula required

Velocity of transverse wave in wire, when tension (T) is applied,

$$v = \sqrt{\frac{T}{m}}$$

where, m is mass of the wire per unit length.

Given: At 10°C, Young's modulus of wire, $Y = 1.3 \times 10^{11}$ N/m², the expansion coefficient $\alpha = 1.7 \times 10^{-5}$ per °C and mass density $\rho = 9 \times 10^3$ kg/m³. Temperature $\theta_2 = 30$°C and $\theta_1 = 10$°C

If T is stress (tension) and A is cross-sectional area of wire, then

$$\frac{T}{A} = Y \times \text{Thermal strain} = Y\alpha\Delta\theta \quad \left[\text{as } \alpha = \frac{\Delta L}{L\Delta\theta}\right]$$

So

$$v = \sqrt{\frac{YA\alpha\Delta\theta}{m}} = \sqrt{\frac{Y\alpha\Delta\theta}{\rho}} \quad \left[\text{as } \rho = \frac{M}{A \times L}\right]$$

$$v = \sqrt{\frac{1.3 \times 10^{11} \times 1.7 \times 10^{-5} \times (30 - 10)}{9 \times 10^3}} = 70 \text{ m/s}$$

Ans. Speed of the transverse wave is 70 m/s.

5.4 NEWTON'S FORMULA FOR THE SPEED OF SOUND IN GASES

Newton assumed that when longitudinal waves (sound) are in the gaseous medium, the temperature of the medium remains constant. Hence, the formula for the speed of sound in gases is given by

$$v = \sqrt{\frac{B_T}{\rho}} \quad ...(5.6)$$

where, B_T is isothermal bulk modulus of the gas. The bulk modulus at constant temperature is equal to pressure (P) of the gas as shown below.

Let and pressure of an ideal gas is P and volume is V. At constant temperature the gas equation is

$$PV = \text{constant} \quad ...(5.7)$$

It is called Boyle's law. Differentiating Eq. (5.7), we get

$$P\,dV + V\,dP = 0$$

or

$$P = -\frac{V\,dP}{dV} = -\frac{dP}{\frac{dV}{V}} = B_T \quad ...(5.8)$$

That is, isothermal bulk modulus (B_T) is equal to the pressure of thegas. Now the velocity of sound wave in the gas is written as

$$v = \sqrt{\frac{P}{\rho}} \quad ...(5.9)$$

Formula given by Eq. (5.9) is called Newton's formula for the speed of sound in gases.

Example 5.6 Determine the change in volume of 7 litres of alcohol if the pressure is decreased from 205 cm of Hg to 80 cm. Velocity of sound in alcohol is 1285 m/s, density of alcohol is 0.84 gm/cc, density of Hg is 13.6 gm/cc and g = 9.81 m/s^2.

Solution: Formula for speed of sound in gases is given by

$$v = \sqrt{\frac{B_T}{\rho}} \quad ...(5.10)$$

where, B_T is isothermal bulk modulus of the gas.

Given: Volume alcohol, $V = 7$ litres

Speed of sound in alcohol, $v = 1285$ m/s $= 1.285 \times 10^5$ cm/s

Density of alcohol, $\rho_{al} = 0.84$ gm/cc

Density of Hg, $\rho_{Hg} = 13.6$ gm/ cc

Acceleration due to gravity, $g = 9.81$ m/s^2

Pressure of Hg decreases from $h_1 = 205$ cm to $h_2 = 80$ cm

Calculation: Isothermal bulk modulus from Eq. (5.10) can be written as

$$B_T = \rho_{al}\, v^2$$

Bulk modulus is defined as

$$B_T = -\frac{\Delta P}{\frac{\Delta V}{V}} = -V\frac{\Delta P}{\Delta V} = \rho_{al}\, v^2$$

or

$$\Delta V = \frac{V(-\Delta P)}{\rho_{al}\, v^2}$$

$$\Delta P = h_2\, \rho_{Hg}\, g - h_1\, \rho_{Hg}\, g = (72 - 200)$$

$$\rho_{Hg}\, g = (80 - 205) \times 13.6 \times 9.81 = -1.667 \times 10^6 \text{ dynes/cm}^2$$

$$\Delta V = \frac{7 \times 10^3 \times 1.667 \times 10^6}{0.84 \times (1.285 \times 10^5)^2} = 0.84 \text{ cc}$$

Ans. Change in volume is 0.84 cc.

5.5 LAPLACE CORRECTION

After about 100 years of establishing the formula of speed of sound in gases, in 1816 Laplace found that the assumption of Newton that temperature of gas remains constant is not correct. When sound waves travel in a gas then at any point in the medium, compression and rarefaction occur alternately. At the time of compression some heat is developed at the point and at the time of rarefaction some heat is lost from the medium at that point. The compressions and rarefactions occur so rapidly that heat produced during compression cannot go out into the surroundings and the heat disappeared during rarefaction cannot come in from the surroundings. Besides this, the exchange of heat does not take place because gases are bad conductors of heat. Hence, the temperature at a point in the medium rises during compression and falls during rarefaction.That is, during propagation of the sound waves, temperature varies in the gas medium. Hence, in the formula, $v = \sqrt{B/\rho}$, B should the adiabatic bulk modulus, B_S (the bulk modulus at constant entropy, S) rather than isothermal bulk modulus, B_T (the bulk modulus at constant temperature). Therefore, the speed of sound in a gas is written as

$$v = \sqrt{\frac{B_S}{\rho}} \qquad ...(5.11)$$

Under adiabatic conditions, the gas obeys Poisson's law

$$PV^\gamma = \text{constant} \qquad ...(5.12)$$

Differentiating Eq. (5.12), we get

$$P(\gamma\, V^{\gamma-1}\, dV) + V^\gamma\, dP = 0$$

or

$$P(\gamma\, V^{\gamma-1}\, dV) = -V^\gamma\, dP$$

or
$$\gamma P = -\frac{dP}{\frac{dV}{v}} = B_S \qquad ...(5.13)$$

From Eq. (5.13), the adiabatic bulk modulus (B_S) or bulk modulus at constant entropy is equal to γP. Therefore, speed of sound in adiabatic approximation is written as

$$v = \sqrt{\frac{\gamma P}{\rho}} \qquad ...(5.14)$$

where, γ is defined as the ratio of specific heat of a gas constant pressure and specific heat of a gas at constant volume, i.e.,

$$\gamma = \frac{\text{Specific heat of a gas at constant pressure}}{\text{Specific heat of a gas at constant volume}} = \frac{C_P}{C_T}$$

For monoatomic gases (such as helium, argon, neon, etc.) the value of γ is 5/3 = 1.67, for diatomic gases (such as oxygen, nitrogen, hydrogen, etc.), the value of γ is 7/5 = 1.4. Atmospheric air around us contains manly nitrogen and oxygen, so the value of γ for air around us is 1.4. According to Laplace, the speed of sound in air is

$$v = \sqrt{1.4} \times 280 = 331 \text{ m/s}.$$

This value agrees with the experimental value. Hence, formula for speed [given by Eq. (5.13)] of sound obtained by Laplace is correct.

Since γ depends on the medium, so speed of sound is different for different media. The speed of sound is minimum in gases (such as oxygen, nitrogen, hydrogen, etc.), some what more in gases (such as helium, argon, neon, etc., and maximum in solids. If a person hammers a railway line at some point and another person at some distance hears two sounds at a small interval of time. The first intense sound reaches through the rail line and the second sound reaches through the air, because the speed of sound through rail line is more than the speed of sound in air.

5.6 EFFECT OF PRESSURE, TEMPERATURE, HUMIDITY, WIND AND FREQUENCY

(i) Effect of pressure

The formula of speed of sound in a gas is given by Eq. (5.14)

$$v = \sqrt{\frac{\gamma P}{\rho}} \qquad ...(5.15)$$

From this formula the speed of sound, $v \propto \sqrt{P}$, so it appears that speed of sound increases with increase in pressure. But it is not true.

Let 1 mole of a gas has a pressure (P) and a volume, V. If T is the absolute temperature of the gas, the gas equation is written as

$$PV = \text{RT} \qquad ...(5.16)$$

where, R is gas constant. Ifthe molecular mass (mass of 1 mole of the gas) of the gas is M, then density, $\rho = M/V$, so gas equation becomes

$$P\frac{M}{\rho} = \text{RT}$$

or
$$\frac{P}{\rho} = \frac{\text{RT}}{M} = \text{constant, if } T \text{ is constant.} \qquad ...(5.17)$$

In Eq. (5.17) the ratio $\left(\frac{P}{\rho}\right)$ remains constant. This is because if P changes, then ρ also chanes in such a way that ratio P/ρ remains constant. Therefore, in the formula given by Eq. (5.15), the speed (v) of sound does not change if pressure changes as the ratio P/ρ remains constant. From this it is concluded that there is no effect of pressure change on the speed of sound, if temperature is constant.

(ii) Effect of temperature

For a gas, value of the ratio (P/ρ) given by Eq. (5.16) is proportional to temperature (T) of the gas. If we heat the gas, there are two possibilities:

(i) If the gas is free to expand, then on heating it will expand, so its density (ρ) will decrease and hence its pressure will remain the same. Thus, the ratio (P/ρ) will increase.

(ii) If the gas is closed in a vessel, then on heating its pressure will increase, but its density will not change. So the value of ratio (P/ρ) will increase.

Therefore, in both cases, the speed of the sound will increase with increase in temperature of the gas.

For 1 mole of gas, the gas equation is

$$PV = \text{RT}$$

or
$$P\frac{M}{\rho} = \text{RT}$$

or
$$\frac{P}{\rho} = \frac{\text{RT}}{M}$$

where, M is the molecular mass of the gas and T is the absolute temperature of the gas. Now the formula for speed of sound given by Eq. (5.15) becomes

$$v = \sqrt{\frac{\gamma\,\text{RT}}{M}} \qquad ...(5.18)$$

It is evident from the formula given by Eq. (5.18) that the of speed of sound wave in a gas is directly proportional to the square root of absolute temperature.

Now, if v_0 is speed at 0°C and v_t is the speed at t°C, then from the Eq. (5.18), we have

$$v_0 = \sqrt{\frac{\gamma R}{M}}\sqrt{273}\text{ , } T = 0°\text{C} = 273\text{ °K} \qquad ...(5.19)$$

and $$v_t = \sqrt{\frac{\gamma R}{M}}\sqrt{273+t} \qquad ...(5.20)$$

Dividing v_t by v_0, we get

$$\frac{v_t}{v_0} = \frac{\sqrt{273+t}}{\sqrt{273}} = \sqrt{1+\frac{t}{273}}$$

or $$v_t = v_0\left(1+\frac{t}{273}\right)^{1/2}$$

Since $t >> 273$, expanding right-hand side by binomial theorem and retaining only dominant term, above equation reduces to

$$v_t = v_0\left(1+\frac{1}{2}\times\frac{t}{273}\right)$$

If the speed of sound in air at 0°C is taken as 332 m/s, then

$$v_t = 332\left(1+\frac{1}{2}\times\frac{t}{273}\right) = (332 + 0.61\ t)\ \text{m/s}$$

Thus, the velocity of speed in air increases roughly by 0.61 m/s per degree centigrade rise in temperature.

(iii) Effect of humidity

The density of moist air, that is, air mixed with water vapour, is less than the density of dry air. Therefore, assuming the value of γ for moist air same as that of dry air, it is clear from the formula, $v = \sqrt{\gamma P/\rho}$, where, ρ is the density of moist air, that the speed of sound in the moist air is slightly greater than the speed of sound in dry air. Thus, speed of sound (v) increases with increase in humidity. That is why the sound of sirens of mills and sound of whistle of trains are heard up to longer distance in the rainy season as compared to summers.

According to the formula, $v = \sqrt{\gamma\, RT/M}$, where, M is the effective molar mass of moist air. The speed of sound depends on the molar mass (M) of the mixture. So, the speed of sound in water vapour is more that its speed in nitrogen gas or in oxygen gas, because molecular mass of water vapour is less than that of nitrogen or oxygen. The molecular mass of water vapour (H_2O) is 18, molecular mass of nitrogen (N_2) gas is 28 and that of oxygen (O_2) is 32.

Note: Actually the value of γ for moist air is slightly less than for dry air, but this difference is generally negligible.

(iv) Effect of wind

In the presence ofblowing wind the speed of sound changes. If the speed of wind is w and the speed of sound in absence of wind is v, then the speed of sound in the direction wind will be $(v + w)$ and in the opposite direction it will be $(v - w)$.

Let us assume that sound is moving in the direction of observer (O) with speed, v. If the wind is blowing with velocity (w) along a direction making an angle (θ) with the direction of sound, as shown in Fig. 5.7, then the component of wind velocity along the direction of observer, i.e., $SO = w \cos \theta$. This component is added to the speed v of sound. Therefore, the resultant speed of sound in the direction SO is $(v + w \cos \theta)$. If the wind is blowing at right angle to the direction of sound, i.e., $\theta = 90°$, then the resultant speed of the sound will remain v. That is, the wind will not affect the speed of sound.

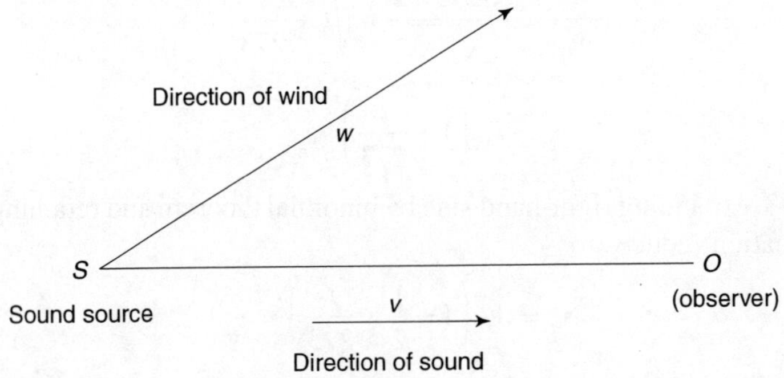

Fig. 5.7

(v) Effect of frequency

There is no effect of frequency on the speed of sound, normally. Sound of different frequencies travels with the same speed in air even if their wavelengths are different in air. If the speed of sound were dependent on the frequency then orchestra were not able to produce musical sound.

5.7 COMPARISON BETWEEN TRANSVERSE AND LONGITUDINAL WAVES

Both transverse and longitudinal waves may be represented by sine waves. Both types of waves can

(a) be reflected,

(b) be refracted,

(c) interfere, and

(d) be diffracted.

In general, what applies to one set of waves allies equally well to other. But there is one major difference in the properties of two. The transverse waves can be polarized whereas longitudinal waves cannot.

Example 5.7 (a) Speed of sound in air is 343 m/s at NTP. What will be the speed of sound in hydrogen at NTP if the density of hydrogen at NTP is (1/9) that of air. [Assume $P_{air}/P_H = 1$]

(b) Calculate the ratio of the speed of sound in neon to that in water vapours at any temperature. [molecular weight of neon = 2.02×10^{-2} kg/mol and for water vapours = 1.8×10^{-2} kg/mol]

Solution: Formula for velocity of sound in gas at pressure (P) and temperature, T

$$v = \sqrt{\frac{\gamma P}{\rho}} = \sqrt{\frac{\gamma RT}{M}}$$

where, M and ρ are molecular weight and density of gas, respectively. R is gas constant. γ is defined as the ratio of specific heat of a gas constant pressure and specific heat of a gas at constant volume, i.e.,

$$\gamma = \frac{\text{Specific heat of a gas at constant pressure}}{\text{Specific heat of a gas at constant volume}} = \frac{C_P}{C_T}$$

Given: Speed of sound in air 342 m/s

Density of hydrogen, $\dfrac{\rho_H}{\rho_{air}} = \dfrac{1}{9}$

Molecular weight of neon, $M_N = 2.07 \times 10^{-2}$ kg/mol

Molecular weight of water vapour, $M_W = 1.9 \times 10^{-2}$ kg/mol

Pressure of air and hydrogen are equal, $P_{air} = P_{air}$

For neon, $\gamma_N = \dfrac{5}{3}$, for water vapour, $\gamma_N = \dfrac{4}{3}$

(a) The velocity of sound in air and hydrogen are written as

$$v_{air} = \sqrt{\frac{\gamma P_{air}}{\rho_{air}}} \quad \text{and} \quad v_H = \sqrt{\frac{\gamma P_H}{\rho_H}}$$

$$\frac{v_H}{v_{air}} = \sqrt{\frac{P_{Hr}}{P_{air}} \times \frac{\rho_{air}}{\rho_H}} = \sqrt{\frac{\rho_{air}}{\rho_H}} \quad \text{as } P_{air} = P_{air}$$

$$v_H = 342\sqrt{9} = 1026 \text{ m/s}$$

(b) The velocity of sound in neon and water at same temperature (T) are written as

$$v_W = \sqrt{\frac{\gamma_W RT}{M_W}} \quad \text{and} \quad v_N = \sqrt{\frac{\gamma_N RT}{M_N}}$$

$$\frac{v_N}{v_W} = \sqrt{\frac{M_W}{M_N} \times \frac{\gamma_N}{\gamma_W}}$$

or

$$\frac{v_N}{v_W} = \sqrt{\frac{1.9 \times 10^{-2}}{2.07 \times 10^{-2}} \times \frac{5}{4}} = 1.071$$

Ans. (a) Speed of hydrogen at NTP is 1026 m/s.

(b) The ratio of the speed of sound in neon to that in water vapour at any temperature is 1.071.

Example 5.8 (a) Find the speed of sound in a mixture of 1 mol of helium and 2 mol of oxygen at 27°C. (b) If the temperature is raised by 1 K from 300 K, find the percentage change in the speed of sound in the gaseous mixture. [R = 8.31 J/mol K].

Solution: Formula for speed of sound in gas at pressure (P) and temperature (T) is

$$v = \sqrt{\frac{\gamma RT}{M}}$$

where, M is molecular weight of gas and R is gas constant. γ is defined as the ratio of specific heat of a gas constant pressure and specific heat of a gas at constant volume, i.e.,

$$\gamma = \frac{\text{Specific heat of a gas at constant pressure}}{\text{Specific heat of a gas at constant volume}} = \frac{C_P}{C_T}$$

Given: Mixture of 1 mol of helium and 2 mol of oxygen at 27°C

Gas constant, R = 8.31 J/mol K

Calculation: Molecular weights of one molecule of helium and 1 molecule of oxygen are $M_H = 4$ and $M_0 = 32$, respectively. The mixture of $n_H = 1$ mol of helium and $n_0 = 2$ mol of oxygen will be

$$M_{\text{mix}} = \frac{n_H M_H + n_O M_O}{n_H + n_O} = \frac{1 \times 4 + 2 \times 32}{1+2}$$

$$= \frac{68}{3} \times 10^{-3} \text{ kg/mol}$$

Helium is monoatomic so $C_v^H = \frac{3}{2} R$ and oxygen is diatomic so $C_v^O = \frac{5}{2} R$

The specific heat of mixture of $n_H = 1$ mol of helium and $n_O = 2$ mol of oxygen at constant volume will be

$$C_v^{\text{mix}} = \frac{n_H C_v^H + n_O C_v^O}{n_H + n_O} = \frac{1 \times \frac{3}{2} R + 2 \times \frac{5}{2} R}{1+2} = \frac{13}{6} R$$

Since the temperature is raised by 1 K, so the specific heat of mixture of $n_H = 1$ mol of helium and $n_O = 2$ mol of oxygen at constant pressure will be

$$C_p^{\text{mix}} = C_v^{\text{mix}} + R = \frac{13}{6} R + R = \frac{19}{6} R$$

Now $$\gamma^{\text{mix}} = \frac{C_p^{\text{mix}}}{C_v^{\text{mix}}} = \frac{19}{13}$$

Speed of sound in mixture of gases is

$$v = \sqrt{\frac{\gamma^{\text{mix}}\,\text{RT}}{M_{\text{mix}}}} = \sqrt{\frac{19}{13} \times \frac{8.31 \times 300}{\frac{68}{3} \times 10^{-3}}} = 400.9 \text{ m/s}$$

(b) Speed of sound in gas at pressure (P) and temperature (T) is

$$v = \sqrt{\frac{\gamma R}{M}}$$

Differentiating v w.r.t. T, we get

$$\Delta v = \sqrt{\frac{\gamma R}{M}} \times \frac{1}{2} T^{-1/2} \, \Delta T = \frac{1}{2}\sqrt{\frac{\gamma \text{RT}}{M}} \frac{\Delta T}{T} = \frac{1}{2} v \frac{\Delta T}{T}$$

$$\frac{\Delta v}{v} = \frac{1}{2}\frac{\Delta T}{T} = \frac{1}{2} \times \frac{1}{300}$$

Percentage change in the speed of sound by raising 1 K to 300 K

$$\frac{\Delta v}{v} \text{ is } = \frac{1}{2} \times \frac{1}{300} \times 100 = 0.167\%$$

Ans. (a) Speed of sound in mixture is 400.9 m/s.

(b) Percentage change of speed of sound by raising 1 K from 300 K is 0.167%.

5.8 COMPARISON BETWEEN SOUND WAVES AND ELECTROMAGNETIC WAVES

(a) Sound waves are longitudinal and electromagnetic waves are transverse waves.

(b) Sound waves travel at a speed of 340 m/s whereas electromagnetic waves travel at a speed of 3×10^8 m/s.

(c) Sound waves do not pass through a vacuum, i.e., sound waves need medium to travel, but electromagnetic wave travesl through vacuum, i.e., electromagnetic waves do not need any medium to travel.

5.9 DOPPLER EFFECT FOR SOUND (FIRST METHOD)

When a source of sound (say radio) and detector (usually listener) of sound are at a large distance from each other and they are stationary relative to the medium (usually air). The detector receives the same number of waves per second as that emitted by the source. However, if source or detector or both are in motion, the number of waves received per second is usually different from that emitted by the source The apparent change of frequency or pitch due to motion of the source and detector is called Doppler effect.

Let us consider the case when both source and detector are moving with speeds, v_S and v_D, respectively, in the direction in which the sound is moving. The detector is moving ahead of the source as shown in Fig. 5.8. Let us consider that the frequency of sound source is f.

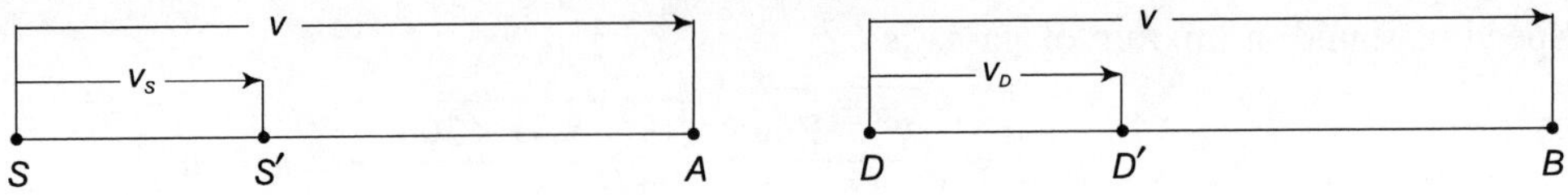

Fig. 5.8

In one second the first wave emitted by the source at S travels a distance $v \times 1 = v$ and reaches A. Mean while the source S moves a distance v_S and reaches S' emitting f^{th} wave. If $v_S < v$, the f number of waves emitted by the source in one second contained in the distance $S'A = v - v_S$.

The first wave received by the detector at D covers a distance v in one second and reaches at B, mean while the detector itself covers a distance v_D and reaches at D', where it receives the last wave emitted in one second. If $v_D < v$ the waves received by the detector in one second are contained in a distance $D'B = v - v_D$.

The waves contained in a distance $v - v_S$ are f, so the number of waves contained in a distance $v - v_D$ will be

$$= f \times \frac{(v - v_D)}{(v - v_S)}$$

Which is the apparent frequency f_{AP} as perceived by the detector, i.e.,

$$f_{AP} = \left[\frac{v - v_D}{v - v_S}\right] \qquad ...(5.21)$$

In this formula detection of sound from source (S) to detector (D) is taken positive, so speeds, v_D and v_S, are positive because they are in the direction from S to D.

[A] Only source is in motion

Let us assume that the source is moving towards the detector, so v_S is considered +ve, i.e., $+ v_S$ as is moving in the direction of travelling sound wave.

In this case $v_D = 0$, so,

$$f_{AP} = f\left[\frac{v - 0}{v - (+v_S)}\right] = f\left[\frac{v}{v - v_S}\right]$$

i.e., the apparent frequency or pitch has increased as $f_{AP} > f$.

If the source moves away from the detector, then v_S is to considered –ve , i.e., $- v_S$, because it is moving in opposite direction of sound wave. Herefore

$$f_{AP} = f\left[\frac{v - 0}{v - (-v_S)}\right] = f\left[\frac{v}{v + v_S}\right]$$

i.e., the apparent frequency or pitch has decreases as $f_{AP} < f$.

[B] Only detector is in motion

In this case $v_S = 0$ and detector is moving away from the source, v_D is taken as +ve, i.e., + v_D, because it is moving in the direction of sound wave. Therefore,

$$f_{AP} = f\left[\frac{v-(+v_D)}{v-0}\right] = f\left[\frac{v-v_D}{v}\right]$$

i.e., the apparent frequency or the pitch increases as $f_{AP} < f$.

If the detector is moving towards the source, v_D will be $-\ v_D$, because the detector is moving in opposite direction of sound waves. Therefore,

$$f_{AP} = f\left[\frac{v-(-v_D)}{v-0}\right] = f\left[\frac{v \pm v_D}{v}\right]$$

i.e., the apparent frequency or the pitch increases as $f_{AP} > f$.

[C] Source and detector both in motion

If the source and detector both move towards each other, i.e., v_S is moving in the direction of v, while v_D is moving opposite to v so that v_S is taken + v_S and v_D is taken $-\ v_D$. Therefore,

$$f_{AP} = f\left[\frac{v-(-v_D)}{v-(+v_S)}\right] = f\left[\frac{v+v_D}{v-v_S}\right]$$

i.e., the apparent frequency or the pitch increases as $f_{AP} < f$.

If the source and detector both move away from each other, i.e., v_D is moving in the direction of v, while v_S is moving opposite to v so that v_D is taken as + v_D and v_S is taken as $-\ v_S$. Therefore,

$$f_{AP} = f\left[\frac{v-(+v_D)}{v-(-v_S)}\right] = f\left[\frac{v-v_D}{v+v_S}\right]$$

i.e., there is maximum decrease in apparent frequency or pitch.

All cases of Doppler effect can be written in a single formula

$$f_{AP} = f\left[\frac{v \pm v_D}{v \mp v_S}\right] \qquad ...(5.22)$$

where, v_S, v_D and v are the speeds of source, detector and sound. The upper sign on v_S (or v_D), is used when the source (detector) moves towards the detector (source) while lower sign is used when it moves away.

5.9.1 Doppler Effect for Light

Doppler effect occurs not only for sound (mechanical waves) but also occur for electromagnetic waves (non-mechanical waves) including microwaves, radio waves and visible light. However, as electromagnetic waves do not require a medium for their propagation, so the speed of light is independent of relative motion between source and observer. Therefore, the formulas for apparent change in frequency or wavelength of light are different from that of sound. When either source or detector or both are in motion, only two cases are possible.

In this section a short introduction of Doppler effect of light is given. Another method for Doppler effect of light is given in Section 5.10.

Let c be the velocity of light and v is the relative speed of approach or recession. For approach and recession, the apparent frequency is given by,

$$f_A = f\sqrt{\frac{c+v}{c-v}} \quad \text{and} \quad f_R = f\sqrt{\frac{c-v}{c+v}} \qquad \text{...(5.23)}$$

or

$$f_A = f\left[\frac{1+\frac{v}{c}}{1-\frac{v}{c}}\right]^{1/2} \quad \text{and} \quad f_R = f\left[\frac{1-\frac{v}{c}}{1+\frac{v}{c}}\right]^{1/2}$$

$$f_A = f\left[1+\frac{v}{c}\right]^{1/2}\left[1-\frac{v}{c}\right]^{-1/2}$$

and

$$f_R = f\left[1-\frac{v}{c}\right]^{1/2}\left[1+\frac{v}{c}\right]^{-1/2}$$

For $v << c$, the above formulas with the help of bionomial theorem reduce to

or

$$f_A = f\left[1+\frac{1}{2}\frac{v}{c}\right]\left[1+\frac{1}{2}\frac{v}{c}\right]$$

and

$$f_R = f\left[1-\frac{1}{2}\frac{v}{c}\right]\left[1-\frac{1}{2}\frac{v}{c}\right]$$

or

$$f_A = f\left[1+\frac{v}{c}\right] \quad \text{and} \quad f_R = f\left[1-\frac{v}{c}\right] \qquad \text{...(5.24)}$$

These formulas are same as those sound (for $u << c$). So at low speed of observer and detector,formulas for Doppler effect in light and sound are the same.

For electromagnetic waves, $c = f\lambda$ and c is independent of relative motion between the source and the observer, the above formulas in terms of λ will become

$$\frac{c}{\lambda_A} = \frac{c}{\lambda}\left[1+\frac{v}{c}\right] \quad \text{and} \quad \frac{c}{\lambda_R} = \frac{c}{\lambda}\left[1-\frac{v}{c}\right]$$

or

$$\lambda_A = \lambda\left[1+\frac{v}{c}\right]^{-1} \quad \text{and} \quad \lambda_R = \lambda\left[1-\frac{v}{c}\right]^{-1}$$

Using binomial theorem again and simplifying, we get

$$\lambda_A = \lambda\left[1-\frac{v}{c}\right] \quad \text{and} \quad \lambda_A = \lambda\left[1+\frac{v}{c}\right]$$

$$\text{or} \qquad \lambda_A - \lambda = -\frac{v}{c}\lambda \quad \text{and} \quad \lambda_R - \lambda = +\frac{v}{c}\lambda$$

$$\text{or} \qquad (\Delta\lambda)_A = -\frac{v}{c}\lambda \quad \text{and} \quad (\Delta\lambda)_R = +\frac{v}{c}\lambda$$

$$\text{or} \qquad |\Delta\lambda| = \frac{v}{c}\lambda \qquad ...(5.25)$$

i.e., when the observer or source or both are moving towards each other, frequency increases while wavelength decreases, i.e., shift $\Delta\lambda$ is towards blue end of the spectrum ($\lambda_A = \lambda - \Delta\lambda$). When the observer or source or both are moving away from each other, frequency decreases and wavelength increases, i.e., shift is towards red end ($\lambda_R = \lambda + \Delta\lambda$). Doppler effect is used by astronomers to determine the speeds of stars and galaxies relative to the earth by studying the wavelength (or frequency) of radiations (light) coming from them.

5.10 DOPPLER EFFECT FOR SOUND (SECOND METHOD)

Doppler effect is the apparent change of frequency or wavelength of a wave when source or observer move relative to each other. The Austrian physicist Johann Christian Doppler first proposed the effect in 1842. Buys Ballot in Holland tested it experimentally in 1845. Doppler effect is awave phenomenon. It holds not only for sound waves, but also for electromagnetic waves. However, here we shall study only sound waves.

The sound of whistle of a fast moving train decreases as it moves away from the listener. When, we approach astationary source of sound with high speed, the pitch of the sound heard appears to be higher than that of the source. As the observer moves away from the source, the observed pitch (or frequency) becomes lower than that of the source. We shall analyze changes in frequency under three different situations: (1) observer is stationary but the source is moving, (2) observer is moving but the source is stationary, and (3) both the observer and the source are moving. The situations (1) and (2) differ from each other because of the absence or presence of relative motion between the observer and the medium. Sound waves require a medium for their propagation. But, electromagnetic waves do not require any medium for propagation. If there is no medium present, the Doppler effects are same irrespective of whether the source moves or the observer moves. Since, there is no way of distinction between the two situations.

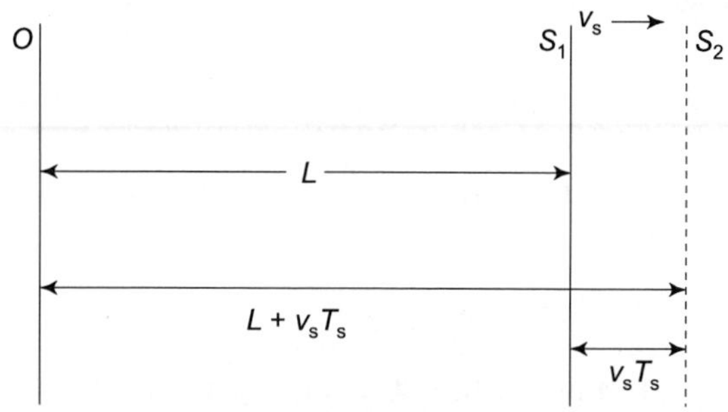

Fig. 5.9 Doppler effect (change in frequency of wave) detected when the source is moving away from observer, while observer is at rest in the medium

Case 1. Source is moving and observer is stationary

For the present treatment the direction from the observer to the source is taken as positive direction of velocity. Let us consider a source S is moving with velocity v_s, and an observer who is stationary and the medium is also at rest. Let the frequency of a sound wave be f, time period be T_s and velocity, v. As shown in Fig. 5.9, at time $t = 0$ the source is at point S_1, located at a distance L from the observer, and emits one sound frequency. One sound frequency reaches from S_1 to the observer in time, $t_1 = L/v$. In the time T_s the source has moved away a distance $v_s\ T_s$ and reaches at the point S_2, located at a distance $(L + v_s\ T_s)$ from the observer. The sound frequency emitted by the source from the point S_2 reaches the observer at time

$$t_2 = T_s + \frac{(L + v_s\ T_s)}{v}$$

The time interval between receiving the first and second frequency is

$$t_2 - t_1 = T_s \frac{(L + v_s\ T_s)}{v} - \frac{L}{v}$$

Since it is time to receive one frequency, so $T = t_2 - t_1$ is the apparent time period observed by the detector. So the apparent time period is given by

$$T = \left[T_s + \frac{v_s\ T_s}{v}\right]$$

or
$$T = T_s\left(1 + \frac{v_s}{v}\right) \qquad ...(5.26)$$

If f' is the apparent frequency observed when the source is moving away from source, as

$$f = f_0\left(1 + \frac{v_s}{v}\right)^{-1} \qquad ...(5.27)$$

If v_s is very small compared with the wave speed (v), taking binomial expansion to terms in the first order in v_s/v and neglecting higher power, Eq. (5.27) can be written as

$$f = f_0\left(1 - \frac{v_s}{v}\right) \qquad ...(5.28)$$

For a source is moving towards the observer, v_s is replaced by $- v_s$ to get

$$f = f_0\left(1 + \frac{v_s}{v}\right) \qquad ...(5.29)$$

The stationary observer receives a lower frequency when the source moves away from him. He receives a higher frequency when the source moves towards him.

Case 2. Observer is moving and source is stationary

Let the observer be moving with velocity (v_0) towards the source and the source is at rest. We work in the reference frame of the moving observer. In this case, the source is approaching at speed v_0 and the speed with which the wave approaches is $v_0 + v$. Following a similar procedure as in the previous case, we find that the time interval between the arrival of the first and second frequency is

$$t_2 - t_1 = T_0 - \frac{v_0 T_0}{v_0 + v}$$

The apparent period of sound wave observed by observer thus is

$$T = t_2 - t_1 = T_0 \left(1 - \frac{v_0}{v_0 + v}\right)$$

or $$= T_0 \left(\frac{v}{v_0 + v}\right)$$

or $$= T_0 \left(1 + \frac{v_0}{v}\right)^{-1} \qquad ...(5.30)$$

Let the apparent frequency of source is f', so $f' = \frac{1}{T}$. Since real frequency of source, $f = \frac{1}{T_0}$, so Eq. (5.32) becomes

$$f = f_0 \left(1 + \frac{v_0}{v}\right) \qquad ...(5.31)$$

Case 3. Both source and observer are moving

As before, let us consider the direction from the observer to the source as the positive direction. Let the source and the observer are moving with velocity v_s and v_0, respectively as shown in Fig. 5.10. Suppose at time $t = 0$, the observer is at O_1 and the source is at S_1. O_1 is in the left of S_1. Let the v, f and T_0 be velocity, frequency and time period wave emitted by source. Let L be the distance between O_1 and S_1 at $t = 0$, when the source emits the first frequency. Since the observer is moving, so the velocity of the wave relative to the observer is $v + v_0$. Therefore, the first frequency reaches the observer at time $t_1 = L/(v + v_0)$. At time $t = T_0$, both the observer and the source reach to the positions O_2 and S_2, respectively by travelling the distance $v_0 T_0$ and $v_s T_0$. The new distance between the observer and the source, $O_2 S_2 = L + (v_s - v_0) T_0$. At S_2, the source emits a second frequency. The second frequency reaches the observer at time,

$$t_2 = T_0 + \frac{[L + (v_s - v_0) T_0]}{(v + v_0)}$$

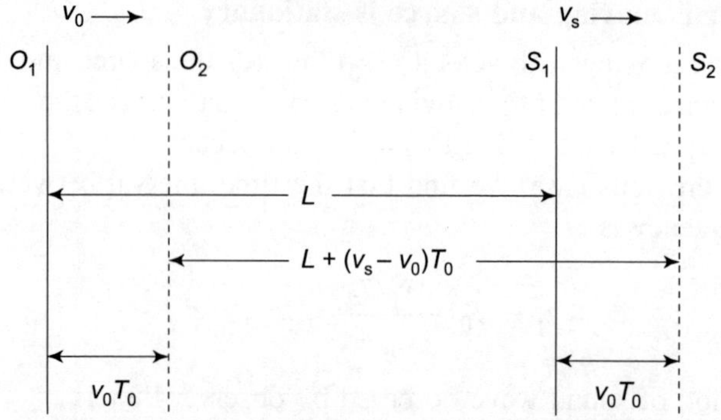

Fig. 5.10 Doppler effect (change in frequency of wave) detected when both source and observer are moving with different velocities

Hence, in a time interval

$$t_2 - t_1 = T_0 + \frac{[L + (v_s - v_0)\, T_0]}{(v + v_0)} - \frac{L}{v_0 + v}$$

one frequency is received by the observer. Let the time period recorded by the observer be T. So

$$T = T_0 \left(1 + \frac{(v_s - v_0)}{(v + v_0)}\right)$$

or

$$T = T_0 \left(\frac{v + v_s}{v + v_0}\right)$$

The frequency f' observed by the observer is given by

$$f' = f\left(\frac{v + v_0}{v + v_s}\right) \qquad ...(5.32)$$

This is a general expression for Doppler shift when both the source and the observer are moving.

Consider a passenger sitting in a train moving on a straight track. Suppose, a passenger hears a whistle sounded by the driver of the train. What frequency will the passenger hear? Here both the observer and the source are moving with the same velocity, so there will be no shift in natural frequency. But an observer outside who is stationary with respect to the track will hear a higher pitch, if the train is approaching him and will hear a lower pitch when train moves away from him.

Note that we have considered the direction from the observer to the source as the positive direction.

1. If the observer is moving towards the source, v_0 has a positive (numerical) value, i.e., v_0 is replaced by $+ v_0$ in Eq. (5.32).

2. If O is moving away from S, v_0 has a negative value, i.e., v_0 is replaced by $-v_0$ in Eq. (5.32).
3. If S is moving away from O, v_s has a positive value, v_0 is replaced by $+v_0$ in Eq. (5.32).
4. If it is moving towards O, v_s has a negative value, i.e., v_0 is replaced by $-v_0$ in Eq. (5.32).

The sound emitted by the source travel in all directions. So a part of sound coming towards the observer is received and detected by the observer. Therefore, the relative velocity of sound with respect to the observer is $v + v_0$ in all cases.

We have considered four different methods to derive the formulas for Doppler effect of light. Three methods are given in Appendix A. One method is given in this section.

A second version of this method is given in Appendix B.

Example 5.9 A rocket is moving at a speed of 210 m/s towards a target, which is stationary. While moving, it emits a wave of frequency 1100 Hz. Some of the sound reaching the target gets reflected back to the rocket as an echo. Calculate (i) the frequency of the sound detected by the stationary target and (ii) the frequency of the echo as detected by the rocket.

Solution: (i) The source (rocket) is moving towards the observer with a speed of 200 m/s and the observer (target) is at rest. The velocity of rocket is comparable with the speed of sound 320 m/s, so the formula used is

$$f = f_0\left(1 - \frac{v_s}{v}\right)^{-1}$$

Given: f_0 = 1100 Hz, v_s = 210 m/s, v = 320 m/s

$$f = 1100 \times \left(1 - \frac{210}{320}\right)^{-1} = 3200 \text{ Hz}$$

(ii) The target is now the source (because it is the source of echo) and the rocket's detector is now the observer (because it detects the echo). Thus, $v_s = 0$ and v_0 has a positive value. The frequency of the sound emitted by the source (the target) is v, the frequency intercepted by the target and not v_0. Therefore, the frequency as registered by the rocket is

$$f' = f\left(\frac{v + v_0}{v}\right)$$

$$= 3200 \times \left(\frac{210 + 320}{320}\right)$$

$$= 5300 \text{ Hz}$$

Example 5.10 When a train is approaching the observer, the frequency of the whistle heard by the observer is 200 cps. When it passed the observer, the frequency heard by the observer is 100 cps. Calculate the frequency when the observer moves with the speed of the train.

Solution: If v and f are velocity and frequency of source (train whistle), respectively.

When observer approaching the train, frequency, $f' = f\dfrac{v}{v - v_s}$

When observer moving away from the train, frequency, $f' = f\dfrac{v}{v + v_s}$

Where v_s is velocity of the source (train whistle)

Given: When train approaching the observer, frequency heard by observer, $f' = 200$ Hz

When train moving away from the observer, frequency heard by the observer, $f' = 100$ Hz

So $$200 = f\frac{v}{v - v_s} \quad \text{or } v - v_s = \frac{fv}{200}$$

$$100 = f\frac{v}{v + v_s} \quad \text{or } \quad v + v_s = \frac{fv}{100}$$

Adding the two equations

$$2v = \frac{3fv}{200} \quad \text{or} \quad f = \frac{400}{3} = 133.33 \text{ Hz}$$

Ans. The actual frequency of whistle is 133.33 Hz.

Example 5.11 A whistle of frequency 530 Hz rotates in a circle of radius 2.5 m at an angular speed of 16 rad/sec. What is the lowest and highest frequency heard by a listener a long distance away at rest with respect to the centre of the circle (v = 320 m/s). Can the apparent frequency be ever equal to actual?

Solution: The detector is rest.

Let f and v be the frequency and velocity of the source (whistle).

When source is approaching the detector, frequency, $f' = f\dfrac{v}{v - v_s}$.

When source moving away from the detector, frequency, $f' = f\dfrac{v}{v + v_s}$.

where, v_s is velocity of the source (whistle)

Given: Radius (r) of circle 2.5 m, frequency of whistle, f = 540 Hz,

Angular velocity of whistle, ω = 16 rad/s, velocity of sound, v = 320 m/s

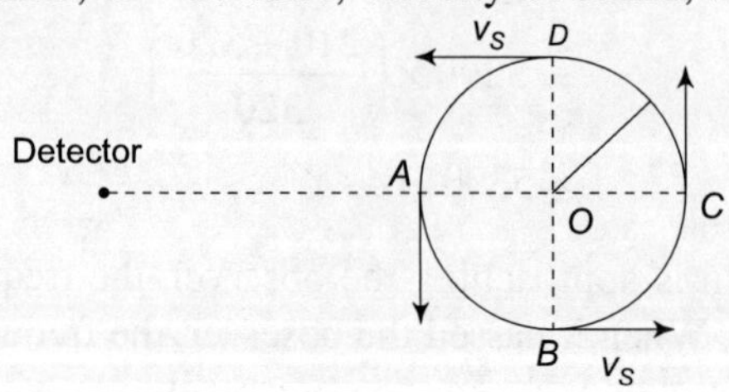

Fig. 5.11

Calculation: In case of circular motion of an object, velocity of object, $v = \omega r$

So the velocity of whistle, $v_s = 16 \times 2.5 = 40$ m /s

The frequency will be minimum, when whistle is at B

$$f' = f\frac{v}{v + v_s} = 530 \times \frac{320}{320 + 40} = 471 \text{ Hz}$$

The frequency will be maximum, when whistle is at D

$$f' = f\frac{v}{v - v_s} = 530 \times \frac{320}{320 - 40} = 606 \text{ Hz}$$

The component of velocity of whistle A and C is $v_s \cos 90 = 0$.

So the velocity at A and C is

$$f' = f\frac{v}{v \pm v_s} = f\frac{v}{v \mp 0} = f = 530 \text{ Hz}$$

Ans. Minimum frequency heard is 471 Hz,

Maximum frequency heard is 606 Hz

Yes, apparent frequency can be equal to actual frequency sometimes.

Example 5.12 A source of sound is moving along a circle, i.e., at A and C, with an angular velocity of 12 rad/s. A sound detector located far away from the source is executing linear simple harmonic motion along the line BD with amplitude $BC = CD = 5$ m. The frequency of oscillation of the detector is $(6/\pi)$ per sec. The source is at the point A when the detector is at the point B. If the source emits a continuous sound wave of frequency 330 Hz, find the maximum and the minimum frequencies recorded by the detector [velocity of sound = 320 m/s].

Solution: Let source and detector are move with velocities v_s and v_D, respectively. Let f be the frequency of the source. When both the source and observer are moving, then formula for frequency (f_{ap}) observed by the detector,

$$f_{ap} = f\left[\frac{v \pm v_D}{v \mp v_s}\right]$$

Given: The angular velocity of source at A and C is 12 rad/s

The detector moving with harmonic around point C with amplitude $A = 5$ m or $BC = CD = 5$ m

Frequency of oscillation of detector is $(6/\pi)$ per second.

Speed of sound, $v = 320$ m/s

Frequency of source, $f = 330$ Hz

Radius of circle, $r = 2$ m

Calculation: Time period of simple harmonic motion of detector, $T = \frac{2\pi}{\omega} = \frac{1}{v} = \frac{\pi}{5}$ or $\omega = 12$ m/s

The speed of source in circular motion, $v_s = \omega r = 12 \times 2 = 24$ m/s

As detector is performing simple harmonic motion, so

Speed of detector, $v_D = \omega\sqrt{A^2 - y^2} = 12\sqrt{5^2 - 0} = 60$ m/s

The formula when both source and detector move towards each other,

$$f_{max} = f\left[\frac{v + v_D}{v - v_S}\right]$$

f_{max} will be maximum, whenthe numerator is maximum and denominator is minimum. This is possible when both v_D and v_S are maximum. v_D is maximum at point C and v_S is maximum at point M and both are moving towards B. So

$$f_{max} = 330\left[\frac{320 + 6}{320 - 2}\right] = 424 \text{ Hz}$$

The formula for frequency, when both move away from each other is

$$f_{max} = f\left[\frac{v - v_D}{v + v_S}\right]$$

f_{ap} will be minimum, when the numerator is minimum and denominator is maximum. This is possible when both v_D and v_S are maximum. v_D is maximum at point C and v_S is maximum at point N and both are moving away from B. So

$$f_{max} = 330\left[\frac{320}{320 + 24}\right] = 249 \text{ Hz}$$

Ans. Maximum frequency detected by detector is 442 Hz.

Minimum frequency detected by detector is 255 Hz.

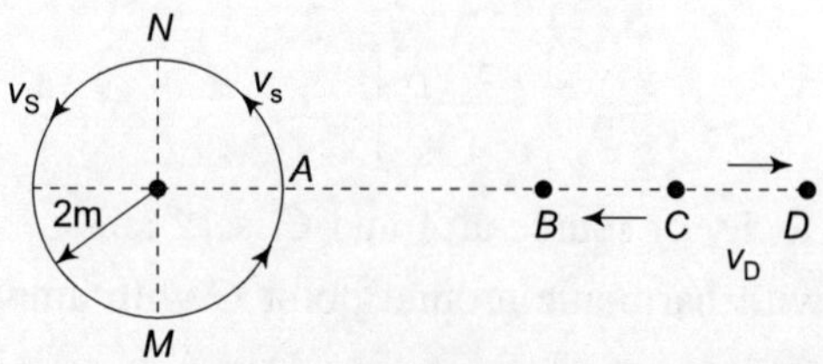

Fig. 5.12

Example 5.13 A train approaching a crossing at a speed of 100 miles/hr sounds a whistle of frequency 500 Hz when 2 mile from the crossing. The speed of sound in air is 0.1 mile /s.

Calculate the frequency is heard by an observer 0.5 miles from the crossing on the straight road which crosses the railroad at right angles, when there is no wind.

Solution: Let v_s be speed of train, v and f are speed and frequency of sound wave. The observer is stationary. When the train is approaching the observer, the formula for frequency received by observer is

$$f' = f\frac{v}{v - v_s}$$

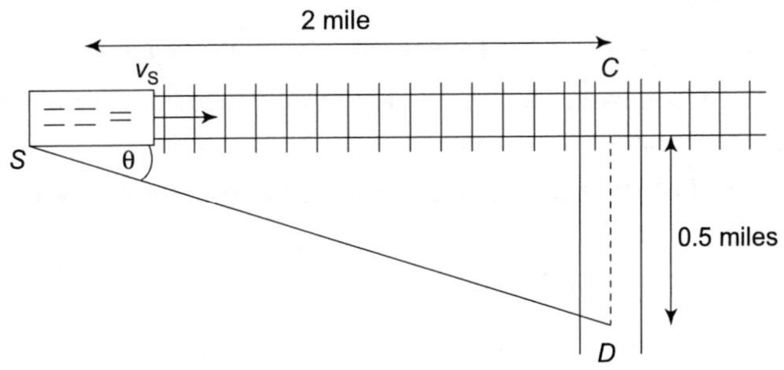

Fig. 5.13

Given: Speed of the train, v_s = 100 mile/hr = 100/(60 × 60) miles per second

Frequency of the sound, f = 500 Hz

Speed of sound, v = 0.1 miles/s

Distance of observer from train, SC = 2 miles

Distance of observer from the track, CD = 0.5 miles

Calculation: Distance SD of the observer,

$$SD = \sqrt{SC^2 + CD^2} = \sqrt{2^2 + 0.5^2} = 2.062 \text{ miles}$$

The component of v_s along SD is $v_s \cos\theta = v_s \cos\theta$

$$= v_s \frac{SC}{SD} = v_s \frac{1}{2.062} = \frac{100}{60 \times 60} \times 0.485 = 0.0135 \text{ miles/s}$$

Frequency heard by the observer will be

$$f' = f\frac{v}{v - v_s \cos\theta} = 500 \times \frac{0.1}{0.1 - 0.0135} = 578 \text{ Hz}$$

Ans. The frequency heard by the observer is 578 Hz.

Example 5.14 Two tuning forks with natural frequencies of 320 Hz each move relative to a necessary observer. One fork moves away from the observer, while the other moves towards him at the same speed. The observer hears beats of frequency 4 Hz. Find the speed of the tuning fork. Velocity of sound in air is 320 m/s.

Solution: The observer is stationary and sources are moving with same speed (v_s) according to this problem.

Let the speed and frequency of sound be v and f, respectively. So the formula for frequency heard by the observer is

$$f' = f\frac{v}{v \pm v_s}$$

Plus sign when source is moving away from the observer and minus sign when the source is approching the observer.

Given: Frequency of fork, $f = 320$ Hz, velocity of sound in air, $v = 320$ m/s

If f_A and f_R and the frequencies for approaching and receding the observer by the forks, the number of beats heard, $f_A - f_R = 4$ per second.

Calculation: If speed of the fork, $v_s = u$, the formula for frequency is

$$f_R = f\frac{v}{v+u}$$

When fork is receding, while the formula for frequency is

$$f_A = f\frac{v}{v-u}$$

When fork is approaching.

Since, $u << v$, the formulas can be written as

$$f_R = f\frac{v}{v+u} = f\left[1+\frac{u}{v}\right]^{-1} = f\left[1-\frac{u}{v}\right]$$

$$f_A = f\frac{v}{v-u} = f\left[1-\frac{u}{v}\right]^{-1} = f\left[1+\frac{u}{v}\right]$$

The beat frequency is

$$f_A - f_R = \Delta f = f\frac{2u}{v}$$

or $$u = \frac{v}{2}\left[\frac{\Delta f}{f}\right] = \frac{320}{2}\left[\frac{4}{320}\right] = 2 \text{ m/s}$$

Ans. The velocity of fork is 2 m/s.

Note: The above problem can also be solved without taking the approximation as done above. In that case one will get a second order equation. Solving of the second order equation makes it time complicated and more time consuming.

Example 5.15 A sonometer wire under tension of 256 N vibrating in its fundamental mode is in resonance with a vibrating tuning fork. The vibrating portion of the sonometer wire has a length of 20 cm and mass of 2 gm. The vibrating tuning fork is now moved away from the vibrating wire at a constant speed and an observer standing near the sonometer hears one beat per second. Calculate the speed with which the tuning fork is moved. The speed of sound in air is 300 m/s.

Solution: The observer is stationary and source (fork) is moving away with speed u m/s.

Let the frequency and speed of sound are f and v, respectively.

The formula for apparent frequency is

$$f_R = f\frac{v}{v+u}$$

When the fork is receding from the source.

If L is length of the wire, M is the mass of the wire and T is the tension of the wire, then frequency of wire is given by formula

$$f_s = \frac{1}{2L}\sqrt{\frac{T}{m}} = \frac{1}{2}\sqrt{\frac{TL}{ML}}, \text{ where } m = \frac{M}{L}$$

Given: Tension of wire, $T = 256\ N$, length of sonometer wire, $L = 20$ cm $= 2 \times 10^{-1}$ m, mass of the wire, $M = 2$ gm $= 2 \times 10^{-3}$ kg

Speed of sound, v = 320 m/s.

Number of beats, $f_R - f = 2$ per second

Calculation: Frequency of the sonometer wire,

$$f_s = \frac{1}{2}\sqrt{\frac{T}{ML}} = \frac{1}{2}\sqrt{\frac{256}{2\times10^{-3}\times2\times10^{-1}}} = 400 \text{ Hz}$$

Initially, the sonometer wire is in resonance with the tuning fork, the frequency of tuning fork,

$$f = f_s = 400 \text{ Hz}$$

Since $u << v$, so we can write $f_R = f\dfrac{v}{v+u}$ as

$$f_R = f\frac{v}{v+u} = f\left[1+\frac{u}{v}\right]^{-1} = f\left[1-\frac{u}{v}\right]$$

or $$f - f_R = \Delta f = f\frac{u}{v}$$

or $$u = v\frac{\Delta f}{f} = 320 \times \frac{2}{400} = 1.6 \text{ m/s}$$

Ans. The speed of fork is 1.6 m/s

Example 5.16 A train approaching a hill at a speed of 70 km/hr sounds a whistle of frequency 570 Hz when it is at a distance of 1 km from the hill. A wind with speed 40 km/hr is blowing in the direction of motion of the train. Find (a) the frequency of the whistle as heard by an observer on the hill, and (b) the distance from the hill at which the echo from the hill is heard by the observer and its frequency. Velocity of sound in air is 1210 km/hr.

Solution: (a) For observer is at rest at hill and source (engine) is moving towards the observer with speed, v_s.

Let v and f be the speed and frequency, respectively. So the formula for approaching observer is

$$f_R = f\left[\frac{v}{(v - v_s)}\right]$$

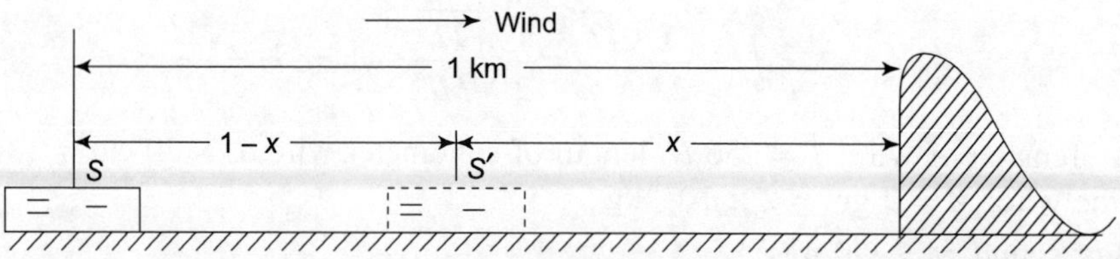

Fig. 5.14

The wind is blowing from source to observer $v \to v + w$, where w is speed of wind. In presence wind, the formula becomes

$$f_R = f\frac{(v + w)}{(v + w) - v_s}$$

Given: Speed of the train, v_s = 70 km/hr

Frequency of sound, f = 570 Hz

Distance from train to hill = 1 km

Speed of sound in air, v = 1210 km/hr

Speed of wind, w = 70 km/hr

Calculation: The frequency heard by the observer at hill will be

$$f_R = 570 \times \frac{(1210 + 70)}{(1210 + 70) - 70} = 603 \text{ Hz}$$

(b) If x km is the present distance of the train from the hill. Thus the distance moved by the train is $(1 - x)$ km. The time taken by the train to move the distance $(1 - x)$ is $\frac{1 - x}{70}$ hour.

In time $\frac{1-x}{40}$, sound travels 1 km distance with speed (1210 + 70) = 1280 km/hr and after reflecting from the hill it travels x distance with speed (1210 – 70) = 1140 km/hr, so

$$\frac{1-x}{70} = \frac{1}{1280}+\frac{x}{1140}$$

or
$$\frac{x}{70}+\frac{x}{1140} = \frac{1}{70}-\frac{1}{1280}$$

or
$$x = \frac{1140\times 70}{1210}\times\frac{1210}{1280\times 70}=\frac{57}{64}\text{ km} = 890.6\text{ m}$$

When the sound is reflected from the hill, the train will be the observer. The speed of observer (train) $v_0 = v_s$ = 40 km/hr, the frequency (f_2) noted by detector at train is given by the formula

$$f_2 = f_A\left[\frac{v+v_0}{v}\right]$$

Now the speed is blowing opposite tothe wind, so

$$v \rightarrow v - w$$

Then formula becomes

$$f_2 = f_A\left[\frac{(v-w)+v_0}{(v-w)}\right] = 603\left[\frac{(1210-70)+70}{(1210-70)}\right]$$

$$= 570\times\frac{1280}{1210}\times\frac{1210}{1140} = 640\text{ Hz}$$

Ans. (a) Apparent frequency heard by the observer at hill is 603 Hz.

(b) The distance from which the echo is heard 890.6 m.

Apparent frequency heard by the observer at train from the echo is 640 Hz.

Example 5.17 The wavelength of light coming from a distant galaxy is found to be 0.4% more than that coming from a source on the earth. Calculate the velocity of the galaxy.

Solution: Formula for frequency shift, $\Delta = \frac{v}{c}\lambda$

Where, v is the velocity of galaxy and c is the speed of light.

Given: Speed of light, $c = 3 \times 10^8$ m/s.

Let the wavelength of light be λ. The wavelength of the light coming from a distant galaxy is 0.4% more than the wavelength of light coming from the earth, that is, Δ = 0.4% of λ, i.e., $\frac{\Delta\lambda}{\lambda}=\frac{0.5}{100}$.

Calculation: The velocity of galaxy is

$$v = c\left(\frac{\Delta\lambda}{\lambda}\right) = 3\times 10^8\times\frac{0.4}{100} = 1.2\times 10^6\text{ m/s}$$

Ans. The velocity of galaxy is 1.2×10^6 m/s

Example 5.18 An astronaut is approaching the moon. He sends out a radio signal of frequency 5×10^9 Hz and finds that the frequency shift in echo received is 10^3 Hz. Find his speed of approach.

Solution: Let the speed of source (astronaut) be v_D and frequency of radio signal be f. The frequency observed by detector (moon) is

$$f_1 = f\sqrt{\frac{c+v_D}{c-v_D}} \quad \text{...(5.33)}$$

Now this will be reflected back by the moon (now source) towards the astronaut (now observer) and moving towards the moon. So the astronaut will hear the frequency

$$f_2 = f_1\sqrt{\frac{c+v_D}{c-v_D}} \quad \text{...(5.34)}$$

Given: Frequency of radio signal, $f = 5 \times 10^9$ Hz
Shift in echo frequency, $\Delta f = 10^3$ Hz
Speed of light, $c = 3 \times 10^8$ m/s
Calculation: Substituting the value of f_1, Eq. (5.34) becomes

$$f_2 = f\left[\frac{c+v_D}{c-v_D}\right] = f\left[\frac{1+\dfrac{v_D}{c}}{1-\dfrac{v_D}{c}}\right] = f\left[1+\frac{v_D}{c}\right]\left[1-\frac{v_D}{c}\right]^{-1}$$

or

$$f_2 = f\left[1+\frac{v_D}{c}\right]\left[1+\frac{v_D}{c}\right] = f\left[1+\frac{2\,v_D}{c}\right] \quad \text{as } v_D << c$$

$$\frac{f_2-f}{f} = \frac{2\,v_D}{c} \quad \text{or} \quad \frac{\Delta f}{f} = \frac{2\,v_D}{c} \quad \text{or} \quad v_D = \frac{c}{2}\frac{\Delta f}{f}$$

The speed of the astronaut,

$$v_D = \frac{3\times 10^8}{2}\times\frac{10^3}{5\times 10^9} = 30 \text{ m/s}$$

Ans. The speed of astronaut is 30 m/s

APPENDIX A

Three methods are considered for Doppler effect of light in this chapter. One method is given in Section 5.10 and two methods are given in this Appendix.

Doppler Effect for Light (Second Method)

The Doppler effect is the apparent change of frequency or wavelength of a wave when source or observer moves relative to each other. The Austrian physicist Johann Christian Doppler first proposed the effect in 1842. Buys Ballot in Holland tested it experimentally in 1845. Doppler effect is awave phenomenon. It holds not only for sound waves, but also for electromagnetic waves. However, here we shall only sound waves.

I. When observer is stationary

Case 1 (a). When source is moving towards object

Consider a source S moving at velocity v_s towards an observer O exactly one wavelength away. The velocity of light source is c. The source emits a continuous series of waves of wavelength λ or frequency f.

In the time t, the wave travels from S to O, the source travels from S to S', a distance $v_s t$ as shown in Fig 5.15. Therefore, the observed wavelength λ' is

$$\lambda' = \lambda - v_s t$$

But $t = \dfrac{\lambda}{c}$, therefore

$$\lambda' = \lambda\left(1 - \frac{v_s}{c}\right)$$

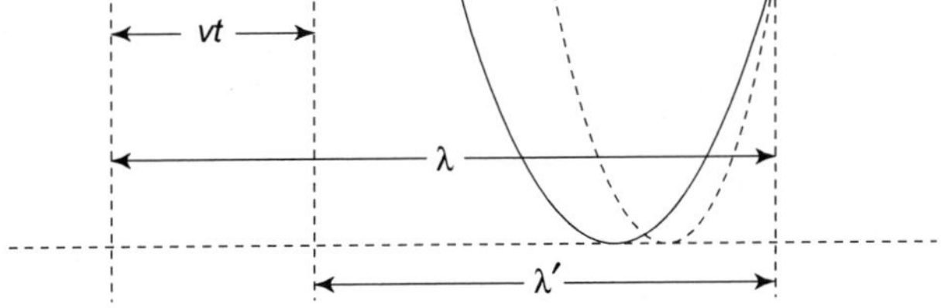

Fig. 5.15

If f and f' are initial and final frequencies, then we can write

$$\frac{c}{f'} = \frac{c}{f}\left(1 - \frac{v_s}{c}\right)$$

$$f' = f\left(1 - \frac{v_s}{c}\right)^{-1}$$

or
$$f' = f\left(1 + \frac{v_s}{c}\right)$$

Case 1 (b). When source is moving away fromthe object

In the time t, the wave travels from S to O, the source travels from S to S', a distance $v_s t$ as shown in Fig. 5.16. Therefore, the observed wavelength λ' is

$$\lambda' = \lambda + v_s t$$

But $t = \dfrac{\lambda}{c}$, therefore

$$\lambda' = \lambda\left(1 + \frac{v_s}{c}\right)$$

In both cases the wavelength change ($\Delta\lambda$) is given by

$$\Delta\lambda = \lambda\frac{v_s}{c}$$

If f and f' are initial and final frequencies, then we can write

$$\frac{c}{f'} = \frac{c}{f}\left(1 + \frac{v_s}{c}\right)$$

$$f' = f\left(1 + \frac{v_s}{c}\right)^{-1}$$

or
$$f' = f\left(1 - \frac{v_s}{c}\right)$$

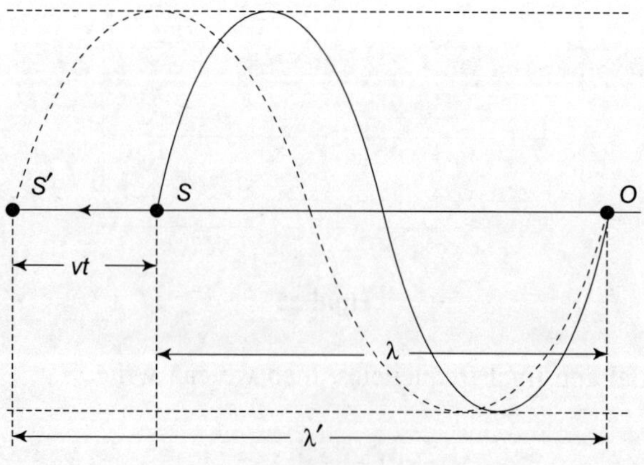

Fig. 5.16

II. When source is stationary

Case 2 (a). When the observer is moving towards the source

Let v be the velocity of observer. The position of observer after t second is O'. So the distance, $OO' = v_0 t$. Let $t = 1$, then $OO' = v_0$ is the distance moved by the observer towards the source.

Let velocity of light be c. The wavelength light source is λ and frequency is f. In one second f waves pass through the stationary observer, so the observer observes f frequency in one second, if observer is stationary. But the observer is moving towards source, so it will observe more waves then f waves. If v_0 is velocity of the observer, the extra waves observed by the observer are calculated as follows:

In c distance number of waves is f.

In unit distance number of waves is $\frac{f}{c} = \frac{1}{\lambda}$.

In $v_0 t$ distance number of waves will be $\frac{v_0 f t}{c}$.

Thus observer will observe $\frac{v_0 f t}{c}$ extra waves in t seconds.

If $t = 1$ second, then observer will observe $\frac{v_0 f}{c}$ extra waves in 1 second.

So, the apparent frequency of the source observed by observer will be

$$f' = f + \frac{v_0 f}{c}$$

or

$$f' = f\left(1 + \frac{v_0}{c}\right)$$

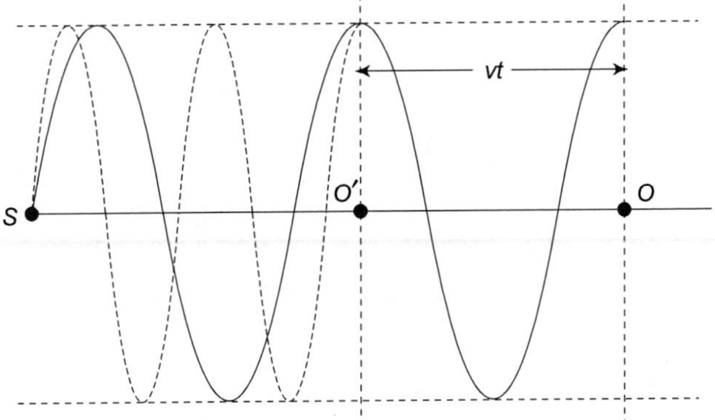

Fig. 5.17

Case 2 (b). When the observer is moving away from source

Let v_0 be the velocity of the observer. The position of the observer after t second is O'. So the distance

$OO' = v_0 t$. Let $t = 1$, then $OO' = v_0$ is the distance moved by the observer away from the source.

Let velocity light be c. Let us assume that the wavelength of source λ and frequency is f. In one second f waves pass through the stationary observer, i.e., the observer observes v frequency in one second, if the observer is stationary. But the observer is moving away from the source, so it will observe less waves then v waves. The less waves observed by the observer are calculated as follows:

In c distance number of waves is f.

In unit distance number of waves is $\frac{f}{c} = \frac{1}{\lambda}$.

In $v_0 t$ distance number of waves will be $\frac{v_0 ft}{c}$.

Thus, observer will observe $\frac{v_0 ft}{c}$ less waves in t seconds.

If $t = 1$ second, then observer will observe $\frac{v_0 f}{c}$ less waves in 1 second.

So, the apparent frequency of the source observed by the observer will be

$$f' = f - \frac{v_0 f}{c}$$

or

$$f' = f\left(1 - \frac{v_0}{c}\right)$$

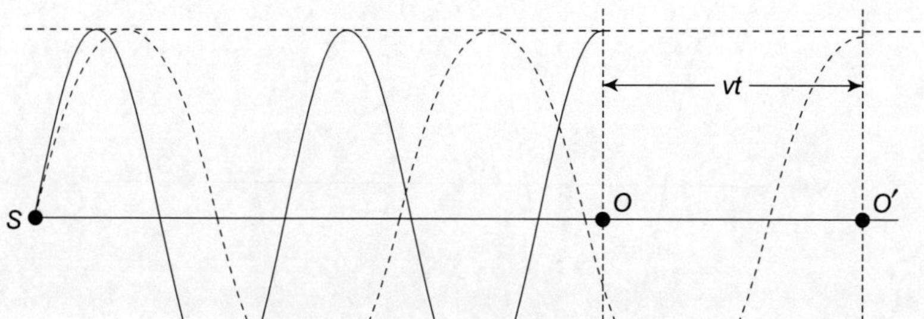

Fig. 5.18

Doppler Effect for Light (Third Method)

Doppler effect is the apparent change of frequency or wavelength of a wave when the source or the observer moves relative to each other.

When the observer is stationary

Case 1 (a). When the source is moving towards object

Consider a source S moving at velocity v_s towards an observer O exactly one wavelength away. Let us assume that velocity light is c. The source emits a continuous series waves of wavelength λ or frequency f.

In time t, the wave travels distance from S to O. If v_s is the velocity of the source, the source travels a distance $v_s t$ while moving from S to S', as shown in Fig. 5.19. Therefore, the apparent distance observed by the observer is $S'O = \lambda'$, i.e.,

$$\lambda' = \lambda - v_s t$$

But $t = \dfrac{\lambda}{c}$, therefore,

$$\lambda' = \lambda\left(1 - \frac{v_s}{c}\right)$$

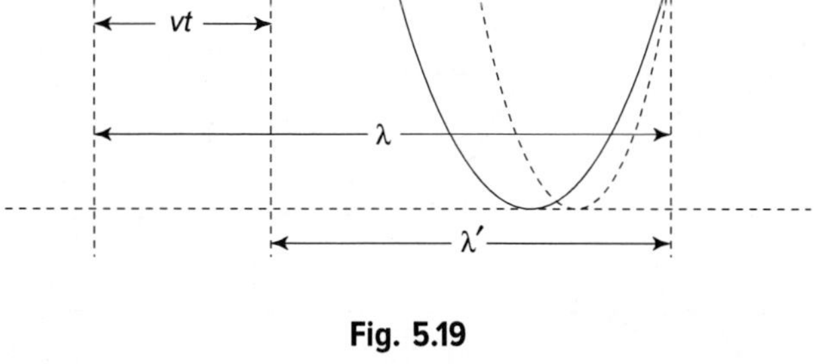

Fig. 5.19

If f' is the apparent frequency observed by observer, then we can write

$$\frac{c}{f'} = \frac{c}{f}\left(1 - \frac{v_s}{c}\right)$$

$$f' = f\left(1 - \frac{v_s}{c}\right)^{-1}$$

or
$$f' = f\left(1+\frac{v_s}{c}\right), \quad \text{as} \quad v_s << c.$$

Case 1 (b). When the source is moving away fromthe object

In the time t, the wave travels a distance from S to O. If v_s is the velocity of the source, the source travels a distance $v_s t$ from S to S', as shown in Fig. 5.20. Therefore, the apparent distance observed by the observer is $S'O = \lambda'$, so

$$\lambda' = \lambda + v_s t$$

But $t = \dfrac{\lambda}{c}$, therefore,

$$\lambda' = \lambda\left(1+\frac{v_s}{c}\right)$$

In both the cases the wavelength change ($\Delta\lambda$) is given by

$$\Delta\lambda = \lambda\frac{v_s}{c}$$

If f' is the apparent frequency observed by observer, then we can write

$$\frac{c}{f'} = \frac{c}{f}\left(1+\frac{v_s}{c}\right)$$

$$f' = f\left(1+\frac{v_s}{c}\right)^{-1}$$

or
$$f' = f\left(1-\frac{v_s}{c}\right), \quad \text{as} \quad v_s << c$$

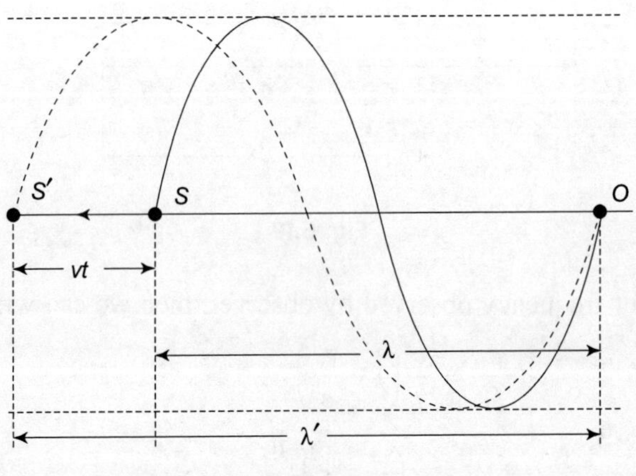

Fig. 5.20

When source is stationary

Case 2 (a). When the observer is moving towards the source

Let v be the velocity of the observer. The position of the observer after t second is O'. If v_0 is the velocity of the observer, the distance travelled by the observer towards the source is $OO' = v_0 t$. Let $t = 1$, then $OO' = v_0$ is the distance moved by the observer towards the source.

Let the velocity of light be c. The source emits a continuous series waves of wavelength (λ) or frequency, f. In one second f waves pass through the stationary observer, i.e., the observer observes f frequency in one second, if the observer is stationary. But the observer is moving towards source, so it will observe more waves than v waves. If v_0 is velocity of observer, the extra waves observed by the observer are calculated as follows:

In distance c number of waves is f.

In unit distance number of waves is $\frac{f}{c} = \frac{1}{\lambda}$.

In $v_0 t$ distance the number of waves will be $\frac{v_0\, ft}{c}$.

Thus, the observer will observe $\frac{v_0\, ft}{c}$ extra waves in t seconds.

If $t = 1$ second, then the observer will observe $\frac{v_0 t}{c}$ extra waves in 1 second.

So, the apparent frequency of the source observed by the observer will be

$$f' = f + \frac{v_0 f}{c}$$

or

$$f' = f\left(1 + \frac{v_0}{c}\right)$$

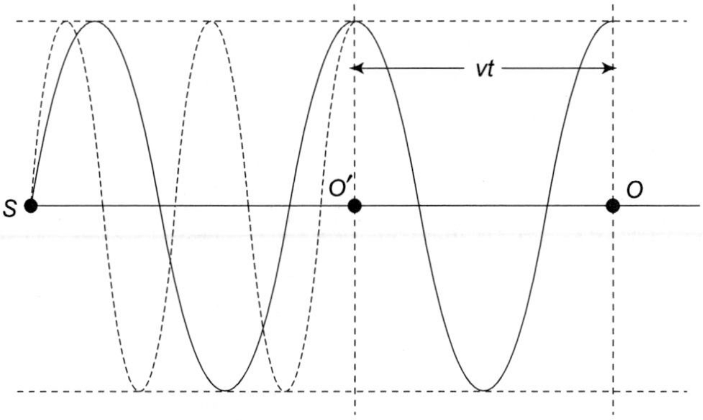

Fig. 5.21

Case 2 (b). When the observer is moving away from the source

Let v_0 be the velocity of the observer. The position of the observer after t second is O'. So the distance $OO' = v_0 t$. Let $t = 1$, then $OO' = v_0$ is the distance moved by the observer away from the source.

Let the velocity of light be c. The source emits a continuous series of waves of wavelength or frequency, f. In one second f waves will pass through the stationary observer, i.e., the observer observes f frequency in one second, if the observer is stationary. But the observer is moving away from the source, so it will observe lesser waves then v waves. The less frequencies observed by the observer are calculated as follows:

In c distance number of waves is f.

In unit distance number of waves is $\frac{f}{c} = \frac{1}{\lambda}$.

In $v_0 t$ distance number of waves will be $\frac{v_0\, ft}{c}$.

Thus, the observer will observe $\frac{v_0\, ft}{c}$ less waves in t seconds.

If $t = 1$ second, then the observer will observe $\frac{v_0 f}{c}$ less waves in 1 second.

So, the apparent frequency of the source observed by the observer will be

$$f' = f - \frac{v_0\, f}{c}$$

or

$$f' = f\left(1 - \frac{v_0}{c}\right)$$

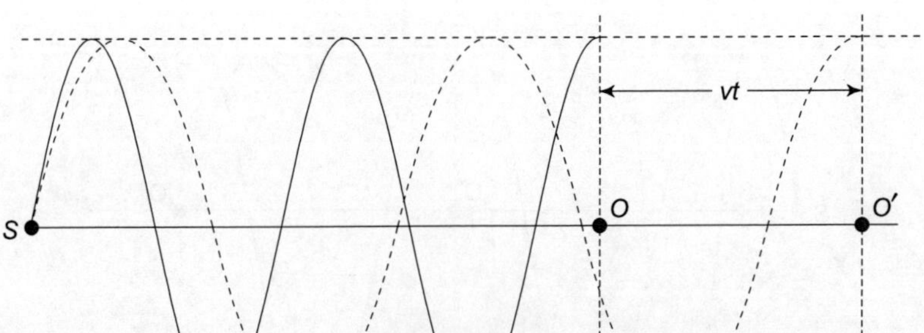

Fig. 5.22

APPENDIX B

Doppler Effect for Sound (Extended Method)

Doppler effect is the apparent change of frequency or wavelength of a wave when the source or the observer moves relative to each other.The Austrian physicist Johann Christian Doppler first proposed the effect in 1842. Buys Ballot in Holland tested it experimentally in 1845. Doppler effect is awave phenomenon. It holds not only for sound waves, but also for electromagnetic waves. However, here we shall study only sound waves.

The sound of whistle of a fast moving train decreases as it moves away from the listener. When, we approach astationary source of sound with high speed, the pitch (or frequency) of the sound heard appears to be higher than that of the source. As the observer moves away from the source, the observed pitch (or frequency) becomes lower than that of the source. We have considered three cases for finding changes in frequency: (1) The observer is stationary but the source is moving, (2) the observer is moving but the source is stationary, and (3) both the observer and the source are moving. Sound waves require a medium for their propagation. But, electromagnetic waves do not require any medium for propagation.

Case 1. Source is moving and the observer is stationary

For present treatment the direction from the observer to the source is takenas positive direction of velocity. Let us consider a source S is moving with velocity v_s, and an observer who is stationary and the medium is also at rest. Let the frequency of a sound wave be f, time period, T_s and velocity, v. As shown in Fig. 5.23, at time $t = 0$ the source is at point S_1, located at a distance L from the observer, and emits one sound frequency. One sound frequency reaches from S_1 to the observer in time, $t_1 = L/v$. In the time T_s the source has moved away a distance $v_s T_s$ and reaches at the point S_2, located at a distance $(L + v_s T_s)$ from the observer. The second sound frequency emitted by the source from the point S_2 reaches the observer at time

$$t_2 = T_s + \frac{(L + v_s T_s)}{v}$$

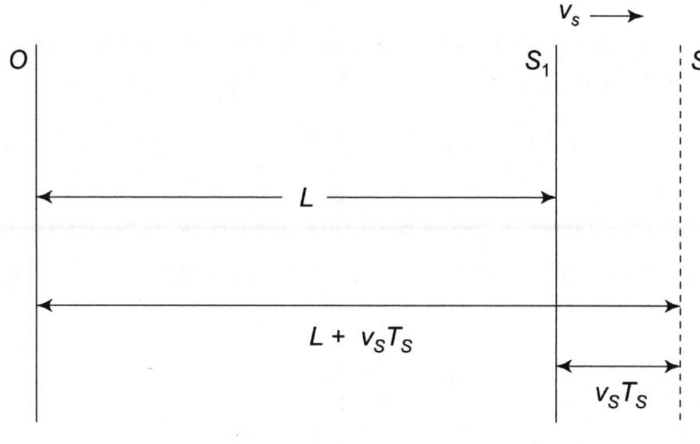

Fig. 5.23 Doppler effect (change in frequency of wave) detected when the source is moving away from the observer, while the observer is at rest in the medium

In time nT_s, the source emits its $(n + 1)^{th}$ frequency. In the time nT_s the source has moved away a distance $n v_s T_s$ and reaches at the point S_2, located at a distance $(L + n v_s T_s)$ from the observer.

In time nT_s, the source emits its $(n + 1)^{th}$ frequency. However, in time nT_s, the source has moved away a distance $n v_s T_s$, making the distance $L + n v_s T_s$ of the source from the observer.

So the time for reaching $(n + 1)^{th}$ frequency to the observer will be

$$t_{n+1} = nT_s + \frac{(L + nv_s T_s)}{v}$$

Hence, in a time interval in reaching n frequencies to the observer is

$$\left[n T_s + \frac{(L + nv_s T_s)}{v} - \frac{L}{v}\right] = \left[n T_s + \frac{nv_s T_s}{v}\right]$$

If the apparent time period is T, then the detector counts n frequencies in time given by

$$nT = \left[nT_s + \frac{nv_s T_s}{v}\right]$$

or
$$T = \left[T_s + \frac{v_s T_s}{v}\right]$$

or
$$T = T_s\left(1 + \frac{v_s}{v}\right) \qquad \text{...(B1)}$$

If f' is the apparent frequency observed when the source is moving away from source, as

$$f = f_0\left(1 + \frac{v_s}{v}\right)^{-1} \qquad \text{...(B2)}$$

If v_s is very small compared to the wave speed v, taking binomial expansion to terms in first order in v_s/v and neglecting higher power, Eq. (A2) can be written as

$$f = f_0\left(1 - \frac{v_s}{v}\right) \qquad \text{...(B3)}$$

For a source is moving towards the observer, v_s is replaced by $-v_s$ to get

$$f = f_0\left(1 + \frac{v_s}{v}\right) \qquad \text{...(B4)}$$

The stationary observer receives a lower frequency when the source moves away from him. He receives a higher frequency when the source moves towards him.

Case 2. Observer is moving and the source is stationary

Let the observer move with velocity v_0 towards the source and the source is at rest. We work in the reference frame of the moving observer. In this case, the source is approaching at speed v_0 and the speed with which the wave approaches is $v_0 + v$. Following a similar procedure as in the previous case, we find that the time interval between the arrival of the first and the $(n + 1)^{th}$ crest is

$$t_{n+1} - t_1 = n\,T_0 - \frac{nv_0\,T_0}{v_0 + v}$$

The apparent period of sound wave observed by the observer thus is

$$T = \frac{t_{n+1} - t_n}{n} = T_0\left(1 + \frac{v_0}{v_0 + v}\right)$$

$$= T_0\left(\frac{v}{v_0 + v}\right)$$

$$= T_0\left(1 + \frac{v_0}{v}\right)^{-1} \qquad \text{...(B5)}$$

Let the apparent frequency of source be f', so $f' = \dfrac{1}{T}$. Since real frequency of the source,

$f = \dfrac{1}{T_0}$, so Eq. (A5) becomes

$$f = f_0\left(1 + \frac{v_0}{v}\right) \qquad \text{...(B6)}$$

Case 3. Both the source and the observer are moving

As before, let us consider the direction from the observer to the source as the positive direction. Let the source and the observer move with velocities v_s and v_0, respectively as shown in Fig. 5.24. Suppose at time $t = 0$, the observer is at O_1 and the source is at S_1. O_1 is in the left of S_1. Let the v, f and T_0 be velocity, frequency and time period wave emitted by source. Let L be the distance between O_1 and S_1 at $t = 0$, when the source emits the first frequency. Since the observer is moving, so the velocity of the wave relative to the observer is $v + v_0$. Therefore, the first frequency reaches the observer at time $t_1 = L/(v + v_0)$. At time $t = T_0$, both the observer and the source reach to the positions O_2 and S_2, respectively, by travelling the distance $v_0\,T_0$ and $v_s\,T_0$. The new distance between the observer and the source, $O_2\,S_2 = L + (v_s - v_0)\,T_0$. At S_2, the source emits the second frequency. The second frequency reaches the observer at time,

$$t_2 = T_0 + \frac{[L + (v_s - v_0)\,T_0]}{(v + v_0)}$$

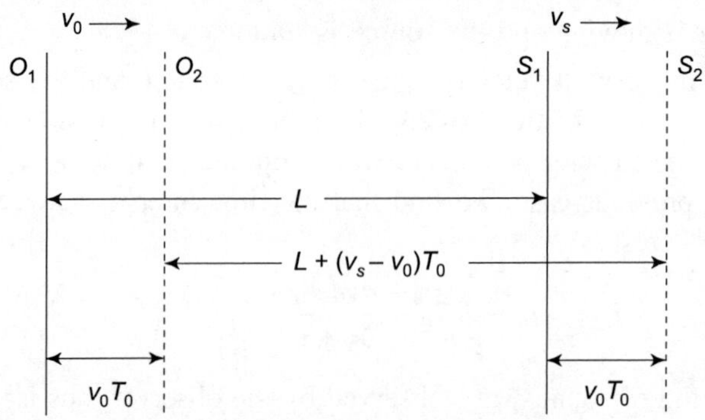

Fig. 5.24 Doppler effect (change in frequency of wave) detected when both source and observer are moving with different velocities

At time nT_0 the source emits its $(n + 1)^{\text{th}}$ frquency and this reaches the observer at time

$$t_{n+1} = nT_0 + \frac{[L + n(v_s - v_0)T_0]}{(v + v_0)}$$

Hence, in a time interval

$$t_{n+1} - t_1 = n\,T_0 + \frac{[L + n(v_s - v_0)T_0]}{(v + v_0)} - \frac{L}{v_0 + v}$$

The observer counts n frequency. Let the time period recorded by the observer be T. So n frequencies reach in time

$$nT = n\,T_0 + \frac{n(v_s - v_0)T_0}{(v + v_0)}$$

or
$$T = T_0\left(1 + \frac{(v_s - v_n)}{(v + v_0)}\right)$$

or
$$T = T_0\left(\frac{v + v_s}{v + v_0}\right)$$

The frequency f'observed by the observer is given by

$$f' = f\left(\frac{v + v_0}{v + v_s}\right) \qquad \text{...(B7)}$$

This is a general expression for Doppler shift when both the source and the observer are moving.

Consider a passenger sitting in a train moving on a straight track. Suppose the passenger hears a whistle sounded by the driver of the train. What frequency will the passenger hear? Here both the observer and the source are moving with the same velocity, so there will be no shift in natural frequency. But an observer outside who is stationary with respect to the train will note a higher pitch, if the train is approaching him and will note a lower pitch when it moves away from him.

Note: That we have considered the direction from the observer to the source as a positive direction.

1. If the observer is moving towards the source, v_0 has a positive (numerical) value, i.e., there is no change in Eq. (A7).
2. If O is moving away from S, v_0 has a negative value, i.e., v_0 is replaced by $-v_0$ in Eq. (B7).
3. If S is moving away from O, v_s has a positive value, i.e., there is no change in Eq. (B7).
4. If it is moving towards O, v_s has a negative value, i.e., v_0 is replaced by $-v_0$ in Eq. (B7).

The sound emitted by the source travels in all directions. So only a fraction of sound coming towards the observer is received and detected by the observer. Therefore, the relative velocity of sound with respect to the observer is $v + v_0$ in all cases.

Example B1. A rocket is moving at a speed of 210 m/s towards a target, which is stationary. While moving, it emits a wave of frequency 1100 Hz. Some of the sound reaching the target gets reflected back to the rocket as an echo. Calculate (i) the frequency of the sound detected by the stationary target, and (ii) the frequency of the echo as detected by the rocket.

Solution: (i) The source (rocket) is moving towards the observer with a speed of 210 m/s and the observer (target) is at rest. The velocity of rocket is comparable with the speed of sound 320 m/s, so the formula used is

$$f = f_0\left(1 - \frac{v_s}{v}\right)^{-1}$$

Given: $f_0 = 1100$ Hz, $v_s = 210$ m/s, $v = 320$ m/s

$$f = 1100 \times \left(1 - \frac{210}{320}\right)^{-1} = 3200 \text{ Hz}$$

(ii) The target is now the source (because it is the source of echo) and the rocket's detector is now the observer (because it detects the echo). Thus, $v_s = 0$ and v_0 has a positive value. The frequency of the sound emitted by the source (the target) is v, the frequency intercepted by the target and not v_0. Therefore, the frequency as registered by the rocket is

$$f' = f\left(\frac{v + v_0}{v}\right)$$

$$= 3200 \times \left(\frac{210 + 320}{320}\right)$$

$$= 5300 \text{ Hz}$$

Formulas at a Glance

	Subject	Formula	Interpretation
1.	Velocity of transverse wave in string	$v = \sqrt{\frac{T}{m}}$	T is tension of the string, m is mass per unit length
2.	Velocity of sound wave in a solid	$v = \sqrt{\frac{\eta}{\rho}}$	η is the bulk modulus (modulus of rigidity) of the solid and ρ is the mass per unit volume of the solid
3.	Velocity of sound wave in a gas	$v = \sqrt{\frac{P}{\rho}}$	P is the pressure of an ideal gas. ρ is the mass per unit volume of the gas
4.	Velocity of sound wave in a gas	$v = \sqrt{\frac{B_T}{\rho}}$	B_T is the isothermal bulk modulus of the gas. ρ is the gas density
5.	Speed of sound in a gas	$v = \sqrt{\frac{\gamma P}{\rho}}$	γ is defined as the ratio of the specific heat of a gas at constant pressure, i.e., $\gamma = C_p/C_v$.
6.	Speed of sound in a gas	$v = \sqrt{\frac{\gamma RT}{M}}$	R is the gas constant. If(that is, mass of 1 mole of the gas) the molecular mass of the gas is M, then density is $\rho = M/V$.
7.	The velocity (v) of a vehicle moving towards or away from a radar speed trap is given by the formula	$v = \frac{c\Delta f}{2f}\ \frac{v}{c} = \frac{1}{2}\frac{f' - f}{f}$ $f' = f\left(1 + \frac{2v}{c}\right)$	where, f is the frequency of the transmitter and Δf is the observed frequency change or Doppler shift.
8.	When the observer is stationary (i) Frequency shift when the source is moving away from the observer (ii) Frequency shift when the source is moving towards the observer	(i) $f = f_0\left(1 - \frac{v_s}{v}\right)$ (ii) $f = f_0\left(1 + \frac{v_s}{v}\right)$	v_s is the speed of source and v is the wave speed of wave. f_0 is the initial frequency of the f is the final frequency of wave.
9.	When the source is stationary (i) Frequency shift when the observer is moving away from the source (ii) Frequency shift when the observer is moving towards source	(i) $f' = f\left(1 - \frac{v_0}{v}\right)$ (ii) $f' = f\left(1 + \frac{v_0}{v}\right)$	v_0 is the speed of the observer and v is the wave speed of the wave. f_0 is the initial frequency of the wave. f is the final frequency of the wave.

10.	Both the source and the observer are moving to the right side	$f = f_0\left(\dfrac{v+v_0}{v+v_s}\right)$	v_s and v_0 are speeds of the source and the observer, respectively. v is the wave speed of the wave. f_0 is the initial frequency of the wave. f is the final frequency of the wave.

SHORT ANSWER QUESTIONS

5.1 Which characteristics of medium determines the speed of longitudinal waves through it?

Ans. Elasticity and density

5.2 Which characteristics of solid determines the speed of transverse waves through it?

Ans. Modulus of rigidity and density pressure

5.3 How does the variation of pressure affect the speed of sound in air at constant temperature?

5.4 At normal temperature and pressure, the speed of sound in air is 330 m/s. What is the speed of sound when pressure doubles?

Ans. 330 m/s, because the density of air is proportional to pressure.

5.5 What is effect of humidity on the speed of sound in air?

Ans. The speed of sound in moist air increases slightly more than in dry air, as the density of moist air decreases slightly in moist air than dry air.

5.6 What is effect of humidity on the speed of sound in dry hydrogen?

Ans. The speed of sound in moist air increases slightly more than in dry hydrogen, as the density of moist hydrogen decreases slightly in moist air than in dry hydrogen.

5.7 When the ear of a person is placed at one end of a long tube of iron and another person strikes at the other end, the sound of striking is heard two times, why? Which sound is heard first and why?

Ans. The person hears the striking sound two times, because one sound travels through the iron rod and another travels through air. Since the speed of sound travelling through the rod is more than the sound travelling through air. So the sound of striking through rod is heard first.

5.8 What will be the speed of sound in a perfectly rigid rod?

Ans. Infinite

5.9 What is the change in the speed of sound in air during rainy season as compared to the dry season?

Ans. Sound increases.

5.10 What type of hearing aids are required by astronauts on the surface of the moon and why?

Ans. There is no atmosphere on the moon and so sound cannot travel from one place to the other at the surface of the moon. Hence, hearing aids capable of transmitting and detecting electromagnetic waves, which can travel in vacuum, are required.

5.11 Why the sound produced in air not heard by a person deep inside the water?

Ans. Density of water is more than the density of air. Hence, most of the sound is reflected from the surface of water and there a little refraction of sound from the air into the water. When there is a glass wall in between two rooms, then conversation in one room cannot be heard in the other room, because sound waves get reflected from the glass wall.

5.12 The speed of sound does not depend upon its frequency. Can you verify this statement from some experience of your daily life?

Ans. In musical concerts, musical sounds of different frequencies produced by different musical instruments simultaneously reach a listener, through the intervening medium, at the same time. Such a medium is called a non-dispersive medium.

5.13 What is dispersive medium?

Ans. A medium is said to be dispersive if the speed of a wave in that medium depends on the frequency of the wave. Waves of different frequencies travel with different speeds in a dispersive medium.

5.14 On what factors does the speed of a transverse wave in a stretched string depend?

5.15 Why do the musicians playing sitar, violin, etc., press the wires at different places by their left hand?

Ans. By doing so, the length of the vibrating wire is changed and notes of different frequencies are produced.

5.16 The weight suspending from the sonometer wire is made four times. Will the frequency of the wire become exactly twice?

Ans. Yes, provided the length of the sonometer wire remains same. By increasing weight four times, it is expected that the length of the wire slightly increases, so the frequency decreases slightly.

5.17 Two strings of equal lengths are stretched by equal forces. The strings are made of the same material. If the diameters are in the ratio 3 : 2, what will be the ratio of their fundamental frequencies?

Ans. 2 : 3.

5.18 Draw a diagram showing the formation of the second overtone in a stretched string.

[Hint: The string producing the second overtone vibrates in three segments.

5.19 The source of energy of the sun is fusion of hydrogen which provides energy in the form of heat, light and sound. Explain why sound from the sun does not reach while light reaches.

Ans. Sound is a mechanical wave, so requires medium for propagation. Since there is no medium between the sun and the earth, sound cannot reach the earth. Heat and light are electromagnetic waves, so do not require any medium for propagation and reach the earth.

5.20 The speed of sound waves in air depends on temperature but that of light does not, why?

Ans. Sound waves are mechanical waves, so depend on a medium for propagation, Formula for speed of sound is: $v = \sqrt{\gamma RT/M}$. Light waves are electromagnetic waves, so do not need any medium for propagation. Formula for speed of light is: $c = \sqrt{1/\mu_0\mu_0}$.

5.21 Light waves can be polarized while sound waves cannot be. Why?

Ans. Light waves are transverse waves, while sound waves are longitudinal waves.

5.22 When sound travels from air into water, does the frequency, wavelength and speed of the wave change?

Ans. No, yes, yes.

5.23 Explain why

(a) The speed of sound is generally more in solids than in gases.

(b) The speed of sound in oxygen is less than in hydrogen.

5.24 Why the speed of sound in air (a) is independent of the pressure, and (b) increases with temperature.

5.25 Calculate the temperature at which the speed of sound in air becomes double of its value at 0°C.

Ans. 819°C.

5.26 Distinguish between beats and interference.

Ans. In beats at a given position, intensity varies periodically with time with periodicity, $T = [1/(f_1 - f_2)]$, while in interference at a given time, intensity varies periodically with position with periodicity (λ).]

5.27 A 1 m long steel wire has mass 20 gm and is stretched with a force of 800 N. What are the frequencies of its fundamental mode and next three higher modes.

Ans. 100, 200, 300 and 400.

QUESTIONS FOR EXAMINATION

5.1 Write the formula for the fundamental frequency of a transverse wave in a stretched string and explain with its help how the frequency of the string depends on the length and radius of the string. Write the frequency of the first overtone produced along with fundamental frequency (n) of the string.

5.2 Explain the formation of the standing wave in a string clamped at both ends and discuss the various modes of vibrations.

5.3 How does the frequency of the transverse oscillations of the stretched string depend on the tension and density of the material of the string? Give the necessary formula. Show in a diagram the position of nodes and antinodes on the string when it vibrates to produce the third harmonic of its fundamental frequency.

5.4 What do you understand by harmonics? Explain, giving necessary sketch, that both type of harmonics, even and odd, are produced in a stretched string.

5.5 Describe a sonometer. How is the frequency of a tuning fork is determined?

5.6 Show that the resultant of two travelling waves of the same amplitude and frequency that travel in opposite direction is a stationary wave.

5.7 What are beats? How are they produced? Find the expression for beat frequency.

5.8 Sketch standing wave patterns for fundamental and second overtones for the vibration of air column in a tube (a) closed at one end, and (b) open at both ends.

5.9 Write the Newton's formula for the speed of sound waves in a gaseous medium. Explain Laplace correction in this formula for an ideal gas.

5.10 Explain, with reason, whether the propagation of longitudinal waves in air is an isothermal process or adiabatic.

5.11 What is the effect of temperature on the speed of sound in air? Derive the necessary expression.

5.12 What is Doppler effect? Derive the formula for frequency observed by the observer when both are moving in the same direction.

5.13 What is Doppler effect? Derive the formula for frequency observed by the observer when they are moving in opposite direction.

5.14 What is Doppler effect? Derive the formula for frequency shift, when the source is moving away from the observer and the observer is stationary.

5.15 What is Doppler effect? Derive the formula for frequency shift, when source is moving towards observer and observer is stationary.

5.16 What is Doppler effect? Derive the formula for frequency shift, when the observer is moving towards the source and the source is stationary.

5.17 What is Doppler effect? Derive the formula for frequency shift, when the observer is moving away from the source and the source is stationary.

5.18 What is Doppler effect? Derive the formula for frequency shift, when both the observer and the source are moving towards right side.

OBJECTIVE QUESTIONS

5.1 At resonance

(a) The amplitude of displacement is maximum.

(b) The amplitude of velocity is maximum.

(c) The amplitude of displacement is minimum.

(d) The amplitude of velocity is minimum.

5.2 Which of the following waves are mechanical?

(a) Heat waves (b) Microwaves (c) Seismic waves (d) γ-rays

5.3 Transverse mechanical waves can travel in

(a) Iron rod (b) Hydrogen gas (c) Water (d) Stretched string

5.4 Which of the following phenomenon cannot take place with sound (in air)?

(a) Interference (b) Diffraction (c) Polarization (d) Scattering

5.5 Which of the following cannot travel through vacuum?

(a) Light waves (b) Heat waves (c) X-rays (d) Sound waves

5.6 A big explosion on the moon cannot be heard on the earth because

(a) The explosion produces high frequency sound waves which are inaudible.

(b) Sound waves require a material medium for propagation.

(c) Sound waves are absorbed in the atmosphere of the moon.
(d) Sound waves are absorbed in the atmosphere of the earth.

5.7 An astronaut cannot hear his companion at the surface of the moon because
(a) Produced frequencies are above the audio frequencies.
(b) There is no medium for sound propagation.
(c) Speed of sound is too low to reach the companion.
(d) Sound waves are absorbed in the atmosphere of moom.

5.8 It is possible to distinguish between transverse and longitudinal waves by studying the property of
(a) Interference (b) Diffraction (c) Superposition (d) Polarization

5.9 Mechanical (sound) waves in a gas are
(a) Transverse (b) Longitudinal (c) Either transverse or longitudinal
(d) Neither transverse nor longitudinal

5.10 Mechanical waves on the surface of a liquid are
(a) Transverse (b) Longitudinal (c) Torsional
(d) Both transverse and longitudinal (ripples)

5.11 Which of the following property of a wave does not change with a change in medium?
(a) Frequency (b) Wavelength (c) Velocity (d) Amplitude

5.12 Velocity of sound in air is nearly
(a) 332 m/s (b) 332 miles/s (c) 332 ft/s (d) 332 km/hr

5.13 In which medium sound has maximum velocity.
(a) Gases (b) Liquids (c) Solids (d) Vacuum

5.14 Velocity of sound in any gas depends on
(a) Intensity (b) Amplitude (c) Density and elasticity
(d) Volume and temperature

5.15 Velocity of sound is proportional to
(a) Square root of isothermal elasticity (b) Adiabatic elasticity
(c) Square root of adiabatic elasticity (d) Isothermal elasticity

5.16 The speed of sound in a gas at NTP is 300 m/s. If the pressure in increased four times, without change in temperature, the velocity of sound will be
(a) 150 m/s (b) 300 m/s (c) 600 m/s (d) 1200 m/s

5.17 If at a place the speed of a sound wave of frequency 300 Hz is V, of the another wave of frequency 150 Hz at the same place will be
(a) V (b) $V/2$ (c) $2V$ (d) $4V$

5.18 The variation of the speed of sound with temperature is greatest in
(a) Metals (b) gases (c) Solids (d) Liquids

5.19 At displacement node
(a) Displacement is minimum (b) Pressure is maximum
(c) Sound is maximum (d) Speed is minimum

5.20 Echoes are produced due to

(a) Reflection (b) Refraction (c) Resonance (d) Beat production

5.21 To demonstrate the phenomenon of interference we require

(a) Two sources which emit radiations of the same frequency.

(b) Two sources which emit radiations of nearly the same frequency.

(c) Two sources which emit radiations of the same frequency and have a definite phase relationship.

(d) Two sources which emit radiation of different frequency.

5.22 Beats are the result of

(a) Diffraction

(b) Destructive interference

(c) Constructive and destructive interference

(d) Superposition of two waves of nearly equal frequencies

5.23 Two tuning forks are sounded together and beats are heard. We can infer that

(a) The pitches of the two notes are nearly the same.

(b) The intensities of the two notes are nearly the same.

(c) The quality of the two notes are nearly the same.

(d) The velocities of the two sounds are nearly the same.

5.24 The frequencies emitted by two adjacent pianos simultaneously are f_1 and f_2. The number of beats heard per second are

(a) $\frac{1}{2}(f_1 - f_2)$ (b) $\frac{1}{2}(f_1 + f_2)$ (c) $(f_1 - f_2)$ (d) $(f_1 + f_2)$

5.25 If a tuning fork of frequency 512 Hz is sounded with a vibrating string of frequency 505.5 Hz the beats produced per second will be

(a) 6 (b) 7 (c) 6.5 (d) 3

5.26 A source (x) of unknown frequency produces 8 beats with a source of 250 Hz and 12 beats with a source of 270 Hz. The frequency of the source is

(a) 258 Hz (b) 242 Hz (c) 262 Hz (d) 282 Hz

5.27 A tuning fork whose frequency is given by the manufacturer as 512 Hz is being tested with an accurate oscillator. It is found that the fork produces a beat of 2 Hz when the oscillator reads 514 Hz but produces a beat of 6 Hz when the oscillator reads 510 Hz. The actual frequency of the tuning fork is

(a) 508 Hz (b) 512 Hz (c) 516 Hz (d) 518 Hz

5.28 If two tuning forks, *A* and *B*, are sounded together they produce 4 beats per second. A is lightly loaded with wax, they produce 2 beats when sounded again. The frequency of *A* is 256. The frequency of *B* will be

(a) 250 Hz (b) 252 Hz (c) 260 Hz (d) 262 Hz

5.29 A fork produces 6 beats per second with another fork of frequency 384 Hz. If the prongs of first fork are slightly filed, 4 beats per second are produced. The frequency of the first fork after filing is

(a) 390 Hz (b) 378 Hz (c) 380 Hz (d) 388 Hz

5.30 The number of beats produced per second by the vibrations Y_1 = A sin (320 πt) and Y_2 = A sin (326 πt) is

(a) 6 (b) 4 (c) 3 (d) 2

5.31 Two vibrating tuning forks producing progressive waves given by

$$Y_1 = 4 \sin (320\,\pi t) \text{ and } Y_2 = 2 \sin (326\,\pi t)$$

are held near the ear of a person. The person will hear

(a) 3 beats with intensity ratio between maxima and minima equal to 2.

(b) 3 beats with intensity ratio between maxima and minima equal to 9.

(c) 6 beats with intensity ratio between maxima and minima equal to 2.

(d) 6 beats with intensity ratio between maxima and minima equal to 9.

5.32 The apparent change in the pitch of a sound source due to relative motion of the source and the observer is called

(a) Doppler effect (b) Huygen's law (c) Kepler's law (d) Gallileo law

5.33 An engine is approaching a stationary observer. It sounds a whistle. Indicate whether

(a) The observer hears the same frequency

(b) The observer hears a slightly higher frequency

(c) The observer hears a slightly lower frequency

(d) The observer hears the same frequency but of different quality

5.34 A train giving a whistle of frequency 450 Hz approaches a stationary observer at a speed of 33 m/s. The frequency heard by the observer is

(a) 409 Hz (b) 429 Hz (c) 519 Hz (d) 495 Hz

5.35 A train giving a whistle of frequency 450 Hz moving away from a stationary observer at a speed of 33 m/s. The frequency heard by the observer is

(a) 409 Hz (b) 429 Hz (c) 519 Hz (d) 495 Hz

5.36 The frequency of the sound of a car horn as heard by an observer towards whom the car is moving differs from the frequency of the horn by 2.5%. Assuming that the velocity of sound in air is 320 m/s, the velocity of car is

(a) 8 m/s (b) 800 m/s (c) 7.5 m/s (d) 6 m/s

5.37 An object producing a pitch of 400 Hz approaches a stationary person in a straight line with a velocity of 200 m/s. Velocity of sound is 300 m/s. The person will note a change in frequency, as the object flies past him equal to

(a) 1440 Hz (b) 240 Hz (c) 1200 Hz (d) 960 Hz

[**Note:** Speed of sound, v = 300 m/s, speed of object v_s = 200 m/s, object producing frequency, f = 400 Hz

When the object approaches the person, the apparent frequency,

$$f'_{ap} = f\frac{v}{v - v_s} \quad \text{or} \quad f' = 400 \times \frac{300}{300 - 2} = 1200 \text{ Hz}$$

When object pasts away the person, the apparent frequency,

$$f'_{ap} = f\frac{v}{v + v_s} \quad \text{or} \quad f' = 400 \times \frac{300}{300 + 2} = 240 \text{ Hz}$$

Change in frequency is 1200 – 240 = 960 Hz.

5.38 A train blowing a whistle moves with a constant velocity (v) away from an observer on the ground. The ratio of the natural frequency of the whistle to that measured by the observer is found to be 0.75. If the train is at rest and the observer moves away from it at the same velocity, this ratio would be given by

(a) 0.5 (b) 0.75 (c) 1.5 (d) 1.2

5.39 Light shows many properties like those of sound. A star, which emits a line of wavelength 5000 Å is receding from you in the line of sight with a velocity (1/100) of speed of light. The wavelength of this line as observed by you will be

(a) 4900 Å (b) 4950 Å (c) 5050 Å (d) 5100 Å

[**Note:** speed of light, $v = c$ m/s, speed of star $v_s = (1/100)\, c$ m/s, the initial wavelength of the star $\lambda = 5000$ Å. Let the initial frequency is f.

When star is moving away from the observer, the apparent frequency,

$$f_{ap} = f\frac{v}{v+v_s} \quad \text{or} \quad \frac{c}{\lambda_{ap}} = \frac{c}{\text{Å}}\frac{v}{v+v_s} \quad \text{or} \quad \lambda_{ap} = \lambda\frac{v+v_s}{v}$$

or $$\lambda_{ap} = 5000 \times \frac{c + \frac{1}{100}c}{c} = 5000 \times \frac{101}{100} = 5050 \text{ Å}$$

5.40 A radio wave of frequency 840 MHz is sent towards an aeroplane. The frequency of radio echo has a frequency 2.8 kHz more than the original frequency. The velocity of the aeroplane is

(a) 3 km/s (b) 2 km/s (b) 4 km/s (d) 0.5 km/s

5.41 When the observer is stationary and the source is moving towards the observer with velocity v_0, v and f are the velocity and initial frequency of the source, the frequency f' noted by the observer will be

(a) $f_0\left(1-\frac{v_s}{v}\right)$ (b) $f_0\left(1+\frac{v}{v_s}\right)$ (c) $f_0\left(1+\frac{v_s}{v}\right)$ (d) $f_0\left(1-\frac{v}{v_s}\right)$

5.42 When the observer is stationary and the source is moving away from the observer with velocity v_0, v and f is the velocity and initial frequency of source, the frequency (f') noted by the observer will be

(a) $f_0\left(1-\frac{v_s}{v}\right)$ (b) $f_0\left(1+\frac{v}{v_s}\right)$ (c) $f_0\left(1+\frac{v_s}{v}\right)$ (d) $f_0\left(1-\frac{v}{v_s}\right)$

5.43 When the source is stationary and the observer is moving towards object with velocity v_0, v and f are the velocity and initial frequency of source, the frequency (f') noted by the observer will be

(a) $f_0\left(1-\frac{v_s}{v}\right)$ (b) $f_0\left(1+\frac{v}{v_s}\right)$ (c) $f_0\left(1+\frac{v_s}{v}\right)$ (d) $f_0\left(1-\frac{v}{v_s}\right)$

5.44 When the source is stationary and the observer is moving away from the source with velocity v_0, v and f are the velocity and initial frequency of source, the frequency (f') noted by the observer will be

(a) $f_0\left(1-\frac{v_s}{v}\right)$ (b) $f_0\left(1+\frac{v}{v_s}\right)$ (c) $f_0\left(1+\frac{v_s}{v}\right)$ (d) $f_0\left(1-\frac{v}{v_s}\right)$

5.45 v_s and v_0 are speeds of the source and the observer, respectively, v is the wave speed of wave, f_0 is the initial frequency of the wave, then final frequency of the wave noted by the observer will be.

(a) $f_0\left(\frac{v-v_s}{v-v_0}\right)$ (b) $f_0\left(\frac{v+v_s}{v+v_0}\right)$ (c) $f_0\left(\frac{v-v_0}{v-v_s}\right)$ (d) $f_0\left(\frac{v+v_0}{v+v_s}\right)$

Answers

5.1 (b)	5.2 (c)	5.3 (d)	5.4 (c)	5.5 (d)	5.6 (b)	5.7 (b)	5.8 (d)
5.9 (b)	5.10 (d)	5.11 (a)	5.12 (a)	5.13 (c)	5.14 (c)	5.15 (c)	5.16 (b)
5.17 (a)	5.18 (b)	5.19 (all)	5.20 (a)	5.21 (c)	5.22 (d)	5.23 (a)	5.24 (c)
5.25 (c)	5.26 (a)	5.27 (c)	5.28 (b)	5.29 (a)	5.30 (c)	5.31 (b)	5.32 (a)
5.33 (b)	5.34 (d)	5.35 (a)	5.36 (a)	5.37 (d)	5.38 (b)	5.39 (c)	5.40 (d)
5.41 (c)	5.42 (a)	5.43 (c)	5.44 (a)	5.45 (d)			

CHAPTER 6

Fluids

6.1 INTRODUCTION

In this chapter, we will study the viscosity and surface tension of liquids. Different types of pumps to produce low pressure and gauges to measure low pressure are also studied.

Viscosity is internal friction of the liquid by which it opposes relative motion between its different layers of liquid.

Surface tension is the force per unit length acting on the free surface of the liquid due to attractive forces among molecules of the liquid. Due to surface tension free surface of a liquid tries to have minimum surface area.

Different types of pumps are used to produce low pressure as low as 10^{-7} mm of mercury or to create vacuum, like rotary pump, diffusion pump, molecular pump. Ordinary manometer cannot measure low pressures. To measure the low pressure functioning of Knudsen absolute gauge and Penning and Pirani gauge are used.

6.2 SURFACE TENSION

The liquid in small quantity, free from any external force, like gravitational force, always takes the spherical drop shape. For example, rain drop, small quantity of mercury when placed on a clean glass surface.

For a given volume, a sphere has the least surface area. That is why, a liquid always takes the spherical shape. The following experiments show the tendency of a liquid to decrease its surface area.

Experiment 6.1 Let us dip a wire ring in a soap solution. When we take it out, we find that a thin film is formed across the ring. If we place a moistened cotton loop gently on the film, it will remain in position in which it is placed as shown in Fig. 6.1(a). This is because equal and opposite forces act on every point of cotton loop due to inside soap film and outside the soap film cancel each other. If the film inside the loop is pricked, it disappears and the loop at once takes the form of a circular ring, as shown in Fig. 6.1(b), because the inside forces acting at all points of cotton loop vanish, and only the outward forces are acting normally at every point of the loop. For a given perimeter, a circle encloses the largest area; so that the circular loop now encloses the maximum area. In other words, the area of the film between the loop and the wire ring is now reduced to minimum, which clearly shows that the film has a tendency to contract or shrink. As a result the tension is developed on the loop.

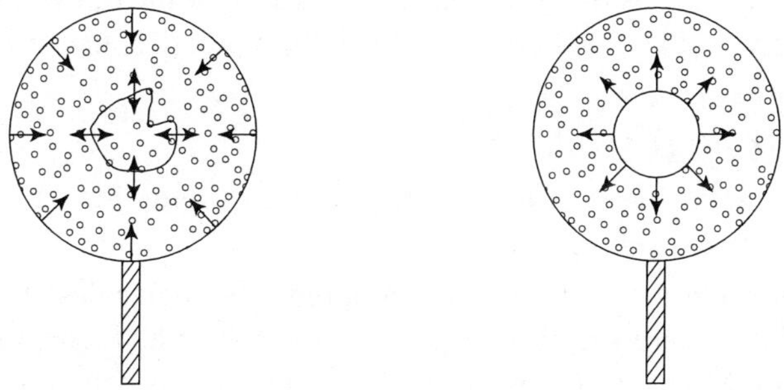

(a) (b)

Fig. 6.1

Experiment 6.2 The mixture of alcohol and water has almost equal density that of olive oil. Now, let us put a drop of olive oil with the help of a pipette inside a mixture of alcohol and water. Olive does not mix in the mixture of water and alcohol, though the density of the mixture is same as that of olive oil. The oil drop will keep on floating in the mixture with a perfectly spherical form. The reason is that the weight of the drop has been exactly balanced by the upthrust of the displaced mixture.The force of surface tension compels the loop to acquire the spherical shape having the least surface area. The action of gravity on oil drop is completely ignored. For a given volume, a sphere has the minimum surface area, therefore, the loop acquires spherical shape. This experiment is known as Plateau's experiment.

Experiment 6.3 Let us place a greased needle (N) on a piece of blotting paper, and then put the paper gently on the surface of water, the blotting paper sinks to bottom, but the needle remains floating on the surface [Fig. 6.2]. If we observe carefully, we find that a small depression is formed below and around the needle and that the free surface of water is slightly extended. The weight (mg) of the needle is supported by the resultant tension in the depression acting upwards, as shown in Fig. 6.2. If one end of the needle is forced to pierce the surface of water, it rapidly goes down slanting to the bottom of the water.

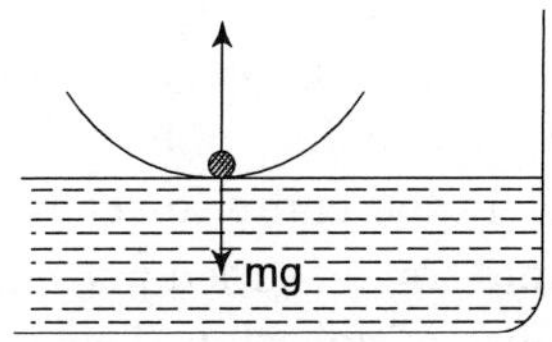

Fig. 6.2

The above experiments clearly show that the surface of a liquid behaves as though it were covered with an elastic skin or membrane, having a natural tendency to contract. Whereas Hooke's law states that in the case of membrane or skin, the tension increases as the skin is stretched or its surface area is increased, but it is quite independent of the area of the surface in the case of a liquid, unless the liquid film is reduced in thickness to less than 10^{-7} cm, when the tension in it decreases rapidly. This tension or pull in the surface of a liquid is called its surface tension. It is defined as the force per unit length of a line drawn in the liquid surface, acting perpendicularly to it at every point and tending to pull the surface apart along the line.

Explanation of the surface tension

By extending the kinetic theory of gases to the case of liquids Laplace explained the phenomenon of surface tension as follow.

Consider four molecules, *A*, *B*, *C* and *D* of a liquid. The molecules attracting them, called their spheres of influence, are shown in Fig. 6.3. Molecule *A* and its sphere of influence are well inside the liquid, Molecule *B* lies inside the liquid, but its sphere of influence is partly out of the liquid. Molecule *C* lies on the surface of the liquid, so one-half of the sphere of influence (*C*) is inside the liquid. Molecule *D* lies completely outside the liquid, however its sphere of influence partly lies inside the liquid.

Since the sphere of influence of *A* lies wholly inside the liquid, it is attracted by the other molecules lying within its sphere of influence equally in all directions, so that there is no resultant cohesive force on it, and therefore, it possesses only *thermal velocity*.

The sphere of influence of molecule *B* lies partly outside the liquid. The portion, which is outside the liquid, contains relatively a few molecules of the gas or vapour above the liquid, so that the upper half of the sphere *B* contains fewer molecules attracting it upwards, than the molecules of the lower half, attracting it downwards. Therefore, a resultant downward force is acting on molecule *B*.

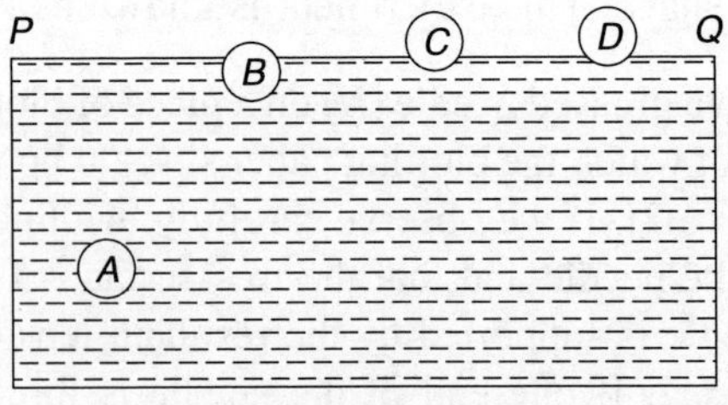

Fig. 6.3

The molecule *C* lies on the surface of the liquid, so that one-half of the sphere of influence lies above the surface of the liquid, containing only a few molecules of the gas or vapour, whereas there are liquid molecules in the entire lower half, and thus, the resultant force is downward. This downward or inward force exerted per unit area of a liquid surface is called its *cohesion pressure*, and is the cause of cohesion. This cohesion pressure is represented by the term a/v^2 in a van der Waals modified gas equation.

Since molecule D lies outside the liquid surface, and only a part of its sphere of influence lies inside the liquid, so the downward force on it is very small. When the sphere of influence lies entirely outside the liquid surface, there will be no downward force on the molecule, and it will be free to wander about as a molecule of the vapour or gas. In conclusion, we can say that there is a downward pull due to attraction between the molecules all over the surface of the liquid.

6.3 PRESSURE DIFFERENCE ACROSS A LIQUID SURFACE

(i) Plane surface of liquid: If the surface of the liquid is plane as shown in Fig. 6.4(i). The molecule at the surface is attracted equally by other nearby molecules at the surface in all directions. If we ignore the gravitational pull, then, the resultant force due to surface tension on a molecule on its surface is almost zero or negligible, and the cohesion-pressure, (i.e., the downward force per unit area of the liquid surface) is, therefore, almost zero or negligible.

(ii) Concave surface of liquid: If the free surface of the liquid is concave as shown in Fig. 6.4(ii), then the resultant component of all forces on a molecule acts normal to the surface and directed upwards. The pressure on the concave side is more than the pressure on the convex side. At equilibrium, the excess pressure due to surface tension, which is the difference of pressure between its two sides, is $2T/r$. Here, T is surface tension and r is radius of curvature of concave surface.

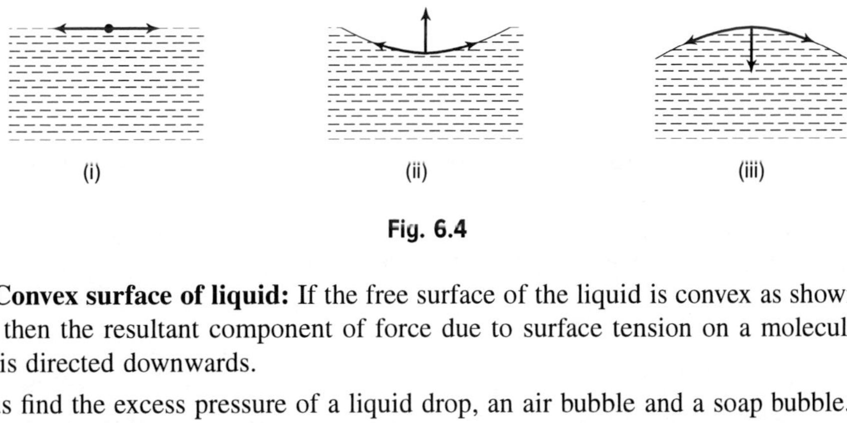

Fig. 6.4

(iii) Convex surface of liquid: If the free surface of the liquid is convex as shown in Fig. 6.4(iii), then the resultant component of force due to surface tension on a molecule on the surface is directed downwards.

Let us find the excess pressure of a liquid drop, an air bubble and a soap bubble.

(a) Excess pressure of a liquid drop: The pressure inside it, therefore, is more than the pressure outside it. Let r be the radius of the drop and T, its surface tension. The molecules near the surface of the drop have convex surface, therefore, they experience a resultant pull inwards. The inside pressure $P + p$ at the surface of the drop is more than its outside pressure Let excess pressure (p) push the drop to outside a distance, ΔR, where ΔR is so small that pressure inside the drop remains unchanged. The work done by the excess pressure is

$$W = \text{force} \times \text{displacement}$$

$$= (\text{Excess pressure} \times \text{Area}) \times \text{Displacement}$$

$$= (p \times 4\pi r^2) \times \Delta R$$

Increase in the surface area of the drop is

$$\Delta A = 4\pi\,(r + \Delta R)^2 - 4\pi\, r^2$$
$$= 4\pi\,[r^2 + 2r\,\Delta R + (\Delta\text{R})^2 - r^2]$$
$$= 8\,\pi r\,(\Delta R) \text{ neglecting } (\Delta R)^2$$

The increase in the surface energy is

$$(p \times 4\pi r^2) \times \Delta R = \text{Increase in the surface area} \times \text{Surface tension}$$
$$= 8\pi r\,(\Delta R) \times T$$

The excess pressure inside the drop will be

$$p = \frac{2T}{r}$$

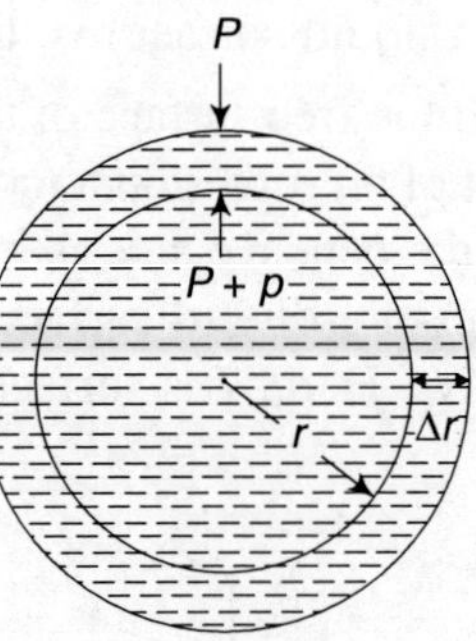

Fig. 6.5

(b) Excess pressure of an air bubble inside the water: If an air bubble of radius R is formed in a liquid of surface tension (T) as shown in Fig. 6.6, then the air bubble has one surface in contact with liquid, so it behaves just like a drop. Proceeding in the same way as in case of a liquid drop, the excess pressure inside the air bubble will be

$$p = \frac{2T}{r}$$

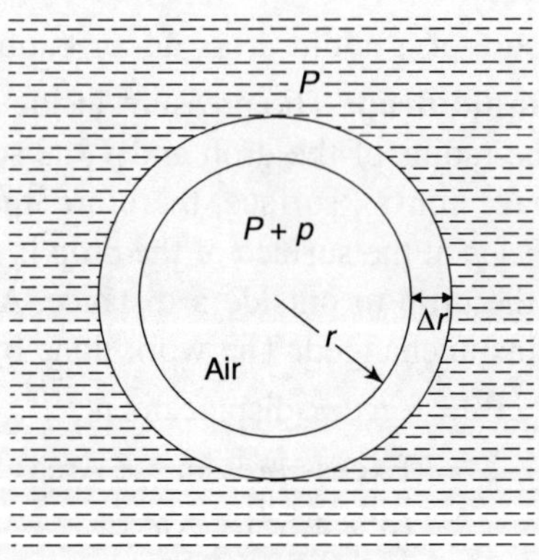

Fig. 6.6

(c) Excess pressure inside a soap bubble: A soap bubble is a thin hollow sphere. So it has two surfaces an inner surface and an outer surface. Therefore, in case of a bubble, we have to consider two surfaces. Let the radii of inner and outer spheres be r and $r + \Delta r$, respectively. If the surface tension on the airbubble is T, then increase of surface energy due to excess pressure will be $2 \times 8\pi r\, T\Delta r$ ($8\pi r\, T\Delta r$ of each surface). The work done due to excess pressure (p) on increasing the radius of bubble by Δr will be $p \times 4\pi r^2$. Therefore, in equilibrium the work done due to excess pressure for increasing the radius Δr will be equal to the increase in surface energy, i.e.,

$$p \times 4\pi r^2\, \Delta r = 2 \times 8\pi r\, T\Delta r$$

The excess pressure inside the soap bubble will be

$$p = \frac{16\,\pi\, rT}{4\,\pi\, r^2} = \frac{4T}{r}$$

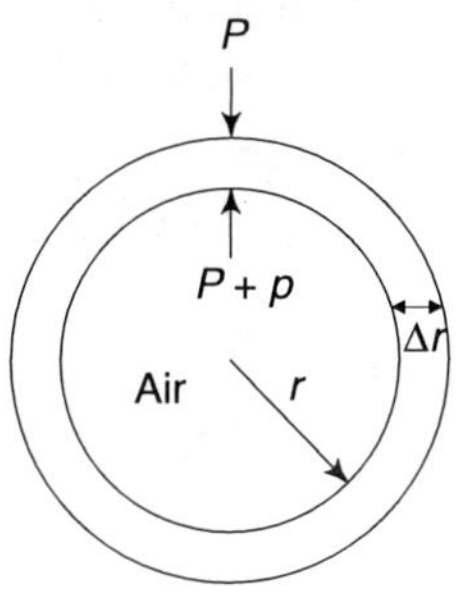

Fig. 6.7

(d) Case of a cylindrical surface: In this case, one of the radii is infinite, whereas the other is equal to the radius r of the cylinder so that

(i) for one single surface, i.e., for cylindrical column of liquid or for a cylindrical bubble in a liquid, excess pressure will be $p = T/r$.

(ii) for two surfaces, i.e., for cylindrical bubble or film, $p = 2\,T/r$, which is clearly half of the spherical bubble of the same radius.

In other words, the excess pressure inside a cylindrical bubble or film is same as that inside a spherical bubble of twice its radius.

This can be demonstrated by the following experiment.

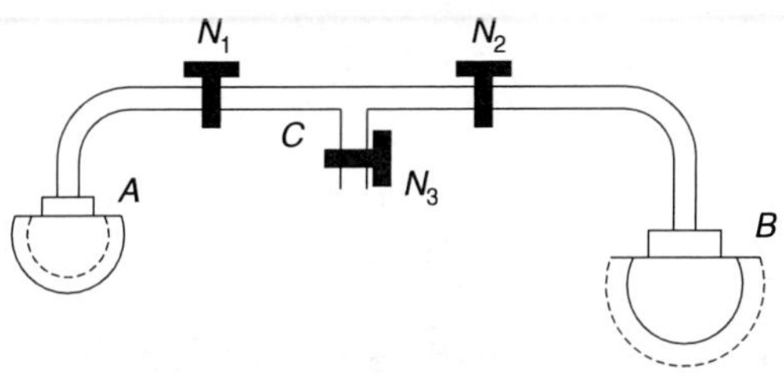

Fig. 6.8

Let us dip two ends of the tube T [Fig. 6.8)] having small cups at their ends in some soap solution and then taken out. Let us open stopcock N_2 and open N_1 and N_3. When the air is blown through tube C, a small bubble is formed at A. First of all close N_1 and open N_3, then blow air through C. A larger bubble is formed at B. Now, close N_3, and open N_1 and N_2. As a result, the two bubbles are put in direct contact with each other. Air at once starts passing from the smaller to the larger bubble, because the pressure inside the former is greater than that inside the latter. As a result, the smaller bubble goes on shrinking and the larger one goes on swelling until smaller one is reduced to a hemisphere. Equilibrium between the two bubbles is attained when the curvature of the two becomes the same.

6.4 EXPERIMENTAL STUDY OF THE VARIATION OF SURFACE TENSION WITH TEMPERATURE

The variation of surface tension can be studied with help of an experiment suggested by Jaeger. This method is applicable only for a small range of temperature.

In Jaeger's experiment a glass capillary tube AB is connected to another glass tube C, of a comparatively wider bore, to form a U-tube as shown in Fig. 6.9. This U-tube has two limbs of different bores.

The required liquid is then poured into this U-tube. Since, the angle of contact for the liquid and glass is less than 90°, the level of liquid is much higher in the narrower limb AB of the tube than its wider limb, CD.

Let r_1 and r_2 be the radii of the bores of the two limbs, T is the surface tension of the liquid, P is atmospheric pressure, and θ is the angle of contact with glass.

Pressure just below the liquid meniscus in AB limb will be

$$= P - 2T \cos \theta / r_1$$

and pressure just below the liquid meniscus in CD limb will be

$$= P - 2T \cos \theta / r_2$$

The difference of pressure in the two limbs is

$$= 2T \cos \theta \left(\frac{1}{r_1} - \frac{1}{r_2} \right) \quad \text{...(6.1)}$$

If h is the difference of liquid levels in the two limbs, and d is the density of liquid, this difference of pressure is balanced by the hydrostatic pressure hdg due to the liquid column in the U-tube, i.e.,

$$2T \cos \theta \left(\frac{1}{r_1} - \frac{1}{r_2} \right) = hdg \quad \text{...(6.2)}$$

The surface tension of the liquid can be calculated from the equation

$$T = \frac{hdg}{2 \cos \theta \left(\frac{1}{r_1} - \frac{1}{r_2} \right)} \quad \text{...(6.3)}$$

A suitable temperature bath is used to find the surface tension at different temperatures. A graph is then plotted between the temperatures ($t°$) versus surface tension (T) to show the variation of the surface tension of the given liquid with temperature.

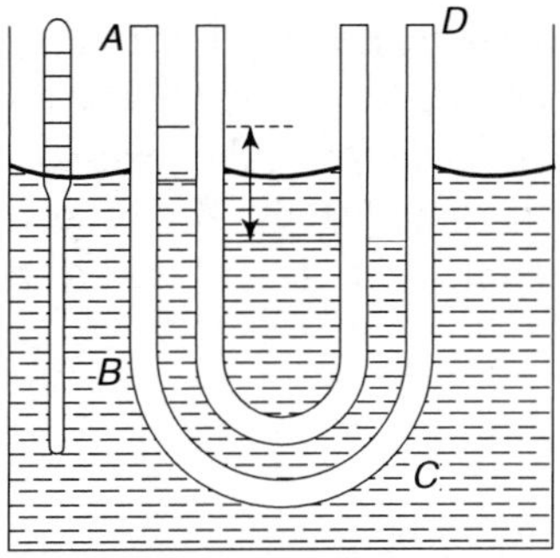

Fig. 6.9

6.5 VISCOSITY

When the flow is streamlined, i.e., when a liquid flows steadily and slowly over a fixed horizontal surface, its layer in contact with the fixed surface remains stationary. However, the velocity of the other layers increases with the distance from the fixed surface. That means that a regular velocity gradient is set up in the liquid, i.e., velocity of liquid increases as distance of a layer increases from the fixed surface. Thus, the topmost layer is moving fastest. In Fig. 6.10 the length of arrow shows that velocity of liquid increases as distance of a layer increases from the fixed surface. The velocity of the liquid versus distance (from the fixed surface) curve is, therefore a straight line inclined to the fixed horizontal surface as shown in Fig. 6.10. Such a flow is also referred to as a laminar flow, i.e., one layer of the liquid slides over another without any two layers getting mixed up.

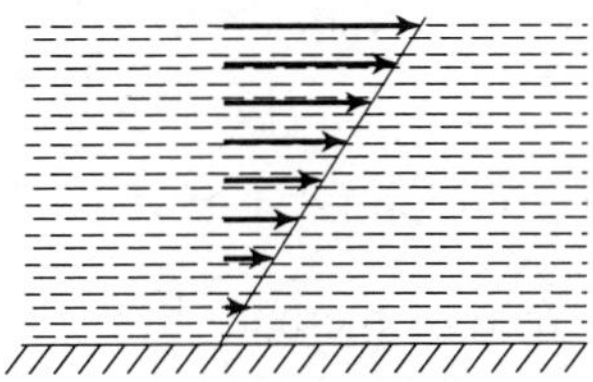

Fig. 6.10

If the liquid flows through a tube, its layers form concentric cylindrical shells coaxial with the tube. The layer in contact with the wall of the tube remains at rest and the velocity increases as we move away from the wall towards the axis of the tube. The velocity versus distance curves will be parabolic as shown in Fig. 6.11.

Now, let us consider a particular layer of the liquid in Fig. 6.10. One can see that there is a layer immediately below it moving slower than it, and the layer immediately above it is moving faster than it. This means that the layer below it tries to retard its motion and the layer above it tries to accelerate its motion. In other words, the two layers oppose their relative motion, as a frictional force acts tangentially between the layers. To maintain the relative velocity between the two layers, an external force must be applied to overcome this frictional force. In the absence of any such outside force, the relative motion between the layers slows down and ultimately the flow of the liquid stops. The property of a liquid by virtue of which it opposes relative motion between adjacent layers is known as internal friction or viscosity of the liquid.

SOLVED EXAMPLES

Example 6.1 The pressure of air in a soap bubble of 0.6 cm diameter is 7 mm of water above the atmospheric pressure. Calculate the surface tension of the soap solution.

Solution: Formula required

Pressure, $P = \dfrac{4T}{r}$

Given: Diameter of soap bubble, $R = 0.6$ cm, pressure, $P = 0.7$ mm of water column

Calculation:

$$\begin{aligned} P &= 0.7 \text{ cm of water column} \\ &= 0.7 \times 1 \times 980 \text{ dynes/cm}^2 \\ &= 686 \text{ dynes/cm}^2 \end{aligned}$$

Radius of soap bubble, $r = 0.3$ cm

The surface tension

$$T = \frac{pr}{4} = \frac{686 \times 0.3}{4} = 51.45 \text{ dynes / cm}$$

Ans. Surface tension of the soap solution is 51.45 dynes/cm.

Example 6.2 Whyis the upper surface of mercury in a glass capillary tube convex upward, while for water it is concave? Assuming the surface tension of rainwater to be 70 dynes/cm, find the difference of pressure inside and outside a rain drop of diameter 0.02 cm. What would be this difference of pressure amount to, if the drop were to be decreased by evaporation to a diameter of 0.0002 cm?

Solution: Formula required

Excess pressure inside a liquid drop over that outside is given by

$$p = \frac{2T}{r}$$

Given: Surface tension of rainwater, $T = 70$ dynes/cm, diameter of rain drop, $d = 0.03$ cm, diameter of evaporated drop, $d_1 = 0.00003$ cm

Calculation: Radius of rain drop, r = 0.015 cm

$$p = \frac{2T}{r} = \frac{2 \times 70}{0.015} = 0.93 \times 10^3 \text{ dynes/cm}^2$$

The radius of reduced drop, $r = 0.000015$ cm

$$p = \frac{2T}{r} = \frac{2 \times 70}{0.000015} = 0.93 \times 10^8 \text{ dynes/cm}^2$$

Ans. The excess pressures in two cases are 0.92×10^3 dynes/cm^2 and 0.93×10^8 dynes/cm^2, respectively.

Example 6.3 What will be the pressure in a spherical cavity within mass of paraffin oil? The cavity is at a depth of 25 cm below the surface and has diameter of 0.0028 cm. The specific gravity of the oil is 0.85 and its surface tension is 26 CGS units. The pressure of the air over the liquid surface is equal to 76 cm of the mercury.

Solution: Formula required

The pressure inside the spherical bubble in excess of the pressure outside it is given by

$$p = \frac{2T}{r}$$

Given: Depth of cavity, $h = 25$ cm, diameter of cavity, $d = 0.0028$ cm, specific gravity, $s = 0.86$, surface tension, $T = 27$ CGS units, pressure over the liquid surface, $p = 76$ mm of mercury.

Calculation: The pressure over the liquid surface, $p = 76$ cm of the mercury.

$$hdg = 76 \times 13.6 \times 980 = 1.014 \times 10^6 \text{ dynes/cm}^2$$

Pressure due to 20 cm oil column, $hdg = 25 \times 0.86 \times 980 = 2.107 \times 10^4$ dynes/cm^2

Total pressure on the spherical cavity, i.e., on the bubble

$$= 1.014 \times 10^6 + 2.107 \times 10^4 = (101.4 + 2.107) \times 10^4$$

$$= 103.507 \times 10^4 \text{ dynes/cm}^2$$

The pressure inside the spherical bubble in excess of the pressure outside it is given by

$$p = \frac{2T}{r} = \frac{2 \times 27}{0.0014} = 3.85 \times 10^4 \text{ dynes/cm}^2$$

Total pressure inside the cavity = $103.507 \times 10^4 + 3.85 \times 10^4 = 107.357$ dynes/cm^2

Ans. Total pressure inside the cavity is 107.357 dynes/cm^2.

Example 6.4 Aminute spherical air bubble is rising slowly through a column of mercury contained in a deep jar. If the radius of the bubble at a depth of 105 cm is 0.2 mm, calculate the depth when its radius is 0.226 mm; given that the surface tension of mercury is 568. Assumethat atmospheric pressure is 760 of mercury.

Solution: Formula required

The excess pressure inside the bubble over the excess of the pressure outside is given by

$$p = \frac{2T}{r}$$

Given: Radius of bubble at depth 100 cm, $r = 0.2$ mm

Calculate the depth (h) when its radius is 0.226 mm

Surface tension of mercury, $T = 568$

Atmospheric pressure is 760 of mercury.

Calculation: Pressure on mercury surface in the jar = 760 mm

$$= 76 \text{ cm of mercury} = 76 \times 13.6 \times 980 \text{ dynes/cm}^2$$

Pressure due to 100 of mercury column = $105 \times 13.6 \times 980$ dynes/cm^2

Total pressure on air bubble = $76 \times 13.6 \times 980 + 105 \times 13.6 \times 980$

$$= (76 + 105) \times 13.6 \times 980 = 2.412 \times 10^6 \text{ dynes/cm}^2$$

Excess pressure inside the air bubble

$$p = \frac{2T}{r} = \frac{2 \times 568}{0.02} = 0.0568 \times 10^6 \text{ dynes/cm}^2$$

Total pressure inside the bubble at a depth of 100 cm

$$P_1 = 2.412 \times 10^6 + 0.0568 \times 10^6 = 2.4688 \times 10^6$$

Volume of the bubble at depth 100 cm, $V_1 = \frac{4}{3}\pi r^3 = \frac{4}{3}\pi (0.02)^3 \text{ cm}^2$

Pressure due to mercury column at depth, $h = h \times 13.6 \times 980$ dynes/cm^2

Total pressure on the bubble at depth, $h = (h + 76) \times 13.6 \times 980$

$$= 13328\,(h + 76) \text{ dynes/cm}^2$$

Excess pressure inside the bubble at depth (h) over this pressure on its outside

$$= \frac{2T}{r} = \frac{2 \times 567}{0.0226} = 50177 \text{ dynes/cm}^2$$

Total pressure inside the bubble at depth h

$$P_2 = 13328\,(h + 76) + 50177 = 13328\,h + 1012928 + 50177$$

$$= 13328\,h + 1063105$$

Volume of the bubble at depth h, $V_2 = \frac{4}{3}\pi r^3 = \frac{4}{3}\pi (0.0226)^3 \text{ cm}^2$

According to Boyle's law, product of pressure and volume in two cases are equal, i.e.,

$$P_1 V_1 = P_2 V_2$$

or $$2.4688 \times 10^6 \times \frac{4}{3}\pi\,(0.02)^3 = [13328\,h + 1063105] \times \frac{4}{3}\pi\,(0.0226)^3$$

or $$2.4688 \times 10^6 \times (0.02)^3 = [13328\, h + 1063105] \times (0.0226)^3$$

or $$19.7504 = 0.15384\, h + 12.27161$$

or $$0.15384\, h = 7.47879$$

or $$h = 48.61408 \text{ cm}$$

Ans. The bubble of radius is 0.226 mm is at a depth 48.61408 cm.

Example 6.5 If a number of little droplets of water, all of the same radius, coalesce to form a single drop of radius R, show that rise in temperature of water will be given by

$$\theta = \frac{2T}{J}\left(\frac{1}{r} - \frac{1}{R}\right)$$

where, T is the surface tension of water and J, the mechanical equivalent of heat.

Let the number of little droplet be n and the radius of each droplet is r. Then, surface area of all the droplets = $n\, 4\, \pi r^2$. If R is the radius of single drop formed after coalescing n drops, then the surface area of the single drop is = $4\, \pi R^2$.

The decrease in surface area = $n\, 4\, \pi r^2 - 4\, \pi R^2$

Decrease in surface energy = $(n\, 4\, \pi r^2 - 4\, \pi R^2) \times T$ ergs

Heat produced = $(n\, 4\, \pi r^2 - 4\, \pi R^2) \times \frac{T}{J}$ calories.

The heat is absorbed by asingle drop of volume $\frac{4}{3}\pi R^3$. If the density of water, $d = 1$ gm/cm^3, then mass of the drop, = $\frac{4}{3}\pi R^3 \times 1$. If θ is rise in temperature of drop, then

$$\left(\frac{4}{3}\pi R^3 \times 1\right) \times 1 \times \theta = (n\, 4\, \pi r^2 - 4\, \pi R^2) \times \frac{T}{J}$$

$$\theta = 3\frac{T}{J}\left(\frac{nr^2}{R^3} - \frac{R^2}{R^3}\right)$$

Volume of a single coalesced drop is equal to the volume of n drops, i.e.,

$$\frac{4}{3}\pi R^3 = n\frac{4}{3}\pi r^3 \quad \text{or} \quad R^3 = n\, r^3, \text{ so}$$

$$\theta = 3\frac{T}{J}\left(\frac{1}{r} - \frac{1}{R}\right)$$

This is the required formula.

Example 6.6 A capillary tube of 0.6 mm bore stands vertically in a wide vessel containing a liquid of surface tension 0.04 N/m. The liquid wets the tube and has a specific gravity 0.7. Calculate the rise of the liquid in the tube.

Solution: Formula required

Pressure on the meniscus, $p = \dfrac{2T\cos\theta}{r}$

Given: Diameter of the capillary tube = 0.6 mm

Surface tension of liquid, $T = 0.04$ N/m

Specific gravity = 0.7

Calculation: Radius of the tube, $r = 0.30$ mm = 0.035 cm = 0.00035 m

Density of liquid, $d = 0.7 \times 10^3$ kg/m^3 as density of water = 1 gm/cm^2 = 10^3 kg/m^3

Acceleration due to gravity, $g = 9.80$ m/sec^2

Angle of contact, $\theta = 0$, because liquid is in the tube, $\cos\theta = 1$, so

$$p = \frac{2T}{r} = hdg$$

or

$$h = \frac{2T}{rdg} = \frac{2 \times 0.04}{0.00035 \times 0.7 \times 10^3 \times 9.80} = 0.033320 \text{ m}$$

Ans. The liquid will rise by 0.033320 m.

Example 6.7 A capillary tube of internal diameter 2 mm and external diameter 6 mm hangs vertically from the arm of a balance, the lower end of the tube is in the liquid of surface tension 45 dynes/cm. assuming that the liquid wets the tube, what is the change in apparent weight of the tube due to surface tension? Acceleration due to gravity, $g = 980$ cm/sec^2

Solution:

Given: Internal diameter of tube, $d = 2$ mm

External diameter of tube, $D = 6$ mm

Surface tension of liquid, $T = 45$ dynes/cm

Acceleration due to gravity, $g = 980$ cm/sec^2

Calculation:

Inner radius of tube, $r = 1.0$ mm = 0.1 cm

Outer radius of tube, $R = 3$ mm = 0.3 cm

The inner circumference of tube = $2\pi r = 2\pi \times 0.1$ cm

The outer circumference of tube = $2\pi R = 2 \times \pi \times 0.3$ cm

Sum of two circumferences = $2\pi \times (0.1 + 0.3)$ cm = $2\pi \times 0.40$ cm = 0.80π

Since the capillary tube is dipped in liquid, so the force of surface tension downwards in the capillary tube along the inner as well as the outer circumference of its lower end dipping into the liquid, i.e., along the length is 0.80π.

Force acting downwards on the tube due to surface tension

$$\text{mg} = 0.80\pi \times T \text{ dynes}$$

$$\text{Weight of liquid, } m = \frac{0.80\,\pi \times T}{g} = \frac{0.80\,\pi \times 45}{980} = 0.11{,}540 \text{ gm}$$

Ans. Change in weight due to surface tension is 0.11, 540 gm.

Example 6.8 Water rises to a height of 6 cm in a certain capillary tube. In the same tube the level of mercury surface is depressed by 1.55 cm. Compare the surface tension of water and mercury, the specific gravity of mercury is 13.6, the angle of contact for water is 0° and for mercury, 120°.

Solution: Formula required

$$\text{Pressure due to surface tension, } P = \frac{2T\cos\theta}{r}$$

Given:

Water height rises to $h = 6$ cm

Mercury surface depressed by $h = -\,1.55$ cm

Specific gravity of mercury is 13.6

Angle of contact $\theta = 0°$ for water

Angle of contact $\theta = 120°$ for mercury

Calculation: If d is the density of liquid, h is depression and g is acceleration due to gravity, then

$$P = \frac{2T\cos\theta}{r} = hdg$$

or

$$T = \frac{r\,hdg}{2\cos\theta}$$

In case of mercury, the surface tension,

$$T_m = \frac{r(-1.55)\times 13.6\times 980}{2\cos 120}$$

In case of water, the surface tension,

$$T_w = \frac{r\times 6\times 13.6\times 980}{2\cos 0}$$

The ratio of surface tensions

$$\frac{T_m}{T_w} = \frac{r(-1.55)\times 13.6\times 980}{2\cos 1}\times\frac{2\cos 0}{r\times 6\times 1\times 980}$$

$$= \frac{(-1.55)\times 13.6}{6\times(-0.5)} = \frac{1.55\times 13.6}{6\times 0.5} = 7.02667$$

Ans. The ratio of surface tensions of mercury and water is 7.02667 : 1.

6.6 COEFFICIENT OF VISCOSITY

Newton showed that the backward dragging, or viscous force (F) acting tangentially on any liquid layer is directly proportional to its surface area, A, and –ve of velocity (v_x) of liquid in a layer and the distance of layer y from the stationary layer, i.e,

$$F \propto A \; ; F \propto -v_x \; ; F \propto \frac{1}{y}$$

Combining these three equations, we can write

$$F \propto -\frac{Av_x}{y}$$

Writing $\frac{v_x}{y} = \frac{dv_x}{dy}$, the above equation becomes

$$F \propto -\frac{Av_x}{y} \quad \text{or} \quad F = -\eta \frac{Adv_x}{dy} \qquad ...(6.4)$$

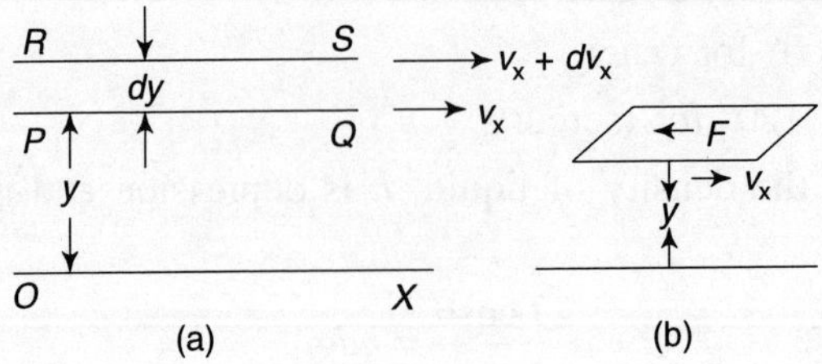

Fig. 6.11

We can also get this equation as follows:

Newton showed that the backward dragging, or viscous force (F) acting tangentially on any liquid layer is directly proportional to its contact surface area (A), and velocity gradient (dv_x/dy), i.e.

$$F \propto A \; ; F \propto \frac{dv_x}{dy}$$

Combining these two equations, we can write

$$F \propto -\frac{Adv_x}{dy}$$

or

$$F = -\eta \frac{Adv_x}{dy} \qquad ...(6.5)$$

where, η is proportionality constant. It depends on the nature of the liquid, and is called its coefficient of viscosity. Negative sign in Eq. (6.5) indicates that the direction of the viscous force is opposite to that of velocity of liquid.

Equation (6.5) is known as Newton's law of viscous flow in streamline motion.

If we consider $A = 1$, $dv_x = 1$, $dy = 1$, so $\frac{dv}{dy} = 1$, then from Eq. (6.5), we have $F = \eta$

Thus, the coefficient of viscosity of a liquid may be defined as the tangential force required per unit area to maintain the unit relative velocity between two layers of unit distance apart.

In CGS system, $\eta = 1$, if one dyne tangential force is required per cm^2 to maintain the relative velocity of 1 cm/s, between two layers 1 cm apart. It is called Poise, after Poiseuille whose has done a lot of work on viscosity of liquids.

In MKS (SI) system, $\eta = 1$, if one newton tangential force per square metre is required to maintain the relative velocity of 1 m/s, between two layers 1 m apart.

In FPS system, $\eta = 1$, if one pound tangential force per square metre is required to maintain a relative velocity of 1 ft/s, between two layers 1 feet apart.

Viscosity in liquids is similar to friction in solid. The relative motion of a solid on solid surface opposes by friction, in the same way the relative motion between two layers is opposed by viscosity. Friction depends on the contact surface area of two solids. On the other hand, the viscosity depends upon: (i) contact surface area of the liquid layers, (ii) the distance between two layers, and (iii) relative velocity of the liquid in two layers.

Dimensions of η

From Eq. (6.5),

$$\eta = \frac{F}{A\frac{dv}{dx}}$$

The dimension of η is obtained by putting the dimensions of area ($A \sim L^2$) and gradient $\left(\frac{dv}{dx} \sim \frac{L/T}{L}\right)$ in the following expression

$$\frac{\text{Force}}{\text{Area} \times \text{Velocity gradient}}$$

Thus, dimension of η is $\frac{\text{Force}}{L^2 \times \frac{L/T}{L}} = \frac{\text{Force}}{L^2\, T^{-1}}$ ...(6.6)

Putting the dimension of force MLT^{-2}, the dimension of η is

$$\frac{MLT^{-2}}{L^2 \times \frac{L/T}{L}} = \frac{MLT^{-2}}{L^2\, T^{-1}} = ML^{-1}\, T^{-1} \quad ...(6.7)$$

In CGS system, η may be expressed in dynes sec cm^{-2} (or Poise) according to Eq. (6.6) or in gm cm^{-1} sec^{-1} in accordance with Eq. (6.7).

In MKS system, η may be expressed in newton m^{-2} sec or in kg m^{-1} sec^{-1} according to Eq. (6.7).

In FPS system, η may be expressed in lb (pound) ft^{-2} sec or in Engineer's units, slug ft^{-1} sec^{-1} where, slug is the unit of mass, the weight of mass of 1 slug being 32 lb or mass of 1 lb weight = $\frac{1}{32}$ slug, i.e., weight in pounds/g (in ft^2/sec) = slugs.

6.7 POISEUILLE'S EQUATION FOR FLOW OF A LIQUID THROUGH A HORIZONTAL CAPILLARY TUBE

Let us consider that a liquid is flowing through a horizontal capillary tube of length l and radius r due to a constant pressure difference, P across its two ends. To drive an expression for the rate of flow of a liquid through the tube, we make the following assumptions:

(i) The tube is horizontal, so that gravity does not affect the flow of liquid.

(ii) The liquid layer in contact with the sides of the tube remains stationary.

(iii) The pressure is constant over every cross section of the tube, so there is no radial flow.

(iv) All stream lines are parallel to the axis of the tube, the flow of the liquid is steady and streamline.

(v) The liquid yields a small shearing stress.

Now, imagine a cylindrical layer or shell of liquid, coaxial with the tube, and of radius, x. The velocity (v) of flow is same at all points on this cylindrical shell. The velocity of the layer in contact with the wall of the tube is zero and goes on increasing towards the axis, that is liquid inside the imaginary cylinder is moving faster than that outside it.

If x and l are radius length of the cylindrical shell, then its the surface area, $A = 2\pi\, xl$, and the backward tangential frictional force due to the outer, slower moving liquid on the inner faster moving liquid, according to Eq. (6.5), is given by

$$F = -\,\eta \cdot 2\pi\, xl \cdot \frac{dv_x}{dy} \quad \text{...(6.8)}$$

If the pressure difference of at two ends of the cylindrical liquid shell is P, the forward force on the cylindrical liquid shell, which accelerates the motion of the liquid in the direction of flow

$$F = P \times \pi x^2 \quad \text{...(6.9)}$$

In steady state of the liquid the motion

$$\eta \cdot 2\,\pi\, x.l.dv/dx = -\,P \times \pi x^2 \quad \text{...(6.10)}$$

the –ve sign indicates that the two forces act in opposite direction. Rewriting Eq. (6.10) as

$$dv = \frac{P\,\pi x^2}{\eta \cdot 2\pi\, x \cdot l}\, dx = -\frac{Pxdx}{2\eta l} \quad \text{...(6.11)}$$

Integrating both sides of Eq. (6.11), we get

$$v = -\frac{P}{2\eta l}\int x\, dx$$

or
$$v = -\frac{P}{2\eta l}\frac{x^2}{2} + C_1 = -\frac{Px^2}{4\eta l} + C_1 \qquad \text{...(6,12)}$$

where, C_1 is the constant of integration.

Since, the layer in contact with sides of the tube is stationary, i.e., $v = 0$, when $x = r$, so Eq. (6.12) becomes

$$0 = -\frac{Pr^2}{4\eta l} + C_1 \quad \text{or} \quad C_1 = \frac{Pr^2}{4\eta l}$$

Putting for C_1 in Eq. (6.12)

$$v = -\frac{Px^2}{4\eta l} + \frac{Pr^2}{4\eta l} = \frac{P}{4\eta l}(r^2 - x^2) \qquad \text{...(6,13)}$$

In Eq. (6.13), v is the velocity of flow of the liquid at a distance x from the axis of the tube. The velocity distribution curve of the liquid motion in the tube is a parabola [Fig. 6.12]. The velocity of the liquid increases from zero (at the wall of the tube) to a maximum along its axis.

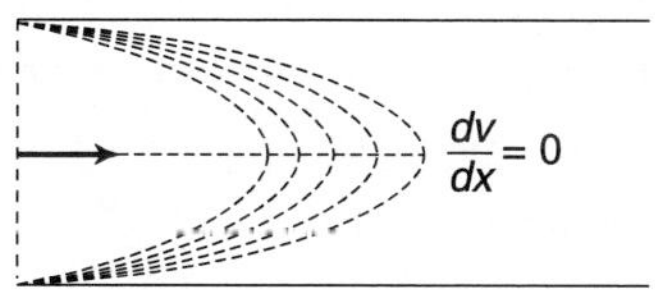

Fig. 6.12

Consider another coaxial cylindrical shell of radius $(x + dx)$ enveloping the first. Then, the cross-sectional area between the two shells is $2\pi x dx$. If the velocity of the liquid is v, then volume of the liquid flowing through this cross-sectional area in 1 sec is $dV = 2\pi x dx v$.

Assume that whole of the capillary tube to consist of such the coaxial cylindrical shells, the volume of the liquid flowing through the entire tube in one second will be

$$V = \int_0^r 2\pi x\, dxv = \int_0^r 2\pi x \cdot \frac{P}{4\eta l}(r^2 - x^2)\, dx$$

$$= \frac{\pi P}{2\eta l}\int_0^r (xr^2 - x^3)\, dx = \frac{\pi P}{2\eta l}\left[\frac{x^2 r^2}{2} - \frac{x^4}{4}\right]_0^r$$

$$= \frac{\pi P}{2\eta l}\left[\frac{r^4}{2} - \frac{r^4}{4}\right] = \frac{\pi Pr^4}{8\eta l}$$

So $$\eta = \frac{\pi P r^4}{8 V l} \qquad ...(6,14)$$

Equation (6.14) is known as *Poiseuille's equation.* If we know pressure difference at the two ends of capillary tube (P), radius of capillary tube (r), rate of flow of volume through capillary tube (V) and length of the capillary tube (l), the coefficient of viscosity of the liquid (η) can be easily calculated.

Alternative derivation of Poiseuille's equation

Consider a cylindrical layer or shell of the liquid, coaxial with the tube and of radius x and thickness dx. According to Newton's law of viscous flow given by Eq. (6.5), tangential viscous force acting on the inner side of the layer

$$F = -\eta \cdot 2\pi x l \frac{dv}{dx}$$

The tangential viscous force acting on the outer side of the layer,

$$= F + \frac{dF}{dx} dx$$

where, $\frac{dF}{dx}$ is the change of force with distance, x.

Since the layer on the inner side of the cylindrical shell is moving faster, and layer on the outer side moving slower, than the shell experience a forward push due to inner side of layer and a backward pull due to outer side of layer.

Therefore, the resultant drag experienced by the shell,

$$\left(F + \frac{dF}{dx} dx\right) - F = \frac{dF}{dx} dx$$

As the liquid is neither accelerated nor retarded, i.e., it continues to flow with same steady velocity, it is clear thatthis backward dragging force on the shell is just balanced by force $P \times 2\pi x dx$ due to the pressure difference (P) across the two ends.

We thus have $\frac{dF}{dx} dx = P \times 2\pi x dx$ or $\frac{dF}{dx} = P \times 2\pi x$.

Putting the tangential dragging forces, $F = -\eta \cdot 2\pi x l \frac{dv}{dx}$, we get

$$\frac{dF}{dx} = \frac{d}{dx}\left(-\eta \cdot 2\pi x l \frac{dv}{dx}\right) = P \times 2\pi x \qquad ...(6.15)$$

Iintegrating both sides of Eq. (6.15) with respect to x, we get

$$-\eta \cdot 2\pi x l \frac{dv}{dx} = \pi x^2 \cdot P + C$$

where, C is a constant of integration.

The velocity is maximum along the axis of the tube, so $\frac{dv}{dx} = 0$ when $x = 0$, so $C = 0$ and hence we get

$$-\eta \cdot 2\pi\, xl \frac{dv}{dx} = \pi\, x^2 \cdot P$$

or

$$\frac{dv}{dx} = -P\, x/2\, \eta l \qquad \text{...(6.16)}$$

Integrating both sides of Eq. (6.16), we get

$$v = -\frac{Px^2}{2\,\eta\, l} + C_1$$

where, C_1 is another constant of integration. At $x = r$, $v = 0$, so

$$C_1 = \frac{Pr^2}{2\,\eta\, l}$$

Putting the value of C_1, we get equation

$$v = \frac{P}{4\,\eta l}(r^2 - x^2) \qquad \text{...(6.17)}$$

as in the first case.

6.8 CRITICISM OF POISEUILLE'S EQUATION

For deriving Poiseuille's equation in Article 6.7 it was assumed that fluid flows through a horizontal capillary tube. This assumption is more or less valid if the velocity of the flow is small and the tube is narrow, because in that case a streamline motion of the liquid is possible. The critical velocity of liquid flow decreases as bore size of the capillary increases, i.e., $v_c \propto l/r$. Therefore, for tubes of wider bores, the value of critical velocity is small, turbulence is set up soon and the relation given by Eq. (6.17) is not valid. This may easily be seen by plotting a graph between the pressure difference (P) across the two ends of the tube and the rate of flow (v) of the liquid through the tube, as shown in Fig. 6.13.

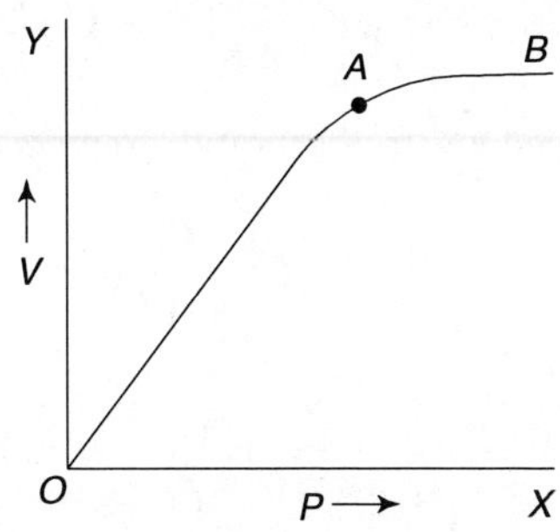

Fig. 6.13

From the Fig. 6.13 it is inferred that the rate of flow (v) is proportional to the pressure difference (P) and thus depends mainly on the viscosity (η) of the liquid. The curved part (AB) corresponds to the velocities above the critical values. In this curved part the rate of flow is no longer proportional to the pressure difference. The kinetic energy gained by increasing the pressure is now almost wholly used up in combating the turbulence set up in the liquid. That is, the rate of flow no longer depends on η, but is found to depend mainly on the density (ρ) of the liquid. Thus, rate of flow is approximately proportional to $\sqrt{P}$.

The assumption for the layer of the liquid in contact with the wall of the tube remains stationary and is found valid for really small velocities of the flow.

Thus, Poisseuille's equation is applicable, if the tube is narrow and the velocity of flow of the liquid is small and the flow of liquid is of almost a thin stream of drops.

However, two factors have not been taken into account in deriving this equation: (i) Apart of the thrust due to the pressure difference across the capillary tube imparts kinetic energy to the liquid and only the rest of it is used for overcoming the viscous resistance of the liquid, hence the effective value of the pressure difference is less than P, (ii) Motion of the liquid where it enters the capillary tube is some what accelerated, as a result the velocity of the flow is not uniform over the first short length of the tube. Proper corrections are necessary for these factors.

Correction for effective pressure

It is also known as the kinetic energy correction and is obtained as follows:

Consider a cylindrical layer or shell of the liquid of radius x and thickness dx, coaxial with the tube. If v is the steady velocity of the liquid flow at distance x from the axis of the tube, volume of liquid flowing along it in l sec $= 2\pi\, x dx\, v$.

If ρ is the density of the liquid, then mass of the liquid $= 2\pi\, x dx\, v\rho$.

$$\text{Hence, the kinetic energy of this liquid} = \frac{1}{2} 2\,\pi\, xdx\, v\, \rho v^2$$

$$= \pi\, xdx\, \rho v^3$$

Putting the value of v from Eq. (14),

$$\text{Kinetic energy of this liquid} = \pi\, xdx\, \rho\left[\frac{P}{4\,\eta\, l}(r^2 - x^2)\right]^3$$

Kinetic energy imparted per second to the liquid flowing through the tube

$$= \int_0^r \pi\, xdx\, x\rho\left[\frac{P}{4\,\eta\, l}(r^2 - x^2)\right]^3$$

$$= \pi\,\rho\left[\frac{P}{4\,\eta\, l}\right]^3 \int_o^r ((r^2 - x^2))^3\, x\, dx$$

$$= \pi \rho \left[\frac{P}{4\eta l}\right]^3 \frac{r^8}{8} = \left[\frac{\pi P r^4}{8\eta l}\right]^3 \frac{\rho}{\pi^2 r^4}$$

As $\frac{\pi P r^4}{8\eta l} = V$, the rate of flow of the liquid.

Kinetic energy imparted per second to the liquid flowing out $= \frac{V^3 \rho}{\pi^2 r^4}$ ...(6.18)

This energy is imparted to the liquid due to a part of pressure difference (P) across the two ends of the tube. Let the remaining pressure difference used for overcoming the viscous resistance of the liquid be P_1. Then, work done per second for overcoming viscous resistance $= P_1 V$.

Total work done $= P_1 V + \frac{V^3 \rho}{\pi^2 r^4}$

is equal to PV, so that

$$PV = P_1 V + \frac{V^3 \rho}{\pi^2 r^4} \quad \text{or} \quad P_1 = P - \frac{V^3 \rho}{\pi^2 r^4} \qquad \text{...(6.19)}$$

This effective value of pressure P_1 mustbe substituted for P in Eq. (6.14) for getting .

(ii) Correction for accelerated imparted to the liquid at the end of the capillary tube: The correction due to acceleration imparted to the liquid at the end of the capillary tube is included by assuming that the length (l) of the tube is increased by1.64 r. In other words, the effective length of the tube is taken equal to (1 + 1.64 r), where r is the radius of the tube.

Taking the value of effective pressure given by Eq. (6.19) and the effective length, expression for η given by Eq. (6.14) becomes

$$\eta = \frac{\pi \left(P - \frac{v^3 \rho}{\pi^2 r^4}\right) r^4}{8V(1 + 1.64r)} = \frac{\pi P r^4}{8V(1 + 1.64r)} - \frac{V^2 \rho}{8\pi(1 + 1.64r)} \qquad \text{...(6.20)}$$

6.9 EXPERIMENTAL DETERMINATION OF η FOR LIQUID – POISEUILLE'S METHOD

In this experimental setup a capillary tube (T) of length (l) and radius (r) is fixed horizontally near the bottom of the vessel (A) as shown in Fig. 6.14. The level of liquid in vessel A is kept constant at any desired height by over flowing the liquid though capillary tube, O. A clean and dry beaker is placed below the outer end of the tube (T) to collect the liquid flowing out of it. The liquid is allowed to flow out in a slow thin stream and collected in the beaker for a known time, and the beaker is then weighed again. The difference of the two weights gives the mass of the liquid flowing out in that time. If density of the liquid is ρ, the volume of the liquid will be m/ρ. Dividing this volume by the time for which liquid flows, we get the volume (V) of the liquid flowing out per second. Substituting the value of V, so obtained, by using Eq. (11), the coefficient of viscosity (η) of the liquid can be calculated.

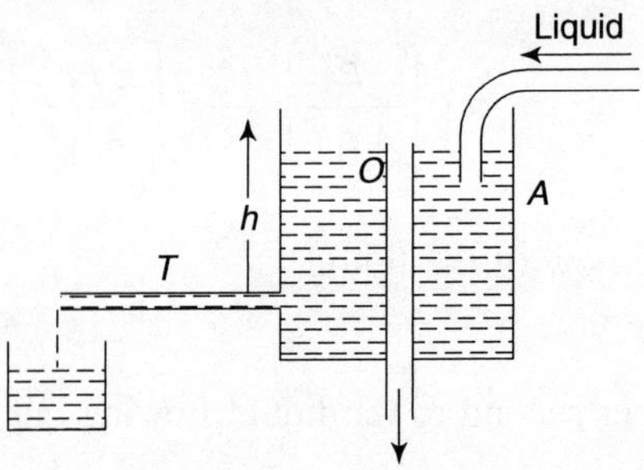

Fig. 6.14

For getting better accuracy of coefficient of viscosity, the corrected formula givenby Eq. (17)

$$\eta = \frac{\pi P r^4}{8V(1+1.64\,r)} - \frac{V\rho}{8\,\pi\,(1+1.64\,r)} \text{ is used .}$$

A much better experimental setupfor finding η with better accuracy is given in Fig. 6.15. In this apparatus, the flow tube (*F*) is a long one, and of a uniform circular cross section. The difference of pressure for a length (*AB*) is measured by a manometer (*M*), whose limbs are connectedto two fine holes at *A* and *B*, as shown in Fig. 6.15, where both *A* and *B* lie at a distance of at least of 10 cm from the two ends of the flow tube, so that the velocity of the out flowing liquid is uniform near them. This arrangement minimizes very much the two sources of error: (i) flow of the liquid is stream lined between two points, *A* and *B*; (ii) the pressure difference is constant at the points *A* and *B*. So, with a slow rate of flow of the liquid and a fairly small size of the holes at *A* and *B*, no correction is necessary.

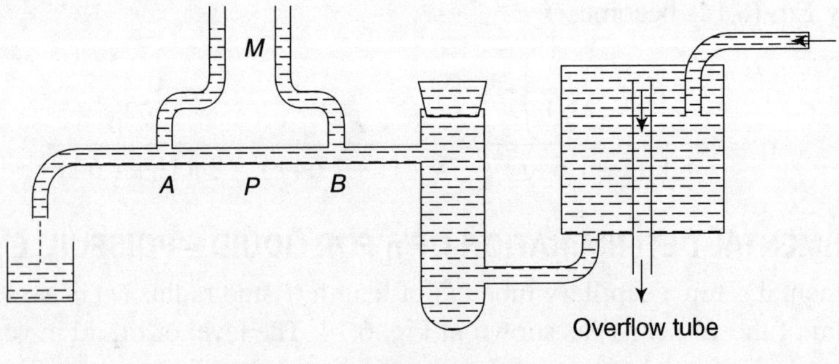

Fig. 6.15

6.10 MOTION IN A VISCOUS MEDIUM: STOKES' LAW

The viscous force is similar to frictional force faced by a moving body on solid surface. When a body falls into a viscous medium, its motion is opposed by thelayer of the liquid medium in immediate contact with it. The property of liquid, by virtue of which it drags the moving body is called the dragging force or viscous force of the liquid. When the motion of the body slows down continuously, its energy is absorbed continuously by the medium and is converted into heat. If the body moving is a fast one, like high speed cars and aeroplanes, eddy currents and waves are set up in the medium. Even if the body is moving so slowly that no eddy currents or waves are set up, energy is still wasted to overcome the viscous drag.

Now, this opposing force increases as the velocity of the body increases and finally attains a constant velocity, called its *terminal velocity.*

Stoke showed that retardation force (F) due to the viscous drag, for a spherical body of the radius r, moving with velocity v, in a medium having coefficient of the viscosity η, is given by

$$F = 6\pi V r\eta \quad ...(6.21)$$

This relation is known as Stokes' law.

If the density of the spherical body is ρ.

Its weight is = Volume × ρ × g = $\frac{4}{3}\pi r^3 \times \rho \times g$

If σ is the density of the medium, then upward thrust on the body due to the displaced medium is = $\frac{4}{3}\pi r^3 \times \sigma \times g$.

Therefore, the resultant downward force on the body will be = $\frac{4}{3}\pi r^3 \rho g - \frac{4}{3}\pi r^3 \sigma g$

$$= \frac{4}{3}\pi r^3 g(\rho - \sigma) \quad ...(6.22)$$

The downward force is balanced by the upward thrust given by Eq. (6.21), i.e.,

$$6\pi v r\eta = \frac{4}{3}\pi r^3 g(\rho - \sigma) \quad ...(6.23)$$

From this equation, the velocity of spherical body (of radius r and density ρ) moving down in the liquid (of density σ) and viscosity coefficient η will be

$$v = \frac{2}{9}\frac{r^2 g(\rho - \sigma)}{\eta} \quad ...(6.24)$$

Equation (6.24) indicates that the terminal velocity of the body of small size falling through a viscous medium is (i) directly proportional to the square of radius (r^2), (ii) directly proportional to the difference in the densities of the body and the medium, ($\rho - \sigma$), and inversely proportional to coefficient of the viscosity of the medium (η).

For deriving Eq. (6.24), stokes made the following assumptions:

(a) The medium through which the body falls is infinite in extent.

(b) The spherical body is perfectly rigid and smooth.

(c) There is no slip between the spherical body and the medium.

(d) The medium is homogeneous, and the diameter of the spherical body is large compared with the spaces between the molecules of the medium.

(e) No eddy currents or waves are set up in the medium due to the motion of the body in the medium. This is possible when the body moves very slowly through it and motion of the medium is smooth and not turbulent.

A good example of a body falling through a viscous medium is that of a tiny drop forming clouds. These tiny drops of water of radius as small as 0.001 cm, when fall through air having $\eta = 0.00018$, their terminal velocity comes to be 1.2 cm/sec using Eq. (6.24). That is why they remain suspended in the air and appear floating about as clouds.

Bigger rain drops have radius about 0.01 cm, which is about 10 times that of cloud drop. Their terminal velocity comes to be 120 cm/sec, so they fall through the air, instead of floating in it.

Also, if the density of the medium in which the body falls is greater than that of the body itself, i.e., if $\sigma > \rho$, then the terminal velocity (v) will have a negative value. That is, the body will move upwards and it will have an upward terminal velocity. That is why bubbles of air or gas rise up through water or anany other liquid. The smaller bubble has smaller velocity.

Stoke's law has several applications: For example, it can be used to determine: (i) the coefficient of viscosity of a liquid, (ii) the radius of the spherical objects like rain drops and ball bearings, etc., and (iii) the electronic charge by Millikan's falling oil drop method.

6.11 DETERMINATION OF COEFFICIENT OF VISCOSITY OF A LIQUID (STOKE'S FALLING BODY VISCOMETER)

The formula for v given by Eq. (6.24) is used to determine the viscosity of a liquid. In this method the time and distance of fall of small spheres, such as ball-bearings, etc., in the liquid are noted down. These values are used to find the terminal velocity of the sphere. Putting this velocity in Eq. (6.24), the coefficient of viscosity

$$\eta = \frac{2}{9}\frac{r^2 \cdot g(\rho - \sigma)}{v} \qquad \text{...(6.25)}$$

can be calculated. One should note that the Eq. (6.25) is valid only when the velocity of the fall is less than $6\eta/\sigma r$. To meet this condition sphere of 1 mm radius or less is used in case of thin machine oil, which has coefficient of viscosity 1 cgs unit or 1 poise nearly. In case of thick oils, or liquids like glycerine, whose coefficient of viscosity is greater than 10 cgs units, steel ball-bearings of maximum $\frac{1}{4}$, diameter can beused. For liquids which are less viscous, spheres of much smaller radii are required. The spheres of radii of 0.0002 to 1 cm can be obtained from flowing melted metal into cold water through capillary tubes of different bores. The diameters of these spheres are measured accurately by means of a microscope fitted with a micrometer eyepiece.

The liquid is filled in a tall jar of a large diameter. A sphere of suitable size is dropped in the centre of the liquid. The time taken by the sphere is noted at three points at different levels. The velocity is calculated for each of the two distances between the three points. If the calculated velocity over each path is the same, it means that the sphere has acquired constant velocity, or terminal velocity, v. If the velocities are different, that is, it is greater over the lower segment than over the upper one, a smaller ball-bearing must be tried. The procedure is repeated until the velocities in the two segments is equal, within the experimental error.

In practice, we make two marks, A and B, at distance (say, 10 to 12 cm) below the top and above the bottom of the jar as shown in Fig. 6.16(a). Now allow to fall small spheres of different known radii in the centre of the liquid. Note down the time taken by each sphere when it crosses the two marks, A and B. We assume that the sphere is of very small size so that it has attained the terminal velocity before crossing the mark A. If, the distance between marks A and B is S and time taken to pass this distance is t, then the terminal velocity of the sphere, $v = S/t$.

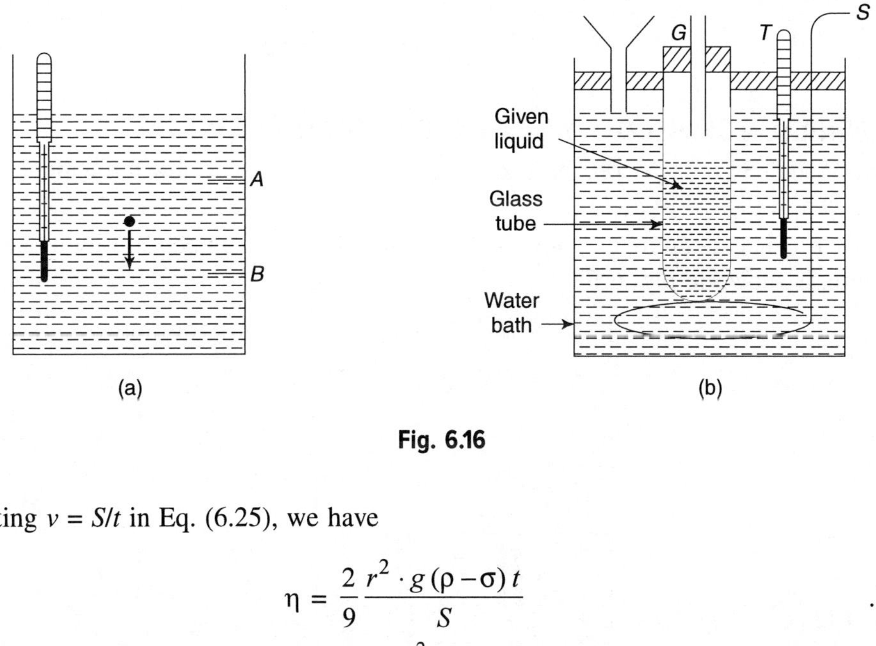

Fig. 6.16

Putting $v = S/t$ in Eq. (6.25), we have

$$\eta = \frac{2}{9}\frac{r^2 \cdot g(\rho - \sigma)t}{S} \qquad ...(6.26)$$

For a given sphere and a given liquid, r^2t is constant, provided its temperature remains constant. So, we get straight line curves, when we plot r^2 versus $1/t$ for a number of spheres. From the slope of the straight-line curves, the mean value of r^2t is obtained. Substituting this value in Eq. (6.26) coefficient of viscosity (η) is obtained.

Further, it is to be noted that viscosity of a liquid depends on its temperature [see Section 6.12], and the rate of change is fairly rapid in the case of liquids like oils. Therefore, a sensitive thermometer must also be put into the jar of the liquid to indicate its temperature, when the experiment is performed with the different spheresin quick succession.

Corrections for wall effect : For greater accuracy, a proper correction must be applied for the finite size of the vessel containing the liquid, because the formula for η is derived by Stoke by assuming that the medium is infinite.

Ladenberg has shown that if the liquid column in the tall jar is divided into three equal parts, lengthwise, and the sphere is dropped into the central compartment, let the velocity of sphere be v. If the liquid were unbounded by the walls of the containing jar, the velocity (v_∞) of the sphere is given by the relation $v_\infty = v\,(1 + 2.4\, r/R)$, where, R is the radius of the bore of jar. This is called the correction 'wall effect'.

Correction for end effect : Similarly, Ladenberg has shown that if the liquid medium is not bounded by the bottom of the containing jar, the velocity of the sphere is given by

$$v_\infty = v\,(1 + 3.3\, r/\mu)$$

where, μ is the full depth of the liquid column in the jar. This is called the correction for the 'end effect'.

Combining the two corrections, relation for η becomes:

$$\eta = \frac{2}{9}\frac{r^2 \cdot g\,(\rho - \sigma)}{9\, v\,(1 + 3.3\, r/\mu)\,(1 + 3.3\, r/\mu)} \qquad \text{...(6.27)}$$

6.12 VARIATION OF VISCOSITY OF A LIQUID WITH TEMPERATURE

To determine the viscosity of a liquid one has to measure its rate of flow through a capillary tube as shown in Figs. 6.14 and 6.15.

The experimental setup to see, how the viscosity of a liquid is affected by a change in temperature, is shown in Fig. 6.17.

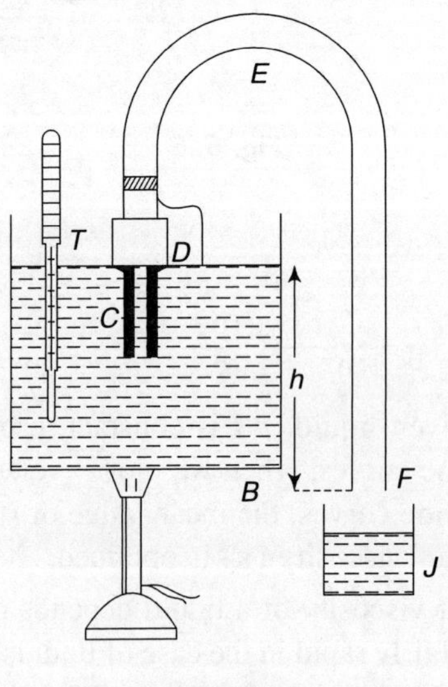

Fig. 6.17

The liquid whose coefficient of viscosity is to be determined is taken in beaker, *B*. The desired temperature is maintained by sprit lamp. A thermometer (*T*) is immersed in the liquid to note its temperature. A capillary tube (*C*) is set vertically, completely immersed in the liquid, and is connected to a tube (*DEF*) of wider bore. The tube is bent as shown in Fig. 6.17 to siphon the liquid into a vessel *J*, whose weight is known.

The method of procedure is the same as in the case of the horizontal capillary tube given in Fig. 6.16, except that the flow of the liquid is through *C DEF* tube. The height of liquid injar is *h*. The liquid will not flow if the end (*D*) of the capillary tube were open to the atmosphere. The pressure difference, P (= $h\rho g$) at *D* and *F* is responsible for the flow of the liquid through capillary tube, C · ρ is the density of the liquid.

As the liquid flows through the tube, the liquid level falls in jar *B*, thus altering the height of liquid, *h*. To maintain height (*h*) constant throughout the experiment in *B*, the liquid flows through capillary (*C*) and *DEF*, so that the height of liquid level in *B* is always maintained at *D*. This is done by attaching a bent pin is to the tube, as shown in Fig. 6.17, to check and ensure that the liquid level in *B* always touches its tip at *D*.

The liquid is thus to flow into vessel (*J*) for a known time, *t*. The mass of the liquid collected in the beaker is determined weighing beaker with collected water. Dividing the mass of liquid collected by its density and the time, *t* (in seconds), the volume blowing out per second, i.e., its rate of flow of volume of liquid (*V*) is determined. Substituting the values of *V*, *P*, *r* (radius of the capillary tube, *C*) and l (length of the tube *C*) in Poiseuille's formula, $\eta = \dfrac{\pi P r^4}{8V}$, one can calculate the value of η for the liquid at the temperature of the bath.

The experiment is repeated with the liquid at different constant temperatures and a graph is plotted between temperature (*T*) and η. It is found that usually η falls with a rise in temperature of the liquid. However, decrease of η with rise in temperature is not universal.

Example 6.9 A flat of area 15 cm^2 is separated from a large plate by a layer of glycerine 1.5 mm thick. If the viscous coefficient of glycerine is 25 gm/cm/sec, what force will be required to keep the plate moving with a velocity of 2 cm/second?

Solution: Formula required

Viscous force, $F = \eta Av/x$

Given: Viscous coefficient, η = 25 gm/cm, area of plate A = 15 cm^2, glycerine layer, x = 1.5 mm = 0.15 cm

Calculation:

$$F = \frac{25 \times 15 \times 2}{0.15} = 25 \times 200 = 5000 \text{ dynes.}$$

Ans. Force required is 5000 dynes.

Example 6.10 Water is escaping from a cistern by way of a horizontal capillary tube, 15 cm long and 0.5 mm in diameter, at a distance of 60 cm below the free surface of water in the cistern. Calculate the rate at which the water is escaping.

Solution: Formula required

$$\text{Coefficient of viscosity, } \eta = \frac{\pi P r^4}{8 V l}$$

Given: Length of capillary tube (l) = 15 cm, diameter of capillary tube, R = 0.5 mm, height of escaping water from surface, h = 60 cm, gravity, g = 980 cm/s^2, density of water, d = 1 gm/cm^2.

Calculation: Pressure of water at escaping point, $P = hdg = 60 \times 1 \times 980$ dynes/cm^2

Radius of capillary tube, r = 0.025 cm

The volume of water flowing per second

$$V = \frac{\pi P r^4}{8 \eta l} = \frac{3.14 \times 60 \times 980 \times (0.025)^4}{8 \eta \times 1.5}$$

$$= 6.01015 \times 10^{-4} / \eta \text{ cm}^3/\text{s}$$

Ans. The rate at which the water is flowing from the capillary tube is $3.0772 \times 10^{-4}/\eta$ cm^3/s.

Example 6.11 Write down Poiseuille's formula for the rate of flow of a fluid through a capillary. From this show that if two capillaries of radii r_1 and r_2 having length l_1 and l_2, respectively are set to series, the rate of flow v is given by

$$V = \frac{\pi P r^4}{8 \eta}\left[\frac{l_1}{r_1^4} + \frac{l_2}{r_2^4}\right]^{-1}$$

where, P is the pressure across the arrangement and η is the coefficient of viscosity of the liquid.

Solution: *Formula required*

$$\text{Coefficient of viscosity, } \eta = \frac{\pi P r^4}{8 V l}$$

Given: Length two capillary tubes are l_1 and l_2, their radii are r_1 and r_2, respectively.

Derivation: Let P' be the average pressure within two capillaries. If P' is the pressure across the first capillary, then the pressure across the second capillary will be $P - P'$.

The rate of flow of the liquid through first capillary will be

$$V_1 = \frac{\pi P' r_1^4}{8 \eta l_1}$$

The rate of flow of liquid through second capillary will be

$$V_2 = \frac{\pi (P - P') r_2^4}{8 \eta l_2}$$

Two capillaries are connected in series, so their rate of flow will be equal, i.e., $V = V_1 = V_2$

or

$$\frac{\pi P' r_1^4}{8 \eta l_1} = \frac{\pi (P - P') r_2^4}{8 \eta l_2}$$

or
$$\frac{P' r_1^4}{l_1} = \frac{P r_2^4}{l_2} - \frac{P' r_2^4}{l_2} \quad \text{or} \quad \frac{P' r_1^4}{l_1} + \frac{P' r_1^4}{l_2} = \frac{P r_2^4}{l_2}$$

or
$$P' = \frac{\dfrac{P r_2^4}{l_2}}{\dfrac{r_1^4}{l_1} + \dfrac{r_2^4}{l_2}} \quad \text{or} \quad P' = \frac{\dfrac{P r_2^4}{l_2}}{\dfrac{1}{l_1 l_2}(l_2 r_1^4 + l_1 r_2^4)}$$

or
$$P' = \frac{P l_1 r_2^4}{(l_2 r_1^4 + l_1 r_2^4)} \quad \text{or} \quad P' = \frac{P l_1}{\left(l_2 \dfrac{r_1^4}{r_2^4} + l_1 \right)}$$

or
$$P' = \frac{P l_1}{r_1^4 \left(\dfrac{l_2}{r_2^4} + \dfrac{l_1}{r_1^4} \right)}$$

Substituting, the value of P',

$$V = \frac{\pi r_1^4}{8 \eta l_1} \times \frac{P l_1}{r_1^4 \left(\dfrac{l_2}{r_2^4} + \dfrac{l_1}{r_1^4} \right)} = \frac{\pi p}{8 \eta} \left(\frac{l_2}{r_2^4} + \frac{l_2}{r_1^4} \right)^{-1}$$

Ans. Rate of flow of liquid, $V = \dfrac{\pi P}{8 \eta} \left(\dfrac{l_2}{r_2^4} + \dfrac{l_1}{r_1^4} \right)^{-1}$

Example 6.12 A gas bubble of diameter 2 cm, rises steadily through a solution of density 1.75 gm/s at the rate of 0.35 cm/sec. Calculate the coefficient of viscosity of the solution. Neglect the density of gas.

Solution: Formula required

Velocity of fluid, $v = \dfrac{2}{9} \dfrac{r^2 \cdot g (d - \sigma)}{\eta}$

Given: Diameter of gas bubble, $R = 3$ cm, density of solution, $d = 1.95$ gm/s, velocity of bubble, $v = 0.40$ cm/s, density of gas bubble, $\sigma = 0$.

Calculation: Coefficient of viscosity, $\eta = \frac{2}{9}\frac{r^2 \cdot gd}{v}$

Radius of gas bubble, $r = 1.5$ cm, $g = 980$ cm/cm^2

$$\eta = \frac{2}{9}\frac{1.5 \times 1.5 \times 980 \times 1.95}{0.40} = \frac{2 \times 2.25 \times 980 \times 1.95}{9 \times 0.40}$$

$$= 2.38875 \times 10^3 \text{ poise}$$

Ans. Coefficient of viscosity of solution is 2.38875×10^3 poise.

Example 6.13 A cylindrical vessel of radius 8 cm is filled with water to a height of 60 cm. It has a capillary tube 15 cm long and 0.25 mm radius protruding horizontally at its bottom. If the viscosity of water is 0.02 poise (poise is a CGS unit, poise – dyne –s/cm^2) CGS units and $g = 980$ cm/sec^2, find the film in which the level will fall to the height at 30 cm.

Solution: Coefficient of viscosity, $\eta = \frac{\pi P r^4}{8 V l}$

where, pressure, $P = hdg = hg$ as density of water, $d = 1$ gm/cm^2.

Given: Radius of cylindrical vessel, $R = 8$ cm, height of water, $h_1 = 60$ cm, length of capillary, $l = 15$ cm, radius of capillary, $r = 0.25$ cm, $g = 980$ cm/sec^2, height of fall, $h_2 =$ 30 cm, coefficient of water viscosity, $\eta = 0.02$ poise. Poise is a CGS unit, 1 poise $\equiv$ 1 dyne – s/cm^2.

Let $A = \pi R^2$ is area of cross section and height of fall is dh in time dt, then rate of flow through capillary tube is

$$-A\frac{dh}{dt} = \frac{\pi P r^4}{8 \eta l} \quad \text{or} \quad -A\frac{dh}{dt} = \frac{\pi h g r^4}{8 \eta l}$$

The negative sign shows that as time increases, the height decreases. Rewriting the above equation, we get

$$dt = -\frac{8\eta l A}{\pi g r^4}\frac{dh}{h}$$

Integrating this equation, one gets

$$\int_0^t dt = -\frac{8 \eta l A}{\pi g r^4}\int_{h_1}^{h_2}\frac{dh}{h} \quad \text{or} \quad t = \frac{8 \eta l A}{\pi g r^4}\log_e \frac{h_1}{h_2}$$

Putting $A = \pi R^2$, $t = \frac{8 \eta l R^2}{g r^4}\log_e \frac{h_1}{h_2}$

Putting the values of various quantities

$$t = \frac{8 \times 0.02 \times 1.5 \times 8 \times 8}{980 \times (0.25)^4} \times 2.3026 \log_{10} \frac{60}{30}$$

$$= 40.12408 \times 0.69315 = 27.8120 \text{ sec}$$

Ans. The water falls by 30 cm in 27.812 seconds.

6.13 PRODUCTION AND MEASUREMENT OF LOW PRESSURE (VACUUM PUMPS)

First time in 1744 Toricelli's famous experiment in connection with the mercury barometry showed the possibility of empty space. Later the experiment was repeated by Pascal using red wine instead of mercury. It motivated the then scientific community the world over, to take up the challenge and devise ways and means to create a perfect vacuum. In 1750, Otto von Guericke, in Germany, and in 1760, Robert Boyle, in England, succeeded in producing their exhaust pumps. Though, these pumps were crude devices, but they broke the myth of impossibility of creating vacuum. However, invention of the barometer and the exhaust pump were hailed with the same importance as it was given to the invention of the telescope by Galileo in the same century.

From then on, continuous efforts have been made to improve upon those earlier devices by developing newer and better techniques. Today pumps have made which can produce a pressure as low as 10^{-9} mm of the mercury, a near perfect vacuum.

With the development of high vacuum pumps, the development of gauges to measure the low pressures was done simultaneously. We will now study these devices in the following sections.

6.14 DIFFERENT TYPE OF PUMPS

The following table shows the classification of the different types of exhaust or vacuum pumps.

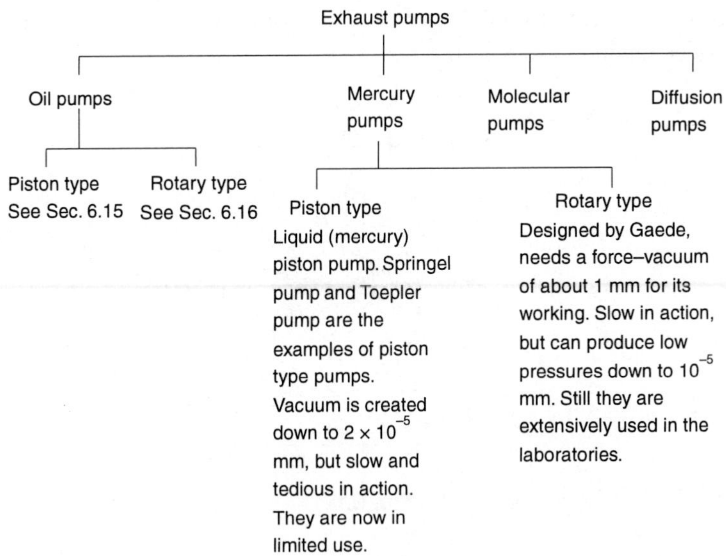

6.15 COMMON AIR PUMP

An *air pump* is a pump for pushing air. For example, a bicycle pump, pumps that are used to aerate an aquarium; a gas compressor used to drive air horn or pipe organ; a pipe used to encourage a fire; a vacuum cleaner and a vacuum pump.

We will study in detail some of the simple and important pumps. For example, rotary pump, diffusion pump, and molecular pump.

6.16 ROTARY OIL PUMPS

Rotary pumps initially were devised by Gaede. These pumps are of two types: (i) the rotary vane oil pump, and (ii) the stationary vane oil pump. The principle of working of both is the same. Both have a massive cylindrical shaft or 'rotor' revolving eccentrically inside a hollow strong steel cylinder, or , 'stator' , which compresses the gas or vapour entering it, and ejecting it finally through a 'non-return' outlet valve. Whole pump is kept immersed in oil. The oil serves three purposes: (i) it provides automatic lubrication, (ii) it prevents the leakage of gas or vapour into the high vacuum created, and (iii) it does the efficient cooling of the pump. We will study both type of pumps in a little more detail.

6.16.1 Rotary Vane Oil Pump, or the Gaede Rotary Oil Pump

Rotary pumps was initially made by Gaede so it is also called, so Gaede rotary oil pump. A rotary vane vacuum pump is an oilsealed rotary displacement pump. The diagram of rotary oil pump is shown Fig. 6.18. The pumping system consists of a (1) container, (2) hollow, cylindrical steel chamber called stator (*S*), (3) an eccentrically installed strong cylindrical shaft called rotor, *R*, (4) vanes (discs) that move radially and keep pressed against the wall of the stator due to spring force, and (5) the inlet (*I*) and outlet (*O*). The outlet valve (*V*) is oilsealed. The inlet valve is designed as a vacuum safety valve that is always open during operation. The stator is located inside the container. Rotor and vanes divide the working chamber into two separate spaces having variable volumes.

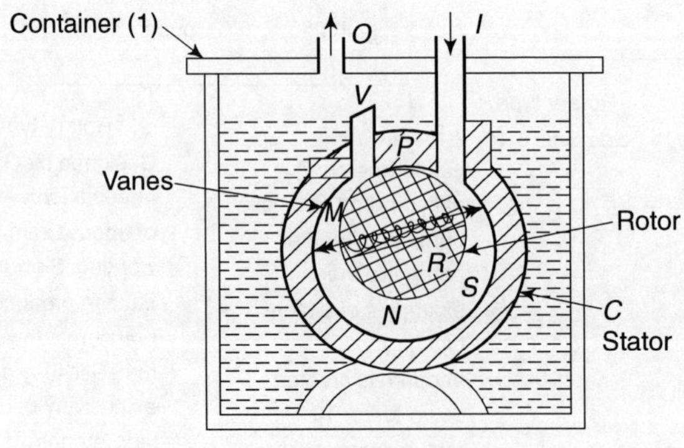

Fig. 6.18

The rotor moves with the help of a motor attached to it. It rotates eccentrically (varying centre) about its own axis by means of an electric motor, parallel to the axis of the chamber, so that it is always in contact with the chamber at a fixed peripheral point, such as *P*.

Two vanes (or discs), V_1 and V_2 are fixed right across the diameter of the rotor in such a way so that they divide the space between the stator and the rotor into two separate compartments (or slots), *M* and *N*. These discs are kept apart from each other and pressed against the walls of the stator by means of one or more springs in between them. The compartments along with discs inside it also rotate with the rotor, hence the pump is called rotary vane oil pump. The whole pump is kept immersed in oil for the reasons mentioned above.

As the rotor rotates in the anticlockwise direction, the space between the rotor and the stator, on the inlet side, goes on increasing, while the space between the rotor and the outlet side of the stator goes on decreasing, so that the gas or vapour from the vessel connected to *I* is continually drawn into the inlet side, and that in the outlet side of rotor gets slowly compressed, until when its pressure becomes significantly high, which forces valve (*V*) to open and the gas or vapours escape out of the outlet, *O*. This process goes on repeating until a pressure as low as 10^{-3} mm is produced in the vessel connected to the pump. Even when the pump stops working, a special self-sealing oil valve is attached to the exhausted vessel to prevent the gas or vapour from being sucked back into it. In the beginning no vacuum is required for the working of this pump, therefore, it can be used directly from the atmospheric pressure.

6.16.2 Stationary Vane or the Cenco-Hyvac Rotary Oil Pump

This oil pump is preferred because of its better performance. It consists of a strong outer cylinder (*C*), inside which cylinder, *R*, called the rotor, is mounted eccentrically as shown in Fig. 6.19. A partition, called the vane (disc) (*V*) is kept pressed against the rotor with the help of a spring, *S*. This disc remains fixed or stationary in the position to the inside cylinder, hence this oil pump is called stationary vaneoil pump. Two tubes, *O* and *I*, are provided to the outer cylinder. An inlet tube is connected to the vessel to be exhausted, and an outlet

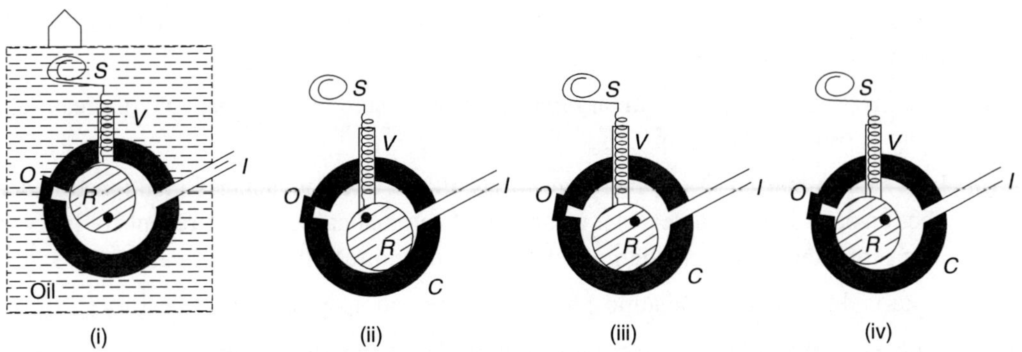

Fig. 6.19

tube (O) is provided with an opening valve outwards. The whole pump is immerged in oil to prevent any leakage as shown in Fig. 6.19 (i). A special type of valve is fixed to prevent the oil from getting into the vessel being exhausted, when the pump is stopped. The rotor is driven at a very high speed by means of a separate electric motor in the direction shown by the arrowheads.

Figure (i) shows starting condition. When the inlet tube is connected to the vessel to be exhausted. The gas or air from the exhausted vessel is pushed into the space in between the cylinder and the rotor, R.

Figure (ii) shows the condition when the rotor has started rotating eccentrically, the gas or air is compressed. The fresh gas or air comes into the cylinder through the inlet tube, I, behind the rotor. The fresh air is kept apart from that already present by the disc V, as explained above.

Figure (iii) shows a step taken further for the process of compression.

Figure (iv) shows the final stage of compression, when due to increased pressure of the gas in C the valve at the mouth of the outlet tube O is forced open, and the gas is driven out.

The gas or air behind the rotor is similarly compressed and forced out in the next cycle. The process is repeated until a high vacuum is produced in the vessel connected to it

In practice, the pumping system consists of two such units, mounted side by side in series with each other, and run by the same motor. The first pump works directly from the atmosphere, as explained above, and the second pump works from the forced vacuum created by the first. The maximum speed of the working exhausts is about 6 litres per minute, and the vacuum created is about 10^{-3} mm. This pump is suited for exhausting vessel of the large capacities.

If a vessel is connected to the outlet tube, O, the gas or air will be compressed into it, and, therefore, this pump can also be used as a compression pump.

Advantages of the pump

(i) It is run by an electric motor, so that the whole operation is quick and automatic, with no manual labour involved.

(ii) It is much faster than an ordinary exhaust pump, therefore, it is capable of creating a vacuum of about 10^{-3} mm.

(iii) It is kept immersed in oil, therefore it has all the three advantages mentioned under Section 6.15. For example: (a) it is lubricated automatically, (b) oil prevents of leakage of gas or vapour into the high vacuum created, and (c) oil cooling is efficient.

(iv) It starts directly from the atmospheric pressure and no vacuum is necessary.

(v) It occupies much less space compared with the other exhaust pumps.

(vi) It can also be used as a compression pump.

6.17 MOLECULAR PUMP

The earliest molecular drag pump was designed by Wolfgang Gaede in1905. The first prototype device was made in 1910, which can achieve a pressure of less than 10^{-6} mbar. A molecular pump is a mechanical kinetic vacuum pump that operates in the area of molecular flow, meaning that molecules are relatively free, and the molecules do not interact with each other, which is acquired at reduced pressure of order 0.1 mbar to 0.01 mbar. The gas particles/molecules when collide with the surface of the rotors moving at high speed receives momentum. The surface of the stator is designed in such a way that it promotes the movement of gas particles/molecules in the direction of forceline (outlet of pump).

Working Principle

A molecular drag pump is a type of vacuum pump that utilizes the drag of air molecules against a rotating surface. The molecular pump is made of spinning cylinder called rotor. The gas in the chamber enters from inlet to one side of a rapidly spinning cylinder. Collisions between the gas and the spinning cylinder give momentum to the molecules of gas in the same direction because the surface of the cylinder is designed to get the gas from the vacuum chamber and to push it towards a forceline (output of pump).

The oil pumps start working straightway from the atmospheric pressure to achieve the vacuum of 10^{-1} to 10^{-2} mm. On the other hand, the molecular pumps (and also the diffusion pumps) cannot work at atmospheric pressure. They operate only at a reduced pressure or fore/vacuum of the vessel to be exhausted. The reason is that the force exerted by the rotor to a molecule is not significant compared to the force of the collision between the molecules, and so the flow of gas remains unaffected by viscous force. A separate backing pump, for example, a rotary pump, at forceline (output) is used to lower the pressure to about 10^{-2} torr. At this pressure, the flow is called "molecular flow", meaning that molecules are relatively free, and the molecules do not interact with each other.

At this enough low pressure the flow of gas is called "molecular flow", meaning that molecules are relatively free and, to a good approximation, do not interact with each other. In other words, at low pressure, the mean free path (λ) of the molecules of the gas or vapour is greater than the gap between the rotor and the stator, so there are too few collisions among molecules as compared with the collisions between molecules and the walls of this gap. So at low pressure, the molecules acquire the 'drift velocity' of the rotating surface they strike upon and rebound from, and this velocity is not altered by any subsequent collisions among themselves. In other words at this pressure, the viscous force imparts momentum to the individual gas molecule. One should note that the dragging force due to viscosity of gas in the direction of motion of rotor is exerted when the space between a rapidly rotating surface (called the rotor) adjacent to a stationary surface (called the stator) is as small as 0.03 mm.

This "Gaede molecular pump" was used in an early experiment testing vacuum gauges.

Figure 6.20 shows a block diagram of the pump, where A represents the rotor revolving around eccentrically and closely adjacent to the walls of the stator, C. There are an inlet and an outlet in the stator at P and Q, respectively. A manometer (M) is also attached at the inlet and outlet points to measure the pressure difference at these points.

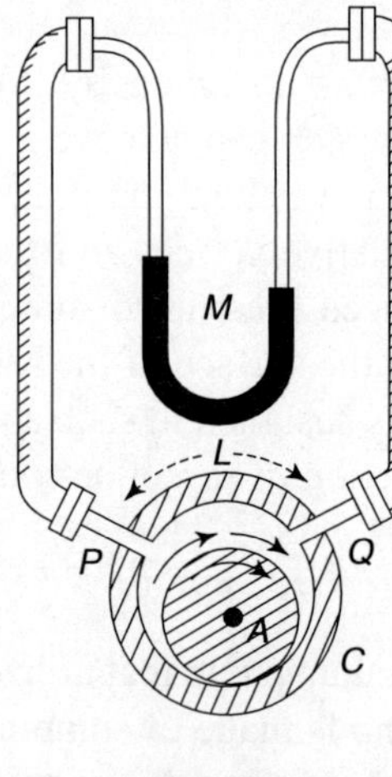

Fig. 6.20

6.18 TURBO-MOLECULAR PUMP (TMP)

A turbo-molecular pump (TMP) or turbo pump can achieve the vacuum pressure of the chamber in the range from about 10^{-2} torr to about 10^{-8} torr. A TMP consists of a stack of rotors with blades, or slots. In between the rotor disks there are stators or fixed discs. The number of blades or slots of each disk and the rotor are equal, but are oriented in the opposite direction. The motor is used to rotate the rotor about the axis. O is outlet for exhausting the gas to the atmosphere. Through inlet there is flow of gas molecules from the attached chamber.

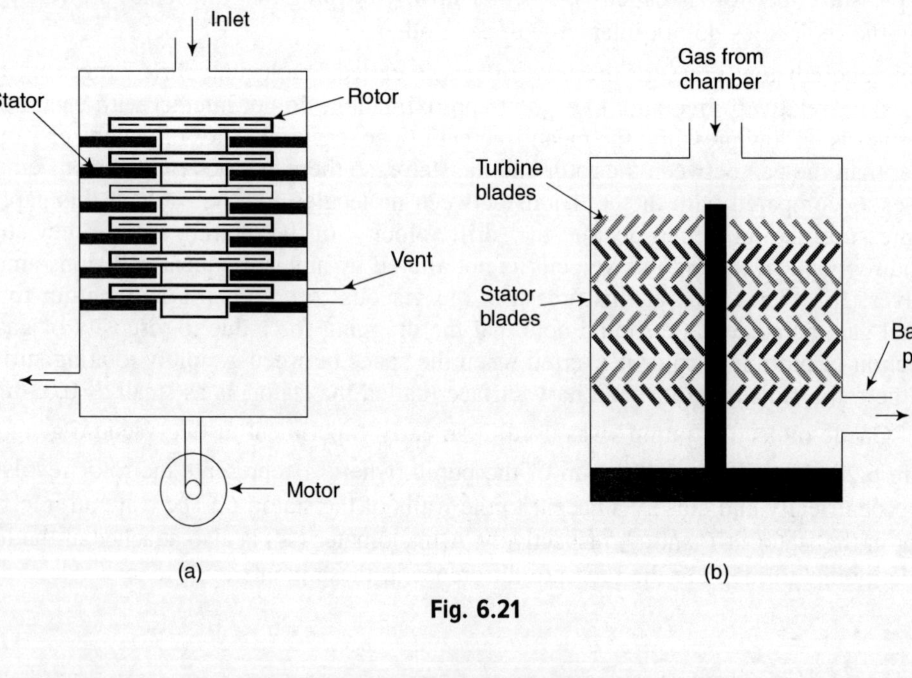

Fig. 6.21

When the blades, spinning at 20,000 rpm (about 2100 rad/s), hit the molecules, they get momentum. The angle of orientation of the blades of the rotor and the stator is such that the molecules move down to the next rotor when they are hit. The successive rotor-stator pairs drive the molecules towards the exhaust, where they are collected by a backing pump.

When the chamber is at atmospheric pressure, the flow of gas is described as "viscous flow", where the molecules move in bulk and interact with each other much more significantly than they do with the walls of the container. Most turbo pumps cannot work at atmospheric pressure. The reason is that the force exerted by a blade to a molecule is not significant compared to the force of the collision between the molecules, and so the flow of gas remains unaffected when blades hit its molecules. For this reason, a rotary pump has to remove the gas molecules before the turbo pump is turned on, down to pressure of about 10^{-2} torr. At this pressure, the flow is called "molecular flow", meaning that molecules are relatively free, and the molecules do not interact with each other. At this pressure, the turbo pump imparts momentum to the individual gas molecule, so they properly go downwards. Once turned on, the turbo pump will work for quite a long time reducing the chamber pressure, until at some point the pressure will not go down significantly any longer. This happens between 10^{-6} and 10^{-8} torr of pressure.

6.19 DIFFUSION PUMP

The principle of diffusion of one gas into another has been used to create rapid vacuum in vessels. The method of diffusion was first used by Gaede in the year 1915. In a mixture of gases, the diffusion of a gas takes place from a region, where concentration is greater to the one where its concentration is smaller. On the basis of pressure, the diffusion of a gas takes place from a region where its partial pressure is higher, to the one, where it is lower. The diffusion of a gas depends on the pressure of a gas rather than on the total pressure in the two regions. The working diagram of the pump is shown in Fig. 6.22.

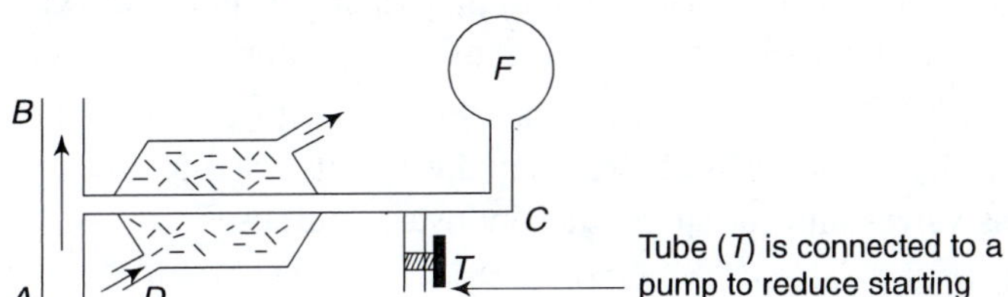

Fig. 6.22

The bulb (F) to be exhausted from air or gas is connected to a wider vertical tube (AB) by means of a tube, C. The tube (C) is kept at low temperature, by circulating cold water round it. A regular stream of air-free liquid vapour is maintained through tube (AB) in the direction shown in the figure.

From the diagram, it is clear that the concentration or the pressure of air is greater at *C* than at the other end connected to tube, *AB*. There is little or no air at this end because the liquid vapour is air-free, therefore, diffusion of air takes place from *C* into *AB*, where it is swept out by the stream of the liquid vapour. As a result the concentration and partial pressure of air are greater at *C* than at the end opposite and hence more air diffuses from *C* into *AB*. This process goes on until the pressure of air at *C* is greater than its partial pressure in *AB*.

The liquid vapour also tends to diffuse from *AB*, where its concentration and, therefore, its pressure is greater, towards *C*, where its concentration and pressure are lower, thus preventing the air diffusion from *C* into *AB* backwards. The diffusion of liquid vapour from tube *AB* to *C* is prevented by (i) making the aperture of *C*, opening into *AB*, narrow, so that its dimensions becomes smaller than the mean free path of the molecules of the gas or air in *AB*, then there will be much fewer vapour/gas or vapour/air molecular collisions and the diffusion velocity of mercury vapour towards *C* is smaller than the diffusion velocity of air from *C* into *AB*, and (ii) condensing the mercury vapour by cooling the tube (*C*), thus not only preventing it from proceeding further and entering the vessel, *F*, but also reducing the residual vapour pressure there less than 10^{-3} mm, thereby ensuring an uninterrupted diffusion of the gas from *C* into *AB*. The condensed mercury is then conveyed back to the boiler (as shown in Fig. 6.23) for using it again.

This pump can be used successfully only when the pressure in the vessel (*F*), is about or less than 1 cm of mercury. This can be achieved by connecting rotary vacuum pump to the vessel through the tube, *T*. Maximum effect can be obtained when the pressure of the mercury vapour is just above the starting vacuum of vessel, *F*.

Such pumps may be used in series, so that each pump, in turn, carries the evacuation a step further than the last. In this way a pressure as low as millionth of a centimetre of mercury column can be rapidly attained. Theoretically speaking, there is no limit to the vacuum produced by them. Under the ideal conditions, mentioned above, there will be a continuous diffusion of air or gas from *C* into *AB*, until the air or gas pressure is reduced to zero.

Since diffusion as well as condensation both play their part in the working of these pumps, diffusion pumps are also called diffusion condensation pumps, or even condensation pumps.

The vacuum order 10^{-6} mm or less can be obtained with the help of these pumps, with a starting vacuum of 10^{-1} mm. Therefore, they are mainly used for the evacuation of X-ray tubes and wireless valves and for such other industrial purposes.

Drawbacks of diffusion pump

(i) Their exhaust speedis comparatively slow.

(ii) A regular vigilance of the vapour/temperature is required to operate them.

(iii) Diffusion pumps cannot discharge directly into the atmosphere, so a mechanical forepump is typically used to maintain an outlet pressure around 0.1 mbar.

6.20 LANGMUIR'S IMPROVED VERSION OF DIFFUSION PUMP

In the year 1916, Langmuir removed these drawbacks and improved the working of the pump greatly by (a) replacing the narrow slit of the diffusion pump by a wider slit, which increases the speed of the pump greatly, (b) making arrangement so that the stream of vapour sweeps out the air molecules more efficiently, and (c) making the speed of the pump almost independent of the liquid vapour pressure, which can be done by increasing the speed of the vapour stream.It removes the necessity of a vigilant and tiresome regulation of the vapour/ temperature.

In the year 1923, Waran designed a simple and more efficient form of diffusion pump as shown in Fig. 6.23. It consists of a conical glass vessel (*A*) containing mercury. When mercury is heated then its vapours move through *B* tube having a nozzle or a 'parallel jet' of about 8 mm in diameter at the end. The heated vapours pass through the jet of tube *B* into a wider tube, *C*, which is surrounded with a cold water jacket. The tube (*C*) is connected to the vessel to be exhausted through an inlet tube, *I*. The tube *B* is covered with asbestos to prevent any condensation of the mercury vapour in it.

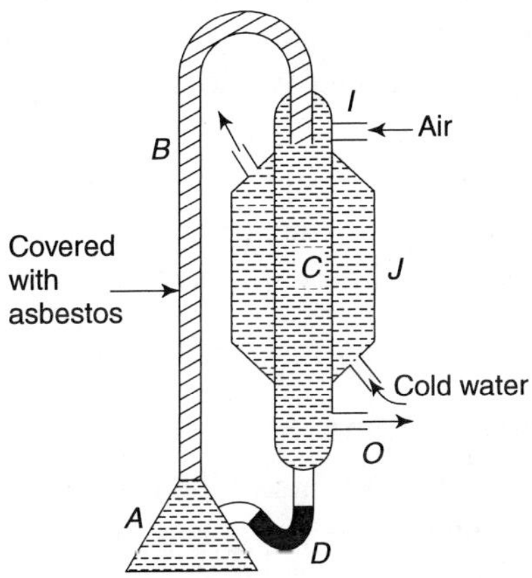

Fig. 6.23

As the mercury vapour sprays out of the nozzle, it carries along with it the air diffusing through (*I*) from the vessel to be exhausted. The diffusion of the mercury vapour into the vessel is prevented because it is almost condensed on entering *C* immediately. The air is discharged through the outlet tube (*O*) and the condensed mercury vapours are collected in the bend (*D*) below *C* and returned back to *A*. The process goes on, until a high vacuum, of the order of about 10^{-3} mm of the mercury is produced.

Drawbacks of the diffusion pump: If vapours of mercury once get into the vacuum system, then mercury becomes a source of trouble as the working substance in a diffusion pump.

6.21 OIL DIFFUSION VACUUM PUMPS

Diffusion pump operates quite differently from a gas displacement pump, such as a mechanical pump. A mechanical pump scoops up air and pushes it out of the chamber by mechanical means such as pistons, fins, or lobes. The oil diffusion pump works in a completely different manner. The diffusion pump uses the vapour of a boiling fluid to capture air molecules. The fluid is then moved to another location and cooled. The cooling forces the air molecules to be released. The combination of gravity and the downward direction of the vapours move the air molecules toward the bottom of the pump.

In the year 1928, Burch suggested that the some high-boiling point oils can be used in place of mercury, because, they have much lower vapour pressure, so no liquid-air trap is possible (unless the operation happens to be a prolonged one) and the maximum speed of the pump can thus be easily utilized. High-boiling point oils are obtained during distillation of petroleum.An oil diffusion pump has invariably a much higher speed than a mercury diffusion pump. One serious drawback of oil diffusion pump is that, unlike mercury, an organic oil dissolves gases, so it requires periodical cleaning.

The most common silicone oils used in oil diffusion pumps are trisiloxanes (tri silicon oxides), which contain the chemical group Si-O-Si-O-Si, to which various phenyl groups or methyl groups are attached. These are made up of the isomers of tetraphenyl tetramethyl trisiloxane and pentamethyl trimethyl trisiloxane, which were formerly manufactured by Dow Corning. These hydrocarbons are inflammable and most likely burn or explode if exposed to oxidizers.

By doing suitable modification in the design of mercury diffusion pumps, the oil diffusion pumps can, with a starting vacuum of 0.1 mm, produce as high a vacuum as 10^{-6} mm. Other liquids like Apiezon oils can be used in place of mercury, mostly in diffusion pumps with an all-metal construction.

6.21.1 Working of Oil Diffusion Pump

At the bottom just below the oil there is a heater. When the oil becomes hot enough to vaporize, the oil vapours rise up at the center of the pump and exit the nozzles at a downward angle. These nozzles are in a ring form and make a curtain or skirt of vapour that extends from the nozzles to the pump wall. The air molecules, which come from the inlet, are captured by the vapours. When the oil vapour hits the water cooled walls of the pump, the oil cools as it runs down the sides of the pump. By the time the oil reaches the reservoir once again, it has given off its trapped gas and is ready to begin the cycle a new. Any gas molecule that tries to go upward is caught by the vapour skirt above it and forced downward again. By continually forcing the air molecules downward, higher in pressure is created in the area near the bottom of the pump than the top of the pump. In other words, when the pump is in operation the pressure is higher below each oil skirt than it is above that skirt. At the bottom of the pump, the pressure is high enough for the gas to be pumped out by a standard mechanical pump. Diffusion pumps cannot discharge directly into the atmosphere, so a mechanical pump is used to maintain an outlet pressure around 0.1 mbar.

Although oil vapour is directed downwards, it is possible for some of it to go to the top of the pump. To avoid migration of oil back into the chamber, a cold cap is fitted at the top of the nozzle assembly to condense vapour in that area. Nowadays, silicone oil is used frequently as the diffusion pump fluid because of its high vapour point and stability. Though it is expensive, but silicone oil performs well for long periods of time with little or no degradation. The oil is also relatively safer because it is not toxic and is inflammable. The only caution when using silicone oil is that it must not be exposed to air when it is heated because it gets converted into gel form.

Unlike turbo molecular pumps, diffusion pumps have no moving parts, so they are quite durable and reliable. They can create vacuum pressure upto 10^{-10} mbar. They are driven only by convection and thus have a very low energy efficiency.

One major disadvantage of diffusion pumps is the tendency to diffuse oil vapours into the vacuum chamber. These oil vapours can contaminate surfaces inside the chamber. Due to diffusion of oil vapours in the vacuum chamber, oil diffusion pumps are not suitable for use with highly sensitive analytical equipment or other applications which require an extremely clean vacuum.

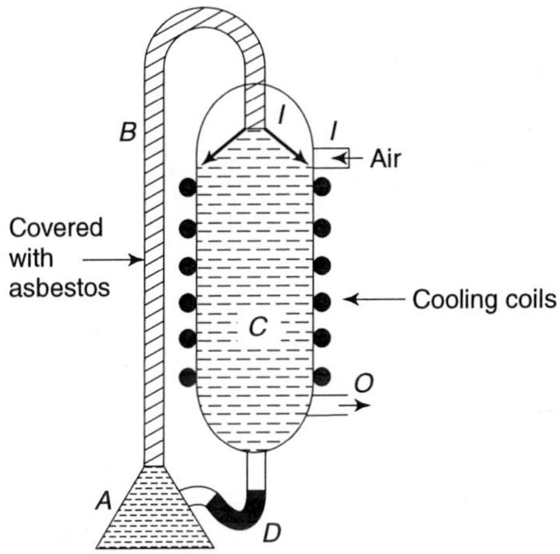

Fig. 6.24

6.22 OTHER METHODS OF PRODUCING VACUA

With the help of a diffusion pump or rotary oil pump the lowest pressure attainable is 10^{-3} mm. Following methods are used to produce still lower pressures:

1. Sorption process
2. Chemical process
3. Thermal process
4. Electrical process

6.23 PENNING AND PIRANI GAUGE

At high pressures, the thermal conductivity (K) of a gas is quite independent of pressure. At pressures below 10^{-2} mm of mercury column, when the mean free path of gas molecules is of the same order of the magnitude as the diameter of the containing vessel, it is directly proportional to the pressure (p), i.e.. K is a linear function of p, or $K = \alpha \cdot p$, where, α is constant.

This fact first used by Warburg in 1907, for the measurement of low pressure and is the basis of the Pirani Gauge (P.G.) as shown in Fig. 6.25(a). It consists of a tungsten or platinum filament, F, and is enclosed in a small detachable glass bulb, B. It is maintained at a temperature higher than the surroundings. The bulb is open at the lower end. It can be connected to the vessel in which the pressure is to be determined.

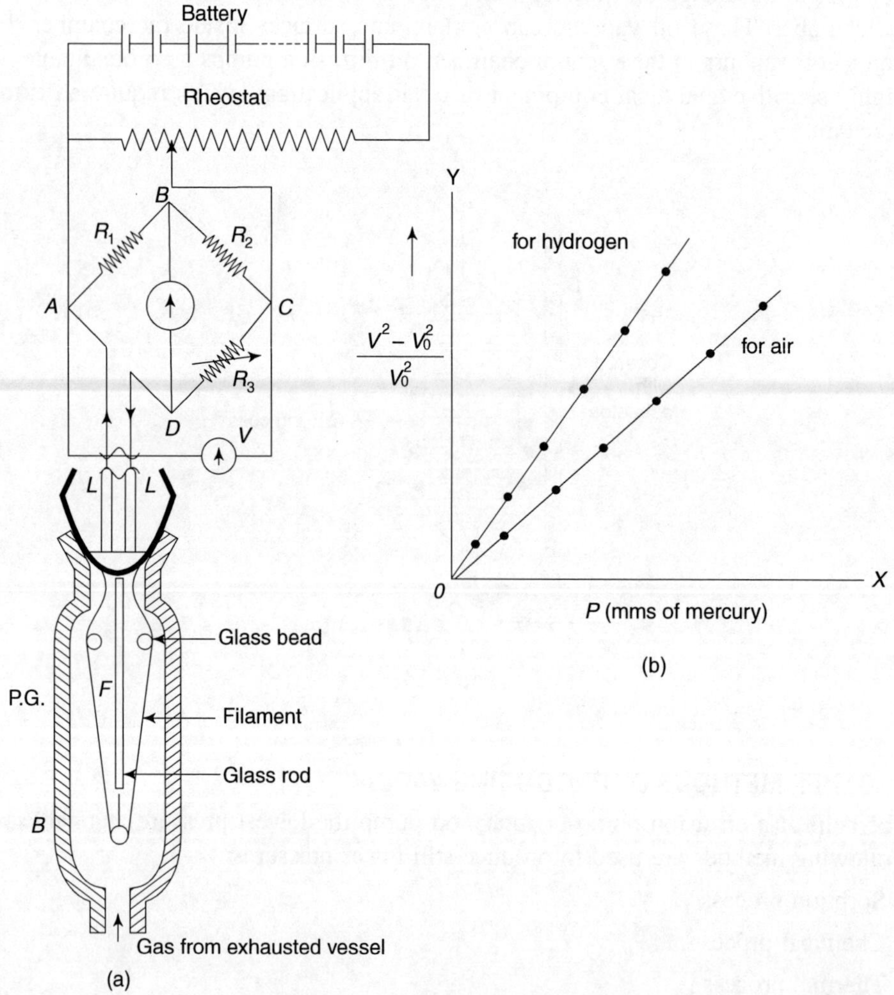

Fig. 6.25

The rate of heat conduction across the gas changes with change in the pressure of the gas in between the filament and the walls of the bulb. As a result, there is a change in the temperature of the filament and hence change in its resistance. Since, thermal conductivity is proportional to the pressure, so by measuring the change in the resistance of the filament, the change in the thermal conductivity and hence, the pressure of the gas can be obtained indirectly. For measuring the change in the resistance, a constant potential difference is applied to one arm of Wheatstone's bridge, so that it can heat the filament to a temperature of about 120°C (but never to red heat). A curve for the gauge is plotted by measuring simultaneously the resistance of the filament from Whetstone's bridge and the pressure of the gas around it by McLeod gauge. The pressure, corresponding to any value of the resistance of the filament, can then be read directly from this curve.

The resistance versus pressure curve is a straight line, so long as the pressure of the gas is below 10^{-2} mm. But at higher pressure, the relation between K and p no longer remains linear and K varies in a complicated manner with both the pressure of the gas and the temperature of the surroundings. In fact, the curve between K and p can not be straight line, rather it is a complicated curve. To remove this difficulty, Campwell suggested that instead of keeping the voltage across the bridge constant and measuring the resistance of the filament, the temperature, and hence the resistance of the filament must be kept constant and the potential difference, applied to the bridge, must be measured.

Accordingly, the Pirani gauge (P.G.) is connected in one arm of the bridge as shown in Fig. 6.25(a), together with fixed resistances (R_1 and R_2) and the variable resistance (R_3) in other third arm. All these resistances are made of an alloy like 'manganin' and 'minalpha', having an almost zero coefficient of temperature. The potential difference is applied to the bridge at A and C, by means of a potential divider, i.e., through a rheostat included in the battery circuit and its value is read with the voltmeter (V) connected across the bridge terminals A and C as shown. To ensure constancy of the temperature of the surroundings of the filament, the bulb of the gauge is placed in a thermostat at 0°C.

Procedure

(i) With a known potential difference across A and C, the bridge is balanced by adjusting the variable resistance, R_3, so that the deflection in the galvanometer is zero and there is sufficient current through the filament of the gauge to raise its temperature to about 100°C. Now, every time the pressure changes, the voltage across A and C is adjusted to restore the balance of the bridge and make the galvanometer deflection zero. If θ is the excess temperature of the filament over that of the surroundings, the heat loss along the leads L and L, assumed small, will be proportional to θ, say equal to $\beta\theta$, where, β is a constant. and, if V is the voltage applied across the bridge, the heat dissipated per second in the filament is equal to αV^2, where, α is another constant. If p is pressure of the gas around the filament, the heat lost per second by conduction across it is $f(p)$, where, $f(p)$ is some function of pressure. We, therefore, have

$$\alpha V^2 = \beta\theta + f(p) \quad \text{...(6.28)}$$

For the same value of θ, but with $p = 0$, [i.e., with vacuum around the filament], if the voltage applied to the bridge is V_0, then

$$\alpha V_0^2 = \beta\theta \quad \text{...(6.29)}$$

Dividing Eq. (6.28) by Eq. (6.29), we get

$$\frac{V^2}{V_0^2} = 1 + \frac{f(p)}{\beta\theta} \quad \text{or} \quad \frac{V^2 - V_0^2}{V_0^2} = k\,f(p) \qquad ...(6.30)$$

where, $\frac{1}{\beta\theta} = k$, a constant, which is almost quite independent of the material and the length of the filament and varies only with the nature of the gas.

We have seen that the pressure of the gas measured by means of a McLeod (or any other absolute) gauge shows that $f(p) = \gamma p$, where, γ is also constant. So that

$$\frac{V^2 - V_0^2}{V_0^2} = k\gamma p \qquad ...(6.31)$$

From Eq. (6.31) it clear that $\frac{V^2 - V_0^2}{V_0^2}$ is directly proportional to the pressure. Therefore, if we plot a graph between $\frac{V^2 - V_0^2}{V_0^2}$ and p for the gas, we get a straight line. The slope of the straight line with the axis of p depends on the nature of the gas in question, as is clear from the two curves drawn by Campbell for hydrogen and air [Fig. 6.25(b)].

(ii) In actual practice, it is difficult to use Campbell's method. Therefore, the following simplified procedure is adopted.

The bridge is first balanced with only vacuum about the filament and then, keeping the voltage across the bridge constant at this value, the gas or air is allowed into the gauge. The balance of the bridge is thus naturally disturbed. The current passes through the galvanometer, at this point is out of the balance, the deflection of galvanometer scale divisions is noted. Corresponding value of pressure of the gas is measured by a McLeod gauge. By allowing more gas or air into the gauge in steps, deflection of galvanometer scale divisions and corresponding pressure are noted.

If N is the number of scale divisions of the deflected galvanometer needle for pressure (p) of the gas surrounding the filament, then we can write $N = f(p)$. The graph for $f(p)$ versus p will be a straight line and gives the required calibration curve for the gauge, from which the pressure of the gas for any deflection in the galvanometer can be noted.

Essential points of the gauge

For the success of the gauge for measurement of a low pressure, the essential points are that (i) the material of the filament must have a high coefficient of temperature, so that the change in its resistance must be appreciable for a small change in its temperature. Thus, it should be made of a tungsten or platinum wire, of a diameter of about 0.06 mm; (ii) the filament must throughout be tightly stretched, so that the distance between it and the walls of the enclosing bulb remains unaltered; (iii) the heat losses through support of the filament must be as small as possible; and (iii) the galvanometer should have a high current sensitivity.

To ensure both the conditions, (i) and (ii), a poor conductor of heat like a glass rod is used by way of support for the filament and it is taken around glass rod, as shown in Fig. 6.25, with its longer positions equidistant from the walls of the bulb on either side.

Range, Utility and Drawback of the Gauge

The range of pressure measurement is small, for example, between 10^{-2} to 10^{-4} mm of mercury. However, it gives almost instantaneous reading, therefore, it is extremely useful for the measurement of pressure fluctuations.

Its main drawbacks are that (i) it is unsuitable for the use with organic vapours, as its filament gets 'poisoned by them'; (ii) it is too sensitive to sudden or accidental thermal or mechanical shocks and vibrations, which must, therefore, be avoided as far as possible. For safeguard against the mechanical shock, a shock absorbing mounting is provided for it; (iii) it is not an absolute gauge and has to be calibrated against a McLeod or some other absolute gauge; (iv) at pressures below 10^{-3} mm of mercury column, the heat loss occurs more by radiation than conduction; and (v) pressures below 10^{-3} mm of mercury cannot be measured with it by any reasonable amount of certainty.

Finally, it is mentioned that, like most other gauges in the pressure range 10^{-3} to 10^{-5} mm of mercury, this too requires some manual adjustments, which are not as reliable as mechanical or automatic ones. This problem was resolved in 1939 by Scott by introducing a triode valve in the gauge circuit, making the working of the gauge quick, smooth and automatic.

6.24 KNUDSEN ABSOLUTE GAUGE

Knudsen gauge gives the pressure directly in dynes . cm^2, and there is no need to calibrate against a McLeod gauge, etc., so it is an absolute gauge. The gauge works properly at low pressure when the mean free path of the gas molecules is longer as compared to the dimensions of the gauge, i.e., when the molecules are free to move. The gauge ceases to function in a proper way, when the pressure has become so high and the mean free path becomes so short that convection currents are setup causing erratic behaviour of the vane (plate). Knundsen absolute gauge enables us to measure pressure in an evacuated vessel, from 10^{-3} mm down to 10^{-7} mm, by simply noting the deflection of a cold plate suspended in the evacuated vessel.

Working Principle

The Knudsen gauge consists of alight suspended vane (plate) in a vacuum chamber whose pressure is to be measured. One side of the vane is exposed to a source of heat at temperature (T_1) while the other side is exposed merely to the walls of the chamber at temperature, T_0. The "hot" gas molecules, leaving the heater, transport more momentum to one side of the vane than the cold molecules from the walls of the chamber transport to the other side of the vane. As a result the vane is deflected because it experiences a net displacing force per unit area given by

$$F = \frac{p}{2} \cdot \left(\left[\sqrt{\frac{T_1}{T_0}} - 1 \right] \right)$$

Here, p is the gas pressure to be measured in the vacuum chamber, T_1 is the temperature of the heater and T_0 the temperature of the walls. It is to be noted that the formula is independent of the molecular weight of the gas.

Figure 6.26 shows diagram of Knudsen absolute gauge. P_1 and P_2 are two fixed plates, kept electrically heated and suspended (in the evacuated vessel) by means of a quartz fibre, carrying a mirror (M) to measure its deflection by the usual 'lamp and scale' arrangement.

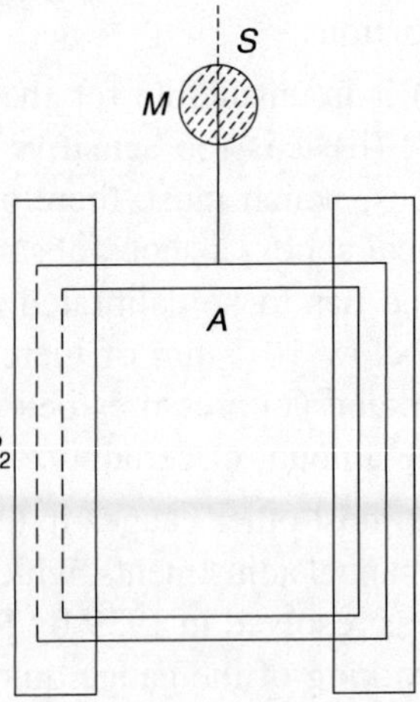

Fig. 6.26

Let the temperature of P_1 and P_2 be T_1 and that wall A of the evacuated vessel, in which it is suspendedis T_0. Let n_1 and n_0 be the number of molecules per cubic centimetre travelling from P_1 to A and from A to P_1, with root mean square velocities, $\bar{c}_1$ and $\bar{c}_0$, respectively. In the equilibrium state, the rate of molecular collisions per square centimetre is same, so we have

$$n_1 \bar{c}_1 = n_0 \bar{c}_0 \qquad \text{...(6.32)}$$

And, if n is the number of molecules per cubic centimetre (cc) in the space between P_1 and A, the number of molecules flowing from P_1 to A and from A to P_1 must be same, therefore, we have

$$n \bar{c}_0 = n_1 \bar{c}_1 + n_0 \bar{c}_0 = 2 n_1 \bar{c}_1 = 2 n_0 \bar{c}_0 \qquad \text{...(6.33)}$$

If m is the mass of each molecule, n is the number of molecules per cc and $\bar{c}_0$ is the root mean square velocity of each molecule away from plate P_1 in the side of A, then pressure (p) exerted by the gas away from the plates, P_1 and P_2, on either sides of A, will be

$$p = \frac{m n \bar{c}_0^2}{3} \qquad \text{...(6.34)}$$

Total pressure exerted by n_1 and n_0 between the plates P_1 and P_2 and A will be

$$= \frac{m\, n_1\, \overline{c}_1^2}{3} + \frac{m\, n_0\, \overline{c}_0^2}{3}$$

Therefore, excess pressure nearer to P_1 and P_2 over that away from wall A, is given by

$$= \frac{m\, n_1\, \overline{c}_1^2}{3} + \frac{m\, n_0\, \overline{c}_0^2}{3} - \frac{m\, n\, \overline{c}_0^2}{3}$$

Substituting $\frac{n\, \overline{c}_0}{2}$ for $n_1\, \overline{c}_1$ and for $n_0\, \overline{c}_0$, excess pressure nearer to P_1 and P_2 is

$$\frac{m\, n\, \overline{c}_1\, \overline{c}_0}{6} + \frac{m\, n\, \overline{c}_0^2}{6} - \frac{m\, n\, \overline{c}_0^2}{3} = \frac{m\, n\, \overline{c}_1\, \overline{c}_0}{6} - \frac{m\, n\, \overline{c}_0^2}{6} = \frac{m\, n\, \overline{c}_0^2}{6}\left[\frac{\overline{c}_1}{\overline{c}_0} - 1\right]$$

If m is the mass of a gas molecule, then at temperature T, it gains kinetic energy $\frac{1}{2}\text{mv}^2$ $= \text{KT}$, which yields the velocity, $v = \sqrt{\frac{2\,\text{KT}}{m}}$ of the molecule.

Assuming that the molecules striking P_1 and A to take up their temperatures T_1 and T_0, respectively, then

$$\frac{\overline{c}_1}{\overline{c}_0} = \sqrt{\frac{T_1}{T_0}} \qquad \text{...(6.35)}$$

Thus, excess pressure on the side of frame nearer to P_1 and P_2 is

$$= \frac{m\, n\, \overline{c}_0^2}{6}\left[\sqrt{\frac{T_1}{T_0}} - 1\right]$$

$$= \frac{1}{2} \cdot \frac{m\, n\, \overline{c}_0^2}{3}\left[\sqrt{\frac{T_1}{T_0}} - 1\right]$$

The pressure inside the vessel, $p = \frac{m\, n\, \overline{c}_0^2}{3}$.

The vane of the heater side experiences the excess pressure or force per sq. cm

$$F = \frac{p}{2} \cdot \left[\sqrt{\frac{T_1}{T_0}} - 1\right] \qquad \text{...(6.36)}$$

If the temperature difference $(T_1 - T_0)$ is not very large (i.e., not more than 250°C), we may consider the flow of molecules in between the space P_1A and rest of the vessel as a thermal transpiration effect between the temperature $\left(\frac{T_1 + T_0}{2}\right)$ inside P_1A and T_0 outside; so that, if p' and c' are the pressure and velocity of molecules inside P_1A, putting the values of c_1 bar $= c'$ and $T_1 = (T_1 + T_0)/2$ in Eq. (6.35), we have

$$\frac{p'}{p} = \frac{c'}{\overline{c}_0} = \sqrt{\frac{\left(\frac{T_1 + T_0}{2}\right)}{T_0}} = \sqrt{\frac{T_1 + T_0}{2\,T_0}} = \sqrt{\frac{2\,T_0 + T_1 - T_0}{2\,T_0}}$$

$$= \left(1 + \frac{T_1 - T_0}{2\,T_0}\right)^{1/2} = 1 + (1/4)\left[\frac{(T_1 - T_0)}{T_0}\right]$$

or $$p' = p + \frac{p}{4}\frac{T_1 - T_0}{T_0} \quad \text{...(6.37)}$$

Since $F = p' - p$ is force per unit area, therefore

$$F = \frac{p}{4}\frac{T_1 - T_0}{T_0} \quad \text{...(6.38)}$$

Equation (6.38) indicates that the force exerted on the frame is quite independent of the nature of the gas in the vessel.

Now, if a is the area of the plate P_1 and P_2, then

force acting on frame = Fa.

Since these forces act in opposite directions on the two sides of frame, constitute a couple, is equal to $F \cdot a \cdot 2b$, where, b is the width of the plate suspended.

The frame is thus deflected, giving rise to a restoring torsional couple in the suspension wire and therefore, comes to rest, when the two couples just balance each other, say, when it has deflected through an angle θ.

Then if c is the torsional couple, set up the suspension wire per unit twist in it (or per unit deflection of the frame) the total torsional couple tending to restore the plate back to the original position is $c\theta$; and we, therefore, have

$$c\theta = 2\,F\cdot a\cdot b \quad \text{or} \quad F = \frac{c\,\theta}{2\,ab} \quad \text{...(6.39)}$$

Substituting the value of F in relation (6.38), we have

$$\frac{c\theta}{2\,ab} = \frac{p}{4}\frac{T_1 - T_0}{T_0}$$

or $$p = \frac{4\,T_0\,c\theta}{2\,ab\,(T_1 - T_0)}$$

or $$p = \frac{2\,c\theta}{ab}\frac{T_0}{(T_1 - T_0)} \quad \text{...(6.40)}$$

Further, if I is the moment of inertia of the frame about the suspension wire as axis and t as time period of the vibration, we have

$$t = 2\,\pi\sqrt{\frac{I}{c}} \quad \text{or} \quad c = \frac{4\,\pi^2 I}{t^2}$$

Substituting the value of c in Eq. (6.40), we get

$$p = \frac{8\pi^2 I\theta}{ab^2t^2}\frac{T_0}{(T_1 - T_0)} \quad ...(6.41)$$

From Eq. (6.41), the pressure (p) of the gas in the vessel, can be easily calculated.

It is to be noted from Eq. (6.41) that the value of p is independent of the molecular weight of the gas.

Advantages of Knudsen absolute gauge

(i) It gives the pressure directly in dynes · cm^2, and is not to be calibrated against a McLeod gauge, etc., so it is an absolute gauge.

(ii) It gives a continuous indication of pressure in the vessel.

(iii) It is unaffected by outside influences.

(iv) It is stable and yet very sensitive at low pressure, down to 10^{-7} mm.

(v) It can be used to measure the pressure of all kinds of gases and vapours, irrespective of their mass or condensability.

(vi) It does not require the use of objectionable liquids, like mercury.

(vii) It is simple, cheap and easy to construct and work with, though obviously, it cannot be used above a pressure of 10^{-3} mm; because, then the mean free path of the molecules becomes comparable with the distance between the plates.

6.25 IONIZATION GAUGE (IG)

The ions can be produced by following two methods:

(i) by knocking atoms the with electrons, which can be obtained by heating a wire like tungsten or molybdenum.

(ii) by ionization of gas by α particles obtained from radioactive substance. α particle is doubly ionized He atom.

Production of electrons by two methods leads to two different types of ion gauges. First type of gauge, where ions are produced by heating, are called hot-cathode ionization gauge (HCIG). Second type of gauge, where ions are produced without heating, are called cold-cathode ionization gauge (CCIG).

6.25.1 Hot Cathode Ionization Gauge

First hot cathode ionization gauge was invented by Dushman and Found. Ionization is a process of knocking out an electron from the outer shell of a gas atom. The gas atoms are neutral. When a fast moving electron collides with the gas atoms, an electron may be knocked out from the outer shell of a gas atom. This process is known as ionization by electron collision. Thus, ionization process produces positive ions and electrons having negative charge. For ionizing a gas an atom, an electron must possess a certain minimum amount of energy, depending on the gas in question. For acquiring certain minimum amount of energy, the electronis accelerated through a certain minimum potential difference, called the ionization potential for that particular gas. The energy acquired by the electron is measured in terms of electron volts, where one electron volt is the energy acquired by the electron when accelerated by a potential difference of 1 volt.

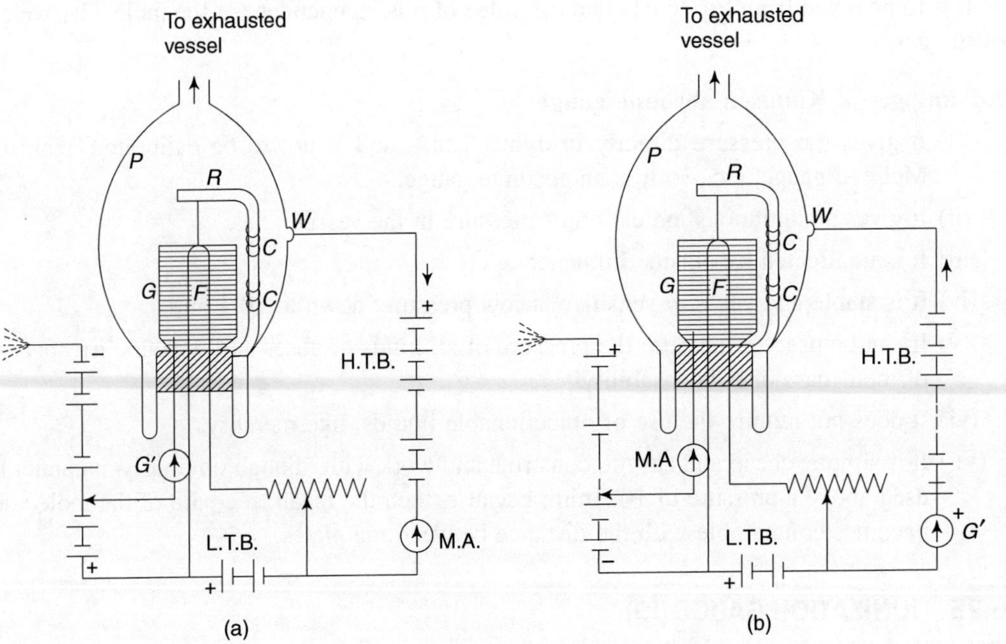

Fig. 6.27

In this gauge, the electrons which ionize the gas atoms, are emitted on heating the filament or the cathode, therefore this gauge is called a *hot cathode ionization gauge*. It consists of cathode filament, screening grid in the form of gauge and +ve plate enclosing the filament and grid acts as anode. Therefore, basically it is a triode. The tungsten filament (F) is supported on a glass rod (R) with a co-axial grid (G) of tungsten or molybdenum around it as shown in Fig. 6.27(a). The nickel or silver is coated on the interior of the glass bulb, enclosing filament and grid and acts as the plate, P. It is sealed with a platinum wire (W) to enable it to connect to the external electrical circuit. In Fig. 6.27(b), a little modified gauge is shown.

The glass rod (R) is provided with loose glass collars C, C in order to prevent deposition of any metal film on the filament or plate to stop inter-electrode leaks.

The positive ions may be collected either on the grid or on the plate, as desired. Necessary electrical connections for the functioning of the gauge are shown in Figs. 6.27(a) and 6.27(b). The grid is given a negative potential in the first case and a positive potential in the second case, with respect to the filament. The plate is connected to the positive terminal in Fig. 6.27(a) and to the negative terminal of high tension battery (H.T.) of about 120 volts in Fig. 6.27(b). A milliameter (MA) is included in the plate circuit and galvanometer (G') in the grid circuit, in the first case, while the galvanometer (G') is included in the plate circuit and the milliameter (MA) in the grid circuit, in the second case.

Working of ionization gauge

(i) First case: In this case, the plate is kept at a positive potential and grid at negative potential with respect to the filament. The electrons emitted by the filament (when heated by the current from the low voltage battery (LTB) are attracted towardsthe plate and pass through the mesh in the grid. On their way to the plate, they ionize the gas between the grid and the plate. The positive ions formed are collected by the grid which is at a negative potential with respect to the filament. A small positive ion current, thus flows through the grid-filament circuit and can be easily read from the galvanometer (G'), included in the circuit, the electron current is read from the milliameter (MA) included in the plate circuit.

(ii) Second case: In this case the grid is kept at a positive potential and the plate at a negative potential with respect to the filament. The electrons emitted by the filament or the hot cathode, are strongly attracted by the grid (as it is +ve with respect to filament and also nearer to it), but quite a number of them get through it on account of their momentum and thus ionize the gas in the space between the grid and the plate (in case, applied accelerating voltage is higher than the ionization potential of the gas). The positive ions thus created are collected at the plate, P. The electrons staying into the region are repelled back by it. The current due to positive ion is read on galvanometer (G') included in the plate circuit and the electron current is read on the milliameter included in grid-filament circuit.

Of the two arrangements shown in Figs. 6.27(a) and (b), the second case, where the plate is given –ve voltage, is far the more sensitive, than the first one, where the plate is given +ve voltage, is simpler to work with.

As already indicated, the relation between the ionization current and pressure is a linear one only at pressure below 10^{-3} mm of the mercury. If the emission of electrons from the filament is relatively small, the relation also becomes linear at higher pressures, but, the sensitivity of the gauge is considerably impaired.

This gauge is not an absolute one. So it has to be calibrated against a McLeod gauge, with the same gas in it, the pressure of which is to be determined by the ionization gauge. Once calibrated, the galvanometer G' is replaced by a micrometer, graduated in pressure units.

The gauge can measure the pressure in the range 10^{-3} mm to 10^{-7} mm. It has some advantages over the McLeod gauge, for example: (i) it can be used to measure pressures of both vapours and gases, and (ii) being smaller in size, so it can be located in close proximity with the vessel being exhausted.

Drawbacks of hot cathode ionization gauges

1. Its operation is some what complicated, because it requires a lot of extra electrical equipment with it.

2. Its sensitivity depends on the nature of the gas, the arrangement of the electrodes and the electric circuits employed.
3. Reactions of the gas molecules with the hot filament can seriously affect the composition of the gas, and the reliability of the pressure measurements becomes questionable.
4. Glass envelope gauges can break and/or implode violently resulting in the danger of flying glass.
5. Electrons hitting the grid produce X-rays that produce photoelectric noise in the ion collector. This limits the range of older hot cathode gauges to 10^{-8} torr.
6. Organic vapours poison its filament, because the emission of the electrons from filament is reduced by organic vapours. Therefore, as a precaution, a 'cold trap' of the carbon dioxide snow or acetone is made in between the gauge and the exhaust of the vessel, some sort of a screen must be used to prevent any oil molecules streaming back into the gauge which weaken its working.

6.25.2 Cold Cathode Ionization Gauge (CCIG) or Crossed-Field Ionization Gauge or Penning Gauge

An important class of gauge in the medium to high vacuum ranges is based on a cold gas discharge in crossed electric and magnetic fields. In such discharges, free electrons are accelerated by the electric field and are trapped by the magnetic field so that they have very long path lengths–much longer than the gauge dimensions. This means that even at low pressures, these electrons have a good chance of ionising a gas molecule.

In 1936, Penning invented the cold-cathode discharge gauge as a method to measure pressures below 10^{-3} torr. Penning increased the probability of ionization by placing a magnetic field parallel to the paths of ions and electrons, which forces electrons into a helical trajectory.

A high voltage, up to 3 kV, was used for generating a discharge between the cathode and the anode. At low pressures (< 1 Pa or so) this discharge could only be maintained, if a magnetic field crosses the electrical field. It turned out that the discharge current was almost linearly proportional to the pressure in the gauge from 1 MPa to 0.1 Pa. Due to the magnetic field the electron is prevented from going directly to the anode and moves instead in helical paths through the gauge, therefore, greatly increases the path length of an electron from cathode to anode. The ions, due to their large mass, are virtually unaffected by the magneticfield and travel directly to the cathode. Secondary electrons released from the cathode by ion bombardment serve to build up and maintain the discharge (Fig. 6.28). The discharge is generally not stable in crossed-field gauges. In the early designs the discharge be came erratic below 10^{-3} Pa, and was often extinguished completely at 10^{-4} Pa. Therefore, better designs were invented to reduce discontinuities by increasing the active volume of the discharge. Many configurations are possible for such gauges which are often referred to as *Penning gauges*, since the most popular configurations are based on the Penning discharge.

Penning's version of gauge, invented in 1949, is shown in Fig. 6.28. In the new design, the anode was changed from a ring in his original version to an open cylinder.

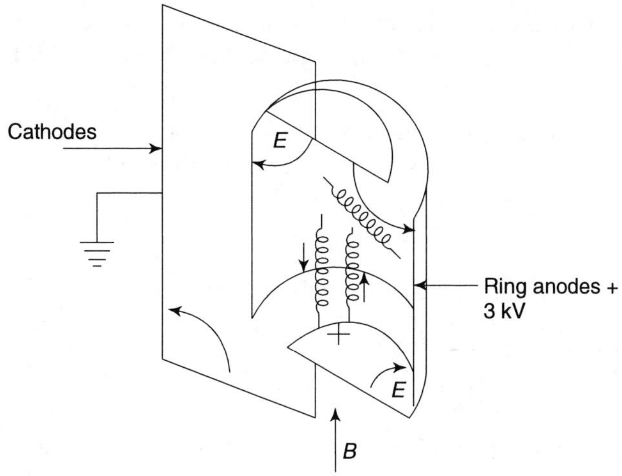

Fig. 6.28 Electrode arrangement, fields, and trajectories in the Penning gauge

Penning gauge measures pressure indirectly. Discharge current is used as a measure of pressure.

The useful range of standard Penning gauges is between 10^{-3} mbar and 10^{-8} mbar, or in special versions, 10^{-9} mbar.

Drawbacks of Penning gauge

1. Difficult to start and maintain discharge $< 10^{-6}$ torr.
2. Indicated value of pressure is gas dependent.
3. Less stable and less accurate than a Bayard Alpert gauge.

6.26 DETECTING METHODS FOR LEAKAGE OF PRESSURE

Several leak testing techniques are available, starting from very simple methods to more sophisticated methods. The most commonly used leak test methods are underwater bubble test, bubble soap paint, pressure and vacuum decay, and tracer gas detectors (helium and hydrogen). The first three techniques, due to their characteristics and sensitivity, can be used only for gross leak detection [300 *g*/*y* (10.5 oz) or more refrigerant leakages]. Tracer gas leak testing methods are much more precise than the other methods. However, theoretical value of sensitivity is important in some methods. The sensitivity is limited by environmental factors and working conditions.

Each method mentioned above along with advantages and drawbacks is discussed briefly in the following sections.

6.26.1 Water Immersion Bubble Test Method

The water immersion bubble test, also called "bubble testing" or "dunking". It is a traditional and relatively primitive technique of leak detection. In this method a pressurized part with high pressure dry air or nitrogen is immersed in a water tank and watched for escaping bubbles. If the bubbles are large and more frequent, the leakage is big. It is very difficult to detect relatively small leaks.

The main limitation of this method is sensitivity, which is the minimum detectable leak rate. If a bubble is spherical of radius ®, its internal volume (V) is

$$V = \frac{4}{3}\pi R^3$$

Let p be the pressure inside the bubble and t the time required to form the first bubble, the leak rate (Q) is written as:

$$Q = \frac{P \times V}{t}$$

The two key parameters determining the sensitivity of this method are the smallest bubble detectable by the operator and the waiting time for generation of bubble. This time must be compatible with the production rate and with operator's attention.

Let us assume that an operator can detect the smallest bubble of radius 0.1 mm and the waiting time is 30 seconds. The volume of bubble is

$$V = \frac{4}{3}\pi R^3 = \frac{4}{3} \times \pi \times (0.01)^3 \approx 4 \times 10^{-6}\ \text{cm}^3 \qquad ...(6.42)$$

Let us assume that the pressure inside the bubble is at atmospheric pressure, therefore the minimum detectable leak rate will be:

$$Q = \frac{P \times V}{t} = \frac{1000 \times 4 \times 10^{-6}}{30} \approx 1 \times 10^{-4}\ \text{mbar/s} \qquad ...(6.43)$$

This is a theoretical value. However, the real sensitivity is strongly influenced by many external, factors, for example, condition of illumination, clean water, location of unit, and movement of water. All these factors and operator dependency limits the practical sensitivity to 5×10^{-4} mbar /s, though 1×10^{-3} mbar/s is normally considered.

For easy detection of a leak following tricks are used:

- By increasing the internal pressure in steps, the probability of finding a leak increases. Using this process less time is required in pinpointing the leak.
- By adding the detergent surface tension of the water is decreased, which helps to prevent the leaking gas from going to the side of the container.
- Hot water in the tank helps to increase the pressure inside the container or tube.

The advantages of this technique for finding leak location are: (i) it is economical in most of the cases, and (ii) its accuracy is about 10^{-3} mbar/s for high volume products.

The disadvantages of this method are: (i) Its sensitivity is relatively low. (ii) its dependency is high on the operator, (iii) wastage of fluid is too much, and (iv) after testing, one has to dry the parts. In addition these drawbacks, this process requires a large amount of space and a certain amount of water, especially for big units, where the tank could be very large and requires a lot of water. To operate and maintain dryers also cost money.

6.26.2 Soap Solution Bubble Method

Instead of submerging the part in water, the soap solution is sprayed on the pressurized unit to be tested. The operator watches the bubbles formed by gas escaping from the point of the leak carefully.

The different types of soap solutions are available for this purpose. Some solutions can be applied with the help of a brush. For covering quickly large areas of tubes, some brands of soap solution have a spray system. The sensitivity of this method is higher than water immersion technique. The sensitivity detection of leaks of this method is up to 10^{-5} mbar/s, therefore, it is more suitable for very large systems.

This soap solution method is less time consuming if the approximate area of a leak location is known. Because, the soap solution is only applied in that specific area where leak is located. It is the simplest, less time consuming and least expensive method. However, if the area of leak location is not easily detectable, then this method is more time consuming and expensive because of labour cost involved.

The probability of finding the leak location increases by raising the gas pressure. But, for safety of operator, the pressure must be limited to 11251 mm or 14996.4 mbar.

Drawbacks of soap solution bubble method

The soap solution bubble method has some drawbacks also. (i) The area at which the soap solution to be sprayed must be a simple and easily accessible surface. (ii) It could be very difficult for the operator to spray that part on the bottom of a large heat exchange and to watch a bubble there. (iii) This method is not suitable for high productivity units.

6.26.3 Pressure Decay Method

High pressure gas, normally dry air or nitrogen, is used for pressurizing the component of the system in this method. After pressurizing the component, it is isolated from the gas supply. After a stabilizing period, its internal pressure is monitored for some time. If the pressure in this component drops fast, then a large leak is present in that part or section of the system. If the pressure of the component drops slowly, then a small leak is present. If the pressure remains the same, then the component is leakfree. Let Δp be the pressure drop in the time, Δt, V is the volume of the component, then the leak rate (Q) from the component is written as

$$Q = \frac{\Delta p \times V}{\Delta t}$$

From this equation it is clear that the leak detection sensitivity depends on the time required for testing, the change in pressure and the volume. In most of the components the pressure variation is up to 70 Pa (0.010 psi) at test pressure, the leak detection cycle time is at least 30 seconds. Let us consider that V = 1.5 dm^3 (0.4 gal) of internal component, Δp = 70 Pa (0.010 psi) at 3450 kPa (500 psi) test pressure and Δt = 60 seconds, then the leak rate will be:

$$Q = \frac{\Delta p \times V}{\Delta t} = \frac{0.7 \times 1.5}{60} = 1.7 \times 10^{-2} \text{ mbar/s}$$

This test method is affected by several external factors, like temperature variation and mechanical deformation of the component. The internal pressure depends on temperature, so it may be changed by thermal fluctuations, altering the results. The pressure change of dry nitrogen is very little when it is exposed to small temperature changes.

The sensitivity of this method depends on measurement of pressure change, testing time and test pressure. The longer testing time makes the test more time consuming, because some low-level leaks may require even several hours. It is faster to determine a leak present at higher pressure. However, safety concern of the operator puts the limit on the maximum admissible pressure value without safety precautions. The leak test of components can be done at low pressures, at less than 2 MPa (290 psi) without protection, and at higher pressures, 7 MPa (1000 psi), by using safety protection hoods.

Usually, the limit of sensitivity for pressure decay test is about 10^{-3} mbar/s without pressure change and 10^{-4} mbar/s for pressure decay test with pressure change.

This leak testing method has some advantages. By monitoring the pressure drop, it is certain in this method to identify whether leak exists or not. If any pressure drop occurs, it means a leak is definitely present. To avoid errors of the operator, this method is completely automatic. This procedure is a preliminary leak test that detects large leaks before the final automatic leak test operation using a tracer gas, like helium. This test can detect over 90%, of the defective parts, especially those not welded correctly. If the test unit has a large leak (i.e., an overlooked welded joint) without doing the pressure decay test first, large quantities of tracer gas can find a leak. During leak test helium may contaminate the component. The resulting helium contamination, of a component of large volume may render the system inoperable for hours. The advantage of using nitrogen instead of helium is that it cleans the component and lower its humidity.

The major drawback of this method is that it does not identify the exact potion of the leakage.

6.26.4 Tracer Gas Leak Testing

The term "tracer gas leak testing" is used to a group of test methods, which are used to detect the flow of a tracer gas through a leak. These techniques differ for different tracer gases.

The most common tracer gases are helium, a mixture of nitrogen 95% and hydrogen 5% and halogen gas (CFC, HCFC and HFC refrigerant). The electronic devices used for detection of a leak using halogen gases are simple. Despite the simplicity of their electronic detection devices, the use of halogens as a tracer gas is decreasing, due to environmental hazards. On the other hand, helium and especially hydrogen/nitrogen mixture are gaining more utility. Due to physical properties of helium, it has been used successfully as a tracer gas for a long time. It is neither toxic nor flammable. It is an inert gas and does not react with other compounds. It has low viscosity and relative molecular mass, so it easily passes through cracks. In the same environmental conditions, it flows through the outlet 2.7 times faster than air. Because of its low concentration (5 ppm) in air, it is easy to detect an increment of helium concentration. However, it has some short comings. Helium disperses slowly into the atmosphere, so, in the case of big leaks, its high concentration will contaminate the area for a long time, even hours. Also, helium is costly, but it is less expensive than halogen gases. The most suitable helium detector is based on a mass spectrometer, which is an expensive and delicate apparatus, requires more care and maintenance, and is more suitable for a laboratory than for the manufacturing industry.

The mixture of nitrogen 95% and hydrogen 5% is a relatively new tracer gas. Hydrogen has a number of properties that make it an excellent tracer gas for leak testing. It is easy to fill, evacuate and dissipate, because it is the lightest element, with higher molecular speed and lower viscosity than any other gas. It is easy to find a leak using tracer hydrogen gas, because it passes through a leak faster, i.e., it flushes out and flushes in faster. Its molecules do not stick to surfaces as easily as helium atoms. It is environment friendly and renewable. More important property of this gas is that it has the highest leakage rate than any other gas. Moreover, the normal background concentration of hydrogen (0.5 ppm) is ten times less than helium. Hydrogen tracer gas detectors use a semiconductor sensor and have no moving parts, so they are completely maintenance free. These devices are not affected by the presence of other gases. Pure hydrogen gas should never be used as tracer gas because it is inflammable and expensive, but a standard industrial grade mix of 5% hydrogen in nitrogen is inexpensive, non-flammable and readily available from industrial gas suppliers.

It is important to remember that background concentration of tracer gas in air is a limiting factor for any tracer gas detector. There are two ways to carry out leak testing with tracer gas: external detection of tracer gas escaping from leaks of a filled unit (inside-out method), and internal detection of tracer gas entering from leaks (outside-in method). For each of these two methods two techniques are used. The inside-out methods can be executed by sniffing the gas in the atmosphere or with vacuum chamber detection, while the outside-in method is generally implemented by putting the unit to be tested in a room containing the tracer gas or, very rarely, spraying the tracer gas on the unit surface.

6.26.5 Vacuum Chamber Inside-out Leak Testing

Vacuum chamber inside-out leak testing is the most complex system of leak detection, but it is theoretically suitable to find very small leakages, using the proper tracer gas. The equipment is composed of one or more vacuum chambers, large enough to house the unit to be tested. For chamber evacuation and gas detection, the chamber is connected to a vacuum pumping unit equipped with the tracer gas detector. A second vacuum unit is required to evacuate the unit under test before filling it with gas. A tracer gas filling device is used to fill gas in the testing apparatus. The unit to be tested is put into the vacuum chamber and connected to proper hoses. Then the vacuum chamber and the unit are evacuated. During evacuation, the part is pressurized with the tracer gas. After a stabilization time, the detector is linked to the vacuum line to detect the tracer gas flow through a leak. In this way, the leakage is detected, but leak position is not pinpointed. For finding the leak location other techniques are used.

This method has some advantages. This technique is fully automatic, so it depends very little on an operator. Its sensitivity depends on the tracer gas and test time, and can reach 10^{-10} mbar/s. However, for a practical application in the refrigeration industry the limit is only upto 10^{-6} mbar/s.

There are also some drawbacks. Depending on the vacuum chamber dimensions, the evacuation unit may need a high pumping speed. Some gas detectors require periodic maintenance to ensure proper performance, because they are composed of complex systems, like vacuum pumps, a mass spectrometer and vacuum fittings. Besides this, the cost of the tracer gas may be significant. In case of a particularly expensive gas, the use of a suitable gas recovery system should be used. In the case of big leak in the part under test, a lot of tracer gas escapes in comparison to the tracer gas used. A long pumping time is required to lower the tracer gas in the detector to the required level needed to function the unit. The unit is unusable during this time. To avoid this, it is a good practice to use a preliminary leak testing units to reject parts with gross leakages. The preliminary test can be integrated with the vacuum chamber test unit to rectify the problem.

Another major disadvantage is that this method does not identify the position of leak, it can only decide whether a leak is present or not. It detects the total system leak and even though more than one leak may exist. To find leak location other techniques are used.

6.26.6 Outside-in Leak Testing

In this technique, the unit to be tested is put into an enclosure containing the tracer gas. The part is connected to a vacuum group and evacuated. A tracer gas detector (i.e., helium mass spectrometer) is placed in the vacuum line in order to detect the tracer gas flow in a leak and pulled in by the pumping group.

This method has some advantages. This technique is fully automatic, so it is not operator dependent. The sensitivity, depending upon the tracer gas and test time, can reach 10^{-6} mbar/s. The gas containment hood can be used to prevent dispersion, in order to reduce working area pollution and tracer gas consumption, saving money by avoiding the need for a recovery system.

There are also some drawbacks. The difference from inner and outer pressure is limited to values slightly above the atmospheric pressure. In case of a big leak in the part under test, a lot of tracer gas escapes from it. A long pumping time could be required to lower the tracer gas in the detector to an acceptable level compatible with the system function. The system is unusable during this time. It is good practice to use a preliminary leak testing system to reject parts with gross leakages. It is possible to integrate this preliminary test, (i.e., a pressure decay test), in the tracer gas detecting machine, so as to simplify the equipment.

Another disadvantage is that this method does not identify the position of the leak, it only determines whether a leak is present or not. It detects the total system leak and though more than one leak may exist. Leak location requires other techniques.

6.27 APPLICATIONS

In the following sections, the main applications of the different methods are described.

6.27.1 Outside-in Helium Spraying

In the "outside-in" helium spraying technique, the sample to be tested is connected to a vacuum unit equipped with a mass spectrometer. The unit is evacuated and its surface is "probed" with a pointed jet of helium by a suitable diffuser. Passing over a leak, the detector senses the helium entering from the leak, identifying the location and leak size. This method is classified as vacuum test method, though it is an exception.

This technique can be used to locate small leaks, since its sensitivity is not limited by the background helium level. However, it is not free from some drawbacks. This method is operator dependent. Moreover, helium is lighter than air so it rises, therefore helium sprayed in the bottom of the unit may pass through a leak in the upper part. Helium has a tendency to accumulate and then saturate the working area if it is not well ventilated. In this condition, it becomes difficult to find the leak location. In case of a big leak, a lot of helium reaches the mass spectrometer and a long time is needed to reduce the helium level to a value required for testing. Since the unit to be tested is evacuated, its internal pressure is lower than atmospheric pressure. Therefore, if the working condition requires a positive pressure inside the unit, then this leak testing method will stress the part in the opposite way.

This testing technique is, however, widely used in research and in all those applications involving big plants, which cannot use other methods. It is not the most suitable method for testing in high productivity lines or large production manufacturing industries. It is noteworthy that parts to be checked must be kept in an area free from helium contamination before testing. If high pressure air is used for a preliminary pressure decay test, the high pressure air compressor inlet must be kept in a fresh air atmosphere.

6.27.2 Outside-in Helium Leak Testing

In the "outside-in" helium leak testing technique, the part to be tested is placed in a containment hood suitable to contain helium and connected to a vacuum machine equipped with mass spectrometer. The test consists of evacuating the unit and flooding the container with helium. Helium has a high penetration capability due to its atomic characteristics. So a mass spectrometer can detect the helium leaked into the component through cracks and porosities not detectable using other methods. The part to be tested is evacuated to less than 15 Pa (100 μm Hg). The unit may be also evacuated using a vacuum rise method to ensure that it is clean from water vapour contamination, other non-condensable gases, and is free of very large leaks. After evacuation, the internal circuit is connected to the mass spectrometer. If any helium has been leaked into the circuit, the mass spectrometer detects it. In this way, it is possible to find quickly the sample leaks and the total leak rate. This method is able to detect leak rates up to 1.8 × 10^{-5} mbar/s, or 2.5 g (0.1 oz) of [in R134A refrifgerant hydrocarbon gas tetrafluoroetthane, known as norflurane or floron, is used] a refrigerant per year.

There are several practical uses of this testing method. The test unit can be designed with one or multiple stations, normally two, for production rate requirements so that multi-station machines can test several parts, one for each station, simultaneously. The containment hood may be designed so as to reduce helium wastage during loading and unloading operations. Recent advances in vacuum technology provide improved sensitivity and faster test cycles even with low helium concentration, for example, at only 10% helium. Recirculation fans and mixing ducts maintain the preset concentration of helium to air distribution uniform in the internal volume. The hood must be of appropriate size to cover the test unit completely and must be designed to be adjustable for lowering the volume and therefore to minimize the helium consumption. In this way cost of operation of test unit can be reduced.

This method has some advantages. It is a final global leak test and is very useful on production lines where a test piece must be simply accepted or rejected. The test technique is fully automatic and operator independent. The total process time for testing leaks in a part is low, regardless of its volume, even less than 80 seconds for a medium size unit.

This leak testing method has some drawbacks also, so some precautions must be taken for the proper use of it. This method is for leak detection only. To pinpoint the leak location another method is required. If the test unit has a large leak such as an overlooked soldered joint, the escaping helium, at this much volume and helium concentration would leave the unit un-operational for long time. It is better to use a preliminary leak test for gross leakage identification. A pressure decay test can be easily combined to the outside-in leak testing machine. This initial nitrogen filling, besides the mechanical stress and leakage test, cleans the circuit to be tested and lowers its humidity. The component is ready for final test after a successful completion of the pressure decay test. At the end of the cycle, the part may be filled again with nitrogen, at a low pressure to avoid helium being sucked into it when disconnected.

Since the unit to be tested is evacuated, its internal pressure becomes lower than atmospheric pressure. If the working condition requires a positive pressure inside the unit, this leak testing method will stress the part in the opposite way. This problem can be solved using a proper preliminary pressure test.

It is worth noting that parts to be checked must be kept in an area free from helium contamination before outside-in leak testing. If high-pressure air is used for a preliminary pressure decay test, the high-pressure air compressor inlet must also be placed in a fresh air atmosphere. At the end of cycle, the unit may be filled with low-pressure nitrogen to avoid helium getting into the part when disconnected.

6.27.3 Inside-out Helium Vacuum Chamber Leak Testing

In "inside-out" techniques, also known as "global hard-vacuum test", the test configuration is reversed. The component is placed inside an airtight chamber provided with service hoses and a vacuum pumping group with a mass spectrometer. The unit is connected to the service hoses, the chamber is closed, the part is evacuated and then pressurized with helium. The chamber is subsequently evacuated and once a suitable vacuum level is reached, the inlet valve of the leak detector is opened. The leak detector begins to analyze the residual gas molecules present in the chamber. Helium molecules escaping from the component are conveyed and measured by the leak detector. At the end of the cycle, the vacuum chamber is opened to pass out air. The leak detector finds the leak and gives the total measurement. Referring to the complexity of the test and the desired degree of automation, different test systems may be used from the simplest to more complex. As an example, a two-chamber machine can be used so that in one chamber the test cycle is in progress while in the other chamber a part is unloaded and reloaded. In many industrial applications, in order for the test chamber to reach the vacuum level in acceptable times, an auxiliary pumping group is needed. The dimensioning of the pumping group depends on different parameters, such as the body of the expected leaks, the dimensions of the parts to test and the cycle time.

Vacuum chamber helium inside-out leak test method has some advantages. A sniffer is not used; testing is automatic and not operator dependent. The sensitivity achievable with this test is very high. In a laboratory application, it is not impossible to get 10^{-10} mbar/s. However, in practical application in the refrigeration industry, the limit is considered 10^{-6} mbar/s.

Vacuum chamber inside-out leak test method has several drawbacks also. It is an expensive system. While the sniffing test requires only the leak detector and the ability of the operator, the hard-vacuum test requires more complex and expensive equipment, but it enormously limits the incidence of the human factor. This fact makes this test attractive in many industrial contexts, while the sniffing test has its applications in the maintenance and the analysis of the defective parts.

In design and utilization of the vacuum chamber, particular care must be taken to avoid helium contamination. In fact, if some helium remains on the chamber's inner surface or in some internal components, the background helium level will be high and the sensitivity will consequently be reduced.

If the test unit has a large leak, a lot of helium would leak out of the test unit and into the vacuum chamber, the mass spectrometer and the vacuum pumping station. The helium, at this much volume and concentration, would render the system inoperable for a long time. To avoid this occurrence, it is a good practice to use a preliminary leak test to detect big leaks. A pressure decay test can be combined with the vacuum chamber inside-out leak test. This initial nitrogen filling, besides the mechanical stress and leakage test, allows cleaning the circuit to be tested, lowering its humidity. On a successful completion of the pressure decay test the component is ready for final helium test. At the end of cycle, the part may be filled again with nitrogen, even at a low pressure, to clean its internal circuit from the helium residuals before vacuum chamber is opened.

Unlike the sniffing test, the hard-vacuum test is a global method that quantitatively defines the total leak of the piece, but does not determine its position. Other techniques are required for leak location.

Since helium is an expensive gas, and to reduce the helium concentration in the working area, it is not convenient to spread it in the atmosphere, so it is a good practice to use helium recovery/reclaim stations to empty the unit at the end of leak testing. The recovered helium can be reutilized for successive tests. However, helium recovery systems are costly.

6.27.4 Inside-out Hydrogen Sniffer Detectors

As in the other "inside-out" techniques, the unit which is to be tested is evacuated first and then pressurized with the hydrogen/nitrogen mixture. An operator moves the sniffer over the part and tests with the probe around suspected leak sites.

Compared with helium sniffer leak testing, this method is operator dependent, but has many advantages. Hydrogen is ideal for leak testing. It is the lightest element with higher molecular speed and lower viscosity than any other gas. Therefore, it easily fills the unit which is to be tested, mixes quickly with any gas, it is easily evacuated and dissipated. Since hydrogen mixes very quickly in other gases, this method requires a low vacuum level compared with helium filling. As a result, the vacuum group is very simple; Venturi vacuum generator is used to do vacuum by using a fluid or gas chamber meant to move fluids or gases via a narrow section of a pipe. Hydrogen easily disperses into air, so its concentration in the working area falls quickly to the background level even in the case of big leaks.

In addition to all these qualities, hydrogen has the highest leakage rate than any other gas. Hydrogen detectors are based on semiconductor sensors and are tough enough for industrial use and completely maintenance free, because they contain no moving parts. They are also unaffected by the presence of other gases. Moreover, these sensors do not require suction for functioning, so the probes can be used without worrying about the dust. In addition, the probe can be equipped with a protective cover that allows it to be used on wet objects. Hydrogen has a very low background concentration (0.5 ppm) and the detectors are sensitive enough to detect up to 5×10^{-7} mbar/s (or 0.5 g/y or 0.018 oz/y) R134a refrigerant) using a mixture of 5% hydrogen and 95% nitrogen, and are ten times more sensitive compared to helium sniffers. This mixture is standard industrial grade mix, inexpensive, non-flammable and readily available from industrial gas suppliers. In R134A refrifgerant hydrocarbon gas tetrafluoroetthane, known as norflurane or floron, is used.

There are some doubts about for using hydrogen as tracer gas. Hydrogen escapes rapidly from a leak in an almost vertical direction, so a leak can be precisely detected and located only if the detector is directly above it. This might cause someone to think that leak detection with hydrogen is more difficult than with helium. However, even if a skilled operator has no difficulty in managing this problem, detectors can be easily integrated in special designs, for example, funnel-shaped devices, to facilitate leak detection. Another common doubt about using a hydrogen/nitrogen mixture is component separation and hazards related to hydrogen. Even though the hydrogen/nitrogen mixture is widely available, large industries with high production lines think that in big cylinders hydrogen tends to accumulate in higher section of the bottle. Bottle is a plastic jar with a narrow neck. This would cause a problem of working with different hydrogen concentrations using the bottle. On other hand, in-house mixing of hydrogen and nitrogen would assure the proper mixture, but it is very hard to achieve due to the risks related to hydrogen handling.

Hydrogen leak testing is a manual operation, but with a proper sampling probe, it can also be automated. An approach similar to the accumulation method can be favourably applied.

Although hydrogen quickly disperses in air and, even in case of big leak, the working area is not saturated by hydrogen, it is convenient to save tracer gas using a preliminary leak test for gross leakage identification. A pressure decay test can be easily integrated in the hydrogen testing machine.

6.27.5 How to Choose the Test Method

The specification of the acceptable leak and the test method are the first parameters to take in consideration in the plan of a product.

The questions are:

- Which type of total leak can damage the product?
- How many points (welding, joints, etc.) must be tested?
- How much time the product last?

The answer to these questions depends on the volume or gas mass that flows in a definite time. For example, 10 gm (0.35 oz) of R134a refrigerant is leaked in seven years. Sometimes answer to this question is not straightforward. For example, to define the total leak specifications of refrigeration system is simple, but to define the valid specifications for the single components and the single welding point is much more complex. Every leak testing methodology has advantages and disadvantages. For leaks in the range 10^{-1} and 10^{-3} mbar/s all the test methods are able to recognize a leak. Besides sensitivity, in selecting the proper testing method, many parameters have to be considered. Among them are repeatability, accuracy, report capability, operator dependency and, most important, cost of the equipment and the necessary work force. In addition, the cost of tracer gas has an important factor especially in case of series controls. In some cases, a gas recovery system, cutting down consumption, is to be taken into account.

For the sake of clarity, in the following a sample application is treated.

Let 82 gm (3 oz) of R134a in 5 years at internal pressure of 1800 kPa (261 psi) and room temperature an acceptable leakage. This leak corresponds to 16.4 gm/y (0.59 oz/y) of refrigerant

$$F = 82/5 = 16.4 \text{ gm/y} = 16.4/(365 \times 24 \times 3600)$$
$$= 5.20 \times 10^{-7} \text{ gm/s}$$

The equivalent leak rate can be computed by the kinetic theory of perfect gas. Since the molar weight of R134a is 102.03, the equivalent molar flow is:

$$F_m = F/M = 5.20 \times 10^{-7}/102 = 5.10 \times 10^{-9} \text{ mol/s}$$

It is known that a mole or a gas occupies a volume of 22.4×10^{-3} m^3 (0.79 ft^3) at atmospheric pressure and 0°C (32°F). Then the molar volume at room temperature and atmospheric pressure is:

$$V_m = 22.4 \times 10^{-3} \times 300/273 = 24.6 \times 10^{-3} \text{ m}^3/\text{mol}$$

Let p be the atmospheric pressure, the volumetric leak flow will be:

$$Q = F_m \cdot V_m \cdot p = 5.10 \times 10^{-9} \times 24.6 \times 10^{-3} \times 105$$
$$= 1.25 \times 10^{-5} \text{ Pa} \cdot \text{m}^3/\text{s} = 1.25 \times 10^{-4} \text{ mbar/s}$$

The internal pressure has effects on the leak rate. The flow through an orifice (outer opening) depends not only on the pressure difference at the two sides of the orifice, but also on their absolute pressure values. Let P_1 and P_2 be the absolute pressures, to both edges of an orifice, that cause the rate of Q_1 leak, and P_3 and P_4 are the absolute pressures that cause, in the same orifice, the Q_2 leak rate. Then the following relationship will be:

$$\frac{Q_1}{Q_2} = \frac{P_1^2 - P_2^2}{P_3^2 - P_4^2} \qquad \text{...(6.44)}$$

This equation allows computing the equivalent leakage when different pressure values are used. For example, resuming the previous case, the leak of $Q_1 = 1.25 \times 10^{-4}$ mbar/s from $P_1 = 1800$ kPa (261 psia) to $P_2 = 100$ kPa (14.5 psia), corresponds to a flow rate (Q_2) from the atmospheric pressure to vacuum:

$$Q_2 = 1.25 \times 10^{-4} \times \frac{1^2 - 0}{18^2 - 1^2} = 3.8 \times 10^{-7} \text{ mbar/s.}$$

If the leak testing is done in vacuum chamber, pressurizing the unit to 100 kPa (14.5 psia), the limit leak rate should be 3.8×10^{-7} mbar/s.

The flow through a leak depends also on the fluid viscosity; the less the viscosity, the larger the flow rate. Therefore, the flow (Q_1) of a medium of viscosity (v_1) is related to the flow (Q_2) of a medium of viscosity (v_2) by the relationship:

$$\frac{Q_1}{Q_2} = \frac{v_2}{v_1} \qquad \text{...(6.45)}$$

Returning to the previous example, considering that the tracer gas is helium, the R134a leak can be converted in the equivalent of leak for helium, based on various viscosity (v) of two gases

$$\frac{Q_{R134a}}{Q_{He}} = \frac{v_{He}}{v_{R134a}} \qquad \text{...(6.46)}$$

Since the viscosity of the R134a is 0.014 cP and that one of helium 0.018 cP, this follows that the permissible leak for helium is:

$$Q_{He} = 3.8 \times 10^{-7}\ 0.014/0.018 = 2.95 \times 10^{-7} \text{ mbar/s}$$

Even though these considerations are useful to identify the best suited testing techniques, they are not precise. Several approximation and simplification errors are introduced in the various steps. The perfect gas assumption is a useful approximation but, especially for refrigerants that should be more properly considered vapours, perfect gas is not possible. In addition, there is no confidence that if a leak occurs at 1800 kPa (261 psi), this leak happens also to 100 kPa (14.5 psi), especially if the pressure acts in opposite sense. Moreover, the flow computations are applicable accurately in known flow regimen (set of rules for diet), but, obviously, this is not the case. In order to obviate (clear away) to these error sources, it is better to carry out the tests with sensitivity ten times higher than that obtained with theory, in order to create a wide safety margin.

6.27.6 Conclusion

The growing demand for products with fewer weaknesses are acceptable due to several compelling market demands, such as economical requirements, environmental protection specifications, safety constraints and quality products requirement. The strict quality control for leak testing is very important. Researchers, technicians, scientists, producers, etc., are working for completely sealed products, so that air cannot go in or out. For making the products tight sealedone should have knowledge for finding location of leaks of equipment. Surprisingly, this technical field is nearly unknown even in engineering and important project organizations. A brief analysis of some of leak detection techniques used most commonly is given in this chapter. Every methodology has advantages and disadvantages. For right choice of the test methodology used, it is necessary to accurately consider all admitted leak limits and all the other factors, not only the technological requirements, but also the corporate image, regular development and the new requirements of the market.

ANNEXURE A

Pressure Unit Conversion

1 atmos = 760 mm of Hg = 1013 mbar = 101325 Pa = 14.7 psi = 760 Torr

1 mm of Hg = 1.3328947 mbar, 1 Pa = 7.500616×10^{-3} mm

1 psi = 6895 Pa

Pressure unit psi is pound per square inch.

psi and psig are the same units.

Formulas at a Glance

S.No.	Subject	Formula	Interpretation
1.	Pressure on the surface of a hemispherical liquid drop	$p = \frac{2T}{r}$	T is surface tension of the liquid and r is the radius of drop.
2.	Excess pressure inside a soap bubble	$p = \frac{4T}{r}$	
3.	Pressure meniscus surface	$p = \frac{2T \cos\theta}{r}$	θ is made of contact of tube surface.
4.	Temperature of coalesce drop	$\theta = \frac{2T}{J}\left(\frac{1}{r} - \frac{1}{R}\right)$	r and R are the radii of small and coalesce drop, respectively.
5.	Viscous force	$F = -\eta \frac{Adv}{dx}$	v is the velocity of liquid and $\frac{dv}{dx}$ is velocity gradient, η is coefficient of viscosity, a is cross sectional-area of tube

6.	Coeffieicent of viscosity	$\eta = \frac{\pi P r^4}{8Vl}$	P is difference of pressure at two ends of tube, V is liquid volume flow g per second, r and l are radius and length of the tube.
7.	Coefficient of viscosity after correction at two ends	$\eta = \frac{\pi P r^4}{8V(1+1.64r)} - \frac{V\rho}{8\pi(1+1.64r)}$	
8.	Stoke's formula of coefficient of viscosity	$\eta = \frac{2}{8}\frac{r^2 \cdot g(\rho-\sigma)}{v}$	v is velocity of the body falling in a viscous liquid, $(\rho - \sigma)$ is difference between density of body and liquid. r is radius of body and g is acceleration due to gravity.
9.	Viscous force	(i) $F = 6\pi v r \eta$	η is coefficient of viscosity, v and r are velocity and radius of ball falling in a viscous medium.
10.	Rate of liquid volume flow	$V = \frac{\pi P}{8\nu}\left(\frac{l_2}{r_2^4}+\frac{l_1}{r_1^4}\right)^{-1}$	r_1 and r_2 are radii of two capillaries of having length l_1 and l_2, respectively. P is pressure difference at two ends of system.
11.	Grade formula of pressure difference $(p_1 - p_2)$ between inlet and outlet of stator is given by the relation.	$p_1 - p_2 = \frac{\pi . 6 I u}{h^2}$	η is the viscosity of the gas in question, u is the speed of the rotor and l and h are the length and radial length of the slot between inlet and outlet. Thus, at ordinary pressures, since η undergoes no change, the pressure difference $(p_1 - p_2)$ remains constant for a constant speed (u) of the rotor.

SOLVED QUESTIONS

6.1 A block of wood floats in a bucket of water in a lift. Will the block sink more or less, if the lift starts accelerating up?

Ans. It will float at the same level as upward thrust on the wooden block depends upon volume of water displaced or the surface area of block within the water and not on g. However, the weight depend on g, so it will increase as the lift accelerated upward.

6.2 A bottle full of a liquid is fitted with a tight cork. Explain why a slight blow on the cork may be sufficient to break the bottle?

Ans. The blow (force f) that is given to the cork exerts a pressure ($= f/s$) on the liquid. Here, s is the inside surface area of cork. According to Pascal's law the pressure is communicated undiminished to each part of the liquid and the bottle. So the force on the area S of the vessel will be $F = pS = (f/s)\ S$. Now as the area of vessel is much larger than the area of cork, s, F may be large enough to break the bottle.

6.3 Explain why an air bubble in water rises from bottom to top and grows in size.

Ans. We know that fluids form higher pressure to lower and in a fluid pressure increases with depth, so pressure at top ($= p_0$) is lesser than the bottom ($p_0 + hdg$) and so the air bubble will move from bottom to the top. It cannot move sideways as the pressure at the same level in a fluid is same. Furthermore, in coming from bottom to top pressure decreases, so in accordance with Boyle's law PV = RT = constant at constant temperature, volume of bubble V increases, i.e., size of bubble will grow.

6.4 A man is sitting in a floating boat in a pond. If the man drinks some water from the pond, what will happen to the level of water in the pond?

Ans: If man drinks m gm of water from the pond, the weight (boat + man) will increase by mg and so that boat will displace m gram more water for floating. So due to removal of water from the pond, the water level in pond will fall but due to water displaced by the floating boat the water level in the pond will rise and so the water removed from the pond is equal to the water displaced by the boat. Hence, the water level in the pond will remain unchanged.

SOLVED EXAMPLES

6.1 A liquid of density (d) is filled in a beaker of the cross-sectional area (S) to a height H and then a cylinder of mass (m) and cross-sectional area (s) is made to float in it as shown in Fig. 6.29. If the atmospheric pressure is p_0, find the pressure (a) at the top face A of the cylinder, (b) at the bottom face (C) of the cylinder and (c) at the base (B) of the beaker. Can ever these three parameters be equal?

Solution: *Formula required*

Pressure due liquid of density (d) at depth (h) of the beaker is given by

$$P = hdg$$

Given: Density of liquid (d), mass of the cylinder (m), cross-sectional area of cylinder (s), atmospheric pressure, p_0.

Calculation:

(a) The pressure at point A will be equal to the atmospheric pressure, p_0.

(b) The pressure p_B at point B at the bottom of cylinder will be

$$p_B = \text{Atmospheric pressure} + \text{Pressure at point } B \text{ due to cylinder}$$

The volume of the cinder is hs, hence liquid displaced by cylinder is *hsd*, where, *h* is height of the cylinder, the weight of the cylinder $hsdg = mg$. Since the weight of the cylinder is equal to the thrust by the liquid at lower surface (s) of the cylinder, therefore the pressure (p_B) at point B will be

$$p_B = p_0 + hdg = p_0 + \frac{mg}{S}$$

(b) Volume of liquid is HSd, mass of the liquid will be $HSdg$, the pressure at point C at the bottom of beaker will be

$$p_C = p_0 + \frac{\text{Mass of the liquid} + \text{Mass of cylinder}}{\text{Surface area of the bottom of beaker}} = p_0 + \frac{HSdg + mg}{S} = p_0 + Hdg + \frac{mg}{S}$$

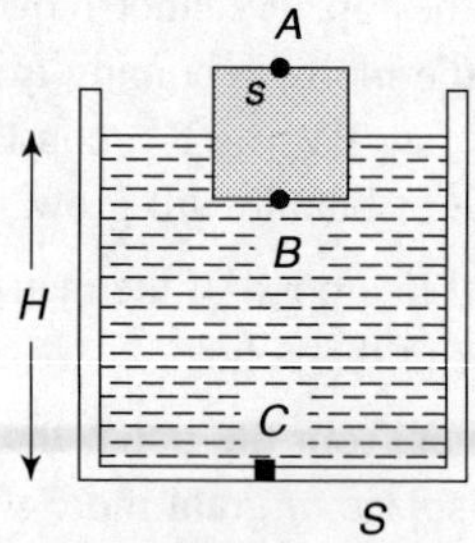

Fig. 6.29

6.2 An open U tube of uniform cross section contains mercury. When 40.8 cm of water is poured into one limb of the tube: (a) How high does the mercury rise in the other limb from its initial level? (b) What is the difference in levels of the liquids of the two sides? [d_w = 1 and d_{Hg} = 13.6]

Solution: *Formula required*

Pressure due liquid of density (d) at depth (h) of the beaker is given by

$$P = hdg$$

Given: Depth of water, h = 40.8 cm, density of water, d_w = 1, and mercury, d_{Hg} = 13.6

Calculation:

(a) If water pushes the mercury in the other limb by y, then the mercury in the other limb will rise by y above its initial level (as fluid is incompressible), so that

$$AA' = BB' = y$$

$$h_2 = B'B + BC = B'B + AA' = 2y$$

If h_1 is the height of water column above A', then the pressure, $p_{A'} = h_1 d_{Hg} g$

If h_2 is the height of water column above C, the pressure, $p_C = h_2 d_w g$

Since points A' and C are at same level, so pressure points A' and C will be same, i.e.,

$$p_{A'} = p_C \quad \text{or} \quad h_1 d_{Hg} g = h_2 d_w g \quad \text{or} \quad h_1 d_{Hg} = h_2 d_w$$

Putting the values of various quantities, we have

$$40.8 \times = 2y \times 13.6$$

which yields $y = 1.5$ cm, i.e., mercury rises by 1 cm from the initial level.

(b) The difference of levels on two sides

$$z = h_1 - h_2 = 40.8 - 3 = 37.8 \text{ cm}$$

Ans. (a) The mercury rise in the other limb from its initial level by 1.5 cm, and (b) difference in levels of the liquids of the two sides is 37.8 cm.

6.3 Upto which height should a cylindrical vessel be filled with a homogenous liquid to make the force with which the liquid presses on the side of the vessel equal to the force exerted by the liquid on the bottom of the vessel?

Solution: Let us consider a cylindrical vessel of radius (r) filled with a liquid of density (d) to a height, h. If p_0 is the atmospheric pressure, the pressure difference inside and outside the base of cylinder is

$$\Delta p_B = (p_0 - hdg) - p_0 = hdg$$

If S is the surface area of base, the extra force inside the base is

$$F_B = S \times \Delta p_B = \text{p } r^3 \times hdg$$

The pressure at depth (y) below the surface of liquid inside the vessel is ($p_0 + ydg$), while it is p_0 outside. So the pressure difference on two sides of the curved surface at depth (y) below the surface will be

$$\Delta p_s = (p_0 - ydg) - p_0 = ydg$$

The force on the strip of curved surface of thickness dy at depth y below the surface of the lquid as shown in Fig. 10.9 will be

$$\Delta F_C = \Delta p_s \times 2\pi rdy = ydg \times 2\pi rdy = 2\pi rdgy\, dy$$

Thus the force on the outside surface will be

$$F_C = 2\pi rdg \int_0^h y\, dy = \pi rdg\, h^2$$

If the force at the base and outside surface are equal, i.e.,

$$F_B = F_C$$

or $\pi r^2 \times hdg = \pi r\, dg\, h^2$ or $h = r$

Ans. If cylinder is filled upto the height equal to its radius, then the force at the base and outside surface are equal.

6.4 Figure 6.30 shows a siphon in action. The liquid flowing through the siphon has a density of 1.5 gm/cc. Calculate the pressure difference between (a) points A and D (b) points B and C. [$g = 980$ cm/s^2]

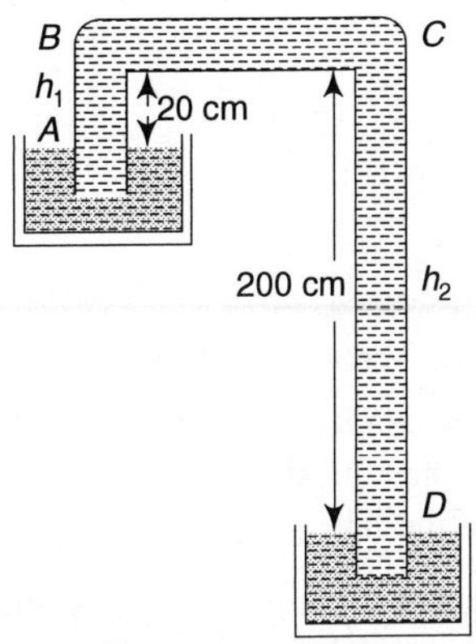

Fig. 6.30 Shows working of a siphon

Solution:

(a) As points A and D are at open surface, $p_A = p_D = p_0$ (atmospheric pressure so $p_A - p_D = 0$.

(b) If p_B is pressure at B,

$$p_A (= p_0) = p_B + h_1\, dg$$

As pressure at local level is same, i.e.,

$$p_B = p_0 - h_1\, dg$$

Similarly, $p_C = p_0 - h_2\, dg$

Therefore, $p_B - p_C = (h_2 - h_1\, dg$

Putting the values of h_1 = 20 cm, h_1 = 200 cm, d = 1.5 gm/cc, g = 980 cm/s^2

$$p_B - p_C = (200 - 20) \times 1.5 \times 980 = 2.65 \times 10^{-5} \text{ dyne/cm}^2$$
$$= 2.65 \times 10^{-4} \text{ N/m}^2$$

Ans. (a) Pressure difference between A and D is zero. (b) Pressure difference between B and C is 2.65×10^{-4} N/m^2.

6.5 An open ended tube is immersed vertically in mercury in such a way that a length 0.04 m extends above the mercury level. The open end of the tube is then closed and the tube is raised further by o.44 m. Calculate the length of the air column above the mercury level in the tube.

Solution: Let us consider that the air in the tube with cross section A when the tube is open, so

$$p_1 = p_0 \quad \text{and} \quad V_1 = A \times 0.04$$

Now when the tube is closed and 0.44 m of it is withdrawn out of mercury, if h is the length of air column above and p_2 is the pressure of air in it, then as at same level pressure is same.

$$p_2 = p_0 + (0.48 - h)\, dg \quad \text{and} \quad V_2 = Ah$$

If the temperature is constant, then according to Boyle's law

$$p_1 V_1 = p_2 V_2$$

or $$p_0 \times A \times 0.04 = Ah \times [p_0 - (0.48 - h)\, dg]$$

or $$0.76\, dg \times A \times 0.04 = Ah \times [0.76\, dg - (0.48 - h)\, dg] \quad \text{as} \quad [p_0 = 0.76\, dg]$$

or $$0.76 \times 0.04 = h \times [0.76 - (0.48 - h)]$$

or $$0.0304 = 0.28\, h + h^2$$

or $$h^2 + 0.28\, h - 0.0304 = 0$$

Solving the quadratic equation, we get

$$h = \frac{1}{2}[-0.28 \pm \sqrt{(0.28)^2 + 4 \times 0.0304}]$$
$$= \frac{1}{2}[-0.28 \pm \sqrt{0.0784 + 0.1216}$$
$$= \frac{1}{2}[-0.28 \pm 0.447]$$
$$= 0.0835 \text{ m}$$

Ans. The length of the air column above the mercury level in the tube is 8.35 cm.

6.6 When a large bubble rises from the bottom of a lake to the surface, its radius becomes double. If the atmospheric pressure is equal to that of a water column of height H, calculate the depth of the lake.

Solution: Let p_1 and p_2 *abe* atmospheric pressure and pressure at the bottom of the lake. V_1 and V_2 are volumes of the drop out of lake and at the bottom of the lake, respectively.

According to Boyle's law at constant temperature

$$p_1 V_1 = p_2 V_2$$

The atmospheric pressure, $p_1 = p$

The atmospheric pressure is equal to the water column of height H, i.e., $p_1 = p = Hdg$

If h is the depth of the lake, the pressure, $p_2 = p + hdg$

If r is the radius of bubble at bottom of the lake, then volume of bubble, $V_2 = \frac{4}{3}\pi r^3$.

Out of lake radius of bubble becomes double, therefore the volume of bubble out of lake, $V_1 = \frac{4}{3}\pi (2r)^3$.

Putting the values of p_1, V_1, p_2 and V_2 in Boyle's equation, we get

$$p \times \frac{4}{3}\pi (2r)^3 = (p + hdg) \times \frac{4}{3}\pi r^3$$

or $$p\,(2)^3 = (p + hdg)$$

or $$8\,p - p = hdg \quad \text{or} \quad hdg = 7p = 7\,Hdg \quad \text{or} \quad h = 7\,H$$

Ans. Depth of the lake is 7 H.

6.7 A piece of ice is floating in water of a beaker. What will happen to the level of water when all ice melts. What will happen if the beaker is filled with liquid (a) denser than water, and (b) lighter than water?

Solution: If m gm ice is floating in a liquid of density, d_L, V_D is the volume of displaced liquid, then for its equilibrium

Weight of ice = Thrust of water, i.e., mg = $V_D\, d_L\, g$,

So the volume displaced by the ice, $V_D = \frac{m}{d_L}$

If m gm ice melts completely, water formed will have mass m gm. If d_w is the density of water, the volume of water formed will be

$$V_F - \frac{m}{d_w}$$

Instead of liquid there is water in beaker, so $d_L = d_w$, and hence $V_D = V_F$, i.e., water displaced by the ice is equal to the water formed due meting of ice. Therefore, level of water remains unchanged.

(a) If $d_L > d_w$, then $\frac{m}{d_L} < \frac{m}{d_w}$, i.e., $V_D < V_F$

i.e., volume of water displaced by the floating ice will be less than water formed on ice melting and so the level of liquid in the beaker will rise.

(a) If $d_L < d_w$, then $\frac{m}{d_L} < \frac{m}{d_w}$, i.e., $V_D > V_F$

i.e., volume of water displaced by floating ice will be more than water formed on ice melting and so the level of liquid in the beaker will fall.

6.8 A boat containing some pieces of material is floating in a pond. What will happen to the level of water in the pond if on unloading the pieces in the pond, the piece (a) floats or (b) sinks?

Solution: If M is the mass of boat and m is the mass material pieces in it, V_D is the volume of displaced water then initially as the system is floating,

$$M + m = V_D\, d_w$$

So volume of displaced water is

$$V_D = \frac{M+m}{d_w} \qquad \text{...(i)}$$

When material pieces are dropped in the pond, the boat will float, so it displaces water, $M = V_1\, d_w$, i.e., now volume of displaced water, $V_1 = \dfrac{M}{d_w}$.

(a) If the unloaded pieces floats in the pond, the water displaced by the pieces, $m = V_2\, d_w$, i.e., the volume of displaced water by pieces, $V_2 = \dfrac{m}{d_w}$

Total water displaced by the boat and the unloaded pieces will be

$$V_1 + V_2 = \frac{M}{d_w} + \frac{m}{d_w} \qquad \text{...(ii)}$$

Thus, water displaced by boat containing some pieces of material and by the boat and unloaded pieces is equal, so level of water in the pond will remain unchanged.

(b) If the unloaded pieces of material sinks, the water displaced by them will be equal to their own volume, i.e.,

$$V_2' = \frac{m}{d} \qquad \left[\text{as } d = \frac{m}{V_2'}\right]$$

Total volume of water displaced by the boat and pieces when sinking will be

$$V_1 + V_2' = \frac{M}{d_w} + \frac{m}{d} \qquad \text{...(iii)}$$

Since pieces are sinking, so $d > d_w$, so this volume will be less than the volume of water displaced by boat containing the pieces, so level of water in the pond will fall.

6.9 A beaker, full of water, has an ice piece floating in it. As the ice melts what happens to the water level if (a) the cube contains an air bubble, (b) the cube contains, and (i) a lead piece and a cork.

Solution:

(a) If m gm of the ice containing an air bubble is floating in water,

$$M = V_D\, d_w, \text{ so displaced volume, } V_D = \frac{M}{d_w}$$

An air bubble encloses air, so it is weightless.

When M gram ice melts, water formed will be M gm and it will occupy volume, $V_F = \dfrac{M}{d_w}$.

As the volume of water formed by melting of ice is equal to the volume of water displaced by the floating ice, the level of water in the beaker will remain unchanged.

(b) Let the ice be M, and the mass of embedded piece be m, then for floating

$$\text{Weight} = \text{upthrust or } M + m = V_D\, d_w$$

So the volume of water displaced by the embedded ice cube when floating

$$V_D = \frac{M}{d_w} + \frac{m}{d_w} \quad \text{...(i)}$$

When ice melts water formed will be M gm and the volume of water formed by melting of ice, $V_1 = \frac{M}{d_w}$.

(i) The ice cube initially contains a piece of lead then $d_{Pb} > d_w$, it will sink when ice melts and the water displaced by it will be equal to its own volume, i.e., $V_2 = \frac{m}{d_{Pb}}$.

Total volume occupied by water formed by melting of ice and lead piece will be

$$V_1 + V_2 = \frac{M}{d_w} + \frac{m}{d_{Pb}} \quad \text{...(ii)}$$

Since $d_{Pb} > d_w$, this volume will be less than the initial water displaced by the floating embedded ice cube, therefore, the water level in the beaker will fall as the ice melts, if ice contains a sinking lead.

(ii) If the ice initially contains a cork, then the cork will float when ice melts as $d_{cork} < d_w$, the cork will float on the water when ice melts, so the water displaced by floating cork piece will be

$$M = V_2' d_w \quad \text{or} \quad V_2' = \frac{m}{d_w}$$

[and not its own volume, $V_1' = \frac{m}{d_{cork}}$].

In case water formed by melting of M gm of ice and water displaced by floating cork,

$$V_1 + V_2' = \frac{M}{d_w} + \frac{m}{d_w} \quad \text{...(iii)}$$

This volume is the same as initial volume displaced by the floating given by Eq. (i), the level of water in the beaker will remain unchanged.

QUESTIONS FOR EXAMINATION

6.1 Will 5 litres of kerosene will weigh more in winter or summer?

Ans. Winter

6.2 Why is a fluid container more likely to burst from the base?

Ans. The hydrostatic pressure is more at base, so fluid container more likely to burst from the base.

6.3 When you drink a liquid through a straw, you reduce the pressure in your mouth and let the atmosphere move the liquid. Could you use the straw to sip a drink on the moon?

Ans. No, as moon has no atmosphere, so there will be no atmospheric pressure to move the liquid.

6.4 Because atmospheric pressure is 10^5 N/m^2 and area of the chest of a person is about 0.1/m^2, the force of the atmosphere on thechest of person is about 10,000 N (or 1000 kg), why our bodies collapse?

Ans. As we have open system to atmosphere, the pressure inside and outside of our body is same due to atmospheric pressure.

6.5 Describe the working of a rotary oil pump for producing low pressure. How are these pressures measured?

6.6 Describe a diffusion pump and explain its working principle. How the vacuum is measured?

6.7 Describe, with a neat diagram, the construction and working of (a) any modern high vacuum pump and (b) a gauge which can measure the low pressure this pump produces.

6.8 Describe in detail the Pirani resistance gauge, illustrating your answer with a neat diagram and explaining the essential points of the gauge, its range, utility and drawbacks.

6.9 Explain the principle and working of the Knundsen gauge. What are its particular advantages over the other forms of gauges.

6.10 Explain the principle and working of a modern high vacuum pump. How can you measure the pressure so obtained?

6.11 Describe a molecular pump and explain its working principle. How the vacuum is measured?

OBJECTIVE QUESTIONS

6.1 Equal masses of water and a liquid of density 2 are mixed together, the density of mixture is

(a) 2/3 (b) 4/3 (c) 3/2 (d) 3/1

6.2 Pressure at a point inside a liquid does not depend on

(a) The the point below the surface of the liquid

(b) The nature of the liquid

(c) The acceleration due to gravity at that point

(d) The shape of the containing vessel

6.3 Two stretched membranes of areas 2 cm^2 and 3 cm^2 are placed in a liquid at the same depth, the ratio of the pressure on them is

(a) 1 : 1 (b) 2 : 3 (c) 3 : 2 (d) 5 : 2

6.4 A dam water reservoir is built thicker at the bottom than at the top because

(a) Pressure of water is very large at the bottom due to its large depth

(b) Water is likely to have more density at the bottom due to its large depth

(c) Quantity of water at the bottom is large

6.5 A vessel contains liquid of density (d) as shown in Fig. 10.50. The pressure of liquid at point (P) is

(a) hdg (b) Hdg (c) $(H - h)\, dg$ (d) $(H - h)\, dg \cos \theta$

6.6 A tank of 5 m high is half filled with water and then rest of the half is filled to the top with oil of density 0.85 g/cm^3. The pressure at the bottom of the tank, due to these liquids, is

(a) 1.85 g/cm^2 (b) 89.25 g/cm^2 (c) 462.5 g/cm^2 (d) 500 g/cm^2

6.7 The height to which a cylinder is filled with a homogeneous liquid, to make the average force with which the liquid presses the side of the cylinder equal to the force exerted by the liquid on the bottom of the cylinder, is equal to

(a) Half of the radius of the cylinder

(b) Radius of the cylinder

(c) One-fourth of the radius of the cylinder

(d) Three-fourths of the radius of the cylinder

6.8 Pressure is applied to an enclosed fluid. It is

(a) Increased and applied to every part of the fluid

(b) Diminished and transmitted to the walls of the container

(c) Increased in proportion to the mass of the fluid and then transmitted

(d) Transmitted unchanged to every portion of the fluid and wall of the container

6.9 Pressure applied to an enclosed fluid is transmitted undiminished to every portion of the liquid and walls of the containing vessel. This law was first formulated by

(a) Bernoulli (b) Archimedes (c) Boyle (d) Pascal

6.10 When a bubble comes from the bottom of a lake to the top, its radius

(a) Increases (b) Decreases (c) Does not change (d) Becomes zero

6.11 When a large bubble rises from the bottom of a lake to the surface, its radius becomes double. The atmospheric pressure is equal to that of a water column of height H, the depth of the lake will be

(a) H (b) 2 H (c) 7 H (d) 8 H

6.12 When a body is weighed in a liquid, then its weight depends upon

(a) Volume of the body (b) Mass of the body

(c) Shape of the body (d) Centre of gravity of the body

6.13 Which of the following would a hydrogen balloon find easier to lift

(a) 1 kg of water (b) 1 kg of steel

(c) 1 kg of light packed feather (d) All of these

6.14 An iron ball is weighed in air and then in water by aspring balance, then

(a) Its weight in air is more than in water

(b) Its weight in water is more than in air

(c) Its weight is same both in air and water

(d) Its weight is zero in water

6.15 Abody weighs 40 gm in air, if its volume is 10 cc, its weight in water will be

(a) 30 gm (b) 40 gm (c) 50 gm (d) 20 gm

6.16 A body of weight (W_1) displaces an amount of water, W_2, when floating

(a) $W_1 > W_2$ (b) $W_1 < W_2$ (c) $W_1 = W_2$ (d) None of these

6.17 When a ship floats in water

(a) The mass of water displaced is two times to the mass of the ship

(b) The mass of water displaced is equal to the mass of the ship

(b) The mass of water displaced is lesser to the mass of the ship

(c) The mass of water displaced is greater to the mass of the ship

6.18 It is easier to swim in the seawater than in river water because

(a) Density of river water is more than that of seawater

(b) Sea is deeper than river

(c) Density of seawater is more than that of river water

(d) Sea has high waves all the times

6.19 When a loaded boat enters the sea from a river, it rises

(a) There is more water in sea than river (b) Seawater is denser the river water

(c) Sea is deeper than river (d) High waves flow in sea

6.20 When a piece of ice floating in a beaker of water completely melts, the level of the water in the beaker

(a) Rises (b) Falls (c) Remain unchanged (d) None of these

6.21 An ice cube containing a lead piece in it is floating in a glass of water. As ice melts the water level will

(a) Fall (b) Rise (c) Remain unchaged (d) None of these

6.22 A man is sitting in a boat which is floating in a pond. If man drinks some water from the pond. The level of water in pond

(a) Rises a little (b) Sinks a little

(c) Remains unchanged (d) None of these

6.23 A boat with wood is floating in a lake. If the wood is thrown in the lake the water level

(a) Goes up (b) Goes down

(c) Remains unchanged (d) None of these

6.24 A boat with scrap of iron is floating in a lake. If the wood is thrown in the lake the water level

(a) Goes up (b) Goes down

(c) Remains unchanged (d) None of these

6.25 A ball floats on the surface of water in a container exposed to atmosphere. If the container is covered and air is compressed, the ball

(a) Rises (b) Goes down

(c) Remains unchanged (d) None of these

6.26 A wooden piece floats half submerged in a trough of water. If the trough is transferred to a lift ascending with acceleration the wooden piece

(a) Sinks a little more
(b) Rises a little
(c) Remains half submerged
(d) Sinks to the bottom.

6.27 When a satellite orbiting the earth, a cork is immersed in a jar of water and release. It

(a) Sinks a little more
(b) Rises a little.
(c) Remains where left
(d) None of these

Answers

6.1 (b)	6.2 (d)	6.3 (a)	6.4 (a)	6.5 (c)	6.6 (c)	6.7 (b)	6.8 (d)
6.9 (d)	6.10 (a)	6.11 (c)	6.12 (a)	6.13 (c)	6.14 (a)	6.15 (a)	6.16 (a)
6.17 (b)	6.18 (c)	6.19 (b)	6.20 (c)	6.21 (a)	6.22 (c)	6.23 (c)	6.24 (b)
6.25 (a)	6.26 (c)	6.27 (c)					

Further Reading

1. *Elements of Properties of Matter* by D.S. Mathur, S. Chand, Daryaganj New Delhi, 1993.

CHAPTER 7

Sound

7.1 INTRODUCTION

Sound waves are longitudinal mechanical waves generated by density or pressure fluctuations. In other words, a sound wave is a pressure disturbance that moves through air or water, or whatever the host medium happens to be. Human vocal sound is generated by the back and forth vibration of the vocal cords. When these cords are extended, they momentarily compress air molecules together creating a high pressure region that is accelerated outwards. As the cords pull back, they generate a momentary vacuum, a low pressure region (in between these two situations, the air pressure obviously passes through a "normal" pressure circumstance). In other words, the vibration of the vocal cords creates regions of high pressure, then normal pressure, then low pressure, then normal pressure, then high pressure, etc., as they vibrate back and forth.

These pressure disturbances move out into the surrounding air at approximately 330 metres per second (i.e., the speed of sound). As they pass a hearing person, the pressure variations motivate tiny hairs in the listener's ears to vibrate generating electrical signals which, upon reaching the brain, are translated into incoming sound.

In brief, sound waves are a disturbance moving through a medium. Without the medium, there can be no sound.

Characteristics of sound

Sound intensity is measured in decibels (dB). The logarithmic scale is used to measure the intensity of sound, so loudness becomes double by increasing the intensity by10 dB.

Sound pitch is measured in hertz (Hz), the standard unit for the measurement for frequency.

Audible sound, that our ears can detect, ranges from about 20 Hz to 20,000 Hz, although, on ageing and exposure to loud sounds the upper limit generally decreases.

Infrasonic: frequencies below 20 Hz, e.g., earthquake, thunder, etc.

Ultrasonic: Frequencies above 20,000 Hz could be heard by cats, dogs, bats, and porpoises. [It is a sea animal with a blunt rounded projected nose and mouth, similar to dolphin or small whale.] Ultrasonic waves are called **SONAR** waves because of their important applications.

RADAR(radio detection and ranging) is an electronic device which detects objects by using ultra-high frequency or microwave segment of the radio frequency spectrum to control the location of the spot or range of an object. It cannot detect submarine under, because, microwaves are strongly absorbed by seawater within feet of their transmission. This renders radar unusable underwater. Hydroacoustic sensors use SONAR technology to detect submarines.

A common use of ultrasound is in SONAR. Ultrasonic waves are used in SONAR instead of audible sound waves.

SONAR (sound navigation and ranging) is the best example that illustrates how a ship on the ocean utilizes the reflecting properties of sound waves to determine the depth of the ocean. A sound wave is transmitted and bounces back from the seabed. Because the speed of sound is known and the time lapse between sending and receiving the sound can be measured. Thus, depth of see can be measured using ultrasonic waves using the technique called **SONAR**.

It is found that the time interval between the sending and receiving of the wave is 1.6 s. If the velocity of ultrasound is 1400 m/s, then the depth of sea is 1120 m.

Acoustics of building

The branch of Physics which deals with the planning, design and construction of a hall, auditorium or a room to provide the best audible sound to every audience is called *acoustics of building*. In this chapter, we will discuss the essential features of acoustics for sound distribution in a hall, like reverberation, time of reverberation, Sabine formula for calculating time of reverberation, coefficient of absorption, etc.

7.2 OSCILLATIONS

The motion that repeats about its mean position is called *oscillation*. For example, boats bobbling at anchor, surging pistons in the engines of cars, swinging chandeliers are oscillatory motions. The oscillation motions are performed by strings in guitars, drums, bells, diaphragms in telephones and speaker systems, pendulum of clocks and quartz crystal in watches. Oscillations of molecules transmit the sensation of sound, oscillations of atoms in the solids convey the sensation of heat.

Oscillations are not confined to material objects such as violin string, beating drums. Light, radio waves, waves from TV antennas, X-rays, gamma rays and motion of electrons in solids are oscillatory phenomena.

Oscillations in the real world are usually damped, i.e., motion dies out gradually because of transferring mechanical energy to thermal energy due to action of frictional forces. It is not possible to eliminate 100% loss of mechanical energy. However, it is possible to reduce the loss of energy of an object by using certain methods. For example, during swinging the children pass mechanical energy by pushing the earth by their legs to maintain swinging (oscillation) motion. Similarly, with help of grease or oil or ball bearing transfer of mechanical energy produced during oscillatory motion to heat energy in fans, piston in engines of vehicles is reduced by minimizing the friction.

7.2.1 Simple Harmonic Motion

The sequence of motion of pendulum bob at equal time intervals about origin is shown in Fig. 7.1(a).

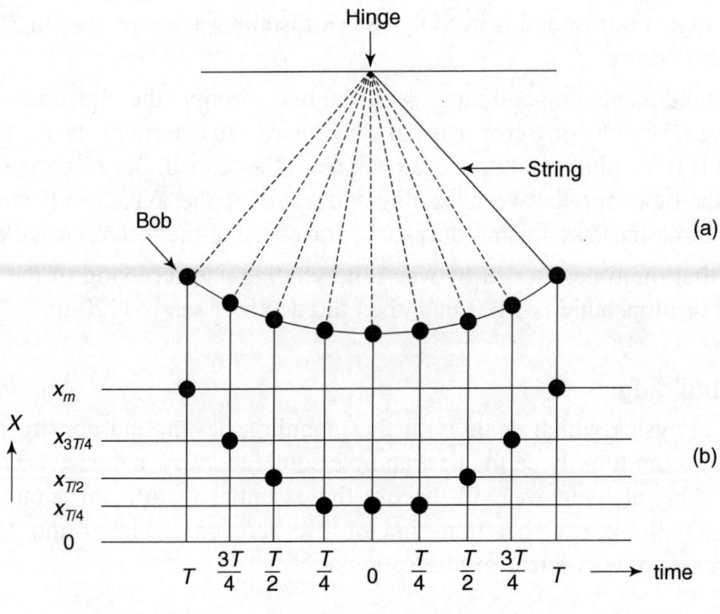

Fig. 7.1 (a) Oscillatory motion of pendulum. (b) Position of bob along X-axis at equal time intervals.

The position of bob along X-axis of the pendulum is shown in Fig. 7.1(b) at equal time intervals by a symmetric curve about the origin.

One important property of oscillatory motion is its frequency. Frequency is equal to the number of oscillations completed in one second. We have chosen the symbol of frequency, f. Its SI unit is hertz (Hz), where,

$$1 \text{ hertz} = 1 \text{ Hz} = 1 \text{ oscillation per second}$$
$$= 1 \text{ cycle per second} \quad ...(7.1)$$

Time required to complete one oscillation is called time period, T. Time period (T) and frequency (f) are related as follows:

$$T = \frac{1}{f} \quad ...(7.2)$$

The bob oscillates about its mean position back and forth about its mean position along the X-axis between maximum limits: – x_m and x_m. The sequence of bob for repeated motion of the pendulum at equal time intervals is shown in Fig. 7.2. The vector arrows are scaled to indicate the speed of the bob. The speed is maximum when the particle is at the origin and zero at ± x_m. Let the time (*t*) is zero when bob is at x_m and the time to return the bob at – x_m is *T*, then *T* for one oscillation is called *time period.*

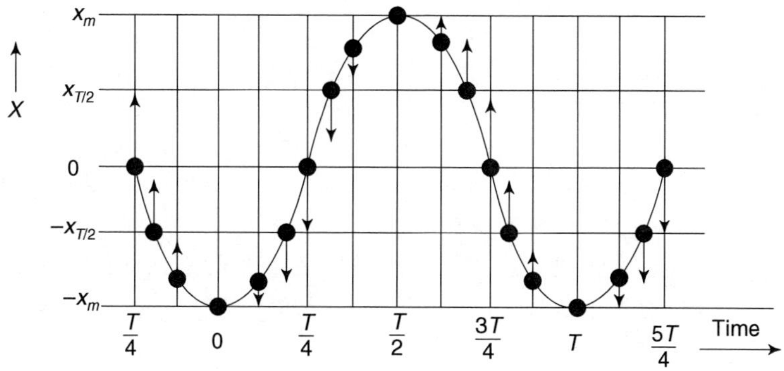

Fig. 7.2 Sequence of position of bob at equal time intervals as it oscillates back and forth about the origin along X-axis between limits

Any motion that repeats at regular intervals is called *periodic motion* or *harmonic motion.* For such motion the displacement of the particle (bob in case of pendulum) from the origin is given as a function of time by

$$x\,(t) = x_m \cos\,(\omega\, t + \varphi) \qquad ...(7.3)$$

where, x_m, ω and φ are constants. This motion is called *simple harmonic motion* (*SMH*). One can infer from Eq. (7.3) that periodic motion is a sinusoidal function of time. The cosine function in Eq. (7.3), varies between the limits ± 1. The maximum displacement of the particle ($|x_m|$) on either directionis called the *amplitude* of the motion of particle in Fig. 7.2.

In Eq. (7.3), the argument of the cosine function (ω *t* + φ) varies with time and is called the phase of the motion. The constant (φ) is called the *phase constant* or *phase angle* of motion. Velocity of the particle can be obtained by differentiating *x* (*t*) with respect to *t*, i.e.,

$$v = \frac{dx\,(t)}{dt} = -\,\omega\, x_m \sin\,(\omega\, t + \varphi) \qquad ...(7.4)$$

at *t* = 0, Eq. (7.4) becomes

$$v = -\,\omega\, x_m \sin\,(\varphi)$$

or

$$\varphi = \sin^{-1}\left[\frac{-\,v}{\omega\, x_m}\right] \qquad ...(7.5)$$

From Eq. (7.5) it is clear that the value of φ depends on the displacement and velocity of the particle at $t = 0$. In Fig. 7.3(a) the displacement is maximum and velocity is zero at $t = 0$, so the phase angle is zero.

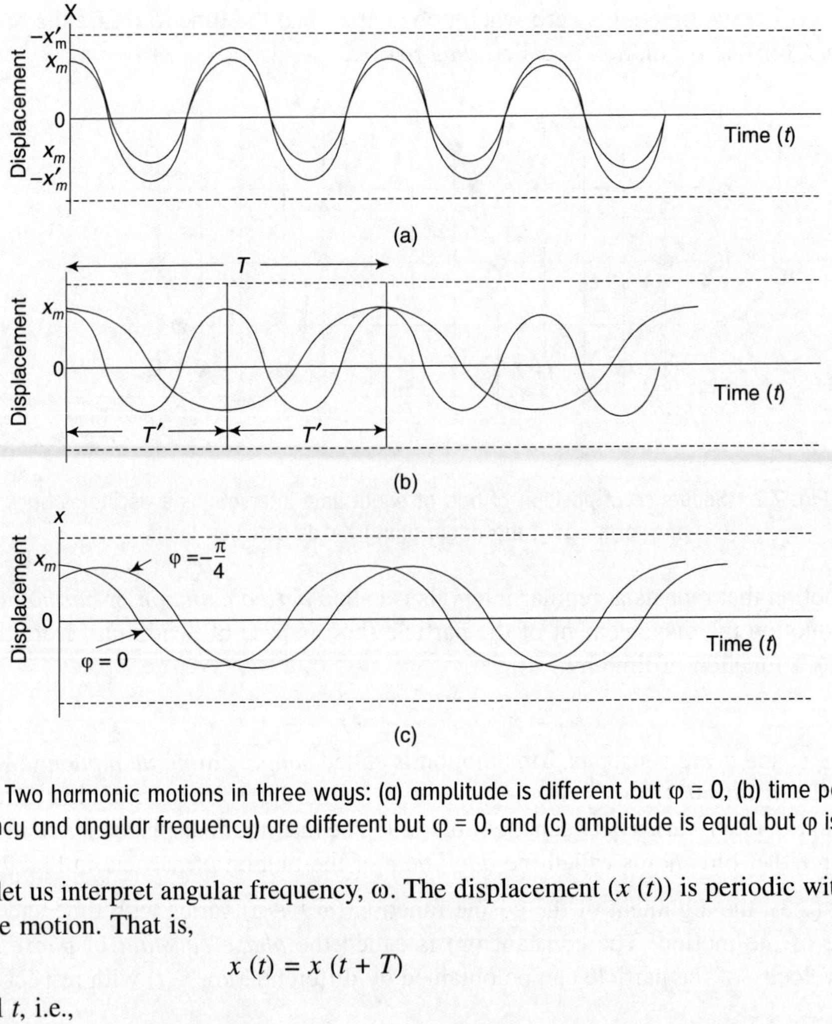

Fig. 7.3 Two harmonic motions in three ways: (a) amplitude is different but φ = 0, (b) time period (i.e., in frequency and angular frequency) are different but φ = 0, and (c) amplitude is equal but φ is different

Now let us interpret angular frequency, ω. The displacement ($x(t)$) is periodic with period (T) of the motion. That is,

$$x(t) = x(t + T)$$

for all t, i.e.,

$$x_m \cos(\omega t) = x_m \cos[\omega(t + T)] \qquad ...(7.6)$$

For simplicity we have taken φ = 0 in writing Eq. (7.6). Since $\cos(A) = \cos(A + 2\pi)$, so Eq. (7.6) can be written as

$$x_m \cos(\omega t + 2\pi) = x_m \cos[\omega(t + T)] \qquad ...(7.7)$$

Equation (7.6) implies that

$$\omega(t + T) = \omega t + 2\pi$$

or

$$\omega T = 2\pi$$

or $$\omega = \frac{2\pi}{T}$$

From Eq. (7.2) $T = \frac{1}{f}$, so

$$\omega = 2\pi f \qquad ...(7.8)$$

The quantity (ω) is called the *angular frequency* of the motion. Its SI unit is the radian per second. In Fig. 7.3, we have plotted two harmonic motions that differ (a) in amplitude (b) in period (i.e., in frequency and angular frequency), and (c) in phase constant.

7.2.2 Simple Harmonic Oscillator

Motion of an object which repeats at definite intervals is called *periodic* or *harmonic.* For example motion of the earth around the sun is said to be periodic or harmonic. If an object undergoes periodic motion between two limits over the same path, the motion of the object is called oscillatory. For example motion of the pendulum in clock is oscillatory. At microscopic level, the motion of atoms in crystalline solids is periodic.

To illustrate the concept of oscillatory motion, let us consider a body of mass m connected to one end of the spring as shown in Fig. 7.4(a). The other end of the spring is fixed with the pole. The body slides on a frictionless, horizontal surface as shown in Fig. 7.4(b).

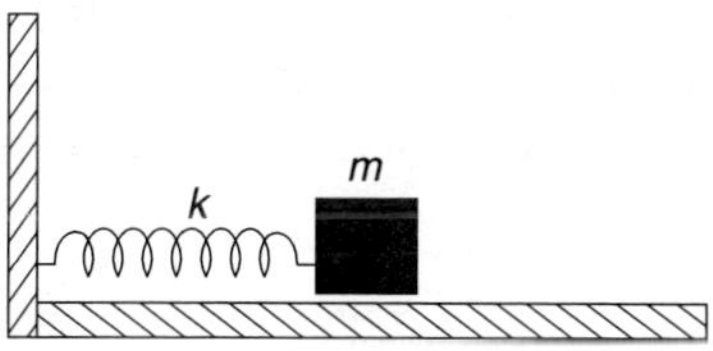

Fig. 7.4 (a) Experimental setup for simple harmonic oscillator

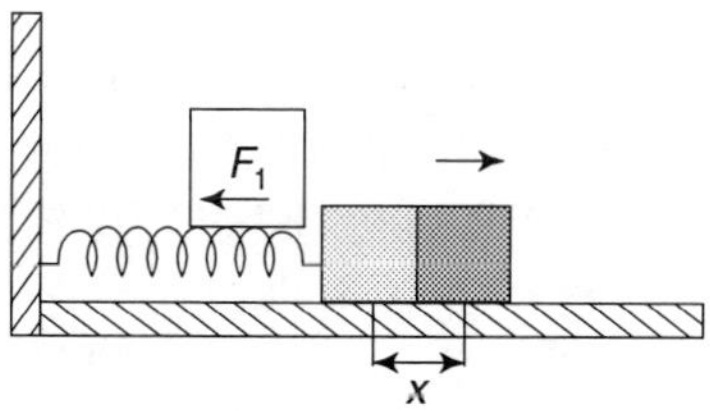

Fig. 7.4 (b) Shows motion of simple harmonic oscillator

Let us consider that the body is at equilibrium at position $x = 0$, that is when the spring is relaxed [Fig. 7.4(a)]. If the body is moved from $x = 0$ and then released, the body oscillates along a horizontal line as shown in Fig. 7.4(b). If the spring is not compressed or stretched too far from $x = 0$, the force applied on the body to move it at position x is proportional to x i.e.,

$$F = -kx \qquad ...(7.9)$$

The proportionality constant (k) measures the stiffness of the spring. The negative sign implies that the force acts to restore the body to its equilibrium position. Newton's equation of motion of the body can be written as

$$m\ddot{a} = -kx \quad \text{or} \quad \ddot{a} = -\frac{k}{m}x$$

Putting the value of acceleration, $\ddot{a} = \dfrac{d^2x}{dt^2}$, we get

$$\frac{d^2x}{dt^2} = -\frac{k}{m}x \quad \text{or} \quad \frac{d^2x}{dt^2} = -\omega_0^2 x \qquad \text{...(7.10)}$$

where, angular frequency of harmonic motion, $\omega_0 = -\sqrt{\dfrac{k}{m}}$. The simple harmonic motion is represented by linear differential Eq. (7.10). The time period of the motion is given by

$$T = \frac{2\pi}{\omega_0} = \frac{2\pi}{\sqrt{k/m}} = 2\pi\sqrt{\frac{m}{k}} \qquad \text{...(7.11)}$$

The frequency (ν) of the motion, i.e., number of cycles per second is given by $\nu = 1/T$. The time period of simple harmonic motion is independent of the amplitude of the motion.

Like acceleration, force is always directed towards the mean position, hence it is sometimes called the restoring force in simple harmonic motion.

Note that force in Eq. (7.9) is linearly proportional to x (t). A particle oscillating under such a force is, therefore, called a *linear harmonic oscillator.* In real conditions, the force may contain small additional terms proportional to x^2, x^3, etc. In that case we call them *non-linear oscillators.*

Example Two identical springs of the spring constant (k) are attached to a block of mass (m) and fixed supports as shown in Fig. 7.5(a). Show that when the mass is displaced from its equilibrium position on either side, it executes a simple harmonic motion. Find period of oscillation.

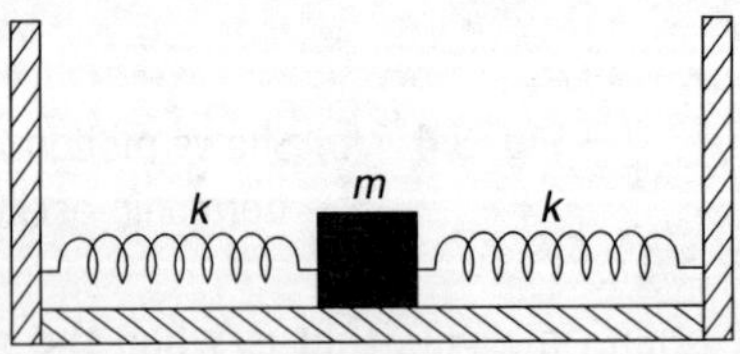

Fig. 7.5 (a) Experimental set up for simple harmonic oscillator

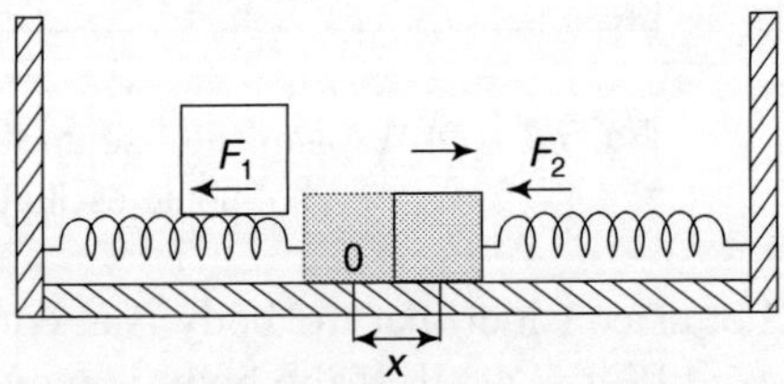

Fig. 7.5 (b) Shows motion of simple harmonic oscillator

Let the mass m is displaced by small distance x to the right side of the equilibrium as shown in Fig. 7.5(b). As a result, the spring on the left side gets elongated by length equal to x and the spring on the right side gets compressed by the same length. The forces acting on the mass are written as

$$F_1 = -kx \quad \text{and} \quad F_2 = -kx$$

The force exerted by the spring on the left side, F_1 is trying to pull the mass towards the mean position.

The force exerted by the spring on the right side, F_2 is trying to pull the mass towards the mean position.

The net force, F, acting on the mass is the sum of these forces, i.e.,

$$F = -2\,kx$$

From this equation it is clear that the force, F, acting on the mass is proportional to the displacement and is directed towards the mean position. Therefore, the motion executed by the mass is simple harmonic. If ω_0 is the angular frequency of simple harmonic motion, then the time period of oscillation will be

$$T = \frac{2\pi}{\omega_0} = \frac{2\pi}{\sqrt{2\,k/m}} = 2\pi\sqrt{\frac{m}{2\,k}} \quad ...(7.12)$$

7.2.3 Energy of Particle in Simple Harmonic Motion

The particle executing simple harmonic motion inherits both kinetic energy and potential energy of a particle. Both of these in simple harmonic motion vary between zero and their maximum value.

The motion of particle is simple harmonic motion with respect to space and with respect to time. So displacement is written as

$$x\,(t) = A\cos(\omega t + \varphi)$$

So velocity of particle will be

$$v = \frac{dx}{dt} = -\,\omega A\sin(\omega t + \varphi) \quad ...(7.13)$$

It is clear from Eq. (7.13) that velocity of particle is also a periodic function of time. Now let us find the velocity of particle at extreme position from Eq. (7.13). For simplicity of calculations, let us consider that phase of wave, $\varphi = 0$.

Let the time period and frequency of simple harmonic motion be T and f, respectively, so that $T = \frac{1}{f} = \frac{2\pi}{\omega}$. We assume that initially the particle is at extreme right position.

(i) So, at $t = 0$ or $\omega t = 0$, $v = 0$.

The time taken by the particle to reach at the mean position is $\frac{T}{4}$.

(ii) So, at $t = \frac{T}{4}$ or $\omega t = \frac{\pi}{2}$, $v = -\,\omega A$. Negative sign of velocity indicates that the particle is moving in the –ve x-direction.

The time taken by the particle to reach at the extreme position in the left will be $\frac{T}{2}$.

(iii) So, at $t = \frac{T}{2} =$ or $\omega t = \pi$, $v = 0$.

After time $\frac{3T}{4}$, when the particle reaches at mean position again.

(iv) So at $t = \frac{3T}{4}$ or $\omega t = \frac{3\pi}{2}$, $v = +\omega A$. Positive sign of velocity indicates that the particle is moving in the +ve x-direction.

Thus, the velocity of particle at extreme displacement is zero and is maximum at the mean position.

Kinetic energy of such particle is defined as

$$K = \frac{1}{2} mv^2$$

Putting the value of v from Eq. (7.13) is

$$K = \frac{1}{2} m\,\omega^2 A^2 \sin^2(\omega t + \varphi)$$

$$= \frac{1}{2} k A^2 \sin^2(\omega t + \varphi) \quad ...(7.14)$$

is also periodic function of time.

It is clear from Eq. (7.14) that kinetic energy of the particle is also a periodic function of time. Now let us find the kinetic energy of particle using Eq. (7.14). For simplicity of calculations, let us consider that phase of wave, $\varphi = 0$.

For calculating kinetic energy, let us write the Eq. (7.14) as

$$K = \frac{1}{2} k A^2 \sqrt{\frac{1 - \cos(\omega t + \varphi)}{2}} \quad ...(7.15)$$

We assume that initially the particle is at extreme right position.

(i) So, at $t = 0$ or $\omega t = 0$ and $K = 0$.

(ii) At mean position, i.e., when $t = t = \frac{T}{4}$ or $\omega t = \frac{\pi}{2}$, $K = \frac{1}{2} k A^2$.

The time taken by the particle to reach at the extreme position in the left will be $\frac{T}{2}$.

(iii) So when $t = \frac{T}{2}$ = or $\omega t = \pi$, $K = 0$.

From extreme left position the pendulum comes back at mean position after time $\frac{3T}{4}$.

(iv) So, when $t = \frac{3T}{4}$ or $\omega t = \frac{3\pi}{2}$, $K = \frac{1}{2} k A^2$

Thus kinetic energy of the particle executing simple harmonic motion is zero when the displacement is maximum and it is maximum when the particle is at the mean position.

The potential energy associated with conservative force,

$$F = -kx$$

will be
$$U = \frac{1}{2}kx^2 = \frac{1}{2}kA^2\cos^2(\omega t + \varphi) \quad ...(7.16)$$

The potential energy of a particle is also periodic function of time. The calculation of potential energy at extreme displacement and mean position can be done easily. It is found that the potential energy of the particle executing simple harmonic motion is maximum when the displacement is maximum and zero when the particle is at the mean position. Therefore, it is concluded that the sum of kinetic energy and potential energy of harmonic oscillator remains constant throughout the motion.

Total energy of the particle executing simple harmonic motion is written as sum of kinetic energy and potential energy, i.e.

$$E = \left(\frac{1}{2}\right)mv^2 + \left(\frac{1}{2}\right)kx^2 = \left(\frac{1}{2}\right)kA^2 \quad ...(7.17)$$

Although the position and velocity of the oscillator are continuously changing, the total energy remains constant.

7.3 DAMPED SIMPLE HARMONIC MOTION

It is difficult for a pendulum to swing in water, because the water exerts a drag force on the pendulum that stops the motion slowly. However, it is easier to swing for pendulum in air, but still the motion dies out after some time, because the air exerts a drag force on the pendulum and friction, if it is there, also acts to slow down the motion by transferring energy from the pendulum.

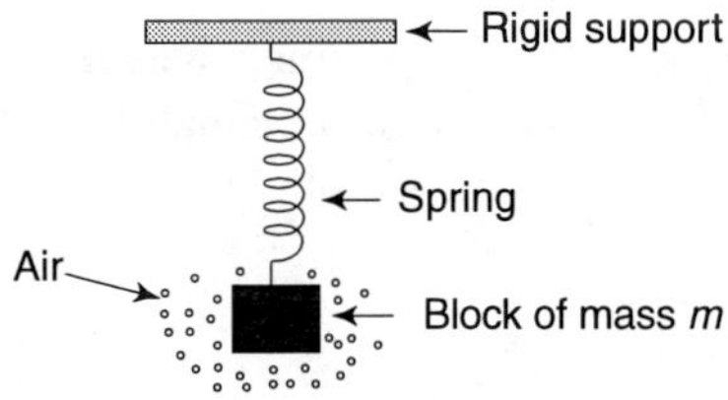

Fig. 7.6 Damped harmonic oscillator. Air exerts a damping force on the oscillating block

When applied external force is withdrawn, the motion of an oscillator slowly and slowly is reduced, then the oscillator and its motion are said to be damped. An ideal example of a damped oscillator is shown in Fig. 7.6, a block with mass m oscillates with the help a spring, which has spring constant, k. A system of block-spring is hanging with help of a rod. As the block moves up and down, the air exerts a drag force on it. With time, the mechanical energy of the block-spring system decreases. The mechanical energy is converted to thermal energy and transferred to the air.

Let a damping force $(\vec{F_d})$ exerted on the liquid is proportional in magnitude to the velocity $(\vec{V})$ of the block. Then

$$F_d \propto v$$

or

$$F_d = -\,bv \qquad ...(7.18)$$

where, b is a damping constant that depends on the characteristics of both, the block and the air, and has the SI unit of metre per second. The minus sign indicates that F_d opposes the motion. The formula given by Eq. (7.18) will be accurate only when v is small, that is, the block moves slowly. The total force acting on the block will be

$$\Sigma F = -\,kx - bv \qquad ...(7.19)$$

If we take $v = \dfrac{dx}{dt}$, then

$$\Sigma F = -\,kx - b\,\frac{dx}{dt} \qquad (7.20)$$

Using Newton's second law of motion, we get the differential equation

$$m\frac{d\;x}{dx} + b\frac{dx}{dt} + kx\; = 0 \qquad ...(7.21)$$

The solution of Eq. (7.21) is

$$x\,(t) = x_m\, e^{-bt/2m} \cos\,(\omega' t + \varphi) \qquad ...(7.22)$$

where, the angular frequency of the damped oscillator, ω', is given by

$$\omega' = \sqrt{\frac{k}{m} - \frac{b^2}{4\,m^2}} \qquad ...(7.23)$$

If we set $b = 0$, i.e., when there is no damping, then Eq. (7.23) reduces to equation of undamped oscillator. The angular frequency of undamped oscillator will be

$$\omega = \sqrt{\frac{k}{m}}$$

If damping constant is very small, i.e.,

$$\frac{b^2}{4\,m^2} << \frac{k}{m}$$

$$b << \sqrt{km}$$

then $\omega' \approx \omega$

Equation (7.22) is considered a cosine function where amplitude, $x_m\ e^{-bt/2m}$, gradually decreases with time, as shown in Fig. 7.7. For an undamped oscillator, the mechanical energy is constant and is given by equation

$$E = \frac{1}{2} k\, x_m^2$$

If the oscillator is damped, the mechanical energy is not constant and decreases with time. If the damping is small, then $E\ (t)$ for damped oscillator can be obtained by replacing x_m in expression $E = \frac{1}{2} k\, x_m^2$ by the effective amplitude ($x_m\ e^{-bt/2m}$) of the damped oscillator, i.e.,

$$E\ (t) = \frac{1}{2} k x_m^2\ e^{-bt/m} \qquad ...(7.24)$$

Equation (7.24) indicates that mechanical energy of the damped oscillator decreases exponentially with time.

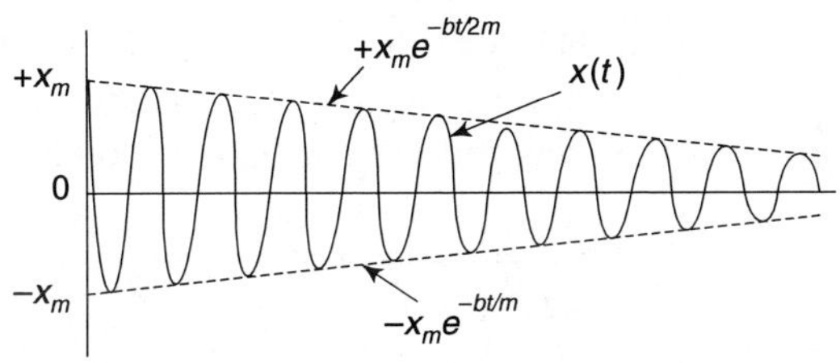

Fig. 7.7 Displacement function (x (t)) for the damped oscillator. The amplitude (+ $x_m\ e^{-bt/2m}$) decreases with time

7.4 FORCED OSCILLATIONS AND RESONANCE

When a block attached to a spring, as shown in Fig. 7.6, is displaced from its equilibrium position and released, it oscillates with natural frequency (ω) and oscillations are called *free oscillations*. All free oscillations slowly and slowly die out because of the ever present damping force. However, an external force can maintain these oscillations. These are called *forced* or *driven oscillations*. The external force, which is periodic with frequencies of oscillation (ω_d), is called the driven frequency. The important fact of force periodic motion is that the system oscillates not with its natural frequency (ω), but at frequency (ω_d) of the driven force. The free oscillations slowly die out due to damping. The familiar example of forced oscillation is that of child enjoying swinging with a rope attached to a branch of a tree. The child maintains oscillations by pushing the ground by his feet perodically.

Let us assume that a periodically varying force with time ($F\ (t)$) of amplitude (F_0) is applied to a damped oscillator. Such deriving forcecan be written as

$$F\ (t) = F_0 \cos\ (\omega_d\ t) \qquad ...(7.25)$$

The oscillatory motion of the block under the two forces, (i) linear restoring force (damping force) and (ii) time dependent driving force given by Eq. (7.25), is represented by equation

$$ma\ (t) = -\ kx\ (t) - bv\ (t) + F_0 \cos\ (\omega_d\ t).$$

Substituting the value of acceleration, $a\ (t) = \dfrac{d^2x}{dt^2}$, we get

$$m\frac{d^2x(t)}{dt^2} + b\frac{dx\ (t)}{dt} + kx\ (t) = F_0 \cos\ (\omega_d\ t) \qquad ...(7.26)$$

where, k is force constant of natural free motion and b is force constant due to linear restoring damped force. ω_d is the angular frequency of applied periodic driven force. Initially, the oscillator oscillates with natural frequency, ω. When an external periodic driven force is applied, oscillations with natural frequency die out, and then the oscillator oscillates with angular frequency (ω_d) of the external periodic force. Its displacement, after the natural oscillations die out, is written as

$$x\ (t) = A \cos\ (\omega_d\ t + \varphi) \qquad ...(7.27)$$

where, t is the time measured from the moment when we apply the external periodic force. The amplitude is the function of forced frequency (ω_d) and natural frequency, ω. After solving Eq. (7.26) and detailed analysis the amplitude of oscillations comes out to be

$$A = \frac{F_0}{\sqrt{m^2\ (\omega^2 - \omega_d^2)^2 + \omega_d^2\ b^2}} \qquad ...(7.28a)$$

and

$$\tan\varphi = \frac{-\ v_0}{\omega_d\ x_0} \qquad ...(7.28b)$$

where, m is the mass of the particle, x_0 and v_0 are displacement and velocity of particle, $t = 0$, the moment when external periodic force was applied. Eq. (7.28a) shows that displacement amplitude (A) depends on angular frequencies, ω_d and ω. Now let us consider two cases:

(i) When damping is small, i.e., $\omega_d << \omega$

In this case the term $m^2\ (\omega^2 - \omega_d^2)^2 >> \omega_d^2\ b^2$, so we can neglect the second term under square root and hence Eq. (7.28a) reduces to

$$A = \frac{F_0}{m\ (\omega^2 - \omega_d^2)} \qquad ...(7.29)$$

Equation (7.29) shows that amplitude of oscillations depends on the applied force, i.e., amplitude is different for different driving forces as shown in Fig. 7.8. Again it is observed that the amplitude is maximum when $\omega_d = \omega$, a position of resonance. Further, one observes from curves in Fig. 7.8 that resonance peak of maxima becomes large and narrow as damping force decreases.

If we go on decreasing the driving frequency, the amplitude increases and becomes infinity when the deriving frequency becomes equal to natural frequency. But this ideal situation of zero damping never arises in real system because the damping never becomes perfectly zero.

You must have experienced during swinging that when the time your push matches with the time of period of the swing, your swing gets the maximum amplitude. This is quite large, but not infinity as there is always some damping in your swing. This will become more clear by point (ii) given below:

(ii) When damping frequency is close to natural frequency, i.e., $\omega_d \approx \omega$

In this case the term, $m^2(\omega^2 - \omega_d^2) << \omega_d^2 b^2$, so we can neglect the first term under square root and hence Eq. (7.28a) reduces to

$$A = \frac{F_0}{\omega_d b} \qquad ...(7.30)$$

Equation (7.30) indicates that maximum amplitude for giving driving frequency depends on deriving frequency and damping, so it will never become infinity. There is maximum amplitude of the oscillating oscillator when the driving force is close to natural frequency of the oscillator. Thus, the phenomenon of maximum amplitude of an oscillator, when the driving frequency is close to natural frequency of the oscillator, is called *resonance*.

One can experience the phenomenon of resonance during swinging. The skill in swinging to achieve greater heights lies in the matching the rhythm of pushing against the ground with the period of swing, i.e., with the natural frequency of the swing.

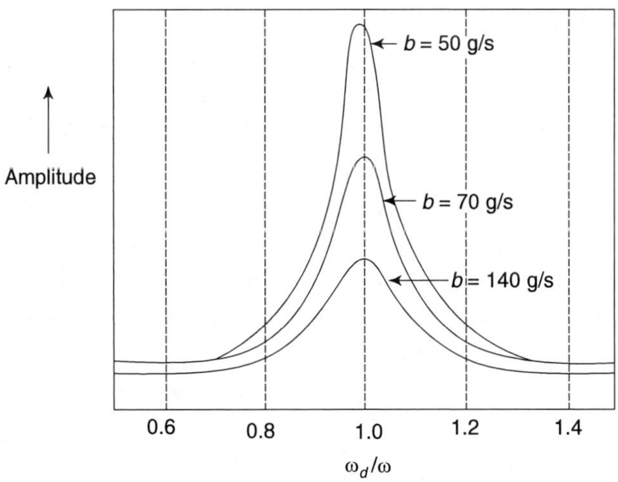

Fig. 7.8 Variation of displacement amplitude (x_m) of a forced oscillator with angular frequency (ω_d) of the driving force for three values of damping constant, *b*. The amplitude is maximum at $\omega = \omega_d$ the resonant frequency

All mechanical structures have one or more natural frequencies, and if a structure is subjected to a strong external deriving force that matches one of these frequencies, the resulting oscillations of the structure may rupture it. Thus, for example, aircraft designer must make sure that none of the natural frequencies at which a wing can vibrate matches the angular frequency of the engines at the cruising speed. It obviously would be dangerous for a wing to flap violently at certain engine speeds.

You might have heard that a large group of people doing march past with a rhythm together like army personnel platoon or light infantry are never allowed to cross through a concrete bridge, because march past may force the bridge to rupture or collapse due to resonance phenomenon, when forced frequency of march past rhythm becomes close to natural frequency of the bridge.

During earthquake many concrete structures either rupture or collapse due to resonance phenomenon. Oscillations of the ground due to the earthquake's seismic waves have their greatest velocity amplitude at an angular frequency of around 9 rad/s, which almost exactly matches a natural angular frequency of the individual horizontal sections of the concrete bridge or building.

7.5 PRINCIPLE OF SUPERPOSITION OF WAVES

When waves travel from one region to another, two or more waves often pass simultaneously from the same region. For example, when we listen a concert, sounds from many instruments fall simultaneously on our eardrums. An array of signals sent by antennas at different broadcasting centres are bound to pass through the same region before reaching our antennas of our radio and TV receivers. The water waves (ripples) in a lake or harbour produced by churning the wakes of many boats may often overlap.

Suppose that two waves travel simultaneously along the same stretched string. Let $Y_1(x, t)$ and $Y_2(x, t)$ be the displacements of two waves passing through a string. In some region when waves overlap the resultant wave $Y(x, t)$ will be algebraic sum of two waves as shown below:

$$Y(x, t) = Y_1(x, t) + Y_2(x, t) \qquad ...(7.31)$$

The summation of displacements along the string means that overlapping waves algebraically add to produce a resultant wave is called the principle of superposition.

Similarly, when several waves overlap simultaneously, the resultant wave will be the algebraic sum of all the waves as follows:

$$Y(x, t) = Y_1(x, t) + Y_2(x, t) + Y_3(x, t) + Y_4(x, t) + ...$$

$$= f_1\left[\omega_1\left(t - \frac{x}{v}\right)\right] + f_2\left[\omega_2\left(t - \frac{x}{v}\right)\right] + f_3\left[\omega_3\left(t - \frac{x}{v}\right)\right] + ... \qquad ...(7.32a)$$

$$= f_1(\omega_1 t - k_1 x) + f_2(\omega_2 t - k_2 x) + f_3(\omega_3 t - k_3 x) + ... \qquad ...(7.32b)$$

as $$\omega = 2\pi f = = 2\pi\frac{v}{A} = kv \quad \text{or} \quad \frac{\omega}{v} = k$$

In Eq. (7.32a), velocity of the wave is v and angular frequency of wave is ω.

Figure 7.9 shows a sequence of steps of two pulses travelling in opposite directions on the same stretched string. When the pulses overlap, the resultant pulse is their algebraic sum. Surprisingly, each pulse moves through the other, as if the other wave were not present.

"Overlapping waves do not in any way alter the travel of each other."

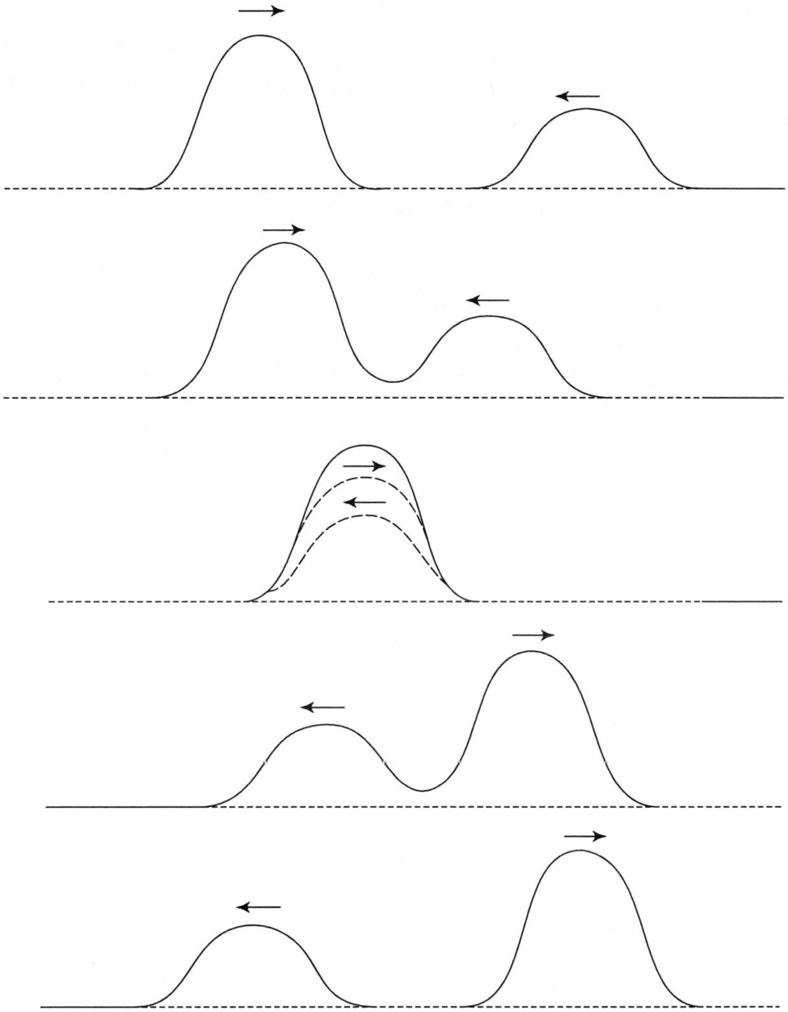

Fig. 7.9 A series of figures show that two pulses travelling in opposite direction along a stretched string. The superposition principle applies as the pulses move through each other

7.6 FOURIER THEOREM AND ANALYSIS

French mathematician Jean Baptiste Fourier (1786–1830) used the principle of superposition to analyse non-sinusoidal or non-harmonic waveforms. He showed that any non-sinusoidal waveform can be represented as the sum of a large number of sinusoidal waves of carefully chosen frequencies and amplitudes.

"Fourier's theorem states that every non-osinusoidal function ($f(x)$) which is reasonably continuous can be exactly reproduced by superposing a sufficient number of simple harmonic [sinusoidal] functions."

Figure 7.10 shows sawtooth curve as an example of a Fourier series. In Fig. 7.10 (a) shows the variation of sawtooth curve with time (at position $x = 0$). The Fourier series for sawtooth curve can be written as

$$Y(t) = -\frac{1}{\pi}\sin\omega t - \frac{1}{2\pi}\sin 2\omega t - \frac{1}{3\pi}\sin 3\omega t \; .... \qquad ...(7.33)$$

in which $\omega = 2\pi/T$, where, T is the period of the sawtooth curve. The black solid curve of Fig. 7.10(a), which represents the sum of the first six terms of Eq. (7.33), matches the sawtooth curve to some extent. Figure 7.10(b) shows individual curve of these six terms. By adding more terms, one can reproduce the sawtooth curve as closely as he/she wants.

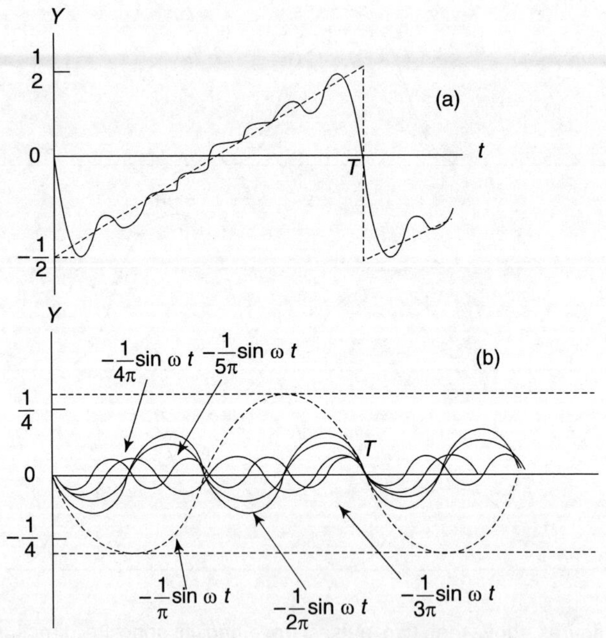

Fig. 7.10 (a) The solid black line curve shows the resultant of first 5 terms in the sum of Eq. (7.21). By including more number (say 50) of terms it will produce a sawtooth curve as shown by dashed line. (b) Individual curves of the first five sinusoidal functions

7.7 INTENSITY AND LOUDNESS LEVEL

Loudness is the sensation of sound received by the ear due to intensity of sound, i.e., loudness depends on intensity. The intensity (I) of a sound wave at a surface is the average rate per unit area at which energy is transferred by the wave through or onto the surface. We can write this as

$$I = \frac{P}{S} \qquad ...(7.34)$$

where, P is the rate of energy transfer (power) of the sound wave and S is the area of the surface facing the sound. The intensity (I) is also related to the displacement amplitude (A) of the sound wave by [Eq. (7.47) of Section 7.7.4]

$$I = \frac{1}{2}\rho\, v\, \omega^2 A^2 \qquad ...(7.35)$$

That is, $I \propto A^2$. Thus larger is the amplitude, larger will be loudness. The loudness being sensation depends on the sensitivity of listener's ear. Therefore, loudness of sound of a given intensity may be different for different listeners. That is, two sounds of equal intensity but different frequency may not appear equally loud to the same listener because the sensitivity of the ear is different for different frequencies. One cannot hear ultrasonic sounds, no matter how intense they are.

The unit of loudness is phon. The ear is most sensitive for the sound frequency 1 kHz. Phon is the intensity level in dB of sound frequency 1 kHz.

7.7.1 Variation of Intensity with Distance

The variation of intensity with distance from a real sound source is not easy due to several barriers like reflection of sound waves from surfaces. For example, some real sources (like loudspeaker) may focus sound in particular directions, and the environment usually produces echoes (reflected sound waves) that overlap with direct sound waves sent by the real source. Also the velocity of air in the atmosphere affects the variation of sound waves drastically. The sound pressure level (SPL) decreases with doubling of the distance by (–) 6 dB. The sound pressure falls 1/2 times (50%) of the sound pressure of the initial value. The sound intensity level decreases with the doubling of the distance by (–) 6 dB.

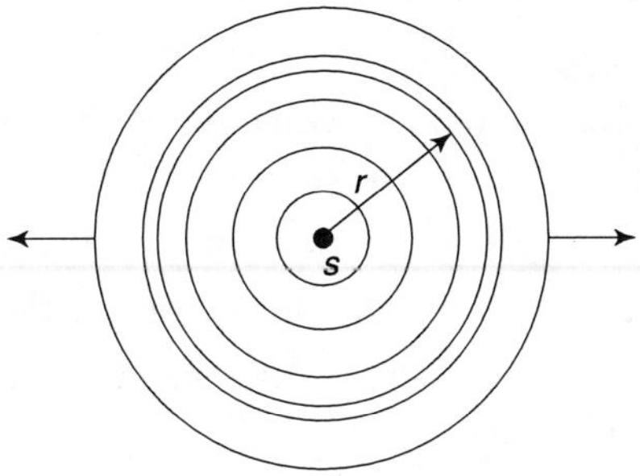

Fig. 7.11 A point source (s) emits sound waves uniformly in all directions. The waves pass through an imaginary sphere of radius →, that is centred on *s*

By making proper conditions we can minimize echoes. In that case we assume that the sound source is a point source that emits the sound isotropically, that is, with equal intensity in all directions. The wavefronts spreading from such an isotropic point source (s) at a particular instant are shown in Fig. 7.11.

Let us assume that the mechanical energy of the sound waves is conserved as they spread from this source. Let us also centre an imaginary sphere of radius r on the source as shown in Fig. 7.11. All the energy emitted by the source must pass through the surface of the sphere. Thus, rate at which the energy is transferred through the surface by the sound waves must be equal to the rate at which energy is emitted by the source (that is, the power (P) of the source). From Eq. (7.34), the intensity (I) at the sphere must then be

$$I = \frac{P}{4\pi r^2} \quad ...(7.36)$$

where, $4\pi r^2$ is the area of the sphere of radius, r. Equation (7.36) indicates that the intensity of sound from an isotropic point source decreases with the square of the distance ® from the source, i.e.,

$$I \propto \frac{1}{r^2}$$

If I_1 and I_2 are intensities at distances d_1 and d_2, then the ratio of intensities will be

$$\frac{I_2}{I_1} = \left(\frac{d_1}{d_2}\right)^2 \quad ...(7.37)$$

This formula can be used to find the intensity I_2 at distance d_2 if intensity I_1 at distance d_1 is known or viceversa.

7.7.2 Decibel scale

The range of audible intensities is from 10^{-12} to 10 W/m^2. The use of a logarithmic scale compresses the range of numbers required to describe this wide range of intensities. We deal with such an enormous range of values by using logarithms. Consider the relation

$$y = \log x,$$

In which y and x are variables. From this equation we can infer that if we multiply x by 10, then y increases by 1. To see this, we write

$$y' = \log (10\,x) = \log 10 + \log x = 1 + y$$

Similarly, if we multiply x by 10^{12}, y increases by only 12.

Thus, instead of speaking of the intensity (I) of a sound wave, it is much more convenient to speak of its sound level (β), defined as

$$\text{SL} = (10\text{ dB}) \log \frac{I}{I_0} \quad ...(7.38)$$

Here, dB is the abbreviation for decibel, unit of sound level, a name that was chosen to recognize the work of Alexander Graham Bell. I_0 in Eq. (7.38) is a standard reference intensity

(= 10^{-12} W/m^2), chosen because it is near the lower limit of the human range of hearing. For $I = I_0$, Eq. (7.38) gives SL = 10 log 1 = 0, so our standard reference level corresponds to zero decibel. Then SL increases by 10 dB every time the sound intensity increases by an order of magnitude (a factor of 10). Thus, SL = 40 corresponds to the intensity, that is, 10^4 times the standard reference level. Table 7.1 lists the sound levels for a variety of environments.

The important point about the decibel is that it is a relative measurement and that each quantity measured in decibels is expressed as a ratio relative to a reference pressure, power or intensity or whatever other quantity is being considered. It is a good policy to employ the word "level" whenever decibel is used, e.g., sound pressure level, sound power level, etc., as a reminder that the measurement is a ratio relative to some reference level.

Table 7.1 Some sound levels (dB)

	SL (dB)
Hearing threshold	0
Rustle of leaves	10
Conversation	60
Rock concert	110
Pain threshold	120
Jet engine	130

SOLVED EXAMPLES

Example 7.1 If the intensity level of audible sound at distance 2 m is 60 dB, what will be intensity level at distance 10 m?

Solution: Formula required

$$\frac{I_2}{I_1} = \left(\frac{d_1}{d_2}\right)^2, \text{SL} = (10 \text{ dB}) \log \frac{I}{I_0}$$

Given: $I_1 = 60$ dB, $d_1 = 2$ m and $d_2 = 10$ m, $I_0 = 10^{-12}$ W/m^2

$$60 = 10 \log \frac{I_1}{I_0}, \frac{I_1}{I_0} = 10^6, \; I_1 = 10^{-6} \text{ W/m}^2$$

$$\frac{I_2}{I_1} = \left(\frac{2}{10}\right)^2 \quad \text{or} \quad I_2 = 10^{-6} \times 0.04 = 4 \times 10^{-8} \text{ W/m}^2$$

$$\text{SL} = (10 \text{ dB}) \log \frac{I_2}{I_0} = 10 \times \log \frac{4 \times 10^{-8}}{10^{-12}}$$

$$= 10 \times \log (4 \times 10^4)$$

$$= 10 \times (4 + \log 4) = 10 \times (4 + 0.6020) = 46.020 \text{ dB}$$

Ans. Sound level at 10 m is 46.020 dB

Example 7.2 If the intensity level of an audible sound at distance 2 m is 60 dB, what will be the intensity level at distance 20 m?

Solution: Formula required

$$\frac{I_2}{I_1} = \left(\frac{d_1}{d_2}\right)^2, \text{ SL} = (10 \text{ dB}) \log \frac{I}{I_0}$$

Given: $I_1 = 60$ dB, $d_1 = 2$ m and $d_2 = 20$ m, $I_0 = 10^{-12}$ W/m^2

$$60 = 10 \log \frac{I_1}{I_0}, \frac{I_2}{I_0} = 10^6, I_1 = 10^{-6} \text{ W/m}^2$$

$$\frac{I_2}{I_1} = \left(\frac{2}{20}\right)^2 \quad \text{or} \quad I_2 = 10^{-8} \text{ W/m}^2$$

$$\text{SL} = (10 \text{ dB}) \log \frac{I_2}{I_0} = 10 \times \log \frac{10^{-8}}{10^{-12}} = 10 \times \log (10^4)$$

$$= 10 \times 4 = 40 \text{ dB}$$

Ans. Sound level at 20 m is 40 dB

Example 7.3 If the intensity of sound level of a rock concert at distance 2 m is 110 dB, what will be intensity level at distance 10 m?

Solution: Formula required

$$\frac{I_2}{I_1} = \left(\frac{d_1}{d_2}\right)^2, \quad \text{SL} = (10 \text{ dB}) \log \frac{I}{I_0}$$

Given: $I_1 = 110$ dB, $d_1 = 2$ m and $d_2 = 10$ m, $I_0 = 10^{-12}$ W/m^2

$$110 = 10 \log \frac{I_1}{I_0}, \quad \frac{I_1}{I_0} = 10^{11}, \; I_1 = 10^{-1} \text{ W/m}^2$$

$$\frac{I_2}{I_1} = \left(\frac{2}{10}\right)^2 \quad \text{or} \quad \text{I}_2 = 10^{-1} \times 0.04 = 4 \times 10^{-3} \text{ W/m}^2$$

$$\text{SL} = (10 \text{ dB}) \log \frac{I_2}{I_0} = 10 \times \log \frac{4 \times 10^{-3}}{10^{-12}}$$

$$= 10 \times \log (4 \times 10^9)$$

$$= 10 \times (9 + \log 4) = 10 \times (9 + 0.6020) = 96.02 \text{ dB}$$

Ans. Sound level at 10 m is 96.02 dB

Example 7.4 If the intensity of sound level of a rock concert at distance 2 m is 110 dB, what will be intensity level at distance 20 m?

Solution: Formula required

$$\frac{I_2}{I_1} = \left(\frac{d_1}{d_2}\right)^2, \text{ SL} = (10 \text{ dB}) \log \frac{I}{I_0}$$

Given: $I_1 = 110$ dB, $d_1 = 2$ m and $d_2 = 20$ m, $I_0 = 10^{-12}$ W/m^2

$$110 = 10 \log \frac{I_1}{I_0}, \frac{I_1}{I_0} = 10^{11}, I_1 = 10^{-1} \text{ W/m}^2$$

$$\frac{I_2}{I_1} = \left(\frac{2}{20}\right)^2 \quad \text{or} \quad I_2 = 10^{-3} \text{ W/m}^2$$

$$\text{SL} = (10 \text{ dB}) \log \frac{I_2}{I_0} = 10 \times \log \frac{10^{-3}}{10^{-12}}$$

$$= 10 \times \log (10^9)$$

$$= 10 \times 9 = 90 \text{ dB}$$

Ans. Sound level at 20 m is 90 dB

7.7.3 Displacement of pressure waves

A sound wave (i.e., longitudinal mechanical wave) can be described either in terms of the longitudinal displacement suffered by the particles of the medium (called *displacement wave*) or in terms of the excess pressure generated due to compression and rarefaction (called *pressure wave*).

Consider a sound wave travelling in the x-direction in a medium. Suppose at time t, the particle at the undisturbed position (x) suffers a displacement (y) in the x-direction. The displacement wave then will be described by

$$y = A \sin (\omega t - kx) \qquad ...(7.39)$$

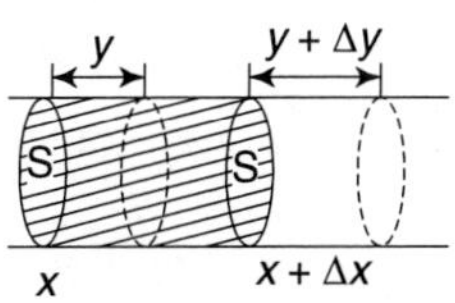

Fig. 7.12

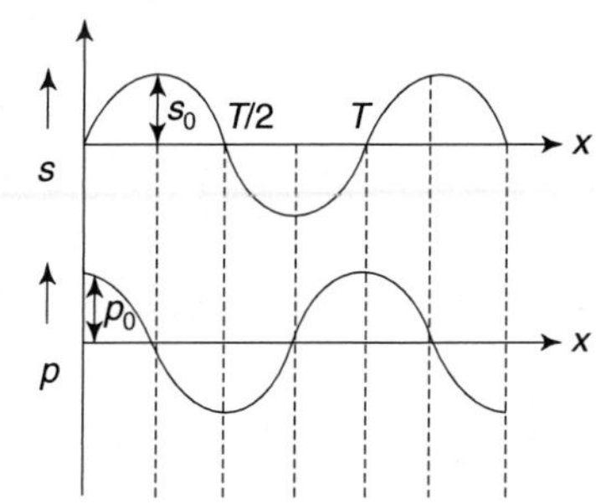

Fig. 7.13

Now consider the element of medium which is confined within x and $x + dx$ in the undisturbed state. If A is the cross-sectional area, the volume element in undisturbed state will be $V = S\Delta x$. As the wave passes, the ends at x and $x + \Delta x$ are displaced by amounts y and $y + \Delta y$ so that increase in volume of the element will be $\Delta V = S\Delta y$ as shown in Fig. 7.12. This, in turn, implies that volume strain for the element under considerations,

$$\frac{\Delta V}{V} = \frac{A\,\Delta y}{A\,\Delta x} = \frac{dy}{dx} \qquad \text{...(7.40)}$$

So the corresponding stress, i.e., excess pressure

$$P = B\left[-\frac{\Delta V}{V}\right] \quad \text{as } B = -V\frac{\Delta P}{\Delta V} = -V\frac{p}{\Delta V}$$

or

$$p = -B\frac{\Delta Y}{\Delta x} \qquad \text{...(7.41)}$$

From Eq. (7.39), we have

$$\frac{\Delta Y}{\Delta x} = -k\,A\cos(\omega t - kx)$$

or

$$p = B\,k\,A\cos(\omega t - kx)$$

or

$$p = p_0\cos(\omega t - kx) \qquad \text{...(7.42)}$$

with

$$p_0 = B\,k\,A \qquad \text{...(7.43)}$$

From Eqs. (7.39) and (7.43) it is clear that

1. A sound wave may be considered either displacement wave, $y = A\sin(\omega t - kx)$, or a pressure wave, $p = p_0\cos(\omega t - kx)$.
2. The pressure wave is 90° out of phase with respect to displacement wave, i.e., displacement will be maximum when pressure is minimum and vice versa. This is shown in Fig. 7.13.
3. The amplitude of pressure wave,

$$p_0 = B\,k\,A = A\,k\,\rho\,v^2 \quad \text{as} \quad v = \sqrt{\frac{B}{\rho}}$$

or

$$p_0 = A\,\rho\,v\,w \quad \text{as} \quad v = \frac{\omega}{k} \qquad \text{...(7.44)}$$

7.7.4 Energy, Power and Intensity of Sound Wave

If a sound wave given by, $y = A\sin(\omega t - kx)$, is propagating through a medium, the particle velocity will be

$$v_{pa} = \frac{dy}{dt} = \omega\,A\cos(\omega t - kx)$$

If ρ is the density of the medium, kinetic energy (KE) of the wave per unit volume will be

$$\text{KE per unit volume} = \frac{1}{2}\rho\left(\frac{dy}{dt}\right)^2 = \frac{1}{2}\rho\,\omega^2 A^2 \cos^2(\omega t - kx)$$

Average KE per unit volume is $\frac{1}{4}\rho\,\omega^2 A^2$, because average of $\overline{\cos^2(\omega t - kx))} = \frac{1}{2}$.

Average potential energy is equal to KE, so total energy of the particle wave per unit volume is the sum of kinetic and potential energies, i.e.,

$$U = KE + PE = \frac{1}{2}\rho\,\omega^2 A^2 \qquad \text{...(7.45)}$$

Let S be the cross-sectional area of the pressure wave as shown in Fig. 7.12, and dx be the displacement in x-direction, then the energy produced due to compression in the volume Sdx will be

$$E = U\,Sdx = \frac{1}{2}\rho\,\omega^2 A^2\,Sdx$$

The power (rate of transmission of energy) will be

$$P = \frac{\Delta E}{\Delta t} = \frac{1}{2}\rho\,\omega^2 A^2\,S\frac{dx}{dt} = \frac{1}{2}\rho\,\omega^2 A^2\,Sv \qquad \text{...(7.46)}$$

where, $v = \frac{dx}{dt}$ is wave velocity.

The intensity is defined as the average energy transmitted per unit area per second normal to the area, i.e., intensity is power per unit area, or

$$I = \frac{\Delta E}{S\Delta t} = \frac{P}{S} = \frac{1}{2}\rho\,v\,\omega^2 A^2 \qquad \text{...(7.47)}$$

From Eq. (7.44), the pressure amplitude, $p_0 = A\,\rho\,v\,\omega$. Removing the wave amplitude A, Eq. (7.48) becomes

$$I = \frac{\Delta E}{S\Delta t} = \frac{P}{S} = \frac{1}{2}\rho\,v\,\omega^2\left(\frac{p_0}{\rho\,v\,\omega}\right)^2 = \frac{1}{2}\frac{p_0^2}{\rho\,v} \qquad \text{...(7.48)}$$

Equations (7.47) and (7.48) represent the intensity of sound in terms of amplitude and pressure amplitude, respectively. For a given source and medium, it is clear from Eqs. (7.47) and (7.48) that

$$I \propto A^2 \quad \text{or} \quad I \propto p_0^2 \qquad \text{...(7.49)}$$

Example 7.5 A sound has an intensity of 4×10^{-8} W/m². What is its sound level?

Solution: Formula required

$$\text{Sound level} = (10\text{ dB})\log\frac{I}{I_0}$$

Given: Intensity of sound, $I = 4 \times 10^{-8}$ W/m^2

Calculation: Sound level = (10 dB) log $\dfrac{4\times10^{-8}}{10^{-12}}$ = (10 dB) log [4 × 10^4] = (10 dB) [log 4 + 4] = (10 dB) [0.6020 + 4] ≈ 46 dB

Ans. Sound level is 46 dB.

Example 7.6 The loudest sound, that the ear can tolerate at a frequency of 1 kHz corresponds to intensity of 1 W/m^2 (the so called threshold pain). Determine (a) the pressure amplitude (b) displacement amplitude and (c) sound level. The density of air is 1.3 kg/m^3 and speed of sound in air is 330 m/s.

Solution: Formula required

Intensity, $I = \dfrac{1}{2}\rho\, v\, \omega^2 A^2$; $I = \dfrac{p_0^2}{2\rho v}$, where, p_0 is the pressure amplitude.

Given: Density of air, ρ = 1.3 kg/m^3, frequency of sound wave, f = 1 khz, Intensity, I = 1 W/m^2, speed of sound, v = 330 m/s

Calculation: Angular velocity, $\omega = 2\pi f = 2\pi \times 10^3$Hz

$$I = \frac{1}{2}\rho\, v\, \omega^2 A^2$$

$$I = \frac{1}{2}\times 1.3\times 330\times(2\,\pi\times10^3)^2\, S^2$$

$$A^2 = \frac{2}{4\,\pi^2\times1.3\times330\times10^6} = \frac{2\times10^{-8}}{4\times9.8596\times1.3\times3.30}$$

$$= 1.1821\times10^{-10}$$

or displacement amplitude, $A = 1.0872 \times 10^{-5}$ m

Let p_0 be the pressure amplitude

$$I = \frac{p_0^2}{2\rho v} \quad \text{or} \quad p_0 = \sqrt{2\rho v I} = \sqrt{2\times1.3\times330\times1}$$

$$= 29.29 \text{ N/m}^2$$

Sound level = (10 dB) log $\dfrac{I}{I_0}$ = (10 dB) log $\dfrac{1}{10^{-12}}$ = 120 dB

Ans. (a) p_0 = 30 N/m^2, (b) $S = 1.08 \times 10^{-5}$ m, and (c) 120 dB

Example 7.7 What is the maximum possible sound level in dB of sound waves in air? Given that density of air is 1.4 kg/m^3, speed of sound is 332 m/s and atmospheric pressure is 1.02 × 10^5 N/m^2.

Solution: Formula required

Intensity of sound, $I = \dfrac{p_0^2}{2\rho v}$

Sound level, SL = $10 \log \dfrac{I}{I_0}$

Given: Atmospheric pressure, $p_0 = 1.02 \times 10^5$ N/m^2

Density of air, $\rho = 1.4$ kg/m^3

Speed of sound, $v = 330$ m/s

Calculation:

$$I = \frac{p_0^2}{2\rho v} = \frac{(1.02 \times 10^5)^2}{2 \times 1.4 \times 330} = \frac{1.02 \times 1.02 \times 10^{10}}{2 \times 1.4 \times 330}$$

$$= 1.12597 \times 10^7 \text{ W/m}^2$$

$$\text{SL} = 10 \log \frac{I}{I_0} = 10 \log \frac{1.12597 \times 10^7}{10^{-12}}$$

$$= 10 \log (1.12597 \times 10^{19}) = 10 (19 + \log 1.12597)$$

$$= 10 (19 + 0.051527) = 190.51527 \text{ dB}$$

Ans. Maximum sound level in air is 190.51527 dB.

Example 7.8 (a) The power of sound from the speaker of a radio is 25 mW. By turning the knob of volume control the power of sound is increased to 500 mW. What is the power increase in dB as compared to original power. (b) How much more intense is the75 dB sound than a 25 dB whisper?

Solution: Formula required

Sound level, SL = $10 \log \dfrac{I}{I_0}$

Given: **(a)** Power of sound from speaker, $P_1 = 25$ mW

Power of sound increased to $P_2 = 500$ mW

The intensity is power per unit area. So for a given source, $P \propto I$

Increase in sound level, Δ SL = $\text{SL}_2 - \text{SL}_1 = 10 \log \dfrac{P_2}{P_1} = 10 \log \dfrac{400}{25}$

$$= 10 \log 16 = 10 (1 + \log 1.6)$$

$$= 10 (1 + 0.2041) = 12.041 \text{ dB}$$

(b) Increase of sound level of whisper

$$\Delta\text{SL} = \text{SL}_2 - \text{SL}_1 = 10 \log \frac{P_2}{P_1}$$

$$75 - 25 = 10 \log \frac{I_2}{I_1} \quad \text{or} \quad 50 = 10 \log \frac{I_2}{I_1}$$

or $$5 = \log \frac{I_2}{I_1}$$

or the ratio of intensity of sound, $\frac{I_2}{I_1} = 10^5$

Ans. (a) Increase in sound level is 12.041 dB (b) Ratio of intensity of sound is 10^5.

Example 7.9 A window whose area is 2.5 m^2 opens on a street where the street noises result in an intensity level at the window of 70 dB. How much 'acoustic' enters the window via sound waves. Now if an acoustic absorber is fitted at the window, how much energy from the street will it collect in 4 hours?

Solution: Formula required

Sound level, $\text{SL} = 10 \log \frac{I}{I_0}$

Given: Area of window, $A = 2.5$ m^2, intensity of sound at window, SL = 70 dB

Calculation:

$$10 \log \frac{I}{I_0} = 70 \quad \text{or} \quad \log \frac{I}{I_0} = 7 \quad \text{or} \quad \frac{I}{I_0} = 10^7$$

or $$I = 10^{-12} \times 10^7 = 10^{-5} \text{ W/m}^2$$

Power entered through window, $P = I\,A = 10^{-5} \times 2.5 = 25$ μW

The energy collected from street in 4 hours is $E = P \times t = 25 \times 10^{-6} \times 4 \times 60 \times 60 = 36 \times 10^{-2}$ J

Ans. The energy collected from street in 4 hours is 36×10^{-2} J.

Example 7.10 A dog while barking delivers about 1.2 mW of power. If this power is uniformly distributed over a hemispherical area, what will be the sound level at a distance of 4 m? What would be the sound level if instead of 1 dog, 4 dogs start barking at the same time each delivering 1 mW of power?

Solution: Formula required

$$\text{Intensity} = \frac{P}{S}$$

Sound level, $\text{SL} = 10 \log \frac{I}{I_0}$

Given: Power sound delivered by dog, $P = 1.2$ mW

Radius of sphere, $r = 4$ m

Calculation:

$$I = \frac{P}{S} = \frac{P}{\frac{1}{2}(4\pi r^2)} = \frac{1.2 \times 10^{-3}}{2 \times 3.14 \times 4 \times 4} = 1.36488 \times 10^{-5}$$

$$= 13.6488\ \mu\text{W/m}^2$$

$$\text{SL} = 10 \log \frac{I}{I_0} = 10 \log \frac{13.488 \times 10^{-6}}{10^{-12}}$$

$$= 10 \log (13.6488 \times 10^6) = 10\ (6 + \log 13.6488)$$

$$= 10\ (6 + 1.13509) = 71.13509 \text{ dB}$$

If there are 4 dogs barking at the same time and same level, $I_2 = 4\,I_1$, so

$$\text{SL}_2 - \text{SL}_1 = 10 \log \frac{I_2}{I_1} = 10 \log \frac{4\,I_1}{I_1}$$

$$= 10 \log 4 = 10 \times 0.6020 = 6.02 \text{ dB}$$

or
$$\text{SL}_2 = \text{SL}_1 + 6.02 = 71.13 + 6.02 = 77.15 \text{ dB}$$

Ans. Sound level at a distance of 4 m is 71.13509 dB and 77.15 dB due to barking of one dog and four dogs, respectively.

Example 7.11 Power of a sound from the speaker of a radio is 30 milliwatt. By turning the knob of the volume control the power of the sound is increased to 600 milliwatt. What is the power increase in decibel as compared to the original power?

Solution: For sound level is

$$\text{SL} = (10 \text{ dB}) \log \frac{I}{I_0}$$

Intensity, I is proportional to power P, so formula for sound level is written as

$$\text{SL} = (10 \text{ dB}) \log \frac{P}{P_0}$$

Given: Power of speaker, $P = 30$ milliwatt, power is increased to $P_0 = 600$ milliwatt

Calculation: The power increased as compared to original power

$$\text{SL} = (10 \text{ dB}) \log \frac{600}{30} = (10 \text{ dB}) \log 20$$

$$= (10 \text{ dB})\ [1 + \log 2]$$

$$= (10 \text{ dB})\ [1 + 0.3010] = 13.010 \text{ dB}$$

Ans. The power increased as compared to original power is 13. 010 dB.

7.8 MUSICAL NOTES

The lowest frequency of a musical instrument is called *fundamental frequency*. This fundamental frequency of the musical instrument is called *note* or *tone*. Other higher frequencies of musical instruments are integral multiples of fundamental frequency. The higher frequencies are called *overtones*. The frequency emitted by musical instrument are called notes or overtones.

Any musical note is characterized by its *loudness*, its *pitch* and its *quality* (sometimes called its ***timre*** or its *tone*).

Loudness

Loudness is a sensation received by the ear due to intensity. That is, loudness is determined by intensity of sound. Intensity is proportional to the square of the amplitude of the sound wave. However, loudness is a subjective quantity and depends on the sensitivity of the listener's ear. Therefore, the loudness of a sound of given intensity may be different for different listeners. A sound appears loud to one person may appear less loud to a person whose hearing is poorer. Loudness increases with intensity, but the relationship between the two quantities is not simple. Two sounds of equal intensity but of different frequency (pitch) may not appear equally loud to the same listener, because the sensitivity of the ear is different for different frequencies. One cannot hear ultrasonic sounds, no matter how intense they are.

The amount of energy associated with a sound wave is measured by its intensity, which is proportional to the square of the amplitude of the wave. The intensity and loudness of a note also depends on the mass of the medium or body which is vibrating in order to produce it. The sound from a loudspeaker appears very loud, because loudspeaker cone has a large surface area. A guitar string has only a small surface area, so its vibrations would be unlikely to be heard if it is not connected to the hollow sounding box of the instrument.

As a sound wave travels outward from its source, its energy is spread over the surface of a sphere centred on the source. Therefore, the intensity (I) and the distance from the source (r) are related by

$$I \propto \frac{1}{r^2}$$

Pitch

The pitch is the sensation received by the ear due to frequency. That is, the pitch of a note is determined by its frequency. A note of high pitch is due to a vibration of high frequency. The following points are to be noted about the pitch:

1. The roar of a lion has loudness but low pitch, while the buzzing of bee or humming of a mosquito has high pitch but low loudness.
2. Normally, the pitch of females is higher than males due to more harmonics.
3. In TV or tape recorder, bass and treble refer to low and high pitch, respectively. Woofer is a loudspeaker designed to play low pitch accurately. So at bass (when woofer is on),

i.e., at low pitch, grave sound of 'tabla' and dholak' becomes loud, while at treble, i.e., at high pitch, shrill sounds of 'flute' or 'ghungroo' become dominant.

4. In record player if record of 47 r.p.m. is played at 79 r.p.m., pitch will increase and sound will become shriller. If record of 47 r.p.m. is played sat 34 r.p.m., pitch will decrease and so sound will become grave (or serious)..

Quality (Timbre or Tone)

Quality is the good sensation received by the ear due to 'waveform'. Three waves of same intensity and frequency, shown in Fig. 7.14, will produce different sensation on the ear, because their 'waveforms' are different. The waveforms depend on the number of overtones. Hence, quality of sound depends on the number of overtones, i.e., harmonics present and their relative intensities. The following points are worth noting regarding quality:

1. By listening sound of a person one can recognize a person without seeing him/her, because it has a definite quality.
2. If same note is played on different instruments, say sitar and guitar at same loudness, it produces different sensation in the ear due to their different quality.
3. Due to poor quality of sound, sometimes it becomes difficult to recognize a person by listening his sound on telephone or tape.

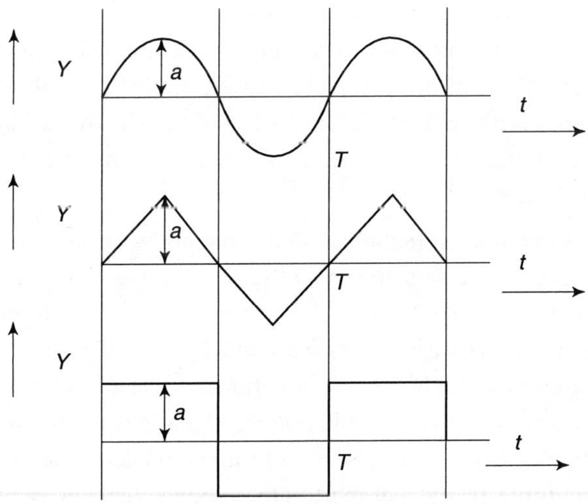

Fig. 7.14

Difference between Tone and Pitch

Human beings can hear a large range of acoustic frequencies, but our perception of sound has a curious periodicity to it. Therein lies the concept of *pitch* – a subjective, perceived (by humans) aspect of sound, associated with musical tones. A musical tone is defined by its frequency. The pitch of a tone is closely related to this frequency. In fact, the frequency and the pitch of a tone can effectively be thought of as being 'proportional', but it is not so.

Let us consider two pure tones having the same frequency. When these two are heard together, the sound is said to be in 'unison' – both the notes are the same. Let us increase one of the frequencies, such that there is now frequency difference ($\delta\nu$) between two notes. The resultant sound wave of base frequency equal to $\nu_a = \nu + \delta\nu/2$, is modulated by a beat frequency of $\delta\nu$. Evidently, the base frequency is very close to the original frequency for small values of $\delta\nu$. For $\delta\nu \leq 12$ Hz, the human ear only detects the sound combination as a single tone (known as a 'fused' tone), because the amplitude modulation for such cases is too low to be detectable by the human ear.

When $\delta\nu \geq 12$ Hz, the human ear begins to perceive two separate tones with some roughness (like 'buzzing'). This perception continues till the difference is large enough for the human ear to perceive two separate tones with clarity. It needs to be noted that the value of $\delta\nu$, at which this clarity is achieved, depends strongly on ν_a. For audio frequencies near the upper limit of human being, $\delta\nu$ could be as large as 400 Hz for the two notes to be perceived separately.

If we now continue to increase the frequency difference, the human ear would continue to perceive the two tones as two separate ones, but only up to a point. As the second tone approaches twice the frequency of the first, its 'pitch' returns to its starting point. This gives rise to the situation when two tones, whose frequencies differ by a factor of two, are perceived to be the same musical 'note', although their frequencies are different. These two notes are said to be one *octave* apart in Western musical tradition. An octave is the interval between one musical note and another with half or double its frequency. In frequency space, the interval between the fundamental and second harmonic (or between any successive harmonics) of a harmonic series, is an octave.

All the notes in between unison [sound of two same notes, when heard together, is said to be in unison] and the octave are said to have different chroma or colour, and notes with the same chroma but separated by one or more octaves are said to have different tone height. While chroma applies to a continuous variation in frequency, pitch refers to tones with specifically defined values of frequency modulo are an integral multiple of octaves. This perception of a unique note across octaves is a natural phenomenon that has been referred to as the 'basic miracle of music', the use of which is common in most musical systems. Because the notes that are an integral multiple of octave apart same 'sound' [except same 'sound'], they are given the same name in musical notations. This is called the *octave equivalence* of musical frequencies and is a part of most contemporary musical cultures.

Other similar relations do exist which are also of great importance to our musical tradition. For example, two notes which differ in frequency by a factor of 3/2 sound extremely consonant together, and this relation is known as the *consonance of the fifth*. We found that this relation is actually a consequence of the octave equivalence, which can be understood in terms of combination tones (or beat frequencies).

A musical fifth comprises of two basic frequencies, ν_1 and ν_2 where, $\nu_2 = 3/2\ \nu_1$. The superposition of these two generates the beat frequency, $\nu_0 = \nu_2 - \nu_1 = \nu_1/2$, which is exactly one octave below ν_1.

Therefore, the tones generated by a musical fifth ($\nu_1/2$, ν_1, $3/2\ \nu_1$) constitute the first three terms in a harmonic series, where, ν_b is the'fundamental', and $\nu_1$and ν_2 are the second and the third harmonics.

Thus, when two tones constituting a musical fifth are played, the periodicity of their superposition curve enables the auditory [hearing] system to sense the presence of a fundamental tone that is not actually present and is, therefore, called a *missing fundamental.* In other words, a musical fifth is perceived as a harmonic series by the human auditory system. Similarly, for a sequence of harmonics consisting of even and odd integral multiples of the fundamental, each pair of successive harmonics creates a combination tone with a periodicity equal to that of the fundamental (which is quite remarkable). Therefore, the fundamental is perceived even when it is absent from the original sequence.

This particular ability of the human auditory system to identify the consonant sound and sense the *pitch* rather than the absolute frequency (tone) of a note is actually at the root of our musical scales.

7.9 MUSICAL SCALE

A pleasant musical sound varies from person to person. Some prefer classical and some Bharat Natyam, while others only prefer to listen western music like rap, pop or jaz. All types of music comprise basically the same set of notes. Because of this reason, more fundamental thing than the type of music is the actual frequencies of the notes that make a musical scale.

Just like an alphabet is the basis of a language, a set of basic notes known as a *scale* forms the basis of music. The term 'scale' comes from the Latin word *scala* meaning *ladder*. Thus, a scale is a set of notes ordered according to their frequency or pitch from which one can build melodies and harmonics. Since scales create various distinctive sensations, atmospheres and tonal flavours, there are as many scales as there are cultures around the world.

A scale is a sequence of tones (whole steps) and semi-tones (half steps). To identify a specific scale, one needs to know its unique sequence of intervals. There exist a number of scales depending on the number of primary notes (or fundamental frequencies) available per octave. However, there are two most common western musical scales: 1) diatonic scale and 2) musician's scale. We will also discuss *shrutis* as third Indian music scale

1. The western traditional classical music, is based on the *diatonic* (in Greek, 'diatonic' means 'progressing through tones') scale. It consists of 5 tones and 2 semi-tones called heptatonic (7 notes per octave). Evidently, in a heptatonic scale, doubling of frequency requires going up by 8 notes. The name octave, for a non-repetitive frequency span, derives from this. The diatonic scale is a scientific scale based on middle note (C) having a frequency of 256 Hz.

2. Another modern musical scale is *musicians's scale,* also known as *equal tempered scale* (ETS). It is developed by subsequent transformation of *diatonic scale* and based on 12 notes per octave, where all the notes are equidistant (in a logarithmic sense) from their neighbouring ones. The musician's scale is based on note *A* having a frequency of exactly 440 Hz.
3. The guiding principle of Indian classical music is to exploit the freedom accorded by the special nature of human sensitivity to the acoustic frequencies. The primary characteristic of Indian classical music is that it is based on a standard set of melodic forms (*ragas*), which are themselves built from a basic set of scales (*thaat*). The *ragas* basically define notes for the music by specifying scales (ascending and descending, which may or may not be the same) and provide the general prescription according to which a piece of music should be composed or performed. As there is no rigidity about a set piece of music, a musician is entirely free to bring her/his individual flavour to the composition as long as the prescription specific to a particular *raga* is concerned.

7.10 BASIC STRUCTURE OF INDIAN MUSIC

In Indian music there are 7 pure notes (*shuddha swara*) – *sa* (*shadaj*), *re* (*rishabh*), *ga* (*gandhar*), *ma* (*madhyam*), *pa* (*pancham*), *dha* (*dhaiwat*) *and ni* (*nishad*). The first and the fifth notes –*sa* and *pa*– have fixed frequencies [shown in Tables 7.2 and 7.3] and are commonly known as *atal swaras* (invariant notes). The other 5 notes are variables, and the variants are known as the *vikrita swaras* or the impure notes. These impure notes are R,G,D,N,M corresponding to the *komal* (flat or lower frequency) variants of *re, ga, dha, ni* and *teevra* (sharp or higher frequency) variant of *ma,* respectively as shown in Table. 7.3. An octave consists of the seven pure notes and is known as a *saptak*, the eighth note having twice the frequency of the first note. In reality though, a *saptak* contains 12 notes – 7 pure and 5 impure. The seven pure notes are obtained according to the ratio–1, 9/8, 5/4, 4/3, 3/2, 5/3, 15/8 between the consecutive notes. The frequency of Indian *shrutis* is compared with the Major scale of the Western tradition in Table 7.2.

In traditional Indian music, a total of 22 microtones or *shrutis* were in use instead of the 12 tones. The practice continues to be the same in the South Indian (Karnatic) music though the north Indian (Hindustani) system is at present more or less based on 12 tones. The division of the *saptak* in 22 *shrutis* exploits the fact that there exists a minimum interval (in pitch or frequency) that can be distinguished by the human ear. [Theoretically, an infinite number of *shrutis* are possible, but any practical division would depend on the actual size of the frequency interval that a listener can sense clearly or a musician can produce.] The list of *shrutis* has been shown in Table 7.2, along with the pure notes and corresponding Western scale. It can also be seen from Table 7.2 that the difference between the frequencies of notes in the Indian system and the equal tempered scale (ETS) is very small. In fact, with the introduction of the reed [strip of metal or cane vibrates to produce sound] instruments (piano, harmonium, etc.) in the Indian music (in particular, the huge popularity of harmonium across musical fraternity) the difference has disappeared. Therefore, for the sake of convenience, we shall use the ETS even while talking about the Indian scales and notes.

As has been mentioned before, a *saptak* corresponds to an octave. Three main *saptaks* are used in Indian music. Unlike Western music, which has an absolute frame of reference, the Indian system changes from instrument to instrument. The three main *saptaks* of the Indian tradition are the *mandra, madhya* and *tar saptak*. The middle register, referred to as the *madhya saptak*, uses a base note that is most comfortable for a particular musician (vocal or instrument). The octave above this base is referred to as the *tar saptak*; the lower one is known as the *mandra saptak*. Additionally, two octaves above the middle, i.e., note D_5 of frequency 587.33 Hz, is called *ati-tar saptak*; three octaves above, note of frequency 622.25 Hz, is called *ati-ati-tar saptak* and so on. The reed instruments also allow us to connect the Indian *saptaks* with the corresponding octaves of an ETS in an easy manner, as shown in Table 7.3. It is also clearly seen that there is one to one correspondence between Indian scale and the 'major' scale of Western tradition, since the pure tones of a *saptak* follows the 'T T S T T T S' pattern.

Table 7.2 The Indian *shrutis* and the Western notes are compared. One should note that the *shuddha swaras* coincide with the pure notes of *C*-major. This is because the Indian base note *sa* has been matched to the Western *C* and Indian *saptak* is intrinsically a major scale. The absolute frequencies have been obtained by setting *A* to 440 Hz.

		Indian		Western
Shruti	**Ratio**	**v(Hz)**	**Note**	**v(Hz)**
Chandovati (sa)	1	261.6256	C	261.6256
Dayavati	256/243	275.6220	C#	277.1826
Raudri (re)	9/8	294.3288	D	293.6648
Krodha	32/27	310.0747	D#	311.1270
Prasarini (ga)	5/4	327.0319	E	329.6275
Marjani (ma)	4/3	348.8341	F	349.2282
Rakta	45/32	367.9109	F#	369.9944
Alapini (pa)	3/2	392.4383	G	391.9954
Madanti	128/81	413.4330	G#	415.3047
Ramya (dha)	5/3	436.0426	A	440.0000
Ksobhini	16/9	465.1121	A#	466.1638
Kumudvati (ni)	15/8	490.5479	B	493.8833
Chandovati (sa)	2	523.2511	C	523.2511

Let there be two tones of equal frequency, equal pitch and equal note in the hearing range. If we now continue to increase the frequency difference, the human ear would continue to perceive the two tones as two separate ones, but only up to a point. As the second tone

Table 7.3 Indian *saptak-mandra-saptak* notes are denoted with *halanta* (·) symbol and *tar-saptak* notes are denoted by ref. (′) symbol.

Key	Note	v (Hz)		Saptak
W	C_3	130.81	sa̖	
B		138.59	R̖	
W	D_3	146.83	re̖	
B		155.56	G̖	
W	E_3	164.81	ga̖	
W	F_3	174.61	ma̖	*Mandra*
B		185.00	M̖	
W	G_3	196.00	pa̖	
B		207.65	D̖	
W	A_3	220.00	dha̖	
B		233.08	N̖	
W	B_3	246.94	Ni̖	
W	C_4	261.63	*sa*	
B		277.18	R	
W	D_4	293.67	*re*	
B		311.13	G	
W	E_4	329.63	*ga*	
W	F_4	349.23	*ma*	
B		369.99	M	*Madhya*
W	G_4	392.00	*pa*	
B		415.30	D	
W	A_4	440.00	*dha*	
B		466.16	N	
W	B_4	493.88	*ni*	
W	C_5	523.25	*sa′*	
B		554.37	*R′*	
W	D_5	587.33	*re′*	
B		622.25	*G′*	
W	E_5	659.26	*ga′*	
W	F_5	698.46	*ma′*	*Tar*
B		739.99	*M′*	
W	G_5	783.99	*pa′*	
B		830.61	*D′*	
W	A_5	880.00	*dha′*	
B		932.33	N′	
W	B_5	987.77	ni′	

approaches twice the frequency of the first, its 'pitch' returns to its starting point. This gives rise to the situation when two tones, whose frequencies are a factor of two apart, are perceived to be the same musical 'note', although their frequencies are different. These two notes are said to be one *octave* apart in Western musical tradition. One octave is the interval between one musical note and another with half or double its frequency. In frequency space, the interval between the fundamental and second harmonic (or between any successive harmonics) of a harmonic series, is equal to one octave. One octave is the interval between one musical note and another with half or double its frequency.

The frequency ratio for two notes, which differ by a factor of 2 or $\frac{1}{2}$, with one octave apart is 2 : 1, while the frequency ratio of other two notes, which differ by a factor of $\frac{3}{2}$ called fifth note, with one octave apart is 1.5 : 1. On the equally tempered scale the frequency ratio for each semitone is 1.0595 : 1. The ratio for the tone interval is the sixth root of 2 [$2^{1/6} = 1.12246$] and that for the semitone is the twelfth root of 2 [$2^{1/12} = 1.05946$].

Fifth note is of great importance to our musical tradition. Two notes which differ in frequency by a factor of 3/2 sound extremely consonant together, and this relation is known as the *consonance of the fifth.* We found that this relation is actually a consequence of the octave equivalence, which can be understood in terms of combination tones (or beat frequencies).

A musical fifth comprises two basic frequencies, ν_1 and ν_2, where, $\nu_2 = 3/2\ \nu_1$. The superposition of these two generates the beat frequency, $\nu_b = \nu_2 - \nu_1 = \nu_1/2$, which is exactly one octave below ν_1.

Therefore, the tones generated by a musical fifth ($\nu_1/2$, ν_1, $3/2\ \nu_1$) constitute the first three terms in a harmonic series, where, ν_b is the 'fundamental', and ν_1 and ν_2 are the second and the third harmonics.

Thus when two tones constituting a musical fifth are played, the periodicity of their superposition curve enables the auditory system to sense the presence of a fundamental tone that is not actually present and is, therefore, called a *missing fundamental.* In other words, a musical fifth is perceived as a harmonic series by the human auditory system. Similarly, for a sequence of harmonics consisting of even and odd integral multiples of the fundamental, each pair of successive harmonics creates a combination tone with a periodicity equal to that of the fundamental (which is quite remarkable). Therefore, the fundamental is perceived even when it is absent from the original sequence.

7.11 ACOUSTICS OF BUILDING

Acoustics of building is scientific method to control the sound in buildings, cinema halls, auditoriums, etc. The purpose of this sound control is to create conditions so that people in the big room or hall can hear the sound clearly at every place of it. This is done by constructing the building by taking various factors that affect the architectural acoustics.

Essential Features of Good Acoustics

W.C. Sabine studied the problem of sound distribution in the buildings in depth. Based on scientific study, he laid down the essential features for good acoustics.

1. The sound must be sufficiently loud so that it is heard in every part of an auditorium or a hall.
2. An auditorium or a hall should be free from echo.
3. For clarity of sound, overlapping of successive echo should not be there.
4. The quality of sound of speech or music must remain same always.
5. A proper (~ 1 to 2 seconds) time of reverberation should be there.
6. Sound should be not concentrated in any part of the hall.
7. The boundaries of the hall should be made sound proof to avoid unwanted noise.
8. No echelon effect should be there.
9. There should not be resonance in the building.

Note: *Echelon* effect. The musical notes produced by regular successive echoes from the reflection of the surfaces of structures, like flight of stairs or a set of railings in the hall (asset of railways), are called echelon effect. In short, a new sound formed by combination of multiple echoes is known as echelon effect.

Reverberation

When a sound is produced in a hall or building, it lasts for a long time. The listener receives the sound directly from the source and the sound reflected from walls, window, ceiling and floor of the hall. Although, the small part of the sound energy is lost at every reflection. As a result, the listener receives series of sounds of diminished intensity, which reduces the clarity of sound.

A sound created in a big hall will persists by repeated reflection from the walls until it is reduced to a value where it is no longer audible. Reverberation is defined as the persistence or prolongation of sound in a hall even though the sound source is stopped. The prolonged reflection of sound from the walls, ceiling, and floor of the room or hall is caused by reverberation. Reverberation can be reduced by covering the roof and walls of the auditorium generally with sound-absorbent materials like compressed fibreboard, rough plaster, draperies or porous tiles, asbestos, etc. The seat materials are also selected on the basis of their sound absorbing quality.

Reverberation time

Repeated reflection of sound from surfaces is called reverberation. Reverberation time in a hall is defined as the time taken by the sound to fall below the minimum audibility after the source of sound is shut off. The minimum audibility of sound is generally considered 60 dB. Therefore, reverberation time or time of reverberation is also defined as the time required for sound to decay to 60 dB from its initial level.

Acoustically good auditorium or a hall is that where time of reverberation is negligibly small. Sabine found that the time of reverberation depends upon the size of the hall, loudness of the sound and the kind of music or sound for which hall is to be used.

It was found that, for a sound of frequency 512 Hz, the time of reverberation is 1 to 1.5 seconds for a capacity of 4500 m^2 hall.

To avoid longer time of reverberation, anti-reverberation materials must be used. Porous tiles, asbestos for plastering can be used to reduce reverberation time. Reverberation can also be reduced by covering the roof and walls of the auditorium generally with sound-absorbent materials like compressed fibreboard, rough plaster, draperies, etc.

Reverberation: Role in acoustics

1. If a hall is to be acoustically satisfactory, it is essential that it should have the right reverberation time.
2. The reverberation time should not be either too long or too short. A very short reverberation time makes a room 'dead'.
3. On the other hand, a long reverberation time destroys clarity of the sound. The optimum value for reverberation time depends on the purpose for which the hall is designed.
4. A reverberation time of 0.5 seconds is acceptable for speeches and lectures, while a reverberation time of 1 to 2 seconds is satisfactory for concerts.
5. In case of theatres, the optimum value varies with volume. 1.1 to 1.5 seconds is suitable for small theatres, whereas for large theatres it may go up to 2.3 seconds.

7.12 SABINE'S EQUATION

(Calculation of reverberation time)

The Sabine equation, named after the scientist Wallace Clement Sabine who established it, is written as

$$RT_{60} = \frac{24(\log 10)V}{C_{20}S_a}$$

RT_{60} is the reverberation time (to drop 60 dB), V is the volume of the room and c_{20} is the speed of sound at 20°C (room temperature). Sa is the total absorption in sabin.

The sabin unit has the same dimension as area (e.g., m^2). A one square metre surface with an absorption coefficient of 0.75 would be considered 0.75 sabins. The absorption coefficient has a range of 0 to 1, where a coefficient of 0 indicates none of the sound is absorbed, and a coefficient of 1 indicates that 100% of it is absorbed.

The speed of sound at 20°C is 343 m/s. Putting these values Sabine's equation becomes 1 m = 3.28084 feet

$$RT_{60} = \frac{24 \times \log 10 \times V}{343 \times S_a} = \frac{24 \times 2.3026 \times V}{343 \times S_a}$$

or
$$RT_{60} = \frac{0.161\,V}{S_a} \qquad ...(7.49a)$$

Note that the factor 0.161 has the units seconds per metre; dimensional analysis on the equation will yield a time in seconds as the volume is measured in cubic metres and sabins in square metres.

If α is the average absorption coefficient and S is the total surface area of the room, then $S_a = S\alpha$. Now eqn. (7.49a) becomes

$$RT_{60} = \frac{0.161\,V}{S_a} \qquad ...(7.50)$$

The derivation of formula given in eqn. (7.50) is given **Appendix A.**

The velocity of sound at room temperature (20°C) is 1125 ft/s. By taking the room dimensions in feet, Sabine's equation is written as

$$RT_{60} = \frac{24 \times \log 10 \times V}{1125 \times S_a} = \frac{24 \times 2.3026 \times V}{1125 \times S_a}$$

or
$$RT_{60} = \frac{0.049V}{S\alpha} \qquad ...(7.51)$$

The value of α is written as:

$$= \frac{S_1\alpha_1 + S_2\alpha_2 + ... + S_n\alpha_n}{S} = \frac{\sum_{i=1}^{n} S_i\alpha_i}{S} = \frac{S_a}{S} \qquad ...(7.52)$$

where, S_α is total absorption in the room and S is sum of s_1, s_2, etc., are surface areas of inside walls, floor and ceiling, etc., of the room, i.e., S is total surface area inside the room. S incudes the surface area of objects or people in the room. α_1, α_2, etc., are the absorption coefficients of the surface area with which they are associated, $S_n\,\alpha_n$ is the total absorption by the people, furniture, etc., present in the room.

APPENDIX A

Sabine's Formula for Reverberation Time

The sound produced by a speaker in a room or hall suffers successive reflections from the walls, the ceiling and the floor of the hall so that in addition to the direct sound a series of sound waves pass the listener's ear called echo. This gives the listener the impression of prolongation or persistence of sound for some time after the original sound has ceased. This is called reverberation.

Causes of reverberation

(i) The intensity of the sound produced by the speaker decays exponentially and takes a long time to fall to zero value. The time depends upon the absorption of sound. It is large if the absorption is small. Hence, the presence of absorbing material like open windows, audience, curtain, etc., reduce reverberation.

(ii) The reverberation is called by reflection of sound waves from the walls, ceiling and floor of the hall because the sound persists for some time due to reflection.

(iii) As all the sound waves reaching a particular point in the hall are not in phase, these produce interference, thereby causing destruction of sound waves. This gives echo effect to sound waves.

Reverberation time

The time interval is taken by a sustained or continuous sound to fall in intensity to one millionth (10^{-6}) of its original value, i.e., fall by 60 decibels in loudness is called reverberation time.

Expressing the reverberation time in terms of volume of the wall, the aera and absorption coefficient of the surface is known as Sabine's formula.

The absorption coefficient for the material of a surface is defined as the ratio of the sound energy absorbed by the surface to the sound energy absorbed by an equal area of perfect absorber, such as an open window. The standard unit of absorption is called a sabine and is the sound energy absorbed by one square foot of an open window. For a surface having a total area S and absorption coefficient α, the effective absorbing area A is given by

$$A = \alpha S$$

If α_1, α_2, α_3, α_4, ... etc., are the absorption coefficients at each reflection of the surfaces having areas S_1, S_2, S_3, S_4, ... etc., then the average value of the absorption coefficient α, is written as

$$\alpha = \frac{\alpha_1 S_1 + \alpha_2 S_2 + \alpha_3 S_3 + \alpha_4 S_4 + \ldots}{S_1 + S_2 + S_3 + S_4 + \ldots}$$

or $\alpha = \dfrac{\sum_{i=1}^{N} \alpha_i S_i}{S}$, where, S is the total area of surface and N is number of reflections from the surfaces.

Let I be the average intensity at any instant and dI the fall in intensity due to absorption in a small interval of time dt, then

$$dI = -\alpha\, n\, I\, dt, \qquad \text{...(A1)}$$

where, α is the mean absorption coefficient of all absorption surfaces in the room and n is number of the reflections of the sound waves per second.

By statistical method it is found that sound travels an average distance $\frac{4V}{S}$ between two successive reflections, where V is the volume of the hall and S the total area of the reflecting surfaces. If v is the velocity of the sound and τ is the time between two successive reflections, then distance between two successive reflections will be $v\tau = \frac{4V}{S}$

So, the time between two successive reflections is $\tau = \frac{4V}{Sv}$

Average number of reflections per second $n = \frac{1}{\tau} = \frac{Sv}{4V}$ ...(A2)

Substituting for n in eqn. (A1), we get

$$dI = -\alpha\frac{Sv}{4V}Idt$$

or
$$dI/I = -\alpha\frac{Sv}{4V}dt$$

If I_0 is the steady value of the intensity at the source is cut off, i.e., at $t = 0$ and I_t is the value at a time t latter, then

$$\int_{I_0}^{I_t}\frac{dI}{I} = -\int_0^t \alpha\frac{Sv}{4V}dt$$

Carrying out the integrations of both sides of this equation, one gets

$$\log_e\left(\frac{I_t}{I_0}\right) = -\alpha\frac{Sv}{4V}dt$$

or
$$\frac{I_t}{I_0} = e^{-\alpha\frac{Sv}{4V}t} \quad \text{(A3)}$$

According to the definition of reverberation time, $\frac{I_t}{I_0} = 10^{-6}$, then we have

$$e^{-\alpha\frac{Sv}{4V}t} = 10^{-6}$$

or
$$e^{\alpha\frac{Sv}{4V}t} = 10^{+6}$$

Taking the log of this equation, we get

$$a\frac{Sv}{4V}t = \log_e(10^6)$$

If $t = T$ is the reverberation time, then

$$T = \frac{\log_e(10^6)}{\alpha\frac{Sv}{4V}} = \frac{6\log_e(10)}{a\frac{Sv}{4V}} = \frac{6\times 2.3026}{v}\frac{4V}{aS}$$

Taking the velocity of sound 344 m/s, reverberation time will be

$$T = \frac{6 \times 2.3026}{344} \frac{4V}{\alpha S}$$

or
$$T = \frac{0.161V}{aS} \tag{A4}$$

This equation is called Sabine's equation.

In the limiting case, the Sabine equation predicts a finite reverberation time in a room with 100% absorption present, and for true absorption values in excess of 0.63, this equation can give α values in excess of 1.0 (100% absorption). The Norris-Eyring equation

$$RT_{60} = \frac{0.049\ V}{-S \ln(1 - \alpha)} \quad ...(7.53)$$

for room dimensions in feet or

$$RT_{60} = \frac{0.161\ V}{-S \ln(1 - \alpha)} \quad ...(7.54)$$

for room dimensions in metres gives α values from 1.0 to 0 for true absorption values when calculated from actual RT_{60}. If the value of RT_{60} is measured and the corresponding value of α is calculated, insertion of this value of α into the expression $-\ln(1 - \alpha)$ converts the α value into a Sabine α. For example, $-\ln(1 - 0.63) = 0.99$.

7.13 ABSORPTION COEFFICIENT AND ITS CALCULATION

The coefficient of absorption of a material is defined as the ratio of sound energy absorbed by the surface to that of total incident sound energy on the surface.

$$\text{Thus, absorption coefficient, } \alpha = \frac{\text{Sound energy absorbed by the surface}}{\text{Sound energy incident on the surface}}$$

The standard unit of absorption is taken as the unit area of open window [called open window unit (OWU)] as all the sound falling on open window passes through it. It is assumed that an open window is a perfect absorber of sound. Thus, absorption coefficient is also defined as "the rate of sound energy absorbed by a certain area of the surface to that of an open window of same area".

Several experimental methods are there to measure absorption coefficient. Reverberation time can be measured with the help of sound meter. By knowing reverberation time, the Sabine equation can be used without and with a standard large sample of material inside the room for calculating absorption coefficient of the desired material.

Let $T_{60}(1)$ be the time of reverberation without putting any absorbing material on the surface area, S_T and absorption coefficient alpha.The reverberation time $T_{60}(1)$ is given by equation.

$$T_{60}\ (1) = \frac{0.161\ V}{S_T \alpha} \quad ...(7.55)$$

If a sample of absorptive material having an area S_1 is placed on the floor and the test repeated, the new reverberation time $T_{60}(2)$ is

$$T_{60}(2) = \frac{0.161}{S_T\alpha - S_1\alpha_0 + S_1\alpha_1} \quad ...(7.56)$$

where, α_0 is the absorption coefficient under the covered portion of the floor, and α_1 is the absorption coefficient of the sample material under test. Equation (7.56) can be rewritten as

$$\frac{1}{T_{60}(2)} = \frac{S_T\alpha - S_1\alpha_0 + S_1\alpha_1}{0.161\,V}$$

$$\frac{1}{T_{60}(2)} = \frac{S_T\alpha}{0.161\,V} + \frac{S_1(\alpha_1 - \alpha_0)}{0.161\,V}$$

Substituting $\frac{1}{T_{60}(1)} = \frac{S_T\alpha}{0.161\,V}$ from Eq. (7.55), we get

or $$\frac{1}{T_{60}(2)} = \frac{1}{T_{60}(1)} + \frac{S_1(\alpha_1 - \alpha_0)}{0.161\,V}$$

This equation leads to equation

$$\alpha_1 = \alpha_0 + \frac{0.161\,V}{S_1}\left[\frac{1}{T_{60}(2)} - \frac{1}{T_{60}(1)}\right] \quad ...(7.57)$$

Absorption of sound by the material depends upon its position in the room or hall. Materials placed in the centere of a surface are more effective absorbers, and yield higher absorption coefficients, than materials located in the corners. This is because the average particle velocity is higher there. There are also *diffraction effects* and *edge absorption* attributable to the sides of the material. For these reasons Sabine absorption coefficients are found greater than one sometimes. A better estimate of absorption coefficient is done using Norris-Eyring equation.

The following table gives absorption coefficient for certain materials.

Table 7.4 Absorption coefficient of certain materials in open window unit (OWU)

Material	Absorption coefficient in OWU
Glass	0.02
Marble	0.01
Wooden floor	0.06 - 0.08
Carpets	0.15 - 0.30
Cushions	0.15 - 0.18
Brick wall	0.03

7.14 ACOUSTIC CONCEPTS OF HALLS AND AUDITORIUM

Building acoustics is the science of controlling sound level and noise in buildings. This includes the maintenance of sound level in a building and minimisation of noise transmission in the building and from one building to another.

Building acoustics plays an important role in the design, operation and construction of most buildings, and can have a significant impact on communication and productivity, health and well-being. The control of sound in buildings, such as concert halls, cinema halls, auditoriums, recording studios, lecture theatres, and so on, is very important for maintaining quality of sound during speech, lecture, showing movie, during presentation of a play, etc.

Building acoustics depends on:

(a) The design and volume of a building.

(b) The sound absorption, transmission and reflection characteristics of surfaces enclosing the hall and in the building.

(c) The generation of sound inside or outside the building.

(d) Impact of noise created inside and outside the building.

In a room, one can measure the reverberation time using a loudspeaker and a sound level meter. One can also calculate the reverberation time using the Sabine's formula established by Wallace Clement Sabine if absorption coefficient is known. Sabine's equation is based on the volume of the space and the total amount of absorption in a space. The total amount of absorption in a space is referred to as sabin. It is important to note that the absorption and surface area must be considered for every material in a space to calculate sabin.

1. To obtain the sabin values, we use the volume (length × width × height) of the room and the materials of the surfaces. Table 7.5 provides the sabins at each octave band frequency for common construction materials. One can calculate the absorption coefficient for absorbing material placed in the hall by measuring the reflected sound by using sound level meter.
2. Then one can get total sound absorption the hall by adding it to the hall coefficient of the empty hall. Thus, by knowing the absorption coefficients before and after the treatment, one can calculate the reverberation time before and after treatment. The knowledge of time of reverberation allows us to understand the kind and amount of material to add to the space to minimize the noise in the hall.

Table 7.5 Absorption coefficient (α) and octave band centre frequencies

		Absorption coefficient (α) octave band centre frequencies					
Location	**Material**	**125 Hz**	**250 Hz**	**500 Hz**	**1000 Hz**	**2000 Hz**	**4000 Hz**
All walls	Gypsum board	0.29	0.10	0.06	0.05	0.04	0.04
Ceiling	Metal deck	0.05	0.04	0.03	0.03	0.02	0.01
Floor	Concrete painted	0.01	0.01	0.02	0.02	0.02	0.02
Ceiling	Acoustic Baffles	0.027	0.65	1.06	1.16	1.14	1.12

Formulas at a Glance

S.No.	Formula	
1.	Equation displacement simple harmonic motion $x(t) = x_m \cos(\omega t + \varphi)$	X_m is maximum displacement of particle, $\omega = \frac{2\pi}{T}$ is angular velocity of particle, φ is phase constant
2.	Time period of harmonic oscillator $T = \frac{2\pi}{\omega_0} = 2\pi\sqrt{\frac{m}{k}}$	m is mass of particle, $k = \frac{2\pi}{\lambda}$ is force constant
3.	Energy of particle of simple harmonic oscillator $E = \frac{1}{2}mv^2 + \frac{1}{2}kx^2$	v is velocity of particle, x is displacement of particle at time, t. Energy of particle in harmonic motion remains conserved.
4.	Displacement of damped harmonic oscillator $x(t) = x_m e^{-b/2m} \cos(\omega' t + \varphi)$ where, $\omega' = \sqrt{\frac{k}{m} - \frac{b^2}{4m^2}}$	B is force constant of damping force
5.	Potential energy of damped harmoci oscillator $E(t) = \frac{1}{2}kx_m^2 e^{-bt/m}$	
	Equation displacement simple harmonic motion $x(t) = x_m \cos(\omega_d t + \varphi)$	ω_d is the angular frequency of article due to periodic driven force, $F_0 \cos(\omega_d t)$. Initially, the oscillator oscillates with natural frequency, ω.
	Amplitude of forced oscillator $A = \frac{F_0}{\sqrt{m^2(\omega^2 - \omega_d^2)^2 + \omega_d^2 b^2}}$	
6.	Intensity of sound, $I = \frac{P}{S}$, The intensity (II) is related to the displacement ampliture (A) of the sound wave as $I = \frac{1}{2}\rho v \omega^2 A^2$ or $I = \frac{1}{2}\frac{p_0^2}{\rho v}$	Where, P is the rate of energy transfer (power) of the sound wave and S is the area of the surface facing the sound.
7.	Intensity, $I = \frac{P}{4\pi r^2}$	$4\pi r^2$ is the area of the sphere of radius, r.

8.	Ratio of intensities	This formula can be used to find the intensity at I_2 at distance d_2 if intensity (I_1) at distance.
	$\frac{I_2}{I_1} = \left(\frac{d_1}{d_2}\right)^2$	(d_1) is known or vice versa.
9.	Decibel level SL = (10 dB) log $\frac{1}{I_0}$	Here, dB is the abbreviation for decibel, unit of sound level, a name that was chosen to recognise the work of Alexander Graham Bell, I_0
10.	Amplitude of pressure wave, $p_0 = A\, k\, \rho\, v^2$ or $p_0 = A\, \rho\, v\, \omega$ as $v = \frac{\omega}{k}$	as $v = \sqrt{\frac{B}{\rho}}$, here, B is bulk modulus and ρ is density of medium
11.	Total energy of particle in sound wave $U + \text{KE} + \text{PE} = \frac{1}{2}\rho\, \omega^2 A^2$	U is sum of KE and PE of particle in sound wave
12.	The Sabine equation for reverberation time $RT_{60} = \frac{0.049\, V}{S\alpha}$ for room dimension in feet $RT_{60} = \frac{0.161\, V}{S\alpha}$ for room dimensions in metres	V is volume of hall, S is durface area of hall and α is absorption coefficient
13.	The Norris-Eyring equation for reverberation time $RT_{60} = \frac{0.049\, V}{-S\ln(1-\alpha)}$ for room dimensions in feet or $RT_{60} = \frac{0.161\, V}{-S\ln(1-\alpha)}$ for room dimensions in meters	V is volume of hall, S is surface area of hall and α is absorption coefficient

Example 7.12 For an empty assembly hall of size 20 × 15 × 12 cubic metre with absorption coefficient 0.107. Calculate reverberation time.

Solution: Formula required

$$\text{Reverberation time, } T = \frac{0.161\, V}{\Sigma\, \alpha S} = \frac{0.161}{\alpha S_T} \text{ in metres}$$

Given:

(i) Size of the room = 20 × 15 × 12 = 3600 m^2

(ii) Absorption coefficient, $\alpha = 0.107$

Calculation: Total surface area of the hall is

$$S_T = 2\,(20 \times 15 + 15 \times 12 + 20 \times 12) = 1440\ m^2$$

$$\text{Reverberation time, } T = \frac{0.161 \times 36000}{0.107 \times 1440} = 3.761\ s$$

Ans. Reverberation time = 3.761 s

Example 7.13 A hall of volume 5600 m^3 is found to have a reverberation time of 2.4 seconds. The sound absorbing surface of the hall has an area of 760 m^2. Calculate the average absorption coefficient.

Solution: Sabine's formula for reverberation time

$$T = \frac{0.161\,V}{\Sigma\,\alpha S} = \frac{0.161\,V}{\alpha S_T} \text{ in metres}$$

Given: $V = 5600\ m^3$, $T = 2.4\ s$, $S_T = 760\ m^2$

Calculation: Let absorption coefficient be α, then

$$2.3 = \frac{0.161 \times 5600}{a \times 760} \quad \text{or} \quad \alpha = \frac{0.161 \times 5600}{2.4 \times 760} = 0.494$$

Ans. Reverberation coefficient is 0.494.

Example 7.14 For an empty hall of size 20 × 11 × 11 cubic metres, the reverberation time is 2.4 seconds. Calculate the average absorption coefficient of the hall. What area of the floor should be covered by carpet so as to reduce the reverberation time to 2.1 seconds. Given that absorption coefficient of carpet is 0.51.

Solution: Sabine's formula for reverberation time

$$T = \frac{0.161\,V}{\Sigma\,\alpha S} = \frac{0.161\,V}{\alpha S_T} \text{ in metres}$$

Given: Size of hall is 20 × 12 × 12 cubic metres

Reverberation time, $T = 2.4$ s

Calculation:

(a) Volume of hall, $V = 20 \times 11 \times 11 = 2420$ cubic metres

Total surface area of hall

$$S_T = 2\,(20 \times 11 + 11 \times 11 + 20 \times 11) = 1122\ m^2$$

Reverberation time

$$T = \frac{0.161\,V}{\alpha\, S_T}, \text{ so absorption coefficient}$$

$$\alpha = \frac{0.161\,V}{TS_T} = \frac{0.161 \times 2420}{2.4 \times 1122} = 0.1446$$

(b) By using the carpet of surface area S_1 whose absorption coefficient is 0.51, the reverberation time is reduced to 2.1 seconds.

So new reverberation time, $T_2 = 2.1$ seconds

Carpet surface area, = S_1, coefficient of absorption of carpet, $\alpha_c = 0.51$

New total surface area of hall will be

$$S_T' = 1122 \text{ m}^2 + S_1$$

$$T_2 = \frac{0.161\,V}{\alpha\, S_T - \alpha\, S_T' + \alpha_c\, S_T'}$$

or $$\frac{1}{T_2} = \frac{\alpha S_T - \alpha\, S_T' + \alpha_c\, S_T'}{0.161\,V}$$

or $$\frac{1}{T_2} = \frac{\alpha S_T}{0.161\,V} + \frac{(\alpha_c - \alpha)\, S_T'}{0.161\,V}$$

or $$\frac{1}{T_2} - \frac{1}{T} = \frac{(\alpha_c - \alpha)\, S_T'}{0.161 V} \quad \text{or} \quad S_T' = \frac{0.161\,V}{(\alpha_c - \alpha)}\left[\frac{1}{T_2} - \frac{1}{T}\right]$$

$$S_T' = \frac{0.161 \times 2420}{(0.51 - 0.1446)}\left[\frac{1}{2.1} - \frac{1}{2.4}\right]$$

or $$S_T' = 1066.2835[0.478 - 0.417] = 65.0433$$

Ans. Carpet area required to reduce reverberation time up to 2 seconds is 65.0433 m^2.

Example 7.15 A hall with dimensions 16 × 12 × 12 cubic metres is found to have reverberation time of 5 seconds. What is the total absorbing power of all the surfaces in the hall?

Solution: Reverberation time

$$T = \frac{0.161\,V}{\Sigma\, aS}$$

Given:

$$V = \text{Volume of a hall} = 16 \times 12 \times 12 = 2304 \text{ m}^3$$

Reverberation time, $T = 5$ seconds

Calculation: Total absorbing power,

$$\Sigma\, aS = \frac{0.161\,V}{T} = \frac{0.161 \times 2304}{5} = 74.19 \text{ sabins}$$

Ans. Total absorbing power of all surfaces of the hall is 74.19 sabins.

UNSOLVED PROBLEMS

7.1 A classroom with a volume of 1500 m^3 has reverberation time of 2.5 seconds. Find the total absorbing power of all the surfaces in the room.

7.2 A cinema hall has a volume of 8000 m^3. It is required to have reverberation time of 1.5 seconds, what should be the total absorption in the hall?

7.3 The volume of the room is 1250 m^3. The wall area is 250 m^2, the floor area is 150 m^2, ceiling area 150 m^2. The average sound absorption coefficient: (i) for wall is 0.04, (ii) for ceiling 0.6, and (iii) for floor is 0.05. Calculate the average sound absorption coefficient and reverberation time.

7.4 For an empty assembly hall of size 20 × 15 × 12 m^3, the reverberation time is 4 seconds. Calculate the average absorption coefficient of the hall. What area of the wall should be covered by the curtain so as to reduce the reverberation time to 2 seconds? Given the absorption coefficient of the curtain cloth is 0.5.

7.5 A classroom has dimensions 20 × 15 × 10 m^3. The reverberation time is 4 seconds. Calculate average absorption coefficient.

7.6 The reverberation time is found to be 2 seconds for an empty hall and it is found to be 1 second when a curtain of 25 m^2 is suspended at the centre of the hall. If the dimensions of the hall are 10 × 10 × 5 m^3, calculate the coefficient of absorption of the curtain.

QUESTIONS FOR EXAMINATION

7.1 Each of the following properties of sound listed in column *A* primarily depends on one of the quantities in column *B*. Write down the matching pairs from the two columns.

Column *A*	Column *B*
Pitch	Waveform
Quality	Frequency
Loudness	Intensity

Ans. Pitch → Frequency, Quality → Waveform, Loudness → Intensity

7.2 In view of characteristics of sound, distinguish between the roar of a lion and humming of a mosquito.

Ans. Lion has large loudness but low pitch, while humming has low loudness but high pitch.

7.3 Why is it sometimes difficult to recognize the voice of a speaker over the telephone?

Ans. Due to poor quality of sound.

7.4 Fill in the blanks

(a) A normal ear cannot listen to mechanical wave of frequency exceeding .... Hz, which are called .... waves.

(b) Phase difference between waves, $y = a \sin(\omega t - kx)$ and $y = a \cos(\omega t - kx)$ is ....

(c) A body vibrating with a certain frequency sends waves of wavelength 15 cm in medium *A* and 20 cm in medium *B*. If the velocity of waves in *A* is 120 m/s, that in *B* will be .... m/s.

(d) A point source emits sound equally in all directions in a non-absorbing medium. Two points, *P* and *Q*, are at distance of 9 m and 25 m, respectively from the source. The ratio of the amplitudes of the waves at *P* and *Q* is ....

(e) The term woofer (or bass) refers to .... frequency part of musical sound.

(f) The quality of note is determined by .... present in it.

Ans. (a) 20 kHz ultrasonic (b) $\pi/2$ (c) 1690 m/s (d) 25/9

(e) low (f) overtones

7.5 What are characteristics of sound?

7.6 Discuss simple harmonic motion.

7.7 Define amplitude, phase angle, time period, angular frequency in simple harmonic motion.

7.8 Discuss the physics of simple harmonic oscillator.

7.9 Find the expression for energy of a particle in simple harmonic motion.

7.10 What is damped harmonic motion? Discuss the physics of damped harmonic oscillator.

7.11 What are forced oscillation and resonance? Discuss the physics of forced harmonic oscillator.

7.12 What is principle of superposition?

7.13 What is Fourier theorem? Discuss the Fourier analysis.

7.14 What is the difference between intensity and loudness?

7.15 How does intensity vary with distance?

7.16 Discuss decibel scale?

7.17 Derive the formula for amplitude of pressure waves.

7.18 Derive the formulas for energy, power and intensity of sound.

7.19 What are loudness , pitch and quality of waves?

7.20 Write differences between tone and pitch.

7.21 What is the purpose of musical scale? Discuss it in detail in western musical scale.

7.22 What is the basic structure of Indian music?

7.23 What is the difference between western and Indian musical scales?

7.24 Define reverberation and reverberation time? Write causes of reverberation. State the method minimizing it.

7.25 Write Sabine's formula for reverberation time of a hall. Write its significance.

7.26 Define absorption coefficient of a material? Establish the relation between absorption coefficient and time of reverberation.

7.27 Write the factors which affect the acoustics of building.

7.28 State the features of acoustically good auditorium?

7.29 Write short note on acoustics of buildings.

OBJECTIVE QUESTIONS

7.1 Which of the following statements are correct about intensity of sound?

(a) It depends only on amplitude of wave.

(b) It depends both on amplitude and frequency of the wave

(c) Its practical unit is decibel

(d) Its practical unit is phon

7.2 In case of sustained forced oscillations the amplitude of oscillations decreases:

(a) Decreases linearly (b) Decreases sinusoidally

(c) Decreases exponentially (d) Always remains constant

7.3 Which of the following cannot travel through vacuum:

(a) Light waves (b) Heat waves (c) *X*-rays (d) Sound waves

7.4 A big explosion on the moon cannot be heard on the earth because:

(a) The explosion produced high frequency sound waves which are inaudible

(b) Sound waves require a material medium for propagation

(c) Sound waves are absorbed in the atmosphere of the moon

(d) Sound waves are absorbed in the earth's atmosphere

7.5 An astronaut cannot hear his companion at the surface of the moon because:

(a) Produced frequencies are above the audio frequencies.

(b) There is no medium for sound propagation

(c) Temperature is too low during night and high during day

(d) There are too many craters on the surface of the moon.

7.6 Mechanical wave (sound wave) in a gas is:

(a) Transverse (b) Longitudinal (c) Neither transverse nor longitudinal

(d) Either transverse or longitudinal

7.7 Mechanical waves on the surface of a liquid are:

(a) Transverse waves (b) Longitudinal

(c) Torsional (d) Both transverse and longitudinal (ripples)

7.8 Which of the following properties of a wave does not change with a change in medium?

(a) Frequency (b) Wavelength (c) Velocity (d) Amplitude

7.9 Which property of a wave is independent of the other?

(a) Amplitude (b) Velocity (c) Wavelength (d) Frequency

7.10 Which of the expression is that of a simple harmonic progressive wave?

(a) $A \sin \omega t$ (b) $A \sin \omega t \cos kx$ (c) $A \sin (\omega t - kx)$ (d) $A \cos kx$

7.11 A simple harmonic wave having amplitude (A) and time period (T) is represented by the equation, $y = 5 \sin \pi (t + 4)$ m. The value of A (in m) and T (in s) are:

(a) $A = 5; T = 2$ (b) $A = 10; T = 1$ (c) $A = 5 ; T = 1$ (d) $A = 10; T = 2$

7.12 A sound wave in air can be represented as

$$Y = 5 \times 10^{-2} \sin [100t - 50 x]\ m,$$

where, t is expressed in seconds. and x in m and y represents displacement. The phase velocity of the wave is

(a) 2 m/s (b) 4 m/s (c) 500 m/s (d) 0.2 m/s

7.13 If the equation of progressive wave is given by

$$Y = 4 \sin \pi \left[\frac{t}{5} - \frac{x}{9} + \frac{\pi}{6}\right] m$$

then which of the following is correct, if x in m and t in seconds:

(a) $v = 5$ cm/s (b) $\lambda = 18$ m (c) $A = 0.04$ cm (d) $f = 50$ Hz

7.14 The velocity of sound in air is nearly:

(a) 332 m/s (b) 3.32 km/s (c) 332 ft/s (d) 332 km/s

7.15 Ultrasonic, infrasonic and audible waves travel through a medium with speeds v_u, v_i and v_2 and v_a, respectively; then

(a) $v_u > v_a > v_i$ (b) $v_a < v_u < v_i$ (c) $v_u = v_i = v_a$ (d) $v_u < v_i < v_a$

7.16 In which medium has sound maximum velocity:

(a) Gases (b) Liquids (c) Solid (d) Vacuum

7.17 In which medium has sound maximum velocity:

(a) Water (b) Steel (c) Air (d) Vacuum

7.18 The velocity of sound in any gas depends on

(a) Intensity (b) Amplitude

(c) Density and elasticity (d) Volume and temperature

7.19 The loudness of sound depends on

(a) Amplitude (b) Pitch (c) Velocity (d) Wavelength

7.20 The bells of a college or temple are made of large size for producing

(a) Sound of high pitch (b) Loud sound

(c) Sound of high quality (d) Pleasant sound

7.21 The intensity of a sound wave gets reduced by 20% on passing through a slab. The reduction in intensity on passing through two slabs is

(a) 40% (b) 36% (c) 30% (d) 50%

7.22 The intensity of a harmonic wave depends on

(a) Frequency (b) Amplitude (c) Frequency and amplitude
(d) Frequency and velocity

7.23 If the ratio of intensity of two waves is 4 : 1, the ratio of their amplitudes is

(a) 2 : 1 (b) 1 : 2 (c) 4 : 1 (d) 1 : 4

7.24 If the pressure amplitude of a sound wave is tripled then the intensity of the sound is increased by a factor

(a) 3 (b) 6 (c) 9 (d) $\sqrt{3}$

7.25 The frequency of a singing note *B* is one-eighth the frequency of singing note *A*. The energy of two sounds are equal. The amplitude of the note of *B* is

(a) Same as that of *A* (b) Is twice that of *A*
(c) Decreased by a factor of 2 (d) Decreased by a factor of 4

7.26 Waves of amplitude (*A*) and frequency (ω) travel in air with the same velocity. Which of the following waves has the highest intensity?

(a) $A = 10 \times 10^{-4}$ m, ω = 500 rad/s (b) $A = 2 \times 10^{-4}$ m, ω = 200 rad/s
(c) $A = 20 \times 10^{-4}$ m, ω = 200 rad/s (d) $A = 20 \times 10^{-4}$ m, ω = 115 rad/s

7.27 Decibel is

(a) A musical measurement (b) A measurement of sound level
(c) A musical note (d) Noise

7.28 Power of a sound from the speaker of a radio is 20 milliwatts. By turning the knob of the volume control the power of the sound is increased to 400 milliwatts. The power increase in decibel as compared to the original power is

(a) 13 dB (b) 10 dB (c) 800 dB (d) 8000 dB

[**Hint:** Sound level, $\beta = (10 \text{ dB}) \log \frac{I}{I_0} = (10 \text{ dB}) \log \frac{400}{20} = (10 \text{ dB}) \log 20 = (10 \text{ dB}) [1 + \log 2]$

$= (10 \text{ dB}) [1 + 0.03010] = 13.\ 010$

7.29 The intensity level of sound *A* is 2 decibel greater than that of *B*. How many times more intense is the sound *A* than *B*?

(a) 10 (b) 100 (c) 1000 (d) 2

7.30 The shrillness (pitch) of a sound note depends on

(a) Amplitude (b) Frequency (c) Wavelength (d) Velocity

7.31 If a music record is made faster in a gramophone, then

(a) Intensity increases (b) Pitch increases
(c) Intensity decreaes (d) Pitch decreases

7.32 The voice of a lion is different from that of a mosquito because
(a) Two animals have different intensities
(b) Two voices travel with different velocities
(c) Sounds of two have different pitch
(d) Sounds of two have different frequencies

7.33 It is possible to recognize a person by hearing his voice without seeing him because voice of each person has
(a) A definite pitch (b) A definite quality (c) A definite loudness (d) A definite intensity

7.34 The quality of a tone
(a) Decreases with loudness (b) Varies directly with pitch
(c) Depends on the overtones (d) Varies inversely with frequency

7.35 The same notebeing played on sitar and veena differs in
(a) Pitch (b) Quality
(c) Both pitch and quality (d) Intensity

7.36 In an orchestra, the musical sounds of different instruments are distinguished from one another by
(a) Pitch (b) Quality (c) Loudness (d) Overtones

7.37 Two wave pulses (one being inverted with respect to the other) travel in opposite direction on a string to approach each other. What will happen to them?
(a) They will vanish after collision.
(b) The pulses will pass through each other without any change in their shape.
(c) The pulses will reflect, that is, the pulse going towards right will move to left after collision and vice versa.
(d) The pulses will pass through each other but their shapes will be modified.

7.38 The best absorber of the sound is
(a) Open window (b) Carpet
(c) Heavy curtain (d) Perforated cellulose fibre files

7.39 Reverberation time is
(a) Directly proportional to volume (b) Inversely proportional to volume
(c) Equal to volume (d) None of the above

7.40 The good choice for acoustically correct auditorium
(a) Parallel walls (b) Splayed walls (c) Curved walls (d) White wall

7.41 The time gap between the initial direct note and the reflected note up to a minimum audibility level is
(a) Deflection time (b) Absorption time (c) Reverberation time (d) Echo time

7.42 The absorption coefficient is measured in.
(a) Closed window unit (b) Open window unit
(c) Open door unit (d) None of these

7.43 Echelon effect is produced by

(a) Floors (b) Ceiling (c) Curved surfaces (d) Staircase

7.44 Use of absorbents .... the reverberation time.

(a) Increases (b) Decreases

(c) Have no effect on (d) Can increase or decrease

7.45 Reverberation of sound in a hall is

(a) A desirable effect in a controlled way (b) Undesirable effect

(c) Sometimes desirable and sometimes undesirable

(d) Essential evil

7.46 Reverberation time is given by

(a) Sabine (b) Weber-Fechner law (c) Weber's law (d) Sabine-Fechner law

7.47 For a given hall absorption coefficient is 0.25 and volume 200 m^3 and absorbing area 840 m^2. The reverberation time is

(a) 2.5 s (b) 2.2 s (c) 1.2 s (d) 1.53 s

7.48 To get good sound effect inside a hall

(a) The reverberation time has to be as large as possible.

(b) The reverberation time has to be zero.

(c) The hall should not have any absorbing material.

(d) The reverberation time has to be optimum.

7.49 Optimum reverberation time for music is

(a) 0.5 to 1 seconds (b) 0 to 1 seconds (c) 1 to 2 seconds (d) Above 5 seconds

7.50 Which one of the following has minimum absorption coefficient?

(a) Glass (b) Felt (c) Open windows (d) Wooden floor

(**Note:** Meaning of felt is compressed and rolled thick cloth of hair, fur or wool.)

7.51 The walls of a hall built for music concerts should

(a) Amplify sound (b) Reflect sound (c) Transmit sound (d) Absorb sound

7.52 The time of reverberation of a hall can be reduced by using

(a) Reflectors (b) Absorbers (c) Domes (d) Arch

7.53 For a uniform distribution of sound intensity is it better to have

(a) Parabolic reflector (b) Concave reflector

(c) Spherical reflector (d) Cylindrical reflector

7.54 The walls of a hall built for music concerts should

(a) Amplify sound (b) Reflect sound (c) Transmit sound (d) Absorb sound

7.55 A cinema hall has a volume of 800 m^3 and the total absorption in the hall is 800 OWU. The time of reverberation should be

(a) 0.161 s (b) 1.61 s (c) 0.48 s (d) 4.8 s

7.56 The Sabine is the unit of

(a) Energy
(b) Reverberation time
(c) Absorption of a hall
(d) Reflection by a wall

Answers

7.1 (b) and (c)		7.2 (d)	7.3 (d)	7.4 (b)	7.5 (b)	7.6 (b)	7.7 (d)
7.8 (a)	7.9 (a)	7.10 (b)	7.11 (a)	7.12 (a)	7.13 (b)	7.14 (a)	7.15 (a)
7.16 (c)	7.17 (b)	7.18 (c)	7.19 (a)	7.20 (b)	7.21 (b)	7.22 (c)	7.23 (a)
7.24 (c)	7.25 (d)	7.26 (c)	7.27 (b)	7.28 (a)	7.29 (b)	7.30 (b)	7.31 (b)
7.32 (c)	7.33 (b)	7.34 (c)	7.35 (b)	7.36 (b)	7.37 (b)	7.38 (a)	7.39 (a)
7.40 (b)	7.41 (c)	7.42 (b)	7.43 (d)	7.44 (b)	7.45 (a)	7.46 (a)	7.47 (d)
7.48 (d)	7.49 (c)	7.50 (a)	7.51 (d)	7.52 (c)	7.53 (a)	7.54 (d)	7.55 (a)
7.56 (c)							

Books Referred for Further Reading

The Sounds of Music : Science of Musical Scales I -Human perception of sound by Sushan Konar
The Sounds of Music : Science of Musical Scales II- Western Classical Music by Sushan Konar
The Sounds of Music : Science of Musical Scales III -Indian Classical Music by Sushan Konar

CHAPTER 8

Wave Optics

8.1 INTRODUCTION

Huygens' principle is based on the travelling of light as wavefronts. In chapter 8, the applications of Huygens' principle such as reflection, refraction and total internal reflection of light are discussed. Diffraction from double slit [Young's experiment], diffraction from single slit using Lloyd's mirror and diffraction from grating, other applications based on Huygens' principle, are also presented in this chapter. An interesting phenomenon, the diffraction of water waves at gap is also given in Section 8.12.

8.2 ELECTROMAGNETIC NATURE OF LIGHT

Properties which are common to all types of electromagnetic waves are listed below.

(i) Electromagnetic waves are associated with varying electric and magnetic fields. The two fields are perpendicular to each other and also perpendicular to the direction of the travelling wave. Each field vibrates at the same frequency. The ratio $E/H = c$, the velocity of light.

(ii) All electromagnetic waves travel at the same speed, $c = 2.998 \times 10^8$ m/s in wavefront.

(iii) Electromagnetic waves remain unaffected by electric and magnetic fields.

(iv) It is assumed that electromagnetic waves travel in straight lines (within the limits set by diffraction).

(v) The electromagnetic waves can be polarized because they are transverse waves.

(vi) Electromagnetic waves produce interference effects.

(vii) Examples of electromagnetic waves are radio waves, microwaves, infrared light, visible light, ultraviolet light, X-rays and γ-rays.

(vii) Order of wavelength and frequency of these electromagnetic waves are given in the following table.

Table 8.1

S.No.	Subject	Wavelength (m)	Frequency (Htz)
1.	Radio wavs		
	Long waves	$> 10^3$	$< 10^5$
	Medium wavs	10^3 to 10	$\sim 10^5$ to 10^7
	Short wavs		
	VHF, UHP	10 to 0.1	$\sim 10^8$ to 10^9
2.	Microwaves	0.1 to 10^{-3}	$\sim 10^9$ to 10^{11}
3.	Infrared light	$\sim 10^{-5}$	$\sim 10^{13}$
4.	Visible light	3.9×10^{-7} to 7.8×10^{-7}	$\sim 10^{15}$
5.	Ultraviolet light	$> 10^{-7}$	$> 10^{15}$
6.	X-rays	$\sim 10^{-9}$ to 10^{-11}	$\sim 10^{17}$ to 10^{19}
7.	γ-rays	$\sim 10^{-13}$ to 10^{-15}	$\sim 10^{21}$ to 10^{23}

8.3 TEMPORAL AND SPATIAL COHERENCE

The temporal or the longitudinal coherence length (L_t) is defined as the distance from the source to the point at which the waves become complexly out of phase.

The transverse spatial coherence length (L_t) is defined as the lateral distance between two points S_1 and S_2 situated on the cardboard placed transverse to the laser beam at which the two waves are out of phase.

For details of temporal and spatial coherence see Section 7.3 of Chapter 7.

8.4 DEFINITION AND PROPERTIES OF WAVEFRONT

A wavefront is a line or surface, in the path of a wave motion, on which the disturbances at every point have the same phase.

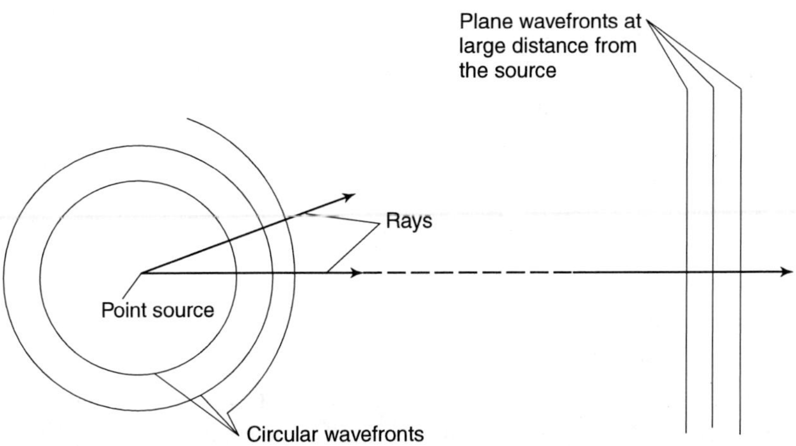

Fig. 8.1 Wavefronts and rays

A medium in which the waves travel is assumed an isotropic medium. All the waves travel with the same speed in all directions. If a point source generates two-dimensional waves in an isotropic medium, the wavefronts are circles centred on the point source as shown in Fig. 8.1. One can see circular wavefronts, when a stone is thrown in pond. The wavefront associated with two-dimensional waves produced by a line source are straight lines. For example, wavefront produced by a straight vibrator in a ripple tank are straight lines. The wavefronts are straight lines at large distances from a source producing two-dimensional waves, as shown in Fig. 8.1.

If a point source in an isotropic medium produces spherical wavefronts centred on the source point in three dimensions. A point source of light produces spherical wavefronts. At large distances from the point source, wavefronts are the surfaces of the spheres of large radii, and therefore, can be considered essentially plane wavefrons. In fact, at large distances from a source of any kind which is generating three-dimensional waves, the wavefronts are plane.

A ray is a line representing the direction of travel of wave is at right angle to the wavefronts as shown in Fig. 8.1.

8.5 HUYGENS' PRINCIPLE AND CONSTRUCTION

Huygens postulated that each point on a wavefront is a secondary source of spherical wavelets and that the surface new wavefront touches all of these secondary wavelets.

Figure 8.2 illustrates the construction to determine subsequent position of a circular wavefront.

Notes:

(i) In this construction, it is assumed that the amplitude of the secondary wavelets is zero in the backward direction. This is done to prevent the existence of an additional wavefront behind the original one, although this assumption is contrary to the observations.

(ii) In two dimensions, the secondary wavelets are circular rather than spherical, due to assumption made in (i), are shown as semicircle.

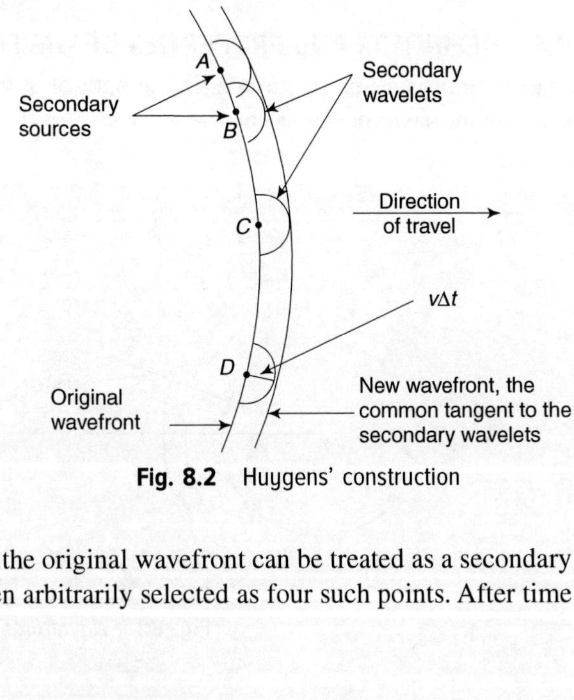

Fig. 8.2 Huygens' construction

According to Huygens, every point on the original wavefront can be treated as a secondary source, therefore, *A*, *B*, *C* and *D* have been arbitrarily selected as four such points. After time

Δt, the position of the wavefront is found by constructing secondary wavelets centred on these points and radius, $v\,\Delta t$. The new wavefront is the common tangent to these wavelets. Thus, according to Huygens' construction, a circular wavefront continues as a circular wavefront. This result is consistent with observation.

8.6 HUYGENS' CONSTRUCTION APPLIED TO REFLECTION

Let us consider that a parallel beam of light incident on a reflecting surface, *XY*, at an angle (i) with respect to the normal to the surface as shown in Fig. 8.3. When the side *A* of plane wavefront *AB* has just reached the surface. The other side, *B*, of the plane wavefront proceeds to *C* as a secondary wavelet of radius *BC*. In the same time, a secondary wavelet of radius *AD* proceeds to *D* in the form of semicircle as shown in Fig. 8.3. Since the speed of light is same in the region above the surface *XY*, so $BC = AD$. Because *XY* is a reflecting surface, the secondary wavelet must be emitted into the region above *XY*. The new plane wavefront (*CE*) is tangent to semicircle at point *D* as shown in Fig. 8.3. The reflected beam must be at right angle to this wavefront and proceed as *AD* and *CF* parallel rays as illustrated in the figure.

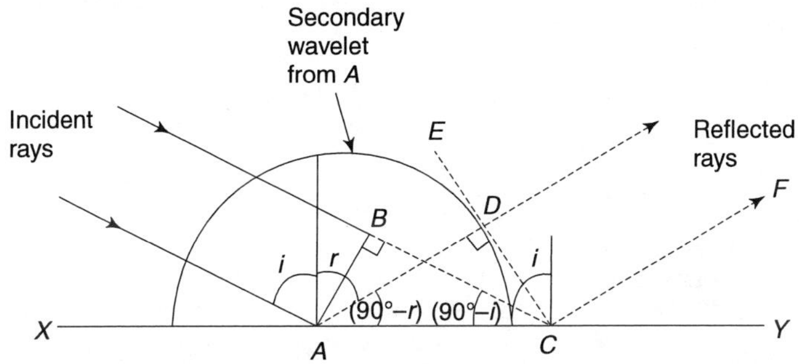

Fig. 8.3 Huygens' construction for reflection at a plane surface

In ΔABC and ΔADC, *AC* is common in both the triangles.

$$BC = AD \quad \text{(by construction)}$$

Right $\angle ABC$ = right $\angle ADC$

The two triangles are congruent, therefore,

$$\angle BCA = \angle DAC$$

$$(90° - i) = (90° - r)$$

or

$$i = r$$

That is, angle of incidence = angle of reflection

This is called the second law of reflection.

Thus, Huygens' construction shows that reflected light is consistent with the observation. Although this result has been obtained by considering the behaviour of light, therefore it applies to any wave motion. No assumption has been made which violate the properties of light.

8.7 HUYGENS' CONSTRUCTION APPLIED TO REFRACTION

Let us consider that a parallel beam of light incident on a refracting surface, *XY*. It incidents at angle θ_1 with respect to normal to the surface as shown in Fig. 8.4. Consider also, that side *A* of an associated wavefront, *AB*, has just reached the surface. If light from the other side, *B*, of the beam subsequently travels to *C*, in time *t*, then $BC = c_1t$, where, c_1 is the velocity of light in the medium above *XY*. If c_2 is the velocity of light in the region below *XY*, then, in the same time, *A* will emit a secondary wavelet of radius c_2t. It is shown by a semicircle centred at *A*. The new plane wavefront (*CE*) will be tanget to this and touches it at *D* as shown in Fig. 8.4. The refracted beam must be at right angle to this wavefront and is bounded by parallel rays (*AD* and *CF*) as illustrated in the figure.

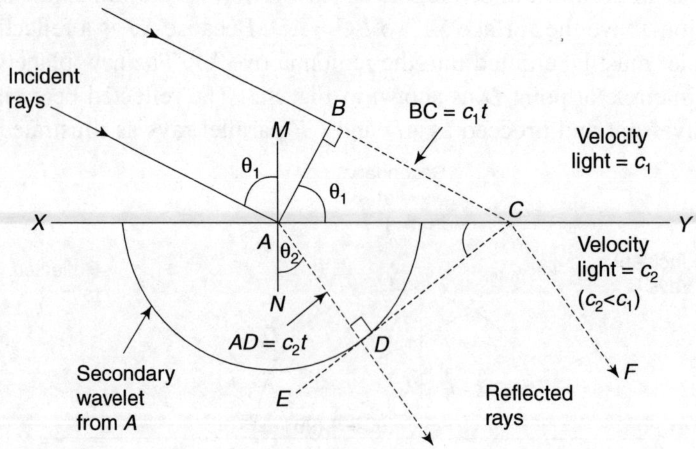

Fig. 8.4 Huygens's construction for refraction at a plane surface

In ΔABC and ΔADC,

$$\frac{\sin \angle CAB}{\sin \angle ACD} = \frac{\frac{BC}{AC}}{\frac{AD}{AC}} = \frac{BC}{AD} = \frac{c_1 t}{c_2 t} = \frac{c_1}{c_2} \quad \text{...(8.1)}$$

$$\angle CAB + \angle BAM = 90° \quad \text{and} \quad \theta_1 + \angle BAM = 90° \quad \text{as} \quad \angle CAB = \theta_1$$

and $$\angle ACD + \angle CAD = 90° \quad \text{and} \quad \theta_2 + \angle CAD = 90° \quad \text{as} \quad \angle ACD = \theta_2$$

Therefore, substituting the values $\angle CAB = \theta_1$ and $\angle ACD = \theta_2$

Equatin (8.1), becomes

$$\frac{\sin \theta_1}{\sin \theta_2} = \frac{c_1}{c_2} \quad \text{...(8.2)}$$

For any given wavelength, c_1 and c_2 are constants, therefore Eq. (8.2) is consistent with Snell's law, i.e., in Huygens' construction one gets a result which is consistent with observations.

There were two views to the light propagation, around the middle of seventeenth century. Huygens tied to establish that light travels in waveform, while according to Isaac Newton theory, light rays travel in streams of particles called corpuscles. Huygens' theory obeys the laws of reflection and refraction. Newton's theory also obeys laws of reflection and refraction. On the basis of Eq. (8.2), the wave theory predicts that light bends towards the normal when it enters a medium where its velocity is smaller, because $\theta_2 < \theta_1$ as $c_2 < c_1$. Newton's corpuscular theory, on the other hand, predicts just opposite to Huygens' theory, i.e., light bends away from the normal on moving into a medium where its velocity is smaller.

In 1950, when two rival theories were proposed, Foucault measured the velocity of light both in air and water. He found that the velocity of light in air is more than it is in water. Foucault's measurement confirmed the prediction of the wave theory, i.e., light bends towards the normal in going from air to water.

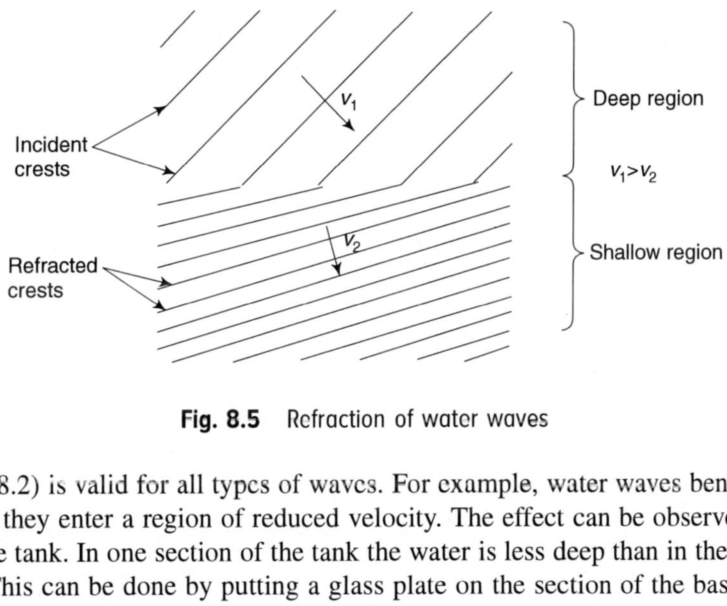

Fig. 8.5 Refraction of water waves

Equation (8.2) is valid for all types of waves. For example, water waves bend towards the normal when they enter a region of reduced velocity. The effect can be observed by making use of a ripple tank. In one section of the tank the water is less deep than in the other section of the tank. This can be done by putting a glass plate on the section of the base of the tank. Since water waves travel more slowly in shallow water, so water waves bend towards the normal as they enter the shallow region as shown in Fig. 8.5.

8.8 HUYGENS' CONSTRUCTION APPLIED TO TOTAL INTERNAL REFLECTION

Let us consider a beam of light incident on boundary, *XY*, between two media, so that it is at an angle θ to the normal to *XY* as shown in Fig. 8.4. Also let us assume that side *A* of the beam has just touched *XY*. At the same time, the light from the other side of the beam, *B*, travels (with velocity c_2) to *C* in time *t*, so that $BC = c_2t$. Let c_1 be the velocity of light in the medium above *XY*. In the same time *t*, *A* emits a secondary wavelet of radius c_1t into the region above *XY* shown by semicircle (*S*) in Fig. 8.4. If $c_1 > c_2$, i.e., the light is travelling from optically dense medium to an optically less dense medium. It is possible that c_1t is greater than AC. In that case, refraction cannot occur because there will be no tangent of wavefront S, which passes through C. The only possibility is that the secondary wavelet from A remains in the region below XY, i.e., reflected in the region below XY. In that case radius of wavefront (CB)will be equal to c_2t. This reflected wavefront leads to total internal reflection. Thus, Huygens' construction applies for reflection at a plane surface also.

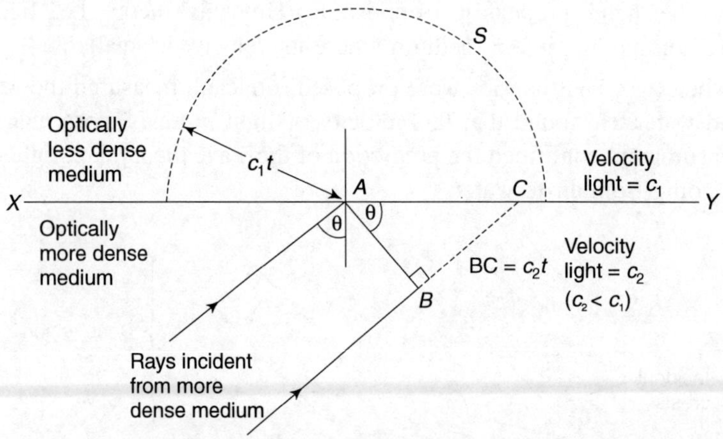

Fig. 8.6 Huygens' construction applied to total internal reflection

In ΔABC, $\sin\theta = \dfrac{BC}{AC}$ and $BC = c_2t$.

Thus, total internal reflection occurs if $c_1t > AC$, i.e., if

$$c_1t > \frac{BC}{\sin\theta}$$

Therefore, total internal reflection occurs if

$$c_1t > \frac{c_2t}{\sin\theta}$$

i.e., if $\sin\theta > \dfrac{c_2}{c_1}$

Therefore, the maximum value of angle of incidence (θ) should be θ_c for occurring total internal reflection. θ_c is called critical angle and can be calculated from the relation

$$\sin\theta_c = \frac{c_2}{c_1} = \frac{n_1}{n_2},$$

where, n_1 and n_2 are the refractive indices of the regions above and below XY, respectively. The refractive index (n_1) is smaller than n_2, i.e., $n_1 < n_2$.

8.9 SECONDARY WAVELET TREATMENT OF YOUNG'S FRINGES

Theory of fringes obtained using Young's method is given in detail in Section 9.5 of Chapter 9. The formation of the fringes is shown in Fig 8.7. S_1 and S_2 are sources which emit light rays in phase with each other. Bright fringes occur in those directions in which the light from S_1 interferes constructively with light from S_2, i.e., in those directions where crests are superimposed. The fringes shown in figure are equally spaced. However, in the diagram they are not appearing equally spaced because of scale problem.

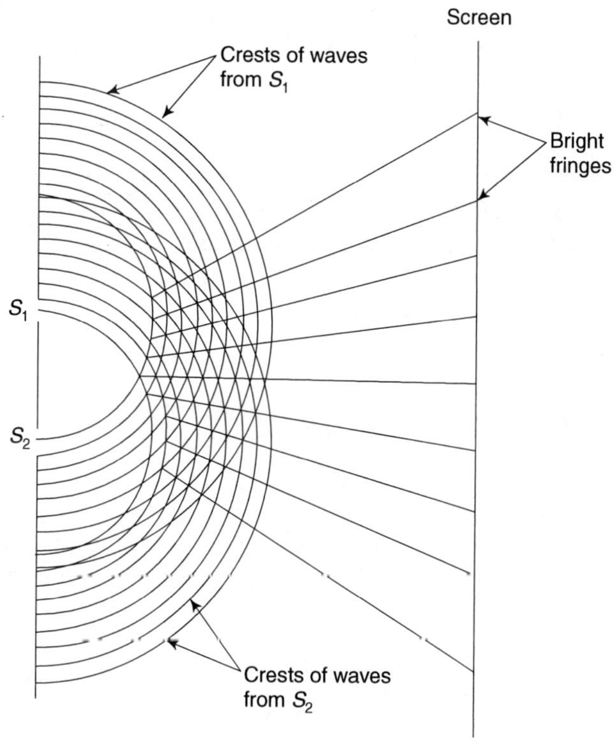

Fig. 8.7 Secondary wavelet treatment of Young's fringes

8.10 SECONDARY WAVELET TREATMENT OF LLOYD'S MIRROR

The theory of diffraction pattern formed by Lloyd's mirror is given in detail in Section 9.6 of Chapter 9. Incident crests on a screen from a point source (S) are shown in Fig. 8.8. The wavefronts undergo a phase change of π radians on reflection. The reflected rays appear to come from S'. The points where crests from S meet the reflector, there are troughs from S', because of the phase change. Destructive interference occurs where crests and troughs are superimposed. The constructive interference occurs where crests from S and S' are superimposed.

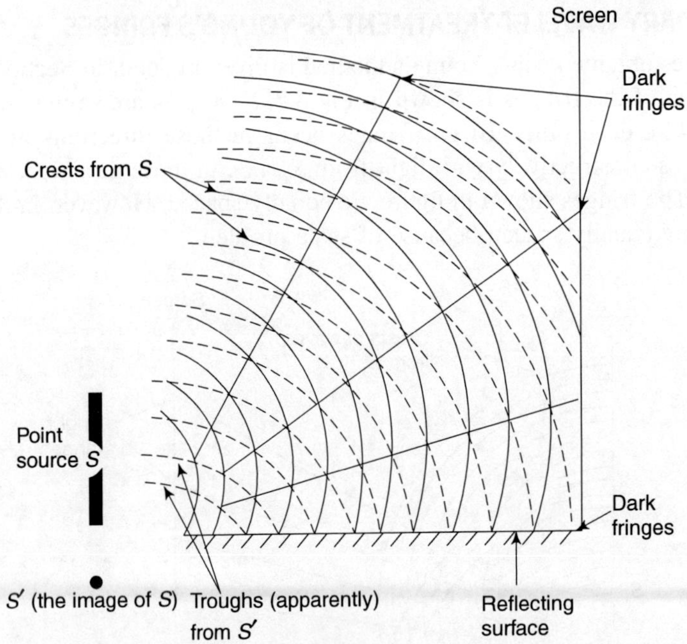

Fig. 8.8 Secondary wavelet treatment of Lloyd's mirror

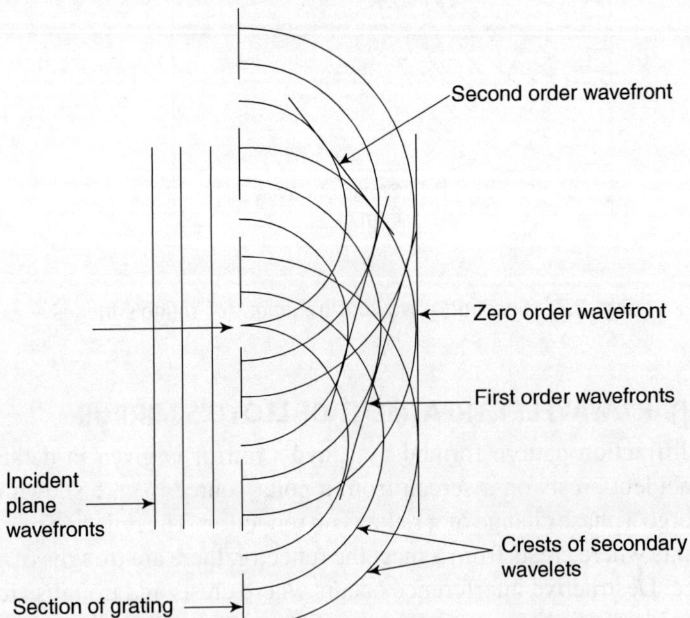

Fig. 8.9 Secondary wavelet treatment of a diffraction grating

8.11 SECONDARY WAVELET TREATMENT OF THE DIFFRACTION GRATING

The diffraction grating is discussed in Section 12.5 of Chapter 12. Diffraction pattern obtained using Huygens' construction is shown in Fig. 8.9.

8.12 DIFFRACTION OF WATER WAVES AT A GAP

Diffraction from a single slit is discussed in detail in Section 12.3 of Chapter 12. The phenomenon of diffraction can be demonstrated using water waves in a ripple tank. Figure 8.10 shows that diffraction occurs when plane waves (produced by a straight vibrator) incident on a gap in a barrier. In Fig. 8.10(a) the width of the gap is smaller as compared with the wavelength of the waves, so there is considerable sideways spreading, i.e., considerable diffraction. In Fig. 8.10(b) the gap is much larger than the wavelength of the waves, so there is very little sideways spreading, i.e., very little diffraction. When the gap is much larger than the wavelength of the waves most of the energy associated with the waves propagates in the direction of incident waves. Therefore, the greater the ratio of wavelength to gap width, the greater the spreading.

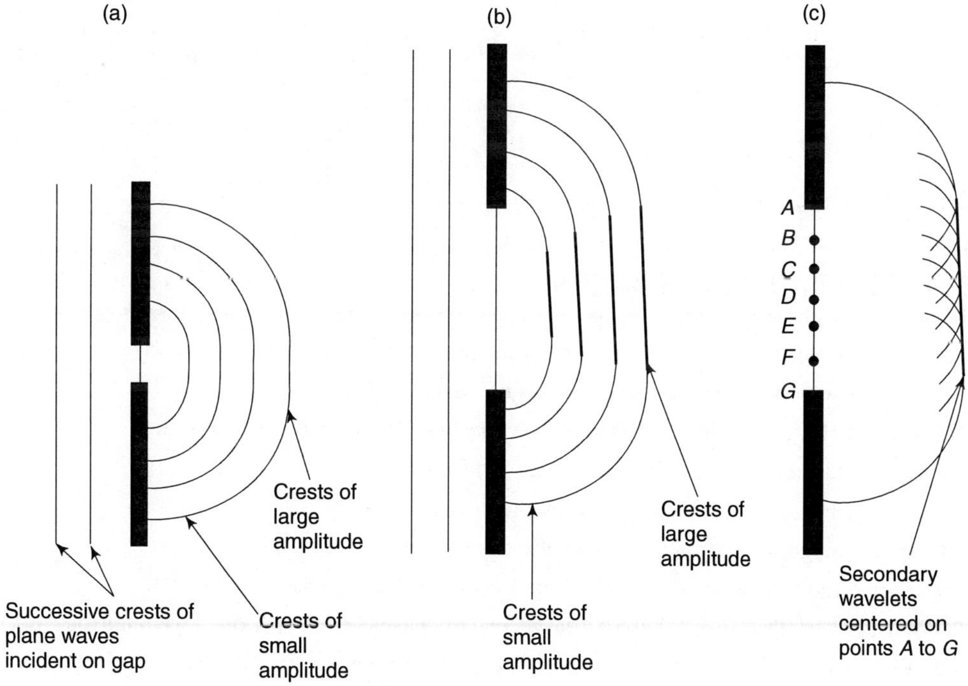

Fig. 8.10 Diffraction of water waves (a) of a narrow gap, (b) of a wide gap (c) Huygens' construction for (b)

The shape and spread of the wavefronts can be predicted using Huygens' construction. Figure 8.10(c) shows the shape and spread for wide gap shown in Fig. 8.10(b). It is to be noted that the construction can show the shape and spread of the wavefronts, but it does not show the reduced amplitude in directions other than the direction of incident wavefront.

Formulas at a Glance

S.No.	Subject	Formula	Interpretation
1.	Formula for refraction	$\frac{\sin\theta_1}{\sin\theta_2} = \frac{c_1}{c_2}$, Eq. (8.2)	θ_1 and θ_2 are angle of incidence and angle of refraction, respectively. c_1 and c_2 are velocities of light in the incident and refracted medium, respectively.
2.	Condition for total internal reflection	$\sin\theta > \frac{c_2}{c_1}$	θ is angle of incidence. c_1 and c_2 are velocities of light in the less dens medium and more dens medium, respectively.
3.	Condition for critical angle θ_c for total internal reflection	$\sin\theta_c = \frac{c_2}{c_1} = \frac{n_1}{n_2}$	n_1 and n_2 are the refractive indices of the less dens medium and more dens medium, respectively. The refractive index n_1 is smaller than n_2, i.e. $n_1 < n_2$. c_1 and c_2 are velocities of light in the less dens medium and more dens medium, respectively.

QUESTIONS FOR EXAMINATION

8.1 What are electromagnetic waves? Write the properties of electromagnetic waves.

8.2 What is wavefront? Write the properties of wavefront.

8.3 What is Huygens' principle? Discuss the construction of circular wavefront.

8.4 Apply Huygens' construction for reflection of rays.

8.5 Apply Huygens' construction for refraction of rays.

8.6 Apply Huygens' construction for total internal reflection of rays.

8.7 Discuss the formation of Young's fringes using secondary wavelet treatment.

8.8 What is Lloyd's mirror? Discuss the secondary wavelet treatment of Lloyd's mirror.

8.9 What is grating? Discuss the secondary wavelet treatment of the diffraction grating.

8.10 What is diffraction? Discuss the diffraction of water waves with a gap.

CHAPTER 9

Interference of Light

9.1 INTRODUCTION

In this chapter phenomena of interference due to monochromatic coherent light are studied. Interference patterns can be obtained by two methods: (i) by division of light wavefront, and (ii) by division of amplitude. The Newton's rings are obtained by interference by reflected or transmitted light waves by division of amplitude. The fringes in double slit experiment and Fresenl's biprism are obtained by division of wavefront. These experiments have several applications. For example, wavelength of monochromatic light source, refractive index of transparent thin films, thickness of thin transparent glass plate or mica sheet can be obtained by studying interference patterns. In the daylight different, colours seen from oil film placed on water and soap bubbles are the phenomena of interference.

9.2 INTERFERENCE

Phenomenon of interference is defined as the interaction between two or more light waves of the same or very close wavelength, where the waves overlap according to the principle of superposition.

Phenomenon of interference can be divided into two categories:

1. Division of wavefront
2. Division of amplitude

Division of Wavefront

When the incident wavefront from a single source is divided into two parts either by slits or prism, then it is called division of wavefront. The intensity of each part of wavefront is equal. The Fresnel's biprism also divides the wavefront of a single source into two parts.

Division of Amplitude

When the amplitude of the incoming beam is divided into two parts either by partial reflection and refraction, then it is called division of amplitude. The intensity of divided parts is sum of the intensity of the original source. The circular rings and other patterns obtained using Michelson interferometer and Newton's rings use the division of amplitude of the beam.

Conditions for interference

In order to get interference pattern of light waves, the following conditions are required:

1. The two sources of light must have the same wavelength, otherwise the maxima, minima will change with time.
2. The two sources must be coherent, i.e., the two sources of light must have either the same phase or a constant difference in phase.
3. The sources should be monochromatic. The width of the beam of sources must be very narrow and constant for observing the maxima with good intensity.
4. Two coherent sources having wavelengths very close to each other may also give rise to interference pattern. If the wavelengths are not very close to each other, then the path difference between the interfering waves will be large and the maxima and minima will be very close to each other or may even overlap.
5. There should be reasonable distance between the sources and screen. The intensity reduces, if the distance of the screen and sources is too large. The gap between maxima and minima will be too small, if the distance between sources and screen is very close.
6. Dark background.
7. In case the sources are obtained by division of amplitude, the amplitude of two waves must be equal or nearly equal. In case, the amplitudes of source nearly equal, the cancellation of amplitude will near zero. While the addition will make the amplitude nearly double and the intensity almost four times.
8. The beam of coherent sources must be narrow otherwise a single source will act as a multi-source. As a result there will be a number of interference patterns and after overlapping they may result in just illumination.

9.3 COHERENCE

Coherence is a property of waves that yields stationary interference. Two waves are said to be coherent if they have a constant relative phase. This also means that they have the same energy or frequency or wavelength. Actually, the coherence is a measure of the correlation that exists between the phases of the wave measured at different points. During interference the waves overlap constructively or destructively, depending on their relative phase. The coherence is of two types: (i) longitudinal coherence or temporal coherence, and (ii) transverse coherence or spatial coherence.

Depending on the distance and time up to which photon waves maintain coherence, the coherence of laser light is of two types: (a) longitudinal or temporal coherence, and (b) transverse or spatial coherence.

9.3.1 Temporal Coherence or Longitudinal Coherence

If the phase difference between moving light waves at one fixed point in a direction is same at any time interval, then it is called temporal coherence. A perfectly coherent source emits light radiations of constant amplitude as shown in Fig. 9.1(a).

This tells that the monochromatic source is coherent. In Fig. 9.1(a) two waves are shown travelling along the positive x-direction, where two points A and B are lying on two waves. Let the phases of the waves at these points at any time t be, φ_A and φ_B, respectively. At a later time t' the phases are φ'_A and φ'_B at the same points, A and B. If the phase difference $\varphi_B - \varphi_A = \varphi'_B - \varphi'_A$, then the wave is said to have temporal coherence.

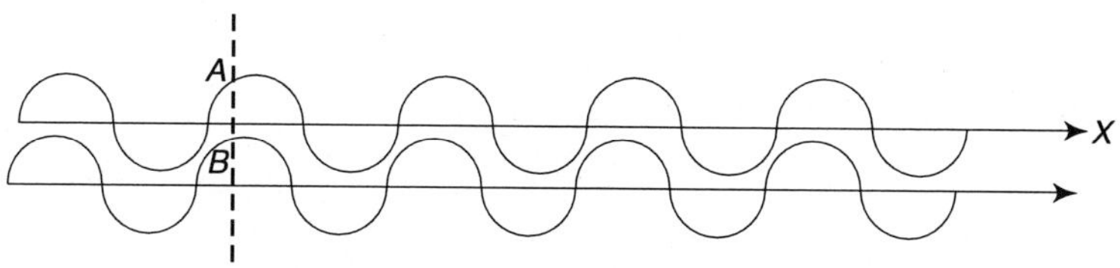

Fig. 9.1 (a) Temporal coherence.

9.3.2 Spatial Coherence or Transverse Coherence

If phase difference between light waves at different points in the direction of propagation is same at any time interval, then it is called spatial coherence.

If the phase difference of the waves crossing the two points lying on a plane perpendicular to the direction of the wave propagation is independent of time, then the waves are said to have spatial coherence. This tells that wave front is coherent. We see that two waves are travelling along the positive x-direction in Fig. 9.1(b), where PQRS is a transverse plane and A and B are the two points situated on this plane within the waveforms. Let the waves crossing these points at any time t have the phases φ_A and φ_B and hence the phase difference is $\varphi_A - \varphi_B$. At a later time t' the phase of the waves are $\varphi_{A'}$ and $\varphi_{B'}$ at other points A' and B'. If the phase difference $\varphi_{A'} - \varphi_{B'} = \varphi_A - \varphi_B$, then waves are said to have spatial coherence.

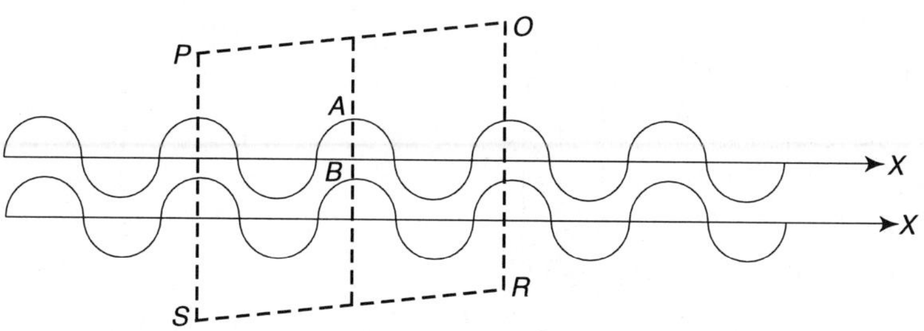

Fig. 9.1 (b) Spatial coherence

9.3.3 Longitudinal or Temporal Coherence Length

Let us consider two sinusoidal photon waves overlapping each other as shown in Fig. 9.2(a). Very close to the source, the two waves are almost in phase. As the distance of propagation increases, the phase difference of the two waves increases slowly and at a certain distance (L_t) the two waves become completely out of phase. Two waves become out of phase because there is difference in frequencies of two waves, even if it is too small. So the longitudinal coherence length (L_t) is defined as the distance from the source to the point at which the waves become complexly out of phase. Let λ_1 and λ_2 be the wavelengths of the two waves and average value of these wavelengths,

$$\lambda = \frac{\sqrt{\lambda_1 \times \lambda_2}}{2},$$

longitudinal coherence length (L_t) is given by

$$L_t = \frac{\lambda^2}{\Delta\lambda}$$

where, $\Delta\lambda = |\lambda_1 - \lambda_2|$. From this equation it is clear that if the difference between wavelengths of the two waves decreases, then longitudinal coherence increases.

The longitudinal coherence is also called temporal coherence. The temporal coherence time (τ) is defined as the time from the source to the point at which the waves become completely out of phase. Let v_1 and v_2 be the frequencies of the two waves. The temporal coherence time is the inverse of the difference, $\Delta v = |v_1 - v_2|$, of the frequencies of the two waves as

$$\tau = \frac{1}{\Delta v}$$

Longitudinal coherence length, $L_t = c\tau$, where, c is velocity of light. After twice of coherence length $2L_t$, the two waves again become in same phase. The derivation of longitudinal coherence length (L_t) is given in **Appendix B of chapter 15.**

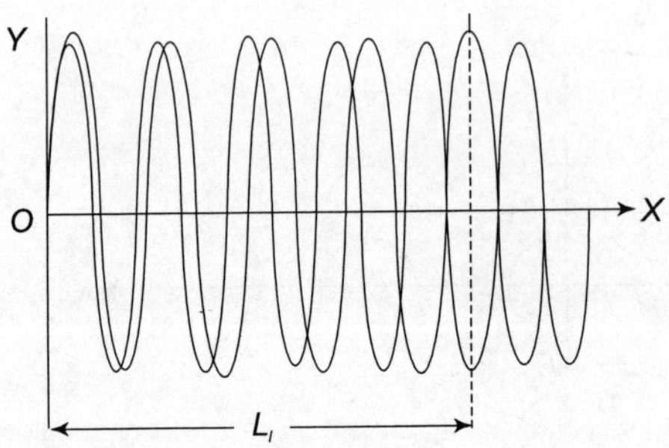

Fig. 9.2(a) Longitudinal coherence or temporal coherence

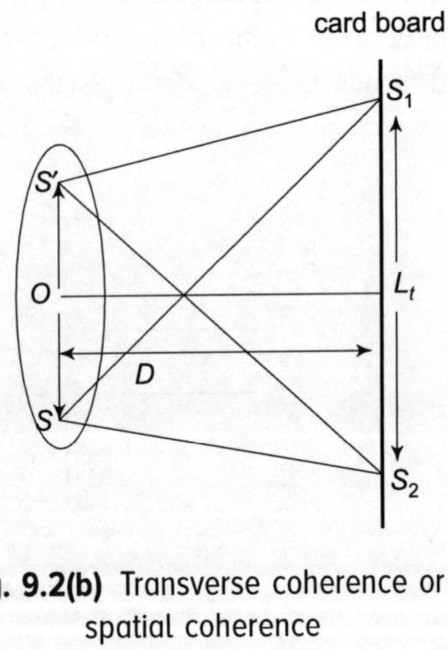

Fig. 9.2(b) Transverse coherence or spatial coherence

9.3.4 Transverse Coherence or Spatial Coherence

Let us consider two waves emitted from two points, S' and S, of a single source as shown in Fig. 9.2(b). The points S' and S are situated on a single plane perpendicular to the laser beam. Very close to the source at points S' and S the two waves are almost in phase, i.e., phase difference is zero or nearly zero. As two wave propagate towards the cardboard the phase difference between the two waves increases. After travelling a distance (D) equal to coherence length (L_{t-}), two waves are out of phase laterally, i.e., the two waves at points S_1 and S_2 are out of phase. The transverse coherence length (L_t) is defined as the lateral distance between two points, S_1 and S_2, situated on the cardboard placed transverse to the laser beam at which the two waves are out of phase. If diameter of beam is $S'\,S = a$ and distance between source and cardboard is D, then the transverse coherence length is given by

$$L_t = \frac{\lambda D}{a}$$

where, λ is the wavelength of the laser source.

9.3.5 Coherence Time and Coherence Length

For a sinusoidal wave of single wavelength, the coherence time is infinite because the phase remains constant throughout. However, practically a monochromatic source of light emits radiation of certain range of wavelengths. For a monochromatic wave, the time interval over which the phase remains constant is called the *coherence time*. The coherence time, generally represented by τ, is practically finite. The distance travelled by the light pulses during this coherence time is known as coherence length, $L_t = c \times \tau$, where, c is velocity of light.

9.4 THEORY OF INTERFERENCE

(Conditions for constructive and destructive interference)

Consider a monochromatic source of light S.

$$Y_1 = a_1 \sin(\omega t - kx) \quad \text{...(9.1a)}$$

and

$$Y_2 = a_2 \sin(\omega t - kx + \delta) \quad \text{...(9.1b)}$$

The light wave passes from two narrow slits, A and B, as shown in Fig. 9.3. Here, a_1 and a_2 are the amplitudes of the waves. ω, and x are angular frequency and position of the waves. k is the momentum of the photon and δ is the phase difference between two waves. According to the principle of superposition of waves, the resultant intensity at point P on the screen XY is given by

$$\begin{aligned} Y &= Y_1 + Y_2 \\ &= a_1 \sin(\omega t - kx) + a_2 \sin(\omega t - kx + \delta) \\ &= a_1 \sin(\omega t - kx) + a_2 \sin(\omega t - kx)\cos(\delta) + a_2 \cos(\omega t - kx)\sin(\delta) \\ &= \sin(\omega t - kx)\,[a_1 + a_2 \cos(\delta)] + \cos(\omega t - kx)\,[a_2 \sin(\delta)] \quad \text{...(9.2)} \end{aligned}$$

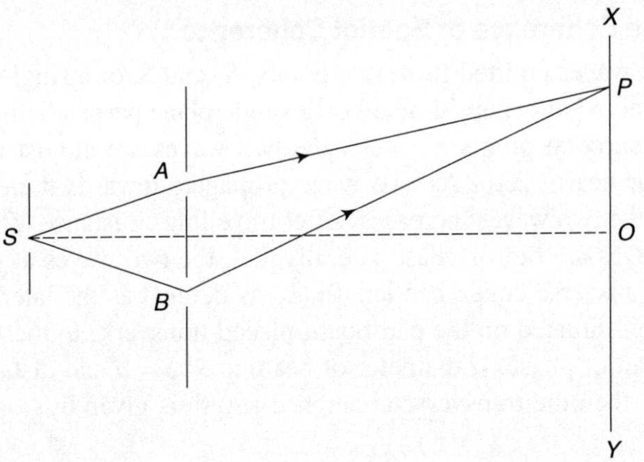

Fig. 9.3

Taking

$$a_1 + a_2 \cos(\delta) = R \cos \varphi \qquad ...(9.3a)$$

and $$a_2 \sin(\delta) = R \sin \varphi \qquad ...(9.3b)$$

Equation (9.2) becomes

$$Y = R \sin(\omega t - kx) \cos \varphi + R \cos(\omega t - kx) \sin \varphi$$
$$= R \sin(\omega t - kx + \varphi) \qquad ...(9.3)$$

The resultant wave is a harmonic wave having amplitude, R. Squaring and adding Eqs. (9.3a) and (9.3b), we get

$$R^2 [\sin^2 \varphi + \cos^2 \varphi] = [a_1 + a_2 \cos(\delta)]]^2 + a_2^2 \sin^2 \delta$$

or $$R^2 = a_1^2 + 2\, a_1\, a_2 \cos(\delta) + a_2^2 \cos^2 \delta + a_2^2 \sin^2 \delta$$
$$= a_1^2 + 2\, a_1\, a_2 \cos(\delta) + a_2^2$$

or $$R = \sqrt{a_1^2 + 2\, a_1\, a_2 \cos(\delta) + a_2^2} \qquad ...(9.4)$$

The resultant intensity at P is proportional to the square of the amplitude of the resultant wave given by Eq. (9.3), i.e.,

$$I \propto R^2$$

or $$I = CR^2$$

or $$I = C\,[a_1^2 + 2\, a_1\, a_2 \cos(\delta) + a_2^2]$$

where, C is the constant of proportionality. If $I_1 = a_1^2$ and $I_2 = a_2^2$, then

$$I = I_1 + 2\sqrt{I_1\, I_2} \cos(\delta) + I_2 \qquad ...(9.5)$$

For constructive interference

Intensity (I) should be maximum, i.e., $\cos(\delta) = 1$, which gives the phase difference,

$$\delta = 2\, n\pi \qquad ...(9.6)$$

where, $n = 0, 1, 2, 3, ...$

The corresponding path difference (Δ) for maximum intensity at point P is

$$\Delta = n\lambda \quad ...(9.6a)$$

Thus for constructive interference, the phase difference between two waves should be zero, or even integral multiple of π, while the path difference should be integral multiple of wavelength, λ.

For destructive interference

Intensity (I) should be minimum, i.e., cos (δ) = 0, which gives the phase difference,

$$\delta = (2n - 1)\pi \quad ...(9.7)$$

where, $n = 1, 2, 3, ...$

The corresponding path difference (Δ) for minimum intensity at point P,

$$\Delta = (2n - 1)\frac{\lambda}{2} \quad ...(9.7a)$$

Thus, for destructive interference, the phase difference between two waves should be an odd integral multiple of π, while the path difference should be odd integral multiple of wavelength, $\frac{\lambda}{2}$.

9.5 YOUNG'S DOUBLE SLIT EXPERIMENT

To determine the fringe width, consider two slits, A and B, at equal distance from source slit, S. The monochromatic light falling on the two slits will be in the same phase at any instant, i.e., the light passing through slits A and B is coherent.

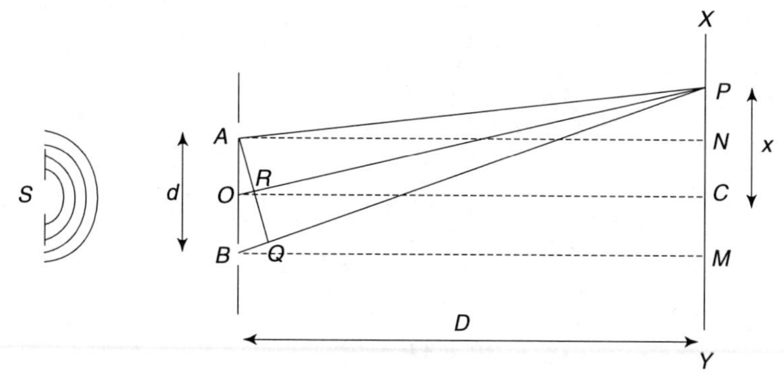

Fig. 9.4

The distance between two slits is $AB = d$. The screen is set at a distance $OC = D$ from the slits. Let x be the position of point P from the centre of the screen, i.e., $CP = x$. Now drop a perpendicular (AR) from point A on line OP and extend it to point Q on line BP. Triangle AQP is an equilateral triangle, so $AP = QP$. The paths traversed by light source from slits

A and B to point P are AP and BP, respectively. So, the path difference traversed by two sources from the slits to point P is

$$BQ = \Delta = BP - AP \qquad \text{...(9.8)}$$

In right angle ΔANP

$$AP^2 = AN^2 + NP^2 = D^2 + \left(x - \frac{d}{2}\right)^2$$

In right angle ΔBMP

$$BP^2 = BM^2 + MP^2 = D^2 + \left(x + \frac{d}{2}\right)^2$$

or

$$AP = \sqrt{D^2 + \left(x - \frac{d}{2}\right)^2} = \left(D^2 + \left(x - \frac{d}{2}\right)^2\right)^{1/2}$$

$$= D\left(1 + \frac{\left(x - \frac{d}{2}\right)^2}{D^2}\right)^{1/2}$$

and

$$BP = \sqrt{D^2 + \left(x + \frac{d}{2}\right)^2} = \left(D^2 + \left(x + \frac{d}{2}\right)^2\right)^{1/2}$$

$$= D\left(1 + \frac{\left(x + \frac{d}{2}\right)^2}{D^2}\right)^{1/2}$$

or

$$AP = D\left(1 + \frac{1}{2}\frac{\left(x - \frac{d}{2}\right)^2}{D^2}\right) \qquad \text{...(9.9)}$$

and

$$BP = D\left(1 + \frac{1}{2}\frac{\left(x + \frac{d}{2}\right)^2}{D^2}\right) \qquad \text{...(9.10)}$$

Thus, the path difference

$$\Delta = BP - AP$$

$$= D\left(1 + \frac{1}{2}\frac{\left(x + \frac{d}{2}\right)^2}{D^2}\right) - D\left(1 + \frac{1}{2}\frac{\left(x - \frac{d}{2}\right)^2}{D^2}\right)$$

$$= D\,\frac{1}{2}\frac{1}{D^2}\left(\left(x + \frac{d}{2}\right)^2 - \left(x - \frac{d}{2}\right)^2\right)$$

$$= \frac{1}{2D}\left(x^2 + \frac{d^2}{4} + xd - x^2 - \frac{d^2}{4} + xd\right)$$

$$= \frac{1}{2\,D} \times 2\,xd$$

$$= \frac{xd}{D} \qquad \text{...(9.11)}$$

The corresponding phase difference,

$$\varphi = \frac{2\,\pi}{\lambda} \times \text{path difference} = \frac{2\,\pi}{\lambda}\frac{xd}{D} \qquad \text{...(9.12)}$$

For bright fringes, the path difference,

$$\Delta = n\lambda = \frac{xd}{D} \quad \text{or} \quad x = \frac{D}{d}n\lambda$$

Thus, distance of nth bright fringe from C is

$$x_n = \frac{D}{d}n\lambda, \qquad \text{where, } n = 0, 1, 2, 3, \ldots \qquad \text{...(9.13)}$$

For dark fringes, the path difference

$$\Delta = (2\,n + 1)\,\frac{\lambda}{2} = \frac{xd}{D} \quad \text{or} \quad x = \frac{D}{d}(2\,n+1)\frac{\lambda}{2}$$

Thus distance of nth dark fringe from C is

$$x_n = \frac{D}{d}(2\,n+1)\frac{\lambda}{2}, \qquad \text{where } n = 0, 1, 2, 3, \ldots. \qquad \text{...(9.14)}$$

The distance of two consecutive bright fringes is given by

$$x_{n+1} - x_n = \frac{D}{d}(n+1)\,\lambda - \frac{D}{d}n\,\lambda = \frac{D\lambda}{d} \qquad \text{...(9.15)}$$

The distance of two consecutive dark fringes is given by

$$x_{n+1} - x_n = \frac{D}{d}(2\,(n+1)+1)\frac{\lambda}{2} - \frac{D}{d}(2\,n+1)\frac{\lambda}{2}$$

$$= \frac{D}{d}(2n+2+1-2\,n-1)\frac{\lambda}{2} = \frac{D\lambda}{d} \qquad \text{...(9.16)}$$

Thus, the distance between two consecutive bright or dark fringes is the same. The distance between any two successive bright or dark fringes is called fringe width and is denoted by

$$\beta = \frac{D}{d}\lambda \qquad \text{...(9.17)}$$

The fringe width is independent of the order of the fringe.

The breadth of a bright or dark fringe

$$\frac{1}{2}\beta = \frac{D}{2d}\lambda \qquad ...(9.18)$$

is equal to half of the fringe width.

Interpretation of fringe width given by Eq. (9.17)

1. The fringe width (β) is directly proportional to the wavelength of light, λ, i.e., $\beta \propto \lambda$.
2. The fringe width (β) is directly proportional to the distance between the slit and screen, D, i.e., $\beta \propto D$.
3. The fringe width (β) is inversely proportional to the distance between the two slits, d i.e., $\beta \propto \frac{1}{d}$.
4. The wavelength (λ) can be calculated by putting the values of fringe width, β, d and D.

9.6 LLOYD'S MIRROR

The experimental setup of Lloyd's mirror is shown in Fig. 9.5. S is a monochromatic light source. It passes through the slit which is parallel to both the glass plate and the screen, and is very close to the plane of the plate. Light waves diffracted by slit travel to a point P on the screen directly along SP and after reflected by the mirror along SAP. The reflected waves appear to come from S', the virtual image of S. The interference can be thought of due to coherent sources at S and S'. The path difference between the two sets of waves reaching P,

$$\Delta = SA + AP - SP$$

Since $SA = S'A$, so path difference can be written as

$$\Delta = S'A + AP - SP$$

$$= S'P - SP \qquad ...(9.19)$$

It is to be noted that the light undergoes a phase change of π radians or introduced an extra path difference $\left(\frac{\lambda}{2}\right)$ whenever it is reflected at a more dens medium, for example, a glass plate as is the present case, and therefore the fringe at N is dark and path difference will be

$$\Delta = S'P - SP + \frac{\lambda}{2}$$

The condition for bright fringes is

$$S'P - SP + \frac{\lambda}{2} = n$$

or

$$S'P - SP = (2n - 1)\frac{\lambda}{2}, \quad \text{where, } n = 1, 2, 3, ... \qquad (9.20)$$

Thus, there are bright fringes where the geometrical path difference is odd number of half-wavelengths.

The condition for dark fringes is

$$S'P - SP + \frac{\lambda}{2} = (2\,n + 1)\,\frac{\lambda}{2}$$

or $$S'P - SP = n\lambda, \quad \text{where } n = 0, 1, 2, 3, \ldots \qquad ...(9.21)$$

Thus, there are dark fringes where the geometrical path difference is whole number of wavelengths. For $n = 0$, there will be dark fringe. That is, there is no (geometrical) path difference between waves arriving at N, so there will be a dark fringe at point N.

Calculation of Fringe Width

The situation is equivalent to that of double-slit experiment. The fringes are therefore equally spaced and the separation, β, of adjacent bright (or dark) fringes is given by Eq. (9.10) as

$$\beta = \frac{\lambda D}{a}$$

where, λ is wavelength of the light being used, D is the distance between source slit and screen and a is the separation between the slits S and S'.

$$\Delta = S'P - SP \qquad ...(9.22)$$

In right angle ΔSQP

$$SP^2 = SN^2 + NP^2 = D^2 + \left(\frac{d}{2} - x\right)^2$$

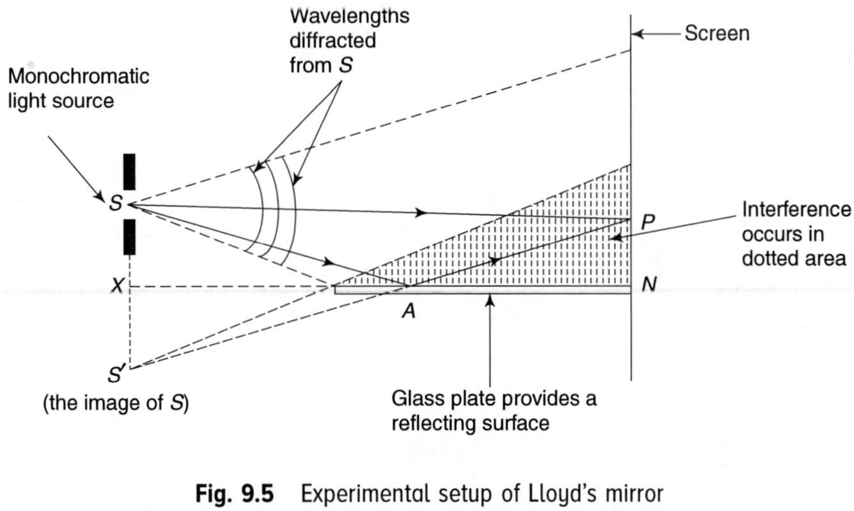

Fig. 9.5 Experimental setup of Lloyd's mirror

In right angle $\Delta S'MP$

$$S'P^2 = S'M^2 + MP^2 = D^2 + \left(\frac{d}{2} + x\right)^2$$

or $$SP = \sqrt{D^2 + \left(\frac{d}{2} - x\right)^2} = \left(D^2 + \left(\frac{d}{2} - x\right)^2\right)^{1/2}$$

$$= D\left(1 + \frac{\left(\frac{d}{2} - x\right)^2}{D^2}\right)^{1/2}$$

and $$S'P = \sqrt{D^2 + \left(\frac{d}{2} + x\right)^2} = \left(D^2 + \left(\frac{d}{2} + x\right)^2\right)^{1/2}$$

$$= D\left(1 + \frac{\left(\frac{d}{2} + x\right)^2}{D^2}\right)^{1/2}$$

or $$SP = D\left(1 + \frac{1}{2}\frac{\left(\frac{d}{2} + x\right)^2}{D^2}\right) \qquad ...(9.23)$$

and $$S'P = D\left(1 + \frac{1}{2}\frac{\left(\frac{d}{2} + x\right)^2}{D^2}\right) \qquad ...(9.24)$$

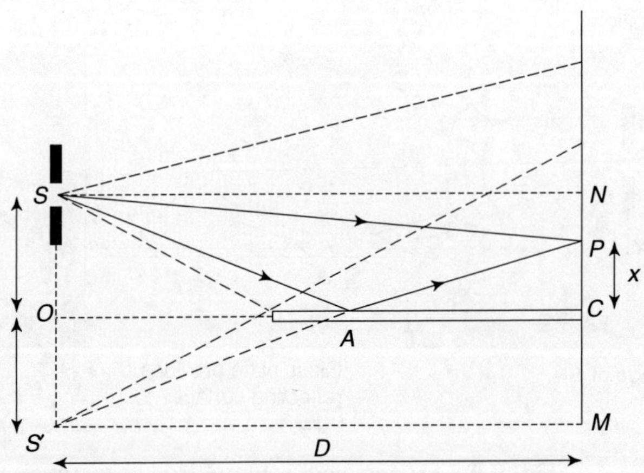

Fig. 9.6 Experimental setup of Lloyd's mirror for calculating path difference

Thus, the path difference = $SP - S'P$

$$= D\left(1+\frac{1}{2}\frac{\left(\frac{d}{2}+x\right)^2}{D^2}\right) - D\left(1+\frac{1}{2}\frac{\left(\frac{d}{2}-x\right)^2}{D^2}\right)$$

$$= D\frac{1}{2}\frac{1}{D^2}\left(\left(\frac{d}{2}+x\right)^2 - \left(\frac{d}{2}-x\right)^2\right)$$

$$= \frac{1}{2D}\left(x^2+\frac{d^2}{4}+xd-x^2-\frac{d^2}{4}+xd\right)$$

$$= \frac{1}{2D}\times 2\,xd$$

$$= \frac{xd}{D} \qquad \text{...(9.25)}$$

The corresponding phase difference is

$$\varphi = \frac{2\pi}{\lambda}\times \text{path difference} = \frac{2\pi}{\lambda}\frac{xd}{D} \qquad \text{...(9.26)}$$

For bright fringes, the path difference

$$\Delta = n\lambda = \frac{xd}{D}+\frac{\lambda}{2} \quad \text{or} \quad x = \frac{D}{d}(2n-1)\frac{\lambda}{2}$$

Thus, distance of nth bright fringe from C is

$$x_n = \frac{D}{d}(2n-1)\frac{\lambda}{2}, \text{ where, } n = 1, 2, 3, \ldots \qquad \text{...(9.27)}$$

For dark fringes, the path difference

$$D = (2n+1)\frac{\lambda}{2} = \frac{xd}{D}+\frac{\lambda}{2} \quad \text{or} \quad x = \frac{D}{d}n\lambda$$

Thus, distance of nth dark fringe from C is

$$x_n = \frac{D}{d}n\lambda, \text{ where } n = 0, 1, 2, 3, \ldots \qquad \text{...(9.28)}$$

Thus, for $n = 0$, there is dark fringe, i.e., at the centre (C) there will be a dark fringe.

The distance of two consecutive dark fringes is given by

$$x_{n+1} - x_n = \frac{D}{d}(n+1)\lambda - \frac{D}{d}n\lambda = \frac{D\lambda}{d} \qquad \text{...(9.29)}$$

The distance of two consecutive bright fringes is given by

$$x_{n+1} - x_n = \frac{D}{d}[2(n+1)-1]\frac{\lambda}{2} - \frac{D}{d}(2n-1)\frac{\lambda}{2}$$

$$= \frac{D}{d}(2n + 2 - 1 - 2n + 1)\frac{\lambda}{2} = \frac{D\lambda}{d} \quad ...(9.30)$$

Thus, the distance between two consecutive bright or dark fringes is the same. The distance between any two successive bright or dark fringes is called fringe width and is denoted by

$$\beta = \frac{D}{d}\lambda \quad ...(9.31)$$

The fringe width is independent of the order of the fringe.

The breadth of a bright or dark fringe

$$\frac{1}{2}\beta = \frac{D}{2d}\lambda \quad ...(9.32)$$

is equal to half of the fringe width,

9.7 FRESNEL'S BIPRISM

Fresnel's biprism consists of two right angled prisms of very small refracting angle α (nearly 30°) joined base to base. In actual practice it is made from a single glass plate by cutting its faces so that two angles are of the order of 30° and the third angle is of 119°.

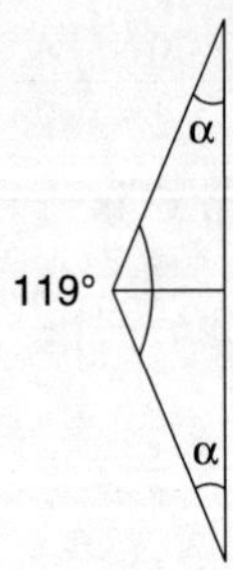

Fig. 9.7 Fresnel's biprism

When a monochromatic light source is incident on it, it produces two coherent sources of light and hence it is used to show the phenomenon of interference.

9.7.1 Formation of Coherent Sources and Interference Fringes using Biprism

When a monochromatic light source (S) is incident on biprism, it shows interference phenomenon after refracting from two halves of it as shown in Fig. 9.8. The incident light refracts from two halves of the biprism in different directions. The waves between AG and BH appear as coming from S_1 and waves between CH and BG appear as coming from S_2. These two virtual images, S_1 and S_2, of source S act as coherent sources and form interference pattern on the screen only in portion GH of the screen.

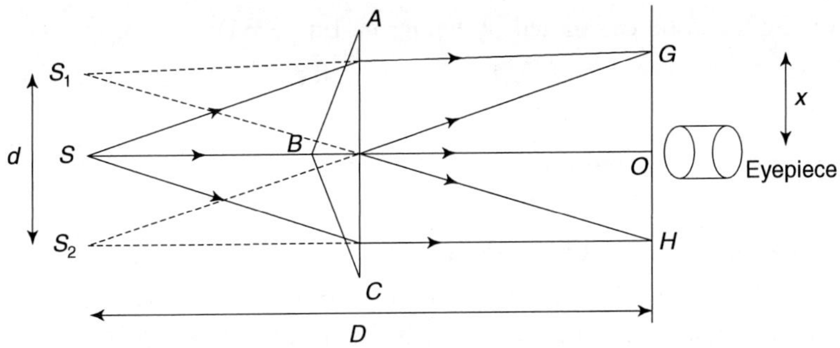

Fig. 9.8

Theory : Let the distance between virtual sources S_1 and S_2 be d and distance between sources and the screen be D. There will be central bright maximum at point O on the screen, because it is at equal distance from S_1 and S_2. On both sides of O, bright and dark fringes will be produced alternately by the principle of superposition of waves. These bright and dark fringes or maxima and minima will depend on path difference of the overlapping rays. The condition for constructive interference, i.e., for maxima or bright fringes is [Eq. (9.13)].

$$\frac{xd}{D} = n\lambda, \text{ where } n = 0.\ 1, 2, 3, \ldots \text{ etc.}$$

The fringe width is [Eq. (9.15)]

$$\beta = \frac{x}{n} = \frac{\lambda D}{d} \qquad \ldots(9.33)$$

The condition for destructive interference for minima or dark fringes is [Eq. (9.14)]

$$\frac{xd}{D} = (2n+1)\frac{\lambda}{2}, \text{ where, } n = 0.\ 1, 2, 3, \ldots \text{ etc.}$$

The fringe width is [Eq. (9.16)]

$$\beta = \frac{x}{n} = \frac{\lambda D}{d} \qquad \ldots(9.34)$$

Here x is the distance of nth fringe from the centre.

Applications of biprism

(a) Wavelength of light can be found.

(b) Fringe width, β, can be calculated.

(c) Thickness of transparent sheet can determined.

9.7.2 Determination of Wavelength of Light

By knowing the fringe width, one can find the wavelength of monochromatic light. The fringe width can be obtained from Eq. (9.34)

$$\beta = \frac{x}{n}$$

and then wavelength can be calculated by using the Eq. (9.34)

$$\beta = \frac{\lambda D}{d} \quad \text{or} \quad \lambda = \frac{\beta d}{D}. \qquad ...(9.35)$$

by putting the values of d, β and D.

Experimental setup of Fresnel's biprism

For setting Fresnel biprism experiment, following componets are required: Biprism, lens, eyepiece, source of light, slit and optical bench. Using Fresnel's biprism, the measurement of β, d and D can be done.

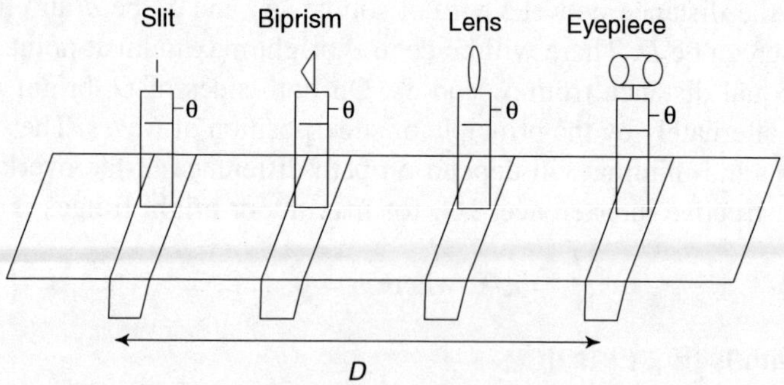

Fig. 9.9 Experimental setup of Fresnel's biprism

Setting of experiment for getting fringes

(i) First of all level the optical bench with a spirit level and then mount the slit, biprism, lens and eyepiece on holders.

(ii) Now make the slit narrow and allow monochromatic light to pass through slit and biprism. Blurred fringes will appear on the screen.

(iii) Adjust the biprism by using tangent screw till distinct fringes are seen through the eyepiece.

(iv) Clear fringes can be obtained by adjusting the width of slit.

9.7.3 Determination of Fringe Width, β

First of all note down the least count of micrometer of the eyepiece. Now fix vertical cross wire of the eyepiece at any bright fringe. When we move the micrometer, a number of fringes cross the field of view. Count the number of fringes crossed, say n and calculate the distance moved by the micrometer, say x. By putting the value x and n in the equation

$$\beta = \frac{x}{n}$$

one can calculate the fringe width.

Now measure the distance between slits and the eyepiece, D and the separation between slits, d.

9.7.4 Measurement of the Separation between the Slits, *d*, using Displacement Method

Take a convex lens of focal length less than one-fourth of distance between biprism. Mount the lens between biprism and the eyepiece without disturbing their positions. Now fix the lens at two positions as shown in Figs. 9.10(a) and 9.10(b), so that sharp images of S_1 and S_2 are formed in the eyepiece. Let d_1 and d_2 be the distances between the images of S_1 and S_2 in the eyepiece at two positions as shown in Figs. 9.10(a) and 9.10(b).

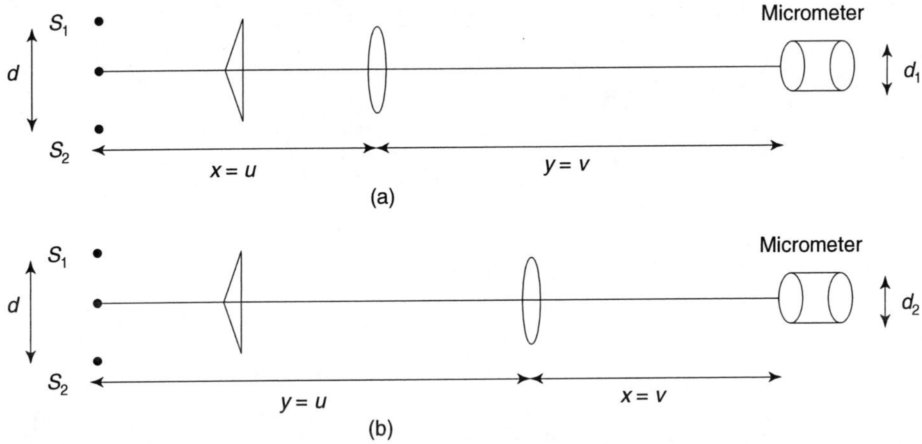

Fig. 9.10 Lens is fixed at two positions so that sharp images of S_1 and S_2 are formed in eyepiece

Using the laws of image formation through lens, one can find the distance between the two slits.

For position of lens shown in Fig. 9.10(a)

$$\frac{d_1}{d} = \frac{v}{u} = \frac{x}{y} \qquad ...(9.36)$$

and for position of lens shown in Fig. 9.10(b)

$$\frac{d_2}{d} = \frac{v}{u} = \frac{y}{x} \qquad ...(9.37)$$

Multiplying Eqs. (9.36) and (9.37), we get

$$\frac{d_1 d_2}{d^2} = 1 \quad \text{or} \quad d = \sqrt{d_1 d_2}$$

Now, λ can be calculated by substituting β, D, and d in relation [Eq. (9.35)]

$$\lambda = \frac{\beta d}{D}$$

9.7.5 Measurement of the Separation between the Slits, *d*, using Refracting Angle (δ) of the Biprism

As angle α is small so deviation (δ) produced by biprism can be given as

$$\delta = (\mu - 1)\,\alpha, \text{ [see Appendix A]} \qquad ...(9.38)$$

where, μ is the refractive index of the material of biprism and the acute angle of the biprism (α) is in radians. Deviation (δ) is defined as the angle between the directions of incident and emergent rays.

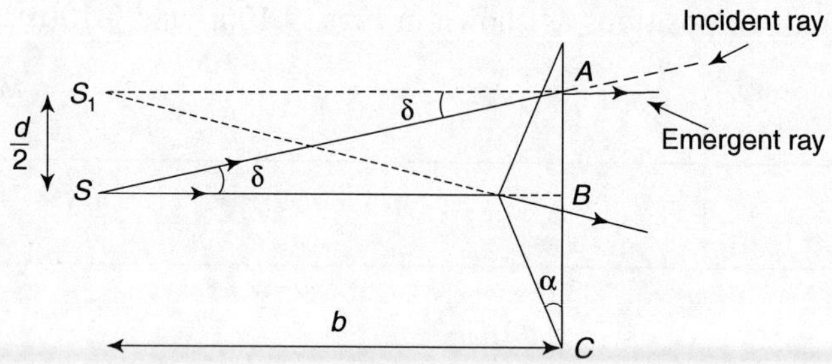

Fig. 9.11

In Δ*SAB*

$$\tan\delta = \frac{AB}{SB} = \frac{\frac{d}{2}}{b} \quad \text{or} \quad \delta = \frac{\frac{d}{2}}{b} \text{ for small } \delta$$

or
$$\delta = 2\,b\delta \qquad ...(9.39)$$

Here, b is the distance between slit source and biprism. Putting the value of δ from Eq. (9.38).

$$\delta = 2b\,(\mu - 1)\,\alpha, \qquad ...(9.40)$$

where, the acute angle of biprism (α) is in radians. Thus, by the values of b, μ and α, one can also calculate the distance between the slits, S_1 and S_2.

9.7.6 Measurement of Thickness of Transparent Sheet

Biprism can be used to measure the thickness of a thin transparent sheet (or mica sheet). Consider two coherent sources, S_1 and S_2, of monochromatic light interfere and produce fringes on a screen (*XY*) placed at a distance (*D*) from the coherent sources. In the absence of thin sheet bright fringe is formed at *C*, because optical paths, S_1C and $S_2\,C$, are equal. But when mica sheet of thickness (c_t) and refractive index (μ) is introduced in the path of ray $S_1\,C$, then ray $S_1\,C$ travels thickness (t) in mica and rest distance ($S_1C - t$) in air while ray $S_2\,C$ travels whole distance through air.

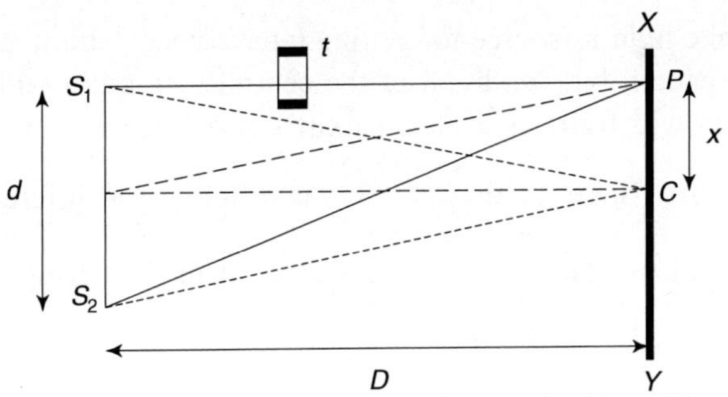

Fig. 9.12

Since the velocity of light is differentin air and mica, the optical path of two rays will not be equal and position of central fringe will shift upwards from C to point P as shown Fig. 9.12.

For optical equal paths

$$\frac{S_2 P}{c/\mu_{air}} = \frac{S_1 P - t}{c/\mu_{air}} + \frac{t}{c/\mu}, \qquad ...(9.41)$$

where, refractive index of air, $\mu_{air} = 1$, c is velocity of light and μ is refractive index of thin sheet. Simplifying Eq. (9.41), we get

$$S_2 P = (S_1 P - t) + \mu t$$

or $$S_2 P - S_1 P = (\mu - 1)\, t \qquad ...(9.42)$$

In the absence of mica sheet, let P be the position of nth bright fringe, which is at a distance x from central bright fringe, then

$$S_2 P - S_1 P = \frac{xd}{D} = n\lambda \quad \text{[see Eq. (9.11)]} \qquad ...(9.43)$$

Comparing Eqs. (9.42) and (9.43), we get $(\mu - 1)\, t = n\lambda$

or $$(\mu - 1)\, t = \frac{xd}{D}$$

or $$x = \frac{D}{d}(\mu - 1)\, t \qquad ...(9.44)$$

Putting the value of $\frac{D}{d}$ from Eq. (9.35), we obtain

$$x = \frac{\beta}{\lambda}(\mu - 1)\, t \qquad ...(9.45)$$

If x, β, λ and μ are known, thickness (t) of the thin transparent sheet can be calculated. Here x is the distance shifted by introducing the thin plate.

Note: If we use white light as source for getting interference pattern, each colour will form its own interference pattern but condition of the central maximum will be same for all the colours, so all colours will form its central maxima at the same point and will combine to produce white fringe. As fringe width, $\beta = \dfrac{\lambda D}{d}$, is different for different colours so pattern of different colours will overlap each other. As a result distinct fringes will not be visible with white light.

9.8 INTERFERENCE DUE TO REFLECTED LIGHT

Let us consider that a monochromatic light of wavelength (λ) is incident on a thin transparent film of refractive index (μ) with incident angle, i. Partially, it is reflected along AB and partially it is refracted along AC at angle r. The refracted ray (AC) is partially reflected along CD at point C and partially transmitted along CE. The reflected ray (CD) is partially refracted along DF at point D and partially reflected. These reflection and refractions continue further. The rays, AB and DF, are produced from the same incident ray SA, so they are coherent. The reflected rays, AB and DF, interfere and form interference pattern according principle of superposition.

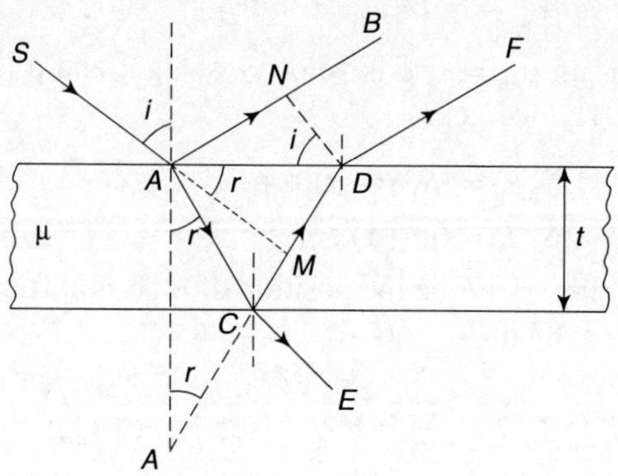

Fig. 9.13

To calculate path difference between interfering rays AB and DF, let us draw perpendicular (DN) from D on AB. Similarly, draw another perpendicular (AM) on CD from A.

The optical path difference (Δ) between interfering rays, AB and DF,

$$\Delta = \mu\,(AC + CD) - AN \qquad \text{...(9.46)}$$

If incident angle of ray SA is i and refracted angle of refracted ray AC is r, then according to Snell's law

$$\mu = \frac{\sin i}{\sin r}$$

In Δ *CND*, $\sin i = \dfrac{AN}{AD}$

In Δ *AMD*, $\sin r = \dfrac{MD}{AD}$

Putting the values of sin *i* and sin *r*, the refractive index is

$$\mu = \frac{\sin i}{\sin r} = \frac{AN/AD}{MD/AD} = \frac{AN}{MD}$$

or $$AN = \mu MD \qquad \text{...(9.47)}$$

Putting the value of *AN* in Eq. (9.46), we get

$$\Delta = \mu\,(AC + CD) - \mu MD$$
$$= \mu\,(AC + CD - MD)$$

In triangle *ACA′*, *AC* = *A′ C*, so, path difference becomes

$$\Delta = \mu\,(A'\,C + CD - MD)$$
$$= \mu\,(A'D - MD) \quad \text{as} \quad A'\,C + CD = A'\,D$$
$$= \mu\,A'\,M \qquad \text{...(9.48)}$$

In triangle *AA′ M*

$$\cos r = \frac{A'M}{AA'} = \frac{A'M}{2t}$$

or $$A'M = 2\,t\cos r \qquad \text{...(9.49)}$$

Putting the value of *A′M* from Eq. (9.49) in Eq. (9.48), we get

$$\Delta = 2\,\mu t\cos r$$

An extra path difference $\left(\dfrac{\lambda}{2}\right)$ is introduced by *AB* ray due to reflection from denser medium at the surface point *A*, while no path difference is introduced by *DF* ray as it is reflected from rarer medium at point *C* on the surface. Thus actual path difference between *AB* and *DF* is

$$\Delta = 2\,\mu t\cos r \pm \frac{\lambda}{2} \qquad \text{...(9.50)}$$

Using negative sign, the condition for maxima or bright fringe is

$$2\,\mu t\cos r - \frac{\lambda}{2} = n\lambda$$

or $$2\,\mu t\cos r = (2\,n + 1)\,\frac{\lambda}{2}, \quad \text{where } n = 0, 1, 2, 2, 3, \ldots \qquad \text{...(9.51)}$$

Using positive sign,the condition for maxima or bright fringe is

$$2\,\mu t\cos r + \frac{\lambda}{2} = n\lambda$$

$$2\,\mu t\cos r = (2\,n - 1)\,\frac{\lambda}{2},\ \text{where } n = 1, 2, 2, 3, \ldots \qquad \ldots(9.52)$$

Using negative sign, the condition for minima or dark fringe is

$$2\,\mu t\cos r - \frac{\lambda}{2} = (2\,n - 1)\,\frac{\lambda}{2}$$

or
$$2\,\mu t\cos r = n\,\lambda,\ \text{where } n = 0, 1, 2, 3, \ldots \qquad \ldots(9.53)$$

Using positive sign, the condition for minima or dark fringe is

$$2\,\mu t\cos r + \frac{\lambda}{2} = (2\,n + 1)\,\frac{\lambda}{2}$$

or
$$2\,\mu t\cos r = n\,\lambda,\quad \text{where } n = 0, 1, 2, 3, \ldots \qquad \ldots(9.54)$$

From Eqs. (9.51), (9.52), (9.53) and (9.54), one should note that the condition of maxima and minima depends on two factors:

(i) Thickness of the film

(ii) Cosine of refracting angle, r

Suppose thickness of film is initially zero and increased gradually. For zero thickness ($t = 0$), the path difference between interfering rays will be $\frac{\lambda}{2}$ and condition of maxima is satisfied. Hence, fillm will appear dark. If thickness is increased gradually, path difference will also increase and will become 0, λ, $2\,\lambda$, $3\,\lambda$, ... for minima and $\frac{\lambda}{2}$, $\frac{3\,\lambda}{2}$, $\frac{5\,\lambda}{2}$, ... for maxima. Thus, alternate bright and dark fringes occur.

If thickness is kept constant and r is varied slowly, again series of maxima and minima are formed.

9.9 INTERFERENCE DUE TO TRANSMITTED LIGHT

Let us consider that a monochromatic light of wavelength λ is incident on the parallel surface of film of refractive index (μ) at angle (I) as shown in Fig. 9.14. It is partly reflected at point A along AB and partly refracted along AC at angle r. At point C, it is partly reflected along CD and partly transmitted along CE. The ray CD is partly reflected along DF at point A. The ray DF is partly transmitted along FG at point F. The rays CE and FG are produced from original incident ray, SA, so they are coherent. The transmitted rays CE and FG form the interference pattern by following the rules of superposition.

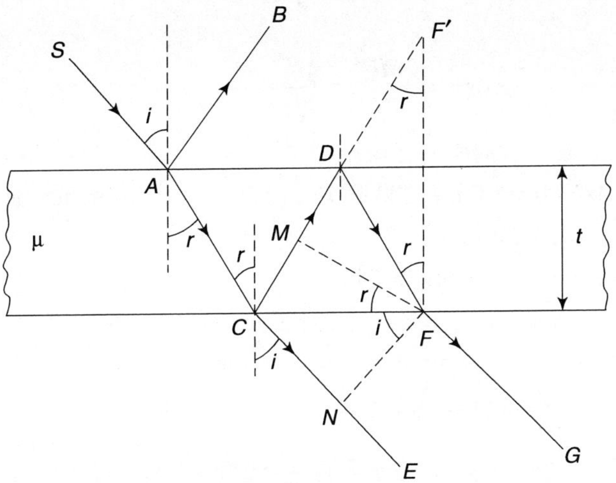

Fig. 9.14

Calculation of path difference between interfering rays, *CE* and *FG*

Let us draw perpendicular *FN* and *FM* from point *F* on *CE* and *CD*, respectively. The optical path difference between rays, *CE* and *FG*, will be

$$\text{Path difference} = \mu\,(CD + DF) - CN \qquad ...(9.55)$$

If incident angle of ray *SA* is *i* and refracted angle of refracted ray *AC* is *r*, then according to Snell's law,

$$\mu = \frac{\sin i}{\sin r}$$

In ΔCNF, $\sin i = \dfrac{CN}{CF}$

In ΔCMF, $\sin r = \dfrac{CM}{CF}$

So, $$\mu = \frac{\sin i}{\sin r} = \frac{CN/CF}{CM/CF} = \frac{CN}{CM}$$

or $$CN = \mu\, CM \qquad ...(9.56)$$

Putting for *CN* from Eq. (9.56) in Eq. (9.55), we get

$$\text{Path difference} = \mu\,(CD + DF) - \mu CM$$
$$= \mu\,(CD + DF - CM)$$

In $\Delta DFF'$, $DF = DF'$, So

$$\text{Path difference} = \mu\,(CD + DF' - CM)$$
$$= \mu\,(CF' - CM)$$
$$= \mu\, MF' \qquad ...(9.57)$$

In $\Delta\, MFF'$

$$\cos r = \frac{MF'}{FF'} = \frac{MF'}{2t}$$

or

$$MF' = 2\, t \cos r \qquad \text{...(9.58)}$$

Substituting for MF' from Eq. (9.58) in Eq. (9.57), path difference becomes

$$\text{Path difference} = 2\, \mu t \cos r \qquad \text{...(9.59)}$$

The condition for maxima or bright fringe is

$$2\, \mu t \cos r = n\lambda, \quad \text{where, } n = 0, 1, 2, 2, 3, ... \qquad \text{...(9.60)}$$

Thus, for $n = 0$, there will be bright fringe, i.e., the central fringe will be bright.

The condition for minima or dark fringe is

$$2\, \mu t \cos r = (2\, n - 1)\, \frac{\lambda}{2}, \quad \text{where } n = 1, 2, 2, 3, ... \qquad \text{...(9.61)}$$

Here one should note that wave FG is reflected at points C and D, where it undergoes reflection more dense medium on two occasions. The phase change introduced at the second point cancels the phase change introduced at the first point. Wave CE does not undergo phase change. Therefore, the fringes formed by transmitted light are complementary to fringes formed by the reflected light. That is, the condition given by Eq. (9.52) is for bright fringes, while the condition given by Eq. (9.61) is for dark fringe. Similarly, the condition given by Eq. (9.54) is for dark fringes, while the condition given by Eq. (9.60) is for bright fringe. The contrast in the intensity of wave CE is considerable more than the twice reflected FG wave in case of fringes formed by transmitted light in comparison to the contrast in the intensity of reflected CE and FG waves in case of fringes formed by reflected light, because in the later case each is reflected only once. Therefore, fringes formed by transmitted light have much less contrast that those formed by reflected light.

9.10 COLOURS IN THIN FILMS

Different colours are seen, when white light is incident on thin films of parallel sides like thin oil films on water and in soap bubbles. This is due to superposition of reflected waves of different colours (wavelengths) obtained from different points. This can be explained as follows.

White light consists of many colours (wavelengths). White light X and Y are incident at different angles at two different points of the film. If the angle of incident of X light is largeer than that of Y light, the light that enters the eye from A travels a greater distance in the film than from B. The refractive index is different for each colour (wavelength) and hence angle of refraction will also be different for each colour (wavelength). Due to refraction the different colours (wavelengths) are separated. The colours that superimpose each other in the light in A are most likely different from those that superimpose in the light from B as angles of incident are different in both cases. The light from A is brought to focus at different point on the eye from that coming from B, therefore, the eye sees the two regions of being different colours.

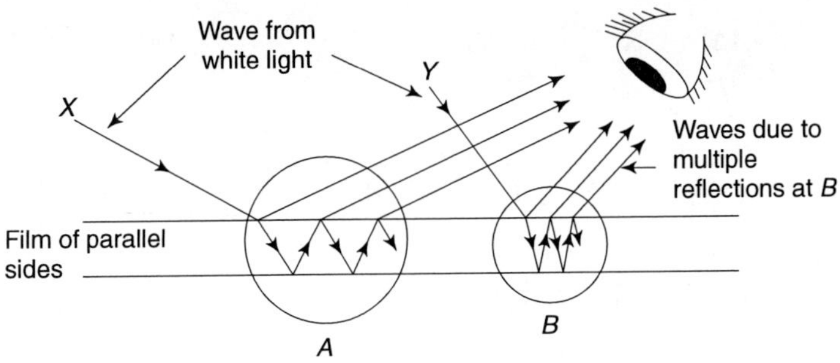

Fig. 9.15

Note: In case of a thick film colours are not observed, because in case of thick film (having thickness of even a few wavelengths), no interference is observed. This is due to the fact that, in case of thick film condition of maxima [$2\mu t \cos r = (2n + 1)\frac{\lambda}{2}$ given by Eq. (9.51)] at a given point is satisfied by a large number of wavelengths (n is different for different wavelengths) and at the same time condition of minima [$2\mu t \cos r = n\lambda$ given by Eq. (9.53)], is also satisfied by another set of wavelengths at the same point. Hence, no colours are seen.

9.11 FRINGES PRODUCED BY WEDGE SHAPED THIN FILMS

Two thin surfaces of glass slightly inclined to each other at an angle θ enclosing a liquid or some other material of refractive index μ is called wedge shaped film. Let us consider two surfaces, *OP* and *OQ*, inclined at an angle (θ), which enclose a wedge shaped film of refractive index (μ) as shown in Fig. 9.16. When the wedge film is illuminated by monochromatic light, the rays fall at a point that is partially reflected and partially refracted. The refracted ray is again reflected from point *C*. Both reflected rays at *A* and *C* act as coherent sources which have been produced by division of amplitude. The reflected wave from point *A* travels in air and the reflected wave from point *Y* travel in medium of refractive index (μ) as well as in air after refraction. So a path difference is created. When they overlap, the bright and dark fringes are formed alternately, which can be observed from the travelling microscope.

To find the path difference between two rays, let us drop a perpendicular (*DK*) on on the ray, *AB*. The path difference between the rays *AB* and *DE* originated from incident ray (*SA*),

$$\mu\,(AC + CD) - AK \qquad \text{...(9.62)}$$

If incident angle of ray *SA* is *i* and refracted angle of refracted ray *AC* is *r*, then according to Snell's law

$$\mu = \frac{\sin i}{\sin r}$$

In ΔAKD, $\sin i = \dfrac{AK}{AD}$

In ΔAND, $\sin r = \dfrac{AN}{AD}$

So, $$\mu = \frac{\sin i}{\sin r} = \frac{AK/AD}{AN/AD} = \frac{AK}{AN}$$

or $$AK = \mu AN \qquad ...(9.63)$$

Putting the value of $AK = \mu AN$ from Eq. (9.63),

Path difference $= \mu\,(AC + CD - AN)$ ...(9.64)

In $\Delta DCD'$, $CD = CD'$, so putting for CD in Eq. (9.64)

Path difference $= \mu\,(AC + CD' - AN)$

$$= \mu\,(AD' - AN) \quad \text{as} \quad AC + CD' = AD'$$

$$= \mu\, ND' \qquad ...(9.65)$$

In $\Delta NDD'$, $\cos(r + \theta) = \dfrac{ND'}{DD'}$

or $$ND' = DD' \cos(r + \theta) = 2t\cos(r + \theta) \qquad ...(9.66)$$

Substituting the value of $ND' = 2t\cos(r + \theta)$ from Eq. (9.66) in Eq. (9.65),

Path difference $= 2\mu t\cos(r + \theta)$ ...(9.67)

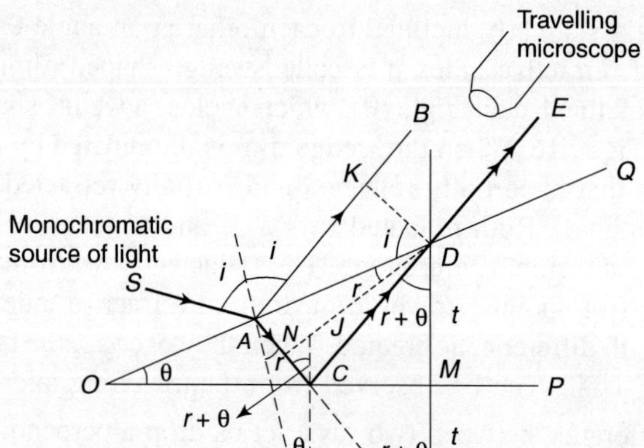

Fig. 9.16

An extra path difference $\left(\dfrac{\lambda}{2}\right)$ is introduced in ray AB due to reflection at A on the surface of denser medium. Therefore, the actual path difference between waves AB and DE becomes

$$2\mu t\cos(r + \theta) \pm \frac{\lambda}{2}$$

Using +ve sign, the condition for maxima is

$$2\mu t \cos(r+\theta) + \frac{\lambda}{2} = n\lambda \quad \text{or} \quad 2\mu t \cos(r+\theta) = (2n-1)\frac{\lambda}{2}, \qquad ...(9.68)$$

where $n = 1, 2, 3, ...$

Using –ve sign, the condition for maxima is

$$2\mu t \cos(r+\theta) - \frac{\lambda}{2} = n\lambda \quad \text{or} \quad 2\mu t \cos(r+\theta) = (2n+1)\frac{\lambda}{2}, \qquad ...(9.69)$$

where $n = 0, 1, 2, 3, ...$

Using +ve sign, the condition for minima is

$$2\mu t \cos(r+\theta) + \frac{\lambda}{2} = (2n+1)\frac{\lambda}{2} \quad \text{or} \quad 2\mu t \cos(r+\theta) = n\lambda, \qquad ...(9.70)$$

where $n = 0, 1, 2, 3, ...$

Using –ve sign, the condition for minima is

$$2\mu t \cos(r+\theta) - \frac{\lambda}{2} = (2n-1)\frac{\lambda}{2} \quad \text{or} \quad 2\mu t \cos(r+\theta) = n\lambda, \qquad ...(9.71)$$

where $n = 0, 1, 2, 3, ...$

When $n = 0$, $t = 0$, the fringe at point O will be dark. In case of wedge shaped film illuminated by parallel beam of monochromatic light, angle of incident and hence angle of refraction are same at every point of film. Thus, in wedge shaped film, μ, r and θ are constant so the path difference, $2\mu t \cos(r+\theta)$, will change due to change in thickness (t) of film from point to point. Thus, the film will appear bright for thickness given by Eq. (9.63),

$$t = \frac{(2n+1)\lambda}{4\mu \cos(r+\theta)}, \quad \text{where, } n = 0, 1, 2, 3, ... \qquad ...(9.72)$$

So, t can have values

$$\frac{\lambda}{4\mu\cos(r+\theta)}, \quad \frac{3\lambda}{4\mu\cos(r+\theta)}, \quad \frac{5\lambda}{4\mu\cos(r+\theta)} \text{ etc.}$$

The dark fringes will appear for thickness given by Eq. (9.64)

$$t = \frac{n\lambda}{2\mu\cos(r+\theta)}, \quad \text{where, } n = 0, 1, 2, 3, ... \qquad ...(9.73)$$

That is, dark fringes will appear for thickness

$$0, \frac{\lambda}{2\mu\cos(r+\theta)}, \quad \frac{2\lambda}{2\mu\cos(r+\theta)}, \quad \frac{3\lambda}{2\mu\cos(r+\theta)} \text{ etc.}$$

It is clear from Eqs. (9.72) and (9.73) that dark and bright fringes are formed due to the wedge of film alternately in the direction of increasing thickness.

Spacing between two consecutive dark bands

Let x_1 and x_2 be the distance of nth and $(n + m)^{th}$ dark ring from the edge for thickness t_1 and t_2, respectively, as shown in Fig. 9.17. Then for n^{th} dark fringe,

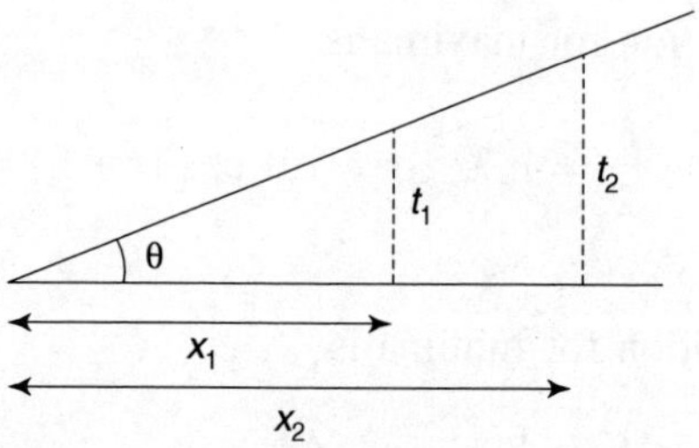

Fig. 9.17

$$t_1 = \frac{n\lambda}{2\mu\cos(r+\theta)} \quad \text{...(9.74)}$$

For $(n + m)^{th}$ dark fringe

$$t_2 = \frac{(n+m)\lambda}{2\mu\cos(r+\theta)} \quad \text{...(9.75)}$$

From Fig. 9.14, we obtain

$$t_1 = x_1 \tan\theta \quad \text{and} \quad t_2 = x_2 \tan\theta$$

Substituting for t_1 and t_2, one gets

$$x_1 = \frac{n\lambda}{2\mu\tan\theta\cos(r+\theta)}$$

and
$$x_2 = \frac{(n+m)\lambda}{2\mu\tan\theta\cos(r+\theta)}$$

Subtracting x_1 from x_2, we get

$$x_2 - x_1 = \frac{m\lambda}{2\mu\tan\theta\cos(r+\theta)} \quad \text{...(9.76)}$$

Thus, the fringe width (β) come out to be

$$\beta = \frac{x_2 - x_1}{m} = \frac{\lambda}{2\mu\tan\theta\cos(r+\theta)} \quad \text{...(9.77)}$$

If θ is small, then $\beta = \dfrac{\lambda}{2\mu\theta\cos(r)}$ as $\tan\theta \approx \theta$ and $r > \theta$

If the light falls normal to the wedge surface, then angle of refraction, $r = 0$, therefore, the fringe width reduces to

$$\beta = \frac{\lambda}{2\mu\theta} \qquad ...(9.78)$$

Note: If wedge shaped film is illuminated by white light, then each colour will form its own system of interference bands. The fringe width for violet colour will be minimum and that for red colour will be maximum. At the edge each colour gives minimum intensity so there will be dark fringe at this point. In the direction of increasing thickness, at certain thickness a few coloured fringes of mixed colours are obtained initially and then at little more thickness overlapping of fringes starts which gives rise to uniform illumination.

Important points for getting clear and bright fringes from wedge shaped film

(i) The fringes are localized to the upper surface of the wedge film, so the travelling microscope must be focused close to it.

(ii) To get the fringes sufficiently far apart so that they are observable, angle θ of the wedge has to be small (~ 10^{-3} rad). From Eq. (9.78), we see that fringe width is in inversely proportional to the angle of wedge film.

(iii) The fringes are clearer at the apex than at the thick end of the wedge. This is because the spread of monochromatic light near the thick end increases as path difference becomes large. For example, the sodium light is not truly monochromatic. It has two waves of wavelengths, $\lambda_1 = 5890 \times 10^{-8}$ cm and $\lambda_2 = 5896 \times 10^{-8}$ cm. It behaves as monochromatic when the path difference is small. The fringes become blurred when path differences are large. A path difference,which is equal to 500 λ_2 for longer wavelengths, may be $500.5\,\lambda_1 = 500\,\lambda_1 + \frac{\lambda_1}{2}$ for shorter wavelengths. Therefore, when one component is producing a bright fringe, the other is producing a dark fringe and vice versa. This will give rise to uniform illumination.

9.12 NEWTON'S RINGS

When a plano-concave lens is placed over a glass plate as shown in Fig. 9.18, the thickness of air film enclosed is minimum at the point of contact and increases gradually from the centre to onwards. If we allow to fall a monochromatic light on plano-concave lens vertically, circular interference fringes are observed by superposition of reflected and transmitted waves.

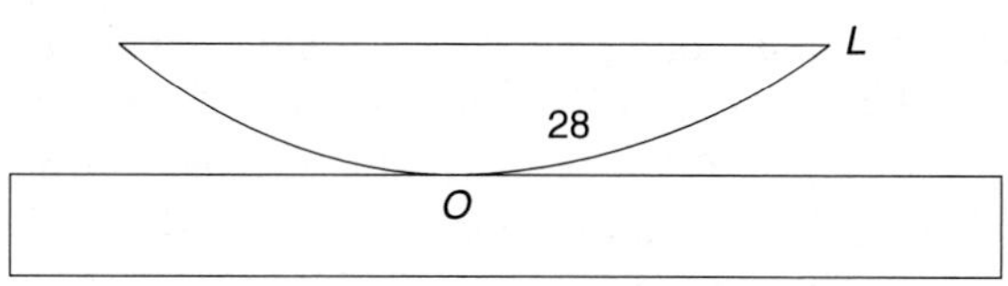

Fig. 9.18

The fringes formed are concentric circles of uniform thickness with centre at the point of contact, are known as Newton's rings. The rings gradually become narrower as their radii increase. In case of monochromatic light alternate bright and dark rings are formed whereas in case of white light coloured rings are formed. Rings formed are used to measure wavelength of monochromatic light and refractive index of liquids.

Experimental Arrangement

The experimental setup is shown in Fig. 9.19. A plano-concave lens (*L*) of large radius of curvature is placed on a glass plate. A monochromatic light from source (*S*) is made parallel by passing through lens (L_1) is allowed to incident on a glass plate (*G*) inclined at 45° to the incident rays.

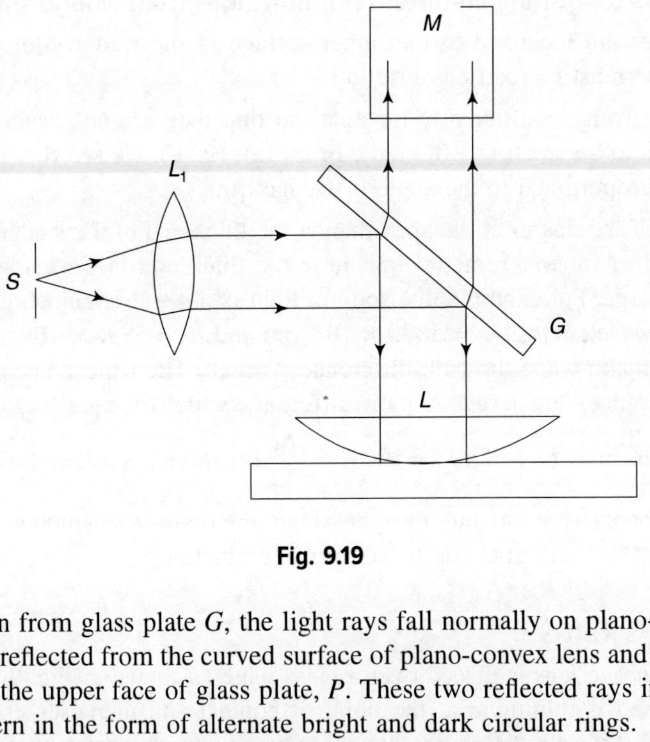

Fig. 9.19

After reflection from glass plate *G*, the light rays fall normally on plano-convex lens (*L*). A part of light is reflected from the curved surface of plano-convex lens and transmitted light is reflected from the upper face of glass plate, *P*. These two reflected rays interfere and give interference pattern in the form of alternate bright and dark circular rings.

9.13 NEWTON'S RINGS BY REFLECTED LIGHT

Suppose *R* is the radius of curvature of a plano-convex lens, *L*. Thickness of air film at point *F* of incident light is *t* (i.e., $DP = FQ = t$). Let the radius of the nth ring be r_n (i.e., $DE = EF = r_n$). When light falls normal to plano-convex lens, it is partially reflected from point *F* and partially refracted from it. The refracted ray again is reflected from glass plate *P* and introduces the additional path difference, $\frac{\lambda}{2}$. The two reflected rays – 1 and – 2 interfere according to principle of superposition. Similarly, the incident light falls at all point of circle of radius (r_n) and interfere. As a result bright rings are formed when path difference between the two rays is odd integral multiple of $\frac{\lambda}{2}$ and dark rings are formed when path difference between the two rays is integral multiple of λ.

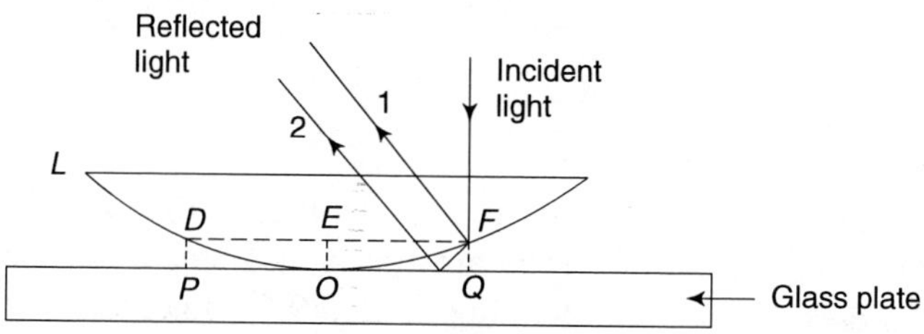

Fig. 9.20

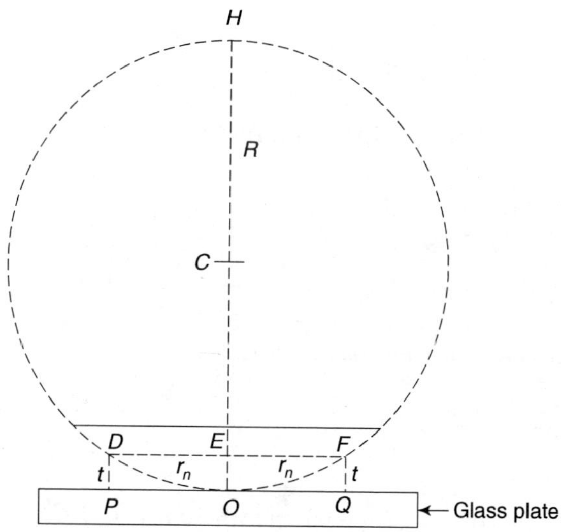

Fig. 9.21

According to theorem of intersecting chords

$$HE \times EO = DE \times EF \quad ...(9.79)$$

or $$(HO - EO) \times EO = DE \times EF$$

or $$(2R - t) \times t = r_n \times r_n$$

or $$(2R - t) \times t = r_n^2 \quad ...(9.80)$$

where, r_n is radius of the nth ring formed and t is the thickness of air film at points D and F.

The radius of curvature of lens R is very large in comparison to the thickness (t) of the air film, so t^2 can be neglected in comparison to $2\,Rt$. Using this approximation, Eq. (9.80) reduces to

$$2\,Rt = r_n^2 = \frac{d_n^2}{4}$$

or
$$2\,t = \frac{d_n^2}{4\,R}, \qquad \text{...(9.81)}$$

where, d_n is the diameter of nth ring.

For small value of thickness of air, t, one can consider the air film as wedge film. For small value of t, the angle of inclination of wedge air film is also small. So, the path difference between two reflected rays is $2\mu t \cos r$ according to Eq. (9.67) in case transmitted light from a thin wedge film. Here, μ is the refractive index of the air film and r is the angle of refraction. But one of the two rays (ray-2) is reflected from denser medium, so an extra path difference of $\frac{\lambda}{2}$ is introduced between interfering ray-1 and ray-2. Therefore, the actual path difference between two interfering rays will be: $2\mu t \cos r \pm \frac{\lambda}{2}$.

Considering –ve sign, the condition for bright ring is

$$2\mu t \cos r - \frac{\lambda}{2} = n\lambda$$

or
$$2\,\mu t \cos r = (2\,n + 1)\,\frac{\lambda}{2}, \text{ where, } n = 0, 1, 2, 3, \ldots \qquad \text{...(9.82)}$$

Considering +ve sign, the condition for bright ring is

$$2\,\mu t \cos r + \frac{\lambda}{2} = n\lambda$$

or
$$2\,\mu t \cos r = (2\,n - 1)\,\frac{\lambda}{2}, \text{ where } n = 1, 2, 3, \ldots \qquad (9.83)$$

That is, bright rings are formed when the path difference between two rays is $\frac{\lambda}{2}, \frac{3\lambda}{2}, \frac{5\lambda}{2}, \ldots$

Considering –ve sign, the condition for dark ring is

$$2\mu t \cos r - \frac{\lambda}{2} = (2\,n - 1)\frac{\lambda}{2}$$

or
$$2\mu t \cos r = n\lambda, \quad \text{where, } n = 1, 2, 3, \ldots \qquad \text{...(9.84)}$$

That is, dark rings are formed when path difference between two rays is 0, λ, 2λ, 3... .Thus, for $n = 0$, there will be dark spot, i.e., at the centre it will be a dark circle.

Considering +ve sign, the condition for dark ring is

$$2\mu t \cos r + \frac{\lambda}{2} = (2n + 1)\frac{\lambda}{2}$$

or $$2\mu t \cos r = n\lambda, \quad \text{where, } n = 0, 1, 2, 3, \ldots \qquad \ldots(9.85)$$

For normal incidence of light on the plano-convex lens, the angle of refraction is zero, i.e., $\cos r = 1$, so bright ring given by Eq. (9.84) reduces to

$$2\mu t = (2n - 1)\frac{\lambda}{2}$$

or $$2t = \frac{(2n-1)\frac{\lambda}{2}}{\mu} \qquad \ldots(9.86)$$

Putting the value of $2t$ from Eq. (9.81), Eq. (9.86) becomes

$$\frac{d_n^2}{4R} = \frac{(2n-1)\frac{\lambda}{2}}{\mu}$$

or $$d_n^2 = \frac{(2n-1)\lambda \times 2R}{\mu}$$

or $$d_n = \sqrt{\frac{2(2n-1)\lambda R}{\mu}} \qquad \ldots(9.87)$$

or $$d_n \propto \sqrt{(2n-1)} \quad \text{as} \quad R, \lambda \text{ and } \mu \text{ are constant.}$$

For normal incidence, the angle of refraction is zero, i.e., $\cos r = 1$, so Eq. (9.85) reduces to equation

$$2\mu t = n\lambda$$

or $$2t = \frac{n\lambda}{\mu} \qquad \ldots(9.88)$$

Putting value of $2t$ from Eq. (9.81), in Eq. (9.88) becomes

$$\frac{d_n^2}{4R} = \frac{n\lambda}{\mu}$$

or $$d_n^2 = \frac{4n\lambda R}{\mu}$$

or $$d_n = \sqrt{\frac{4n\lambda R}{\mu}} \qquad \ldots(9.89)$$

or $$d_n \propto \sqrt{n} \quad \text{as} \quad R, \lambda \text{ and } \mu \text{ are constants.}$$

From Eq. (9.89) it is clear that there will be a dark spot at the centre (for which $t = 0$). From Eq. (9.87) one can see that when $n = 1$, the diameter of the bright ring is $\sqrt{\frac{2\lambda R}{\mu}}$, i.e., the first ring is bright. Taking $n = 1$, radii of the first dark ring and the second bright ring are $\sqrt{\frac{4\lambda R}{\mu}}$ and $\sqrt{\frac{6\lambda R}{\mu}}$, respectively. After that alternate bright and dark rings are formed.

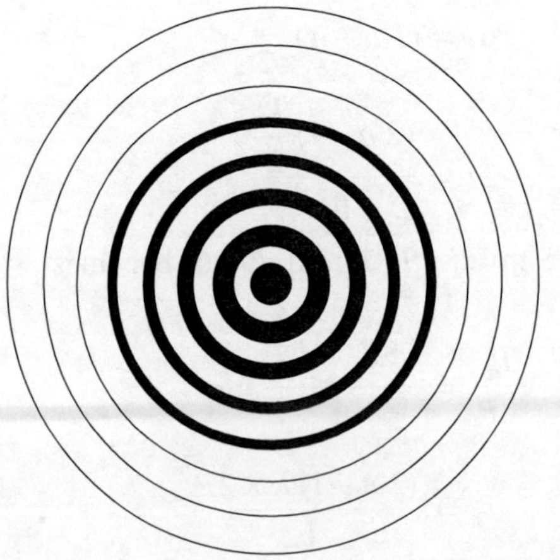

Fig. 9.22 Newton's rings from reflected light

The radius of the:

first dark ring is $d_1 = 2\sqrt{\frac{\lambda R}{\mu}}$, for $n = 1$

second dark ring, $d_2 = 2\sqrt{\frac{2\lambda R}{\mu}}$, for $n = 2$

third dark ring, $d_3 = 2\sqrt{\frac{3\lambda R}{\mu}}$, for $n = 3$

fourth dark ring, $d_4 = 2\sqrt{\frac{4\lambda R}{\mu}}$, for $n = 4$

and so on.

Differences in diameters of two consecutive dark rings are

$$d_2 - d_1 = 2\sqrt{\frac{\lambda R}{\mu}}[\sqrt{2} - 1] = 0.414 \times 2\sqrt{\frac{\lambda R}{\mu}}$$

$$d_3 - d_2 = 2\sqrt{\frac{\lambda R}{\mu}}[\sqrt{3} - \sqrt{2}] = 0.318 \times 2\sqrt{\frac{\lambda R}{\mu}}$$

$$d_4 - d_3 = 2\sqrt{\frac{\lambda R}{\mu}}[\sqrt{4} - \sqrt{3}] = 0.268 \times 2\sqrt{\frac{\lambda R}{\mu}}$$

From the above it is clear that as the radius of dark rings increases, the width of rings decreases. Similar results can be obtained for bright rings.

9.13.1 Determination of Wavelength (μ) of Light

Let us use monochromatic light to get Newton's ring. To determine the wavelength of light, measure the diameter of the nth and $(n + m)^{th}$ dark ring with the help of a travelling microscope. Let diameters of nth and $(n + m)^{th}$ dark rings are

$$d_n^2 = \frac{4\,n\,\lambda\,R}{\mu} \qquad ...(9.90)$$

$$d_{n+m}^2 = \frac{4\,(n+m)\,\lambda\,R}{\mu} \qquad ...(9.91)$$

Subtracting Eq. (9.91) from Eq. (9.92), we get

$$d_{n+m}^2 - d_n^2 = \frac{4\,m\,\lambda\,R}{\mu}$$

or

$$\lambda = \frac{\mu\,(d_{n+m}^2 - d_n^2)}{4\,m\,R} \qquad ...(9.92)$$

For air, μ = 1, therefore, Eq. (9.92) reduces to

$$\lambda = \frac{d_{n+m}^2 - d_n^2}{4\,m\,R} \qquad ...(9.93)$$

The radius of curvature (*R*) of the plano-convex lens (*L*) can be measured with the help of a spherometer. By putting the value of *R* in formula given by Eq. (9.93), wavelength of light used can be calculated.

9.13.2 Determination of Refractive Index (μ) of a Liquid

Place the liquid, whose refractive index (μ) is to be determined between the plano-convex lens (*L*) and the glass plate. Then Newton's rings are obtained by using monochromatic light, whose wavelength is known. Now measure diameter of nth and $(n + m)^{th}$ dark ring with the help of a travelling microscope. Let the diameters of nth and $(n + m)^{th}$ ring are

$$d_n^2 = \frac{4\,n\,\lambda\,R}{\mu} \qquad ...(9.94)$$

$$d_{n+m}^2 = \frac{4\,(n+m)\,\lambda\,R}{\mu} \qquad ...(9.95)$$

Subtracting Eq. (9.94) from Eq. (9.95), we get

$$d_{n+m}^2 - d_n^2 = \frac{4\,m\,\lambda\,R}{\mu}$$

or

$$\mu = \frac{4\,m\,\lambda\,R}{d_{n+m}^2 - d_n^2} \quad ...(9.96)$$

The radius of curvature (R) of the plano-convex lens (L) can be measured with help of a spherometer. By putting the value of R and wavelength λ in the formula given by Eq. (9.96), the refractive index of a liquid can be calculated.

Special Case

If wavelength of light used is not known, then diameter of a particular ring (say n^{th}) is first measured with air in between the plano-convex lens and the glass plate and then measure the diameter of the nth ring by replacing the air with the given liquid. Let diameters d_{air} and d_{liquid} be

$$d_{air}^2 = \frac{4\,n\,\lambda\,R}{1} \quad ...(9.97)$$

and

$$d_{liquid}^2 = \frac{4\,n\,\lambda\,R}{\mu} \quad ...(9.98)$$

Then the formula

$$\mu = \frac{d_{air}^2}{d_{liquid}^2}$$

can be used to find the refractive index.

9.14 NEWTON'S RINGS BY TRANSMITTED LIGHT

Newton's rings are also formed by interference of transmitted light as shown in Fig. 9.23.

Ray 1 is produced from the incident ray by refraction. Ray 2 is produced from the incident ray by reflection at two points of the denser medium. So the path difference introduced at one point is cancelled by the path difference at the second point. So effectively no additional path difference is introduced. The two transmitted rays, 1 and 2, interfere according to the principle of superposition. Similarly, the incident light falls at all points of the circle of radius (r_n) and interfere. As a result dark rings are formed when the path difference between two rays is odd integral multiple of $\frac{\lambda}{2}$ and bright rings are formed when the path difference between two rays is an integral multiple of λ.

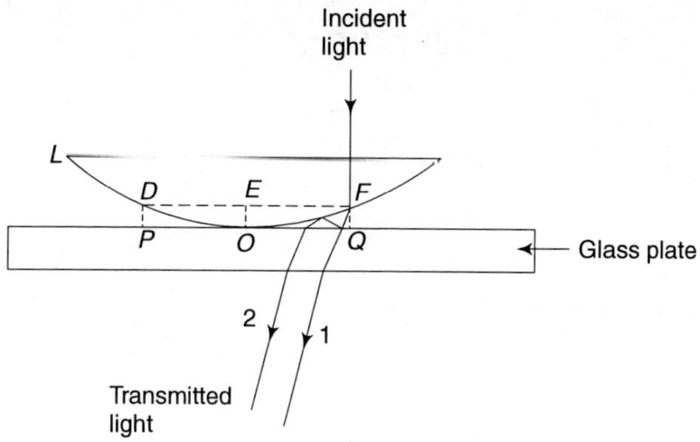

Fig. 9.23

The path difference between two transmitted rays is $2\mu t \cos r$ according to Eq. (9.67) from a thin wedge film. Here, μ is the refractive index of the air film and r is the angle of refraction.

Condition for bright ring is

$$2\mu t \cos r = n\lambda, \quad \text{where, } n = 0, 1, 2, 3, \ldots \qquad \ldots(9.99)$$

That is, bright rings are formed when the path difference between two rays is $0, \lambda, 2\lambda, 3\lambda, \ldots$. Thus, for $n = 0$, there will be a bright spot, i.e., at the centre it will be a dark circle.

Condition for dark ring is

$$2\mu t \cos r = (2n - 1)\frac{\lambda}{2}, \quad \text{where, } n = 1, 2, 3, \ldots \qquad \ldots(9.100)$$

That is, dark rings are formed when the path difference between two rays is $\frac{\lambda}{2}, \frac{3\lambda}{2}, \frac{5\lambda}{2}, \ldots$

Thus, for $n = 1$, there will be a dark ring.

For normal incidence of light on the plano-convex lens, the angle of refraction is zero, i.e., $\cos r = 1$, so condition dark ring given by Eq. (9.100) reduces to

$$2\mu t = (2n - 1)\frac{\lambda}{2}$$

or

$$2t = \frac{(2n-1)\frac{\lambda}{2}}{\mu} \qquad \ldots(9.101)$$

Putting the value of $2t$ from Eq. (9.81) in Eq. (9.101), it becomes

$$\frac{d_n^2}{4R} = \frac{(2n-1)\frac{\lambda}{2}}{\mu}$$

or
$$d_n^2 = \frac{(2n-1)\lambda \times 2R}{\mu}$$

or
$$d_n = \sqrt{\frac{2(2n-1)\lambda R}{\mu}} \qquad ...(9.102)$$

or
$$d_n \propto \sqrt{(2n-1)} \quad \text{as} \quad R, \lambda \text{ and } \mu \text{ are constant.}$$

For normal incidence, the angle of refraction is zero, i.e., $\cos r = 1$, so Eq. (9.99) reduces to equation

$$2\mu t = n\lambda.$$

or
$$2t = \frac{n\lambda}{\mu} \qquad ...(9.103)$$

Putting value of $2t$ from Eq. (9.81), in Eq. (9.103), it becomes

$$\frac{d_n^2}{4R} = \frac{n\lambda}{\mu}$$

or
$$d_n^2 = \frac{4n\lambda R}{\mu}$$

or
$$d_n = \sqrt{\frac{4n\lambda R}{\mu}} \qquad ...(9.104)$$

or
$$d_n \propto \sqrt{n} \quad \text{as} \quad R, \lambda \text{ and } \mu \text{ are constants.}$$

From Eq. (9.104) it is clear that there will be a bright circle at the centre (for which $t = 0$). From Eq. (9.102) one can see that when $n = 1$, the diameter of the dark ring is $\sqrt{\frac{2\lambda R}{\mu}}$, i.e., first ring is dark. Taking $n = 1$, radius of first bright ring is $\sqrt{\frac{4\lambda R}{\mu}}$. After that alternate bright and dark rings are formed.

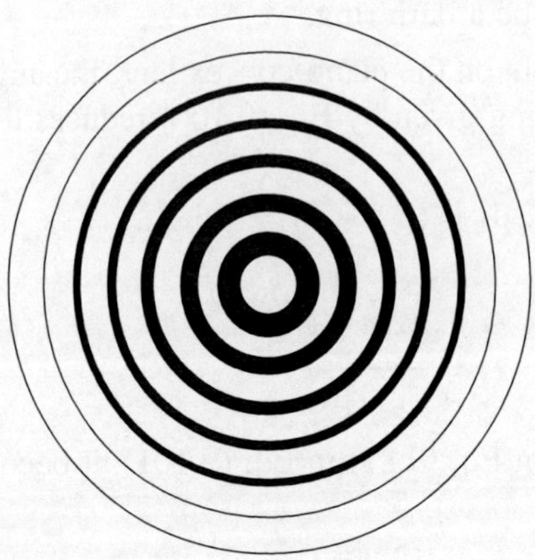

Fig. 9.24 Newton's rings from transmitted light

The radius of the:

first bright ring, $d_1 = 2\sqrt{\dfrac{\lambda R}{\mu}}$, for $n = 1$

second bright ring, $d_2 = 2\sqrt{\dfrac{2\lambda R}{\mu}}$, for $n = 2$

third bright ring, $d_3 = 2\sqrt{\dfrac{3\lambda R}{\mu}}$, for $n = 3$

fourth bright ring, $d_4 = 2\sqrt{\dfrac{4\lambda R}{\mu}}$, for $n = 4$

and so on.

Differences in diameters of two consecutive bright rings are:

$$d_2 - d_1 = 2\sqrt{\frac{\lambda R}{\mu}}[\sqrt{2} - 1] = 0.414 \times 2\sqrt{\frac{\lambda R}{\mu}}$$

$$d_3 - d_2 = 2\sqrt{\frac{\lambda R}{\mu}}[\sqrt{3} - \sqrt{2}] = 0.318 \times 2\sqrt{\frac{\lambda R}{\mu}}$$

$$d_4 - d_3 = 2\sqrt{\frac{\lambda R}{\mu}}[\sqrt{4} - \sqrt{3}] = 0.268 \times \sqrt{\frac{\lambda R}{\mu}}$$

From the above it is clear that as radius of bright rings increases, width of rings decreases. Similar results can be obtained for dark rings.

Note:

(1) The Newton's rings in case of transmitted light are not as distinct as in case of reflected light. Because in case of transmitted light the difference in the intensity of two rays is large. In this case, the intensity of ray 1 is far more than the intensity of ray 2, as ray 1 is obtained after refractions and ray 2 is obtained after reflection at two points of dense medium. The Newton's rings in case of reflected of light are distinct as both the rays are obtained by reflection at one point. So in case of reflected light the difference in the intensities of two rays is large and is almost equal.

(2) In case of monochromatic light bright and dark rings are formed alternately.

(3) The diameter of rings depends on the wavelength.

(4) In case of white light, coloured rings are formed, because the diameter of rings depends on the wavelength. Only the first few rings will be clear and after that either the rings will be blurred or a uniform illumination will be observed because of overlap.

APPENDIX A

Deviation Produced by Prism

$$\delta = (\mu - 1) A, \quad ...(A1)$$

where, A is the angle of prism and μ is refractive index of prism material.

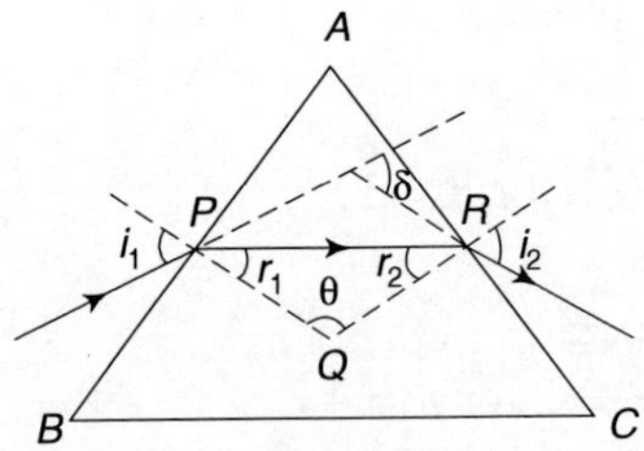

Fig. 9.25

If light is incident on face AB of a prism of prism angle (A) at an angle of incident (i_1) and emerges from the face (AC) at an angle (i_2) as shown in Fig. 9.25, the angle of deviation (δ) will be

$$\delta = (i_1 - r_1) + (i_2 - r_2)$$

or
$$\delta = (i_1 + i_2) - (r_1 + r_2) \quad ...(A2)$$

In ΔPQR

$$r_1 + r_2 + \theta = 180° \quad ...(A3)$$

In quadrilateral $PQRA$

$$A + 90° + \theta + 90° = 360°$$

or
$$A + \theta = 180° \quad ...(A4)$$

Equating Eqs. (A3) and (A4), one gets

$$r_1 + r_2 = A \quad ...(A5)$$

Substituting, for $r_1 + r_2 = A$ from Eq. (A5) in Eq. (A2), we get

$$\delta = (i_1 + i_2) - A \quad ...(A6)$$

Deviation by a Prism at Small Prism Angle

If angle A is small, r_1 and r_2 will be small as $r_1 + r_2 = A$ [Eq. (A2)]. Therefore, i_1 and i_2 will also be small. The Snell's law at the first and second surfaces of the prism are written as

$$\mu = \frac{\sin i_1}{\sin r_1} \quad \text{and} \quad \mu = \frac{\sin i_2}{\sin r_2}$$

Since i_1, i_2, r_1 and r_2 are small , so refractive index is written as

$$\mu = \frac{i_1}{r_1} \quad \text{and} \quad \mu = \frac{i_2}{r_2}$$

or
$$\mu r_1 = i_1 \quad \text{and} \quad \mu r_2 = i_2$$

so that
$$\mu (r_1 + r_2) = i_1 + i_2$$

or
$$\mu A = i_1 + i_2 \quad \text{as} \quad r_1 + r_2 = A \qquad \text{...(A7)}$$

Putting the value of $i_1 + i_2 = \mu A$ from Eq. (A7), Eq. (A6) becomes

$$\delta = \mu A - A$$

or
$$\delta = (\mu - 1) A \qquad \text{...(A8)}$$

Thus, deviation by prism can be obtained from Eq. (A8), by putting the value of refractive index of prism and the value of angle of prism.

Formulas at a Glance

S.No.	Subject	Formula	Explanation
1.	Intensity of two coherent light sources	$I = I_1 + 2\sqrt{I_1 I_2}\cos(\delta) + I_2$ see Eq. (9.5)	$I_1 = a_1^2$ and $I_2 = a_2^2$,
2.	Maximum intensity of two coherent light sources	$I_{max} = I_1 + I_2 + 2\sqrt{I_1 I_2}$ $= \left(\sqrt{I_1} + \sqrt{I_2}\right)^2$	
3.	Minimum intensity of two coherent light sources	$I_{min} = I_2 + I_2 - 2\sqrt{I_1 I_2}$ $= \left(\sqrt{I_1} - \sqrt{I_2}\right)^2$	
4.	Maximum intensity of n coherent waves	$(I_{max})_{co} =$ $\left(\sqrt{I_0} + \sqrt{I_0} + \sqrt{I_0} + \ldots\right)^2 = n^2 I_0$	
5.	Average intensity due to two incoherent waves	$(I_R)_{inco} = I_1 + I_2$	
6.	Intensity of identical incoherent n waves	$(I_R)_{inco} = I_0 + I_0 + \ldots = n I_0$	
	Young's double slit experiment	(1) Interference due to division of wavefront	
7.	For bright fringes, the path difference	$\Delta = n\lambda = \frac{xd}{D}$ or $x = \frac{D}{d} n\lambda$ x-distance of fringe from the centre λ-wavelength of light source	D-distance of screen from the slit d-distance between two slits

8.	Distance of nth dark fringe from the centre	$x_n = \frac{D}{d} n\lambda$, where $n = 0, 1, 2, 3, \ldots$ see Eq. (9.13)	x_n-distance of nth bright fringe from the centre
9.	Distance of nth dark fringe from the centre	$x_n = \frac{D}{d}(2n+1)\frac{\lambda}{2}$, where $n = 0, 1, 2, 3, \ldots$ see Eq. (9.14)	x_n-distance of $(n + 1)^{th}$ dark fringe from the centre
10.	Width of bright or dark fringe	$\beta = \frac{D}{d}\lambda$, see Eq. (9.17)	
11.	Thickness of transparent slab	$t = \frac{x\lambda}{\beta(\mu-1)}$ see Eq. (9.45) μ-refractive index of slab	Here x is moved by n fringe after introducing the thin plate
		Lloyd's Mirror	
12.	Distance of nth bright fringe from centre	$x_n = \frac{D}{d}(2n-1)\frac{\lambda}{2}$, see Eq. (9.27)	where, $n = 1, 2, 3, \ldots$
13.	Distance of nth dark fringe from the centre	$x_n = \frac{D}{d} n\lambda$, see Eq. (9.28)	where, $n = 0, 1, 2, 3, \ldots$
14.	Width of bright or dark fringe	$\beta = \frac{D}{d}\lambda$, see Eq. (9.31) Interference due to division of amplitude	
15.	Condition for bright fringes	$2\mu t \cos r = (2n + 1)\frac{\lambda}{2}$, see Eq. (9.81)	$n = 0, 1, 2, 3, \ldots.$
16.	Condition for dark fringes	$2\mu t \cos r = n\lambda$, see Eq. (9.83)	$n = 1, 2, 3, \ldots$
		Newton's rings due to reflected light	
17.	Condition for bright rings	$2\mu t \cos r = (n + 1)\frac{\lambda}{2}$, see Eq. (9.82)	where $n = 0, 1, 2, 3, \ldots$
18.	Diameter of bright fringes	$d_n = \sqrt{\frac{2(2n+1)R\lambda}{\mu}}$ where $n = 0, 1, 2, 3, \ldots\ldots$ see Eq. (9.87)	R is radius of curvature of plano-convex lens. μ is refractive index of lens.

19.	Diameter of dark fringes	$d_n = \sqrt{\frac{4\,n\,R\,\lambda}{\mu}}$ $n = 1, 2, 3, ...$ see Eq. (9.90)	
20.	Wavelength	$\lambda = \frac{4\,m\,R\lambda}{(d_{n+m}^2 - d_n^2)}$ see Eq. (9.82)	
		Newton's rings due to transmitted light	
21.	Bright fringes	$2\mu t \cos r = n\lambda$, see Eq. (9.99)	where, $n = 1, 2, 3, ...$
22.	Dark fringes	$2\mu t \cos r = (2\,n - 1)\,\frac{\lambda}{2}$, see Eq. (9.100)	where, $n = 1, 2, 3, ...$ r is the angle of refraction from normal of slab.
23.	Deviation of prism	$\delta = (\mu - 1)\,A$ see Eq. (A8) μ is refractive index of slab	Here, A is the angle of prism (179°)
		Fringes produced by reflected light from transparent thin slab	
24.	Condition for maxima or bright fringe	$2\mu t \cos r = (2\,n - 1)\,\frac{\lambda}{2}$, see Eq. (9.52) μ is refractive index of slab.	where, $n = 1, 2, 2, 3, ...,$ t is thickness of slab, r is the refractive angle
25.	Condition for minima or dark fringe	$2\mu t \cos r = n\lambda$, see Eq. (9.53)	where, $n = 0, 1, 2, 3, ...$
		Fringes produced by transmitted light from transparent thin slab	
26.	Condition for maxima or bright fringe	$2\mu t \cos r = n\lambda$, see Eq. (9.60)	where, $n = 0, 1, 2, 2, 3,$ ... Thus, for $n = 0$, there will be bright frings, i.e., the central fringe will be bright
27.	Condition for minima or dark fringe	$2\mu t \cos r = (2\,n - 1)\,\frac{\lambda}{2}$, see Eq. (9.61)	where, $n = 1, 2, 2, 3, ...$

		Fringes produced by wedge shaped thin film	
28.	Condition for bright fringes	$2\mu t \cos(r+\theta) = (2n+1)\dfrac{\lambda}{2}$, see Eq. (9.69)	$n = 0, 1, 2, 3, \ldots$ θ is inclination angle of thin surface
29.	Condition for dark fringes	$2\mu t \cos(r+\theta) = n\lambda$, see Eq. (9.70)	$n = 1, 2, 3, \ldots$
30.	Width of m number of fringes	$x_2 - x_1 = \dfrac{m\lambda}{2\mu \tan\theta \cos(r+\theta)}$, see Eq. (9.76)	
31.	Fringe width	$\beta = \dfrac{x_2 - x_1}{m}$ $= \dfrac{\lambda}{2\mu \tan\theta \cos(r+\theta)}$, see Eq. (9.77)	$\beta = \dfrac{\lambda}{2\mu\theta}$, for $r = 0$ and small θ

SOLVED EXAMPLES

9.1 A biprism is placed at 0.04 m from a slit illuminated by light, $\lambda = 5890$ Å. The width of fringes obtained on a screen at 0.76 m from the biprism is 9.424×10^{-4} m. What is the distance between the coherent sources.

Solution:

Formula used: Width of fringe, $\beta = \dfrac{\lambda D}{d}$

$$\lambda = 5890 \text{ Å} = 5890 \times 10^{-8} \text{ cm} = 5890 \times 10^{-10} \text{ m}$$

$$D = 0.04 + 0.76 \text{ m} = 0.80 \text{ m}, \quad \beta = 9.4 \times 10^{-4} \text{ m}$$

$$d = \frac{\lambda D}{\beta} = \frac{5890 \times 10^{-10} \times 0.80}{9.4 \times 10^{-4}} = \frac{4712}{9.4} \times 10^{-6} = \frac{47.12}{9.4} \times 10^{-4} = 5 \times 10^{-4} \text{ m}$$

Ans. The distance between coherent sources is 5×10^{-4} m.

9.2 The distance between the slit and biprism is 45 cm and between biprism and screen are 55 cm. Angle of biprism and refractive index are 179° and 1.5, respectively. Calculate the wavelength of the light if the distance between successive fringes is 0.0124 cm.

Solution:

Formula used : Distance of successive fringes, $\beta = \dfrac{\lambda D}{d}$

Given: Angle of biprism = 179°, $\mu = 1.5$, $\beta = 0.0124$ cm,
distance between slit and biprism, $a = 45$ cm,

distance between biprism and screen, b = 55 cm

The distance between slits and screen is

$$D = 45 + 55 \text{ cm} = 100 \text{ cm} = 100 \text{ cm}$$

The refracting angle, A

$$A = \frac{180 - 179}{2} = \frac{1}{2} = 0.5° = 0.5 \times \frac{\pi}{180} \text{ radians}$$

The separation of two slits is

$$d = 2\,b\,(\mu - 1)\,A$$

$$= 2 \times 55\,(1.5 - 1) \times 0.5 \times \frac{\pi}{180}$$

$$= 2 \times 55 \times 0.5 \times 0.5 \times \frac{3.14}{180}$$

$$= \frac{27.5 \times 3.14}{180} = 0.480 \text{ rad}$$

The wavelength of source is

$$\lambda = \frac{\beta d}{D} = \frac{0.0124 \times 0.480}{100} = 0.5952 \times 10^{-4}$$

$$= 5952 \times 10^{-8} \text{ cm} = 5952 \text{ Å}$$

Ans. The wavelength of source is 5952 Å

9.3 When a thin piece of glass 3.4×10^{-4} cm thick is placed in the path of one of the interfering beams in a biprism experiment, it is found that central fringe shift through a distance equal to width of four fringes. Find the refractive index of the piece of glass. (λ = 5896 Å).

Solution:

The refractive index of glass plate, μ = ?

Formula for central fringe shift is

$$(\mu - 1)\,t = n\lambda \quad \text{or} \quad (\mu - 1) = \frac{n\lambda}{t} \quad \text{or} \quad \mu = 1 + \frac{n\lambda}{t}$$

Given:

The thickness of glass plate $t = 3.7 \times 10^{-4}$ cm, λ = 5893 Å = 5896 10^{-8} cm, $n = 4$

$$\mu = 1 + \frac{n\lambda}{t} = 1 + \frac{4 \times 5896\,10^{-8}}{3.7 \times 10^{-4}}$$

$$= 1 + \frac{4 \times 0.5896}{3.7} = 1 + 0.64 = 1.64$$

Ans. Refractive index of the glass, μ = 1.64

9.4 A parallel beam of sodium light (A = 5896 Å) strikes the film of oil floating on water. When viewed at angle 30λ from the normal, 6th dark band is seen. What is film's thickness? (μ_{oil} = 1.5)

Solution:

Thickness of film, t = ?

Formula for dark band in reflected system is

$$2\mu t \cos r = n\lambda$$

or $$t = \frac{n\lambda}{2\,\mu\cos r}$$

Given: The dark band seen is $n = 6$

$$\mu = 1.5,\quad \lambda = 5890\text{ Å} = 5890 \times 10^{-8}\text{ cm}$$

Angle of incident, $i = 30° = 30 \times \frac{\pi}{180} = \frac{\pi}{6}$

$$\mu = \frac{\sin i}{\sin r} \quad \text{or} \quad \sin r = \frac{\sin i}{\mu} = \frac{\sin 30}{1.5} = \frac{\frac{1}{2}}{1.5} = \frac{1}{3}$$

$$\cos r = \sqrt{1 - \sin^2 t} = \sqrt{1 - \frac{1}{9}} = \sqrt{\frac{8}{9}} = \frac{2}{3}\sqrt{2} = 0.94$$

Thickness oil film, $t = \frac{n\lambda}{2\,\mu\cos r} = \frac{6 \times 5896 \times 10^{-8}}{2 \times 1.5 \times 0.94} = 1.25 \times 10^{-4}$ cm

Ans. Oil film's thickness is 1.25×10^{-4} cm.

9.5 White light is reflected from an oil film of thickness 0.011 mm and refractive index 1.41 at an angle of 45° to vertical. If the reflected light falls on the slit of a micrometer, calculate the number of dark bands seen between wavelengths 4500 Å and 5500 Å.

Solution:

Number of dark bands seen, n = ?

Formula for dark fringes is

$$2\mu t \cos r = n\lambda$$

Given: Angle of incidence, $i = 45°$, $\mu = 1.41$

$$\lambda_1 = 4000\text{ Å} = 4500 \times 10^{-8}\text{ cm},$$
$$\lambda_2 = 5000\text{ Å} = 5500 \times 10^{-8}\text{ cm}$$

Thickness of oil film, $t = 0.011\text{ mm} = 0.0011\text{ cm} = 1 \times 10^{-3}\text{ cm} = 0.0000011\text{ m} = 1.1 \times 10^{-5}$ m

Let n_1 and n_2 be the bands seen for wavelengths λ_1 and λ_2, where,

$$n_1 = \frac{2\,\mu\, t\cos r}{\lambda_1}; \qquad n_2 = \frac{2\,\mu\, t\cos r}{\lambda_2}$$

$$\mu = \frac{\sin i}{\sin r} \quad \text{or} \quad \sin r = \frac{\sin i}{\mu} = \frac{\sin 45}{1.41} = \frac{\frac{1}{\sqrt{2}}}{1.41}$$

$$= \frac{1}{1.414 \times 1.41} \approx \frac{1}{2} \quad \text{or} \quad r = 30°,$$

$$\cos r = \cos 30 = \frac{\sqrt{3}}{2} = \frac{1.732}{2} = 0.866$$

$$n_1 = \frac{2\times 1.41\times 1.1\times 10^{-3}\times 0.866}{4500\times 10^{-8}}$$

$$= \frac{2\times 1.41\times 1.1\times 0.866\times 10^{2}}{4.5} = 60$$

$$n_2 = \frac{2\times 1.41\times 1.1\times 10^{-3}\times 0.866}{5500\times 10^{-8}}$$

$$= \frac{2\times 1.41\times 1.1\times 0.866\times 10^{2}}{5.5} = 49$$

Bands between 4500 Å and 5500 Å = $n_1 - n_2 = 60 - 49 = 11$

Ans. 11 bands are there between 4000 Å and 5000 Å..

9.6 If the diameter of the nth dark ring in an arrangement giving Newton's ring changes from 0.32 cm to 0.26 cm as liquid is introduced between the lens and the plate. Calculate the value of the refractive index of the liquid.

Solution:

The diameter of the nth dark ring by putting a liquid of refractive index μ is given as

$$d_{nl}^2 = \frac{4\,n\,\lambda\,R}{\mu} \qquad ...(1)$$

For air medium, μ = 1, so the diameter of the nth ring with air as wedge film is

$$d_{na}^2 = 4\,n\,\lambda R \qquad ...(2)$$

Dividing Eq. (1) by Eq. (2), we have

$$\frac{d_{nl}^2}{d_{na}^2} = \frac{\frac{4\,n\,\lambda\,R}{\mu}}{\frac{4\,n\,\lambda\,R}{1}} = \frac{1}{\mu} \quad \text{or} \quad \mu = \left(\frac{d_{na}}{d_{nl}}\right)$$

Given:

$$d_{na} = 0.32 \text{ cm}, \quad d_{nl} = 0.26 \text{ cm}$$

$$\mu = \left(\frac{d_{na}}{d_{nl}}\right)^2 = \left(\frac{0.32}{0.26}\right)^2 = 1.51$$

Ans. The value of refractive index of liquid is 1.51.

9.7 In a Newton's ring experiment the diameter of the 5th and 20th rings are 0.3 cm and 0.8 cm, respectively. Find the wavelength of light. R = 110 cm.

Solution:

Wavelength of light source in Newton's ring experiment is given by the formula

$$\lambda = \frac{D_{m+p}^2 - D_m^2}{4\,Rp}$$

Given:

$$m = 5, m + p = 20, p = 20 - 5 = 15, R = 110 \text{ cm},$$
$$D_m = 0.3 \text{ cm}, D_{m+p} = 0.7 \text{ cm}$$

$$\lambda = \frac{0.7^2 - 0.3^2}{4 \times 110 \times 15} = \frac{(0.7+0.3)(0.7-0.3)}{4 \times 110 \times 15} = \frac{1 \times 0.4}{4 \times 110 \times 15}$$

$$= 6.061 \times 10^{-5} \text{ cm} = 6061 \times 10^{-8} \text{ cm} = 6061 \text{ Å}$$

Ans. The wavelength of light source is 6061 Å.

9.8 When a liquid is placed between the lens and the plate,the diameter of the nth ring of Newton's ring changes from 0.35 cm to 0.30 cm. Calculate the velocity of light in the fluid when the velocity of light in vacuum is 3×10^8 m/s.

Solution:

Diameter of nth dark ring in the presence of liquid is given by

$$D^2_{n \text{ liq}} = \frac{4 n \lambda R}{\mu} \quad ...(1)$$

Diameter of nth dark ring in the presence of air is given by

$$D^2_{n \text{ air}} = 4 n \lambda R \quad ...(2)$$

Dividing Eq. (2) by Eq. (1), we get

$$\frac{D^2_{n \text{ air}}}{D^2_{n \text{ liq}}} = \frac{2 n \lambda R}{\frac{4 n \lambda R}{\mu}} = \mu \quad \text{or} \quad \mu = \frac{D^2_{n \text{ air}}}{D^2_{n \text{ liq}}}$$

$$\mu = \frac{\text{Velocity of light in vacuum}}{\text{Velocity of light in liquid}} = \frac{D^2_{n \text{ air}}}{D^2_{n \text{ liq}}}$$

Given: Diameter of nth ring with air $D_{air} = 0.35$ cm

Diameter of nth ring with air $D_{liq} = 0.30$ cm

Let the velocity of light in liquid be v_1. The velocity of light, $c = 3 \times 10^8$ m/s

Now, putting the values of various variables, we get

$$\frac{3 \times 10^8}{v_1} = \frac{(0.35)^2}{(0.30)^2} = \frac{12.25}{9}$$

or
$$v_1 = \frac{3 \times 10^8 \times 9}{12.25} = 2.20 \times 10^8 \text{ m/s}$$

Ans. Velocity of light in liquid is 2.20×10^8 m/s.

9.9 White light is incident on a soap film at an angle $\sin^{-1}(4/5)$ and the reflected light shows dark bands. The consecutive dark bands correspond to wavelength 6.2×10^{-5} cm and 6.1×10^{-5} cm. If $\mu = 1.33$ for the film, calculate the thickness.

Solution:

Condition for dark fringes, $2\mu t \cos r = n\lambda$

If n and $n + 1$ are successive dark bands corresponding to wavelengths λ_1 and λ_2, then

$$2\mu t \cos r = n\lambda_1 \quad ...(1)$$

$$2\mu t \cos r = (n+1)\lambda_2 \quad ...(2)$$

Comparing Eqs. (1) and (2), we find

$$n\lambda_1 = (n+1)\lambda_2 \quad \text{or} \quad n(\lambda_1 - \lambda_2) = \lambda_2 \quad \text{or} \quad n = \frac{\lambda_2}{\lambda_1 - \lambda_2}$$

Substituting the value of n in Eq. (1), the thickness of film is

$$t = \frac{\lambda_1}{2\mu \cos r} \times \frac{\lambda_2}{\lambda_1 - \lambda_2} = \frac{1}{2\mu \cos r} \frac{\lambda_1 \lambda_2}{\lambda_1 - \lambda_2}$$

Given:

$$\lambda_1 = 6.2 \times 10^{-5} \text{ cm}$$

$$\lambda_2 = 6.1 \times 10^{-5} \text{cm}$$

$$\mu = 1.33, \cos r = ?$$

The incident angle, $i = \sin^{-1}(4/5)$ or $\sin i = \frac{4}{5}$.

$$\mu = \frac{\sin i}{\sin r} = \frac{\frac{4}{5}}{\sin r} = 1.33 = \frac{4}{3} \quad \text{or} \quad \sin r = \frac{4}{5} \times \frac{3}{4} = \frac{3}{5}$$

$$\cos r = \sqrt{1 - \sin^2 r} = \sqrt{1 - \frac{9}{25}} = \frac{4}{5}$$

Now the thickness (t) of the film,

$$t = \frac{1}{2 \times \frac{4}{3} \times \frac{4}{5}} \times \frac{6.2 \times 10^{-5} \times 6.1 \times 10^{-5}}{6.2 \times 10^{-5} - 6.1 \times 10^{-5}}$$

$$= \frac{3 \times 5}{2 \times 4 \times 4} \times \frac{38.0 \times 10^{-5}}{0.1} = 1.78 \times 10^{-3} \text{ cm}$$

Ans. Thickness of the film is 1.78×10^{-3} cm

9.10 In Young's experiment the interference pattern is found to have an intensity ratio between the bright and dark fringes equal to 9. (a) What is the ratio of intensities and (b) amplitudes of the two interfering waves?

Solution: The formula used

$$I = I_1 + I_2 + 2\sqrt{I_1 I_2} \cos \delta$$

Here, I_1 and I_2 are intensities of two waves and δ is the phase difference between two waves.

(a) For I to be maximum $\cos \delta = 1$ for I to be minimum $\cos \delta = -1$

Putting the value of cos δ, we have

$$I_{max} = I_1 + I_2 + 2\sqrt{I_1 I_2} = \left(\sqrt{I_1} + \sqrt{I_2}\right)^2$$

$$I_{min} = I_1 + I_2 - 2\sqrt{I_1 I_2} = \left(\sqrt{I_1} - \sqrt{I_2}\right)^2$$

Given: $I_{max} = 9$, $I_{min} = 1$

$$\frac{I_{max}}{I_{min}} = \frac{\left(\sqrt{I_1} + \sqrt{I_2}\right)^2}{\left(\sqrt{I_1} - \sqrt{I_2}\right)^2} = \frac{9}{1}, \quad \text{i.e.,} \quad \frac{\sqrt{I_1} + \sqrt{I_2}}{\sqrt{I_1} - \sqrt{I_2}} = \frac{3}{1}$$

or $$\sqrt{I_1} + \sqrt{I_2} = 3\sqrt{I_1} - 3\sqrt{I_2}$$

or $$4\sqrt{I_2} = 2\sqrt{I_1} \quad \text{or} \quad \frac{I_1}{I_2} = 4$$

(b) For a wave, $I \propto A^2$

So $$\frac{I_1}{I_2} = \left(\frac{A_1}{A_2}\right)^2 = 4 \quad \text{or} \quad \frac{A_1}{A_2} = 2$$

Ans. The ratio of intensity is 4 and ratio of amplitude is 2.

9.11 Find the maximum intensity in case of interference of n identical waves each of intensity, I_0, if the interference is (a) coherent (b) incoherent.

Solution: In case of interference of two waves

$$I = I_1 + I_2 + 2\sqrt{I_1 I_2}\cos\delta$$

Here, I_1 and I_2 are intensities of two waves and δ is the phase difference between two waves.

(a) In case of coherent interference at a given point, δ does not vary with time, so I will be maximum when cos δ is maximum, i.e., cos δ = 1, so

$$I_{max} = I_1 + I_2 + 2\sqrt{I_1 I_2} = \left(\sqrt{I_1} + \sqrt{I_2}\right)^2$$

For n identical waves intensity of each wave is I_0. Now maximum intensity of n coherent waves is

$$(I_{max})_{co} = \left(\sqrt{I_0} + \sqrt{I_0} + \sqrt{I_0} + \ldots\right)^2 = n^2 I_0$$

(b) In case of incoherent interference at a given point, δ varies randomly with time, so average value of cos δ is zero, i.e., cos δ = 0, so

$$(I_R)_{inco} = I_1 + I_2$$

In case of identical incoherent n waves

$$(I_R)_{inco} = I_0 + I_0 + \ldots = n\, I_0$$

Ans. Maximum intensity of n identical coherent waves is $n^2 I_0$ and intensity of identical incoherent n waves is $n\, I_0$.

QUESTIONS FOR EXAMINATION

9.1 Describe the difference between phase difference and path difference.

9.2 Explain the phenomenon of interference of light.

9.3 What is coherent source?

9.4 Discuss the conditions for interference of light.

9.5 What is fringe width?

9.6 Discuss the difference between division of wavefront and division of amplitude.

9.7 What is Fresnel's biprism?

9.8 Discuss the effect of putting a very thin film in the path of one of the interfering beams.

9.9 Explain, why the central spot is dark in Newton's rings case of reflected light white spot in case of transmitted light?

9.10 Why Newton's rings are circular?

9.11 Describe the theory of interference due to two slits and find the formula for finge width.

9.12 Show that the area enclosed between consecutive rings does not depend on the number of rings.

9.13 How the coherent sources are produced with the help of Fresnel biprism.

9.14 What is interference? Describe and explain Newton's rings of reflected monochromatic light.

9.15 What is coherence? Describe and explain Newton's rings of transmitted monochromatic light.

9.16 Describe the formation of Newton's rings by reflected monochromatic light. Derive the formula for finding the diameter of nth dark ring in this case.

9.17 Describe Newton's experiment to find the wavelength of light.

9.18 Obtain interference fringes using Fresnel biprism.

9.19 Describe the method to find the wavelength of monochromatic light using Fresnel's biprism.

9.20 Derive the formula to find the diameter of Newton's rings. Show that width of rings decreases as order of rings increases.

9.21 Why do we see the colours from oil film on water? Under what conditions the film may appear dark.

9.22 Describe the method to find the refractive index of a liquid using Newton's rings.

OBJECTIVE QUESTIONS

9.1 Two coherent light waves produce destructive interference, when the phase difference between them is

(a) $\frac{3\pi}{2}$ (b) 2π (c) $\frac{\pi}{2}$ (d) π

9.2 The diameter of Newton's rings is proportional to

(a) $\frac{1}{\lambda}$ (b) $\sqrt{\lambda}$ (c) λ (d) $\frac{1}{\sqrt{\lambda}}$

9.3 The phase difference (π) between two light waves at a point is equal to the path difference

(a) $\frac{\lambda}{4}$ (b) 2λ (c) λ (d) $\frac{\lambda}{2}$

9.4 The refracting angle of Fresnel biprism is

(a) 179° (b) 0° (c) 180° (d) $\frac{1}{2°}$

9.5 The diameter of Newton's rings is proportional to

(a) $\frac{1}{\mu}$ (b) $\sqrt{\mu}$ (c) μ (d) $\frac{1}{\sqrt{\mu}}$

9.6 In case of reflected light the central spot in Newton's rings is

(a) dark (b) bright (c) first bright and then dark
(d) first dark and then bright

9.7 In case of reflected light the central spot in Newton's rings is

(a) dark (b) white (c) coloured (d) none of them

9.8 When order increases, the intensity of dark fringes

(a) none of these (b) remain constant (c) decreases (d) increases

9.9 When order increases, the width of dark fringes

(a) none of these (b) decreases (c) remains constant (d) increases

9.10 The path difference introduced by reflection of light from a dense medium is

(a) $\frac{3\lambda}{2}$ (b) $\frac{\lambda}{2}$ (c) λ (d) 2λ

9.11 The path difference introduced by reflection of light two times from dense medium is

(a) $\frac{\lambda}{4}$ (b) $\frac{\lambda}{2}$ (c) λ (d) 2λ

9.12 The area enclosed between consecutive rings.

(a) depends on the number of rings
(b) does not depend on the number of rings
(c) depends on wavelength only
(d) depends on radius of curvature of a plano-convex lens only

9.13 When light wave is reflected at the interface of air and glass, the change of phase of the reflected wave is equal to

(a) 0 (b) 2π (c) $\frac{\pi}{2}$ (d) π

9.14 In Fresnel biprim experiment, the central ring is

(a) dark (b) bright
(c) first bright and then dark (d) first dark the bright

9.15 In Fresnel biprim experiment, the fringe width of which colour will be minimum

(a) red (b) violet (c) yellow (d) green

9.16 Newton's rings are
   (a) locus of points of equal thickness
   (b) locus of points of equal inclination
   (c) locus of points of equal thickness and equal incilnation
   (d) Neither (a) nor (b)

9.17 In Fresnel biprism experiment, the distance between the slits is gradually increased. The fringe width
   (a) first increases and then decreases (b) remains constant
   (c) decreases (d) increases

9.18 In the interference experiment monochromatic light is replaced by white light, we will see
   (a) Uniform illumination on the screen
   (b) Uniform darkness on the screen
   (c) Equally spaced white and dark fringes
   (d) A few coloured fringes and then uniform illumination

9.19 In Young's double slit experiment the separation between the slits is halved and the distance between the slits and screen is doubled. The fringe width becomes
   (a) halve (b) 2 times (c) 4 times (d) one-fourth

9.20 In the double slit experiment, the intensity at the site of the central maxima is I. When either of the two slits is closed, the intensity at the same spot is I_0. The ratio between I and I_0 will be
   (a) $\frac{1}{2}$ (b) 2 (c) 4 (d) 6

9.21 The contrast in the fringes in any interference pattern depends on
   (a) Distance between slits and screen (b) Wavelength
   (c) Intensity ratio of sources (d) Distance between the sources

Answers

9.1 (d)	9.2 (b)	9.3 (d)	9.4 (a)	9.5 (d)	9.6 (a)	9.7 (b)	9.8 (c)
9.9 (b)	9.10 (b)	9.11 (c)	9.12 (b)	9.13 (d)	9.14 (a)	9.15 (b)	9.16 (a)
9.17 (c)	9.18 (d)	9.19 (c)	9.20 (c)	9.21 (c)			

CHAPTER 10

Interferometers

10.1 INTRODUCTION

In this experiment the beam of one light source is divided into two parts of equal intensity by partial reflection and refraction. These two beams after reflection from two mirrors overlap and show interference patterns. These patterns are analyzed with the help of a telescope. Michelson interferometer has several applications. It can be used to do accurate measurement of refractive index of thin transparent plates and wavelength of monochromatic light source. It can be used to find the thickness of thin plates of material (transparent) by putting it in the path of one of the interfering beams. The difference between wavelengths of two neighbouring spectral lines can be obtained with good accuracy by using the Michelson interferometer.

10.2 MICHELSON'S INTERFEROMETER

Experimental setup: Experimental setup of Michelson's interferometer has two highly polished optically plane mirrors, two optically flat glass plates of equal thickness cut from same material, one monochromatic light source and a telescope. The polished optically plane mirrors, M_1 and M_2, placed perpendicular to each other as shown in Fig. 10.1. The glass plates, P and Q, are placed at an angle 45° with incident light beam. The rear face of glass plate (P) is semi-silvered. The mirror (M_1) is mounted on such a stand so that it can be moved forward and backward with the help of micrometer screw. The labelling screws attached at the back of both the mirrors are used to make them exactly perpendicular. The telescope (T) is used to see interference rings made by light rays reflected from the mirrors, M_1 and M_2.

Working: The monochromatic light source (S) is placed at the focus of the collimating lens, L. The parallel beam emerging from collimating lens is allowed to fall on the glass plate, P. As the rear surface of glass plate (P) is semi-silvered, so it is divided into two parts (say, beam 1 and beam 2) of nearly same intensity. As the plate (P) is inclined at an angle 45° with incident light beam, the reflected beam 1 falls normally on mirror (M_1), while the refracted beam 2 falls also falls normally on mirror (M_2) as this mirror is set perpendicular

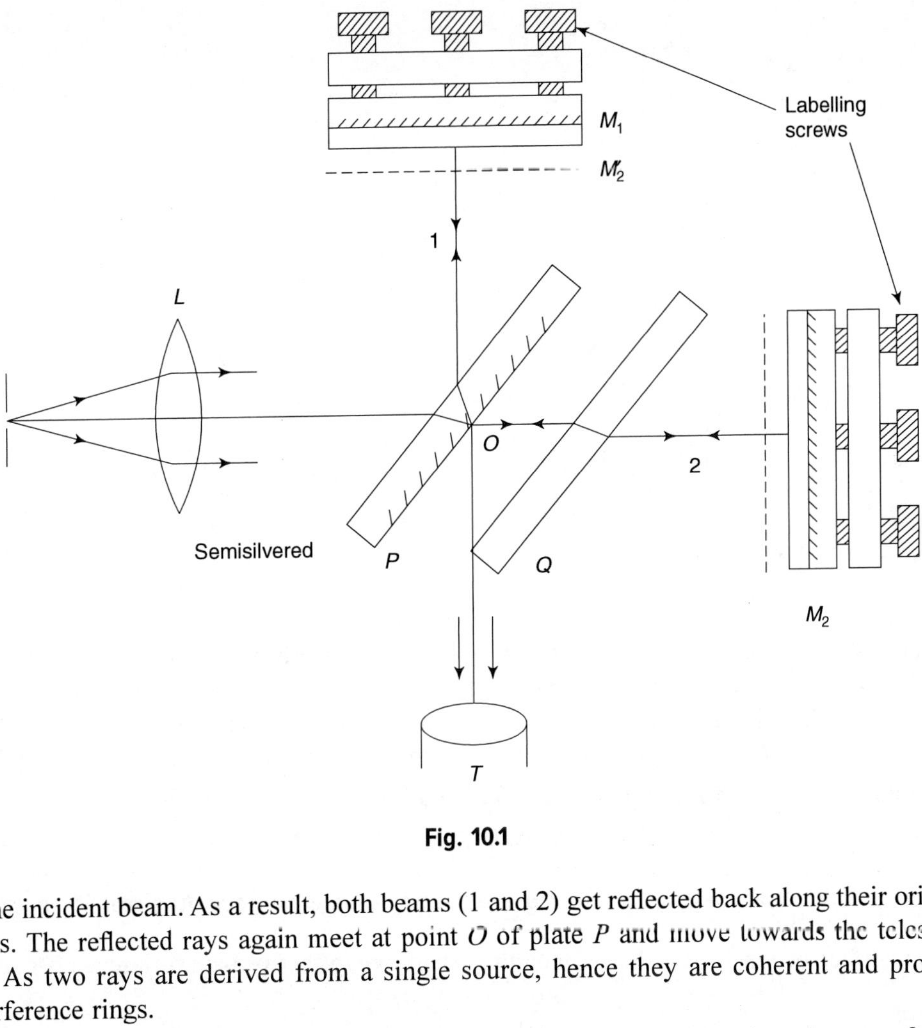

Fig. 10.1

to the incident beam. As a result, both beams (1 and 2) get reflected back along their original paths. The reflected rays again meet at point O of plate P and move towards the telescope (T). As two rays are derived from a single source, hence they are coherent and produce interference rings.

Purpose of plate Q : The reflected ray1 crosses the glass plate (P) twice whereas refracted ray 2, in the absence of the glass plate, Q, moves in air. Thus, in the absence of glass plate (Q), the optical paths, OM_1 and OM_2 are not equal. To equalize the two paths, a glass plate (Q) of same material and same thickness as that of P is introduced in the path of ray 2. The plate (Q) is also known as *compensating plate*.

10.3 TYPES OF FRINGES

10.3.1 Circular Fringes

In Michelson's interferometer, if mirror M_1 is parallel to virtual image M'_2 of mirror M_2, then circular fringes are produced with monochromatic light. As shown in Fig. 10.2, the source (S) produces two virtual sources, S_1 and S_2, due to mirror M_1 and virtual mirror M'_2, respectively.

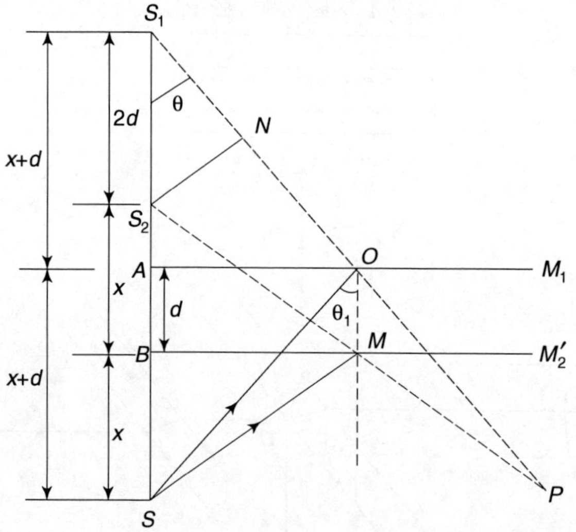

Fig. 10.2

SMS_1 and SOS_2 are equilateral triangles. Let d be the distance between mirrors M_1 and M'_2, and $SB = S_2B = x$, then $SA = S_1A$ will be $x + d$, and the distance between virtual sources, S_1 and S_2, will be $2d$. The rays from S_1 and S_2 form maxima or minima at point P by travelling the paths S_1P and S_2P. The maxima or minima of fringes (seen in telescope, T) will depend on

$$\text{Path difference, } \Delta = S_1P - S_2P = S_1N = S_1 S_2 \cos\theta = 2d\cos\theta$$

an extra path difference of $\frac{\lambda}{2}$ is introduced at reflecting silvered surface of mirror M_2, because silvered surface of mirror is a denser medium. Therefore, effective path difference becomes

$$\Delta = 2d\cos\theta - \frac{\lambda}{2}$$

For bright fringes the path difference must be integral multiple of wavelength λ, i.e.,

$$2d\cos\theta - \frac{\lambda}{2} = n\lambda$$

or
$$2d\cos\theta = (2n+1)\frac{\lambda}{2} \qquad ...(10.1)$$

For dark fringes the path difference must be half integral multiple of wavelength λ, i.e.,

$$2d\cos\theta - \frac{\lambda}{2} = (2n-1)\frac{\lambda}{2}$$

or
$$2d\cos\theta = n\lambda \qquad ...(10.2)$$

For given value of d, n and λ, the value of θ will be constant, that is circular fringes are formed. The fringes of equal inclination or cicular fringes are known as Haldinger's fringes.

When M_1 and M'_2 coinside, the path difference becomes zero and the field of view is completely dark as shown Fig. 10.3(b).

When M'_2 is moved away from M_1, then circular fringes are produced as shown in Figs. 10.3(a) and 10.3(c).

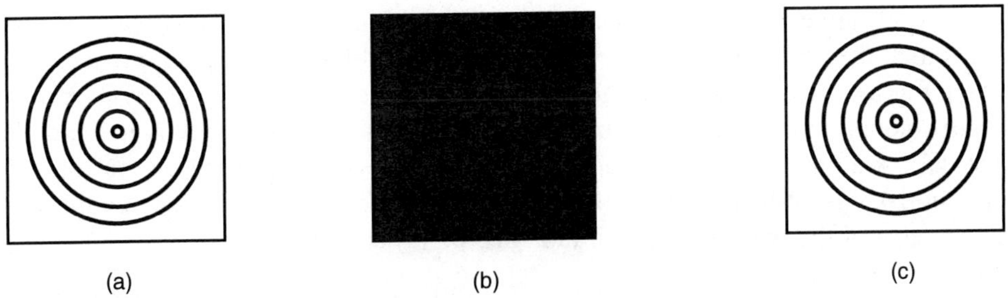

Fig. 10.3 Circular fringes from Michelson interferometer

10.3.2 Localized Fringes

When M_1 and M_2, are not exactly at right angle to each other, then M_1 and M'_2 (image of mirror M_2) will not be parallel. The air between them forms a wedge shaped film. In that case we get localized fringes, which are parallel to the line of intersection of M_1 and M'_2 (image of mirror M_2) as shown in Fig. 10.4(b). The locus of the points of equal thickness is a straight line parallel to the edge of the wedge shaped film. When the M'_2 is inclined to M_1 as shown in Figs. 10.4(a) and 10.4(c), the locus of the points of equal thickness is a straight line parallel to the line of intersection of M_1 and M'_2 (image of mirror M_2) or the edge of the wedge shaped film.

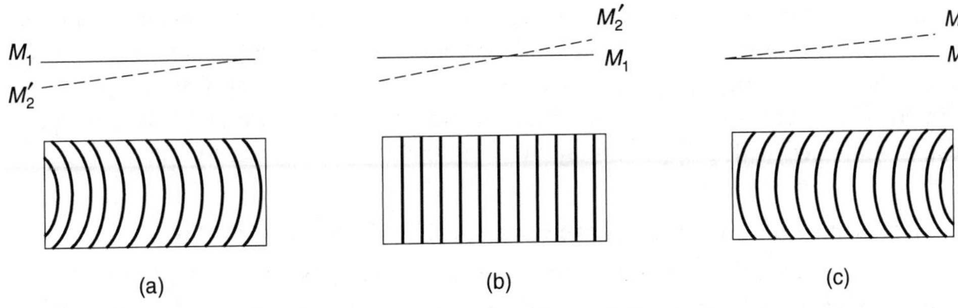

Fig. 10.4 Localised fringes

When M'_2 intersects M_1 in the middle, the fringes are parallel to the line of intersection of M_1 and M'_2 (image of mirror M_2) as shown in Fig. 104(b). When M'_2 intersects M_1 at one end, the shape of fringes is curved with convex towards the straight fringe at the point of intersection as shown in Figs. 10.4(a) and 10.4(c). Such fringes are not observed when the path difference is large.

10.4 WHITE LIGHT FRINGES

When M_1 and M_2 are exactly at right angle to each other, then M_1 and M'_2 (image of mirror M_2) are parallel, then the dark central fringe is surrounded by 8 to 10 colour circular fringes as shown in Fig. 10.5(d), when white light is used in Michelson's interferometer. Overlapping of different colours starts as order of fringe increases. When M'_2 intersects M_1 in the middle, a distinct dark straight fringe is surrounded by colour fringes on either side are formed as shown in Fig. 10.5.(b). When M'_2 is inclined to M_1, the shape of fringes is curved with convex towards the edge of the wedge as shown in Figs. 10.5(a) and 10.5(c).

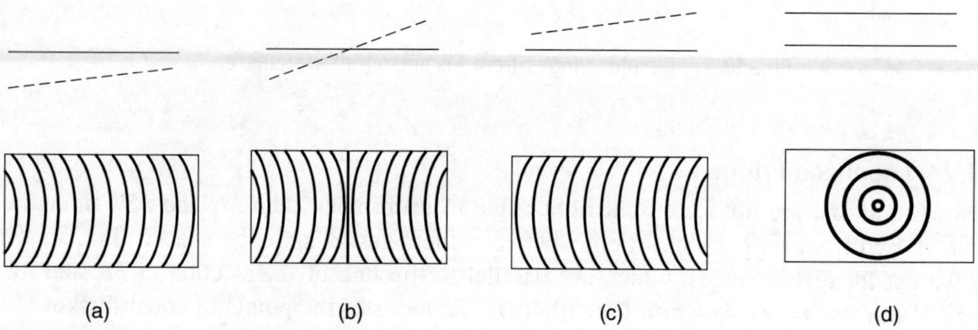

Fig. 10.5 Fringes obtained using white light

10.5 APPLICATIONS OF MICHELSON INTERFEROMETER

Michelson interferometer has several applications. It can be used to do accurate measurement refractive index of thin transparent plates and wavelength of the light source. It can be used to find the thickness of thin plates of material (transparent) by putting it in the path of one of the interfering beams. The difference between wavelengths of two neighbouring spectral lines can be obtained with good accuracy by using the Michelson interferometer.

10.5.1 Determination of Refractive Index of a Thin Transparent Plate

For measuring the refractive index of a thin transparent plate, the cross-wire of the telescope on the centre of fringes obtained using the Michelson interferometer using white light.When a thin transparent plate of thickness (t) and refractive index (μ) is introduced into the path of one of the interfering beams in the interferometer, the optical path in this beam is increased from t to μt. The beam traverses the thin film two times: once when it is incident on it and secondly when it is reflected. This introduces an extra path difference of 2 (μ – 1μ)t [= 2 (μt – t)] between the two beams. The introduction of a thin transparent plate shifts the fringes from their position. M_1 is moved towards or backwards so that cross-wire again coincides with central fringe. If x is the distance moved, then

$$2(\mu - 1)t = 2x \qquad \text{...(10.3)}$$

or

$$\mu - 1 = \frac{x}{t}$$

or

$$\mu = 1 + \frac{x}{t} \qquad \text{...(10.4)}$$

Here, x is distance moved by M_1 for bringing the central fringe again at the cross-wire after introducing the thin plate of thickness, t.

When m fringes are shifted by introducing a thin transparent of thickness (t), then

$$2(\mu - 1) = m\lambda \qquad \text{...(10.5)}$$

or

$$\mu - 1 = \frac{m\lambda}{2t}$$

or

$$\mu = 1 + \frac{m\lambda}{2t} \qquad \text{...(10.6)}$$

Here, m is number of fringes shifted after introducing the thin plate of thickness t and λ is the wavelength of source light.

10.5.2 Determination of Thickness of Thin Transparent Film

For measuring the thickness of a thin transparent plate, set the cross-wire of the telescope on the centre of fringes obtained using the Michelson interferometer using monochromatic light of wavelength, λ. When a thin transparent plate of thickness (t) and refractive index (μ) is introduced into the path of one of the interfering beams in the interferometer, the optical path in this beam is increased from t to μt. The beam traversed the thin film two times: once when it is incident on it and secondly when it is reflected. This introduces an extra path difference of 2 (μ – 1)t [= 2 (μt – t)] between the two beams. The introduction of a thin transparent plate shifts the fringes from their positions. M_1 is moved towards or backwards so that cross-wire again coincides with the central fringe. If m is the number of fringes shifted by introducing the thin plate, then from Eq. (10.5), we have

$$2(\mu - 1)t = m\lambda \qquad \text{...(10.7)}$$

The thickness of the plate is given by

$$t = \frac{m\lambda}{2(\mu - 1)} \qquad \text{...(10.8)}$$

Here, λ is the wavelength of light source, m is the number of shifted bright fringes by introducing a thin transparent film, and μ is the refractive index of plate.

10.5.3 Determination of Wavelength of Monochromatic Light

The monochromatic light, whose wavelength is to be determined, is used as a source of light, while getting the fringes using the Michelson interferometer. The position of M_1 is adjusted till bright fringes appear. The cross-wire of the telescope is adjusted on a particular bright fringe. When M_1 is moved slowly, each fringe gets displaced parallel to itself in the field of view. The number of fringes which cross the centre of the field of view gives the measure of the distance the mirror has moved in terms of wavelength.

If d is the distance between M_1 and M'_2 (image of mirror M_2), then the condition for interference for n^{th} bright fringe is

$$2d \cos \theta = (2n + 1) \frac{\lambda}{2} \quad ...(10.9)$$

Since the rays incident normal to the mirrors, i.e., the incident angle of the rays, $\theta = 0°$, so $\cos \theta = 1$, so

$$2d = (2n + 1) \frac{\lambda}{2} \quad ...(10,10)$$

When M_1 is moved through a distance x and m fringes are moved through the cross-wire, then the condition (8.10) for $(n + m)^{th}$ fringe will be

$$2(d + x) = (2n + 2m + 1) \frac{\lambda}{2} \quad ...(10.11)$$

Subtracting Eq. (10.10) from Eq. (10.11), we get

$$2x = m\lambda$$

or

$$\lambda = \frac{2x}{m} \quad ...(10.12)$$

Measuring the distance x moved by M_1 when m bright fringes are shifted, the wavelength (λ) can be determined.

This is the most accurate method for finding the wavelength of monochromatic source, because x can be measured to an accuracy of 10^{-4} mm by taking m sufficiently large.

10.5.4 Determination of Difference in Wavelengths between Two Neighbouring Spectral Lines

Let a source of light (S) emit the light consisting of two wavelengths, λ_1 and λ_2 ($\lambda_1 > \lambda_2$), which differ slightly. For example, sodium emits the light consisting of wavelengths $\lambda_1 = 5896$ Å and $\lambda_1 = 5896$ Å. Each wavelength gives rise to its own system of fringes. Due to the superposition phenomenon, the fringes become distinct and give rise to positions of maximum and minimum intensity in the field of view. If d is the separation between M_1 and M'_2 (image of mirror M_2), then the conditions for interference for bright fringes for light of wavelengths, λ_1 and λ_2, are

$$2d \cos \theta = (2n_1 + 1) \frac{\lambda_1}{2} \quad ...(10.13a)$$

$$2d \cos \theta = (2 n_2 + 1) \frac{\lambda_2}{2} \quad \text{...(10.13b)}$$

where, n_1 and n_2 are the order of bright fringes corresponding to the wavelengths λ_1 and λ_2, respectively. Since the rays incident normal to the mirrors, i.e., the incident angle of the rays $\theta = 0°$, so $\cos \theta = 1$ and hence Eqs. (10.13a) and (10.13a) can be written as

$$2d = (2 n_1 + 1) \frac{\lambda_1}{2} \quad \text{...(10.14a)}$$

$$2d = (2 n_2 + 1) \frac{\lambda_2}{2} \quad \text{...(10.14b)}$$

At the position of a maximum, a bright fringe of one system (say with λ_1) falls over a bright fringe of the other system (say with λ_2) and a fringe of intensity I_{max} is observed. Similarly, at the position of minimum a dark fringe of one system (say with λ_1) falls over the dark fringe of the other system (say with λ_2) giving rise to intensity, I_{min}. [The visibility of fringe is defined as $\frac{I_{max} - I_{min}}{I_{max} + I_{min}}$, where, I_{max} is the brightness of the bright fringe at the centre and I_{min} is the brightness of dark fringe on either of this bright fringe. The derivation this formula of visibility of fringe is given in Section 10.6. As the M_1 is moved, the brightness of bright fringes starts decreasing and that of dark fringes starts increasing. A position is reached when bright fringes due to λ_1 fall on the dark fringes due to λ_2 or vice versa. In this position the distinction between bright and dark fringes disappears and uniform illumination is observed in the telescope. On further moving M_1, a position is reached when bright fringes of both the wavelengths again fall over each other and clear distinct bright and dark fringes start appearing on the screen.

Let a distance x be moved by M_1 for two consecutive positions for clear visibility of bright and dark fringes, then there will be new path difference, $2 (d + x)$. During this movement, let $(n_1 + m)^{th}$ bright fringe due to λ_1 falls on the $(n_2 + m + 1)\, t^h$ bright fringe due to λ_2. Now the conditions for bright fringes given by Eqs. (10.14b) and (8.14b) can be written as

$$2 (d + x) = (2 n_1 + 2 m + 1) \frac{\lambda_1}{2} \quad \text{...(10.15a)}$$

$$2 (d + x) = (2 n_2 + 2 m + 2 + 1) \frac{\lambda_2}{2} \quad \text{...(10.15b)}$$

Subtracting Eq. (10.14a) and Eq. (10.14b) from Eq. (10.15a) and Eq. (10.15b), respectively, we get

$$2 x = m\lambda_1 \quad \text{and} \quad 2 x = (m + 1) \lambda_2 \quad \text{...(10.16)}$$

Equating both these equations, we get

$$m (\lambda_1 - \lambda_2) = \lambda_2$$

or
$$m = \frac{\lambda_2}{(\lambda_1 - \lambda_2)} \qquad ...(10.17)$$

Substituting for m in Eq. (10.16), we get

$$2\,x = \frac{\lambda_1\,\lambda_2}{(\lambda_1 - \lambda_2)}$$

or
$$\lambda_1 - \lambda_2 = \frac{\lambda_1\,\lambda_2}{2x} = \frac{\lambda^2}{2x}, \qquad ...(10.18)$$

where, λ is the average root mean square wavelength, $\lambda = \sqrt{\lambda_1\,\lambda_2}$ or the wavelength as measured by other experiments without resolving the pair of lines.

The position of one bright fringe is recorded when the centre of maximum is on the cross-wires. On moving M_1, visibility decreases, becomes almost zero and again improves till the centre of next maximum is brought on the cross-wires. If x is the distance through which M_I is moved for getting two successive positions of the maximum intensity of bright fringe, then by putting the value of x and λ_{av} in Eq. (10.5), one can calculate the difference between the wavelengths, $\lambda_1 - \lambda_2$.

10.6 VISIBILITY OF FRINGE

In linear optical interferometer like Michelson's interferometer, interference waves give rise to maxima and minima as intensity oscillation over time or space, called fringes. Under these circumstances, the interferometric visibility is also known as the "Michelson visibility" or the "fringe visibility." For this type of interference, the sum of the intensities (powers) of the two interfering waves equals the average intensity over a given time or space domain. The visibility is written as

$$v = \frac{A}{\overline{I}} \qquad ...(10.19)$$

in terms of the amplitude envelope of the oscillating intensity (A) and the average intensity (I) can be written as

$$A = \frac{I_{max} - I_{min}}{2} \qquad ...(10.20)$$

$$\overline{I} = \frac{I_{max} + I_{min}}{2} \qquad ...(10.21)$$

Now the visibility of fringe is written as

$$v = \frac{I_{max} - I_{min}}{I_{max} + I_{min}} \qquad ...(10.22)$$

The intensity (I) after overlapping of two oscillations is written as [Eq. 9.5 of Chapter 9]

$$I = I_1 + I_2 + 2\sqrt{I_1 I_2}\cos\delta \qquad ...(10.23)$$

Here, I_1 and I_2 are intensities of two waves and δ is the phase difference between two waves.

In case of coherent interference at a given point, δ does not vary with time, so I will be maximum (I_{max})

(a) When cos δ is maximum, i.e., cos δ = 1, so intensity of two fringes is maximum

$$I_{max} = I_1 + I_2 + 2\sqrt{I_1 I_2} = \left(\sqrt{I_1} + \sqrt{I_2}\right)^2 \qquad ...(10.24)$$

I will be minimum (I_{min})

(b) When cos δ is minimum (I_{min}), i.e., cos δ = – 1, so intensity of dark fringes on either side of bright fringe is minimum

$$I_{min} = I_1 + I_2 - 2\sqrt{I_1 I_2} = \left(\sqrt{I_1} - \sqrt{I_2}\right)^2 \qquad ...(10.25)$$

If the two optical fields are ideally monochromatic (consist of only single wavelength) point sources of the same polarization, then the predicted visibility will be

$$v = \frac{2\sqrt{I_1 I_2}}{I_1 + I_2} \qquad ...(10.26)$$

where, I_1 and I_2 indicate the intensity of the respective wave. Any dissimilarity between the optical fields will decrease the visibility from the ideal. In this sense, the visibility is a measure of the coherence between two optical fields. This definition of interference directly applies to the interference of water waves and electric signals also.

10.7 FABRY–PEROT INTERFEROMETER

Objectives:

I. To observe concentric circular fringes

II. Measurement of the wavelength of a diode Laser

Introduction

The Fabry–Perot interferometer uses the phenomenon of multiple beam interference when light falls on two reflective parallel surfaces bounded by a gap. Each time the light falls on the surfaces, a portion of it is transmitted out, and remaining part is reflected back. The net effect is to break a single beam into multiple beams which interfere with each other. If the path difference of the reflected beam (due to multiple reflections) is an integral multiple of the light wavelength, then the reflected beams will interfere constructively. More is the number of reflection inside the gap, sharper is the interference maximum.

Principle of Working

The basic principle of working of the Fabry – Perot interferometer is pictorially shown in Fig. 10.6.

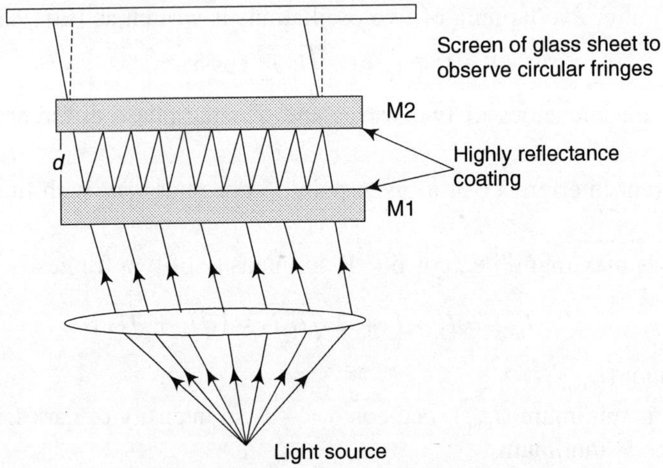

Fig. 10.6 Febry – Perot interferometer

Febry–Perot interferometer consists of two mirrors M1 and M2 with partially reflecting surfaces. They can be either curved or flat. The mirrors are aligned parallel to one another at distance, *d*. The distance between two semi–silvered glass can be varied. One plate remains stationary, while the other is mounted using an accurate screw. When a monochromatic light incidence on reflecting surface of mirror M1 with angle φ from the normal, multiple reflections and refractions take place within the gap. Part of the light is transmitted each time the light reaches the second reflecting surface of mirror, M2. The ray diagram of this phenomenon is shown in Fig. 10.7. All such transmitted light rays interfere with each other and produce a maxima or minima depending upon path difference between them. Let μ be the refractive index of the medium in gap (normally it is air), then the path difference between two neighbouring rays is:

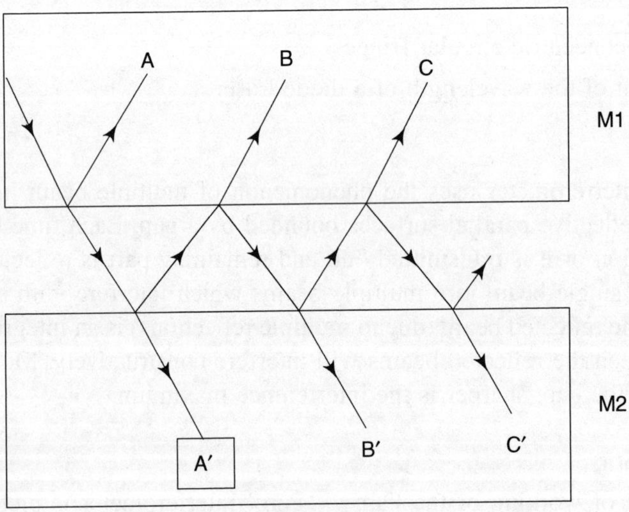

Fig. 10.7

$$\Delta = 2\mu d \cos \vartheta \qquad ...(10.27)$$

Then the phase difference is given by $\delta = \left(\frac{2\pi}{\lambda}\right)\Delta$ ...(10.28)

The path difference given by eqn. (10.27) can be obtained as follows:

For finding the path difference the ray diagram is shown by Fig. 10.8. The light falls at point A on the first partially reflecting surface of mirror M1 at angle φ from the normal. A part of it is reflected in the direction AD and another part is refracted. When the light travels from denser medium (mirror M1) to rarer medium (normally it is air in the gap) it bends away normal. The refracted ray AB makes an angle ϑ from the normal BN. Ray AB is partially reflected at point B in the direction BC.

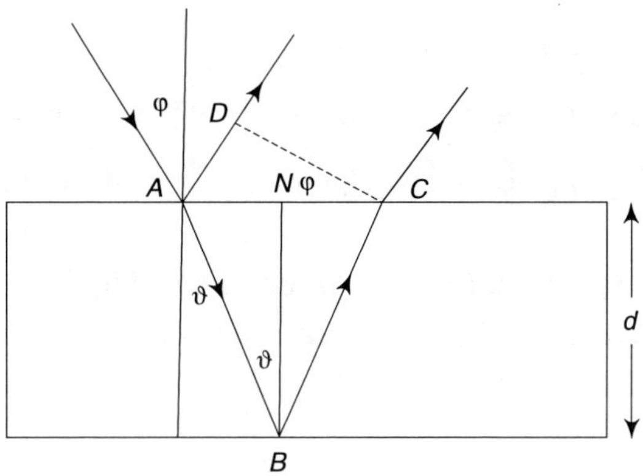

Fig. 10.8

The path difference Δ is written as

$$\Delta = \mu\,(AB + BC) - AD = 2\,\mu\, AB - AD \quad \text{as} \quad AB = BC$$

From the $\Delta\, ABN$ $\quad BN = AB \cos \vartheta \quad$ or $\quad d = AB \cos \vartheta$

From the $\Delta\, ANB$ $\quad AN = \frac{AC}{2} = d \tan \vartheta \quad$ or $\quad AC = 2d \tan \vartheta$

From the $\Delta\, ADC$ $\quad AD = AC \sin \varphi \quad$ or $\quad AD = 2d \tan \vartheta \sin \varphi$

or $\quad AD = 2d \tan \vartheta\,(\mu \sin \vartheta) = 2\mu d \tan \vartheta \sin \vartheta \quad$ as $\quad \mu = \frac{\sin \varphi}{\sin \vartheta}$

Putting the values of AB and AD, the path difference can be written as

$$\Delta = 2\mu \frac{d}{\cos \vartheta} - 2\mu d \tan \vartheta \sin \vartheta$$

or
$$\Delta = 2\mu d \left[\frac{1}{\cos \vartheta} - \tan \vartheta \sin \vartheta\right]$$

or
$$\Delta = 2\mu d \left[\frac{1 - \sin^2 \vartheta}{\cos \vartheta}\right] = 2\mu d \cos \vartheta$$

The resultant transmitted light intensity I_T is:

$$I_T = I_0 \frac{1}{1 + \frac{4R}{(1-R^2)} \sin^2 \frac{\delta}{2}} \qquad \text{...(10.29)}$$

where, I_0 is the incident intensity, R is the reflectivity of the mirrors. It can be noticed that I_T varies with δ.

The condition for maximum intensity is

$$\Delta = m\lambda \quad \text{or} \quad 2\mu d \cos \vartheta = m\lambda, \quad \text{where } m = 0, 1, 2, 3, \text{----} \qquad \text{...(10.30)}$$

The condition for minimum intensity is

$$\Delta = (2m+1)\frac{\lambda}{2} \quad \text{or} \quad 2\mu d \cos \vartheta = (2m+1)\frac{\lambda}{2}, \quad \text{where } m = 0, 1, 2, 3, \text{----} \qquad \text{...(10.31)}$$

Complete interference pattern appears as a set of concentric rings. The sharpness of the rings depends on a parameter called coefficient of finesse, F, defined as $F = \frac{4R}{(1-R^2)^2}$

10.7.1 Determination of Wavelength λ of Diode Laser Light

Let the initial gap between the mirrors be d_1. The condition for the maxima at the centre for $\vartheta = 0$ is

$$2\mu d_1 = m_1\lambda$$

Let us displace the movable mirror. Now the gap between the mirrors is d_2. The condition for the maxima at the centre for $\vartheta = 0$ will be

$$2\mu d_2 = m_2\lambda$$

Then the difference $2\mu d_2 - 2\mu d_1$ is

$$2\mu d_2 - 2\mu d_1 = m_2\lambda - m_1\lambda$$

or

$$2\mu (d_2 - d_1) = (m_2 - m_1)$$

or

$$\lambda = \frac{2\mu(d_2 - d_1)}{(m_2 - m_1)} \qquad \text{...(10.32)}$$

The difference $(m_2 - m_1)$ is the number of fringes (maxima) appearing or disappearing at the centre. When we displace the movable mirror from gap d_1 to d_2, the fringes appear or disappear at the center. By counting the number of fringes (maxima) that appear or disappear at the centre $\vartheta = 0$, one can find the wavelength of the source (say diode laser).

Circular Fringes

Allow to fall the diode laser so that the beam passes from the centre of two mirrors. Place a lens (f = 100 mm) in front of the laser to expand the beam to create a broad source. Adjust the movable mirror to let the multiple reflected beams coincide on the screen. It means the mirrors are now nearly parallel. With all components perfectly set, the observer can find a series of very intense circular interference rings on the glass screen.

Formulas at a Glance

S.No.	Subject	Formula	Explanation
		Michelson Interferometer	
1.	Refractive index of transparent thin plate:	$\mu = 1 + \frac{x}{t}$, see eqn. (10.4) t is the thickness of the transparent thin plat.	x is the distance moved by the mirror.
2.	The thickness of the plate:	$t = \frac{m\lambda}{2(\mu - 1)}$, see eqn. (10.8) μ is the refractive index of plate.	λ is the wavelength of light source, m is number of bright fringes shifted by introducing a thin transparent film.
3.	Wavelength of the light source using Michelson interferometer:	$\lambda = \frac{2x}{m}$, see eqn. (10.12)	x is the distance moved by the mirror M_1 and m is number of bright fringes moved through the cross wire.
4.	Resolution of spectral lines:	$\Delta\lambda = \lambda_1 - \lambda_2 = \frac{\lambda_1 \lambda_2}{2x}$, see eqn. (10.18) x is distance moved by the mirror M_1 for two consecutive positions for clear visibility of bright and dark fringes.	$\Delta\lambda = \frac{\lambda^2}{2x}$, where $\lambda = \sqrt{\lambda_1 \lambda_2}$ is mean square root value of the source light.
		Febry-Perot Interferometer	
5.	Condition for maxima:	$2\mu d \cos \vartheta = m\lambda$ Where $m = 0, 1, 2, 3 \ldots.$	λ is the wavelength of source, d is distance between mirror M_1 and M_2, μ is refractive index of the medium in gap, ϑ is the refractive angle from mirror M_1.
6.	Condition for maxima:	$2\mu d \cos \vartheta = (2m + 1)\lambda/2$ Where $m = 0, 1, 2, 3 \ldots.$	λ is the wavelength of source, d is distance between mirror M_1 and M_2, μ is refractive index of the medium in gap, ϑ is the refractive angle from mirror M_1.
7.	Formula for wavelength of the source:	$\lambda = \frac{2\mu(d_2 - d_1)}{(m_2 - m_1)}$	μ is refractive index of the medium in gap, d_1 and d_2 are gaps between the mirrors $(m_2 - m_1)$ number of maxima appeared or disappeared

8.	Formula for intensity of transmitted light source	$I_T = I_0 \dfrac{1}{1+\dfrac{4R}{(1-R^2)^2}\sin^2\dfrac{\delta}{2}}$	where, I_0 is the incident intensity, R is the reflectivity of the mirrors. It can be noticed that I_T varies with phase δ.

SOLVED EXAMPLES

10.1 A thin film of a plastic of refractive index 1.55 for a wavelength of 5890×10^{-10} m is inserted in one arm of Michelson's interferometer. Find the thickness of the film if a shift of 15 fringes is observed.

Solution: Formula used

$$2(\mu - 1)\, t = m\lambda$$

Given: $\mu = 1.55$, $\lambda = 5890 \times 10^{-10}$ m, $m = 15$

The thickness of the transparent film is

$$t = \frac{n\lambda}{2(\mu-1)} = \frac{15 \times 5890 \times 10^{-10}}{2 \times (1.55 - 1)}$$

$$= \frac{15 \times 5890 \times 10^{-10}}{2 \times 0.55} = 8.03 \times 10^{-6} \text{ m}$$

Ans. The thickness of the film is 8.03×10^{-6} m.

10.2 Calculate the distance between successive positions of the movable mirror of Michelson interferometer for getting successive bright fringes in case of a sodium source having wavelengths 5896 Å and 5890 Å.

Solution:

The formula for distance moved by mirror for getting successive bright fringe is

$$x = \frac{\lambda_1 \lambda_2}{2(\lambda_1 - \lambda_2)}$$

$$\lambda_1 = 5896 \times 10^{-10} \text{ m}, \quad \lambda_3 = 5890 \times 10^{-10} \text{ m}$$

$$x = \frac{5896 \times 10^{-10} \times 5890 \times 10^{10}}{2(5896 - 5890) \times 10^{-10}}$$

$$= \frac{5896 \times 5890 \times 10^{-10}}{2 \times 6} = 2.894 \times 10^{-4} \text{ m}$$

Ans. The distance moved by the mirror for getting successive bright fringes is 2.894×10^{-4} m.

10.3 A transparent film of glass of refractive index 1.5 is introduced normally in the path of one of the interfering beams of Michelson's interferometer which is illuminated with light of wavelength 5000 Å. This causes 600 dark fringes to sweep across the field. Find the thickness of the film.

Solution: Formula used

$$2(\mu - 1)t = m\lambda$$

Given: $\mu = 1.55$, $\lambda = 5800 \times 10^{-8}$ cm, $m = 500$

The thickness of the transparent film is

$$t = \frac{m\lambda}{2(\mu - 1)} = \frac{500 \times 5800 \times 10^{-8}}{2 \times (1.55 - 1)}$$

$$= \frac{500 \times 5800 \times 10^{-8}}{2 \times 0.55} = 2.64 \times 10^{-2} \text{ cm}$$

Ans. The thickness of the film is 2.64×10^{-2} cm.

10.4 When one mirror of a Michelson's interferometer is moved by 0.254 mm, 800 fringes cross the field wire. Find the wavelength of the light used.

Solution:

$$2x = m\lambda$$

Given: $x = 0.0204$ cm, $m = 750$

Wavelength of light is

$$\lambda = \frac{2x}{m} = \frac{2 \times 0.0254}{800} = 6354 \text{ Å}$$

Ans. Wavelength of the source light is 6354 Å.

10.5 In an experiment for determining the refractive index of a gas using Michelson's interferometer a shift of 250 fringes is observed, when all the gas is removed from the tube. If the wavelength of light used is 5430 Å and the length of the tube is 25 cm, calculate the refractive index of the gas.

Solution: Formula used

$$2(\mu - 1)t = m\lambda$$

Given: $x = 25$ cm, $\lambda = 5430 \times 10^{-8}$ cm, $m = 250$

The refractive index of gas is

$$\mu = 1 + \frac{m\lambda}{2t} = 1 + \frac{250 \times 54.30 \times 10^{-8}}{2 \times 25} = 1.0002715$$

Ans. The refractive index of the gas is 1.0002715.

10.6 In an experiment with Michelson's interferometer the distance travelled by the mirror for two successive positions of maxima and distinctiveness was 0.295 mm. If the mean wavelength of the two components of sodium lines is 5893 Å, calculate the difference between then.

Solution: The formula between two wavelengths of sodium D lines is

$$\lambda_1 - \lambda_2 = \frac{\lambda^2}{2x}$$

$$\lambda = 5893 \times 10^{-8} \text{ cm}, \quad x = 0.0295 \text{ cm}$$

$$\lambda_1 - \lambda_2 = \frac{5893 \times 10^{-8} \times 5893 \times 10^{-8}}{2 \times 0.0295}$$

$$= \frac{5.893 \times 5.893 \times 10^{-8}}{2 \times 2.95} = 6.1 \text{ Å}$$

Ans. The difference between two d lines of sodium is 6.1 Å.

QUESTIONS FOR EXAMINATION

10.1 Describe the construction and working of Michelson's interferometer and explain the formation of circular fringes and straight lines.

10.2 Describe the construction and working of Michelson's interferometer. How it is used to find the thickness of a thin transparent plate?

10.3 How will you measure the difference between the wavelength of two d lines of sodium light using Michelson's interferometer.

10.4 Explain the setup and working of Michelson's interferometer. Discuss how it can be used to measure wavelength of light.

10.5 Explain the formation of interference fringes in a Michelson's interferometer. How it is used to find the refractive index of a transparent thin plate?

10.6 Discuss the fringes obtained using white light using Michelson's interferometer.

10.7 What are localized fringes?

OBJECTIVE QUESTIONS

10.1 Which is the correct formula to find the thickness (t) of a thin film using Michelson's interferometer?

(a) $t = \dfrac{x}{4(\mu - 1)}$ (b) $t = \dfrac{x}{(\mu - 1)}$ (c) $t = \dfrac{x}{3(\mu - 1)}$ (d) $t = \dfrac{2x}{(\mu - 1)}$

Here μ is the refractive index of thin plate and x is the distance moved by mirror (M_1) to bring the central fringe at the cross-wire.

10.2 Which is the correct formula to find the thickness (t) of a thin film using Michelson's interferometer?

(a) $t = \dfrac{m\lambda}{(\mu - 1)}$ (b) $t = \dfrac{2m}{(\mu - 1)}$ (c) $t = \dfrac{m\lambda}{2(\mu - 1)}$ (c) $t = \dfrac{m\lambda}{3(\mu - 1)}$

Here, μ is the refractive index of a thin plate and λ is the wavelength of the source light, and m is the number of fringes shifted by introducing the thin plate.

10.3 Which is the correct formula to find the refractive index (μ) of a thin film using Michelson's interferometer?

(a) $\mu = 1 + \dfrac{2x}{t}$ (b) $\mu = 1 + \dfrac{x}{2t}$ (c) $\mu = \dfrac{x}{t}$ (d) $\mu = 1 + \dfrac{x}{t}$

Here, μ is refractive index of thin plate and x is the distance moved by mirror (M_1) to bring the central fringe at the cross-wire.

10.4 Which is the correct formula to find the refractive index (μ) of a thin film using Michelson's interferometer?

(a) $\mu = 1 + \frac{2m\lambda}{t}$ (b) $\mu = 1 + \frac{m\lambda}{2t}$ (c) $\mu = \frac{m\lambda}{t}$ (d) $\mu = 1 + \frac{m\lambda}{t}$

Here, μ is refractive index of thin plate and λ is the wavelength of the source light, and m is the number of fringes shifted by introducing the thin plate.

10.5 Which is the correct formula to find wavelength λ of a monochromatic source light using Michelson's interferometer?

(a) $\lambda = \frac{2x}{m}$ (b) $\lambda = \frac{x}{m}$ (c) $\lambda = \frac{x}{2m}$ (c) $\lambda = 1 + \frac{2x}{m}$

Here, x is the distance moved by mirror (M_1) when m number of bright fringes are shifted.

10.6 What type of fringes are formed by using the monochromatic light when mirrors M_1 and M'_2 (image of mirror M_2) are parallel to each other.

(a) Wedge shaped and convex at the line perpendicular to intersection point of mirrors, M_1 and M'_2.
(b) Circular fringes
(c) Straight line fringes
(d) Colour circular fringes

10.7 What type of fringes are formed by using the monochromatic light when mirrors, M_1 and M'_2 (image of mirror M_2) intersect each other in the middle?

(a) Wedge shaped and convex at the line perpendicular to intersection point of mirrors, $M_1$and M'_2.
(b) Circular fringes
(c) Localized straight lines
(d) Localized colour straight lines

10.8 What type of fringes are formed by using the monochromatic light when mirrors, M_1 and M'_2 (image of mirror M_2), intersect at the right end?

(a) Wedge shaped and convex at the line perpendicular to intersection point of mirrors M_1 and M'_2 at right end.
(b) Circular fringes
(c) Straight line fringes
(d) Wedge shaped and convex at the line perpendicular to intersection point of mirrors, M_1 and M'_2 at the left end.

10.9 What type of fringes are formed by using the monochromatic light when mirrors, M_1 and M'_2 (image of mirror M_2), intersect at the left end?

(a) Wedge shaped and convex at the line perpendicular to intersection point of mirrors M_1 and M'_2 at the right end.

(b) Circular fringes

(c) Straight line fringes

(d) Wedge shaped and convex at the line perpendicular to intersection point of mirrors M_1 and M'_2 at the left end

Answers

10.1 (b)	10.2 (c)	10.3 (d)	10.4 (b)	10.5 (a)	10.6 (b)	10.7 (c)	10.8 (a)
10.9 (d)							

CHAPTER 11

Diffraction Theory

11.1 INTRODUCTION

In this chapter, the review of the articles is presented. Kirchhoff integral theorem and Fresnel-Kirchhoff integral formula are two most important articles of diffraction of light theory. The Fresnel-Kirchhoff diffraction integral formula is also solved using Fresnel approximations for incident spherical wave $\left(\frac{e^{ikr}}{r}\right)$, and diffracted wave $\left(\frac{e^{ikr}}{r}\right)$.

11.1.1 Kirchhoff's Integral Theorem

Gauss theorem states that the total flux enclosed by the closed surface is equal to the charge enclosed by it or Gauss theorem states that the surface integral of electrostatic field $\left(\vec{E_0}\right)$ over a closed surface enclosing volume (V) is equal to the total charge enclosed within that surface.

Similar to Gauss theorem, Helmholtz-Kirchhoff integral theorem is defined as the flux of light rays enclosed by the closed surface is equal to the incident spherical wavefront, $u(x) = e^{ikx}/x$, enclosed within the surface. Mathematically, it is written as [see Eq.11.30)]:

$$\frac{1}{4\pi} \int_{\partial V} (-\hat{n}) \cdot \vec{F}\, dS = u(x) = \frac{e^{ikx}}{x} \qquad ...(11.1)$$

where, $\hat{n}$ is inward normal to surface (∂V) that bounds volume, V. The vector field due to spherical wave emerging from point light source,

$\vec{F} = (u \nabla G - G \nabla u)$. The green function is defined as

$$G\left(\vec{r}, \vec{X}\right) = \frac{e^{ikx|\vec{r}-\vec{x}|}}{|\vec{r}-\vec{x}|}$$

at position $\vec{r}$ due to spherical wave at position $\vec{x}$.

11.1.2 Fresnel-Kirchhoff's Integral Formula

An approximate solution of Helmholtz equation

$$(\nabla^2 + k^2)\, u(\vec{r}) = 0 \qquad \text{...(11.2)}$$

$$u(\vec{x}) = \frac{i\,A}{2\,\lambda} \int_{\partial V} \frac{e^{ikx\,(r+S)}}{r\,S} (\hat{n}\cdot\hat{s} + \hat{n}\cdot\hat{r})\, dS \ \text{[see Eq. (11.38)]} \qquad \text{...(11.3)}$$

is called **Fresnel- Kirchhoff Diffraction formula.** Here, $\hat{n}$ is inward normal to surface (∂V) that bounds volume, $V \cdot k = \dfrac{2\,\pi}{\lambda}$ is wave vector of incident spherical wave $\left(\dfrac{e^{ikr}}{r}\right)$ at point ⓡ and diffracted spherical wave $\left(\dfrac{e^{ikS}}{S}\right)$ at point, S.

11.1.3 Fresnel-Kirchhoff Diffraction Formula, if the Inclination Factor is Nearly Constant over the Aperture

If the inclination factor is nearly constant over the aperture, the Fresnel-Kirchhoff diffraction formula can be written as

$$u(\vec{x}) \approx \frac{ikAI}{4\,\pi} \int_{\partial V} \frac{e^{ik\,(r+S)}}{rS}\, dS \qquad \text{...(11.4)}$$

The Fresnel-Kirchhoff diffraction formulais also solved using Fresnel approximations for incident spherical wave $\left(\dfrac{e^{ikr}}{r}\right)$ and diffracted wave $\left(\dfrac{e^{ikr}}{r}\right)$.

11.1.4 Fresnel-Kirchhoff Formulas for Transmitted Wave

The Fresnel-Kirchhoff formulas for transmitted wave from aperture is written as

$$u(\vec{x}) = \frac{ikI}{4\,\pi} \int_{\partial V} \tau(\vec{r}) \frac{e^{ik\,(r+S)}}{rS}\, dS\,, \qquad \text{...(11.5)}$$

Transmission function of a thin lens of thickness (t) and focal length (f) is defined as

$$\tau(\rho) \approx |\tau(\rho)|\; e^{iknt}\, e^{-ik\rho^2/(2f)} \qquad \text{...(11.6)}$$

Here, ρ is half width of the beam.

11.1.5 Fraunhofer Integral Formula for Rectangular Aperture

For the rectangular aperture having dimensions: $2a$ (along x) × $2b$ (along y), Fraunhofer integral formula in two dimensions is written as

$$u(\vec{x}) = \frac{ikAI}{4\,\pi} \frac{e^{ik\,(r'+S')}}{r'\,S'}\, e^{i\vec{k}_2\cdot\vec{r}_1} \int_{-a}^{a}\int_{-b}^{b} e^{i(q_x\,\xi + q_y\,\eta)}\, d\xi\, d\eta$$

$$= \text{(constant) } [2 \sin c\, (q_x\, a)]\ [2 \sin c\, (q_y\, b)] \qquad \text{...(11.7)}$$

where, $\vec{q} = \vec{k}_1 - \vec{k}_2$, $\vec{k}_1 = k\,\hat{r}'$, $\vec{k}_2 = k\hat{S}' \cdot \vec{q}$ is not necessarily in the plane of the aperture, but we can assume that in this case $\vec{q}$ is in the plane of aperture. The intensity of maxima in case of rectangular aperture is given by

$$S(\vec{q}) = S(0) \sin c^2\, (q_x\, a) \sin c^2\, (q_x\, a) \qquad \text{...(11.8)}$$

11.1.6 Fraunhofer Integral Formula for Circular Aperture

Fraunhofer diffraction integral formula for a circular aperture of radius a is

$$u(\vec{x}) = \frac{ikAI}{4\pi} \frac{e^{ik(r'+S')}}{r'\,S'} e^{i\vec{k}_2 \cdot \vec{r}_1} \int_0^{2\pi} \int_0^{a} e^{iq\rho\cos\phi}\, \rho d\rho d\phi$$

$$= \text{(constant) } (\pi a^2) \left[\frac{2\, J_1\,(qa)}{qa}\right]^2 \qquad \text{...(11.9)}$$

Here, J_1 is Bessel function of the first order. The intensity of maxima is given by

$$S(\vec{q}) = S(0) \left[\frac{2\, J_1\,(qa)}{qa}\right]^2 \qquad \text{...(11.10)}$$

(the airy pattern)

11.1.7 Formula for getting Diffraction Pattern by a Thin Lens, when Incident Source of Light is Coherent and Extended

The fields due to a distribution (ψ) of point sources is written as

$$u(\vec{r}_1) \approx -\frac{i}{\lambda} \iint_{SA} \psi(\vec{r}_0) \frac{e^{ikr}}{r} \alpha(\vec{\rho})\, e^{-ik\frac{\rho^2}{2f}} \frac{e^{ikS}}{S} d^2\rho d^2 r_0 \qquad \text{...(11.11)}$$

The solution of this formula in Fresnel approximation is given in Solved Examples 11.7 and 11.8.

11.2 LIGHT WAVE PROPAGATION EQUATION

Maxwell's equations in dielectric and magnetic medium are

$$\nabla \cdot \vec{D} = \rho \qquad \text{...(11.12a)}$$

$$\nabla \cdot \vec{B} = 0 \qquad \text{...(11.12b)}$$

$$\nabla \times \vec{E} = -\frac{\partial \vec{B}}{\partial t} \qquad \text{...(11.12c)}$$

$$\nabla \times \vec{B} = \mu\left(\vec{J} + \frac{\partial \vec{D}}{\partial t}\right) \qquad \text{...(11.12d)}$$

Putting

$$\vec{D} = \in \vec{E}$$

$$\vec{B} = \mu \vec{H}$$

$$\vec{J} = \sigma \vec{E}$$

Maxwell's equations become

$$\in \nabla \cdot \vec{E} = \rho \quad \text{...(11.13a)}$$

$$\mu \Delta \cdot \vec{H} = 0 \quad \text{...(11.13b)}$$

$$\nabla \times \vec{E} = -\mu \frac{\partial \vec{H}}{\partial t} \quad \text{...(11.13c)}$$

$$\nabla \times \vec{H} = \sigma \vec{E} + \in \frac{\partial \vec{E}}{\partial t} \quad \text{...(11.13d)}$$

In free space, $\rho = 0$, $\vec{J} = 0$, $\mu = \mu_0$, $\in = \in_0$ and $\sigma = 0$, so in free space $\vec{D}, \vec{B}$ and $\vec{J}$ become

$$\vec{D} = \in_0 \vec{E}$$

$$\vec{B} = \mu_0 \vec{H}$$

$$\vec{J} = 0$$

Here $\in_0 = 8.854 \times 10^{-12}\ C^2/\text{Nm}^2$ and $\mu_0 = 4\pi \times 10^{-7}$ weber/amp-m.

Putting the values of $\vec{D}, \vec{B}$ and $\vec{J}$ from Eqs. (11.14a – 11.14c), Maxwell's Eqs. (11.13a – 11.13d) in free space become

$$\in \nabla \cdot \vec{E} = 0 \quad \text{...(11.14a)}$$

$$\mu_0 \nabla \cdot \vec{H} = 0 \quad \text{...(11.14b)}$$

$$\nabla \times \vec{E} = -\mu \frac{\partial \vec{H}}{\partial t} \quad \text{...(11.14c)}$$

$$\nabla \times \vec{H} = \in \frac{\partial \vec{E}}{\partial t} \quad \text{...(11.14d)}$$

Taking curl of (11.14c), we get

$$\nabla \times (\nabla \times \vec{E}) = \nabla \times \left(-\mu \frac{\partial \vec{H}}{\partial t} \right) = -\mu \frac{\partial}{\partial t} (\nabla \times \vec{H})$$

Putting the value of $\nabla \times \vec{H}$ from Eq. (11.13d), we get

$$\nabla \times (\nabla \times \vec{E}) = -\mu \frac{\partial}{\partial t} \left(\in \frac{\partial \vec{E}}{\partial t} \right)$$

or $$\nabla \times (\nabla \times \vec{E}) = -\mu \in \frac{\partial^2 \vec{E}}{\partial t^2}$$

or $$\text{grad}\,(\text{div}\,\vec{E}) - \nabla^2 \vec{E} = -\mu \in \frac{\partial^2 \vec{E}}{\partial t^2}$$

But div $\vec{E} = \nabla \cdot \vec{E} = 0$, so we get

$$\nabla^2 \vec{E} = \mu \in \frac{\partial^2 \vec{E}}{\partial t^2} \qquad ...(11.15)$$

or $$\left(\nabla^2 - \frac{1}{v^2}\frac{\partial^2}{\partial t^2}\right)\vec{E}\,(\vec{r}, t) = 0 \qquad ...(11.16)$$

This is the wave equation for propagation of electromagnetic wave in free space. Here, velocity of light, $v = 1/\sqrt{\mu \in}$. Scalar wave equation can be written as

$$\left(\nabla^2 - \frac{1}{v^2}\frac{\partial^2}{\partial t^2}\right)U\,(\vec{r}, t) = 0, \qquad ...(11.17)$$

where, $U\,(\vec{r}, t) = u\,(\vec{r})\,e^{-i\omega t}$ and ω is the angular frequency of the light wave.

There two methods to solve the light wave propagation Eq. (11.17)

(1) Partial differential equation (PDE) approach
  (i) Analytical solution
    – Can be obtained by solving PDE by separation of variables.
    – Boundary conditions are applied to restrict set of possible solutions.
    – It is useful only for certain highly symmetrical geometries.
  (ii) Numerical solution
    – Can be obtained by discretizing by using finite differences or finite elements.
    – Obtained a system of linear equations by incorporating the boundary conditions
(2) Integral equation approach
  (i) Analytical solution
    – By transforming the PDE into an integral equation.
    – Incorporating analytical solution.
  (ii) Numerical solution

11.3 HELMHOLTZ EQUATION

Let the solution of scalar wave Eq. (11.17) be

$$U\,(\vec{r}, t) = u\,(\vec{r})\,e^{-i\omega t}, \qquad ...(11.18)$$

Here, we have assumed that it is a perfectly monochromatic wave.

Putting $U(\vec{r}, t)$ in Eq. (11.17), we get

$$\left(\nabla^2 + \frac{\omega^2}{v^2}\right) u(\vec{r})\, e^{-i\omega t} = 0, \quad \text{...(11.19)}$$

Taking $k^2 = \frac{\omega^2}{v^2}$, Eq. (11.19) can be written as

$$(\nabla^2 + k^2)\, u(\vec{r}) = 0 \quad \text{...(11.20)}$$

Equation (11.20) is called Helmholtz equation. The boundary conditions for the Helmholtz equation are:

(1) Dirichlet (*u* given on the boundary)

(2) Neumann (*du*/*dn* given on the boundary)

11.3.1 Bessel Beams

The solution of radial part of Helmholtz equation is a Bessel function of the first kind.

A Bessel beam propagates without spreading.

An exact solution of the Helmholtz equation $(\nabla^2 + k^2)\; u(\vec{r}) = 0$ is

$$U(\rho, \phi, z) = J_m(\alpha \rho)\, e^{im\phi}\, e^{i\beta z} \quad \text{...(11.21)}$$

– The condition $k^2 = \alpha^2 + \beta^2$ must be satisfied.

– Most useful for $m = 0$, because J_0 gives the beam a central bright spot.

– Because the radial part is independent of *z*, diffraction does not lead to spreading.

Fig. 11.1 Intensity pattern of a Bessel beam

11.3.2 Dirichlet Boundary Conditions

Approximate scalar boundary conditions

(i) The optical field *u* in an aperture is the same as if the aperture were not present, i.e., $u = f$.

(ii) $u = 0$ at all points on the screen

(iii) $u = 0$ at large distances in image space are obtained from St.Venant's hypothesis and shown in Fig. 11.2. Obviously, these boundary conditions are not entirely correct, but they are useful, because:

(i) They ignore currents in the edges of the aperture.

(ii) Dirichlet or Kirchhoff's boundary conditions apply on $u\ (\vec{r})$.

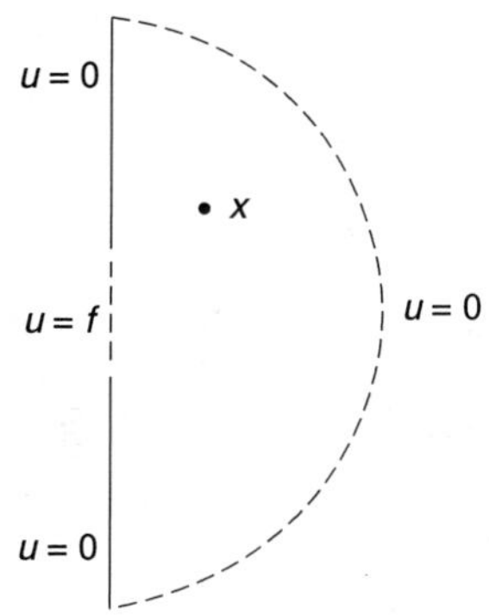

Fig. 11.2 Dirichlet boundary conditions

11.3.3 Solution of Scalar Helmholtz Equation using Green Function

Scalar Helmholtz equation is

$$(\nabla^2 + k^2)\ u = 0$$

In free space, Green function

$$G(\vec{r}, \vec{x}) = \frac{e^{ik|\vec{r} - \vec{x}|}}{|\vec{r} - \vec{x}|} \quad ...(11.22)$$

is the solution of equation

$$(\nabla^2 + k^2)\, G(\vec{r}, \vec{x}) = -4\pi\,\delta(\vec{r} - \vec{x}) \quad ...(11.23)$$

because, the solution of Eq. (11.23) is

$$u(\vec{r}) = \frac{1}{4\pi}\int_0^x G(\vec{r}, \vec{x}) \times 4\pi\,\delta(\vec{r} - \vec{x})\, d^3 r$$

$$G(\vec{r}, \vec{x}) = \frac{e^{ik|\vec{r} - \vec{x}|}}{|\vec{r} - \vec{x}|}$$

$$\text{as Kronecker } \delta(\vec{r} - \vec{x}) = \begin{cases} 1 \text{ for } \vec{r} > \vec{x} \\ 0 \text{ for } \vec{r} < \vec{x} \end{cases} \quad ...(11.24)$$

and outgoing boundary conditions at ∞. G is a spherical wave expanding out from a point source at $\vec{r}$

Similarly, a solution of the equation

$$(\nabla^2 + k^2)\, u(\vec{r}) = -S(\vec{r}) \text{ in a region } V \text{ is}$$

$$u(\vec{x}) = \begin{cases} (4\pi)^{-1} \int_V G(\vec{r}, \vec{x})\, S(\vec{r})\, d^3r & \text{If } (\vec{x}) \in V; \\ 0 & \text{If } \vec{x} \notin V; \end{cases} \qquad \text{...(11.25)}$$

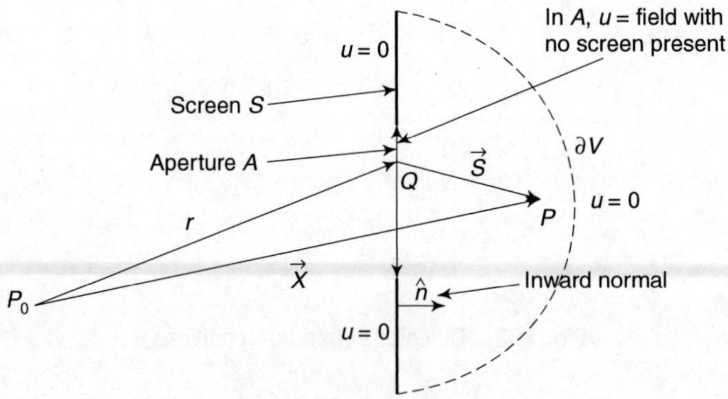

Fig. 11.3 Kirchhoff boundary conditions

11.4 HELMHOLTZ-KIRCHHOFF INTEGRAL THEOREM

Gauss theorem states that total flux enclosed by the closed surface is equal to the charge enclosed by it or Gauss theorem states that the surface integral of electrostatic field ($\vec{E}_0$) over a closed surface enclosing volume (V) is equal to the total charge enclosed within that surface.

Similar to Gauss theorem, Helmholtz-Kirchhoff integral theorem is defined as the total flux of light rays enclosed by the closed surface is equal spherical wavefront, $u(x) = e^{ikx}/x$. Mathematically, it is written as:

$$\int_V \nabla \cdot \vec{F}\, d^3r = \int_{\partial V} (-\hat{n}) \cdot \vec{F}\, dS \qquad \text{...(11.26)}$$

where, ∂V = surface that bounds the volume (V), $\hat{n}$ is inward normal to surface, ∂V. Putting the vector field

$$\vec{F} = u\, \nabla G - G\, \nabla u, \text{ Eq. (11.26) becomes}$$

$$\int_{\partial V} (-\hat{n}) \cdot \vec{F}\, dS$$

$$= \int_V \nabla \cdot (u \nabla G - G \nabla u)\, d^3r$$

$$= \int_V [(\nabla u)\cdot(\nabla G) - (\nabla G)\cdot(\nabla u)]\, d^3r + \int_V (u \nabla^2 G - G \nabla^2 u)\, d^3r$$

$$= \int_V (u \nabla^2 G - G \nabla^2 u)\, d^3r \quad \text{as} \quad (\nabla u)\cdot(\nabla G) - (\nabla G)\cdot(\nabla u) = 0 \qquad \text{...(11.27)}$$

By adding and subtracting $\int_V uG\, k^2\, d^3r$ in right-hand side of Eq. (11.27), one gets

$$\int_{\partial V} (-\hat{n})\cdot \vec{F}\, dS$$

$$= \int_A (-\hat{n})\cdot (u \nabla G - G \nabla u)\, dS$$

$$= \int_V [u(\nabla^2 + k^2) G - G(\nabla^2 + k^2) u]\, d^3r \qquad \text{...(11.28)}$$

Putting $(\nabla^2 + k^2)\, u = 0$ in Eq. (11.28), we get

$$\int_{\partial V} (-\hat{n})\cdot \vec{F}\, dS$$

$$= \int_A (-\hat{n})\cdot (u \nabla G - G \nabla u)\, dS$$

$$= -\int_V [u(\nabla^2 + k^2) G]\, d^3r \qquad \text{...(11.29)}$$

Now substituting $(\nabla^2 + k^2)\ G(\vec{r}, \vec{x}) = -4\pi\, \delta(\vec{r} - \vec{x})$, Eq. (11.29) becomes

$$\int_{\partial V} (-\hat{n})\cdot \vec{F}\, dS = -\int_V [u(-4\pi\, \delta(\vec{r} - \vec{x})]\, d^3r = 4\pi u\,(x)$$

or
$$\frac{1}{4\pi} \int_{\partial V} (-\hat{n})\cdot \vec{F}\, dS = u(x) = \frac{e^{ikx}}{x} \qquad \text{...(11.30)}$$

or
$$\frac{1}{4\pi} \int_A (u\, \hat{n}\cdot \nabla G - G\, \hat{n}\cdot \nabla u)\cdot dS = u(x) = \frac{e^{ikx}}{x} \qquad \text{...(11.31)}$$

as $\vec{F} = u \nabla G - G \nabla u$. Eq. (11.30) or Eq. (11.31) is called Helmholtz-Kirchhoff integral theorem. Here, $\hat{n}$ is inward normal to surface (∂V) that bounds volume, V.

11.5 FRESNEL-KIRCHHOFF DIFFRACTION FORMULA

From Fig. 11.3, we can see that $\vec{r}$ is position vector of incident ray of the point Q at slit from the light source, $P_0 \cdot \vec{S}$ is the position vector of point P at the screen.

In the triangle $P_0\, QP$, $\vec{X} = \vec{r} + \vec{S}$ or $\vec{X} - \vec{r} = \vec{S}$, so that the Green's function, $G(\vec{r}, \vec{x})$, can be written as

$$G(\vec{r}, \vec{x}) = \frac{e^{ik|\vec{r}-\vec{x}|}}{|\vec{r}-\vec{x}|} = \frac{e^{ikS}}{S} \quad ...(11.32)$$

Putting the value of G in Eq. (11.31), we get

$$u(x) = \frac{1}{4\pi} \int\!\!\int_{\partial V} \left[u(\vec{r}) \left\{ \hat{n} \cdot \nabla \frac{e^{ikS}}{S} \right\} - \frac{e^{ikS}}{S} \{ \hat{n} \cdot \nabla u(\vec{r}) \right] dS \quad ...(11.33)$$

Equation (11.33), expresses $u(\vec{r})$ in terms of fields at points $\vec{r}$ on the surface that bounds V. Kirchhoff's boundary conditions are:

(i) $u = 0$ every where except in the aperture

(ii) In the aperture, $u = f = u(\vec{r}) = A \dfrac{e^{ikr}}{r}$ the field due to a point source at P_0,

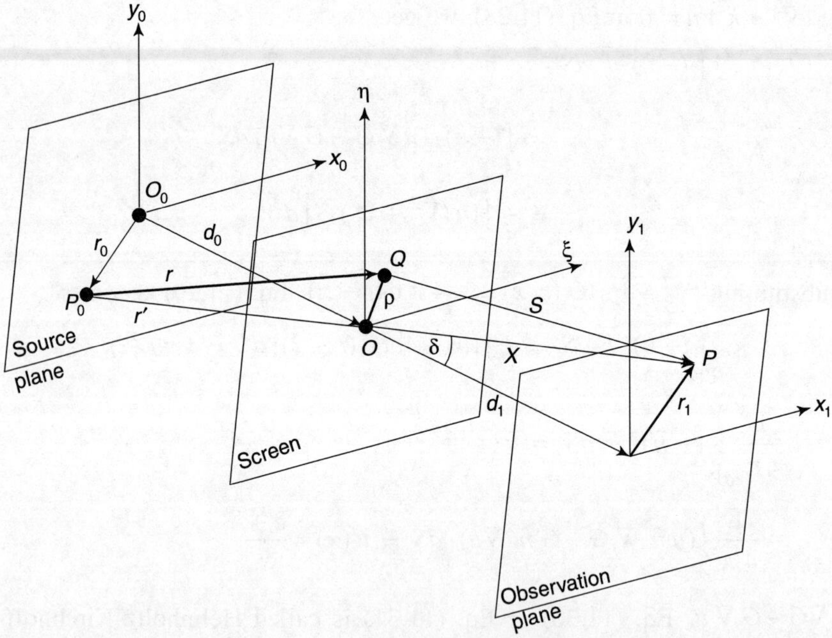

Fig. 11.4 Coordinates for the Fresnel-Kirchhoff diffraction integral

The incident ray at point $\vec{r}$ in the aperture $u(\vec{r}) = A \dfrac{e^{ikr}}{r}$, so

$$\nabla u(\vec{r}) = \nabla A \frac{e^{ikr}}{r} = A\,\hat{\mathbf{r}} \left(\frac{ik}{r} e^{ikr} - \frac{e^{ikr}}{r^2} \right)$$

or
$$\nabla u(\vec{r}) = A\,\hat{\mathbf{r}}\frac{e^{ikr}}{r}\left(ik - \frac{1}{r}\right) \qquad ...(11.34)$$

The emerging ray moving in the direction $\vec{S}$ from the aperture $G(\vec{r}, \vec{x}) = \dfrac{e^{ikS}}{s}$, so

$$\nabla G(\vec{r}, \vec{x}) = \nabla\left(\frac{e^{ikS}}{s}\right) = -\hat{\mathbf{S}}\left(\frac{ik}{s}e^{ikS} - \frac{e^{ikX}}{s^2}\right)$$

$$\nabla G(\vec{r}, \vec{x}) = -\hat{S}\frac{e^{ikS}}{S}\left(ik - \frac{1}{S}\right) \qquad ...(11.35)$$

where $\vec{S} = \vec{X} - \vec{r}$

Putting the value of $\nabla u(\vec{r}) = A\,\hat{\mathbf{r}}\dfrac{e^{ikr}}{r}\left(ik - \dfrac{1}{r}\right)$ and $\nabla\dfrac{e^{ikS}}{s} = -\hat{S}\dfrac{e^{ikS}}{s}\left(ik - \dfrac{1}{s}\right)$ in Eq. (11.33), we get

$$u(\vec{x}) = \frac{1}{4\pi}\int_{\partial V}\left[A\frac{e^{ikr}}{r}\left\{\hat{n}\cdot\nabla\frac{e^{iks}}{s}\right\} - \frac{e^{iks}}{s}\left\{\hat{n}\cdot\nabla\left(A\frac{e^{ikr}}{r}\right)\right\}\right]dS$$

$$= \frac{1}{4\pi}\int_{\partial V}\left[A\frac{e^{ikr}}{r}\left\{\hat{n}\left(-\hat{S}\frac{e^{iks}}{s}\left(ik - \frac{1}{s}\right)\right)\right\} - \frac{e^{iks}}{s}\left\{\hat{n}\cdot\left(A\,\hat{\mathbf{r}}\frac{e^{ikr}}{r}\left(ik - \frac{1}{r}\right)\right)\right\}\right]dS$$

$$= -\frac{ikA}{4\pi}\int_{\partial V}\frac{e^{ik(r+s)}}{rS}(\hat{n}\cdot\hat{s} + \hat{n}\cdot\hat{\mathbf{r}})\,dS + \frac{A}{4\pi}\int_{A}\frac{e^{ik(r+s)}}{rS}\left(\frac{\hat{n}\cdot\hat{r}}{r} + \frac{\hat{n}\cdot\hat{S}}{s}\right)dS$$

$$\approx -\frac{ikA}{4\pi}\int_{\partial V}\frac{e^{ik(r+s)}}{rs}(\hat{n}\cdot\hat{\mathbf{S}} + \hat{n}\cdot\hat{r})\,dS \qquad ...(11.36)$$

Fresnel-Kirchhoff diffraction formula for spherical wave is

$$u(\vec{x}) = -\frac{ikA}{4\pi}\int_{\partial V}\frac{e^{ik(r+s)}}{rS}(\hat{n}\cdot\hat{s} + \hat{n}\cdot\hat{r})\,dS \qquad ...(11.37)$$

or
$$u(\vec{x}) = -\frac{iA}{2\lambda}\int_{\partial V}\frac{e^{ik(r+s)}}{rS}(\hat{n}\cdot\hat{s} + \hat{n}\cdot\hat{r})\,dS \quad \text{as} \quad k = \frac{2\pi}{\lambda} \qquad ...(11.38)$$

is an approximate solution of Helmholtz equation and is called Fresnel-Kirchhoff diffraction formula.

The Fresnel-Kirchhoff diffraction formula given by Eq. (11.37).

(i) is an approximate scalar solution of the Helmholtz equation, valid for wavelengths small. compared to the aperture size,

(ii) is based on deep physical insight,

(iii) is accurate enough to give useful, quantitative results in optics, but

(iv) does not make $u \to 0$ at large distances (as assumed in Kirchhoff's boundary conditions) and

(v) is not the first term in any iterative solution of the boundary value problem.

The Fresnel-Kirchhoff diffraction formula

$$u(\vec{x}) = -\frac{ikA}{4\pi} \int_{\partial V} \frac{e^{ik(r+s)}}{rS} (\hat{n} \cdot \hat{\mathbf{S}} + \hat{n} \cdot \hat{\mathbf{r}})\, dS \qquad \text{...(11.39)}$$

is consistent of with Huyghen's principle, because

(i) It expresses $u(\vec{x})$ as a superposition of spherical waves e^{iks}/s emanating from the wavefront e^{ikr}/r produced by a point source, and

(ii) the inclination factor is

$$I(\hat{\mathbf{r}} \cdot \hat{\mathbf{S}}) = -(\hat{n} \cdot \hat{\mathbf{S}} + \hat{n} \cdot \hat{\mathbf{r}})$$

In the backward direction $(\hat{\mathbf{S}} = -\hat{\mathbf{r}})$,

$$I(\hat{\mathbf{r}} - \hat{\mathbf{r}}) = 0$$

In the forward direction $(\hat{\mathbf{S}} = \hat{\mathbf{r}})$,

$$\begin{aligned} I(\hat{\mathbf{r}} \cdot \hat{\mathbf{r}}) &= -2\cos\delta \\ &= -2 \quad \text{for small } \delta \end{aligned}$$

In the forward direction the integral formula reduces to

$$u(\vec{x}) = \frac{2\,ikA}{4\pi} \int_{\partial V} \frac{e^{ik(r+s)}}{r\,S}\, dS \qquad \text{...(11.40a)}$$

or

$$u(\vec{x}) = \frac{iA}{\lambda} \int_{\partial V} \frac{e^{ik(r+s)}}{rS}\, dS \qquad \text{...(11.40b)}$$

11.6 FRESNEL-KIRCHHOFF DIFFRACTION THEORY

If the inclination factor is nearly constant over the aperture, the Fresnel-Kirchhoff diffraction formula can be written as

$$u(\vec{x}) \approx \frac{ikAI}{4\pi}\int_{\partial V} \frac{e^{ik(r+s)}}{rS}\, dS \qquad \text{...(11.41)}$$

If we replace the point source incident wavefront, e^{ikr}/r, produced by a point source with a general wavefront, $u_{\text{inc}}(\vec{r})$, produced by an extended source (modelled as a collection of point sources), then Eq. (11.41) becomes

$$u(\vec{x}) = \frac{ikI}{4\pi}\int_{\partial V} u_{\text{inc}}(\vec{r})\frac{e^{iks}}{s}\, dS \qquad \text{...(11.42)}$$

By introducing a transmission function (τ) for phase changes due to lenses and diffraction grating in Eq. (11.42), the transmission wave propagation from aperture can be written as

$$u(\vec{x}) = \frac{ikI}{4\pi}\int_{\partial V} \tau(\vec{r})\, u_{\text{inc}}(\vec{r})\frac{e^{iks}}{s}\, dS \qquad \text{...(11.43a)}$$

or

$$u(\vec{x}) = \frac{iI}{2\lambda}\int_{\partial V} \tau(\vec{r})\, u_{\text{inc}}(\vec{r})\frac{e^{iks}}{s}\, dS \qquad \text{...(11.43b)}$$

11.7 APPLICATIONS OF THE FRESNEL-KIRCHHOFF DIFFRACTION FORMULA

$$u(\vec{x}) = \frac{ikI}{4\pi}\int_{\partial V} \tau(\vec{r})\, u_{\text{inc}}(\vec{r})\frac{e^{iks}}{s}\, dS \qquad \text{...(11.44a)}$$

The diffracted spherical wavefront from a point source at position $\vec{r}$ of the aperture is

$$u_{\text{inc}}(\vec{r}) = A\frac{e^{ikr}}{r} \qquad \text{...(11.44b)}$$

In case of Fraunhofer diffraction the incident wave is a plane wave.

In case of Fresnel diffraction, diffraction theory of image formation depends on:

(i) Aberrations

(ii) Focal-plane filtering

(iii) Transmission functions

(iv) Propagation of the mutual coherence function

11.8 TRANSMISSION FUNCTION, τ (ρ), OF A THIN LENS

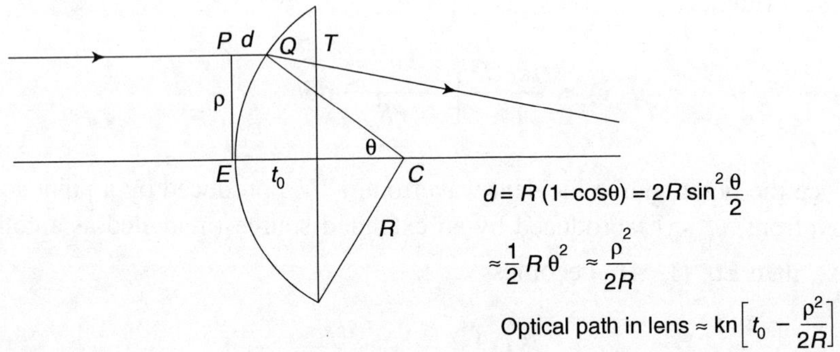

Fig. 11.5 Thin lens

In Fig. 11.5, ρ is half width of the beam, the curvature of plano-convex lens is R and thickness is t_0. Let $PQ = d$, then

$$EC = R = R\cos\theta + PQ = R\cos\theta + d$$

or
$$d = R(1 - \cos\theta) = 2R\sin^2\frac{\theta}{2}$$

$$\approx \frac{1}{2}R\theta^2 \qquad \text{...(11.45)}$$

When $PQ = d$ is small, then the arc $EQ \approx \rho = R\theta$. Putting the value of $\theta = \frac{\rho}{R}$, we get

$$d \approx \frac{\rho^2}{2R} \qquad \text{...(11.46)}$$

The distance travelled by the ray in the lens is

$$QT \approx t_0 - d = t_0 - \frac{\rho^2}{2R} \qquad \text{...(11.47)}$$

Thus the optical path in lens $\approx kn\left[t_0 - \frac{\rho^2}{2R}\right]$

Transmission function of a thin lens of thickness (t) and focal length (f) is defined as

$$\tau(\rho) \approx |\tau(\rho)|\, e^{iknt}\, e^{-ik\rho^2/(2f)} \qquad \text{...(11.48)}$$

Here, ρ is half width of the beam. Let the curvature of plano-convex lens be R and thickness, t_0. If the incident wave is $u_{inc}(\vec{r}) = e^{ikz}$, then from Fig. 11.5, we can see that the transmitted wave from the plano-convex thin lens (neglecting reflections) is written as

$$u_{inc}(\vec{r}) = \exp ik\{z + d + \text{optical path of lens}\}$$

or $$u_{\text{inc}}(\vec{r}) = \exp ik\left\{z + \frac{\rho^2}{2R} + n\left[\tau_0 - \frac{\rho^2}{2R}\right]\right\}$$

or $$u_{\text{inc}}(\vec{r}) = \exp ik\left\{z - (n-1)\frac{\rho^2}{2R} + n\tau_0\right\} \qquad ...(11.49)$$

Therefore, the plano-convex transmission function is

$$\tau_{\text{plano}}(\rho) \approx e^{ikn\tau_0}\, e^{-ik(n-1)/\rho^2}/(2R) \qquad ...(11.50)$$

For a biconvex lens of thickness $(t_1 + t_2)$ and curvatures, R_1 and $-R_2$, the transmission function is

$$\tau_{\text{bioconvex}}(\rho) \approx e^{ik(t_1+t_2)}\, e^{-ik(n-1)\rho^2[(2R_1)^{-1}-(2R_2)^{-1}]} \qquad ...(11.51)$$

Focal length, f = net curvature $\dfrac{1}{f} = (n-1)\left(\dfrac{1}{R_1} - \dfrac{1}{R_2}\right)$

Note: When d and thickness of the plano-convex lens t_0 are very small, then approximation taken in Eq. (11.45) is valid.

11.9 FRESNEL APPROXIMATION TO A SPHERICAL WAVE $\dfrac{e^{ikr}}{r}$

(i) When angles between $\vec{r}$ and $\vec{r}'$ are small angles or when $\vec{r} \approx \vec{r}'$

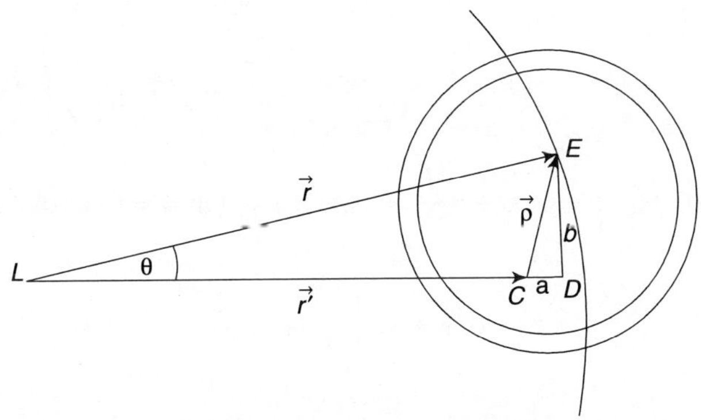

Fig. 11.6 Fresnel approximation

$$a = \hat{r}' \cdot \vec{\rho}$$

$$b^2 = \rho^2 - a^2$$

$$= \rho^2 - (\hat{r}' \cdot \vec{\rho})^2$$

$$r\cos\theta = r' + a$$

$$r = \frac{(r'+a)}{\cos(\theta)} \approx \frac{r'}{\cos(\theta)} + \frac{a}{\cos(\theta)}$$

$$= r'/\cos\theta + a$$

as $$\cos(\theta) = 1 \text{ for small } \theta$$

$$= r'\left(1 - 2\sin^2\frac{\theta}{2}\right)^{-1} + a$$

$$\approx r'\left(1 + \frac{1}{2}\theta^2\right) + a$$

$$\approx r' + a + \frac{b^2}{2r'}$$

In right angle ΔEDC,

$$b^2 = \rho^2 - a^2 = \rho^2 - (\hat{r}' \cdot \vec{\rho})^2 \quad \text{as} \quad CD = a = \hat{r}' \cdot \vec{\rho}$$

In right angle ΔEDL,

$$LD = r\cos\theta = r' + a$$

or

$$r = \frac{r' + a}{\cos\theta} = \frac{r'}{\cos\theta} + \frac{a}{\cos\theta} \approx \frac{r'}{\cos\theta} + a$$

as

$$\cos\theta = 1 \text{ for small } \theta$$

$$= r'\left(1 - 2\sin^2\frac{\theta}{2}\right)^{-1} + a$$

$$\approx r'\left(1 + \frac{1}{2}\theta^2\right) + a, \quad \text{as } \sin\frac{\theta}{2} = \frac{\theta}{2} \text{ for small } \theta$$

From triangle EDL, $\tan\theta = \dfrac{b}{r' + a} \approx \dfrac{b}{r'}$ as $r' >> a$, $\tan\theta = \theta$ so that

$$r \approx r' + a + \frac{b^2}{2r'} \quad \text{...(11.52)}$$

$$\text{Phase} = kr \approx kr' + \vec{k}_1 \cdot \vec{\rho} + k\left[\frac{\rho^2 - (\hat{r}' \cdot \vec{\rho})^2}{2r'}\right] \quad \text{...(11.53)}$$

Putting for kr, the transmitted spherical wave becomes

$$\frac{e^{ikr}}{r} \approx \frac{e^{ikr'}}{r'} e^{i\vec{k}_1 \cdot \vec{\rho}} e^{ik\left|\rho^2 - (\hat{r}' \cdot \vec{\rho})^2\right|/(2r')} \quad \text{...(11.54)}$$

$\dfrac{e^{ikr'}}{r'}$ = spherical wave centered at P_0

$e^{i\vec{k}_1 \cdot \vec{\rho}}$ = plane wave with propagation constant $\vec{k}_1 = k\hat{r}'$

$e^{ik\left|\rho^2 - (\hat{r}'\cdot\vec{\rho})^2\right|/(2r')}$ = lowest correction for wavefront curvature

The factor, $e^{ik\left|\rho^2 - (\hat{r}'\cdot\vec{\rho})^2\right|/(2r')}$, is approximately 1 when the Fresnel number is small, that is, $\dfrac{b^2}{r'\lambda} \ll 1$, where, $b^2 = r^2 - \left(\hat{r}'\cdot\vec{\rho}\right)^2$.

It is to be noted that we get a nearly plane wave if we move far away from a point source. But it is easier to use a lens.

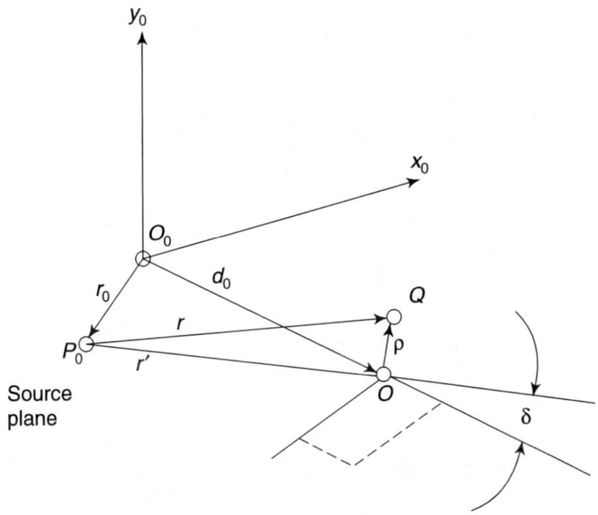

Fig. 11.7

The phase given by Eq. (11. 43) can also be obtained.

(ii) When $\vec{r}'$ is fixed (approximately) and $\vec{\rho}$ varies over aperture as follows:

From the triangle $P_0\,QO$ in Fig. 11.7

$$\vec{r} = \vec{r}' + \vec{\rho}$$

or

$$r^2 = r'^2 + 2\,\vec{r}'\cdot\vec{\rho} + \rho^2$$

or

$$r = r'\left[1 + 2\,\frac{\vec{r}'}{r'^2}\cdot\vec{\rho} + \frac{\rho^2}{r'^2}\right]^{1/2} \qquad \text{...(11.55)}$$

Binomial expansion correct to order $(\rho/r')^2$:

$$r = r'\left[1+\frac{1}{2}\left(2\frac{\vec{r}'}{r'^2}\cdot\vec{\rho}+\frac{\rho^2}{r'^2}\right)+\frac{\frac{1}{2}\times\left(\frac{1}{2}-1\right)}{2!}\left(\left(2\frac{\vec{r}'}{r'^2}\cdot\vec{\rho}+\frac{\rho^2}{2r'}\right)\right)^2+\ldots\right]$$

or $$r = r'\left[1+\frac{\vec{r}'}{r'^2}\cdot\vec{\rho}+\frac{\rho^2}{2r'^2}-\frac{1}{8}\left(2\frac{\vec{r}'}{r'^2}\cdot\vec{\rho}\right)^2+\ldots\right]$$

or $$r \approx r'+\frac{\vec{r}'}{r'}\cdot\vec{\rho}+\frac{\rho^2}{2r'}-\frac{(\vec{r}'\cdot\vec{\rho})^2}{2r'^3}$$

or $$r \approx r'+\hat{r}'\cdot\vec{\rho}+\frac{\rho^2}{2r'}-\frac{(\hat{r}'\cdot\vec{\rho})^2}{2r'} \quad \ldots(11.56)$$

The phase $\mathbf{k}r \approx kr' + k\hat{r}'\cdot\vec{\rho}+k\dfrac{\rho^2-(\hat{r}'\cdot\vec{\rho})^2}{2r'}$

or $$\mathbf{k}r \approx kr' + \vec{k}_1\cdot\vec{\rho}+k\frac{\rho^2-(\hat{r}'\cdot\vec{\rho})^2}{2r'} \quad \ldots(11.57)$$

From the triangle *OQP* in Fig. 11.8

$$\vec{S} = \vec{S}'-\vec{\rho}$$

or $$S^2 = S'^2 - 2\vec{S}'\cdot\vec{\rho}+\rho^2$$

or $$S = S'\left[1-2\frac{\vec{s}'}{s'^2}\cdot\vec{\rho}+\frac{\rho^2}{s'^2}\right]^{1/2}$$

Expanding the right-hand side by binomial theorem, we get

$$S \approx S'-\hat{S}'\cdot\vec{\rho}+\frac{\rho^2}{2s'}-\frac{(\hat{s}'\cdot\vec{\rho})^2}{2s'}$$

The phase $\mathbf{k}S \approx kS' - k\hat{S}'\cdot\vec{\rho}+k\dfrac{\rho^2-(\hat{s}'\cdot\vec{\rho})^2}{2s'}$

$$\mathbf{k}S \approx kS' - \vec{k}_2\cdot\vec{\rho}+k\frac{\rho^2-(\hat{s}'\cdot\vec{\rho})^2}{2s'} \quad \ldots(11.58)$$

where, $\vec{k}_1 = k\hat{r}', \vec{k}_2 = k\hat{S}'$

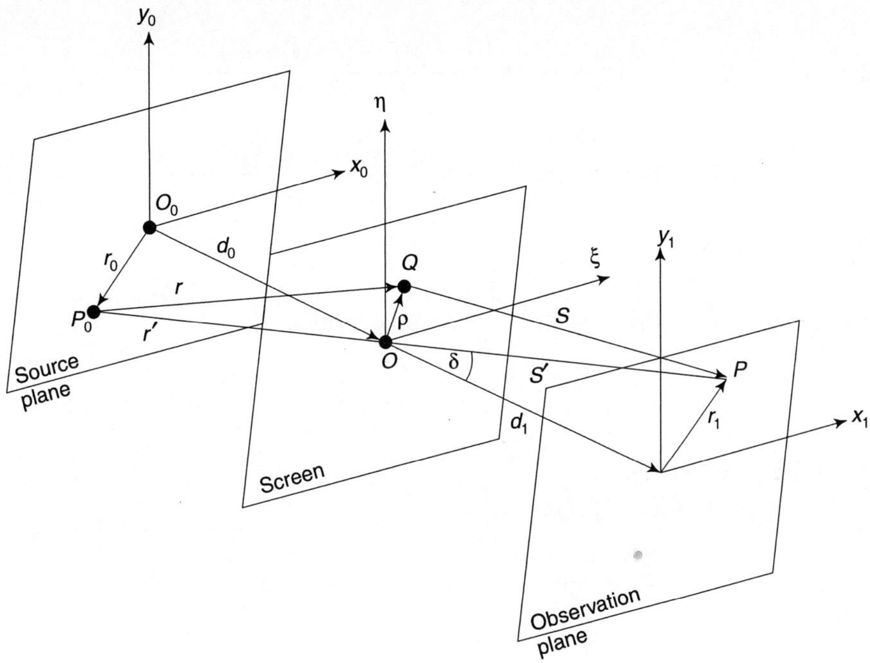

Fig. 11.8 Coordinates for the Fresnel-Kirchhoff diffraction integral

11.10 FRESNEL-KIRCHHOFF'S INTEGRAL FORMULA IN FRESNEL APPROXIMATION

Using Fresnel's approximation, the incident spherical wave, $\frac{e^{ikr}}{r}$, can be written as Eq. (11.54).

$$\frac{e^{ikr}}{r} \approx \frac{e^{ikr'}}{r'} e^{i\vec{k}_1 \cdot \vec{\rho}} \, e^{ik\left|\rho^2 - (\hat{r}' \cdot \vec{\rho})^2\right|/(2r')} \qquad ...(11.59)$$

In the denominator, one can consider $r \approx r'$, if the spread of the beam is small.

From Fig. 11.9, $\vec{S} + \vec{\rho} = \vec{S}' + \vec{r}_1$ or $\vec{S} = \vec{S}' + \vec{r}_1 - \vec{\rho}$

or
$$S_2 = S'^2 + 2\,\vec{S}' \cdot (\vec{r}_1 - \vec{\rho}) + (\vec{r}_1 - \vec{\rho})^2$$

or
$$S = S'\left[1 + 2\,\frac{\vec{s}'}{s'^2} \cdot (\vec{r}_1 - \vec{\rho}) + \frac{(\vec{r}_1 - \vec{\rho})^2}{s'^2}\right]^{1/2}$$

Expanding the right-hand side by binomial theorem, we get

$$S \approx S' + \hat{S}' \cdot (\vec{r}_1 - \vec{\rho}) + \frac{(\vec{r}_1 - \vec{\rho})^2}{2\,s'} - \frac{(\hat{s}' \cdot (\vec{r}_1 - \vec{\rho}))^2}{2\,s'}$$

The phase $\mathbf{k}S \approx kS' + k\hat{S}' \cdot (\vec{r}_1 - \vec{\rho}) + k \frac{((\vec{r}_1 - \vec{\rho}))^2 - (\hat{s}' \cdot (\vec{r}_1 - \vec{\rho}))^2}{2\,r'}$

$$\mathbf{k}S \approx kS' + \vec{k}_2 \cdot (\vec{r}_1 - \vec{\rho}) + k \frac{((\vec{r}_1 - \vec{\rho}))^2 - (\hat{s}' \cdot (\vec{r}_1 - \vec{\rho}))^2}{2\,s'} \qquad ...(11.60)$$

where, $\vec{k}_1 = k\hat{r}', \vec{k}_2 = k\hat{S}'$.

Now the transmitted spherical wave, $\frac{e^{ikS}}{s}$, can be written as

$$\frac{e^{iks}}{s} \approx \frac{e^{iks'}}{s'} e^{i\vec{k}_2 (\vec{r}_1 - \vec{\rho})} e^{ik\left|(\vec{r}_1 - \vec{\rho})^2 - (\hat{s}' \cdot (\vec{r}_1 - \vec{\rho})^2\right|/(2s)} \qquad ...(11.61)$$

Where, the propagation vectors are

$\vec{k}_1 = k\,\hat{r}', \; \vec{k}_2 = k\hat{S}', \; \vec{q} = \vec{k}_1 - \vec{k}_2$

By putting the values of $\frac{e^{ikr}}{r}$ and $\frac{e^{iks}}{s}$, the integral equation of diffracted spherical wave [Eq. (11.43a)]

$$u(\vec{x}) = \frac{ikAI}{4\pi} \int_{\partial V} \tau(\vec{r}) \frac{e^{ikr}}{r} \frac{e^{iks}}{s}\, dS \qquad ...(11.62)$$

in Fresnel's approximation is written as

$$u(\vec{x}) \approx \frac{ikAI}{4\pi} \frac{e^{ik(r'+s')}}{r'\,s'} e^{i\vec{k}_2 \cdot \vec{r}_1} \int_A \tau(\vec{r})\, e^{i\vec{q} \cdot \vec{\rho}}\, e^{ik\left[\frac{\rho^2 - (\hat{r}' \cdot \vec{\rho})^2}{2\,r'} + \frac{(\vec{r}_1 - \vec{\rho})^2 - \left|s'(\vec{r}_1 - \vec{\rho})\right|^2}{2\,s'}\right]} dS$$

...(11.63)

where, $\vec{q} = \vec{k}_1 - \vec{k}_2, \vec{k}_1 = k\hat{r}', \vec{k}_2 = k\hat{S}'$. Equation (11.63) is known as Fresnel-Kirchhoff's integral formula, where,

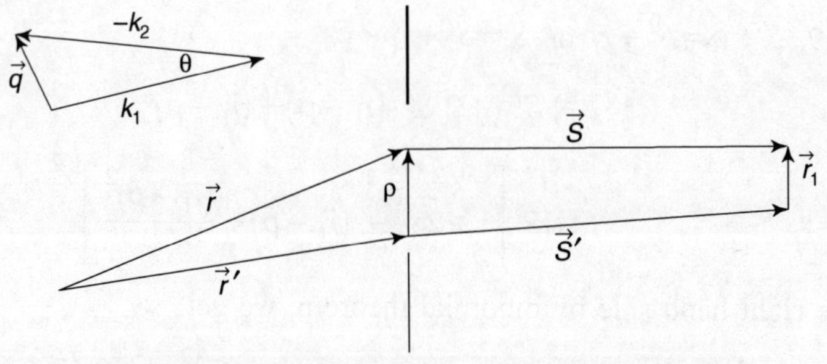

Fig. 11.9 Coordinates for the diffraction integral

– $e^{i\vec{k}_2 \cdot \vec{r}_1}$ is plane wave "carrier" incident at P

– Integral at incident point on the aperture

– $e^{ik\left[\frac{\rho^2 - (\hat{r} \cdot \vec{\rho})^2}{2r'} + \frac{(\vec{r}_1 - \vec{\rho}) - |\hat{s}'(\vec{r}_1 - \vec{\rho})|^2}{2s'}\right]}$ is wavefront curvature correction

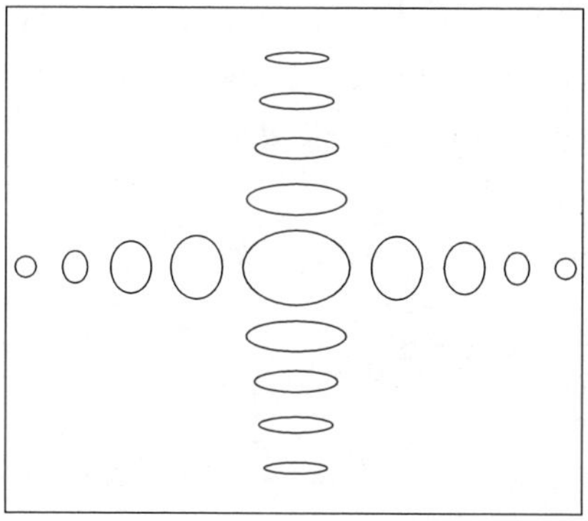

Fig. 11.10 Fraunhofer diffraction pattern of a rectangular aperture 8 mm × 7 mm, magnification 5.0 times, mercury yellow light λ = 5790 Å. To show the existence of the weak secondary maxima the central portion was overexposed

SOLVED EXAMPLES

Example 11.1 Frauanhofer diffraction from a rectangular aperture.

In Fresnel approximation, the wavefront curvature is an essential part of physics, while in Fraunhofer diffraction, no wavefront curvature correction is needed, because, the incident wave in fraunhofer diffraction is plane wave. The diffraction integral formula in case of Fraunhofer diffraction is

$$u(\vec{x}) \approx \frac{ikAI}{4\pi} \frac{e^{ik(r'+s')}}{r'S'} e^{i\vec{k}_2 \cdot \vec{r}_1} \int_{\partial V} \tau(\vec{\rho})\, e^{i\vec{q} \cdot \vec{\rho}}\, dS \qquad ...(11.64)$$

where, $\vec{q} = \vec{k}_1 - \vec{k}_2, \vec{k}_1 = k\hat{r}, \vec{k}_2 = k\hat{S} \cdot u(\vec{x})$ is the Fourier transform of $\tau(\vec{\rho})$, where, $\tau(\vec{\rho})$ is given by Eq. (11.50) or (11.51).

Let the aperture have dimensions: $2a$ (along x) × $2b$ (along y). $\vec{q}$ is not necessarily in the plane of the aperture, but we can assume that in this case $\vec{q}$ is in the plane of aperture, because its z-component is small. Integral formula in 2-d over the components of $\vec{\rho}$ is written as

$$u(\vec{x}) = \frac{ikAI}{4\pi}\frac{e^{ik(r'+S')}}{r'S'}e^{i\vec{k}_2\cdot\vec{r}_1}\int_{-a}^{a}\int_{-b}^{b} e^{i(q_x\xi+q_y\eta)}\,d\xi d\eta$$

$$= (\text{constant})\ [2 \sin c\ (q_x a)]\ [2 \sin c\ (q_y b)] \qquad ...(11.65)$$

The intensity of the pattern is given by

$$S(\vec{q}) = S(0) \sin c^2 (q_x a) \sin c^2 (q_y b) \qquad ...(11.66)$$

The Fraunhofer diffraction pattern in case of two dimensions is shown in Fig. 11.11.

The dimensions of rectangular aperture = $2a \times 2b$

For a small diffraction angle, $q_x \approx k\dfrac{\xi}{S'},\ q_y \approx k\dfrac{\eta}{S'}$

First zero of sinc function is at $x = \pi$

At the first zero in ξ is at $q_x a = k\dfrac{\xi_0}{S'}a = \dfrac{2\pi\xi_0}{\lambda S'}a = \pi$ or $\xi_0 = \dfrac{\lambda S'}{2a}$ ...(11.67)

At the first zero in η is at $q_y b = k\dfrac{\eta_0}{S'}b = \dfrac{2\pi\eta_0}{\lambda S'}b = \pi$ or $\eta_0 = \dfrac{\lambda S'}{2b}$ ...(11.68)

Example 11.2 Frauanhofer diffraction from a circular aperture.

Frauanhofer diffraction integral formula is

$$u(\vec{x}) \approx \int_{\partial V} \tau(\vec{\rho})\, e^{i\vec{q}\cdot\vec{\rho}}\, dS \qquad ...(11.69)$$

where $\vec{q} = \vec{k}_1 - \vec{k}_2, \vec{k}_1 = k\hat{r}, \vec{k}_2 = k\hat{S}'$

Let a be the radius of aperture. The integral over the components of $\vec{\rho}$ is

$$u(\vec{x}) = \frac{ikAI}{4\pi}\frac{e^{ik(r'+s')}}{r'S'}e^{i\vec{k}_2\cdot\vec{r}_1}\int_0^{2\pi}\int_0^{a} e^{iq\rho\cos\phi}\,\rho d\rho d\phi$$

$$= (\text{constant})\ (\pi a^2)\left[\frac{2J_1(qa)}{qa}\right] \qquad ...(11.70)$$

Here, J_1 is Bessel function of the first order. The intensity of maxima is given by

$$S(\vec{q}) = S(0)\left[\frac{2J_1(qa)}{qa}\right]^2 \qquad ...(11.71)$$

(the airy pattern)

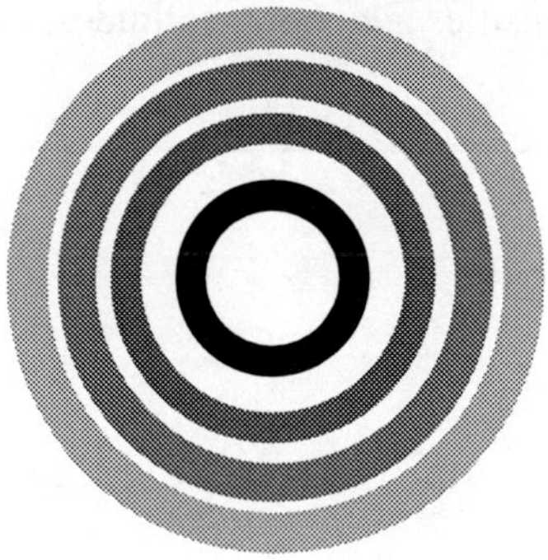

Fig. 11.11 Fraunhofer diffraction pattern of a circular aperture (the AIRY pattern) 6 mm in diameter, magnification 50 times, mercury yellow light λ = 5790 Å.

Fraunhofer diffraction pattern through circular aperture is shown in Fig. 11.11 for mercury yellow light of wavelength, λ = 5790 Å.

The radius of circular aperture is a

The diameter of circular aperture, $d = 2a$

For a small diffraction angle θ, $q \approx k\,\theta$ [Fig. 11.9]

First zero of J_1 is $j_{1,1}\,(qa) = 3.832.... = 1.220\,\pi$

At the first zero, $qa \approx k\theta_0\, a \approx \dfrac{2\pi}{\lambda}\,\theta_0\, a \approx 1.220\,\pi$ or $\theta_0 \approx 0.610\,\dfrac{\lambda}{a} \approx 1.220\,\dfrac{\lambda}{d}$

Example 11.3 Fraunhofer diffraction by N randomly arranged identical apertures.

Let aperture A_m be displaced from aperture A_1 by a random vector, a_m. Amplitude of diffraction field is proportional to

$$\int_A e^{i\vec{q}\cdot\vec{\rho}}\,dS = \sum_{m=1}^{N} e^{i\vec{q}\cdot\vec{a}_m} \int_{A_1} e^{i\vec{q}\cdot\vec{\rho}}\,dS \qquad ...(11.72)$$

where, $\int_{A_1} e^{i\vec{q}\cdot\vec{\rho}}\,dS$ is amplitude of diffraction field due to single aperture, A_1. The intensity of diffracted field is

$$S(\vec{q}) = (\text{const.}) \left| \sum_{m=1}^{N} e^{i\vec{q}\cdot\vec{a}_m} \int_{A_1} e^{i\vec{q}\cdot\vec{\rho}}\,dS \right|^2 \qquad ...(11.73)$$

$$(\text{const.}) \sum_{m=1}^{N} \sum_{n=1}^{N} e^{i\vec{q}\cdot(\vec{a}_m - \vec{q}_n)} \left| \int_{A_1} e^{i\vec{q}\cdot\vec{\rho}}\,dS \right|^2 = N \times (\text{const.}) \times \left| \int_{A_1} e^{i\vec{q}\cdot\vec{\rho}}\,dS \right|^2$$

The intensity of diffraction pattern due to single aperture, $\propto \left| \int_{A_1} e^{i\vec{q}\cdot\vec{\rho}}\,dS \right|^2$

One should note that the intensities add, not amplitudes, when the arrangement is random.

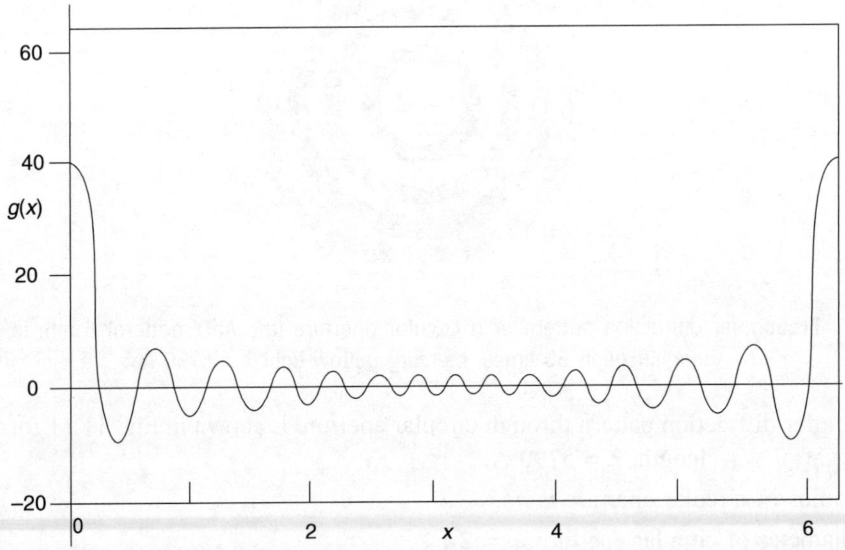

Fig. 11.12 Plotted the grating function, $g(x) = \dfrac{\sin\left[\left(M + \frac{1}{2}\right)x\right]}{\sin\frac{1}{2}x}$ for $M = 20$

Example 11.4 Calculation of Fraunhofer diffraction by periodically arranged identical apertures.

Let $\vec{a}_1$ and $\vec{a}_2$ be lattice vectors in 2-*d* space. The translational vector, $\vec{\rho} = m_1\,\vec{a}_1 + m_2\,\vec{a}_2$, where m_1 and m_2 are integers. The same unit cell is translated through integer multiples of the vectors, $\vec{a}_1$ and $\vec{a}_2$, are not necessarily orthogonal.

Amplitude of diffraction field is proportional to

$$\int_A e^{i\vec{q}\cdot\vec{\rho}}\,dS = \sum_{m=-M_1}^{M_2}\sum_{m=-M_2}^{M_2} e^{i\vec{q}\cdot(m_1\vec{a}_1 + m_2\vec{a}_2)}\int_{A_1} e^{i\vec{q}\cdot\vec{\rho}}\,dS$$

$$= g_1\,(\vec{q}\cdot\vec{a}_1)\,g_2\,(\vec{q}\cdot\vec{a}_2)\int_{A_1} e^{i\vec{q}\cdot\vec{\rho}} \qquad \text{...(11.74)}$$

Grating functions are defined as

$$g\,(\vec{a}\cdot\vec{a}\,) = \sum_{m=-M_j}^{M_j} e^{im_j\vec{q}_j\cdot\vec{q}_j}$$

$$= \frac{\sin\left[\left(M_j + \frac{1}{2}\right)\vec{q}_j \cdot \vec{a}_j\right]}{\sin\frac{1}{2}(\vec{q}_j \cdot \vec{q}_j)} \qquad ...(11.75)$$

Properties of grating function

$$g(\vec{a} \cdot \vec{a}) = \frac{\sin\left[\left(M + \frac{1}{2}\right)\vec{q} \cdot \vec{a}\right]}{\sin\frac{1}{2}(\vec{q} \cdot \vec{a})} \qquad ...(11.76)$$

Peak value of g is $2M + 1$ at $\vec{q} \cdot \vec{a} = 2n\pi$

Half-width of central peak $= \dfrac{\pi}{M + \frac{1}{2}}$

Amplitude of the first side lobe is equal to 0.212 × amplitude of central peak

Intensity of diffracted field is proportional to

$$\left|\int_A e^{i\vec{a}\cdot\vec{\rho}}\, dS\right| = |g_1(\vec{a} \cdot \vec{a}_1)|^2\, |g_2(\vec{q} \cdot \vec{a}_2|^2 \left|\int_{A_1} e^{i\vec{q}\cdot\vec{\rho}}\, dS\right|^2 \qquad ...(11.77)$$

The grating functions are sharply peaked at values of $\vec{q}$ such that

$$\vec{q} \cdot \vec{a}_1 = 2\, n_1\, \pi \quad \text{and} \quad \vec{q} \cdot \vec{a}_2 = 2\pi_2\, \pi$$

These conditions imply that $\vec{q}$ is 2π times a reciprocal lattice vector, i.e.,

$$\vec{q} = 2\pi(n_1 \vec{b}_1 + n_2 \vec{b}_2)$$

The intensity of diffraction pattern due to single aperture is $\propto \left|\int_{A_1} e^{i\vec{q}\cdot\vec{\rho}}\, dS\right|^2$

Thus, peaks are modulated by the intensity of diffraction pattern of a single aperture.

The grating function, $g(x) = \dfrac{\sin\left[\left(M + \frac{1}{2}\right)x\right]}{\sin\frac{1}{2}x}$ for $M = 20$ is plotted in Fig. 11.13.

Example 11.5 Calculation of the field in the image plane.

If an incident beam is normal to the aperture, the inclination factor

$$I = -(\hat{n} \cdot \hat{S} + \hat{n} \cdot \hat{r}) = -2$$

If source of light is a point source and the lens is thin with focal length f, the diffraction integral is [Eq. (11.43a)]

$$u(\vec{r}_1) = \frac{iA}{\lambda}\int_A \frac{e^{ik(r+s)}}{rS}\tau(\vec{\rho})\,dS \qquad \text{...(11.78)}$$

Putting the transmission factor τ $(\vec{\rho})$ from Eq. (11.5), the field at position $\vec{r}_1$ becomes

$$u(\vec{r}_1) \approx \frac{iA}{\lambda}\int_A \frac{e^{ik(r+s)}}{rS}e^{-ik\frac{\rho^2}{2f}}\alpha(\vec{\rho})\,d^2\rho \qquad \text{...(11.79)}$$

Here, τ is transmission function of the lens and aperture and α is transmission function of aperture alone.

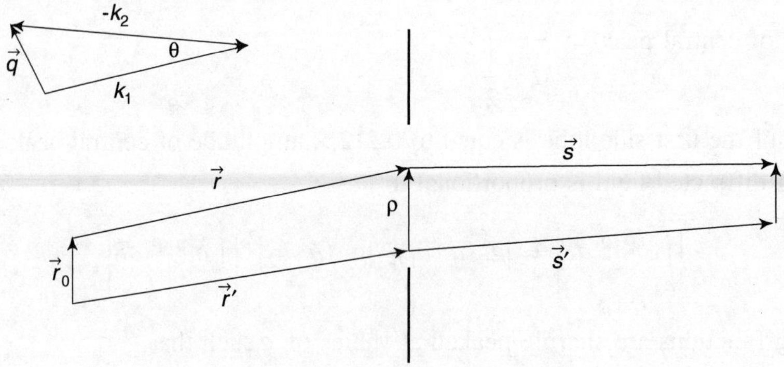

Fig. 11.13 Coordinates for the diffraction integral

Considering the whole beam into account as shown in Fig. 11.13, a correction is needed in the integral formula.

Figure 11.13 shows that

$$\vec{r}_0 + \vec{r} = \vec{r}' + \vec{\rho} \quad \text{or} \quad \vec{r} = \vec{r}' + \vec{\rho} - \vec{r}_0$$

Taking square of $\vec{r}$, we get

$$r^2 = r'^2 + (\vec{\rho} - \vec{r}_0)^2 + 2\,\vec{r}'\cdot(\vec{\rho} - \vec{r}_0)$$

or
$$r = \vec{r}'\left[1 + \frac{(\vec{\rho} - \vec{r}_0)^2}{r'^2} + 2\,\frac{\hat{r}\,(\vec{\rho} - \vec{r}_0)}{\vec{r}'}\right]^{1/2}$$

or
$$r \approx \vec{r}'\left[1 + \frac{(\vec{\rho} - \vec{r}_0)^2}{2\,r'^2} + \frac{\hat{r}\cdot(\vec{\rho} - \vec{r}_0)}{\vec{r}'} + \frac{\frac{1}{2}\times\left(\frac{1}{2} - 1\right)}{2!}\left(2\,\frac{\hat{r}\cdot(\vec{\rho} - \vec{r}_0)}{\vec{r}'}\right)^2\right]$$

or
$$r \approx \vec{r}'\left[1 + \frac{(\vec{\rho} - \vec{r}_0)^2}{2\,r'^2} + \frac{\hat{r}\cdot(\vec{\rho} - \vec{r}_0)}{\vec{r}'} - \frac{1}{2}\left(\frac{\hat{r}\cdot(\vec{\rho} - \vec{r}_0)}{\vec{r}'}\right)^2\right]$$

or
$$r \approx \vec{r}' + \frac{(\vec{\rho} - \vec{r}_0)^2}{2\vec{r}'} + \hat{r} \cdot (\vec{\rho} - \vec{r}_0) - \frac{(\hat{r} \cdot (\vec{\rho} - \vec{r}_0))^2}{2\vec{r}'}$$

or
$$r \approx \vec{r}' + \frac{(\vec{\rho} - \vec{r}_0)^2}{2\,\vec{r}'} \qquad \text{...(11.80)}$$

Figure 11.13 also shows that

$$\vec{\rho} + \vec{S} = \vec{S}' + \vec{r}_1 \quad \text{or } \vec{S} = \vec{S}' + \vec{r}_1 - \vec{\rho}$$

Taking square of $\vec{S}$, we get

$$S^2 = S'^2 + (\vec{r}_1 - \vec{\rho})^2 + 2\,\vec{S}' \cdot (\vec{r}_1 - \vec{\rho})$$

or
$$S = \vec{S}' \left[1 + \frac{(\vec{r}_1 - \vec{\rho})^2}{s'^2} + 2\,\frac{\hat{s}\,(\vec{r}_1 - \vec{\rho})}{\vec{s}'} \right]^{1/2}$$

or
$$S \approx \vec{S}' \left[1 + \frac{(\vec{r}_1 - \vec{\rho})^2}{2\,s'^2} + \frac{\vec{s}' \cdot (\vec{r}_1 - \vec{\rho})}{\vec{s}'} + \frac{\frac{1}{2} \times \left(\frac{1}{2} - 1\right)}{2!} \left(2\,\frac{\vec{s}' \cdot (\vec{r}_1 - \vec{\rho})}{\vec{s}'} \right)^2 \right]$$

or
$$S \approx \vec{S}' \left[1 + \frac{(\vec{r}_1 - \vec{\rho})^2}{2\,s'^2} + \frac{\vec{s}' \cdot (\vec{r}_1 - \vec{\rho})}{\vec{s}'} - \frac{1}{2} \left(\frac{\vec{s}' \cdot (\vec{r}_1 - \vec{\rho})}{\vec{s}'} \right)^2 \right]$$

or
$$S \approx \vec{S}' + \frac{(\vec{r}_1 - \vec{\rho})^2}{2\,\vec{s}'} + \vec{s}' \cdot (\vec{r}_1 - \vec{\rho}) - \frac{(\hat{r} \cdot (\vec{r}_1 - \vec{\rho})^2}{2\,\vec{S}'}$$

or
$$S \approx \vec{S}' + \frac{(\vec{r}_1 - \vec{\rho})^2}{2\,\vec{s}'} \qquad \text{...(11.81)}$$

Example 11.6 Calculation of the field in the image plane.

If the source of light is a point source and a thin lens with focal length f, diffraction integral is

$$u(\vec{r}_1) \sim -\frac{iA}{\lambda} \int_A \frac{e^{ik(r+s)}}{rS}\, e^{-ik\frac{\rho^2}{2f}}\, \alpha(\vec{\rho})\, d\rho^2 \qquad \text{...(11.82)}$$

The sum of exponents is (in Fresenel approximation) using Eqns. (11.80) and (11.81) is

$$ik\left[r + S - \frac{\rho^2}{2f} \right] \approx ik\left[r' + S' - \frac{\rho^2}{2f} + \frac{(\vec{\rho} - \vec{r}_0)^2}{2r'} + \frac{(\vec{r}_1 - \vec{\rho})^2}{2S'} \right]$$

$$\approx ik\left[r' + S' - \frac{\rho^2}{2f} + \frac{(r_0^2 - 2\vec{r}_0\,\vec{\rho} + \rho^2)}{2r'} + \frac{(\rho^2 - 2\vec{r}_1 . \vec{\rho} + r_1^2}{2S'} \right]$$

$$\approx ik\left[r' + S' + \frac{\rho^2}{2}\left(-\frac{1}{f} + \frac{1}{r'} + \frac{1}{S'}\right) + \frac{r_0^2}{2r'} + \frac{r_1^2}{2S'} - \vec{\rho}\cdot\left(\frac{\vec{r}_0}{r'} + \frac{\vec{r}_1}{S'}\right)\right] \qquad ...(11.83)$$

At the Gaussian focal point of a thin lens

$$-\frac{1}{f} + \frac{1}{r'} + \frac{1}{s'} = 0$$

This makes the coefficient of ρ^2 vanish in the exponent.

The diffraction integral [eqn. 11.82] reduces to a phase factor times the Fourier transform of the aperture function, α:

$$u(\vec{r}_1) \approx -\frac{iA}{\lambda r' s'} e^{ik\left[\left(\frac{r_0^2}{2r'} + \frac{r_1^2}{2s'}\right) + r' + S'\right]} \int_A \alpha(\vec{\rho})\, e^{-ik\vec{\rho}\cdot\left(\frac{\vec{r}_0}{r'} + \frac{\vec{r}_1}{s'}\right)} d^2\rho \qquad ...(11.84)$$

The image is spread out by diffraction.

The peak amplitude occurs at the stationary-phase point

$$\frac{\vec{r}_0}{r'} + \frac{\vec{r}_1}{s'} = 0 \;\Rightarrow\; \vec{r}_1 = -\frac{s'}{r'}\vec{r}_0 \text{ (Gaussian image displacement)}$$

Example 11.7 Diffraction pattern by a thin lens, when incident source light is coherent and extended

The fields due to a distribution (ψ) of point sources is written as [see Eq. 11.102)]

$$u(\vec{r}_1) \approx -\frac{i}{\lambda}\int_S\int_A \psi(\vec{r}_0)\frac{e^{ikr}}{r}\alpha(\vec{\rho})\, e^{-ik\frac{\rho^2}{2f}}\frac{e^{iks}}{s} d^2\rho\, d^2 r_0 \qquad ...(11.85)$$

Instead of performing two 2-*d* integrations simultaneously, we can evaluate a diffraction integral to go from the source to the lens and another to go from the lens to the observation plane.

In Fresnel approximation $\mathbf{r} \approx \vec{\mathbf{r}}' + \frac{(\vec{\rho} - \vec{r}_0)^2}{2\vec{r}'}$ [Eq. (11.80)]. Putting it and taking $\mathbf{r} \approx \vec{\mathbf{r}}'$ in the denominator, the integral Eq. (11.85) becomes

$$u(\vec{r}_1) \approx -\frac{i}{\lambda}\int_S\int_A \psi(\vec{r}_0)\frac{e^{ikr'}}{r'} e^{ik\frac{(\vec{\rho}-\vec{r}_0)^2}{2\vec{r}'}} \alpha(\vec{\rho})\, e^{-ik\frac{\rho^2}{2f}}\frac{e^{iks}}{s} d^2\rho\, d^2 r_0 \qquad ...(11.86)$$

$$\approx \int_A u_{\text{inc}}(\vec{\rho})\,\alpha(\vec{\rho})\, e^{-ik\frac{\rho^2}{2f}}\frac{e^{iks}}{s} \qquad ...(11.87)$$

Field incident at lens is written as follows:

$$u_{\text{inc}}(\vec{\rho}) \approx -\frac{i}{\lambda}\int_S \psi(\vec{r}_0)\frac{e^{ikr}}{r}d^2 r_0$$

In Fresnel approximation, $r = r' + \frac{(\vec{\rho}-\vec{r}_0)^2}{2r'}$ [see Eq.(11.80)] and in denominator taking approximation $r = r'$, u_{inc}(row) is written as

$$u_{\text{inc}}(\vec{\rho}) \approx -\frac{i}{\lambda}\frac{e^{ikr'}}{r'}\int_S \psi(\vec{r}_0)\, e^{ik\frac{(\rho-r_0)^2}{2r'}} d^2 r_0$$

$$\approx -\frac{i}{\lambda}\frac{e^{ikr'}}{r'}\int_S \Psi(\vec{r}_0)F_{k/r'}(\vec{\rho}-\vec{r}_0)d^2 r_0 \qquad \text{...(11.88)}$$

u_{inc} is a convolution of the source distribution (ψ) with the Fresnel function, $F_{k/r'}$, where,

$$F_{\alpha}(\vec{r}) = e^{i\alpha r^2} \qquad \text{...(11.89)}$$

here, $\vec{r}$ lies in the $x-y$ plane.

The 2-d Fourier transform of $F_{\alpha}(\vec{r})$ is

$$F_{\alpha}(\vec{r}) = \int e^{-iqr} F_{\alpha}(\vec{q})\, d^3 r$$

$$= \frac{2i\pi}{\alpha}F_{-1/\alpha}(\vec{q})$$

$$= \frac{2i\pi}{\alpha}e^{-\frac{iq^2}{2\alpha}} \qquad \text{...(11.90)}$$

or

$$F_{k/r'} = \frac{2i\pi\vec{r}'}{k}e^{-\frac{i\vec{r}'q^2}{2k}} = i\lambda\vec{r}'e^{-\frac{i\vec{r}''q^2}{2k}} \qquad \text{...(11.91)}$$

Imaging by a thin lens of an extended, coherently illuminated object, part 2:

The field incident on the lens, u_{inc}, is a convolution of the source distribution (ψ) with the Fresnel function $F_{k/r'}$:

$$u_{\text{inc}}(\vec{\rho}) = -\frac{i}{\lambda}\frac{e^{ikr'}}{r'}\int_S \Psi(\vec{r}_0)F_{k/r'}(\vec{\rho}-\vec{r}_0)d^2 r_0$$

$$= -\frac{i}{\lambda}\frac{e^{ikr'}}{r'}\frac{1}{(2\pi)^2}\int_S \Psi(\vec{q})F_{k/r'}(\vec{q})e^{i\vec{q}\vec{\rho}}d^2 q \qquad \text{...(11.92)}$$

Putting $u_{\text{inc}}(\vec{\rho})$, in the diffraction integral given by Eq.(11.87), we get

or $$u(\vec{r}_1) \approx -\frac{i}{\lambda}\frac{e^{ikr'}}{r'}\frac{1}{(2\pi)^2}\int_S\int_A \Psi(\vec{q})F_{k/r'}(\vec{q})e^{i\vec{q}\cdot\vec{\rho}}\alpha(\vec{\rho})e^{-ik\frac{\rho^2}{2f}}\frac{e^{ikS}}{S}d^2\rho\, d^2q \qquad ...(11.93)$$

Putting the exponent from Eq. (11.81) in Fresnel approximation,

$S \approx \left[S' + \frac{\rho^2}{2S'} + \frac{r_1^2}{2S'} - \vec{\rho}\cdot\left(\frac{\vec{r}_1}{S'}\right)\right]$ and considering $S \approx S'$ in denominator, Eq. (11.93) becomes

$$u(\vec{r}_1) \approx -\frac{i}{\lambda}\frac{e^{ikr'}}{r'}\frac{1}{(2\pi)^2}\int_S\int_A \Psi(\vec{q})F_{k/r'}(\vec{q})e^{i\vec{q}\vec{\rho}}\alpha(\vec{\rho})e^{-ik\frac{\rho^2}{2f}}\frac{e^{ik\left[s'+\frac{\rho^2}{2s'}+\frac{r_1^2}{2s'}-\vec{\rho}\cdot\left(\frac{\vec{r}_1}{s'}\right)\right]}}{S'}d^2\rho d^2 q$$

$$\approx -\frac{l}{\lambda}\frac{e^{ik(r'+S')}}{(2\pi)^2 r' S'}\int_S\int_A \psi(\vec{q})\,F_{k/r'}(\vec{q})\,\alpha(\vec{\rho})\,e^{ik\left[-\frac{\rho^2}{2f}+\frac{\rho^2}{2S'}+\frac{r_1^2}{2S'}\right]+1\rho\cdot\left(q-k\frac{\vec{r}_1}{S'}\right)}d^2\rho\, d^2 q \qquad ...(11.94)$$

or $$u(\vec{r}_1) = \left(-\frac{i}{\lambda}\right)^2\frac{e^{ik(r'+s')}}{(2\pi)^2 r'S'}\int_S\int_A \Psi(\vec{q})F_{k/r'}(\vec{q})\alpha(\vec{\rho})e^{E(\vec{q},\vec{\rho},\vec{r})}d^2\rho d^2\vec{q} \qquad ...(11.95)$$

where,

$$E(\vec{q},\vec{\rho},\vec{r}) = ik\left[-\frac{\rho^2}{2f}+\frac{\rho^2}{2S'}+\frac{r_1^2}{2S'}\right]+i\left(\vec{q}-k\frac{\vec{r}_1}{S'}\right)\cdot\vec{\rho} \qquad ...(11.96)$$

In the back focal plane, $S' = f$, then

$$E(\vec{q},\vec{\rho},\vec{r}) = ik\frac{r_1^2}{2S'}+i\left(\vec{q}-k\frac{\vec{r}_1}{S'}\right)\cdot\vec{\rho}$$

or $$E(\vec{q},\vec{\rho},\vec{r}) = ik\frac{r_1^2}{2S'}+i\vec{q}'\cdot\vec{\rho} \qquad ...(11.97)$$

where, $\vec{r}_1 = \vec{r}_1 \big|_{S'=f}$ and $\vec{q}' = \vec{q} - k\frac{\vec{r}_1}{S'}$.

Putting the value of $E(\vec{q},\vec{\rho},\vec{r})$, the diffraction integral for the field in the back focal plane is

$$u(\vec{r}_1') = \left(-\frac{i}{\lambda}\right)^2\frac{e^{ik(r'+S')}}{(2\pi)^2 r'S'}e^{ik\frac{r_1'^2}{2f}}\int_S \Psi(\vec{q})F_{k/r'}(\vec{q})\alpha(\vec{q}')\,d^2q \qquad ...(11.98)$$

The Fourier transform of the pupil function, α, is very sharply peaked at $\vec{q}' = 0$ (for a circular pupil, α is the airy disk)

For an infinite aperture, $\alpha\,(\vec{q}') = (2\pi)^2\delta(\vec{q}') \Rightarrow \vec{q}' = \vec{q} - k\frac{\vec{r}_1}{s'} = 0$

or
$$\vec{q} = k\frac{\vec{r}_1}{f} \text{ so that } E(\vec{q},\vec{\rho},\vec{r}) = ik\frac{r_1^2}{2S'}.$$

Substituting
$$F_{k/r'}(\vec{q}) = i\lambda r' e^{-\frac{i\vec{r}' q^2}{2k}},\ E(\vec{q},\vec{\rho},\vec{r}) = ik\frac{r_1^2}{2S'}$$

and
$$\alpha(\vec{q}') = (2\pi)^2\delta(\vec{q}') = \int_A \alpha(\vec{\rho})e^{i\vec{\rho}\cdot\vec{q}'}d^2\rho,\ \text{Eq. (11.95) becomes}$$

$$u(\vec{r}_1) = \left(-\frac{i}{\lambda}\right)^2 \frac{e^{ik(r'+s')}}{(2\pi)^2 r' s'} e^{ik\frac{r_1^2}{2f}} \int_S \Psi(\vec{q}) \times i\lambda\cdot\vec{r}' e^{-\frac{i\vec{r}' q^2}{2k}} \times (2\pi)^2 \delta(\vec{q}') d^2q$$

Putting $\vec{q} = k\frac{\vec{r}_1}{f}$, $e^{-\frac{i\vec{r}' q^2}{2k}} = e^{-\frac{ikr' r_1^2}{2f^2}}$, now

$$u(\vec{r}_1') = -\frac{i}{\lambda f}\frac{e^{ik(r'+S')}}{(2\pi)^2 f} e^{ik\frac{r_1'^2}{2f}} \int_S \psi(\vec{q}) \times e^{-\frac{ikr' r_1^2}{2f^2}} \times (2\pi)^2 \delta(\vec{q}') d^2q$$ as $S' = f$ in the back

focal plane

$$= -\frac{i}{\lambda f}\frac{e^{ik\left(r'+S'+\frac{r_1'^2}{2f}-\frac{r' r_1^2}{2f^2}\right)}}{(2\pi)^2 f} \int_S \psi(\vec{q}) \times (2\pi)^2\, \delta(\vec{q}')\, d^2q$$

$$\approx -\frac{i}{\lambda f} e^{ik\frac{r_1'^2}{2}\left(\frac{1}{f}-\frac{r'}{f^2}\right)} \psi\left(k\frac{\vec{r}_1}{f}\right) \qquad \text{...(11.99a)}$$

$$\approx -\frac{i}{\lambda f}\psi\left(k\frac{\vec{r}_1}{f}\right) \qquad \text{...(11.99b)}$$

The back plane contains the Fourier transformation of coherently illuminated object.

Sum of exponents, $\frac{1}{f} - \frac{r'}{f^2} = 0$ or $r' = f$

In Fresnel approximation, the phase factor, $e^{ik\left(r'+s'+\frac{r_1'^2}{2f}-\frac{r' r_1^2}{2f^2}\right)} \approx 1$ as shown below:

$$e^{\frac{ikr'r_1^2}{2f^2}} = e^{\frac{ikr'r_1'^2}{2f^2}} \quad \text{as} \quad \vec{r}_1' = \vec{r}_1 \big|_{S'=f}, \ \vec{r}_1 = \frac{S'}{r'}\vec{r}_0 = \frac{f}{r'}\vec{r}_0, \text{ if } \vec{r}_1 \approx \vec{r}_0, r' = f = S'$$

$$e^{ik\left(\frac{r_1'^2}{2f} - \frac{r' r_1'^2}{2f^2}\right)} = e^{ik\left(\frac{r_1'^2}{2f} - \frac{r_1'^2}{2f}\right)} = 1, \ e^{ik(r' + S')} = 1$$

Example 11.8 Diffraction pattern by a thin lens, when incident light is coherent and extended: Focal plane filtering.

Alter the image by (intentionally or unintentionally) filtering the Fourier transform in the focal plane, propagate the field from the back focal plane to the image plane using yet another diffraction integral.

Field in the image plane is

$$u(\vec{r}_1) = -\frac{1}{\lambda}\int_F \frac{e^{iku}}{u}\sigma(\vec{r}_1')\,u(\vec{r}_1')\,d^2 r_1'$$

Here, σ is focal plane filter transmission function.

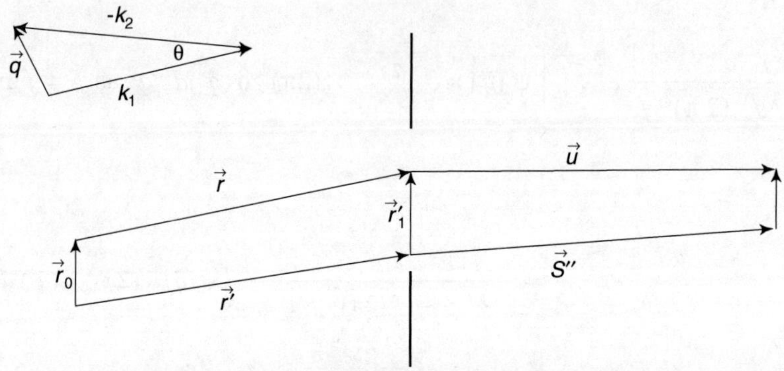

Fig. 11.14 Coordinates for the diffraction integral

Vector from point in focal plane to point in image plane is

$$\vec{u} = \vec{S}'' + \vec{r}_1 - \vec{r}_1'$$

$$u^2 = (\vec{S}'' + \vec{r}_1 - \vec{r}_1')^2$$

$$= S''^2 + (\vec{r}_1 - \vec{r}_1')^2 + 2\vec{S}'' \cdot (\vec{r}_1 - \vec{r}_1')$$

$$u = S''\left[1 + \frac{(\vec{r}_1 - \vec{r}_1')^2}{S^{2''}} + 2\frac{\vec{s}'' \cdot (\vec{r}_1 - \vec{r}_1')}{S^{2'}}\right]^{1/2}$$

$$\approx S''\left[1+\frac{(\vec{r}_1-\vec{r}_1')^2}{2S''^2}+\frac{\vec{S}''\cdot(\vec{r}_1-\vec{r}_1')}{S''^2}+\frac{\frac{1}{2}\times\left(\frac{1}{2}-1\right)}{2\,!}\left(2\frac{\vec{S}''\cdot(\vec{r}_1-\vec{r}_1')}{S''^2}\right)^2\right]$$

$$\approx\left[S''+\frac{(\vec{r}_1-\vec{r}_1')^2}{2\,S''}+\frac{\vec{S}''\cdot(\vec{r}_1-\vec{r}_1')}{S''}+\frac{\frac{1}{2}\times\left(\frac{1}{2}-1\right)}{2!}S''\left(2\frac{\vec{S}''\cdot(\vec{r}_1-\vec{r}_1')}{S''^2}\right)^2\right]$$

$$\approx\left[S''+\frac{(\vec{r}_1-\vec{r}_1')^2}{2S''}+\hat{S}''\cdot(\vec{r}_1-\vec{r}_1')-\frac{\left(\hat{S}''\cdot(\vec{r}_1-\vec{r}_1')\right)^2}{2\hat{S}''}\right]$$

$$\approx S''+\frac{(\vec{r}_1-\vec{r}_1')^2}{2\,S''}$$

$$\approx S''+\frac{(r_1^2+r_1'^2-2\,\vec{r}_1\cdot\vec{r}_1')}{2\,s''} \qquad \text{...(11.100)}$$

$$\vec{r}_0+\vec{r}\;=\;\vec{r}'+\vec{r}_1' \quad \text{or} \quad \vec{r}\;=\;\vec{r}'+\vec{r}_1'-\vec{r}_0$$

Taking square of $\vec{r}$, we get

$$r^2 = r'^2+(\vec{r}_1'-\vec{r}_0)^2+2\,\vec{r}'\cdot(\vec{r}_1'-\vec{r}_0)$$

or $$r = \vec{r}'\left[1+\frac{(\vec{r}_1'-\vec{r}_0)^2}{r'^2}+2\frac{\hat{r}\,(\vec{r}_1'-\vec{r}_0)}{\vec{r}'}\right]^{1/2}$$

or $$r\approx\vec{r}'\left[1+\frac{(\vec{r}_1'-\vec{r}_0)^2}{2\,r'^2}+\frac{\hat{r}\cdot(\vec{r}_1'-\vec{r}_0)}{\vec{r}'}+\frac{\frac{1}{2}\times\left(\frac{1}{2}-1\right)}{2\,!}\left(2\frac{\hat{r}\cdot(\vec{r}_1'-\vec{r}_0)}{\vec{r}'}\right)^2\right]$$

or $$r\approx\vec{r}'\left[1+\frac{(\vec{r}_1'-\vec{r}_0)^2}{2\,r'^2}+\frac{\hat{r}\cdot(\vec{r}_1-\vec{r}_0)}{\vec{r}'}-\frac{1}{2}\left(\frac{\hat{r}\cdot(\vec{r}_1'-\vec{r}_0)}{\vec{r}'}\right)^2\right]$$

or $$r\approx\vec{r}'+\frac{(\vec{r}_1'-\vec{r}_0)^2}{2\,\vec{r}'}+\hat{r}\,.\,(\vec{r}_1'-\vec{r}_0)-\frac{(\hat{r}\,(\vec{r}_1'-\vec{r}_0))^2}{2\,\vec{r}'}$$

or $$r\approx\vec{r}'+\frac{(\vec{r}_1'-\vec{r}_0)^2}{2\,\vec{r}'} \qquad \text{...(11.101)}$$

The fields due to a distribution (ψ) of point sources is written as [see Eq. (11.85)]

$$u(\vec{r}_1)\;\approx\;-\frac{i}{\lambda}\iint_{SA}\psi(\vec{r}_0)\frac{e^{ikr}}{r}\sigma(\vec{r}_1')\,e^{-ik\frac{r_1'^2}{2f}}\frac{e^{iku}}{u}d^2\,\vec{r}_1'\,d^2\,r_0 \qquad \text{...(11.102)}$$

Here, σ is the focal-plane filtering function. It propagates the field from the back focal plane to the image plane.

Instead of performing two 2-*d* integrations simultaneously, we can evaluate a diffraction integral to go from the source to the lens and another to go from the lens to the observation plane.

$$u(\vec{r}_1) \approx \int_A u_{\text{inc}}(\vec{r}_1')\,\sigma(\vec{r}_1')\,e^{-ik\frac{r_1'^2}{2f}}\,\frac{e^{iku}}{u}\,d^2\vec{r}_1' \qquad \text{...(11.103)}$$

Field incident at lens is written as follows:

$$u_{\text{inc}}(\vec{r}_1') \approx -\frac{i}{\lambda}\int_S \psi(\vec{r}_0)\,\frac{e^{ikr}}{r}\,d^2 r_0 \qquad \text{...(11.104)}$$

In Fresnel approximation, $r = r' + \frac{(\vec{r}_1' - r_0)^2}{2r'}$ [Eq. (11.101)] and taking, $r = r'$ in the denominator, we get

$$u_{\text{inc}}(\vec{r}_1') \approx -\frac{i}{\lambda}\frac{e^{ikr'}}{r'}\int_S \psi(\vec{r}_0)\,e^{ik\frac{(\vec{r}_1' - r_0)^2}{2r'}}\,d^2 r_0$$

$$\approx -\frac{i}{\lambda}\frac{e^{ikr'}}{r'}\int_S \psi(\vec{r}_0)\,F_{k/r'}(\vec{r}_1' - r_0)\,d^2 r_0 \qquad \text{...(11.105)}$$

u_{inc} is a convolution of the source distribution (ψ) with the Fresnel function, $F_{k/r'}$, where,

$$F_\alpha(\vec{r}) = e^{i\frac{\alpha}{2}r^2} \qquad \text{...(11.106)}$$

$\vec{r}$ lies in the $x - y$ plane.

The 2-*d* Fourier transform of $F_\alpha(\vec{r})$ is

$$F_\alpha(\vec{r}) = \int e^{-iqr}\,F_\alpha(\vec{q})\,d^3 r$$

$$= \frac{2i\pi}{\alpha}F_{-1/\alpha}(\vec{q})$$

$$= \frac{2i\pi}{\alpha}e^{-\frac{iq^2}{2\alpha}} \qquad \text{...(11.107)}$$

or

$$F_{k/\vec{r}'}(\vec{r}) = \frac{2i\pi\vec{r}'}{k}e^{-\frac{i\vec{r}'q^2}{2k}} = i\lambda\vec{r}'\,e^{-\frac{i\vec{r}'q^2}{2k}} \qquad \text{...(11.108)}$$

Imaging by a thin lens of an extended, coherently illuminated object, part 2:

The field incident on the lens, u_{inc}, is a convolution of the source distribution (ψ) with the Fresnel function, $F_{k/r'}$.

$$u_{inc}(\vec{r}_1') = -\frac{i}{\lambda}\frac{e^{ikr'}}{r'}\int_S \psi(\vec{r}_0)\, F_{k/r'}(\vec{r}_1' - \vec{r}_0)\, d^2 r_0$$

$$= -\frac{i}{\lambda}\frac{e^{ikr'}}{r'}\frac{1}{(2\pi)^2}\int_S \psi(\vec{q})\, F_{k/r'}(\vec{q})\, e^{i\vec{q}\cdot\vec{r}_1'}\, d^2 q \quad ...(11.109)$$

Putting $u_{inc}(\vec{r}_1')$, the diffraction integral for the field in the observation plane becomes

$$\text{or } u(\vec{r}_1) \approx -\frac{i}{\lambda}\frac{e^{ikr'}}{r'}\frac{1}{(2\pi)^2}\int\int_{SA} \psi(\vec{q})\, F_{k/r'}(\vec{q})\, e^{i\vec{q}\cdot\vec{r}_1'}\, \sigma(\vec{\rho})\, e^{-ik\frac{r_1'^2}{2f}}\, \frac{e^{iku}}{u}\, d^2 r_1'\, d^2 q \quad ...(11.110)$$

Putting $F_{k/r'}(\vec{q}) = i\lambda r' e^{-\frac{i r' q^2}{2k}}$, the exponent u in Fresnel approximation, $u \approx \left[S'' + \frac{r_1'^2}{2S''} + \frac{r_1^2}{2S''} - \vec{r}_1' \cdot \left(\frac{\vec{r}_1}{S''}\right)\right]$ Eq. (11.100} and considering $u \approx S''$ in denominator, Eq. (11.110) becomes

$$u(\vec{r}_1) \approx -\frac{i}{\lambda}\frac{e^{ikr'}}{r'}\frac{1}{(2\pi)^2}\int\int_{SA} \psi(\vec{q})\left(i\lambda r' e^{-\frac{i r' q^2}{2k}}\right) e^{i\vec{q}\cdot\vec{r}_1'}\, \sigma(\vec{r}_1')\, e^{-ik\frac{r_1'^2}{2f}}\, \frac{e^{ik\left[S'' + \frac{r_1'^2}{2S''} + \frac{r_1^2}{2S''} - \vec{r}_1'\cdot\left(\frac{\vec{r}_1}{S''}\right)\right]}}{S''}\, d^2\vec{r}_1'\, d^2 q$$

$$\approx -\frac{i}{\lambda}\frac{e^{ik(r'+S'')}}{(2\pi)^2 r' S''}\int\int_{SA} \psi(\vec{q})\,(i\lambda r')\,\sigma(\vec{r}_1')\, e^{ik\left[-\frac{\vec{r}' q^2}{2k^2} - \frac{r_1'^2}{2f} + \frac{r_1'^2}{2S''} + \frac{r_1^2}{2S''}\right] + 1\vec{r}_1'\cdot\left(q - k\frac{\vec{r}_1}{S''}\right)}\, d^2\vec{r}_1'\, d^2 q$$

...(11.111)

$$\text{or } u(\vec{r}_1) = \left(-\frac{i}{\lambda}\right)^2 \frac{e^{ik(r'+s'')}}{(2\pi)^2 r' S''}\int\int_{SA} \psi(\vec{q})\,(i\lambda r')\,\sigma(\vec{r}_1')\, e^{E(\vec{q}\cdot\vec{r}_1', \vec{r})}\, d^2\vec{r}_1'\, d^2\vec{q} \quad ...(11.112)$$

where,

$$E(\vec{q}, \vec{r}_1', \vec{r}) = ik\left[-\frac{\vec{r}' q^2}{2k^2} - \frac{r_1'^2}{2f} + \frac{r_1'^2}{2S''} + \frac{r_1^2}{2S''}\right] + i\left(\vec{q} - k\frac{\vec{r}_1}{S''}\right)\cdot\vec{r}_1' \quad ...(11.113)$$

In the back focal plane, $S'' = f$, then

$$E(\vec{q}, \vec{r}_1', \vec{r}) = ik\left(-\frac{\vec{r}' q^2}{2k^2} - \frac{r_1'^2}{2f} + \frac{r_1'^2}{2f} + \frac{r_1^2}{2f}\right) + i\left(\vec{q} - k\frac{\vec{r}_1}{S''}\right)\cdot\vec{r}_1'$$

or $$E(\vec{q},\vec{r}_1',\vec{r}) = ik\left(-\frac{\vec{r}'q^2}{2k^2}+\frac{r_1^2}{2f}\right)+i\vec{q}'\cdot\vec{r}_1' \qquad ...(11.114)$$

where, $\vec{q}' = \vec{q} - k\dfrac{\vec{r}_1}{S''}$.

Putting the value of $E(\vec{q},\vec{r}_1',\vec{r})$, the diffraction integral, for the field in the back focal plane, given by Eq. (11.112) becomes

or $$u(\vec{r}_1) = \left(-\frac{i}{\lambda}\right)^2 \frac{e^{ik(r'+S'')}}{(2\pi)^2 r'S''}\int_S\int_A \psi(\vec{q})\,(i\lambda r')\,\sigma(\vec{r}_1')\,e^{ik\left(-\frac{\vec{r}'q^2}{2k^2}+\frac{r_1^2}{2f}\right)+i\vec{q}'\cdot\vec{r}_1'}\,d^2\vec{r}_1'\,d^2\vec{q}$$

as $$\vec{r}_1' = \vec{r}_1'\lceil_{S''=f}$$

or $$u(\vec{r}_1') = \left(-\frac{i}{\lambda}\right)^2 \frac{e^{ik(r'+S'')}}{(2\pi)^2 r'S''}e^{ik\left(\frac{r_1'^2}{2f}-\frac{r_1^2}{2f^2}\right)}\int_S \psi(\vec{q})\,(i\lambda r')\,\sigma(\vec{q}')\,d^2q \qquad ...(11.115a)$$

where $$\sigma(\vec{q}\,') = \int_A \sigma(\vec{r}_1')\,e^{i\vec{q}'\cdot\vec{r}_1'}\,d^2\vec{r}_1'$$

The Fourier transform of pupil function, σ, is very sharply peaked at $\vec{q} = 0$ (for a circular pupil, σ is the Airy disk). For an infinite aperture, $\sigma(\vec{q}\,') = (2\pi)^2\,\delta(\vec{q}\,') \Rightarrow \vec{q}\,' = \vec{q} - k\dfrac{\vec{r}_1}{S''} = 0$ or $\vec{q} = k\dfrac{\vec{r}_1}{f}$, so that $\dfrac{i\vec{r}'q^2}{2k} = \dfrac{ikr'\,r_1^2}{2f^2}$.

Putting the value $\dfrac{i\vec{r}'q^2}{2k} = \dfrac{ikr'\,r_1^2}{2f^2}$, Eq. (11.115a) becomes

$$u(\vec{r}_1') = \left(-\frac{i}{\lambda}\right)^2 \frac{e^{ik(r'+S'')}}{(2\pi)^2 r'S''}e^{ik\left(\frac{ikr'r_1^2}{2f^2}+\frac{r_1^2}{2f}\right)}\int_S \psi(\vec{q})\,(i\lambda r')\,\sigma(\vec{q}')\,d^2q \qquad ...(11.115b)$$

Substituting. $\alpha(\vec{q}\,') = (2\pi)^2\,\delta(\vec{q}\,')$, eqn. (11.115b) becomes

$$u(\vec{r}_1') = \left(-\frac{i}{\lambda}\right)^2 \frac{e^{ik(r'+S'')}}{(2\pi)^2 r'S''}e^{ik\left(-\frac{ikr'r_1^2}{2f^2}+\frac{r_1^2}{2f}\right)}\int_F \psi(\vec{q})\times(i\lambda r')\times(2\pi)^2\,\delta(\vec{q})\,d^2q$$

Meaning of $\delta(\vec{q}\,')$ is that for $\vec{q}\,' = 0$, $\vec{q} = k\dfrac{\vec{r}_1}{f}$, integral is removed, then one gets

$$u(\vec{r}_1') = -\frac{1}{\lambda f} e^{ik\frac{r_1'^2}{2}\left(\frac{1}{r} - \frac{r'}{f^2}\right)} \psi\left(k\frac{\vec{r}_1}{f}\right) \quad \text{...(11.116)}$$

$$e^{ik(r' + S'')} = 1$$

Sum of exponents in the diffraction integral is

$$\frac{ikr_1'^2}{2}\left(\frac{1}{f} - \frac{r'}{f^2} + \frac{1}{S''}\right)$$

Using Newton's equation for the focus-to-image at distance r_1, factor

$$\left(\frac{1}{f} - \frac{r'}{f^2} + \frac{1}{S''}\right) = 0$$

or

$$\frac{1}{f} - \frac{r'}{f^2} = -\frac{1}{S''} \quad \text{or} \quad \frac{f - r'}{f^2} = -\frac{1}{S''}$$

Since , $\vec{r}_1' = \vec{r}_1\big|_{S''=f}$ at focus, so $f - r_1 = -S''$

This states that all quadratic terms cancel in the image plane, $f - r_1 = -S''$

UNSOLVED PROBLEM

Example 11.1 Imaging by a thin lens of an extended, coherently illuminated object.

Field amplitude in the image plane is

$$u(\vec{r}_1') = \left(-\frac{i}{\lambda}\right)^2 \frac{1}{(2\pi)^2 s'' f} e^{ik\frac{r_1'^2}{2s''}} \int_F \sigma(\vec{r}_1')\, \psi\left(k\frac{\vec{r}_1}{f}\right) \frac{e^{ik\frac{\vec{r}_1'\cdot\vec{r}_1}{s''}}}{s''} d^2 r_2' \quad \text{...(11.117)}$$

The image-plane field is proportional to the Fourier transform of the product of the focal-plane filtering function (σ) and the Fourier transform of the original field.

APPENDIX A

Two-Dimensional Reciprocal Lattice Vectors

Let $\vec{a}_1$ and $\vec{a}_2$ abe non-parallel lattice vectors in 2-*d* lattice. The reciprocal lattice vector is defined as

$$\vec{R} = m_1 \vec{a}_1 + m_2 \vec{a}_2 \qquad ...(A1)$$

The reciprocal lattice vectors are

$$\vec{b}_1 = \frac{\vec{a}_2 \times \hat{z}}{\hat{\vec{z}}\,(\vec{a}_1 \times \vec{a}_2)}, \quad \vec{b}_2 = \frac{\hat{z} \times \vec{a}_1}{\hat{z}\,(\vec{a}_1 \times \vec{a}_2)}$$

Here, $\hat{z}$ is unit vector perpendicular to $x - y$ plane. The reciprocal lattice vector is now defined as

$$\vec{G} = m_1 \vec{b}_1 + m_2 \vec{b}_2 \qquad ...(A2)$$

where, $\vec{b}_j \cdot \vec{a}_j = \delta_{i,j}$.

Lattice vectors of 2-d hexagonal lattice are: $\vec{a}_1 = \hat{x}$, $\vec{a}_2 = \frac{1}{2}\hat{x} + \frac{\sqrt{3}}{2}\hat{y}$ ...(A3)

Reciprocal lattice vectors of 2-d hexagonal lattice are: $\vec{b}_1 = \hat{x} - \frac{1}{\sqrt{3}}\hat{y}, \vec{b}_2 = \frac{2}{\sqrt{3}}\hat{y}$ ...(A4)

Formulas at a Glance

S.No.	Formulas	Explanation of symbols
1.	Kirchhoff's integral theorem $\frac{1}{4\pi}\int_{\partial V} (-\hat{n})\cdot\vec{F}\,dS = u(x) = \frac{e^{ikx}}{x}$	Incident spherical wavefront $u(x) = e^{ikx}/x$, $\hat{n}$ is inward normal to surface (∂V) that bounds volume V. The vector field due to spherical wave emerging from point light source, $\vec{F} = (u\,\nabla G - G\,\nabla u)$
2.	The Green function $G(\vec{r},\vec{x}) = \frac{e^{ik[\vec{r}-\vec{x}]}}{\lvert\vec{r}-\vec{x}\rvert}$	At position $\vec{r}$ due to spherical wave at position $\vec{x}$
3.	Helmholtz equation $(\nabla^2 + k^2)\,u\,(\vec{r}) = 0$	Spherical wavefront, $u(\vec{r}) = e^{ikr}/r$ at position $r \cdot k = 2\pi/\lambda$
4.	Fresnel-Kirchhoff's integral formula $u(\vec{x}) = -\frac{iA}{2\lambda}\int_{dV} \frac{e^{ik(r+s)}}{rS}(\hat{n}\cdot\hat{s} + \hat{n}\cdot\hat{r})\,dS$	$\hat{n}$ is inward normal to surface (dV) that bounds volume, $V \cdot k = \frac{2\pi}{\lambda}$ is wave vector incident spherical wave $\frac{e^{ikr}}{r}$ at point r on the aperture and diffracted spherical wave $\frac{e^{ik}}{s}$ at point S of the diffraction pattern on screen.
5.	Wave inclimation angle $I = (\hat{n}\cdot\hat{s} + \hat{n}\cdot\hat{r})$	$\hat{n}$ is inward normal to surface dV that bounds volume V. $\hat{r}$ and $\hat{s}$ are unit vectors aling incident direction $\vec{r}$ and $\vec{S}$

6.	Fresnel-Kirchhoff diffraction formula $u(\vec{x}) \approx \frac{ikAI}{4\pi} \int_{\partial V} \frac{e^{ik(r+s)}}{rS} dS$ The inclimation factor, $I = 2$	
7.	Fresnel-Kirchhoff formula for transmitted wave $u(\vec{x}) = \frac{ikl}{4\pi} \int_{\partial V} \tau(\vec{r}) \frac{e^{ik(r+s)}}{rS} dS$	Transmission function of a thin lens of thickness (t) and focal length (f) is defined as $\tau(\rho) \approx \lvert\tau(\rho)\rvert \exp^{(iknt)} \exp\{k \rho^2/2t)$ is half width of the beam.
8.	Fraunhofer integral formula for rectangular aperture in two dimensions $u(\vec{x}) = \frac{ikAI}{4\pi} \times \frac{e^{ik(r'+s')}}{r's'} e^{i\vec{k}_2 \cdot \vec{r}_1}$	Dimensions of the rectangular aperture - $2a$ (along x) × $2b$ (along y), Fraunhofer integral formula
9.	Formula for intensity of maxima in case of rectangular aperture $S(\vec{q}) = S(0) \sin c^2 (q_x a) \sin c^2 (q_x a)$	$\vec{q} = \vec{k}_1 - \vec{k}_2, \vec{k}_1 = k\hat{r}', \vec{k}_2 = k\hat{S}, \ldots \vec{q}$ is not necessarily in the plane of the aperture, but we can assume that in this case $\vec{q}$ is in the plane of aperture.
10.	Frauanhofer diffraction integral formula for a circular aperture of radius a is $u(\vec{x}) = \frac{ikAI}{4\pi} \times \frac{e^{ik(r'+s')}}{r'S'}$ $e^{i\vec{k}_2 \cdot \vec{r}_1} \int_0^{2\pi} \int_0^{a} e^{iq\rho\cos\pi} \rho \, d\rho \, d\phi$ = (constant (πa^2) $\left[\frac{2J_1(qa)}{qa}\right]$	$e^{i\vec{k}_2 \cdot \vec{r}_1}$ is an incident plane wave. A is the radius of the circular aperture.
11.	The intensity of maximum in case of circular aperture $S(\vec{q}) = S(0)\left[\frac{2J_1(qa)}{qa}\right]^2$ (the airy pattern)	J_1 is Bessel function of first order.
12.	Fromula for fields due to a distribution(y) of point source for getting diffraction pattern by a thin lens, when incident source light iscoherent and extended $u(\vec{r}_1) \approx -\frac{i}{\lambda} \int_S \int_A \psi(\vec{r}_0) \frac{e^{ikr}}{r} \alpha(\vec{\rho}) \times$ $e^{-ik\frac{\rho^2}{2f}} \frac{e^{iks}}{S} d^2\rho \, d^2 r_0$	The solution of this formula in Fresnel approximation is given in solved examples 11.7 and 11.8

QUESTIONS FOR EXAMINATION

11.1 Write Maxwell's equations for dielectric medium and free space.

11.2 Write similarties and dissimilarties between Gauss theorem and Kirchhoff integral theorem.

11.3 What is Gauss theorem?

11.4 What is Kirchhoff integral theorem?

11.5 Derive the light propagation equation.

11.6 Derive the Fresnel Kirchhoff integral formula.

11.7 Derive and plot the grating function for Fraunhofer diffraction by periodically arranged identical apertures.

11.8 Solve fields distribution formula of point sources for finding the diffraction pattern by a thin lens, when incident source light is coherent and extended using Fresnel approximation.

11.9 Solve fields distribution formula of point sources for finding the diffraction pattern by a thin lens, when incident source light is coherent and extended for focal plane filtering using Fresnel approximation.

OBJECTIVE QUESTIONS

11.1 Tell which statement is right? Gauss theorem is defined as

(a) Total flux of light rays enclosed by the closed surface is equal to incident spherical wavefront, $u(x) = e^{ikx}/x$.

(b) Total flux of light rays enclosed by the closed surface is twice to incident spherical wavefront, $u(x) = e^{ikx}/x$.

(c) Total flux enclosed by the closed surface is equal to the charge enclosed by it.

(d) The surface integral of vector field $(\vec{E}_0)$ over a closed surface enclosing volume (V) is equal to the total flux of light waves enclosed in that surface.

11.2 Tell which statement is right? Helmholtz-Kirchhoff integral theorem is defined as

(a) Total flux of light rays enclosed by the closed surface is equal to incident spherical wavefront, $u(x) = e^{ikx}/x$.

(b) Total flux of light rays enclosed by the closed surface is twice to incident spherical wavefront, $u(x) = e^{ikx}/x$.

(c) Total flux enclosed by the closed surface is equal to the charge enclosed by it.

(d) The surface integral of electrostatic field $(\vec{E}_0)$ over a closed surface enclosing volume (V) is equal to the total charge enclosed within that surface.

11.3 The inclination factor, $I = -\ (\hat{n}\cdot\hat{s} + \hat{n}\cdot\hat{r})$ for forward direction is nearly

(a) 0 (b) – 1 (c) – 2 (d) none of these

11.4 The inclination factor $I = (\hat{n}\cdot\hat{s} + \hat{n}\cdot\hat{r})$ for backward direction is

(a) 0 (b) – 1 (c) – 2 (d) none of these

11.5 Fresnel-Kirchhoff formula is

(a) $u(\vec{x}) \approx \frac{ikAl}{4\pi} \int_{\partial V} \frac{e^{ikr}}{r} dS$ (b) $u(\vec{x}) \approx \frac{ikAl}{4\pi} \int_{\partial V} \frac{e^{ik(r+s)}}{rS} dS$

(c) $u(\vec{x}) \approx \frac{ikAl}{4\pi} \int_{\partial V} \frac{e^{iks}}{s} dS$ (d) none of these

11.6 Maxwell's equations in dielectric and magnetic medium are

$$\in \nabla\cdot\vec{E} = \rho,\ \nabla\cdot\vec{B} = 0,\ \nabla\times\vec{E} = -\mu\frac{\partial\vec{H}}{\partial t},\ \nabla\times\vec{H} = \sigma\vec{E} + \in\frac{\partial\vec{E}}{\partial t}$$

Maxwell's equations in free space are

$$\in \nabla\cdot\vec{E} = 0,\ \mu_0\nabla\cdot\vec{H} = 0,\ \nabla\times\vec{E} = -\mu\frac{\partial\vec{H}}{\partial t},\ \nabla\times\vec{H} = \in\frac{\partial\vec{E}}{\partial t}$$

Out of the following, which set of Maxwell equation is used to derive light wave propagation equation?

(a) $\in \nabla\cdot\vec{E} = \rho,\ \nabla\times\vec{E} = -\mu\frac{\partial\vec{H}}{\partial t},\ \nabla\times\vec{H} = \sigma\vec{E} + \in\frac{\partial\vec{E}}{\partial t}$

(b) $\nabla\cdot\vec{B} = 0,\ \nabla\times\vec{E} = -\mu\frac{\partial\vec{H}}{\partial t},\ \nabla\times\vec{H} = \in\frac{\partial\vec{E}}{\partial t}$

(c) $\in \nabla\cdot\vec{E} = \rho,\ \nabla\cdot\vec{B} = 0,\ \nabla\times\vec{E} = -\mu\frac{\partial\vec{H}}{\partial t}$

(d) $\in \nabla\cdot\vec{E} = 0,\ \nabla\times\vec{E} = -\mu\frac{\partial\vec{H}}{\partial t},\ \nabla\times\vec{H} = \in\frac{\partial\vec{E}}{\partial t}$

11.7 Which is the formula maximum intensity of diffraction pattern in case of rectangular aperture?

(a) $S(\vec{q}) = S(0)\sin c^2(q_x a)\sin c^2(q_y b)$

(b) $S(\vec{q}) = S(0)\sin c^2(q_x a)\sin c^2(q_y b)$

(c) $S(\vec{q}) = S(0)\left[\frac{2J_1(qa)}{qa}\right]^2$

(d) $S(\vec{q}) = S(0)\sin c^2(q_x q_y a b)$

11.8 Which is the formula maximum intensity of diffraction pattern in case of circular aperture?

(a) $S(\vec{q}) = S(0)\sin c^2(q_x a)\sin c^2(q_y b)$

(b) $S(\vec{q}) = S(0)\sin c^2(q_x a)\sin c^2(q_y b)$

(c) $S(\vec{q}) = S(0)\left[\frac{2J_1(qa)}{qa}\right]^2$

(d) $S(\vec{q}) = S(0)\sin c^2\,(q_x\, q_y\, a\, b)$

Answers

11.1 (c) 11.2 (a) 11.3 (c) 11.4 (a) 11.5 (b) 11.6 (d) 11.7 (a) 11.8 (c)

CHAPTER 12

Fraunhofer Diffraction

12.1 INTRODUCTION

In this chapter the diffraction of light using Fraunhofer diffraction is discussed. However, answers of various questions, like:

- What is diffraction?
- What is Fraunhofer diffraction?
- What is width of central maxima in Fraunhofer diffraction?
- What is resolving power of grating? What is angular dispersive power?
- What is linear dispersive power?

are given in brief.

When, photons are incident on a crystal, the process of emitting secondary waves is called diffraction.

When these secondary (or diffracting) waves overlap, then the distribution of light intensity in a certain manner is called diffraction pattern.

The diffraction phenomena are of two types: (i) Fraunhofer diffraction, and (ii) Fresnel diffraction.

In case of Fraunhofer diffraction the distance of the source and screen is infinite from the slit. So incident wavefront is plane. The secondary wavefront emerging from the slit is plane wavefront and all the wavelets are in same phase. In Fresnel diffraction the source of light or screen or both are at finite distance from the diffracting element (biprism). So incident wavefront is spherical. The secondary wavefront emerging from the biprism is also spherical wavefront and all the wavelets are not in same phase.

In Fraunhofer diffraction pattern, the maxima at the central point has the highest intensity. The maxima at the central point is known as *principal* or *central maxima*. The central maxima is surrounded by the diminishing maxima separated by minima. As we move away from the central maxima, the intensity of each maxima diminishes very fast.If intensity of the central maxima (I_0) is assumed to be unity, then intensity of the first, second, third,.... secondary maxima on either side of central maxima are in the ratio, $\frac{1}{22}:\frac{1}{61}:\frac{1}{121}:....$.

The width of the central maxima is [Eq. 12.10]

$$2x = \frac{2f\lambda}{a},$$

where, x is the half width of central maxima. λ and a are wavelengths of the source and width of the slit, respectively. f is focal length of convex lens used to focus the diffraction pattern.

Resolving power of a grating is referred to as its ability to resolve two close spectral lines, so that they can be viewed or photographed as distinct separate lines.

If a parallel beam of light of wavelength, λ and $\lambda + d\lambda$, is diffracted through the plane transmission grating, then the resolving power of the grating is defined as the ratio of wavelength (λ) to the difference ($d\lambda$) of the wavelengths, i.e., $\frac{\lambda}{d\lambda}$.

Rayleigh's criterion for lines resolution is as follows: If two point sources or two spectral lines of equal intensity can be resolved by an optical instrument when the central maximum of the diffraction pattern due to one falls on the first minimum of the diffraction pattern due to the other.

If $d\vartheta$ is the angular separation between the central points of two lines of wavelength, λ and $\lambda + d\lambda$, [Fig. 12.16], then angular dispersive power $\left(\frac{d\theta}{d\lambda}\right)$ is defined as [Eqs. (12.72) and (12.75)]

$$\frac{d\theta}{d\lambda} = \frac{n}{(a+b)\cos\vartheta} = \frac{n}{\sqrt{(a+b)^2 - n^2\lambda^2}}$$

If dl is the separation between the central points of two lines of wavelengths, λ and $\lambda + d\lambda$ [Fig. 12.16], then linear dispersive power is defined as [Eqs. 12.79 and (12.80)].

$$\frac{dl}{d\lambda} = f\frac{d\theta}{d\lambda} = \frac{fn}{(a+b)\cos\vartheta} = \frac{fn}{\sqrt{(a+b)^2 - n^2\lambda^2}},$$

where, n and $(a + b)$ are order of diffraction and grating element, respectively. ϑ is the dispersive angle [Fig. 12.6] and f is the focal length of convex lens used to focus the diffraction pattern.

12.2 DIFFRACTION

For laymen, the phenomenon of bending of beam around the corners or edges of an obstacle is called diffraction. In fact, when photons are incidenton a crystalthey emit secondary waves of same wavelength. These secondary waves are also called the *diffracting waves*. In short, the diffraction is defined as follows:

When, photons incident on a crystal, the process of emitting secondary waves is called diffraction.

When these secondary (or diffracting) waves overlap, then the distribution of light intensity in a certain manner is called diffraction pattern.

The secondary (or diffracted) waves are emitted as follows:

When photons are incident on the atoms of the crystal, the electrons of atoms in solid are excited by absorbing photon energy. After some time, when excited electrons drop down to parent energy estate, they emit the rays of same energy in all directions. Thus, the process absorption of photons by atoms of solid and subsequently emitting the rays of same frequency in all directions is called diffraction. These diffracted rays after overlapping give rise to diffraction pattern.

Comparison between interference and diffraction

	Interference	Diffraction
1.	Interference is the phenomenon of overlapping of two wavefronts originating from two coherent sources.	Diffraction is the phenomenon of overlapping of secondary waves emerging from different points of the crystal by one source.
2.	In an interference pattern the intensity of all maxima is equal.	In diffraction pattern the intensity the intensity of maxima is not equal. It decrases as the order of it increases.
3.	In an interference pattern, the intensity of dark fringes is almost zero	In diffraction pattern, the intensity of dark fringes is not exactly zero.
4.	In the interference pattern, the spacing between maxima may (double slit) or may not (Newton's rings) be equal.	In the diffraction pattern, the spacing between maxima are not equal.

Types of diffraction phenomena:

1. Fraunhofer diffraction
2. Fresnel diffraction

Fraunhofer diffraction

(i) The diffraction pattern is produced when the wavefront (plane) passes through a slit.

(ii) The source of light and screen are at infinite distance from the diffracting element (slit).

(iii) A convex lens is used to produce plane wavefront from the spherical or cylindrical wavefront.

(iv) Another convex lens is used to focus the diffracting light on the screen.

(v) The secondary waves diffracting from the aperture are in the same phase at every point.

(vi) The secondary waves originating from aperture are focused with a convex lens. The light emerging convex lens interference and form diffraction pattern on the screen.

(vii) In Fraunhofer diffraction, angular inclination are important rather than distance of aperture from source and screen.

(viii) In Fraunhofer diffraction pattern can be reproduced by theoretical analysis almost exactly.

It has a number of important practical applications.

Fresnel diffraction

(i) The diffraction pattern is produced when the wavefront (spherical or cylindrical) passes through a biprism.

(ii) The source of light or screen or both are at finite distance from the diffracting element (biprism).

(iii) No lens is used because in Fresnel diffraction as spherical or cylindrical wavefront is incident on the prism to produce diffraction pattern.

(iv) No lens is used to focus the diffracted wavefront on the screen.

(v) The phase of secondary waves emerging from the prism are not the same.

(vi) The diffraction pattern on the screen is obtained by interference of secondary waves coming from different part of prism.

(vii) In Fresnel diffraction distance from the prism from source and screen is important rather the angular inclination.

(viii) The Fresnel diffraction can be reproduced approximately by theoretical analysis.

12.3 FRAUNHOFER DIFFRACTION AT SINGLE SLIT

In case of Fraunhofer diffraction the distance of source and screen is infinite from the slit. So incident wavefront is plane. The secondary wavefront emerging from the slit is plane wavefront and all the wavelets are in same phase.

In Fraunhofer diffraction method the spherical wavefront emerging from light source is made plane wavefront by putting the source at focus of a convex lens. This plane wavefront is incident on the slit. The diffracted light is focused on the screen with the help of a convex lens.

In Fig. 12.1 the ray diagram of experimental setup is shown. *S* is a narrow source of the monochromatic light of wavelength, λ. Source (*S*) is placed at the focus of the convex lens 1 before a narrow slit *CD* of width '*a*'. When the plane wavefront *WW'* is incident on the slit *CD*, each atom at the edges of slit becomes a source of secondary waves. The diffracted secondary waves are focused on the screen (*YY'*) by using a convex lens 2. The path difference of the waves, emerging from upper half of slit *OC* and lower half of slit *OD* and reaching at point (P_0), is zero, so the secondary waves after overlapping gives rise to a bright spot at central point (P_0) of the screen. Let us now find the intensity at point *P* due to secondary rays diffracted at an angle ϑ. The waves travelling at an angle ϑ with the horizontal axis are focused at point P on the screen. The intensity at point P is maximum or minimum depends on the path difference between secondary waves originating from upper half of slit OC and lower half of slit OD and reaching at point P.

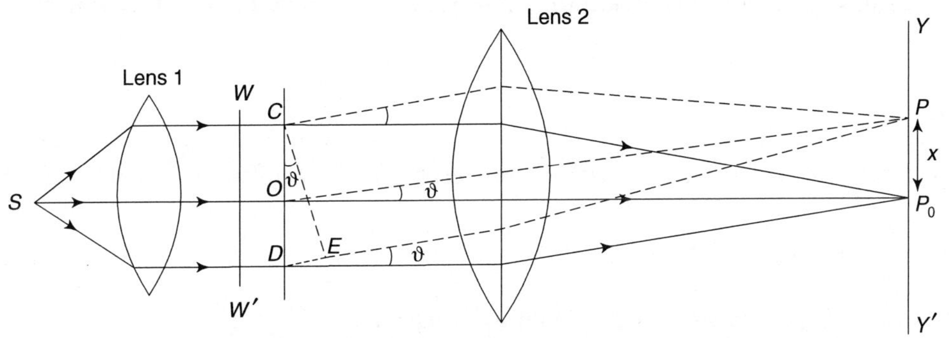

Fig. 12.1 Fraunhofer diffraction at a single slit

In order to find the intensity at point P, let us draw a perpendicular (CE) at line DE. The optical path of the upper half and lower half of the plane wavefront CE to point P are equal. Therefore, the path difference between waves originating from C and D in the direction ϑ is

$$DE = CD \sin \vartheta = a \sin \vartheta \qquad ...(12.1)$$

If this path difference is equal wavelength, λ i.e.,

$DE = a \sin \vartheta = \lambda$ or then there will be minimum intensity point, P. This can be understood as follows:

Let assume that slit CD is divided into two parts, CO and OD. Now Eq. (12.1) is written as

$$\frac{a}{2} \sin \vartheta = \frac{\lambda}{2} \qquad ...(12.2)$$

That is, the secondary waves emerging from each half of CD will differ in path at P by $\frac{\lambda}{2}$, so the destructive interference takes place. Hence, the first minimum on either side of P_0 will occur in a direction ϑ, which can be obtained from the equation

$$DE = a \sin \vartheta = \lambda \quad \text{or} \quad \sin \vartheta = \frac{\lambda}{a}$$

or

$$\vartheta = \sin^{-1} \frac{\lambda}{a} \quad \text{or} \quad \vartheta = \frac{\lambda}{a}$$

as the angle of diffraction (ϑ) is usually very small.

If the direction of secondary waves is such that

$$DE = a \sin \vartheta = 2 \qquad ...(12.3)$$

then there will be minimum intensity point P. This can be understood as follows:

Let assume that slit CD is divided into four parts of $\frac{a}{4}$. Now Eq. (12.3) is written as

$$\frac{a}{4} \sin \vartheta = \frac{\lambda}{2} \qquad ...(12.4)$$

That is, the secondary waves emerging from each part of CD of width $\frac{a}{4}$ will differ in path at P by $\frac{\lambda}{2}$, so the destructive interference takes place. Hence, the second minimum on either side of P_0 will occur in a direction (ϑ) given by equation

$$DE = a \sin \vartheta = 2\lambda \quad \text{or} \quad \sin \vartheta = 2$$

or

$$\vartheta = \sin^{-1} \frac{2\lambda}{a} \quad \text{or} \quad \vartheta = \frac{2\lambda}{a}$$

as the angle of diffraction (ϑ) is usually very small.

In general, the various minima will occur when path difference between extreme waves is an integral multiple of λ or an even multiple of $\frac{\lambda}{2}$, i.e.,

$$BE = a \sin \vartheta_n = 2n \frac{\lambda}{2} = n\lambda, \qquad ...(12.5)$$

where, ϑ_n gives the direction of the nth minimum and n is an integer.

If however, the path difference (DE) is an odd multiple of $\frac{\lambda}{2}$, the direction of various secondary maxima can be obtained from equation

$$DE = a \sin \vartheta_n = (2n - 1) \frac{\lambda}{2}$$

or

$$\sin \vartheta_n = \frac{(2n-1)}{2a} \lambda \qquad ...(12.6)$$

where, $n = 0, 1, 2, ...$ etc. In short, we can say that the condition for n^{th} minimum intensity in Fraunhofer diffraction pattern is given by equation

$$a \sin \vartheta = n\lambda, \quad \text{where,} \quad n = 1, 2, 3, ... \qquad ...(12.7)$$

and the condition for n^{th} maximum intensity in Fraunhofer diffraction pattern is given by equation

$$a \sin \vartheta = (2n - 1) \frac{\lambda}{2}, \text{ where } n = 1,2, 3, ... \qquad ...(12.8)$$

Intensity of the second, third and fourth order of maxima

The secondary maxima lie between the minima on either side of the central maximum and their intensity falls off rapidly as we move away from central maximum. Let I_0 be the intensity of first order or central maximum.

Let us imagine that the slit is divided into three parts, then condition for second order ($n = 2$) maximum is $\frac{a}{3}\sin\vartheta = \frac{\lambda}{2}$. This condition is obtained by overlapping the secondary waves from $\frac{a}{3}$ part of the slit only. The amplitude of the waves corresponding to angle ϑ is approximately $\frac{1}{3}$ times of that at P_0 and hence the intensity (which is equal to the square of the amplitude) of second order maximum will be approximately $\left(\frac{1}{3}\right)^2$ times I_0, i.e. the intensity for second order ($n = 2$) maximum is equal to $\frac{I_0}{9}$.

Let us imagine that the slit is divided into five parts, then condition for third order ($n = 3$) maximum is $\frac{a}{5}\sin\vartheta = \frac{\lambda}{2}$. This condition is obtained by overlapping the secondary waves from $\frac{a}{5}$ part of the slit only. The amplitude of the waves corresponding to angle ϑ is approximately $\frac{1}{5}$ times of that at P_0 and hence the intensity (which is equal to the square of the amplitude) of third order maximum will be approximately $\left(\frac{1}{5}\right)^2$ times I_0, i.e., the intensity for third order ($n = 3$) maximum is equal to $\frac{I_0}{25}$.

Let us imagine that the slit is divided into seven parts, then condition for fourth order ($n = 4$) maximum is $\frac{a}{7}\sin\vartheta = \frac{\lambda}{2}$. This condition is obtained by overlapping the secondary waves from $\frac{a}{7}$ part of the slit only. The amplitude of the waves corresponding to angle ϑ is approximately $\frac{1}{7}$ times of that at P_0 and hence the intensity (which is equal to the square of the amplitude) of fourth order maximum will be approximately $\left(\frac{1}{7}\right)^2$ times I_0, i.e., the intensity for fourth order ($n = 4$) maximum is equal to $\frac{I_0}{49}$.

In general, the intensity of nth order maxima is given by $I_0\left(\frac{1}{2n-1}\right)^2$, where the slit of width a is considered to be divided into ($2n - 1$) parts, so that intensity of nth maximum is reduced by a factor $\left(\frac{1}{2n-1}\right)^2$. Because, we are left with waves from $\frac{1}{2n-1}$ part of the slit only, the amplitude corresponding to this angle is approximately $\frac{1}{2n-1}$ times of that at P_0

and hence the intensity (= Amplitude^2) is approximately $\left(\frac{1}{2n-1}\right)^2$ times I_0, the intensity of central maxima. The intensity distribution curve is shown in Fig. 12.2.

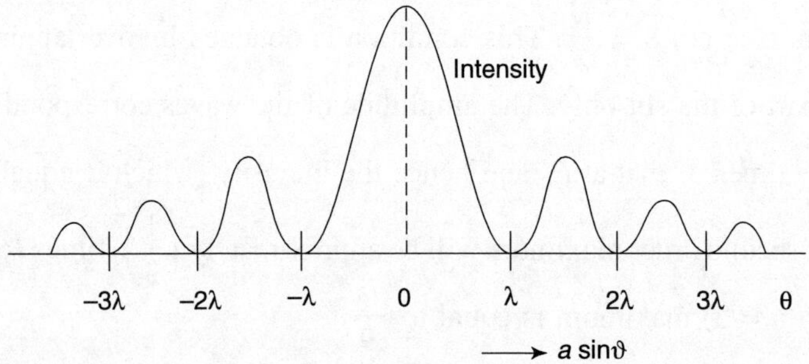

Fig. 12.2 Diffraction pattern of light from single slit

Width of Central Maxima

Let f be the focal length of the convex lens and x be half width of the first minimum. If the lens 2 is very near to the slit or the screen is far away from the lens 2 as shown in Fig. 12.3, then f is almost equal to the distance between the slit and screen. In that case from Fig. 12.1, we have

$$\sin \vartheta = \frac{\lambda}{a} = \frac{x}{f} \quad \text{or} \quad x = \frac{f\lambda}{a} \qquad ...(12.9)$$

Hence, the width of the central maxima will be

$$2x = \frac{2f\lambda}{a} \qquad ...(12.10)$$

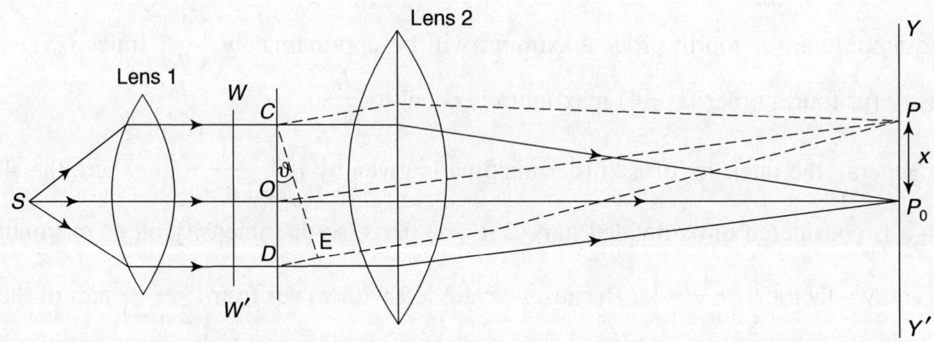

Fig. 12.3 Fraunhofer diffraction at a single slit

From the above equation one can infer that width of central maxima is proportional to the wavelength (λ) of light source, therefore for red light it is roughly twice as great as for violet light. Moreover, if the slit is wide (i.e., a is large), the central maxima will be very narrow. In this case, the minima and maxima lie very close to P_0 so there will not be clear distinction between maxima and minima. On the other hand, if the slit is narrow (i.e., a is small), central maxima is broad resulting in the appearance of distinct minima and maxima in diffraction pattern.

12.4 INTENSITY DISTRIBUTION IN THE DIFFRACTION PATTERN DUE TO A SINGLE SLIT

The total path difference between the waves originating from extreme points, C and D, is $DE = CD \sin \vartheta = a \sin \vartheta$ [Fig. 12.3]. Therefore, the path difference between different waves originating from all the points of the slit CD vary between zero and a sin ϑ. The phase difference corresponding to path difference a sin ϑ will be $\frac{2\pi}{\lambda}$ a sin ϑ. If the slit is considered to be divided into n equal parts as shown in Fig. 12.17, then the phase difference between any two consecutive parts will be $\frac{1}{n}\frac{2\pi}{\lambda}$ a sin ϑ (= φ, say).

For a parallel beam of incident light, the amplitude of the waves from each strip is considered same. Let diffraction angle of secondary wave at point D be ϑ as shown in Fig. 12.3. The phase difference of the wave is proportional to wavelength, i.e., when path difference increase, the phase difference also increases. Let 2 α be the phase difference between the secondary waves emerging from D and C points of the slit as shown in Fig. 12.3.

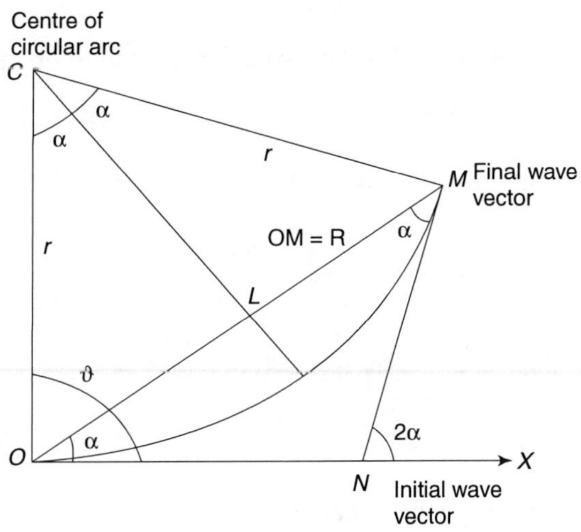

Fig. 12.4

The resultant amplitude of all secondary waves emerging from the slit can be obtained by the vector polygon method as shown in Appendix A. Amplitudes of all the secondary waves are same and small, and phase difference increases by very small amount from wave to wave, so polygon is almost circular arc (*OM*) as shown Fig. 12.4. The chord, $OM = R$, is the resultant amplitude of waves.

In Fig. 12.4, *OX* gives direction of initial vector of secondary waves emerging from point *D* [Fig. 12.3] and *NM* gives direction of vector representing secondary wave emerging from point *C*.

From the geometry of Fig. 12.4 one can see that $\angle MNX = 2\,\alpha$.

If *C* is the centre of circular arc, then $\angle OCM = 2\,\alpha$.

In ΔOCL, $OL = r \sin \alpha$, where, *r* is radius of circular arc.

Therefore, chord $OM = 2\,OL = 2\,r \sin \alpha$ ...(12.11)

If *n* vectors form the polygon in Fig. 12.4 and magnitude each vector is *b*, then

$$\text{Arc } OM = nb = A_0 \text{ (say)}$$

From the geometry of Fig. 12.4 one can see that

$$2\,\alpha = \frac{\text{Arc } OM}{\text{Radius}} = \frac{A_0}{r} \quad \text{or} \quad 2\,r = \frac{A_0}{\alpha} \qquad \text{...(12.12)}$$

Putting the value of $2r$ in Eq. (12.11), we get

$$\text{Chord } OM = A_0 \frac{\sin \alpha}{\alpha}$$

or

$$R = A_0 \frac{\sin \alpha}{\alpha} \qquad \text{...(12.13)}$$

wherer *R* is the resultant amplitude of waves. Another method to obtain Eq. (12.13) is given in Appendix A.

We know that

$$\text{Intensity} = (\text{amplitude})^2$$

or

$$I = R^2 = A_0^2 \left(\frac{\sin \alpha}{\alpha}\right)^2$$

or

$$I = I_0 \left(\frac{\sin \alpha}{\alpha}\right)^2, \qquad \text{...(12.14)}$$

where $I_0 = A_0^2$.

Phase difference (2π) corresponds to path difference λ.

Therefore, phase difference (2α) corresponds to path difference, $2\alpha \times \dfrac{\lambda}{2\,\pi}$.

From Fig. 12.3, we can see that path difference, $DE = a \sin \vartheta$, so diffraction condition can be written as

$$a \sin \vartheta = 2\alpha \frac{\lambda}{2\pi}$$

$$2\alpha = \frac{2\pi}{\lambda} a \sin \vartheta$$

or

$$\alpha = \frac{\pi a \sin \theta}{\lambda} \qquad ...(12.15)$$

From Eq. (12.15) shows that α depends upon ϑ, so value of $\frac{\sin^2 \alpha}{\alpha^2}$ will be different for different values of ϑ and hence intensity at different points in the intensity distribution pattern will be different, therefore a series of maxima and minima will be obtained.

Principal Maxima

Putting the series expansion of $\sin \alpha = \alpha - \frac{\alpha^3}{3!} + \frac{\alpha^5}{5!} - \frac{\alpha^7}{7!} ...$, the chord OM given by Eq. (9.13) becomes

$$R = A_0 \frac{\sin \alpha}{\alpha} = \frac{A_0}{\alpha}\left[\alpha - \frac{\alpha^3}{3!} + \frac{\alpha^5}{5!} - \frac{\alpha^7}{7!} ...\right]$$

or

$$R = A_0 \left[1 - \frac{\alpha^2}{3!} + \frac{\alpha^4}{5!} - \frac{\alpha^6}{7!} ...\right] \qquad ...(12.16)$$

R will be maximum, if each negative term in Eq. (12.16) vanishes. This is possible if $\alpha = 0$, therefore, Eq. (12.15) becomes

$$\alpha = \frac{2\pi a \sin \vartheta}{\lambda} = 0$$

or

$$\sin \vartheta = 0 \qquad ...(12.17)$$

Equation (12.17) implies that $\vartheta = 0$. Thus, for $\vartheta = 0$, the resultant intensity at point P will be maximum. The maximum intensity at point P is known as principal or central maximum.

Diffraction Condition for Minima

The diffraction condition for minima [Eq. (12.7)] is

$$a \sin \vartheta = n\lambda, \quad \text{where } n = 1, 2, 3, ... \qquad ...(12.18)$$

From Eq. (12.18), it is clear that resultant intensity at point P will be zero.

if $\sin \alpha = 0$, for $\alpha \neq 0$. $\sin \alpha = 0$ implies that

$$\alpha = \pm n\pi, \quad \text{where, } n = 1, 2, 3, ...$$

Putting $\alpha = \frac{2\pi a \sin\vartheta}{\lambda}$ from Eq. (12.15), we get

or $$\frac{\pi a \sin\vartheta}{\lambda} = \pm n\pi$$

or $$a \sin\vartheta = \pm n\pi \qquad ...(12.19)$$

where, $n = 1, 2, 3, ...$ gives position of the first, second, third, etc., order minima.

Phase Angles of Secondary Maxima

In diffraction pattern, the principal maximum is surrounded by secondary maxima of weak intensity, whose phase angles can be calculated by using Eq. (12.14)

$$I = I_0\left(\frac{\sin\alpha}{\alpha}\right)^2 \qquad ...(12.20)$$

as follows:

Differentiating intensity (I) w.r.t. α in Eq. (12.20), we get

$$\frac{dI}{d\alpha} = 2I_0\frac{\sin\alpha}{\alpha}\left[\frac{\alpha\cos\alpha - \sin\alpha}{\alpha^2}\right]$$

When I is maximum, $\frac{dI}{d\alpha} = 0$, therefore

$$2I_0\frac{\sin\alpha}{\alpha}\left[\frac{\alpha\cos\alpha - \sin\alpha}{\alpha^2}\right] = 0 \qquad ...(12.21)$$

The left-hand side of Eq. (12.21) will be zero, if

$$\sin\alpha = 0$$

or $$\alpha\cos\alpha - \sin\alpha = 0 \quad \text{or} \quad \alpha = \tan\alpha.$$

Equation $\sin\alpha = 0$ yields the phase angles of secondary minima given by Eq. 12.19), While, equation $\alpha = \tan\alpha$ yields the phase angles of secondary maxima.

The equation $\alpha = \tan\alpha$ can be solved graphically by plotting two curves (i) Y vs. α and (ii) Y vs. $\tan\alpha$ on the same graph as shown in Fig. 12.5(b).

Y vs. α curve is a straight line. Y vs. $\tan\alpha$ curves are continuous curves but become infinite for

$$a = 0, \pm\frac{3}{2}\pi, \quad \pm\frac{5}{2}\pi, \quad \pm\frac{7}{2}\pi, ... \qquad ...(12.22)$$

as shown in Fig. 12.5.

The point of intersection of Y vs. α and Y vs. $\tan\alpha$ is the value of α which satisfies equation $\alpha = \tan\alpha$. The values of α which satisfy equation $\alpha = \tan\alpha$ are given by Eq. (12.22). That is, the values of α for maximum intensity are given by Eq. (12.22). The value of phase angle $\alpha = 0$ corresponds to principal or central maxima.

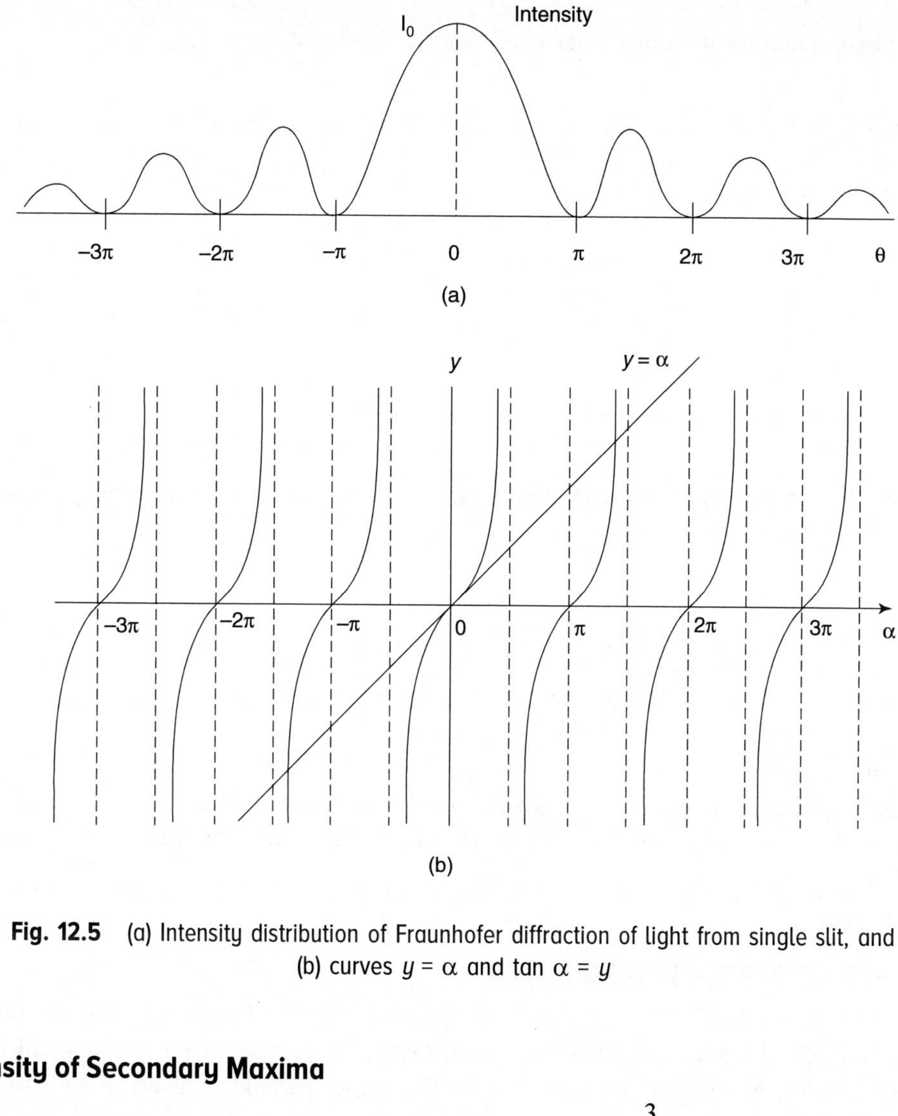

Fig. 12.5 (a) Intensity distribution of Fraunhofer diffraction of light from single slit, and (b) curves $y = \alpha$ and $\tan \alpha = y$

Intensity of Secondary Maxima

The phase angle for first order of secondary maxima is $\alpha = \pm \frac{3}{2}\pi$.

Putting $\alpha = \pm\frac{3}{2}\pi$ in Eq. (12.20), the intensity of second order maximum comes out to be

$$I_1 = I_0 \left(\frac{\sin\left(\pm \frac{3}{2}\pi\right)}{\pm \frac{3}{2}\pi} \right)^2 = I_0 \left(\frac{-1}{\frac{3}{2}\pi} \right)^2 = \frac{4I_0}{9\pi^2} = \frac{I_0}{22} \qquad ...(12.23)$$

The phase angle for the third order maximum is $\alpha = \pm\frac{5}{2}\pi$.

Putting $\alpha = \pm\frac{5}{2}\pi$ in Eq. (12.20), the intensity of third order of secondary maximum comes out to be

$$I_2 = I_0\left(\frac{\sin\left(\pm\frac{5}{2}\pi\right)}{\pm\frac{5}{2}\pi}\right)^2 = I_0\left(\frac{-1}{\frac{5}{2}\pi}\right)^2 = \frac{4I_0}{25\,\pi^2} = \frac{I_0}{61} \qquad \text{...(12.24)}$$

Phase angle for fourth order of secondary maximum is $\alpha = \pm\frac{7}{2}\pi$.

Putting $\alpha = \pm\frac{7}{2}\pi$ in Eq. (12.20), the intensity of the fourth order maximum comes out to be

$$I_3 = I_0\left(\frac{\sin\left(\pm\frac{7}{2}\pi\right)}{\pm\frac{7}{2}\pi}\right)^2 = I_0\left(\frac{-1}{\frac{7}{2}\pi}\right)^2 = \frac{4I_0}{49\,\pi^2} = \frac{I_0}{121} \qquad \text{...(12.25)}$$

and so on.

If intensity of central maxima, I_0, is assumed to be unity, then intensity of the first, second, third, ... order of secondary maxima on either side of central maxima will be in the ratio of $\frac{1}{22} : \frac{1}{61} : \frac{1}{121} : \ldots$

12.5 PLANE TRANSMISSION GRATING

A plane transmission diffraction grating is an optical device in which a very large number of parallel slits of same width, separated by equal opaque spaces, are arranged in small linear region on a plane. In a diffraction grating, slit widths are comparable to the wavelength of light.

The grating has parallel transparent lines of equal width, separated by opaque lines of equal width made by diamond point on a glass surface. The number of transparent lines on grating is quite large, e.g., 17,500 lines per inch or 7000 lines per cm. If lines are ruled on a plane mirror the grating is called *plane reflection grating*.

When a monochromatic wavefront is incident on plane diffraction grating perpendicular to its surface, then light is obstructed by opaque portion, while it is transmitted by the transparent slit. Each slit of grating becomes the source of secondary wavelets. Each secondary wavelet has the same amplitude and almost same initial phase.

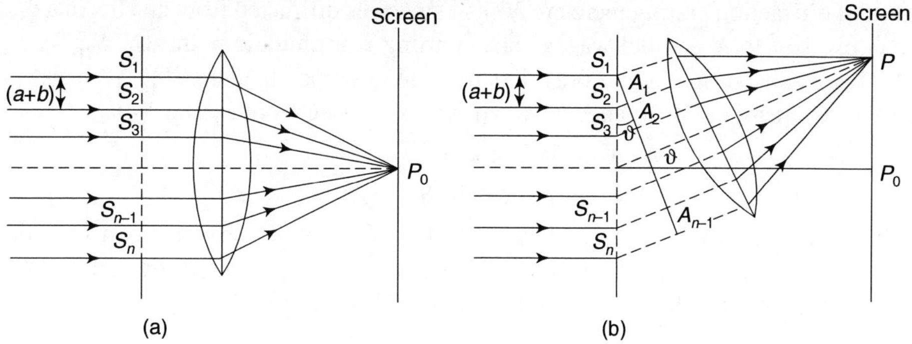

Fig. 12.6

Theory of Plane Transmission Grating

Let a diffraction grating have N slits each of width, a. Each slit is separated by opaque line of width, b. The sum of width of one transparent line and one opaque line $(a + b)$ is called the *grating element.* When light incidents normal on the grating, the diffraction pattern is combined effect of all N slits.

Let a plane wavefront of monochromatic light of wavelength λ be incident normal on grating. According to Huygen's principle, every point on the slit becomes a source of secondary wavelets. All secondary wavelets travel in the direction of incident light. The secondary wavelets emerging from each slit of grating parallel to the incident direction are focused by convex lens at point P_0 as shown in Fig. 12.6(a). A band formed at point P_0, after overlapping by all secondary waves, is called principal or central maxima, because its intensity is largest of all bright principal bands in diffraction pattern.

The secondary waves diffracted in direction other than the incident direction are focused by the convex lens at point P as shown in Fig. 12.6(b). The secondary waves diffracted in other directions diffracted at an angle ϑ (say) form a series of dark and bright bands on both sides of central maxima after overlapping.

The resultant amplitude of the wave formed due to overlapping of all the secondary waves from a single slit is [Eq. (12.13)].

$$R = A_0 \frac{\sin \alpha}{\alpha}, \quad ...(12.26)$$

where, A_0 is the resultant amplitude of wave. The angle [Eq. 12.15]

$$\alpha = \frac{\pi a \sin \vartheta}{\lambda} \quad ...(12.27)$$

is the phase angle of resultant wave achieved on overlapping of all secondary waves diffracted at angle ϑ from middle point of the slit to the end point of slit in any half part of single slit.

In case of diffraction grating there are N slits, so waves diffracted from all slits in a direction (ϑ) are equivalent to N parallel waves, each starting from middle points S_1, S_2, S_3, ..., S_{n-1}, S_n of N slits as shown in Fig. 12.6(a). Let us draw a perpendicular $S_1 A_{n-1}$ on the path of parallel waves, then path difference between waves starting from S_1 and S_2 is

$$S_2 A_1 = S_1 S_2 \sin \vartheta$$

or

$$S_2 A_1 = (a + b) \sin \vartheta \qquad ...(12.28)$$

Similarly, the path difference, $S_3 A_2$ and $S_n A_{n-1}$ will be $(a + b) \sin \vartheta$. In fact, the path difference between any two secondary waves diffracted at angle ϑ from two adjacent slits of grating is $(a + b) \sin \vartheta$. The corresponding phase difference is

$$\frac{2\pi}{\lambda} (a + b) \sin \vartheta = 2\beta \text{ (say)} \qquad ...(12.29)$$

The resultant amplitude of the wave formed due to overlapping of all the secondary waves from single slit diffracted at an angle ϑ is given by Eq. (12.26). The resultant intensity at any point on the screen along a given direction (ϑ) can be calculated by finding the resultant of N waves of equal amplitude, $A_0 \dfrac{\sin \alpha}{\alpha}$. The phase difference of each these wave is 2β with respect to preceding one.

Let us consider a polygon consisting of sides MP_1, $P_1 P_2$, $P_2 P_3$, ..., $P_{n-1} P_N$ representing each wave from each slit of equal amplitude (R) as shown in Fig. 12.7(a). The inclination of vectors increases by an amount 2β as we move from one vector at point M to another vector at point P_N. The chord MP_1 of the polygon in Fig. 12.7(c) gives resultant amplitude (R) of the waves from single slit, while the chord MP_N of the polygon in Fig. 12.7(d) gives resultant amplitude (T) of the waves from n slits of grating. If O is the centre of the polygon, then from the geometry of Fig. 12.7(c) we have

$$MP_1 = 2\, OM \sin \beta \qquad (12.30)$$

and from the geometry of Fig. 12.7(d) we have

$$MP_N = 2\, OM \sin N\beta \qquad ...(12.31)$$

Putting the value of OM from Eq. (12.30), Eq. (12.31) becomes

$$MP_N = MP_1 \frac{\sin N\beta}{\sin\beta} \qquad ...(12.32)$$

If we consider a slit of length, $M_1 = a$. When diffraction takes place from a single slit of length, $MP_1 = a$, and we consider that it is divided into n parts of equal parts (a') as shown in Fig. 12.7(b), then the resultant of all n secondary waves emerging from a single slit is $R = MP_1 = A_0 \dfrac{\sin \alpha}{\alpha}$ as given Eq. (12.26). Substituting this value of $MP_1 = R$, the resultant amplitude N waves, $T = MP_N$ will be

$$T = MP_N = A_0 \frac{\sin \alpha}{\alpha} \frac{\sin N\beta}{\sin \beta} \qquad ...(12.33)$$

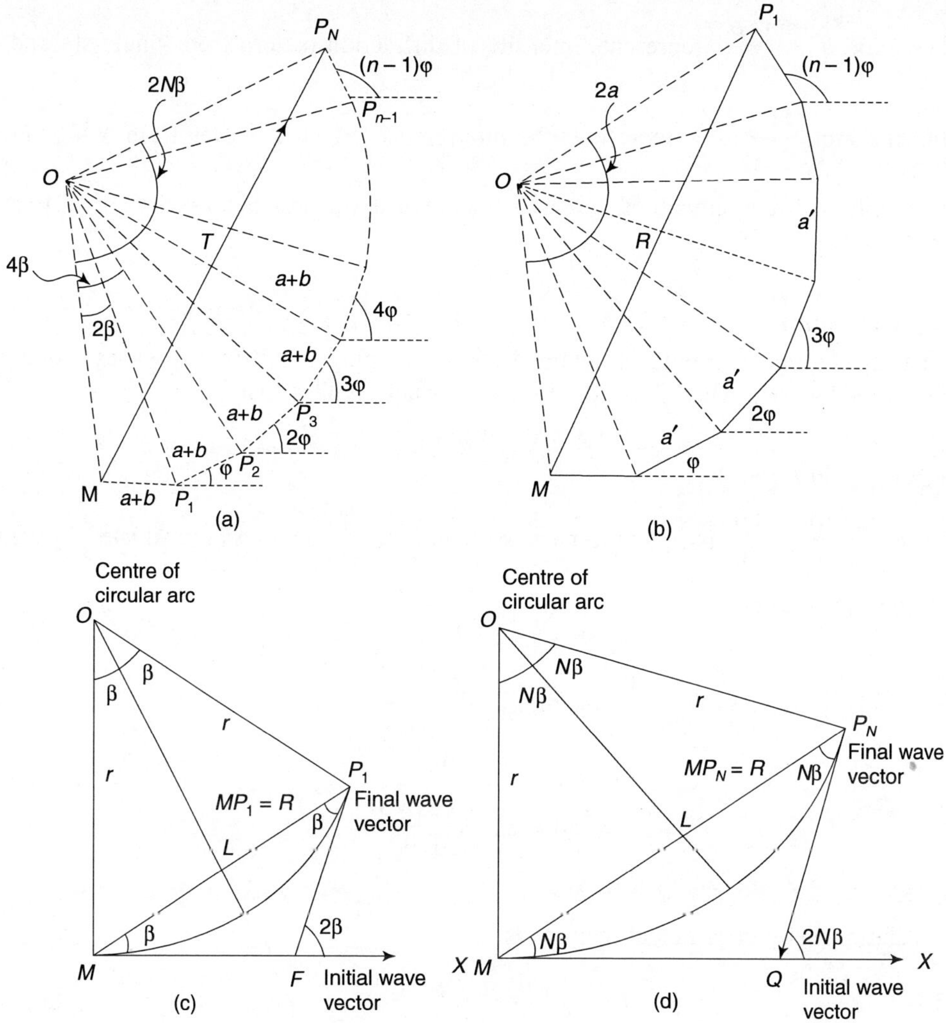

Fig. 12.7 Polygon law of vectors

The intensity (I) is equal to the square of the amplitude of the resultant wave, i.e.,

$$I = T^2 = A_0^2 \left(\frac{\sin\alpha}{\alpha}\right)^2 \left(\frac{\sin N\beta}{\sin\beta}\right)^2$$

or

$$\text{I} = I_0 \frac{\sin^2\alpha}{\alpha^2}\frac{\sin^2 N\beta}{\sin^2\beta} \qquad ...(12.34)$$

The factor, $I_0 \dfrac{\sin^2 \alpha}{\alpha^2}$, represents intensity of diffraction pattern from single slit and the additional factor, $\dfrac{\sin^2 N\beta}{\beta^2}$, represents the interference effect of waves from N slits in the intensity in the diffraction pattern. $I_0 = A_0^2$ is the intensity of principal maxima by diffraction pattern from single slit.

Principal Maxima

Principal or central maximum is obtained by overlapping of diffracted waves moving in incident direction, i.e., when $\beta = 0$ or $\sin \beta = 0$, which implies that

$$\beta = \pm n\pi, \quad \text{where } n = 0, 1, 2, 3,...$$

Therefore, $\sin N\beta = 0$.

Hence, $\dfrac{\sin N\beta}{\sin \beta} = \dfrac{0}{0}$ is indeterminate, so value of $\dfrac{\sin N\beta}{\sin \beta}$ can be evaluated by taking $\beta \to 0$ limit as follows:

$$\lim_{\beta \to 0} \frac{\sin N\beta}{\sin \beta} = \lim_{\beta \to 0} \frac{\dfrac{d}{d\beta} \sin N\beta}{\dfrac{d}{d\beta} \sin \beta}$$

$$= \lim_{\beta \to 0} \frac{N \cos N\beta}{\cos \beta}$$

$$= N$$

Now the intensity of principal maxima is

$$I = I_0 \frac{\sin^2 \alpha}{\alpha^2} N^2 \qquad ...(12.35)$$

Equation (12.35) indicates that intensity of principal maxima increases as number of slits increases in the grating.

Diffraction Condition for Maxima

Putting $\beta = \pm n\pi$ in Eq. (12.29), the direction of maxima is written as

$$\frac{2\pi}{\lambda} (a + b) \sin \vartheta = \pm 2n\pi$$

or

$$(a + b) \sin \vartheta = \pm n\lambda \qquad ...(12.36)$$

When $n = 0$, then $\sin \vartheta = 0$, which implies that $\vartheta = 0$. That is, the principal maximum is formed at the centre point of the diffraction pattern by diffracted waves travelling in the direction of incident light. First, second, third order maxima formed in direction θ obtained for $n = 1, 2, 3, ...$, respectively. The $\pm$ sign indicates that other maxima are formed on either side of principal maxima. $(a + b) \sin \vartheta$ is the path difference between two parallel rays diffracted from any pair of corresponding points on the grating. The number n is the order of maximum. The expression given by Eq. (12.36) is independent of N. Hence, the direction of principal maxima depends only on the grating element $(a + b)$.

Diffraction Condition for Minima

The intensity given by Eq. (12.34) will be zero when $\sin N\beta = 0$ and $\sin \beta \neq 0$

$\sin N\beta = 0$ implies that $N\beta = \pm m\pi$. Putting this value of β in Eq. (12.29), diffraction condition for minimum intensity is written as

$$N \frac{2\pi}{\lambda}(a+b)\sin\vartheta = \pm 2m\pi$$

or

$$N(a + b)\sin\vartheta = \pm m \qquad ...(12.37)$$

where, m can have all integral, values except $0, N, 2N, 3N, ...$, because for these values of m, Eq. (12.37) becomes the diffraction condition for maxima. Putting $m = nN$ into Eq. $N\beta = \pm m\pi$, we get

$\beta = \pm n\pi$, so that $\sin \beta = 0$ and $\sin N\beta = 0$, then we will get the diffraction condition [Eq. (12.35)] for principal maxima. This indicates that $m \neq nN$.

Secondary Maxima

For $m = 0$, the diffraction condition given by Eq. (12.37) yields principal maxima and for $m = 1, 2, 3, ..., (N - 1)$ yields minima. Again for $m = N$, the diffraction condition Eq. (12.37) yields principal maxima. Thus in between two consecutive principal maxima, there are $(N - 1)$ minima. There are $(N - 2)$ other maxima called secondary maxima between two consecutive principal maxima.

For finding the diffraction angle of secondary maxima, $\frac{dI}{d\beta}$ must be zero. This condition can be obtained by equating the differentiation I [Eq. (12.37)] to zero, i.e.,

$$\frac{dI}{d\beta} = I_0 \frac{\sin^2\alpha}{\alpha^2}\frac{d}{d\beta}\left(\frac{\sin N\beta}{\sin\beta}\right)^2 = 0$$

or

$$I_0 \frac{\sin^2\alpha}{\alpha^2} 2 \frac{\sin N\beta}{\sin\beta} \frac{\sin\beta\, N\cos N\beta - \sin N\beta\cos\beta}{\sin^2\beta} = 0$$

The left-hand side is zero when

$$\sin\beta\, N\cos N\beta - \sin N\beta\cos\beta = 0$$

or

$$N\tan\beta = \tan N\beta \qquad ...(12.38)$$

The roots of this equation gives diffraction angle of secondary maxima except for which $\beta = \pm n\pi$, which correspond to principal maxima.

Intensity of Secondary Maxima Triangle

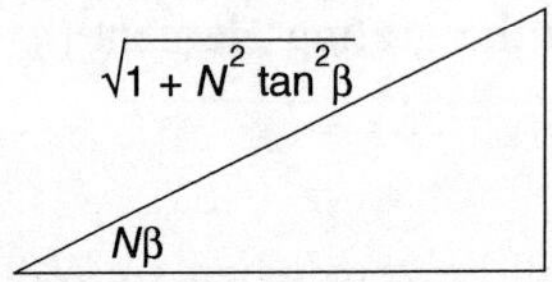

Fig. 12.8

From the triangle shown in Fig. 12.8, we have

$$\sin N\beta = \frac{N \tan\beta}{\sqrt{1 + N^2 \tan^2\beta}} \quad ...(12.39)$$

taking square of $\sin N\beta$ and dividing it by $\sin^2\beta$, we get

$$\frac{\sin^2 N\beta}{\sin^2\beta} = \frac{N^2 \tan^2\beta}{(1 + N^2 \tan^2\beta)\sin^2\beta} = \frac{N^2}{(1 + N^2 \tan^2\beta)\cos^2\beta}$$

$$= \frac{N^2}{(\cos^2\beta + N^2 \sin^2\beta)} = \frac{N^2}{(1 + (N^2 - 1)\sin^2\beta)} \quad ...(12.40)$$

Substituting for $\frac{\sin^2 N\beta}{\sin^2\beta}$, Eq. (12.34) becomes

$$I = I_0 \frac{\sin^2\alpha}{\alpha^2} \frac{N^2}{(1 + (N^2 - 1)\sin^2\beta)} \quad ...(12.41)$$

The intensity of principal maxima defined by Eq. (12.35) is

$$I = I_0 \frac{\sin^2\alpha}{\alpha^2} N^2 \quad ...(12.42)$$

The ratio intensity of secondary maxima to intensity of first principal maxima is

$$\frac{\text{Intensity of secondary maxima}}{\text{Intensity of first principal maxima}} = \frac{1}{(1 + (N^2 - 1)\sin^2\beta)} \quad ...(12.43)$$

Equation (12.18) indicates that as the number of ruling of grating increases, the intensity of secondary maxima decreases. Normally, the value of N is very large, therefore, only a few secondary maxima are visible in the grating spectrum as shown in Fig. 12.9.

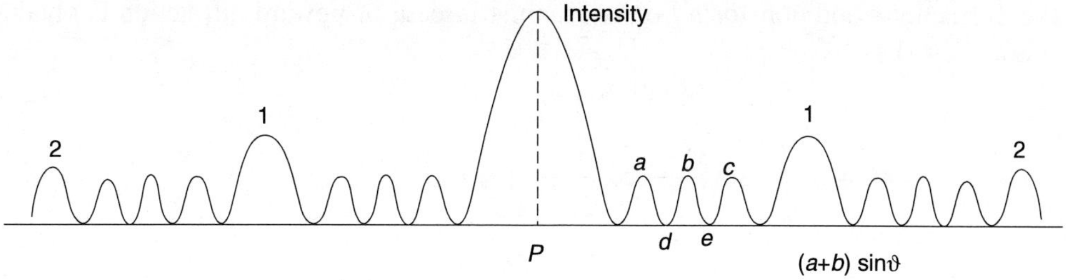

Fig. 12.9 Diffraction pattern of light from grating

12.6 THEORY OF PLANE DIFFRACTION GRATING UNDER OBLIQUE INCIDENT

Let a parallel beam of light be indent at an angle i with respect to horizontal line on a plane diffraction grating as shown in Fig. 12.10(a). The secondary wave is diffracted at an angle ϑ upwards as shown in Fig. 12.10(a).

The path difference between the secondary wavelets emerging from points A and B is

$$DB + BC = (a + b) \sin i + (a + b) \sin \vartheta = (a + b) (\sin \vartheta + \sin i) \quad ...(12.44)$$

Let a parallel beam of light be indent at an angle i with respect to horizontal line on a plane diffraction grating as shown in Fig. 12.10(b). The secondary wave is diffracted at an angle ϑ downwards as shown in Fig. 12.10(b). The path difference between the secondary wavelets passing through the points A and B is

$$DB - AC = (a + b) \sin i + (a + b) \sin \vartheta = (a + b) (\sin i - \sin \vartheta) \quad ...(12.45)$$

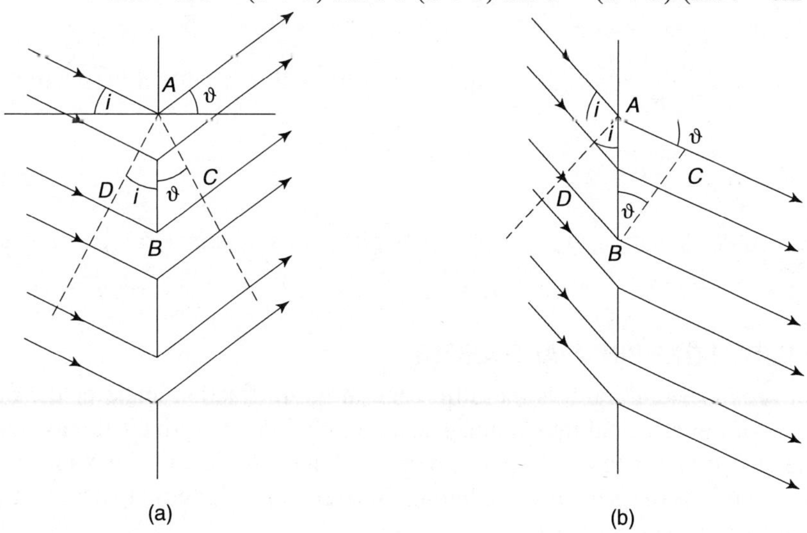

Fig. 12.10 The rays incident obliquely on plane diffraction grating. i is angle of incidence. (a) incident wave is diffracted upward. Here, the path difference is $DB + BC$. (b) incidence wave is diffracted downward. Here, the path difference is $AC - DB$

The diffraction condition for n^{th} order maxima in case of upward diffraction is obtained from Eq. (12.44) as

$$(a + b)\,(\sin \vartheta + \sin i) = n\lambda \qquad ...(12.46)$$

or
$$(a + b)\left(2 \sin \frac{\vartheta + i}{2} \cos \frac{\vartheta - i}{2}\right) = n\lambda$$

or
$$\sin \frac{\vartheta + i}{2} = \frac{n\lambda}{2\,(a + b)\left(\cos \frac{\vartheta - 1}{2}\right)} \qquad ...(12.47)$$

The diffracted beam is deviated from incident beam by an angle $\vartheta + i$. When deviation angle, $\vartheta + i$ is minimum, then sin $\frac{\vartheta + i}{2}$ must be be minimum. According to Eq. (12.47), sin $\frac{\vartheta + i}{2}$ is minimum when cos $\frac{\theta - i}{2}$ is maximum, i.e., cos $\frac{\vartheta - i}{2}$ = 1. cos $\frac{\vartheta - i}{2}$ = 1 is one when

$$\frac{\vartheta - i}{2} = 0 \quad \text{or} \quad \vartheta = i$$

Thus, for $\vartheta = i$, the deviation $(\vartheta + i)$ is minimum. Therefore, the angle of minimum deviation is

$$\delta_m = \vartheta + i = 2\vartheta$$

or
$$\vartheta = \frac{\delta_m}{2}$$

Putting the value of $\vartheta = \frac{\delta_m}{2}$ and $i = \vartheta$, in Eq. (12.46), diffraction condition for maxima is

$$2\,(a + b) \sin \frac{\delta_m}{2} = n\lambda \qquad ...(12.48)$$

The angle of minimum deviation, δ_m, for n^{th} order diffraction maxima can be obtained from Eq. (12.48).

12.7 ANGULAR WIDTH OF PRINCIPAL MAXIMA

Angular width of a principal minimum is the difference in the diffraction angle of the first minima on either side of the principal maximum. The angular width of a principal maximum is a measure of the sharpness of the principal maximum. Lesser is the angular width of a principal maximum, sharper is the principal maximum. The direction of n^{th}principal maximum in case of grating is given as [Eq. (12.36)]

$$(a + b) \sin \vartheta = n\lambda \qquad (12.49)$$

If $\vartheta + d\vartheta$ and $\vartheta - d\vartheta$ are the directions of first maxima on either side of nth principal maximum, then $2d\vartheta$ is the angular width of the principal maximum. The mathematical expression of angular width can be obtained as follows.

In case of grating, the direction of minima is given by Eq. 12.37

$$N(a+b)\sin\vartheta = m\lambda \qquad \text{...(12.50)}$$

where, m can have all integral values except 0, N, $2N$, $3N$, ..., because these values correspond to zero, first, second, third...order principal maxima. If first order minima on either side of nth principal maxima lies in the direction $(\vartheta \pm d\vartheta)$ for which the corresponding value of m is $nN \pm 1$, then, Eq. (12.50) can be written as

$$N(a+b)\sin(\vartheta \pm d\vartheta) = (nN \pm 1)\lambda \qquad \text{...(12.51)}$$

or $$N(a+b)[\sin\vartheta\cos d\vartheta \pm \cos\vartheta\sin d\vartheta] = (nN \pm 1)$$

or $N(a+b)[\sin\vartheta \pm \cos\vartheta\, d\vartheta] = (nN \pm 1)\lambda$, as $\cos d\vartheta = 1$ and $\sin d\vartheta = d\vartheta$ for small $d\vartheta$

or $$N(a+b)\sin\vartheta \pm N(a+b)\,d\vartheta\cos\vartheta = nN\lambda \pm \lambda \qquad \text{...(12.52)}$$

Substituting the value of $(a+b)\sin\vartheta = n\lambda$ from Eq. (12.49), Eq. (12.52) becomes

$$Nn\lambda \pm N(a+b)\,d\vartheta\cos\vartheta = nN\lambda \pm \lambda$$

or $$N(a+b)\,d\vartheta\cos\vartheta = \lambda$$

or $$d\vartheta = \frac{\lambda}{N(a+B)\cos\vartheta} \qquad \text{...(12.53)}$$

Putting the value of $(a+b)$ from Eq. (12.49) in Eq. (12.53), we get

$$2d\,\vartheta = \frac{2}{nN\cot\theta} \qquad \text{...(12.54)}$$

This expression for angular width indicates that angular width $2d\vartheta$ is

(i) $2d\vartheta \propto \frac{1}{N}$, that angular width of nth order of principal maxima is inversely proportional to number lines on the grating.

(ii) $2d\vartheta \propto \frac{1}{n\cos\vartheta}$

When order of principal maxima n increases, the value of ϑ increases. When the value of ϑ increases the value of $\cot\vartheta$ decreases. That is, when n increases the value of $\cot\vartheta$ decreases. However, the rate of increase of n is much more than the decrease of $\cot\vartheta$. Hence, $\frac{1}{n\cos\vartheta}$ decreases with the increase of n. Therefore, we conclude that the angular width of nth order decreases with increase of n, i.e., higher order principal maxima have less angular width.

One can see from equation $(a+b)\sin\vartheta = n\lambda$, ϑ is more for longer wavelength for fixed order n. That is in the diffraction of white light, the diffraction angle for blue band is least, while the diffraction angle for red band is largest.

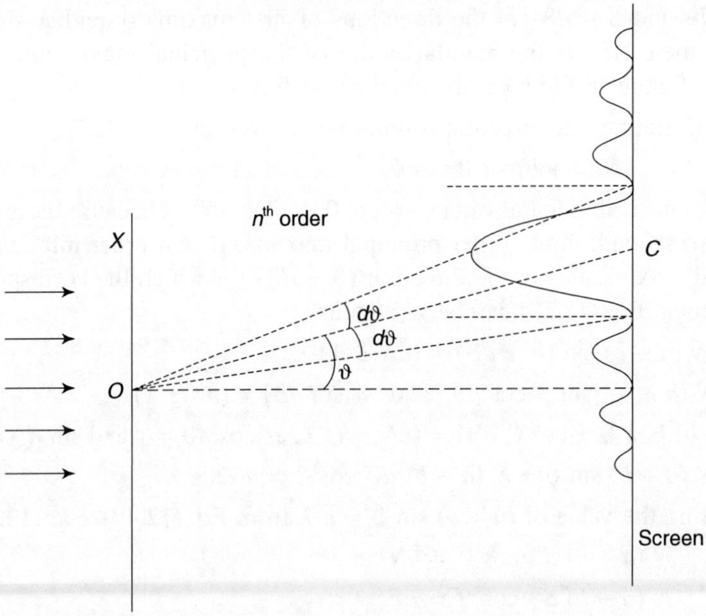

Fig. 12.11 Angular width of principal maxima

Putting the value of n from Eq. (12.49) into Eq. (12.54), we get

$$d\vartheta = \frac{1}{N\cos\vartheta}\frac{\lambda}{(a+b)\sin\vartheta} = \frac{\lambda}{(a+b)N\cos\vartheta} \qquad ...(12.55)$$

N is the number of lines on the grating, so the width of grating is

$$(N-1)(a+b) = N(a+b) \quad \text{as} \quad N >> 1$$

Hence, Eq. (12.55) can be written as

$$d\vartheta = \frac{\lambda}{\text{Width of grating} \times \cos\vartheta}$$

or $$d\vartheta \propto \frac{\lambda}{\text{Width of grating}}, \text{ if } \frac{\lambda}{\cos\vartheta} \text{ is constant}$$

That is, the spectral lines are more sharper in grating having more width.

12.8 ABSENCE OF SPECTRAL LINES FROM DIFFRACTION GRATING

In diffraction pattern, the particular order spectral line for a given diffraction angle ϑ will be absent if path difference between the rays diffracted from two extreme ends of one slit is integral multiple of λ. If path difference is λ and each slit is supposed to consist of two equal parts. Then path difference between corresponding points in two parts of a single slit will be $\frac{\lambda}{2}$. Therefore, the resultant rays from two parts of single slit interfere destructively and give zero intensity. Thus a minimum intensity band is formed in that direction. That is, the diffraction condition of mth minima in case of single slit is

$$a \sin \vartheta = m\lambda, \quad \text{where } m = 1, 2, 3,.... \qquad ...(1256)$$

The diffraction condition for n^{th} principal maximum band in case of grating is

$$(a + b) \sin \vartheta = n\lambda, \quad \text{where } n = 1, 2, 3,... \qquad ...(12.57)$$

If Eqs. (12.56) and (12.57) are to be satisfied simultaneously, then a particular maximum line of order n will be missing in the grating spectrum. Dividing Eq. (12.57) by a sin ϑ, we get

$$\frac{(a+b)\sin\vartheta}{a\sin\vartheta} = \frac{n\lambda}{m\lambda}$$

or

$$\frac{(a+b)}{a} = \frac{n}{m} \qquad ...(12.58)$$

Equation (12.57) is the condition of missing spectral bands.

Case (i): If $b = a$, then from Eq. (12.58), we have

$$n = 2m$$

That is, 2nd, 4th , 6th,..., order of the spectral lines will be missing in grating spectrum when $m = 1, 2, 3,...,$ corresponding to minima due to a single slit.

Case (ii): If $b = 2a$, then from Eq. (9.58), we get

$$n = 3m$$

That is, 3rd, 6th, 9th,..., order of the spectral lines will be missing in the grating spectrum when $m = 1, 2, 3,...$ corresponding to minima due to single slit and so on.

12.9 OVERLAPPING OF SPECTRAL LINES

If incident light has more than one wavelength in case of diffraction grating, then spectral lines of different wavelengths are formed at same space. If light consists of several wavelengths incident normal on the grating, certain spectral lines of higher wavelength in lower order may overlap to spectral lines of lower wavelength in higher order.

If the first order spectral line of wavelength λ_1, second order spectral line of wavelength λ_2 and third order spectral line of wavelength λ_3 overlap in diffraction pattern in case of grating, then the diffraction condition for maxima in a direction ϑ can be written as

$$(a + b) \sin \vartheta = 1 \cdot \lambda_1 = 2 \cdot \lambda_2 = 3 \cdot \lambda_3 \qquad ...(12.59)$$

The equation $1 \cdot \lambda_1 = 2 \cdot \lambda_2 = 3 \cdot \lambda_3$ shows that $\lambda_1 > \lambda_2 > \lambda_3$.

For example, red line of wavelength 7100 Å in the third order, the yellow line of wavelength 5905 Å in fourth order and violet line of wavelength 4300 Å in fifth order will be formed in same direction on the screen because,

$$(a + b) \sin \vartheta = 3 \cdot 7100 \times 10^{-8}$$

$$(a + b) \sin \vartheta = 4 \cdot 5905 \times 10^{-8}$$

$$(a + b) \sin \vartheta = 5 \cdot 4300 \times 10^{-8}$$

So, red line in third order, the yellow line n fourth order and violet line fifth order will overlap each other.

In case of visible spectrum having wavelength range 4300 Å, 7500 Å, there will be no overlapping of the first order and second order spectra, because second order of 4300 Å line will overlap with first order of 8300 Å line which is beyond the visible range.

However, the spectrum in ultraviolet region 2100 may be recorded on photography plate Å. In that case spectral line of wavelength 4000 Å in first order and spectral line of wavelength 2100 Å in second order will overlap. The undesired wavelengths which overlap the region under investigation can be absorbed by using suitable filters and a spectrum having distinct maxima and minima can be observed.

12.10 NUMBER OF ORDER OF SPECTRA WITH GRATIING

The equation of nth principal maximum in case of grating is

$$(a + b) \sin \vartheta = n\lambda, \quad \text{where, } n = 1, 2, 3,... \qquad ...(12.60)$$

From Eq. (12.60) value of n is

$$n = \frac{(a+b)\sin\vartheta}{\lambda} \qquad ...(12.61)$$

where, $(a + b)$ is the grating element, ϑ is diffraction angle, λ is wavelength of the incident light and n is the order of maximum.

For maximum value of $\vartheta = 90°$, $\sin \vartheta = 1$, therefore,

$$n = \frac{(a+b)}{\lambda} \qquad ...(12.62)$$

If grating element $(a + b)$ lies between λ and $2\,\lambda$, i.e., $(a + b) < 2\,\lambda$, then

$$n < \frac{2\lambda}{\lambda} \quad \text{or} \quad n < 2$$

Thus, for normal incidence only first order spectrum is seen.

If grating $(a + b)$ lies between $2\,\lambda$ and $3\,\lambda$ i.e. $(a + b) < 3\,\lambda$, then

$$n < \frac{3\,\lambda}{\lambda} \quad \text{or} \quad n < 3$$

Thus, for normal incidence only second order spectrum is seen and so on.

12.11 DETERMINATION OF WAVELENGTH OF LIGHT

Generally, plane transmission grating is used in laboratories to measure the wavelength of monochromatic light.

When a parallel beam of light is incident normally on a direction a grating, the diffraction angle of principal maxima of the nth order is obtained by equation

$$(a + b) \sin \vartheta = n\lambda, \quad \text{where } n = 1, 2, 3,.... \qquad ...(12.63)$$

where, n is the order of principal maximum, ϑ is diffraction angle, λ is wavelength of the incident light and $(a + b)$ is the grating element. The wavelength of incident light is given by the expression

$$\lambda = \frac{(a+b)\sin\vartheta}{n} \qquad ...(12.64)$$

The grating element $(a + b)$ can be measured directly from the number of rulings per inch (1 inch = 2.54 cm) mentioned by the manufacture. If N is the number of lines per inch, then

$$a + b = \frac{2.54}{N}\text{ cm}$$

By knowing the angle of diffraction (ϑ) for a particular order n, wavelength can be calculated

Apparatus required

Sodium vapor lamp as a source of monochromatic light, collimator to make parallel rays, grating and telescope.

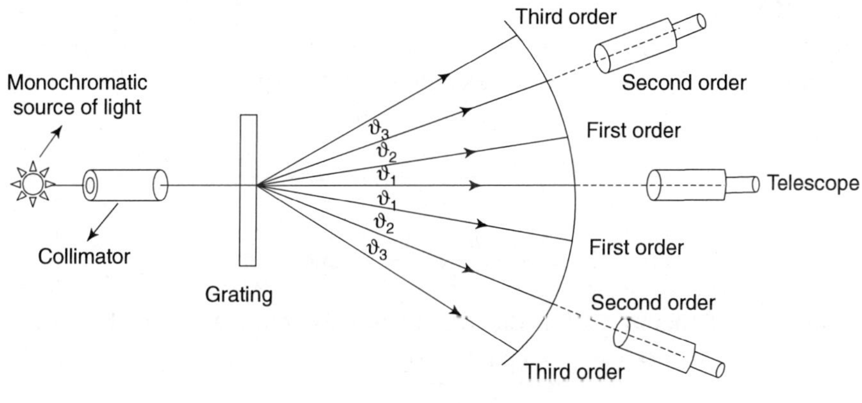

Fig. 12.12 Spectrum of monochromatic light

An ordinary spectrometer is used to measure the angle of diffraction. When a parallel beam of light emerges from the collimator, the telescope is adjusted to view the slit of the collimator. First of all, its vertical cross-wire is focused to view that image of the slit. In this position, telescope and collimator are in the same line. The position of telescope is noted with the help of circular vernier scale attached to the spectrometer. From this position, the telescope is rotated by 90° and fixed with the help of a fixing screw. In this position, the axis of collimator and axis of telescope are perpendicular to each other. Now the grating is placed on the centre of prism table so that its plane is perpendicular to the prism table. After this, the prism table is rotated slowly so that image of collimator slit formed in grating must be seen on vertical cross-wire of telescope. In this position the grating is at 45° to the incident light and telescope and axis of collimator are perpendicular to each other. The prism table is now rotated through 45° in such a way so that grating becomes normal to the incident light.

Measurement of ϑ

The collimator slit illuminated by a monochromatic light source is seen through the telescope. The telescope is moved slowly towards the left. It is stopped when a vertical bright line is visible. Now adjust the vertical cross-wire of the telescope on the bright line. In this position note the reading of vernier scale. This reading (R_{L1}) is the position of telescope for the first order ($n = 1$) spectrum in the left. By moving the telescope further to the left, the readings, R_{L2} and R_{L3}, of second order ($n = 2$) and third order ($n = 3$) bright images are noted. Similarly, by moving the telescope to the right, readings of first order ($n = 1$) and second order ($n = 2$) bright images, R_{R1}, R_{R2} and R_{R3}, are noted. In case of sodium light source, the spectral lines up to third order are visible.

As seen from Fig. 12.12, the angle of diffraction of first order ($n = 1$), second order ($n = 2$) and third order ($n = 3$) are

$$\vartheta_1 = \frac{R_{L1} - R_{R1}}{2},\ \vartheta_2 = \frac{R_{L2} - R_{R2}}{2},\ \vartheta_3 = \frac{R_{L3} - R_{R3}}{2}, \text{ respectively.}$$

Calculation of λ

By putting the values of grating element ($a + b$), the order of bright line (n) and angle of diffraction in Eq. (12.57), we obtain wavelength of first order, second order and third order lines

$$\lambda_1 = \frac{(a+b)\sin\vartheta_1}{1},\ \lambda_2 = \frac{(a+b)\sin\vartheta_2}{2}$$

and

$$\lambda_3 = \frac{(a+b)\sin\vartheta_3}{3}, \text{ respectively.}$$

The wavelength of sodium light is the average of these three wavelengths, i.e.,

$$\lambda = \frac{\lambda_1 + \lambda_2 + \lambda_3}{3}$$

The wavelength of the light source can be measured with good accuracy by using the spectrometer. However, the accuracy of the value of depends on the accuracy of measurement of diffraction angles.

12.12 FORMATION OF SPECTRA WITH GRATING FOR WHITE LIGHT SOURCE

Monochromatic source of light is now replaced with source of white light. As there are different colours in white light, therefore, white light gets dispersed after passing through grating and a spectrum of seven constituent colours is formed as shown in Fig. 12.13. To measure the angle of diffraction (ϑ) for blue, indigo, violet, green, orange, yellow and red colours, setting of spectrometer is done as mentioned in Section 12.11. Now, with the help of the telescope the readings of a particular colour of first order in left position from vernier scale is noted by adjusting vertical cross-wire onbright image. The telescope is then rotated towards right and the reading of first order of that colour is noted by the vernier scale by adjusting vertical cross-wire on bright image. The difference between two readings of the vernier scale is angle 2ϑ. Half of it gives the value of diffraction angle, ϑ. Similarly, the diffraction angles of other colours can be found. The wavelength of line under consideration can be obtained by using the equation

$$\lambda = \frac{(a+b)\sin\vartheta}{n}.$$

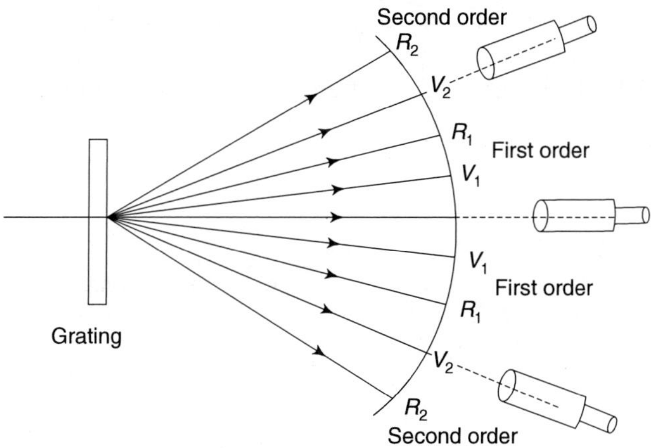

Fig. 12.13 Spectrum of white light

12.13 RESOLVING POWER OF A PLANE DIFFRACTION GRATING

Resolving power of a grating is referred to as its ability to resolve two close spectral lines, so that they can be viewed or photographed as distinctly separate lines.

Let us consider a parallel beam of light of wavelength, λ and $\lambda + d\lambda$, incident normally on the plane transmission grating, having grating element $(a + b)$ and total number of gratings elements, N. Then the resolving power of the grating is defined as the ratio of wavelength (λ) to the difference $(d\lambda)$ of the wavelengths, i.e., $\frac{\lambda}{d\lambda}$.

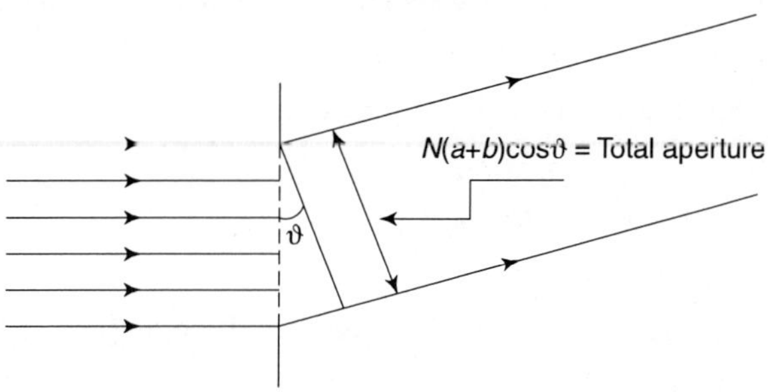

Fig. 12.14

Rayleigh's Criterion for Lines Resolution

Rayleigh suggested that two-point source or two spectral lines of equal intensity can be resolved by an optical instrument when the central maximum of the diffraction pattern due to one falls on the first minimum of the diffraction pattern due to the other. This is called Rayleigh's criterion for lines resolution.

In other words, this criterion states that the least angular separation between two spectral lines (maxima) of wavelength, λ and $\lambda + d\lambda$, (and corresponding phase angles ϑ and $\vartheta + d\vartheta$) in a given diffraction order should be equal to half the angular width ($d\vartheta/2$) of either maximum as shown in Fig. 12.15. If $d\lambda$ is the least difference in wavelength of the two spectral lines, the resolving power is equal to $\dfrac{\lambda}{d\lambda}$.

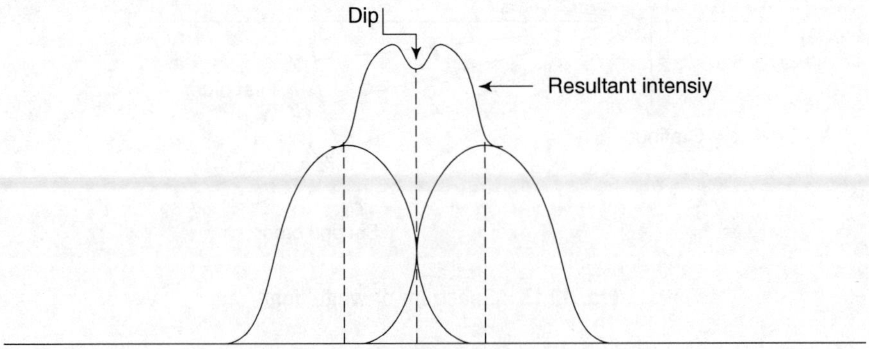

Fig. 12.15

The direction of the nth maximum for a wavelength λ is given as

$$(a + b)\sin\vartheta = n\lambda \qquad ...(12.65)$$

where, $n = 0, 1, 2, 3,..., n,...$

The direction of the nth maxima for a wavelength, $\lambda + d\lambda$, is given as

$$(a + b)\sin(\vartheta + d\vartheta) = n(\lambda + d\lambda) \qquad ...(12.66)$$

Multiplying by N, Eqs. (12.65) and (12.66) become

$$N(a + b)\sin\vartheta = Nn\lambda \qquad ...(12.67)$$

$$N(a + b)\sin(\vartheta + d\vartheta) = Nn(\lambda + d\lambda) \qquad ...(12.68)$$

where, $nN = 0, N, 2N, 3N,..., nN$

the condition for m^{th} minima of wavelength λ can be written as [Eq. (12.50)]

$$N(a + b)\sin\vartheta = m\lambda \qquad ...(12.69)$$

Here, m has integral values except 0, N, $2N$,..., nN, because for these values of m the condition for maxima is satisfied. The condition for first minimum adjacent to n^{th} maxima in the direction $(\vartheta + d\vartheta)$ can be obtained by substituting the value of $m = nN + 1$ in Eq. (12.69). Therefore, first minima adjacent to n^{th} maxima in the direction $(\vartheta + d\vartheta)$ is given by

$$N (a + b) \sin (\vartheta + d\vartheta) = (nN + 1) \qquad ...(12.70)$$

Comparing Eqs. (12.68) and (12.70), we get

$$Nn (\lambda + d\lambda) = (nN + 1) \quad \text{or} \quad Nnd\lambda = \lambda$$

or

$$\frac{\lambda}{d\lambda} = nN \quad \text{or} \quad \frac{\lambda}{d\lambda} \propto N \qquad ...(12.71)$$

This is the required expression for resolving power of the plane diffracting grating.

Equation (12.71) for resolving power of the plane diffracting grating shows that:

(i) The resolving power of grating is independent of the grating element.

(ii) The resolving power of grating is directly proportional to the order of the spectrum.

(iii) The resolving power of grating is directly proportional to the number of elements (N) in the grating. That is the number gratings elements per cm of a grating should be large in order to have large resolving power of the grating.

Angular Dispersive Power

The angular dispersive power $\left(\frac{d\vartheta}{d\lambda}\right)$ is defined as

$$\frac{d\vartheta}{d\lambda} = \frac{n}{(a+b)\cos\vartheta} \qquad ...(12.72)$$

The angular dispersive power $\left(\frac{d\vartheta}{d\lambda}\right)$ can be obtained from Eq. (12.65) by differentiating with respect to λ as

$$(a + b) \cos\vartheta \, d\vartheta = n d\lambda \qquad ...(12.73)$$

Equation (12.73) leads to Eq. (12.72). The value of $\cos \vartheta$ from Eq. (12.65) is

$$\cos \vartheta = \sqrt{1 - \frac{n^2 \lambda^2}{(a+b)^2}}. \qquad ...(12.74)$$

Putting this value of $\cos \vartheta$ in Eq. (12.72), dispersive power of grating becomes

$$\frac{d\vartheta}{d\lambda} = \frac{\quad}{(a+b)\sqrt{1 - \frac{2 \quad 2}{(a \quad b)}}}$$

$$\frac{d\vartheta}{d\lambda} = \frac{n}{\sqrt{(a+b)^2 - n^2 \lambda^2}} \qquad ...(12.75)$$

The dispersive power of grating can be calculated from Eq. (12.75) without knowing the diffraction angle.

Equation (12.75) forangular dispersive power of the plane diffracting grating shows that:

(i) The angular dispersive power of grating is independent of number of grating elements, N.

(ii) The angular dispersive power of grating is directly proportional to the order of the spectrum, n.

(iii) The angular dispersive power of grating is inversely proportional to the element $(a + b)$ of grating. That is, the grating element should be small in order to have large angular dispersive power of the grating.

(iv) The angular dispersive power of grating increases as the angular dispersion increases.

Further resolving power $\left(\frac{\lambda}{d\lambda}\right)$ can also be expressed in terms of angular dispersive power as follows:

$$\frac{\lambda}{d\lambda} = nN = N\,(a + b)\cos\vartheta \times \frac{n}{(a+b)\cos\vartheta} \qquad \text{...(12.76)}$$

The second factor in right-hand side of Eq. (12.76) is angular dispersive power. So Eq. (12.76) reduces to

$$\frac{\lambda}{d\lambda} = A \times \frac{d\vartheta}{d\lambda}$$

$$= \text{Total aperture} \times \text{Angular dispersive power} \qquad \text{...(12.77)}$$

where, total aperture $A = N\,(a + b)\cos\vartheta$ is shown in Fig. 12.12.

Linear Dispersive Power

The linear dispersive power of the grating is defined as the ratio of the linear distance between two spectral lines with the change in wavelength, i.e., linear dispersive power of the grating $= \frac{dl}{d\lambda}$.

Let dl be the linear distance along the screen between two spectral lines differing in wavelength $d\lambda$, then

$$dl = f\,d\vartheta \qquad \text{...(12.78)}$$

where, f is focal length of the lens used to focus the spectra. Linear dispersive power is expressed as $\frac{dl}{d\lambda}$. Putting the value of $d\vartheta$ from Eq. (12.73), we get

$$dl = f \times \frac{n\,d\lambda}{(a+b)\cos\vartheta}$$

or

$$\frac{dI}{d\lambda} = f\frac{d\vartheta}{d\lambda} = \frac{fn}{(a+b)\cos\vartheta} \qquad \text{...(12.79)}$$

Putting this value of $\cos\vartheta$ from Eq. (12.74), linear dispersive power of grating becomes

$$\frac{dI}{d\lambda} = \frac{fn}{(a+b)\sqrt{1-\frac{n^2\lambda^2}{(a+b)^2}}}$$

or

$$\frac{dI}{d\lambda} = \frac{nf}{\sqrt{(a+b)^2 - n^2\lambda^2}} \qquad \text{...(12.80)}$$

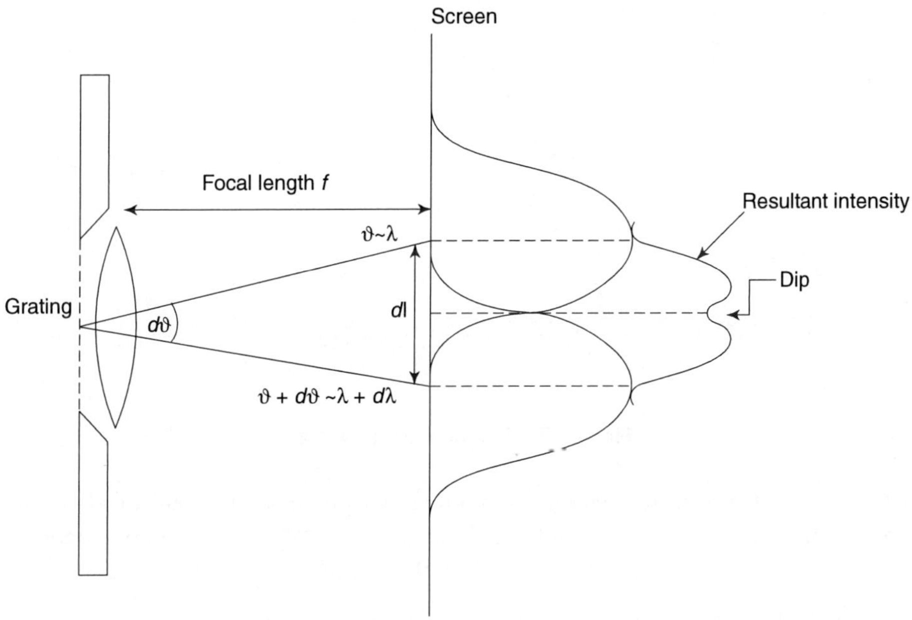

Fig. 12.16

Equation (12.80) indicates that

(i) Linear dispersive power is proportional to the focal length of convex lens. That is, linear dispersive power is large if focal length (f) is large.

(ii) It is inversely proportional to the grating element $(a + b)$. That is linear dispersive power is large grating element is small.

(iii) It proportional to the order of maxima.

(iv) It increases as the angular dispersion increases.

APPENDIX A

Resultant of *n* Simple Harmonic Motions

Let there be *n* secondary waves having the same period, same amplitude (a) and same phase difference (φ) between successive waves. Our aim is to find the resultant amplitude of these waves. For finding the resultant amplitude of all waves, let us consider a polygon of amplitudes as shown in Fig. 12.17.

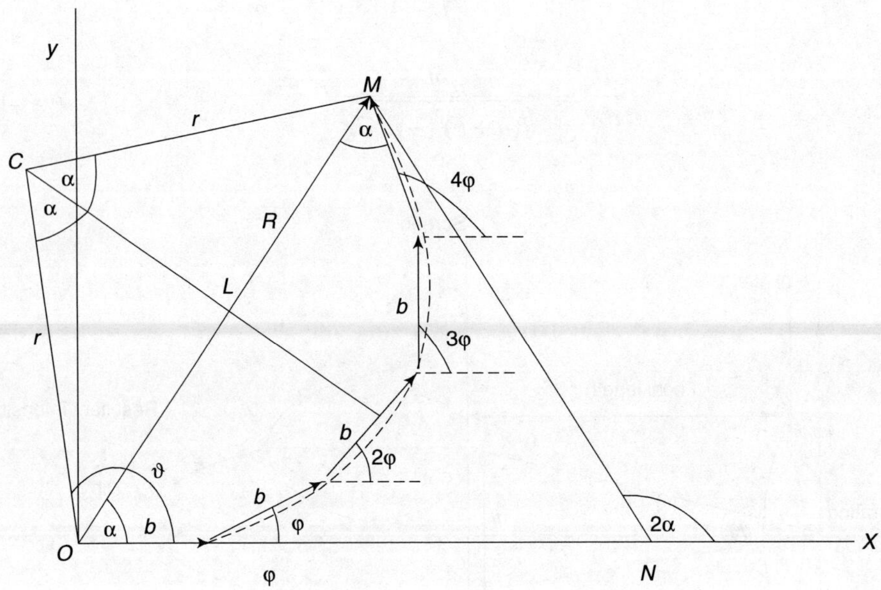

Fig. 12.17 Polygon law of vectors

Let $OP = R$ and ϑ are the resultant amplitude and phase of resultant of all waves. The resultant component of R along X and Y axis are the sum of respective components of the amplitudes of all the waves. If n is the number of secondary waves, then

$$R \cos \vartheta = b\,[1 + \cos \varphi + \cos 2\,\varphi + \cos 3\,\varphi + \ldots + \cos (n-1)\,\varphi] \quad \text{...(A1)}$$

$$R \sin \vartheta = b\,[1 + \sin \varphi + \sin 2\varphi + \sin 3\varphi \ldots. + \sin (n-1)\,\varphi] \quad \text{...(A2)}$$

Multiplying by $2 \sin \dfrac{\varphi}{2}$, we get

$$2R \cos \vartheta \sin \frac{\varphi}{2} = b\left[2 \sin \frac{\varphi}{2} + 2 \cos \varphi \sin \frac{\varphi}{2} + 2 \cos 2\,\varphi \sin \frac{\varphi}{2} + 2 \cos 2\,\varphi \sin \frac{\varphi}{2} + \ldots \right.$$
$$\left. + 2 \cos (n-1)\,\varphi \sin \frac{\varphi}{2}\right]$$

$$= b\left[2\sin\frac{\varphi}{2} + \left\{\sin\frac{3\varphi}{2} - \sin\frac{\varphi}{2}\right\} + \left\{\sin\frac{5\varphi}{2} - \sin\frac{3\varphi}{2}\right\} + \left\{\sin\frac{7\varphi}{2} - \sin\frac{5\varphi}{2}\right\}\ldots \left\{\sin\left(n-\frac{1}{2}\right)\varphi - \sin\left(n-\frac{3}{2}\right)\varphi\right\}\right]$$

$$= b\left[\sin\frac{\varphi}{2} + \sin\left(n-\frac{1}{2}\right)\varphi\right]$$

$$= 2\,b\sin\frac{n\varphi}{2}\cos\frac{(n-1)\,\varphi}{2}$$

$$R\cos\vartheta = b\frac{\sin\frac{n\varphi}{2}}{\sin\frac{\varphi}{2}}\cos\frac{(n-1)\,\varphi}{2} \quad \text{...(A3)}$$

Similarly, multiplying by $2\sin\frac{\varphi}{2}$ in Eq. (A2) and simplifying, one gets

$$R\sin\vartheta = b\frac{\sin\frac{n\varphi}{2}}{\sin\frac{\varphi}{2}}\sin\frac{(n-1)\,\varphi}{2} \quad \text{...(A4)}$$

Squaring Eqs. (A3) and (A4) and adding, we get

$$R^2 = b^2\frac{\sin^2\frac{n\varphi}{2}}{\sin^2\frac{\varphi}{2}}$$

or

$$R = b\frac{\sin\frac{n\varphi}{2}}{\sin\frac{\varphi}{2}} \quad \text{...(A5)}$$

Dividing $R\sin\vartheta$ given in Eq. (A4) by $R\cos\vartheta$ given in Eq. (A3), one gets

$$\tan\vartheta = \tan\frac{(n-1)\,\varphi}{2} \quad \text{...(A6)}$$

or

$$\vartheta = \frac{(n-1)\,\varphi}{2} \quad \text{...(A7)}$$

Since $n >> 1$, so

$$\vartheta = \frac{n\varphi}{2} = \alpha \quad \text{or} \quad n\varphi = 2\alpha$$

In terms of α, Eq. (A5) is

$$R = b = \frac{\sin \alpha}{\sin \frac{\alpha}{n}} \quad ...(A8)$$

Since $\frac{\alpha}{n}$ is extremely small, so

$$R = b\frac{\sin \alpha}{\frac{\alpha}{n}} = nb\frac{\sin \alpha}{\alpha} = A_0 \frac{\sin \alpha}{\alpha} \quad ...(A9)$$

Here, R is the resultant of all the secondary waves and, $A_0 = na$, is the amplitude of the resultant wave.

Formulas at a Glance

S.No.	Subject	Formulas	Explanation of symbols
1.	Resolving power of grating	$\frac{\lambda}{d\lambda} = nN$	$\lambda \to$ Wavelength of source $n \to$ Order of maxima
2.	Angular dispersive power	$\frac{d\vartheta}{d\lambda} = \frac{n}{(a+b)\cos\vartheta}$ $= \frac{n}{\sqrt{(a+b)^2 - n^2\lambda^2}}$	$N \to$ No. of rulings in the grating
3.	Condition for diffraction for maxima with plane grating	$(a + b) \sin \vartheta = n \lambda$	$(a + b) \to$ Grating element $\vartheta \to$ angular dispersion
4.	Half width of central maximum	$x = \frac{f\lambda}{a}$	$f \to$ focal length of lens $a \to$ width of slit
5.	Formulas for the direction of spectral lines due to diffraction from grating	$\tan \vartheta = \frac{x}{f}$	$x \to$ from the central line distance of spectral line
6.	Linear dispersive power	$\frac{dl}{d\lambda} = f\frac{d\vartheta}{d\lambda} = \frac{fn}{(a+b)\cos\vartheta}$ $= \frac{fn}{\sqrt{(a+b)^2 - n^2\lambda^2}}$	$(a + b) \to$ Grating element $J \to$ angular dispersion $f \to$ focal length of lens

SOLVED EXAMPLES

12.1 Calculate the minimum number of lines in a grating which will just resolve the sodium lines in the first under spectrum. Wavelengths are 5890 Å and 5896 Å.

Solution: $\lambda_1 = 5896$ Å and $\lambda_2 = 5890$ Å

Resolving power of grating, $\dfrac{\lambda}{d\lambda} = nN$

where, $\lambda = \dfrac{\lambda_1 + \lambda_2}{2}$ and $d\lambda = \lambda_1 - \lambda_2$, n is the order of the spectrum and N is the total number of lines on the grating.

$$\lambda = \frac{\lambda_1 + \lambda_2}{2} = \frac{5896 + 5890}{2} = 5893 \text{ Å}$$

$$d\lambda = \lambda_1 - \lambda_2 = 589 - 5890 = 6 \text{ Å}$$

Let $n = 1$, $\dfrac{\lambda}{d\lambda} = 1 \times N$ or $\dfrac{5893}{6} = 1 \times N$ or $N = 982$

Ans. Minimum number of lines in grating is 982.

12.2 A diffraction grating which has 5000 lines to a cm is used at normal incidence. Calculate the dispersive power of the grating in the third order spectrum in the wavelength region 6000 Å.

Solution: Formula for dispersive power is

$$\frac{d\vartheta}{d\lambda} = \frac{n}{(a+b)\cos\vartheta} = \frac{n}{(a+b)\sqrt{1-\sin^2\vartheta}}$$

Given number of lines in grating, $N = 5000$, $\lambda = 6000 \times 10^{-8}$ cm

Grating element, $a + b = \dfrac{1}{5000}$ cm

Order of spectrum, $n = 3$

Condition for maxima is

$$(a + b)\sin\vartheta = n\lambda \quad \text{or} \quad \sin\vartheta = \frac{n\lambda}{a+b} = \frac{3 \times 6000 \times 10^{-8}}{\frac{1}{5000}} = 0.9$$

Dispersive power is

$$\frac{d\vartheta}{d\lambda} = \frac{n}{(a+b)\sqrt{1+\sin^2\vartheta}} = \frac{3}{\frac{1}{5000}\sqrt{1-0.9^2}} = \frac{3 \times 5000}{\sqrt{1-0.81}}$$

$$= \frac{3 \times 5000}{\sqrt{0.19}} = \frac{3 \times 5000}{0.436} = 3.44 \times 10^4 \text{ radians/cm}$$

Ans. The dispersive power is 3.44×10^4 radians/cm.

12.3 A plane transmission grating has 15,000 lines to an inch over a length of 5 inches. Find (i) the resolving power of the grating in the second order, and (ii) the smallest wavelength difference that can be resolved for the light of the wavelength 5000 Å.

Solution: Formula for resolving power of the grating is $\dfrac{\lambda}{d\lambda} = nN$

Given: Number of lines 15000 lines per inch, $\lambda = 5000 \times 10^{-8}$ cm

Order of spectrum, $n = 2$

Total number of lines on the grating, $N = 5 \times 15{,}000 = 7{,}50{,}000$

(i) Resolving Power

$$\frac{\lambda}{d\lambda} = nN = 2 \times 75{,}000 = 1{,}50{,}000$$

(ii) Smallest wavelength difference ($d\lambda$) that can be resolved is given by

$$d\lambda = \frac{\lambda}{nN} = \frac{5000 \times 10^{-8}}{2 \times 75000} = 0.0333 \text{ Å}$$

Ans. (i) Resolving power of grating is 1,50,000 (ii) The smallest wavelength difference that can be resolved is 0.0333 Å.

12.4 Light is incident normally on a grating of total ruled width 4×10^{-3} m with 2400 lines in all. Calculate the anguar separation of the two sodium lines in the first order spectrum. Can they be seen distinctly?

Solution: Formula for the direction of spectral lines is

$$(a + b) \sin \vartheta = n\lambda$$

Given: Width of grating, $a = 4 \times 10^{-3}$ m, No. of lines in the grating, $N = 2400$

So sodium light, $\lambda_1 = 5890$ Å $= 5890 \times 10^{-8}$ cm, $\lambda_2 = 5896$ Å $= 5896 \times 10^{-8}$ cm

Order of spectrum, $n = 1$

$$\text{Grating element} = \frac{4 \times 10^{-3}}{2400} \text{ m} = \frac{1}{6000} \text{ cm}$$

Let ϑ_1 and ϑ_2 be the angles of diffraction corresponding to two D lines of sodium. Therefore,

(i) For $\lambda_1 = 5890$ Å $= 5890 \times 10^{-8}$ cm

$$n = 1$$

$$(a + b) \sin \vartheta_1 = n\lambda \quad \text{or} \quad \frac{1}{4000} \sin \vartheta_1 = 1 \times 5890 \times 10^{-8}$$

or $$\sin \vartheta_1 = 0.2356 \; \vartheta_1 = 13°37.6'$$

(i) For $\lambda_2 = 5896$ Å $= 5896 \times 10^{-8}$ cm

$$n = 1$$

$$(a + b) \sin \vartheta_2 = n\lambda \quad \text{or} \quad \frac{1}{4000} \sin \vartheta_2 = 1 \times 5896 \times 10^{-8}$$

or $$\sin \vartheta_2 = 0.23584 \; \vartheta_2 = 13°38.5'$$

Angular separation $\vartheta_2 - \vartheta_1 = 13°38.5' - 13°37.6' \approx 1'°37.6 \approx 1'$

(iii) Condition for just resolution or formula for resolving power of grating is

$$\frac{\lambda}{d\lambda} = nN, \quad \text{total number of lines on grating, } N = ?$$

$$= \frac{\lambda_1 + \lambda_2}{2} = \frac{5896 + 5890}{2} = 5893 \text{ Å}$$

$$d\lambda = \lambda_1 - \lambda_2 = 5896 - 5890 = 6 \text{ Å}$$

For $n = 1$, $\dfrac{\lambda}{d\lambda} = 1 \times N$ or $\dfrac{5893}{6} = 1 \times N$ or $N = 982$

Ans. Since total number of lines on the grating is 2400 which is more than 982 hence, the two lines can be resolved well.

12.5 What is the highest order of spectrum which may be seen with monochromatic light of wavelength 5000 Å by means of a diffraction grating with 6000 lines/cm.

Solution: Diffraction condition for maxima is

$$(a + b)\sin\vartheta = n\lambda$$

Given: $\lambda = 5000 \times 10^{-8}$ cm, Number of lines in grating 6000 lines per cm, $n = ?$

Maximum value of $\sin\vartheta = 1$, so

$$(a + b) = n\lambda$$

$$a + b = \frac{1}{6000}\text{ cm},\quad \lambda = 5000 \text{ Å}$$

$$n = \frac{a+b}{\lambda} = \frac{1}{6000} \times \frac{1}{5000 \times 10^{-8}}$$

$$= \frac{1}{30 \times 10^{6} \times 10^{-8}} = \frac{10}{3} = 3.33$$

Ans. Highest order of spectrum is 3.

12.6 How many orders will be visible if the wavelength of incident radiation is 5000 Å and number of lines on the grating is 2400 per inch.

Solution: Diffraction condition for maxima is

$$(a + b)\sin\vartheta = n\lambda$$

Given: $\lambda = 5000$ Å, No. of lines in grating 2400 per inch

Maximum value of $\sin\vartheta = 1$, so

$$(a + b) = n\lambda$$

$$a + b = \frac{2.54}{2400}\text{ cm},\quad \lambda = 5000 \text{ Å}$$

$$n = \frac{a+b}{\lambda} = \frac{2.54}{2400} \times \frac{1}{5000 \times 10^{-8}}$$

$$= \frac{2.54}{12 \times 10^{6} \times 10^{-8}} = \frac{254}{12} = 21.16$$

Ans. Highest order of spectrum is 21.

12.7 In Fraunhofer diffraction screen is placed at a distance of 2.5 m from the slit. Calculate the slit width if the first minimum lies 6 mm on either side of the central maximum.

Given: $\lambda = 6000$ Å.

Solution: Formula for half width of central maximum, $x = \dfrac{f\lambda}{a}$

Given: Distance between slit and screen, D = 200 cm, $\lambda = 6000 \times 10^{-8}$ cm, x = 0.6 cm

When the convex lens to focus the diffraction pattern is placed near the slit, then the focal length of convex lens, $f \approx D$

$$\text{The width of slit, } a = \frac{D\lambda}{x} = \frac{250 \times 6000 \times 10^{-8}}{0.6} = 2.5 \times 10^{-2} \text{ cm}$$

Ans. The width of slit is 2.5×10^{-2} cm.

12.8 Find the angle of separation of two helium lines (5016 Å and 5048 Å) in the second order spectrum using a plane transmission grating having 5000 lines per cm.

Solution: Formula for the direction of spectral lines is

$$(a + b) \sin \vartheta = n\lambda$$

Given: No. of lines in the grating, 5000 per cm

So sodium light, λ_1 = 5016 Å = 5016×10^{-8} cm, λ_2 = 5048 Å = 5048×10^{-8} cm

Order of spectrum, n = 2

$$\text{Grating element } a + b = \frac{21}{5000} \text{ cm}$$

Let ϑ_1 and ϑ_2 be the angles of diffraction corresponding to two lines of helium. Therefore,

(i) For λ_1 = 5016 Å = 5016×10^{-8} cm

$$n = 2$$

$$(a + b) \sin \vartheta_1 = n\lambda \quad \text{or} \quad \frac{1}{5000} \sin \vartheta_1 = 2 \times 5016 \times 10^{-8}$$

or

$$\sin \vartheta_1 = 0..5016 \Rightarrow \vartheta_1 = 30°6'$$

(i) For λ_2 = 5048 Å = 5048×10^{-8} cm

$$n = 2$$

$$(a + b) \sin \vartheta_2 = n\lambda \quad \text{or} \quad \frac{1}{5000} \sin \vartheta_2 = 2 \times 5048 \times 10^{-8}$$

or

$$\sin \vartheta_2 = 0.5048 \Rightarrow \vartheta_2 = 30°19'$$

Angular separation, $\vartheta_2 - \vartheta_1 = 30°19' - 30°6' \approx 13'$

Ans. Angular separation is $\vartheta_2 - \vartheta_1 = 13'$.

12.9 A mixture of two wavelengths (λ_1 = 5000 Å and λ_1 = 5100 Å) incident normally on a plane transmission grating having 5000 lines per cm. Calculate separation in cm of two lines in the first order if the focal length of lens used for observing the spectrum is 150 cm.

Solution: Formula for the direction of spectral lines is

$$(a + b) \sin \vartheta = n\lambda \quad \text{and} \quad \tan \vartheta = \frac{x}{f}$$

Given: No. of lines in the grating, 5000 per cm

So sodium light, λ_1 = 5000 Å = 5000×10^{-8} cm, λ_2 = 5100 Å = 5100×10^{-8} cm

Order of spectrum, n = 1, focal length of lens, f = 150 cm

Grating element, $a + b = \dfrac{1}{5000}$ cm

Let ϑ_1 and ϑ_2 be the angles of diffraction corresponding to two lines of mixture. Therefore,

(i) For $\lambda_1 = 5000$ Å $= 5000 \times 10^{-8}$ cm

$$n = 1$$

$$(a + b) \sin \vartheta_1 = n\lambda \quad \text{or} \quad \frac{1}{5000} \sin \vartheta_1 = 1 \times 5000 \times 10^{-8}$$

or $$\sin \vartheta_1 = 0.25 \ \vartheta_1 = 14.4775°$$

(i) For $\lambda_2 = 5100$ Å $= 5100 \times 10^{-8}$ cm

$$n = 1$$

$$(a + b) \sin \vartheta_2 = n\lambda \quad \text{or} \quad \frac{1}{5000} \sin \vartheta_2 = 1 \times 5100 \times 10^{-8}$$

or $$\sin \vartheta_2 = 0.255 \ \vartheta_2 = 14.7736°$$

If x_1 and x_2 are distances of two lines from the central line, and ϑ_1 and ϑ_2 are the angles of diffraction corresponding to two lines of mixture are, then

$$\tan \vartheta_1 = \frac{x_1}{f} \quad \text{and} \quad \tan \vartheta_2 = \frac{x_2}{f}$$

Now the separation between two lines is

$$\begin{aligned} \Delta x &= x_2 - x_1 = f(\tan \vartheta_2 - \tan \vartheta_1) \\ &= 150 \times (\tan 14.7736 - \tan 14.4775) \\ &= 150 \times (0.2637 - 0.2582) \\ &= 0.825 \text{ cm} \end{aligned}$$

Ans. The separation between two lines is 0.825 cm.

12.10 Calculate number of lines per cm of a grating having diffraction angle of 35° for wavelength 5500 Å for second order.

Solution: Formula for direction of diffraction line is

$$(a + b) \sin \vartheta = n\lambda$$

Given: Angle of diffraction is $\vartheta = 35°$, $\lambda = 5500$ Å $= 5500 \times 10^{-8}$ cm

Order of spectrum line, $n = 2$

$$(a + b) \sin 35 = 2 \times 5500 \times 10^{-8} \quad \text{or} \quad (a + b) = 1.918 \times 10^{-4} \text{ cm}$$

$$\text{No. of lines per cm} = \frac{1}{a+b} = \frac{1}{1.918 \times 10^{-4}} = 5213$$

Ans. No. of lines per cm is 5213.

QUESTIONS FOR EXAMINATION

12.1 Differentiate between interference and diffraction.

12.2 What is Fresnel diffraction?

12.3 What is Fraunhofer diffraction?

12.4 What is diffraction? Write difference between Fresnel and Fraunhofer diffraction.

12.5 What is diffraction? Find the diffraction conditions to find the diffraction pattern in case of Fraunhofer diffraction with a single slit.

12.6 What is intensity of light? Discuss the method to find the ratio of intensities of the first, second and third order maxima with respect to central maxima in case of Fraunhofer diffraction with single slit.

12.7 Write short notes on the following:

(a) Width of central maxima (b) Angular dispersive power.

(c) Linear dispersive power.

12.8 What are the factors on which the width of principal maxima in case of Fraunhofer diffraction with a single slit depends?

12.9 Write a short note on plane transmission grating.

12.10 Derive the diffraction conditions for diffraction pattern using a plane grating.

12.11 What is resolving power a plane grating? What is Reyleigh criterion to resolve the two lines?

12.12 Derive the diffraction condition for getting principal maxima of the diffraction beam diffracted by a plane grating on normal incident.

12.13 What are the factors on which the dispersive power of a grating depends?

12.14 Derive an expression for dispersive power of a grating?

12.15 What do you mean by dispersive power of a grating?

12.16 Discuss the method to find the wavelength of monochromatic light using plane transmission grating.

12.17 Discuss the spectrum obtained from plane grating when white light is used.

12.18 Discuss the method to find the diffraction angle of a certain line in the spectrum obtained from plane grating.

12.19 How will you find the order of spectrum formed from a grating?

12.20 How will you find missing orders of maxima and minima?

12.21 Discuss the overlapping of spectral lines when the incident beam has more than one wavelength?

12.22 Derive the diffraction condition when a light beam falls obliquely on the grating. Also obtain diffraction condition for n^{th} order diffraction maxima when angle of deviation, δ_m, is minimum.

OBJECTIVE QUESTIONS

12.1 What type of wavefront is incident in case of Fraunhofer diffraction?

(a) elliptical (b) cylindrical (c) spherical (d) plane

12.2 What type of wavefront is incident in case of Fresnel diffraction?

(a) elliptical (b) cylindrical (c) spherical (d) plane

12.3 Who gave the correct interpretation of rectilinear propagation of light?

(a) Fresnel (b) Fraunhoper (c) Maxwell (d) Newton

12.4 In Fraunhofer diffraction, how many times is the intensity of central maxima than that of first maxima?

(a) 20 (b) 21 (c) 22 (d) 23

12.5 In Fraunhofer diffraction with single slit, how many times is the intensity of central maxima than that of second maxima?

(a) 60 (b) 61 (c) 62 (d) 63

12.6 In Fraunhofer diffraction with single slit, how many times is the intensity of central maxima than that of second maxima?

(a) 120 (b) 121 (c) 122 (d) 123

12.7 What is the unit of angular dispersive power of a grating?

(a) degree/radian (b) radian/degree (c) radian/cm (d) radian × cm

12.8 What is the unit of linear dispersive power of a grating?

(a) no unit (b) cm (c) meter (d) nanometer

12.9 Half width of central maxima is

(a) inversely proportional to focal length of the convex lens.

(b) inversely proportional to wavelength of light source.

(c) proportional to length of slit.

(d) proportional to focal length of the convex lens.

12.10 Half width of central maxima is

(a) inversely proportional to focal length of the convex lens.

(b) inversely proportional to wavelength of light source.

(c) proportional to length of slit.

(d) inversely proportional to length of slit..

12.11 In Fraunhofer diffraction with single slit, yellow light is replaced by red light without changing the experimental setup. The diffraction pattern will

(a) disappear (b) be unchanged (c) be wider (d) be narrower

12.12 A CD (Compact Disk) gives a glare of rainbow colour because of

(a) diffraction (b) interference (c) polarization (d) scattering

12.13 When white light is incident on a grating, which of the following colours deviate most?

(a) violet (b) blue (c) yellow (d) orange

12.14 What is the phase difference between between light rays from two half of the slit for getting first minima in case of Fraunhofer diffraction?

(a) $\frac{\pi}{2}$ (b) π (c) 2π (d) $\frac{\pi}{4}$

12.15 When the length of slit is increased in case of Fraunhofer diffraction, then the width of central maxima

(a) increases (b) decreases (c) remains unchanged (d) none of these

12.16 Why the tip of the needle does not give a sharp image on the screen?

(a) polarization (b) interference (c) diffraction (d) scattering

12.17 If N, λ and n are number of elements in the grating, order of fringe and wavelength of light, then the resolving power of plane grating is

(a) nN (b) nN (c) $\frac{N\lambda}{n}$ (d) $\frac{Nn}{\lambda}$

12.18 Which of the following has largest wavelength?

(a) red light (b) blue light (c) X-rays (d) Gamma-rays

Answers

12.1 (d) 12.2 (c) 12.3 (a) 12.4 (c) 12.5 (b) 12.6 (b) 12.7 (c) 12.8 (a) 12.9 (d) 12.10 (d) 12.11 (d) 12.12 (d) 12.13 (d) 12.14 (c) 12.15 (b) 12.16 (c) 12.17 (a) 12.18 (a)

CHAPTER 13

Fresnel Diffraction

13.1 INTRODUCTION

Fresnel assumed that a point light source emits lights in the form of spherical wavefronts. In Fresnel diffraction the source of light and the screen are at finite distance from the biprism, while in Fraunhofer diffraction the source of light and the screen are at infinite distance from the slit. The incident wavefront is considered spherical or cylindrical, because source of light in Fresnel diffraction is placed at finite distance from biprism. The diffraction pattern is formed only when the wavefront of light is incident on a sharp obstacle or passes through slits. A wavefront consists of a large number of elements or zones of small area. Each element sends secondary wavelets continuously. A biprism consists of two right angled prisms of very small refracting angle (α, nearly 30′) joined base to base as shown in Fig. 13.1. In practice it is made from a single glass plate. Its faces are cut in such a way that one angle is 119° and other two angles are nearly 30° each. To prove his theory Fresnel showed that when light is incident on the biprism from a source placed near it, then a diffraction pattern is formed on the other screen placed on the other side. Fresnel diffraction through biprism has a number of applications. For example, wavelength of light source and thickness of transparent sheet can be determined.

In Section 13.7 Fresnel's half period zones are discussed. On the basis of half period zones, Fresnel showed that light travels with rectilinear path. A zone plate is simply a thin parallel glass plate containing concentric circles of radii proportional to the square root of natural numbers. The annular space so created is deeply blackened alternately. Fresnel zone plate behaves like a convex lens. So for getting bright and sharp images a zone plate is good alternative of lens.

13.2 COMPARISON BETWEEN THE FRAUNHOFER DIFFRACTION AND FRESNEL DIFFRACTION

S.No.	Fraunhofer diffraction	Fresnel diffraction
1.	In Fraunhofer diffraction the source of light and the screen are at infinite distance from the slit.	In Fresnel diffraction the source of light and the screen are at finite distance from the biprism.
2.	The incident wave front is plane.	Incident wavefront is either spherical or cylindrical.
3.	A convex lens is used to make plane wavefront from spherical wavefront.	No lens is used as spherical wavefront is used for getting diffraction pattern.
4.	The secondary waves originating from the slit are in the same phase at every point.	The secondary waves originating from the biprism are not in the same phase.
5.	Theoretical analysis of Fraunhofer diffraction yields is almost exact diffraction pattern.	Theoretical analysis of Fresnel diffraction yields approximate dispersion pattern.
6.	In Fraunhofer diffraction, the angular inclination plays an important role rather than distance of slit from the source and the screen.	In Fresnel diffraction, the distance of prism from source and screen plays important rather than angular inclination.

13.3 FRESNEL'S BIPRISM

Fresnel's biprism consists of two right angled prisms of very small refracting angle (α) nearly 30′) joined base to base as shown in Fig. 13.1. In practice, it is made from a single glass plate. Its faces are cut in such a way that one angle is 119° and other two angles are nearly 30° each.

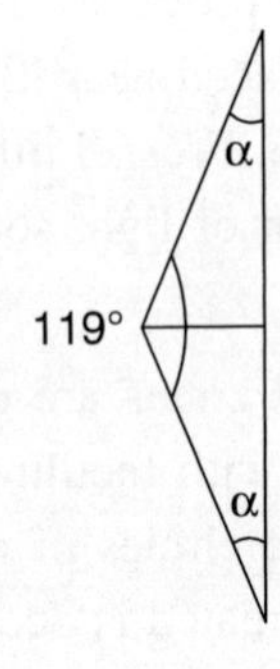

Fig. 13.1 Fresnel's biprism

It is used to show the phenomenon of interference and its function is to produce two images of a given slit which are separated by a certain distance and behave as two coherent sources.

13.3.1 Formation of Coherent Sources and Interference Fringes

Let us consider a monochromatic source (S) incident on a biprism. The waves between GA and OC appear as coming from S_1 and waves between HC and OA appear as coming from S_2 as shown in Fig. 13.2. These two virtual images, S_1 and S_2, of source S act as coherent sources and form an interference pattern on the screen only in AB portion of the screen.

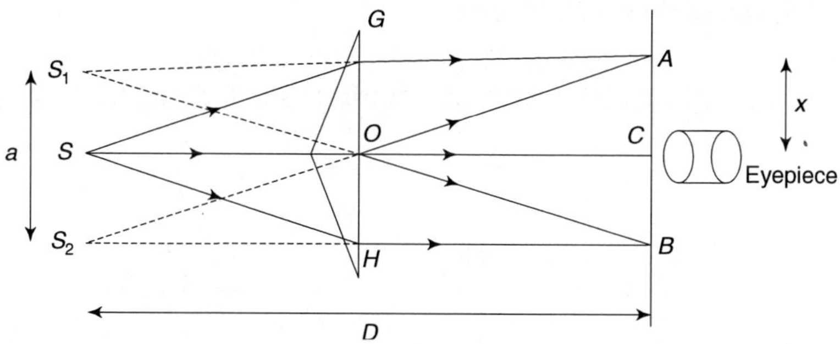

Fig. 13.2

Theory: Let us assume that the distance between virtual sources, S_1 and S_2 is a and distance between source S and the screen is D. The point C on the screen, which is at equal distance from S_1 and S_2, will be a point of central bright maximum. Alternate bright and dark fringes are produced on both sides of C due to interference.

Condition for maxima or bright fringes is [see Appendix A]

$$\frac{xa}{D} = n\lambda, \quad \text{where } n = 0, 1, 2, 3, \ldots \text{ etc.} \qquad \ldots(13.1)$$

Distance of nth and $(n + 1)^{th}$ maxima from central maxima is

$$x_n = \frac{n\lambda D}{a} \quad \text{and} \quad x_{n+1} = \frac{(n+1)\lambda D}{a} \qquad \ldots(13.2)$$

The fringe band width,

$$\beta = x_{n+1} - x_n = \frac{\lambda D}{a} \qquad \ldots(13.3)$$

Condition for minima or dark fringes is

$$\frac{xa}{D} = (2n + 1)\frac{\lambda}{2}, \quad \text{where, } n = 0, 1, 2, 3, \ldots, \text{ etc.} \qquad \ldots(13.4)$$

Distance of nth and $(n + 1)^{th}$ minima from central maxima is

$$x_n = \frac{(2n+1)\lambda D}{2a} \quad \text{and} \quad x_{n+1} = \frac{(2n+3)\lambda D}{2a} \qquad \ldots(13.5)$$

The fringe band width,

$$\beta = x_{n+1} - x_n = \frac{\lambda D}{a} \qquad ...(13.6)$$

Thus, fringe width of bright and dark fringes is equal.

Experimental Setup of Fresnel's Biprism

For getting diffraction pattern from the Fresnel biprism, following components are required: an optical bench with four uprights for holding variable slit, biprism, lens and eyepiece.

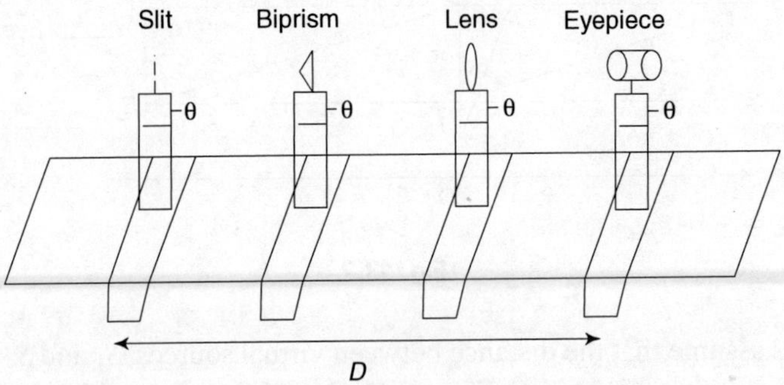

Fig. 13.3 Experimental set up of Fresnel's biprism

Setting the Experiment for Getting Fringes

(i) First of all level the optical bench with spirit level .

(ii) Then mount slit, biprism and the eyepiece on uprights as shown in Fig. 13.3.

(iii) Illuminate the slit by monochromatic light.

(iv) Now make the slit narrow.

(v) Fringes will appear in the eyepiece, which may not be very distinct.

(vi) With the help of a tangent screw adjust the biprism till the fringes seen through eyepiece become most distinct.

(vii) To make fringes more clear adjust width of the slit.

(i) Procedure to Measure Fringe Width, β

Note down the least count of micrometer of the eyepiece. Now set vertical cross-wire of the eyepiece on any bright fringe. Note down the reading of present position of micrometer. Move the micrometer so that n number of fringes cross the field of view. Note down the reading of displaced position of micrometer. Let x be the distance moved by the micrometer then fringe width can be obtained from the relation

$$\beta = \frac{x}{n}$$

(ii) The distance between slits and eyepiece (D) can be measured from the position of upright holders holding them.

(iii) Procedure to Measure of Separation 'a' between Two Virtual Slits

The separation between the virtual slits can be measured using the displacement method as follows:

Mount a convex lens of focal length less than one-fourth of distance between biprism and eyepiece between biprism and the eyepiece without disturbing their positions. Now adjust the lens in two positions as shown in Figs. 13.4(a) and 13.4(b) till sharp images of S_1 and S_2 are formed in the eyepiece. Let a_1 and a_2 be the distance between the images of S_1 and S_2 in two positions of the lens.

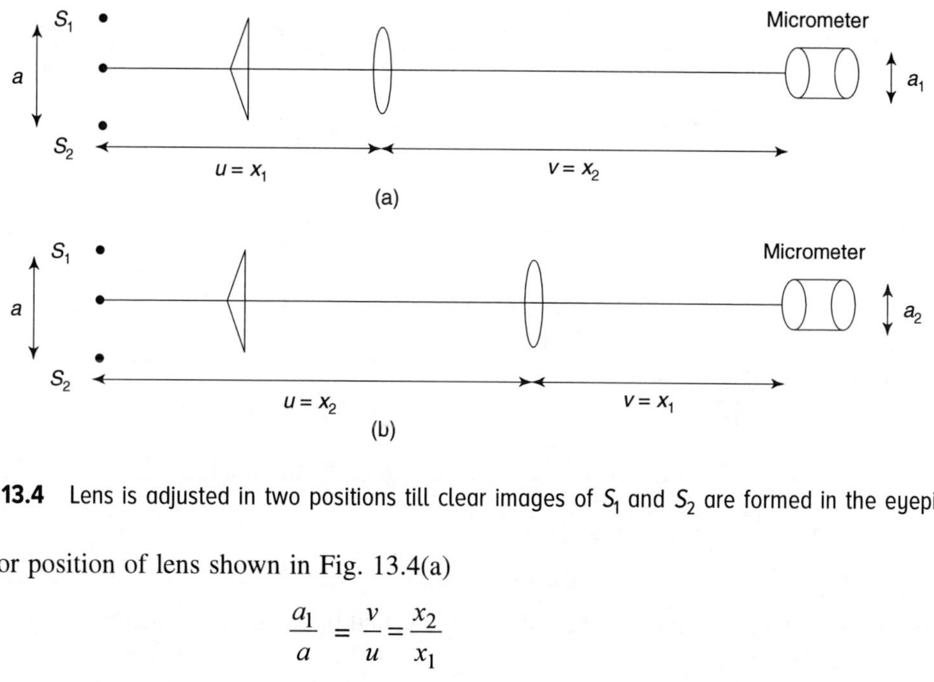

Fig. 13.4 Lens is adjusted in two positions till clear images of S_1 and S_2 are formed in the eyepiece

For position of lens shown in Fig. 13.4(a)

$$\frac{a_1}{a} = \frac{v}{u} = \frac{x_2}{x_1}$$

and for position of lens shown in Fig. 13.4(b)

$$\frac{a_2}{a} = \frac{v}{u} = \frac{x_1}{x_2}$$

The product of $\frac{a_1}{a}$ and $\frac{a_2}{a}$ will be

$$\frac{a_1 a_2}{a^2} = 1$$

So that the distance between two virtual slits, S_1 and S_2, will be given by relation

$$a = \sqrt{a_1 a_2} \quad \text{...(13.7)}$$

Knowing β, D and a, the wavelength of the source (λ) can be obtained from the relation,

$$\lambda = \frac{\beta d}{D} \quad \text{...(13.8)}$$

Note: Separation between two virtual slits, S_1 and S_2, can also be obtained directly if refracting angle (α) of the biprism is known.

Since refracting angle (α) of biprism is small, so angle of deviation (δ) produced by biprism can be written as

$$\delta = (\mu - 1)\,\alpha \quad \text{...(13.9)}$$

where, μ is refractive index of the material of biprism and α is in radians.

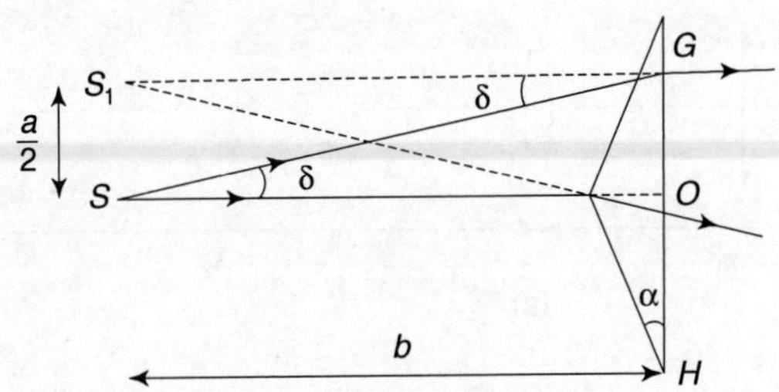

Fig. 13.5

In ΔSGO

$$\tan\delta = \frac{GO}{SO} = \frac{\frac{a}{2}}{b} \quad \text{or} \quad \delta = \frac{\frac{a}{2}}{b} \text{ for small } \delta$$

or $$a = 2b\delta$$

or $$a = 2b\,(\mu - 1)\,\alpha \quad \text{...(13.10)}$$

where, α is in radians. By putting the values of b, α and μ, a can be calculated.

13.4 APPLICATIONS OF FRESNEL BIPRISM

(a) Determination of Wavelength of Light λ

One of the applications of Fresnel biprism is to find the wavelength of a incident light source. Wavelength of a monochromatic light source can be determined by measuring the fringe width. Once we know the fringe width then from the relation

$$\beta = \frac{\lambda D}{a}$$

one can get wavelength, $\lambda = \dfrac{\beta a}{D}$.

Putting the values of β, d and D, λ can be calculated.

(b) Measurement of Thickness of Transparent Sheet

Biprism can also be used to measure thickness of a thin sheet (or mica sheet) of transparent material. Let us consider two coherent sources, S_1 and S_2, of monochromatic light placed at a distance D from the screen, AB. These two coherent sources after overlapping produce interference bands on the screen AB. In the absence of transparent sheet, bright band is formed at point C on the screen, because optical paths, S_1C and S_2C, are equal. Let us now introduce a mica sheet of thickness (t) and refractive index (μ) in the path ray, S_1C. Then, ray S_1C travels a distance equal to thickness (t) through mica sheet and rest distance ($S_1C - t$) it travels in air, while ray S_2C travels whole distance through air.

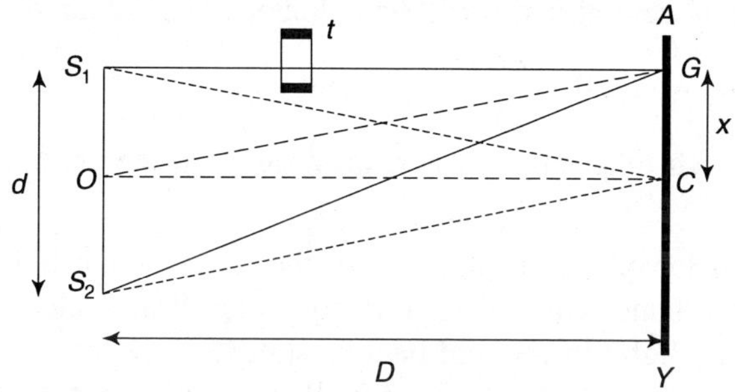

Fig. 13.6

Since velocity of light in air and mica is different, so optical path of two rays will not be equal and position of central band will be shifted upward from point C to point P as shown Fig. 13.6.

According to the format principle

$$S_2G = (S_1G - t) + \mu t$$

or
$$S_2G - S_1G = (\mu - 1)\,t \qquad ...(13.11)$$

In the absence of mica sheet, let G be the position of nth bright band, which is at a distance x from the central bright band, then

$$S_2G - S_1G = \frac{xa}{D} = n\lambda \qquad ...(13.12)$$

Comparing Eqs. (13.11) and (13.12), we get

$$(\mu - 1)\,t = n\lambda \qquad ...(13.13)$$

or $$(\mu - 1)\, t = \frac{xd}{D}$$

or $$x = \frac{D}{d}(\mu - 1)\, t$$

or $$x = \frac{\beta}{\lambda}(\mu - 1)\, t \qquad ...(13.14)$$

If x, β, λ and μ are known, thickness t of the thin transparent sheet can be calculated.

13.5 FRINGES WITH WHITE LIGHT

When white light is used as source of light in the biprism experiment, then the central fringe will be white as all rays of different colours will overlapat central point. Since all the colours of white light have different wavelengths, so path difference will be different for different colours and all colours will form bands at different points thus each colour produces its own bands. All bands, except central band, will be coloured. The coloured fringes will be only a few in numbers as band width, $\beta = \frac{\lambda D}{d}$, is different for different colours so bands of different colours will overlap each other after a short distance from the central white band forming blurred bands.

When mica sheet is introduced in the path of the ray of white light emerging from one of the slits, then central band will not be perfectly white. When mica sheet is introduced in the path of ray of white light, the central band is shifted away from point C, therefore, due to dispersion of light, all bands of all colours will not overlap at the central point and so it will not be perfectly white. It is to be noted that the central band due to white light will be white only when mica sheet introduced in the path of light is very thin.

SOLVED EXAMPLES

Example 13.1 The light of wavelength, $\lambda = 5896$ Å, is incident on a biprism placed at 0.1 m from a slit. The fringe width observed at a screen 0.7 m from the biprism is 9.524×10^{-4} m. What is the distance between coherent sources?

Solution: Formula used

Width of fringe, $\beta = \frac{\lambda D}{a}$

So formula for distance between two coherent sources

$$a = \frac{\lambda D}{\beta}$$

Given:

Wavelength of source, $\lambda = 5890$ Å $= 5896 \times 10^{-8}$ cm $= 5896 \times 10^{-10}$ m

$$D = 0.1 + 0.7 \text{ m} = 0.80 \text{ m},$$

Fringe width, $\beta = 9.524 \times 10^{-4}$ m

Distance between two coherent sources,

$$a = \frac{\lambda D}{\beta} = \frac{5896 \times 10^{-10} \times 0.80}{9.524 \times 10^{-4}} = \frac{4716.8}{9.524} \times 10^{-6}$$

$$= \frac{47.168}{9.524} \times 10^{-4} = 4.95 \times 10^{-4} \text{ m}$$

Ans. The distance between coherent sources is 4.95×10^{-4} m.

Example 13.2 The distances between the slit and biprism and between biprism and screen are 60 cm each. Angle of biprism and refractive index are 179° and 1.5, respectively. Calculate the wavelength of the light, if the distance between successive fringes is 0.0115 cm.

Solution Formula used

Fringe width, $\beta = \dfrac{\lambda D}{d}$

So formula for wavelength of light, $\lambda = \dfrac{\beta d}{D}$

Given:

$$D = 60 + 60 \text{ cm} = 110 \text{ cm} = 1.2 \text{ m},$$

Fringe width, $\beta = 0.0115$ cm

Angle of biprism, 179°, distance between slit and biprism, $b = 50$ cm

Refractive index of prism, $\mu = 1.5$

Refracting angle of biprism, $\alpha = \dfrac{180 - 179}{2} = \dfrac{1}{2} = 0.5° = 0.5 \times \dfrac{\pi}{180}$ radians,

Distance between two coherent sources

$$d = 2(\mu - 1)\alpha b$$

Putting the values of various quantities

$$d = 2(1.5 - 1) \times 0.5 \times \frac{\pi}{180} \text{ 60 cm}$$

$$= 2 \times 0.5 \times 0.5 \times \frac{3.14}{180} \times 60 \text{ cm}$$

$$= \frac{2 \times 0.25 \times 3.14}{3} = 0.523 \text{ cm}$$

The wavelength of source

$$\lambda = \frac{\beta d}{D} = \frac{0.0115 \times 0.523}{100} = 0.6015 \times 10^{-4}$$

$$= 6015 \times 10^{-8} \text{ cm} = 6015 \text{ Å}$$

Ans. The wavelength of source is 6015Å

Example 13.3 When a thin glass sheet of 3.0×10^{-4} cm thickness is placed in the path of one of the interfering beams in a biprism experiment, it is found that the central fringe shifts through a distance equal to the width of three fringes. Find the refractive index of the glass sheet (λ = 5893 Å).

Solution: Formula for thickness of glass sheet is

$$(\mu - 1)t = n\lambda \quad \text{or} \quad (\mu - 1) = \frac{n\lambda}{t}$$

Formula for refractive index of glass sheet is

$$\mu = 1 + \frac{n\lambda}{t}$$

Given: Thickness of glass sheet, $t = 3.0 \times 10^{-4}$ cm

Wavelength of light source, λ = 5890 Å = 5890 10^{-8} cm

Number of fringe shifts, $n = 3$

$$\mu = 1 + \frac{n\lambda}{t} = 1 + \frac{3 \times 5890 \times 10^{-8}}{3.0 \times 10^{-4}}$$

$$= 1 + 0.5890 = 1 + 0.589 = 1.589$$

Ans. Refractive index of the glass, μ = 1.589.

Example 13.4 A beam of sodium light (λ = 5893 Å) falls on the film of oil floating on water. When viewed at angle 30° from the normal, 6th dark band is seen. Find the thickness of the oil film. ($\mu_{oil} = 1.5$)

Solution: Formula used

Condition for dark band in reflected system is

$2\mu t \cos r = n\,\lambda$, here, r is the angle of refraction in the oil film

So, thickness of oil film is given by formula

$$t = \frac{n\lambda}{2\mu \cos r}$$

Given:

$$n = 6,\ \mu = 1.5,\ i = 30° = 30 \times \frac{\pi}{180} = \frac{\pi}{6},$$

$$\lambda = 5893\ \text{Å} = 5893 \times 10^{-8}\ \text{cm}$$

$$\mu = \frac{\sin i}{\sin r} \quad \text{or} \quad \sin r = \frac{\sin i}{\mu} = \frac{\sin 30}{1.5} = \frac{\frac{1}{2}}{1.5} = \frac{1}{3}$$

$$\cos r = \sqrt{1 - \sin^2 r} = \sqrt{1 - \frac{1}{9}} = \sqrt{\frac{8}{9}} = \frac{2}{3}\sqrt{2} = 0.94$$

Thickness of oil film, $t = \dfrac{n\lambda}{2\mu \cos r} = \dfrac{6 \times 5893 \times 10^{-8}}{2 \times 1.5 \times 0.94} = 1.25 \times 10^{-4}$ cm

Ans. Thickness of oil film is 1.25×10^{-4} cm.

13.6 FRESNEL'S EXPLANATION OF RECTILINEAR PROPAGATION OF LIGHT

For the propagation of light wave Huygen's proposed the idea of secondary wavelets. Interference phenomenon due to light is explained by the wave theory. The correct interpretation of the rectilinear propagation of light was given by a French physicist named Augustin-Jean Fresnel in 1815 on the basis of the wave theory of light by combining Huygen's principle of secondary wavelets and the principle of interference. However, it was mentioned by Fresnel that the rectilinear propagation of light is only approximately true.

13.6.1 Fresnel's Assumptions

(i) The diffraction pattern is formed only when the wavefront of light is incident on a sharp obstacle or passes through slits.

(ii) A wavefront consists of a large number of elements or zones of small area. Each element sends secondary wavelets continuously.

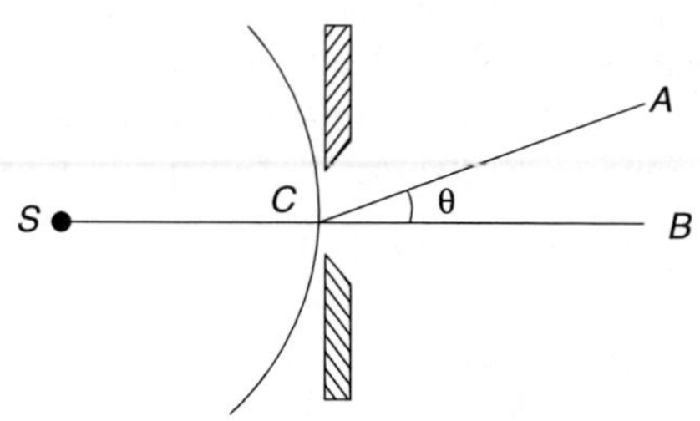

Fig. 13.7

(iii) The intensity due to overlap of wavefront is maximum in the normal direction of wavefront, that is, when $\theta = 0$ [Fig. 13.7]. Intensity decreases as inclination (θ) increases. It becomes zero in opposite direction, i.e., $\theta = 180°$. Mathematically, the intensity is proportional, $\propto$ (amplitude)^2. But amplitude is proportional to $(1 + \cos\theta)$. When $\theta = 0$, that is intensity is maximum, because $(1 + \cos\theta) = 2$ as $\cos\theta = 1$. When $\theta = 180°$, that is, intensity is zero, because $(1 + \cos\theta) = 0$, as $\cos 180° = -1$. Hence, wavefront does not travel in the backward direction. This result removes the weak point of Huygen's principle, which states that each point of secondary wavelets becomes the centre of emitting secondary wavefront of spherical shape, i.e., waves move in all directions including backward direction. As amplitude in the backward direction is zero, waves do not move in backward direction.

(iv) The resultant amplitude at any point is determined by the principle of superposition.

13.7 FRESNEL'S HALF PERIOD ZONES

For the sake of convenience Fresnel divided the wavefront into zones of radii such that the average mean radii of spheres from point P [see Fig. 13.8(a)] increases uniformly in the steps of $\frac{\lambda}{2}$ in going from one zone to another zone. These are called Fresnel's half period zones.

Consider monochromatic light of wavelength (λ) a plane wavefront $ABCD$ perpendicular to the plane of paper as shown in Fig. 13.8(a). Draw, a perpendicular at point C_0 on the wavefront from the point P. The point C_0 is called the pole of the wave with respect to P, where, P is a point situated at a perpendicular distance (b) from the wavefront, i.e., $PC_0 = b$.

Now taking P as the centre, draw concentric spheres of radii

$$PC_1 = b + \frac{\lambda}{2},\quad PC_2 = b + \frac{2\lambda}{2},$$

$$PC_3 = b + \frac{3\lambda}{2}, \ldots,\quad PC_n = b + \frac{n\lambda}{2}. \qquad \ldots(13.15)$$

such that these spheres intercept plane wavefront $ABCD$ in concentric circles C_1, C_2, C_3... C_n, etc, respectively. The circular area enclosed by the circle C_1 is called Fresnel's first half period zone. The circular strip enclosed by the circle C_2 and C_1 is called Fresnel's second half period zone. The circular strip enclosed by the circles, C_3 and C_2, is called Fresnel's third half period zone and so on. Thus, this construction divides the entire wavefront into a large number of half period zones called Fresnel's half period zones.

From the construction of zones given in Fig. 13.8(a), we observe that

$$PC_1 = b + \frac{\lambda}{2} = PC_0 + \frac{\lambda}{2}$$

The radius first half period zone

$$\bar{r}_1 = PC_1 - PC_0 = \frac{\lambda}{2} \qquad \ldots(13.16)$$

The radius of second half period zone,

$$\bar{r}_2 = PC_2 - PC_1 = b + \frac{2\lambda}{2} - \left(b + \frac{\lambda}{2}\right) = \frac{\lambda}{2} \quad ...(13.17)$$

The radius of third half period zone,

$$\bar{r}_3 = PC_3 - PC_2 = b + \frac{3\lambda}{2} - \left(b + \frac{2\lambda}{2}\right) = \frac{\lambda}{2} \quad ...(13.18)$$

The radius of nth half period zone,

$$\bar{r}_n = PC_n - PC_{n-1} = b\ \frac{n\lambda}{2} - \left(b + \frac{(n-1)\lambda}{2}\right) = \frac{\lambda}{2} \quad ...(13.19)$$

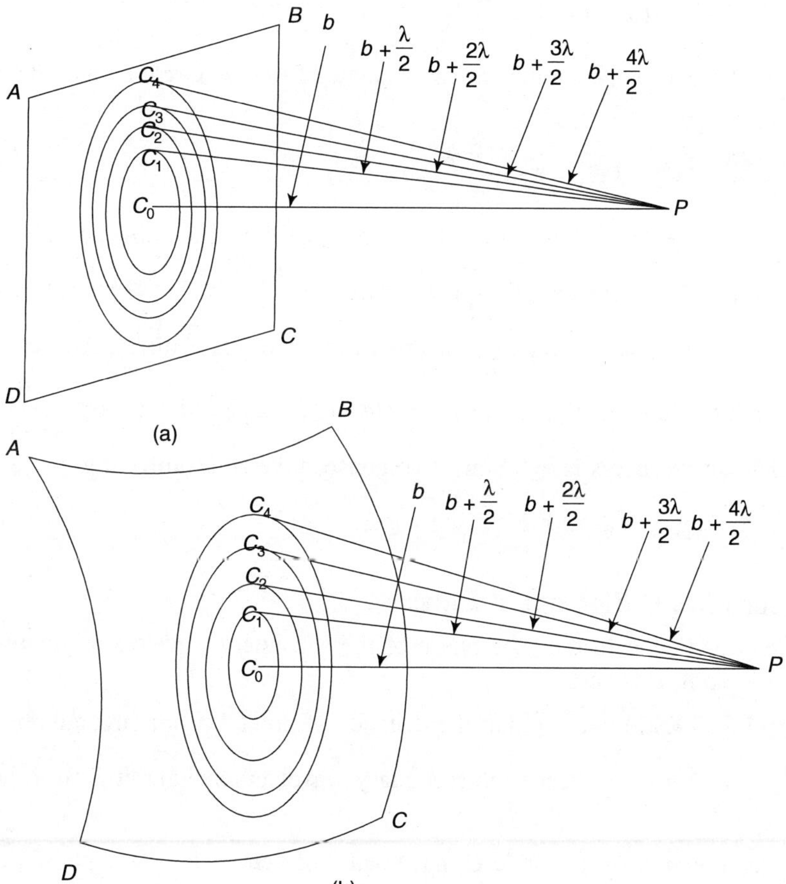

Fig. 13.8 Construction of Fresnel's half period zones in a plane wavefront. The plane wavefront *ABCD* is perpendicular to the plane of the paper. The concentric spheres of the radii $b + \frac{n\lambda}{2}$ with $n = 1, 2, 3,....$ are drawn with *P* as the centre which intersect the wavefront *ABCD* in circles $C_0, C_1, C_2,..., C_n$, respectively. The area of the nth half period zone is A_0 = (area of circle C_0} – (area of the circle C_n).
(b) Construction of the Fresnel half period zones in the spherical wavefront

13.7.1 Phase Difference Between Zones

$PC_1 - PC_0$ is the geometrical path difference between the light waves originating from the circle C_1 and the point C_0. The phase difference between the waves originated from the circle C_1 and the point C_0 will be

$$\frac{2\pi}{\lambda}(PC_1 - PC_0) = \frac{2\pi}{\lambda}\frac{\lambda}{2} = \pi \qquad ...(13.20)$$

The phase difference between the waves originated from the circle C_2 and the circle C_1 will be

$$\frac{2\pi}{\lambda}(PC_2 - PC_1) = \frac{2\pi}{\lambda}\frac{\lambda}{2} = \pi \qquad ...(13.21)$$

The phase difference between the waves originated from the circle C_n and the circle C_{n-1} will be

$$\frac{2\pi}{\lambda}(PC_n - PC_{n-1}) = \frac{2\pi}{\lambda}\frac{\lambda}{2} = \pi \qquad ...(13.22)$$

where, λ is the wavelength of light. If c is the velocity of light, then time period of light wave is $T = \frac{\lambda}{c}$ second. The corresponding phase difference is 2π. Thus, phase difference (π) corresponds to half of the period $\left(\frac{T}{2}\right)$ of the wave. The phase difference between waves originating from two consecutive circles is π. Hence, time period between waves originated from two consecutive circles is $\frac{T}{2}$. Since two consecutive zones differ by phase π or a half period $\left(\frac{T}{2}\right)$, so they are called half period zones.

13.7.2 Mean Phase Difference of Zones

Suppose at any particular instant, the phase of the secondary wavelets originating from C_0 and reaching at point P is zero.

(i) From Fig. 13.8(a), we see that the distance of point P from circumference of circle C_1 is $C_1 P = b + \frac{\lambda}{2}$, hence the secondary wavelets reaching at P from pole C_0 and the circumference of the circle C_1 have path difference of $\frac{\lambda}{2}$ or a phase difference of $\frac{2\pi}{\lambda} \times \frac{\lambda}{2} = \pi$. The phase difference due to other points lying on inside the circle C_1 will vary from zero to π. The mean phase difference of the secondary wavelets originated from the first zone will be $\left(\frac{0+\pi}{2}\right) = \frac{\pi}{2}$.

(ii) From Fig. 13.8(a), we see that the distance of point P from circumference of circle C_2 is $C_2 P = b + \lambda$, hence the secondary wavelets reaching at P from pole C_0 and the circumference of the circle C_2 have path difference of λ or a phase difference of $\frac{2\pi}{\lambda} \times \lambda = 2\pi$. The phase at point P due to wavelets originated from circumference C_1 of first zone is π. So the phase of secondary wavelets reaching at P from the points between circumference of circle C_1 and the circumference of the circle C_2 varies from π to 2π. Therefore, the mean phase difference of secondary wavelets reaching at point P from second zone is $\left(\frac{\pi + 2\pi}{2}\right) = \frac{3\pi}{2}$.

(iii) From Fig. 13.8(a), we see that the distance of point P from circumference of circle C_3 is $C_2 P = b + \frac{3\lambda}{2}$, hence the secondary wavelets reaching at P from pole C_0 and the circumference of the circle C_3 have path difference of $\frac{3\lambda}{2}$ or a phase difference of $\frac{2\pi}{\lambda} \times \frac{3\lambda}{2} = 3\pi$. The phase at point P due to wavelets originated from circumference C_2 of second zone is 2π. So, the phase difference of secondary wavelets reaching at P from the points between circumference of circle C_2 and the circumference of the circle C_3 varies from 2π to 3π. Thus, mean phase of wavelets reaching at point P from the third zone is $\left(\frac{2\pi + 3\pi}{2}\right) = \frac{5\pi}{2}$.

(iv) Similarly, proceeding further, the mean phase difference of all the secondary wavelets originating from the nth Fresnel's half period zone reaching P will be $\frac{(2n-1)\pi}{2}$.

(v) In summary, relative mean phase difference of wavelets reaching at point P from, the first, second, third, ..., n^{th} zones are $\frac{\pi}{2}, \frac{3\pi}{2}, \frac{5\pi}{2}, \ldots, \frac{(2n-1)\pi}{2}$, respectively.

Since two consecutive zones differ by phase π or a half period $\frac{T}{2}$, so they are called half period zones.

Let $R_1, R_2, R_3, \ldots, R_n$ be the amplitudes of the secondary wavelets at P due to the first, second, third Fresnel's half period zone, respectively. From the earlier discussion, the odd numbered Fresnel's half period zones are in phase and the even numbered Fresnel's half period zones are out of phase. Therefore, the resultant amplitude (R) of the secondary wavelets originating from all the Fresnel's half period zones at P will be given by

$$R = R_1 - R_2 + R_3 - R_4 + \ldots, + (-1)^{n-1} R_n \qquad \ldots(13.23)$$

13.7.3 Radii of Zones

From Fig. 13.8(a), let the radii of circles C_0, C_1, C_2,..., C_n are r_1, r_2,..., r_n, respectively.

The radius of Fresnel's first half period zone is

$$r_1 = \left[\left(b + \frac{\lambda}{2}\right)^2 - b^2\right]^{1/2} = \left[b^2 + \frac{\lambda^2}{4} + b\lambda - b^2\right]^{1/2} = \left[\frac{\lambda^2}{4} + b\lambda\right]^{1/2}$$

$$\approx \sqrt{b\lambda} \text{ as } \frac{\lambda^2}{4} << b\lambda, \text{ because } \lambda \text{ is very small.} \quad ...(13.24)$$

The radius of Fresnel's second half period zone is

$$r_2 = [(b + \lambda)^2 - b^2]^{1/2} = [b^2 + \lambda^2 + 2b\lambda - b^2]^{1/2} = [\lambda^2 + 2b\lambda]^{1/2}$$

$$\approx \sqrt{b\lambda} \text{ as } \left(\frac{2}{2}\right)\lambda^2 << 2b\lambda, \text{ because } \lambda \text{ is very small.} \quad ...(13.25)$$

The radius of Fresnel's $(n-1)^{th}$ half period zone is

$$r_{n-1} = \left[\left(b + \frac{(n-1)\lambda}{2}\right)^2 - b^2\right]^{1/2} = \left[b^2 + \frac{(n-1)^2\lambda^2}{4} + (n-1)b\lambda - b^2\right]^{1/2}$$

$$= \left[(n-1)b\lambda + \frac{(n-1)^2\lambda^2}{4}\right]^{1/2}$$

$$\approx \sqrt{(n-1)b\lambda} \text{ as } \frac{(n-1)^2\lambda^2}{4} << (n-1)b\lambda, \text{ because } \lambda \text{ is very small.} \quad ...(13.26)$$

The radius of Fresnel's n^{th} half period zone is

$$r_n = \left[\left(b + \frac{n\lambda}{2}\right)^2 - b^2\right]^{1/2} = \left[b^2 + n^2\frac{\lambda^2}{4} + nb\lambda - b^2\right]^{1/2} = \left[nb\lambda + n^2\frac{\lambda^2}{4}\right]^{1/2}$$

$$\approx \sqrt{nb\lambda} \text{ as } \lambda^2 << nb\lambda, \text{ because } \lambda \text{ is very small.} \quad (13.27)$$

From the radii of Fresnel's half period zones, it is clear that radius of each zone is proportional to the square root of the number of zones.

13.7.4 Area of Zones

The area of the first half period zone

$$A_1 = \pi r_1^2 = \pi\left(\frac{\lambda^2}{4} + b\lambda\right) = \pi b\lambda + \pi\frac{\lambda^2}{4}$$

or

$$A_1 = \pi r_1^2 \approx \pi b\lambda \quad ...(13.28)$$

The area of second half period zone

$$A_2 = \pi\, [r^2_2 - r^2_1] = \pi \left[(\lambda^2 + 2\, b\lambda) - \left(\frac{\lambda^2}{4} + b\lambda\right)\right]$$

$$A_2 = 2\pi b\lambda + \pi\frac{3\lambda^2}{4}$$

or
$$A_2 = \pi r_1^2 \approx \pi b\lambda \qquad ...(13.29)$$

The area, A_n of the nth half period zone will be

$$A_n = \pi\, (r_n^2 - r_{n-1}^2)$$

$$= \pi\left[\left(nb\lambda + n^2\frac{\lambda^2}{4}\right) - \left((n-1)\, b\lambda + \frac{(n-1)^2\,\lambda^2}{4}\right)\right]$$

$$= \pi\left[b\lambda + n^2\frac{\lambda^2}{4} - \frac{(n-1)^2\,\lambda^2}{4}\right]$$

$$= \pi\left[b\lambda + n^2\frac{\lambda^2}{4} - n^2\frac{\lambda^2}{4} - \frac{\lambda^2}{4} + n\frac{\lambda^2}{2}\right]$$

$$= \pi b\lambda + \pi(2n - 1)\frac{\lambda^2}{4} \qquad ...(13.30a)$$

Eqn. (13.30a) can also be obtained as follows:

$$A_n = \pi\left[\left(b + \frac{n\,\lambda}{2}\right)^2 - \left(b + \frac{(n-1)\,\lambda}{2}\right)^2\right]$$

or
$$A_n = \pi\left[\left(b^2 + n^2\frac{\lambda^2}{4} + b\,n\lambda\right) - \left(b^2 + (n-1)^2\frac{\lambda^4}{4}\right) + b\,(n-1)\,\lambda\right]$$

or
$$A_n = \pi\left[\left(b^2 + n^2\frac{\lambda^2}{4} + b\,n\,\lambda\right) - \left(b^2 + n^2\frac{\lambda^2}{4} + \frac{\lambda^2}{4} - \frac{\lambda^2}{2}n + b\,(n-1)\,\lambda\right)\right]$$

or
$$A_n = \pi\left[b\,n\,\lambda - \frac{\lambda^2}{4} + \frac{1}{2}n\lambda^2 - b\,n\,\lambda + b\,\lambda\right]$$

or
$$A_n = \pi b\lambda + \pi\,(2\,n - 1)\frac{\lambda^2}{4} \qquad ...(13.30)$$

Putting $(n - 1)$ for n in Eq. (13.30), the area of $(n - 1)^{th}$ half period zone will be

$$A_{n-1} = \pi b\lambda + \pi\,(2\,n - 3)\frac{\lambda^2}{4} \qquad ...(13.31)$$

Equations (13.30) and (13.31) show that n^{th} half period zones have larger area than $(n-1)^{th}$ half period zone, i.e.,

$$A_n > A_{n-1}$$

This statement indicates that higher half period zones have larger area than that of lower half period zones. The factor, $\frac{\lambda^2}{4}$, in Eq. (13.1) is very small quantity, so that the term, $\pi (2n-1)\frac{\lambda^2}{4}$, can be neglected in comparison to the term, $\pi b\lambda$. Thus, Eq. (13.31) can be written as

$$A_n = \text{p} b\lambda \qquad \text{...(13.32)}$$

Equation (13.32) shows that the area of each half period zone is almost equal.

13.7.5 Average Distance of the nth Fresnel's Half Period Zone from the Pole

The area of the first half period zone is circular. While, the area of the second, third,..., and n^{th} half period zone is a circular strip. The distance of the inner and outer circumference of nth half period zone from the point P is $b+\frac{(n-1)\lambda}{2}$ and $b+\frac{n\lambda}{2}$, respectively. Therefore, the average distance (d_n) of the nth Fresnel's half period zone from point P will be

$$d_n = \frac{b+\frac{(n-1)\lambda}{2}+b+\frac{n\lambda}{2}}{2}$$

or

$$d_n = b + (2n-1)\frac{\lambda}{4} \qquad \text{...(13.33)}$$

Dividing the area of nth half period zone [given From Eq. (13.30] by average distance of half period zone [given by Eq. (13.33)], we have

$$\frac{A_n}{d_n} = \frac{\pi b\lambda + \pi(2n-1)\frac{\lambda^2}{4}}{b+(2n-1)\frac{\lambda}{4}} = \frac{\pi\lambda\left[b+(2n-1)\frac{\lambda}{4}\right]}{b+(2n-1)\frac{\lambda}{4}}$$

or

$$\frac{A_n}{d_n} = \pi\lambda \qquad \text{...(13.34)}$$

13.8 RESULTANT AMPLITUDE OF SECONDARY WAVES AT POINT *P*

Let the transverse displacement at point P, Y_1, due to the secondary wave coming from the first half period zone is

$$Y_1 = R_1 \sin\frac{2\pi}{\lambda}(vt-x) \qquad \text{...(13.35)}$$

where, average distance of the secondary wave originating from is first zone is $d_1 = x = b + \frac{\lambda}{4}$. v is velocity of light wave at time (t) and distance, x. Eq. (13.35) can be written as

$$Y_1 = R_1 \sin\left(\frac{2\pi}{\lambda} vt - \frac{2\pi}{\lambda} x\right)$$

or

$$Y_1 = a_1 \sin\left(2\pi\, vt - \frac{2\pi}{\lambda} x\right)$$

or

$$Y_1 = R_1 \sin(\omega t - \phi) \quad ...(13.36)$$

where, a_1 is the amplitude of the secondary wave originating from the first zone, $\omega = 2\pi v$ is its angular frequency and $\phi = \frac{2\pi}{\lambda} x$ is phase of the wave.

Let the transverse displacement at point P, Y_2, due to the secondary wave coming from second half period zone be

$$Y_2 = R_2 \sin \frac{2\pi}{\lambda}\left(vt - \left(x + \frac{\lambda}{2}\right)\right) \quad ...(13.37)$$

where, average distance of the secondary wave originating from the second zone is $d_2 = x + \frac{\lambda}{2} = b + \frac{3\lambda}{4}$. v is velocity of light wave at time (t) and distance, $x + \frac{\lambda}{2}$. Equation (13.37) can be written as

$$Y_2 = R_2 \sin\left(\frac{2\pi}{\lambda} vt - \frac{2\pi}{\lambda}\left(x + \frac{\lambda}{2}\right)\right)$$

or

$$Y_2 = R_2 \sin\left(2\pi vt - \frac{2\pi}{\lambda} x - \pi\right)$$

$$Y_2 = -R_2 \sin(\omega t - \phi) \quad ...(13.38)$$

where, a_2 is the amplitude of the secondary wave originating from the second zone.

Let the transverse displacement at point P, Y_3, due to the secondary wave coming from the third half period zone be

$$Y_3 = R_3 \sin \frac{2\pi}{\lambda}(vt - (x + \lambda)) \quad ...(13.39)$$

where, average distance of the secondary wave originating from the third zone is $d_3 = x + \lambda = b + \frac{5\lambda}{4}$. v is velocity of light wave at time (t) and distance, $x + \lambda$. Equation (13.39) can be written as

$$Y_3 = R_3 \sin\left(\frac{2\pi}{\lambda} vt - \frac{2\pi}{\lambda}(x + \lambda)\right)$$

or

$$Y_3 = R_3 \sin\left(2\pi v - \frac{2\pi}{\lambda} x - 2\pi\right)$$

$$Y_3 = R_3 \sin(\omega t - \phi) \qquad ...(13.40)$$

where, a_3 is the amplitude of the secondary wave originating from the second zone.

Let the transverse displacement at point P, Y_4, due to the secondary wave coming from the fourth half period zone be

$$Y_4 = R_4 \sin\frac{2\pi}{\lambda}\left(vt - \left(x + \frac{3}{2}\lambda\right)\right) \qquad ...(13.41)$$

where, average distance of the secondary wave originating from the third zone is $d_4 = x + \frac{3}{2}\lambda$ $= b + \frac{7\lambda}{2}$. v is velocity of light wave at time (t) and distance, $x + \frac{3}{2}$. Equation (13.41) can be written as

$$Y_4 = R_4 \sin\left(\frac{2\pi}{\lambda}vt - \frac{2\pi}{\lambda}\left(x + \frac{3}{2}\lambda\right)\right)$$

or

$$Y_4 = R_3 \sin\left(2\pi\, vt - \frac{2\pi}{\lambda}x - 3\pi\right)$$

$$Y_4 = -R_4 \sin(\omega t - \phi) \qquad ...(13.42)$$

where, a_4 is the amplitude of the secondary wave originating from the second zone. The other displacements, Y_5, Y_6, etc. n can be obtained in the same manner.

The resultant displacement (y) at point P due to all half period zones is sum of all the displacements due to all waves originating from all points of all half period zones, i.e.,

$$Y = Y_1 + Y_2 + Y_3 + Y_4 + ...$$

Putting the values of Y_1, Y_2, Y_3, Y_4, etc. from Eqs. (13.36), (13.38), (13.40), (13.42), etc.

$$Y = (R_1 - R_2 + R_3 - R_4 + ...) \sin(\omega t - \phi)$$

or

$$Y = A \sin(\omega t - \phi) \qquad ...(13.43)$$

where, resultant amplitude, $A = R_1 - R_2 + R_3 - R_4 + ...$ (13.44) ...

From the resultant amplitude at point P due to secondary wavelets, it is clear that the odd numbered Fresnel's half period zones are in phase and the even numbered Fresnel's half period zones are out of phase. Therefore, the resultant amplitude (R) of the secondary wavelets originating from all the Fresnel's half period zones at P will be given by

$$R = R_1 - R_2 + R_3 - R_4 + ... + (-1)^{n-1} R_n \qquad ...(13.45)$$

The amplitudes due to secondary wavelets, R_1, R_2, R_3, R_4, ... depend upon the following factors:

(i) Area of the corresponding zone

(ii) Average distance travelled by the wave from the corresponding zone

(iii) Inclination factor $(1 + \cos\theta)$

The amplitude.

(i) $a_n \propto A_n$

Because amplitudes due to secondary wavelets, R_1, R_2, R_3, R_4, ... depend upon the number of secondary wavelets at point P. In turn, number of secondary wavelenths originating from Fresnel's half period zone is proportional to the area of the zone.

(ii) $R_n \propto \dfrac{1}{d_n}$,

i.e., amplitude of secondary wave at point P is inversely proportional to the average distance, because amplitudes to secondary wavelets decreases as the average distance of originating secondary wave from point P increases, i.e., $R_1 > R_2 > R_3 > R_4 > ...$

(iii) $R_n \propto (1 + \cos \vartheta_n)$

i.e., amplitude of secondary wave at point P is proportional to the inclination factor $(1 + \cos \vartheta_n)$ of Fresnel half period zones. As shown in Fig. 13.9, the inclination factor $(1 + \cos \vartheta_n)$ is maximum, when inclination angle, $\vartheta_n = 0$, i.e., the maximum contribution to amplitudes at point P is from secondary waves from those half period zones whose inclination angle is zero or small [Fig. 13.9]. The contribution becomes zero from zones in backward direction, whose inclination angle, $\vartheta_n = 180°$.

By combining factors given by (i), (ii) and (iii), the amplitude

$$R_n = K\frac{A_n}{d_n}(1 + \cos \vartheta_n) \qquad ...(13.46)$$

Putting the value of $\dfrac{A_n}{d_n}$ from Eq. (13.34) into Eq. (13.46), we get

$$R_n = K\,\pi\lambda\,(1 + \cos \vartheta_n)$$

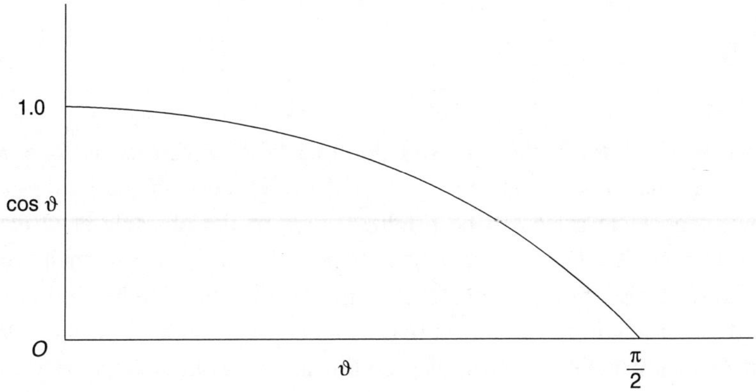

Fig. 13.9 Variation of cos ϑ with ϑ for $0 \le \vartheta \le \dfrac{\pi}{2}$

or $$R_n = k\,(1 + \cos \vartheta_n) \qquad ...(13.47)$$

where, the constant, $k = K\,\pi\lambda$. The optical intensity, $\propto$ (amplitude)^2.

If we place a circular obstacle at O such that amplitude of waves reaching at P due to the first half period zone is $R_1/2$ only, then resultant amplitude at P will be

$$R' \approx \frac{R_1}{2} - R_2 + R_3 - R_4 + R_5$$

$$\approx \frac{R_1}{2} - \frac{R_1 + R_2}{2} + R_3 - \frac{R_3 + R_5}{2} + \frac{R_5}{2}$$

$$\approx 0 \qquad ...(13.48)$$

Equation (13.48) indicates when the amplitude due to first zone at P becomes half by putting a circular object at O, then the resultant amplitude due to wavefront at p is zero.

If we place a circular obstacle at O so that amplitude of waves reaching at P from the first half period zone is zero, i.e., $R_1 = 0$, then resultant amplitude at P due to wavefront will be

$$R' \approx R_2 + R_3 - R_4 + R_5$$

$$\approx -R_2 + \frac{R_2 + R_4}{2} - R_4 + \frac{R_4 + R_5}{2}$$

$$\approx -\frac{R_2}{2} + \frac{R_6}{2}$$

$$\approx -\frac{R_2}{2} \text{ as } \frac{R_6}{2} \text{ is negligible} \qquad ...(13.49)$$

The intensity at point P

$$I' = k\,R_2^2/\,2$$

is almost equal to $I = k\,R_1^2/2$, the intensity due to whole wavefront as $R_2 \approx R_1$. Thus, on placing a small circular obstacle at O [Fig. 13.10], which cuts off the first zone, the centre of the shadow of obstacle at P will be bright. If size of the obstacle at O is increased to an extent that it cuts off a large number of half period zones, then intensity at P becomes negligible. If an obstacle of size comparable with wavelength of light used is placed in the path of light, the illumination is produced in the geometrical shade of obstacle. With the help of half period zones Fresnel established that rectilinear propagation is the only approximation and light bends around the corners of the obstacle.

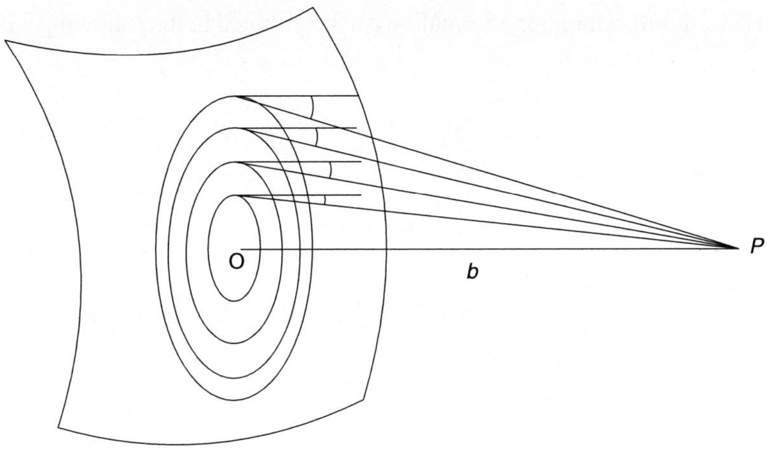

Fig. 13.10

13.9 SCHUSTER'S METHOD OF SUMMING THE SERIES

In this section, the sum of the series given by Eq. (13.45) is done by using Schuster's method. In this method, the terms of series are grouped in certain manner and by taking two assumptions:

(i) The average out of three consecutive amplitudes R_1, R_2 and R_3 due to half period zones at point P, R_2 can be considered the average of R_2 and R_3, because two consecutive zones are close to each other, because wavelength of light is of the order of 10^{-5} cm, i.e.,

$$R_2 = \frac{R_1 + R_2}{2}$$

In general, it can be written as

$$R_i = \frac{R_{i-1} + R_{i+1}}{2}, \; i = 2, 3, 4,... \quad ...(13.50)$$

(ii) For large n, the R_n is negligible in comparison to amplitudes R_1, R_2 and R_3 due to first, second and third half period zones. This is true because of the following reason:

When the number n increases, the inclination angle and distance of each zone increases and hence, the amplitude (R_n) due to half period zones decreases.

Case A: *n* is odd

If n is odd, Eq. (13.45) can be written as

$$R = R_1 - R_2 + R_3 - R_4 + ... + R_n$$

In case, RHS of the aforementioned equation can be grouped in the following two alternative ways; either

$$R = \frac{1}{2}R_1 + \left(\frac{1}{2}R_1 - R_2 + \frac{1}{2}R_3\right) + \left(\frac{1}{2}R_3 - R_4 + \frac{1}{2}R_5\right) + \ldots$$

$$+ \left(\frac{1}{2}R_{n-2} - R_{n-1} + \frac{1}{2}R_n\right) + \frac{1}{2}R_n \qquad \ldots(13.51)$$

or

$$R = R_1 - \frac{1}{2}R_2 - \left(\frac{1}{2}R_2 - R_3 + \frac{1}{2}R_4\right) - \left(\frac{1}{2}R_4 - R_5 + \frac{1}{2}R_6\right) - \ldots$$

$$-\left(\frac{1}{2}R_{n-3} - R_{n-2} + \frac{1}{2}R_{n-1}\right) - \frac{1}{2}R_{n-1} + R_n \qquad \ldots(13.52)$$

Substituting the values R_2, R_4 and R_{n-1} in Eq. (13.51) and R_3, R_5 and R_{n-2} in Eq. (13.52) the bracketed terms become zero. So Eqs. (13.51) and (13.52) reduce to

$$R = \frac{1}{2}R_1 + R_n \qquad \ldots(13.53)$$

and

$$R = R_1 - \frac{1}{2}R_2 - \frac{1}{2}R_{n-1} + R_n \qquad \ldots(13.54)$$

respectively. Considering $R_1 = R_2$ and $R_{n-1} = R_n$, Eq. (13.54) become

$$R = \frac{1}{2}R_1 + \frac{1}{2}R_n \qquad \ldots(13.54a)$$

If n is large, then R_n is negligible due to large inclination angle and distance of nth half period zone, so the resultant amplitude at P due to the entire wavefront will be

$$R = \frac{1}{2}R_1 \qquad \ldots(13.55)$$

when n is odd.

Case B: *n* is even

If n is even, Eq. (13.45) becomes

$$R = R_1 - R_2 + R_3 - R_4 + \ldots. - R_n$$

In case, RHS of the aforementioned equation can be grouped in the following two alternative ways; either

$$R = \frac{1}{2}R_1 + \left(\frac{1}{2}R_1 - R_2 + \frac{1}{2}R_3\right) + \left(\frac{1}{2}R_3 - R_4 + \frac{1}{2}R_5\right) + \ldots$$

$$+ \left(\frac{1}{2}R_{n-3} - R_{n-2} + \frac{1}{2}R_{n-1}\right) + \frac{1}{2}R_{n-1} - \frac{1}{2}R_n \qquad \ldots(13.56)$$

or $$R = R_1 - \frac{1}{2}R_2 - \left(\frac{1}{2}R_2 - R_3 + \frac{1}{2}R_4\right) - \left(\frac{1}{2}R_4 - R_5 + \frac{1}{2}R_6\right) - \ldots$$

$$- \left(\frac{1}{2}R_{n-2} - R_{n-1} + \frac{1}{2}R_n\right) - \frac{1}{2}R_n \qquad \ldots(13.57)$$

Putting the values of R_2, R_4, R_{n-2} and R_3, R_5, R_{n-1} in Eqs. (13.56) and (13.57) bracketed terms become zero and Eqs. (13.56) and (13.57) reduce to

$$R = \frac{1}{2}R_1 - \frac{1}{2}R_n \qquad \ldots(13.58)$$

and $$R = R_1 - \frac{1}{2}R_2 - \frac{1}{2}R_n \qquad \ldots(13.59)$$

Considering, $R_1 = R_2$ and $R_{n-1} = R_n$, Eq. (13.59) becomes

$$R = \frac{1}{2}R_1 - \frac{1}{2}R_n \qquad \ldots(13.60)$$

If n is large, then R_n is negligible due to large inclination angle and distance of nth half period zone, so the resultant amplitude at P due to the entire wavefront will be

$$R = \frac{1}{2}R_1 \qquad \ldots(13.61)$$

when n is even.

That is, whether n is even or odd, the resultant amplitude at P due to the entire wavefront is half of the amplitude due to the first Fresnel's half period zone.

The intensity due to whole wavefront,

$I \propto (\text{amplitude})^2$

$\propto R_1^2/4$

The intensity at P due to light wavefront is calculated by assuming that light propagation is rectilinear. When a small circular obstacle at O is placed, the centre of the shadow of obstacle at P will be bright. If size of the obstacle at O in increased to such an extent that it cuts off large number of half period zones, then intensity at P becomes negligible. If an obstacle of size comparable with wavelength of light used is placed in the path of light, the illumination is produced in the geometrical shade of obstacle. With the help of half period zones Fresnel established that rectilinear propagation is the only approximation and light bends around the corners of the obstacle.

13.10 RECTILINEAR PROPAGATION OF LIGHT

Let us use the result given in Eq. (13.24) to prove that light propagation is rectilinear. Let a plane wavefront of monochromatic light be incident normally on a rectangular aperture *ABCD* as shown in Fig. 13.11(a). Let O be pole with respect to point P on the screen. Let us divide the wavefront at O in Fresnel's half period zones. The outermost zone at pole O is at a large distance (compared to wavelength of light) from the edges of rectangular aperture. Thus, light originating from pole O passes through the aperture giving the resultant amplitude, $R_1/2$, at P. The same result will be obtained if aperture is removed from the screen. As the area of these zones is extremely small, so we can assume that light from pole O to point P travels in straight line. Let O_3 be the pole with respect to point P_3 as shown in Fig. 13.11(b). All the zones at pole O_3 are obstructed by edge of aperture, so resultant amplitude at point P_3 will be zero. That is, there is complete darkness at point P_3, if we assume that light travels in straight line. The poles O_1 and O_2 lie in between two dotted rectangles with respect to points P_1 and P_2 as shown in Fig. 13.11(b). The zones at poles O_1 and O_2 are partially obstructed by the aperture. That means the light from zones at poles O_1 and O_2 is partially obstructed by the aperture. Therefore, at points P_1 and P_2, there is neither complete darkness nor complete illumination, i.e., there is dim light in the shadow region close to the edge of aperture.

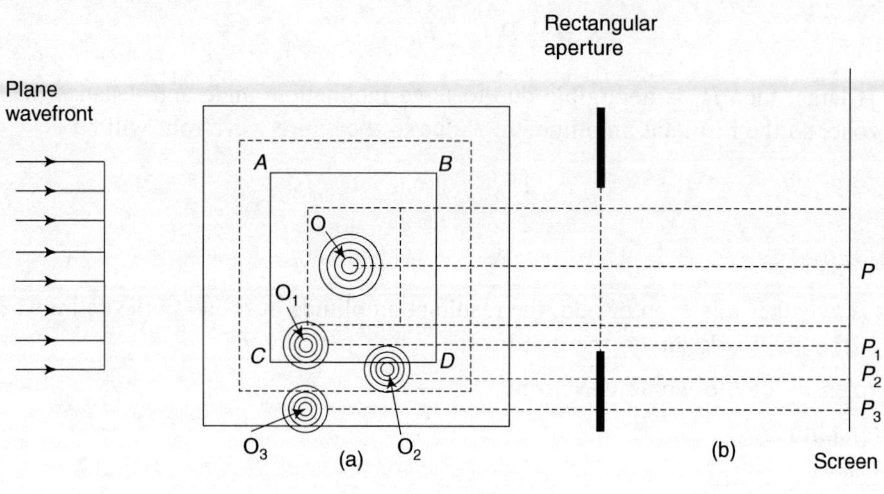

Fig. 13.11

Thus, from the above-mentioned discussion, it is concluded that there is complete illumination at all points on the screen, the corresponding pole which lie well within the inner dotted rectangle. On the other hand, there is complete darkness at all points on the screen corresponding to poles which lie outside the outer dotted rectangle. This conclusion is drawn by assuming that light travels in straight line. The assumption that light travels in straight line is failed in case of poles which lie in between two dotted rectangles. Since the wavelength of light is very small (~ 10^{-5} cm), the sides of two dotted rectangles lie very close to the edges of the aperture. Hence, when light passes close to the edges of the aperture (or an obstacle) there is some deviation from its rectilinear path and there is dim illumination of light in the shadow region. Thus, the propagation of light is approximately linear.

13.11 ZONE PLATE

The period zone theory of Fresnel can be experimentally confirmed with the help of an optical device, called the zone plate. In reality, either all the odd numbered terms or all the even numbered terms of the RHS of Eq. (13.6) can be removed by increasing the resultant amplitude to many times. This concept has been applied practically in constructing a zone plate. A zone plate is simply a thin parallel glass plate containing concentric circles of radii accurately proportional to the square root of natural numbers (recollect Newton's rings). The annular space so created is deeply blackened alternately. Practically, a highly reduced photograph of the Newton's rings is taken on a thin parallel glass plate.

13.11.1 Types of Zone Plate

Either the even numbered annular spaces are blackened having the odd numbered annular space transparent or the odd numbered annular spaces are blackened leaving the even numbered annular space transparent. Depending on this, the zone plates are of two types.

(i) Positive zone plate: When the central circular zone, i.e., the first half period zone on the thin glass plate is transparent, the zone plate is called a positive zone plate. In a positive zone plate, even numbered half period zones, i.e., 2nd, 4th, 6th,..., etc., are blackened to make them opaque for light, leaving odd numbered half period zones, i.e., 1st, 3rd, 5th,..., etc., transparent. A typical positivezone plate is shown in Fig. 13.12(a).

(ii) Negative zone plate: When the central circular zone, i.e., the first half period zone on the thin glass plate is opaque, the zone plate is called a negative zone plate. In negative zone plate, odd numbered half period zones, i.e., 1st, 3rd, 5th,..., etc., transparent. A typical negative zone plate is shown in Fig. 13.12(b).

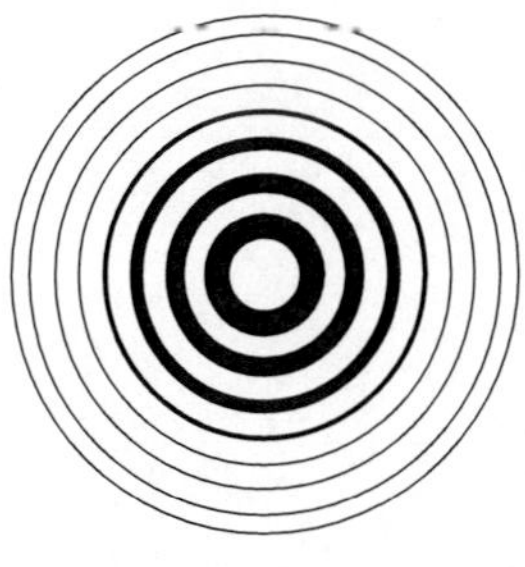

Positive zone plate

(a)

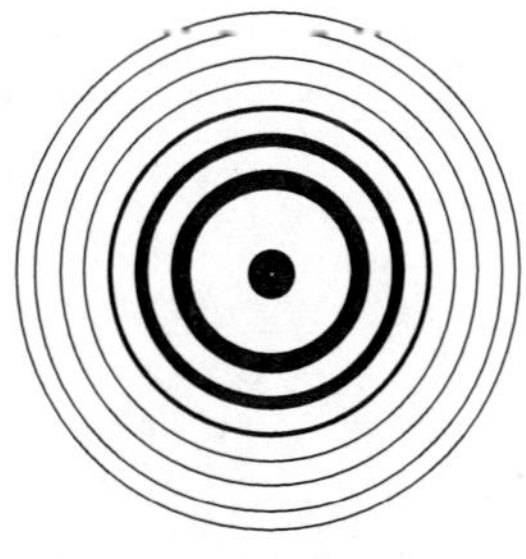

Negative zone plate

(b)

Fig. 13.12 (a) Positive zone plate where the central circular zone is transparent. (b) Negative zone plate where the central circular zone is opaque

13.11.2 Action of the Zone Plate

Let a thin transparent screen (*AB*) is placed perpendicular to the plane of the page as shown in Fig. 13.13. *C* is the central pole of the zone. Let *S* be a point source emitting spherical wavefront of monochromatic light of wavelength (λ) and *P* is a point situated on the other side of the screen. A line (*SP*) perpendicular to the screen passes through the central point *C*. For calculating resultant amplitude at *P* on the screen due to secondary wavelets diffracting from various points of Fresnel's half period zones, we divide the transparent screen into a large number of half period zones as shown in Fig. 13.13.

We place points C_1, C_2, C_3,..., C_n along *AB* on the transparent screen such that

$$SC_1P = SCP + \frac{\lambda}{2}$$

$$SC_2P = SCP + \frac{2\lambda}{2}$$

$$SC_3P = SCP + \frac{3\lambda}{2}$$

$$\text{-------------------------------}$$

$$SC_nP = SCP + \frac{n\lambda}{2} \qquad \text{...(13.62)}$$

The circular area enclosed by the circle C_1 of radius CC_1 is called Fresnel's first half period zone. The circular strip enclosed by the circles C_1 and C_2 is called Fresnel's second half period zone. The circular strip enclosed by the circles C_2 and C_3 is called Fresnel's third half period zone. In the same manner *n* number of zones are formed.

Let the outer radius of the first half period zone, $CC_1 = r_1$

The outer radius of the second half period zone, $CC_2 = r_2$

The outer radius of the nth half period zone, $CC_n = r_n$.

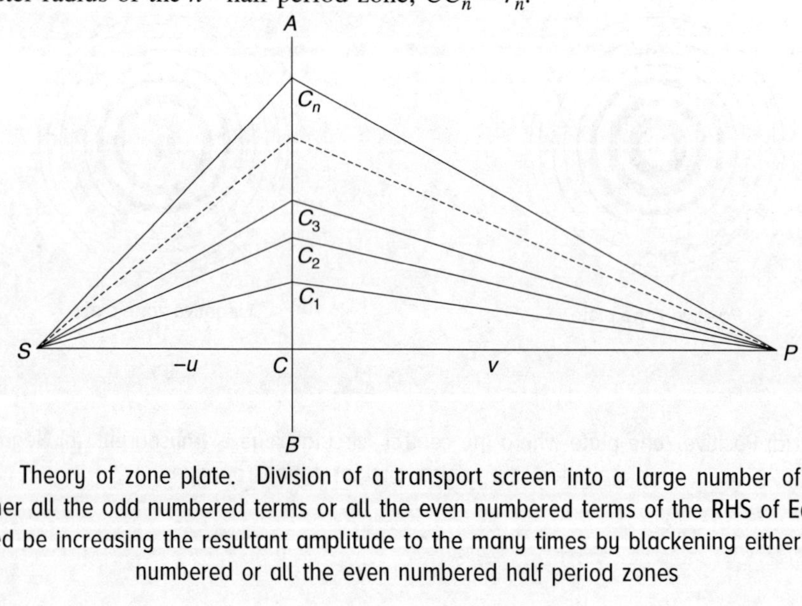

Fig. 13.13 Theory of zone plate. Division of a transport screen into a large number of half period zones. Either all the odd numbered terms or all the even numbered terms of the RHS of Eq. (13.6) can be removed be increasing the resultant amplitude to the many times by blackening either all the odd numbered or all the even numbered half period zones

Let us consider that the distance in right-hand side of C is +ve and the distance in left hand side of C is –ve. Using this sign of convention,

let the distance between the screen and the source, $CS = -u$,

and the distance between the screen and the focal point, $CP = v$

Let $r_1 = CC_1$, $r_2 = CC_2$, $r_3 = CC_3$,..., $r_n = CC_n$ are the radii of the first, second, third,..., n^{th} Fresnel half period zones, respectively. The path difference between two consecutive wavelets reached at P is $\lambda/2$, so that

$$SC_1P = SCP + \frac{\lambda}{2},$$

$$SC_2P = SCP + \frac{2\lambda}{2},$$

$$SC_3P = SCP + \frac{3\lambda}{2},$$

$$SC_nP = SCP + \frac{n\lambda}{2}, \qquad ...(13.63)$$

In the right angled triangle SCC_n, we get

$$SC_n = \sqrt{CS^2 + CC_n^2} = \sqrt{CS^2 + r_n^2}$$

or

$$SC_n = CS\left(1 + \frac{r_n^2}{CS^2}\right)^{1/2}$$

As $CS >> r_n$, so using binomial theorem, we get

$$SC_n = CS\left(1 + \frac{1}{2}\frac{r_n^2}{CS^2}\right)$$

or

$$SC_n = CS + \frac{r_n^2}{2\,CS} \qquad ...(13.64)$$

In the right angled triangle PCC_n, we get

$$PC_n = \sqrt{CP^2 + CC_n^2} = \sqrt{CP^2 + r_n^2}$$

or

$$PC_n = CP\left(1 + \frac{r_n^2}{CS^2}\right)^{1/2}$$

As $CP >> r_n$, so using binomial theorem to this equation, we get

$$PC_n = CP\left(1 + \frac{1}{2}\frac{r_n^2}{CP^2}\right)$$

or
$$PC_n = CP + \frac{r_n^2}{2\,CP} \qquad ...(13.65)$$

Putting $CS = -u$ and $CP = v$, Eqs. (13.64) and (13.65) become

$$SC_n = -u + \frac{r_n^2}{2(-u)} = -u - \frac{r_n^2}{2u} \qquad ...(13.66)$$

$$PC_n = v + \frac{r_n^2}{2v} \qquad ...(13.67)$$

Adding SC_n and PC_n, we get

$$SC_n + PC_n = -u - \frac{r_n^2}{2u} + v + \frac{r_n^2}{2v}$$

or
$$SC_nP = -u - \frac{r_n^2}{2u} + v + \frac{r_n^2}{2v} \qquad ...(13.68)$$

Putting the value of SC_nP from Eq. (13.68) into Eq. (13.63), we obtain

$$SCP + \frac{n\lambda}{2} = -u - \frac{r_n^2}{2u} + v + \frac{r_n^2}{2v}$$

or
$$-u - \frac{r_n^2}{2u} + v + \frac{r_n^2}{2v} - SCP = \frac{n\lambda}{2}$$

or
$$-u - \frac{r_n^2}{2u} + v + \frac{r_n^2}{2v} - (-u + v) = \frac{n\lambda}{2}$$

or
$$\frac{r_n^2}{2}\left(\frac{1}{v} - \frac{1}{u}\right) = \frac{n\lambda}{2}$$

or
$$r_n^2 = \frac{nuv\lambda}{u - v} \qquad ...(13.69)$$

for $n = 1$, we get

$$r_1^2 = \frac{uv\lambda}{u - v} \qquad ...(13.70)$$

Combining Eqs. (13.69) and (13.70), we get

$$r_n^2 = n\,r_1^2 \qquad ...(13.71)$$

or
$$r_n = r_1\sqrt{n} \qquad ...(13.72)$$

Thus, radii of half period zones are directly proportional to the square root of natural numbers.

Putting n = 1, 2, 3, 4,...,. into Eq. (13.72), we can find radii of the first, second and third Freasnel's half period zone, if $r_1$is known.

Area of nth half period zone: From Eq. (13.69), the square of the outer radius of the nth half period zone is

$$r_n^2 = \frac{nuv\lambda}{u-v}$$

The outer radius of the $(n-1)^{th}$ half period zone is

$$r_{n-1}^2 = \frac{(n-1)\,uv\lambda}{u-v}$$

Therefore, area of the nth half period zone A_n will be

$$A_n = \pi\left(r_n^2 - r_{n-1}^2\right)$$

or
$$A_n = \pi\left(\frac{nuv\lambda}{u-v} - \frac{(n-1)\,uv\lambda}{u-v}\right)$$

or
$$A_n = \frac{\pi uv\lambda}{u-v} \qquad ...(13.73)$$

or
$$A_n = \frac{\pi v\lambda}{1-\frac{v}{u}} \qquad ...(13.74)$$

Equation (13.74) indicates that the area of the nth half period zone is independent of n.

Putting n =1, 2, 3,... etc., the area of the first, second, third,..., etc., half period zones will be

$$A_1 = \frac{\pi v\lambda}{1-\frac{v}{u}},\quad A_2 = \frac{\pi v\lambda}{1-\frac{v}{u}},\quad A_3 = \frac{\pi v\lambda}{1-\frac{v}{u}},\ldots, \text{etc.}$$

Therefore, area of each Fresnel half period zone is equal. If, incident monochromatic light is a plane wavefront then light source is considered at infinity, i.e., $u \to \infty$. Putting $u \to \infty$ in Eq. (13.74), the area of the nth half period zone is

$$A_n = \pi v \lambda \qquad ..(13.75)$$

Intensity at Point *P*

The resultant amplitude R at P due to secondary wavelets originating from all the Fresnel's half period zones is given by Eq. (13.45)

$$R = R_1 - R_2 + R_3 - R_4 + \ldots, + (-1)^{n-1} R_n$$

where, R_1, R_2, R_3,..., R_n are the amplitudes at P due to secondary wavelets originating from the first, second, third and n^{th} Fresnel's half period zone, respectively. Summing the series following the procedure given in Section 13.9, the resultant amplitude of secondary wavelets at P is written as

$$R = \frac{1}{2}R_1$$

when $n \to \infty$. Therefore, the intensity (I) at P due to the entire wavefront is

$$I = \frac{k}{4}R_1^2 \quad ...(13.76)$$

where, k is the proportionality constant.

The resultant amplitude at P due to only odd numbered half period zone is

$$R = R_1 + R_3 + R_5 + R_7 + ... = \frac{1}{2}NR_1 \quad ...(13.77)$$

The resultant amplitude at P due to only even numbered half period zone is

$$R = R_2 + R_4 + R_6 + R_8 + ... = \frac{1}{2}NR_1 \quad ...(13.78)$$

where, N is the number of half period zones on the transparent screen AB shown in Fig. 13.11 and $\frac{N}{2}$ is the number of zones that are transparent. Therefore, the intensity at P either due to only all odd or only all even numbered half period zones is given by

$$I = \frac{k}{4}N^2 R_1^2 = k(R_1 + R_3 + R_5 + R_7 + ...)^2 \quad ...(13.79a)$$

due to only odd numbered half period zones,

or

$$I = \frac{k}{4}N^2 R_1^2 = k(R_2 + R_4 + R_6 + R_8 + ...)^2 \quad ...(13.79b)$$

due to only even numbered half period zones.

Thus, the intensity at P due to either only all odd or only all even numbered half period zones is N^2 times than that of the intensity (I) at P due to the entire wavefront given by Eq. (13.76).

Principle behind Zone Plates

From the previous discussion, we conclude that the intensity at point due to only either even numbered half period zones or due to only odd numbered half period zones is many times more than that of the intensity at the same point due to the entire wavefront. If somehow either all the odd numbered half period zones or all the even numbered half period zones on the screen were blocked or made opaque, then the intensity at a point will be many times more than that of the intensity at the same point due to the entire wavefront. In a zone plate, either all the odd numbered half period zones or all the even numbered half period zones are blackened or made opaque. Hence, when light passes through a zone plate, the intensity at a forward point will be maximum and is the image of the light source. The position of the image should satisfy the Eq. (13.69), i.e.,

$$\frac{r_n^2}{2}\left(\frac{1}{v}-\frac{1}{u}\right) = \frac{n\lambda}{2}$$

or

$$\frac{1}{v}-\frac{1}{u} = \frac{n\lambda}{r_n^2}$$

or

$$\frac{1}{v}-\frac{1}{u} = \frac{1}{f_n} \qquad ...(13.80)$$

where, the focal length of the nth zone

$$f_n = \frac{r_n^2}{n\lambda} \qquad ...(13.81)$$

where, r_n is the radius of the nth zone. Equation (13.80) is an equation similar an equation for a thin lens. This equation shows that Fresnel zone plate behaves like a lens. Here u is the distance of the source from the zone plate and v is the distance of the image from the zone plate.

Multiple Foci of a Zone Plate

The focal length of a lens or a zone plate is defined as the distance of image when the distance of object is infinity. Therefore, by putting $u \to \infty$ and $v = f_n$ into Eq. (13.80), we will get the expression for the focal length of a zone plate:

$$\frac{1}{f_n}-\frac{1}{\infty} = \frac{n\lambda}{r_n^2}$$

or

$$f_n = \frac{r_n^2}{n\lambda} \qquad ...(13.82)$$

Putting $r_n = r_1\sqrt{n}$

Focal length of the first zone, $f_n = \dfrac{r_1^2}{\lambda}$

or

$$f_1 = \frac{r_1^2}{\lambda} \qquad ...(13.83)$$

Here, f_1 is called the first order focal length or primary or principal focal length of the zone plate and the corresponding point is called first order focus or primary or principal focus of the zone plate. The intensity of the primary focus is maximum as compared to other foci. The intensity of the other foci goes on decreasing as we move toward the zone plate along the axis. The intensity of the primary focus (I_1) can be obtained from Eq. (13.79a)

$$I_1 = \frac{k}{4} N^2 R_1^2 = k(R_1 + R_3 + R_5 + R_7 + ...)^2 \qquad ...(13.84)$$

where, R_1 is the amplitude of wavelets originating from the first zone and k is proportionality constant.

13.12 PRESENCE OF ODD NUMBERED FOCI

The intensity at odd numbered foci due to plane wavefront on positive zone plate is quite high. So we can say that odd numbered foci are present both for positive and negative zone plates. The presence of odd numbered foci both in case of positive and negative zone plates can be explained in this section. The areas of all the Fresnel's half period zones are approximately equal. The area of a Fresnel's half period zone of a plane wavefront incident on a zone plate is approximately given by Eq. (13.32) as

$$A = \pi \lambda v$$

where, v is the distance of P from C. When incident light is plane wavefront, $v = f_1$, so that we have

$$A = \pi \lambda f_1$$

Dividing this equation by $(2\,m + 1)$, we get

$$\pi \lambda \frac{f_1}{2\,m+1} = \frac{A}{2\,m+1} \qquad ...(13.85)$$

where, m = 0, 1, 2, 3,..., etc.

The intensity of the third order focus

For m = 1 Eq. (13.85) yields

$$\pi \lambda \frac{f_1}{3} = \frac{A}{3} \qquad ...(13.86)$$

We can infer from Eq. (13.86) that when $v = \frac{f_1}{3}$, area of a Fresnel's zone $= \frac{A}{3}$. This result points that each Fresnel's half period zone is divided into three half period zones each having area, $\frac{A}{3}$.

Let us consider the case of positive zone plate. When $v = \frac{f_1}{3}$ each zone is divided into three zones of equal area of $\frac{A}{3}$. In a positive zone plate, the first zone of incident wavefront is transparent and second zone is opaque. When $v = \frac{f_1}{3}$, the first zone is divided into three [1st, 2nd and 3rd] half period zones and the second opaque zone is divided into three [4th, 5th and 6th] opaque half period zones. Similarly, its transparent third zone is divided into three

[7th, 8th and 9th] half period zones and the fourth opaque zone is divided three [10th, 11th and 12th] half period zone of the incident plane wavefront. The odd numbered transparent half period zones of a positive zone plate are transparent, so they allow the secondary wavelets to pass. The even numbered half period zones of positive zone plate are opaque, so they do not allow the secondary wavelets to pass. Hence, the secondary wavelets originating from the 1st, 2nd, 3rd; 7th, 8th, 9th; 13th, 14th, 15th; 19th, 20th, 21st, etc., of the Fresnel's half period zones of the incident plane wavefront can pass through a positive zone plate. The resultants of secondary wavelets originating from these Fresnel's half period zones will have equal amplitudes and even number of zones have opposite phases. Therefore, the resultant amplitude, A_3, when $v = \frac{f_1}{3}$, will be

$$A_3 = (R_1 - R_2 + R_3) + (R_7 - R_8 + R_9) + (R_{13} - R_{14} + R_{15}) + (R_{19} - R_{20} + R_{21}) + \ldots$$

or

$$A_3 = \left\{R_1 - \frac{1}{2}(R_1 + R_3) + R_3\right\} + \left\{R_7 - \frac{1}{2}(R_7 + R_9) + R_9\right\}$$

$$+\left\{R_{13} - \frac{1}{2}(R_{13} + R_{15}) + R_{15}\right\} + \left\{R_{19} - \frac{1}{2}(R_{19} + R_{21}) + R_{21}\right\} + \ldots$$

or

$$A_3 = \frac{1}{2}\ (R_1 + R_3 + R_7 + R_9 + R_{13} + R_{15} + R_{19} + R_{21} + \ldots)$$

The intensity at third order focus I_3 is

$$I_3 = \frac{k}{4}\ (R_1 + R_3 + R_7 + R_9 + R_{13} + R_{15} + R_{19} + R_{21} + \ldots.)^2 \qquad \text{...(13.87}$$

Comparing Eqs. (13.84) and (13.87), we find that

$$I_3 < I_1$$

This inequality indicates that the intensity at the third order focus is less than the intensity at the primary first order focus. The position of the third order focus from the zone plate is

$$f_3 = \frac{f_1}{3} \qquad \text{...(13.88)}$$

The Intensity of the Fifth Order Focus

For $m = 2$, Eq. (13.85) yields

$$\pi\lambda\ \frac{f_1}{5} = \frac{A}{5} \qquad \text{...(13.89)}$$

From Eq. (13.48), we can infer that when $v = \frac{f_1}{5}$, the area of a Fresnel's zone $= \frac{A}{5}$. This indicates that each Fresnel's half period zone is divided into five half period zones with equal area $\frac{A}{5}$.

Let us consider the case of a positive zone plate. When $v = \frac{f_1}{5}$ each zone is divided into five zones of equal area of $\frac{A}{5}$. In a positive zone plate, the first zone of incident wavefront is transparent and the second zone is opaque. When $v = \frac{f_1}{5}$, the first zone is divided into five [1st, 2nd, 3rd, 4th amd 5th] half period zones and second opaque zone is divided into five [6th, 7th, 8th, 9th and 10th] opaque half period zones. Similarly, its transparent third zone is divided into five [11th, 12th, 13th, 14th and 15th] half period zones and fourth opaque zone is divided five[16th, 17th, 18th, 19th and 20th] half period zone of the incident plane wavefront. The odd numbered transparent half period zones of a positive zone plate are transparent, so they allow the secondary wavelets to pass. The even numbered half period zones of positive zone plate are opaque, so they do not allow the secondary wavelets to pass. Hence, the secondary wavelets originating from the 1st, 2nd, 3rd; 4th, 5th,11th, 12th, 13th; 14th, 15th, 21st; 22nd, 23rd, 24th, 25th, etc., of the Fresnel's half period zones of the incident plane wavefront can pass through a positive zone plate. The resultants of secondary wavelets originating from these Fresnel's half period zones will have equal amplitudes and even number of zones have opposite phases. Therefore, the resultant amplitude A_5, when $v = \frac{f_1}{5}$, will be

$$A_5 = (R_1 - R_2 + R_3 - R_4 + R_5) + (R_{11} - R_{12} + R_{13} - R_{14} + R_{15}) + (R_{21} - R_{22} + R_{23} - R_{24} + R_{25})...$$

or
$$A_5 \approx \left\{R_1 - \frac{1}{2}(R_1 + R_3) + R_3 - \frac{1}{2}(R_3 + R_5) + R_5\right\} + \left\{R_{11} - \frac{1}{2}(R_{11} + R_{13}) + R_{13} - \frac{1}{2}(R_{13} + R_{15}) + R_{15}\right\} + \left\{R_{21} - \frac{1}{2}(R_{21} + R_{23}) + R_{23} - \frac{1}{2}(R_{23} + R_{25}) + R_{25}\right\} + ...$$

or
$$A_5 \approx \left\{R_1 - \frac{1}{2}R_1 - \frac{1}{2}R_3 + R_3 - \frac{1}{2}R_3 - \frac{1}{2}R_5 + R_5\right\} + \left\{R_{11} - \frac{1}{2}R_{11} - \frac{1}{2}R_{13} + R_{13} - \frac{1}{2}R_{13} - \frac{1}{2}R_{15} + R_{15}\right\} + \left\{R_{11} - \frac{1}{2}R_{11} - \frac{1}{2}R_{23} + R_{23} - \frac{1}{2}R_{23} - \frac{1}{2}R_{25} + R_{25}\right\} + ...$$

or
$$A_5 \approx \left\{\frac{1}{2}R_1 + \frac{1}{2}R_5\right\} + \left\{\frac{1}{2}R_{11} + \frac{1}{2}R_{15}\right\} + \left\{\frac{1}{2}R_{21} + \frac{1}{2}R_{25}\right\} + ...$$

or
$$A_5 \approx \frac{1}{2}\{R_1 + R_5 + R_{11} + R_{15} + R_{21} + R_{25}\} + ...$$

The intensity of the fifth order focus (I_5) is

$$I_5 = \frac{k}{4} (R_1 + R_5 + R_{11} + R_{15} + R_{21} + R_{25} + ...)^2 \quad ...(13.90)$$

Comparing Eqs. (13.84), (13.87), and (13.90) it is found that

$$I_5 < I_3 < I_1$$

This inequality indicates that intensity due to incident plane wavefront on positive zone plate at the fifth order focus is less than the intensity at the third order focus. In general, we can infer that the intensity at higher order foci is less than the intensity at lower order foci. Closer is the foci to the zone plate, lesser is the intensity. This conclusion is made for foci of positive zone plate. Similar conclusion is applicable to foci of a negative zone plate. The distance of the fifth order focus from the zone plate is

$$f_5 = \frac{f_1}{5} \quad ...(13.91)$$

Following Eqs. (13.83), (13.88) and (13.91), the general expression for odd numbered focal lengths of positive zone plate is written as

$$f_{2m-1} = \frac{f_1}{2m-1} \quad ...(13.92)$$

or

$$f_{2m-1} = \frac{r_1^2}{(2m-1)\lambda} \quad ...(13.93)$$

where, m = 1, 2, 3, ...

For m =1. Equation (13.93) gives the focal length of first zone

$$f_1 = \frac{r_1^2}{\lambda}$$

called primary or principal focal length of zone.

For m = 2, 3, 4,..., Eq. (13.93) gives the length of foci of second, third, ..., etc., zones

$$f_3 = \frac{r_1^2}{3\lambda}, \quad f_5 = \frac{r_1^2}{5\lambda}, \ ..., \text{etc.} \quad ...(13.94)$$

called secondary foci, where, r_1 is the radius of first Fresnel's first half period zone.

13.13 ABSENCE OF EVEN NUMBERED FOCI

The intensity at even numbered foci is almost zero. Therefore, one can say that even numbered foci are are absent. The absence of even numbered foci both in case of positive and negative zone plates is explained in the present section. The area of each the Fresnel's half period zone is approximately equal. The area of a Fresnel's half period zone of a plane wavefront incident on a zone plate is approximately [(13.32)] given by

$$A = \pi v$$

For a plane incident wavefront, the distance of point P from zone plate v = focal length f_1, we have

$$A = \pi\lambda f_1$$

Dividing this equation by $2m$, we have

$$\pi\lambda \frac{f_1}{2\,m} = \frac{A}{2\,m} \qquad ...(13.95)$$

where, m = 1, 2, 3,..., etc.

Intensity at the Second Order Focus

For m =1, Eq. (13.95) becomes

$$\pi\lambda \frac{f_1}{2} = \frac{A}{2} \qquad ...(13.96)$$

Equation (13.96) shows that when $v = \frac{f_1}{2}$, area of a Fresnel's zone $= \frac{A}{2}$, i.e., each Fresnel's half period zone is divided into two half period zones each having area, $\frac{A}{2}$.

Let us consider the case of positive zone plate. When $v = \frac{f_1}{2}$ each zone is divided into two zones of equal area, $\frac{A}{3}$. In a positive zone plate, the first zone of incident wavefront is transparent and the second zone is opaque. When $v = \frac{f_1}{2}$, the first zone is divided into two [1st and 2nd] half period zones and second opaque zone is divided into two [3rd and 4th] opaque half period zones. Similarly, its transparent third zone is divided into two 5th and 6th half period zones and fourth opaque zone is divided two [6th and 7th] half period zones of the incident plane wavefront. The odd numbered transparent half period zones of a positive zone plate are transparent, so they allow the secondary wavelets to pass. The even numbered half period zones of positive zone plate are opaque, so they do not allow the secondary wavelets to pass. Hence, the secondary wavelets originating from the 1st, 2nd;5th, 6th; 9th, 10th, 13th; 14th etc., of the Fresnel's half period zones of the incident plane wavefront can pass through a positive zone plate. The resultants of secondary wavelets originating from these Fresnel's half period zones will have equal amplitudes and even number of zones have opposite phases. Therefore, the resultant amplitude, A_2, when $v = \frac{f_1}{2}$, will be

$$A_2 = (R_1 - R_2) + (R_5 - R_6) + (R_9 - R_{10}) + (R_{11} - R_{14}) + ... \qquad ...(13.97)$$

The magnitudes of the terms written in brackets of Eq. (13.97) are very small, because $R_1 \approx R_2$, $R_5 \approx R_6$, $R_9 \approx R_{10}$ and $R_{11} \approx R_{14}$. Hence, the resultant amplitude A_2 will be very small. Therefore, the intensity I_2 ($\propto A_2^2$) at a distance $\frac{f_1}{2}$ from the zone plate will be very low.

Intensity at the Fourth Order Focus

For $m = 2$, Eq. (13.95) becomes

$$\pi\lambda \frac{f_1}{4} = \frac{A}{4} \qquad \text{...(13.98)}$$

Equation (13.56) shows that when $v = \frac{f_1}{4}$, area of a Fresnel's zone = $\frac{A}{4}$, i.e., each Fresnel's half period zone is divided into four half period zones each having area $\frac{A}{4}$.

Let us consider the case of positive zone plate. When $v = \frac{f_1}{4}$ each zone is divided into four zones of equal area $\frac{A}{4}$. In a positive zone plate, the first zone of incident wavefront is transparent and the second zone is opaque. When $v = \frac{f_1}{4}$, the first zone is divided into four [1st, 2nd, 3rd and 4th] half period zones and second opaque zone is divided into four [5th, 6th, 7th, and 8th] opaque half period zones. Similarly, its transparent third zone is divided into four [9th, 10th, 11th and 12th] half period zones and fourth opaque zone is divided four [13th, 14th, 15th and 16th] half period zones of the incident plane wavefront. The odd numbered transparent half period zones of a positive zone plate are transparent, so they allow the secondary wavelets to pass. The even numbered half period zones of positive zone plate are opaque, so they do not allow the secondary wavelets to pass. Hence, the secondary wavelets originating from the 1st, 2nd; 3rd4th; 9th, 10th, 11th; 12th ; 13th, 17th,18th, 19th, 20th, etc., of the Fresnel's half period zones of the incident plane wavefront can pass through a positive zone plate. The resultants of secondary wavelets originating from these Fresnel's half period zones will have equal amplitudes and even number of zones have opposite phases. Therefore, the resultant amplitude, A_4, when $v = \frac{f_1}{4}$, will be

$$A_4 = (R_1 - R_2 + R_3 - R_4) + (R_9 - R_{10} + R_{11} - R_{12}) + (R_{17} - R_{18} + R_{19} - R_{20}) + \ldots$$

The magnitudes of the terms given by brackets in this equation are very small, because $R_1 \approx R_2$, $R_3 \approx R_4$, $R_9 \approx R_{10}$, $R_{11} \approx R_{12}$, $R_{17} \approx R_{18}$, $R_{20} \approx R_{21}$... etc. Hence, the resultant amplitude (A_4) will be very small. As a result, the intensity I_4 at distance $\frac{f_1}{4}$ from the zone plate will be very low.

Similar, logic can be drawn for all the other even numbered foci. Therefore, we can say that practically, even numbered foci do not exist for positive zone plates. The same logic is applicable to foci of negative zone plates.

13.14 COMPARISON OF A ZONE PLATE WITH A CONVEX LENS

The similarities and dissimilarities between a convex lens and a zone plate are elaborated here.

Similarities

S.No.	Convex lens	Zone plate
1.	Convex lens has focusing property.	Zone plate also has focusing property.
2.	When source is placed on one side of lens, real image is formed on the other side of lens.	When source is placed on one side of zone plate, real image is formed on the other side of the zone plate.
3.	Formula of focal length (f) of convex lens is $\frac{1}{v}-\frac{1}{u}=\frac{1}{f}$, where, v is distance of image and u is distance of object.	Formula for foci $f_n = \frac{r_2^2}{n\lambda}$ of zone plate is $\frac{1}{v}-\frac{1}{u}=\frac{n\lambda}{r_n^2}$, where v is distance of image and u is the distance of object.
4.	Focal length of a convex lens depends on the refractive index of lens material or wavelength, therefore, there is chromatic aberration in the image fromed by convex lens.	Focal length of a zone plate depends on the wavelength and radius of zone, there is also chromatic aberrations in the image formed by zone plate.
	Note: Because of refraction by the two surfaces of lens depends on the wavelength. a lens does not focus light of different wavelengths to the same point, called chromatic aberration.	

Dissimilarties

S.No.	Convex lens	Zone plate
1.	For monochromatic light, a convex lens has a single focus.	For monochromic light, a zone plate has multiple foci.
2.	All light waves refracting through a lens are in phase.	All light waves diffracted through a zone plate are in different phase.
3.	All the monochromatic light waves after refracting through the lens reach simultaneously after traversing sameoptical paths.	The monochromatic secondary wavelets diffracting from different points on the zone plate take different time to reach the focal point, because wavelets from successive transparent zones have λ path difference.
4.	Focusing property is due to refraction of light through the lens.	Focusing property is due to diffraction of light through the zone plate.
5.	The intensity of image formed by convex lens is very high.	The intensity of image at primary focus of a zone plate is less than the intensity of image formed by lens.

6.	In case of convex lens, the focal length of violet light is less than that of red light, because focal length of a convex lens is inversely proportional to refractive index or directly proportional to wavelength, λ.	In case of zone plate, the focal length of violet light is more than that of red light, because the focal length of zone plate is inversely proportional to wavelength, λ.
7.	For a given wavelength of light, a convex lens has a single focal length given by $\frac{1}{r} = (\mu - 1)\left(\frac{1}{R_1} - \frac{1}{R_2}\right)$, R_1 is radius of curvature of surface of lens near the object and R_2 is radius of curvature of lens surface near the image. μ is refractive index of glass.	For a given wavelength of light, a zone plate has multiple focal lengths given by $f_{2m-1} = \frac{r_1^2}{(2m-1)\lambda}$, r_n is radius of the nth zone, λ is wavelength of light source. Here, m = 1, 2, 3, ... The distance of first or primary, third and fifth (secondary) foci from zone is given by $f_1 = \frac{r_1^2}{\lambda}, f_3 = \frac{r_1^2}{3\lambda}, f_5 = \frac{r_1^2}{5\lambda}$.
8.	A convex lens has no virtual focus on the side of the source.	A zone plate has no virtual focus on the side of the source.
9.	A convex lens cannot act like a concave lens in a single medium.	A zone plate can act like a convex lens as well like a concave lens
10.	A concave lens only acts like a converging lens.	A zone plate acts like a converging as well as a like diverging lens.

13.15 FRESNEL DIFFRACTION PATTERN OF A STRAIGHT EDGE

Let FF' be a cylindrical wavefront emerging from monochromatic light source of wavelength (λ) emerging through a narrow slit, S. P is a point on the screen in illuminated region. AB is a straight edge of opaque obstacle like razor blade. It is placed at distance 'a' from the slit. The screen XY is placed at distance b from the straight edge. The ray diagram of experimental set up is shown in Fig. 13.14.

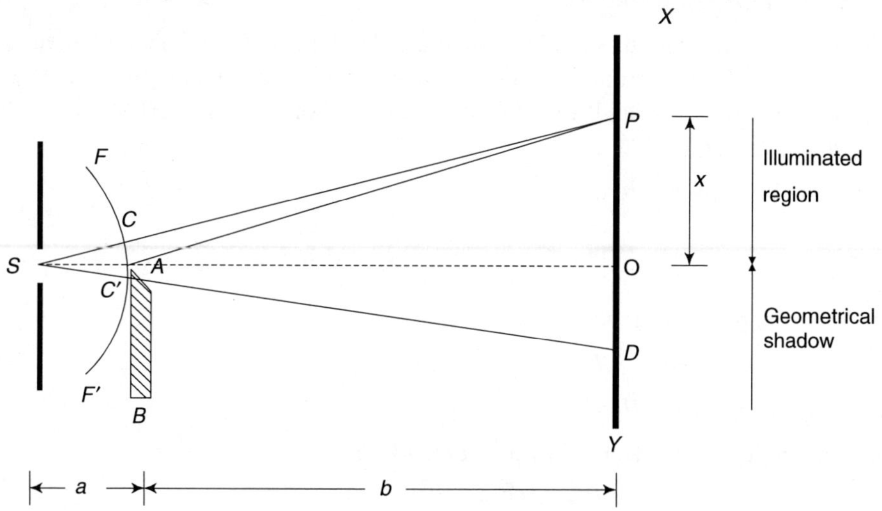

Fig. 13.14 Diffraction at straight edge

If there were no diffraction of light at the straight edge, then there will be uniform illumination on the screen above point O and complete darkness below it. On the other hand because of diffraction of light from the edge, a few unequally spaced bright and dark bands with poor contrast are observed on the screen above point O. In the geometrical shadow below O, the intensity falls rapidly and become zero at point D.

The diffraction pattern observed on the screen is shown graphically in Fig. 13.15. In the geometrical shadow the intensity falls off rapidly, while above point O there is a diffraction pattern consisting of bright and dark bands till uniform illumination is reached.

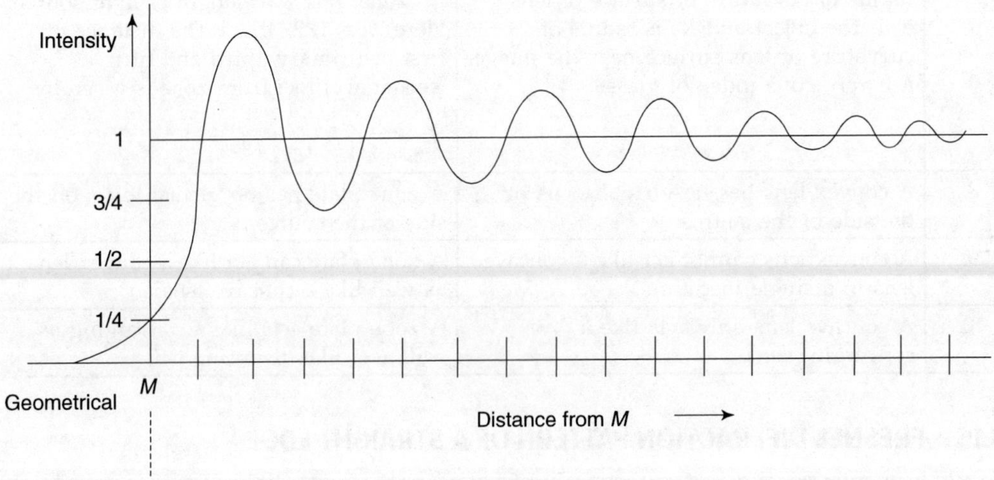

Fig. 13.15 Intensity distribution curve

13.15.1 Mathematical Analysis

From Fig. 13.14, it observed that monochromatic light from source S incident on edge of the obstacle. The FF' is the wavefront of incident light. For getting diffraction pattern theoretically, let us consider two rays, CP and AP, overlaps at point P on the screen. The path difference between two rays is $AP - CP$.

For maxima the path difference

$$AP - CP = (2n + 1)\frac{\lambda}{2}, \text{ where, } n = 0, 1, 2, 3,... \quad ...(13.99)$$

For minima the path difference

$$AP - CP = n\lambda, \text{ where, } n = 0, 1, 2, 3,... \quad ...(13.100)$$

Let $SA = a$, $AO = b$ and $OP = x$

From right angle triangle AOP shown in Fig. 13.14

$$AP = (AO^2 + OP^2)^{1/2}$$

$$= (b^2 + x^2)^{1/2}$$

$$= b\left(1+\frac{x^2}{b^2}\right)^{1/2}$$

Using binomial expansion theorem, we get

$$AP = b\left(1+\frac{1}{2}\frac{x^2}{b^2}\right)$$

$$= b+\frac{x^2}{2b} \qquad \text{...(13.101)}$$

and $$CP = SP - SC = SP - SA$$

From right angle triangle SOP shown in Fig. 13.12

$$SP = (SO^2 + OP^2)^{1/2}$$

$$= ((a + b)^2 + x^2)^{1/2}$$

$$= (a + b)\ b\left(1+\frac{x^2}{(a+b)^2}\right)^{1/2}$$

Using binomial expansion theorem, we get

$$SP = (a + b)\left(1+\frac{1}{2}\frac{x^2}{(a+b)^2}\right)$$

$$= (a + b) + \frac{x^2}{2(a+b)} \qquad \text{...(13.102)}$$

From Fig. 13.14, we see that

$$CP = SP - SC = a + b + \frac{x^2}{2(a+b)} - a = b + \frac{x^2}{2(a+b)} \qquad \text{...(13.103)}$$

Using Eqs. (13.101) and (13.102), the path difference

$$AP - CP = \left(b+\frac{x^2}{2b}\right)-\left(b+\frac{x^2}{2(a+b)}\right)$$

$$= \frac{a x^2}{2b(a+b)} \qquad \text{...(13.104)}$$

Combining Eqs. (13.99) and (13.104), the condition for maxima is

$$\frac{x^2}{2(a+b)} = (2n + 1)\ \frac{\lambda}{2}$$

or
$$x = \sqrt{\frac{b\,(a+b)}{a}\{2\,n+1\}\,\lambda}$$

or
$$x = k\,\sqrt{\{2\,n+1\}} \qquad ...(13.105)$$

where, $k = \sqrt{\frac{b\,(a+b)}{a}}\,\lambda$ is a constant. This expression gives the position of the nth maxima with respect to point O on the screen in the diffraction pattern.

Putting n = 0, 1, 2, 3,... in Eq. (13.105), the distances of the successive maxima from O are

$$\left.\begin{aligned} x_1 &= k \\ x_2 &= k\sqrt{3} \\ x_3 &= k\sqrt{5} \\ x_4 &= k\sqrt{7} \\ &\text{------} \end{aligned}\right\} \qquad ...(13.106)$$

The wdth of successive maxima are:

$$\left.\begin{aligned} (x_2 - x_1) &= k\,\sqrt{(3-1)} = 0.73\,k \\ (x_3 - x_2) &= k\,\sqrt{(5-3)} = 0.50\,k \\ (x_4 - x_3) &= k\,\sqrt{(7-5)} = 0.43\,k \end{aligned}\right\} \qquad ...(13.107)$$

Equation (13.107) indicates that width of successive maxima is in decreasing order. Similarly, one can see that the width successive minima in decreasing order. Thus, as we move from O towards P on the screen, the images come closer and closer. This can be seen from diffraction pattern given in Fig. 13.15.

Example 13.5 In a diffraction pattern due to a straight edge, the distance between the slit and the straight edge is 6.0 metres and that between the straight edge and the eyepiece is 4.0 meters. Calculate the position of the first three maxima and their separation, if the wavelength of light source, λ = 6000 Å.

Solution: Formula used

Position of maxima

$$x_n = \sqrt{\frac{(b\,(a+b)}{a}\{2\,n+1\}\,\lambda}, \quad n = 0, 1, 2, 3,...$$

Given: The distance between the slit and the straight edge, a = 600 cm

The distance between the straight edge and eyepiece, b =300 cm

Wavelength of light source, $\lambda = 6000 \times 10^{-8}$ cm

$$x_n = \sqrt{\frac{400 \times 1000}{600}(2\,n+1) \times 6000 \times 10^{-8}}$$

$$= \sqrt{\frac{(4(2n+1)}{100}} \text{ cm},$$

$$= \frac{1}{5}\sqrt{(2n+1)} \text{ cm}, \quad \text{where, } n = 0, 1, 2, 3,...$$

The position of the first three maxima for $n = 0, 1, 2$ is

$$x_1 = \frac{1}{5} = 0.2 \text{ cm}$$

$$x_2 = \frac{1}{5}\sqrt{3} = 0.346 \text{ cm}$$

$$x_3 = \frac{1}{5}\sqrt{5} = 0.447 \text{ cm}$$

and separations between maxima are

$$x_2 - x_1 = 0.346 - 0.2 = 0.146 \text{ cm}$$

$$x_3 - x_2 = 0.446 - 0.346 = 0.1 \text{ cm}$$

13.15.2 Determination of the Wavelength of Monochromatic Light from the Study of the Diffraction at the Straight Edge

Let the straight edge, AB, be placed at a distance 'a' from the slit and a screen XY is placed at a distance 'b' from the straight edge as shown in Fig. 13.14. If a monochromatic light of wavelength λ passing from a narrow slit (S) incident on the edge, due to the diffraction of light at the straight edge, a few unequally spaced bright and dark bands with poor contrast are observed on screen from point O towards P. In the geometrical shadow below point O, the intensity falls rapidly and becomes zero after a small distance.

The maxima at point P on the screen will be observed, if the wavefront, CA, consists of odd number of half period strips, i.e., path difference,

$$\Delta = PA - PC = (2n - 1)\frac{\lambda}{2} \quad ...(13.108)$$

Putting the value of $PA - CP$ from Eq. (13.104), we get

$$\frac{a x_n^2}{2b(a+b)} = (2n - 1)\frac{\lambda}{2} \quad ...(13.109)$$

Thus the distance of the nth maxima from point O will be

$$x_n^2 = \frac{b(a+b)\lambda}{a}(2n - 1) \quad ...(13.110)$$

The condition for the $(n + m)^{\text{th}}$ maxima is written as

$$x_{m+n}^2 = \frac{b(a+b)\lambda}{a}(2n + 2m - 1) \quad ...(13.111)$$

Subtracting x_n^2 from x_{m+n}^2, we get

$$x_{m+n}^2 - x_n^2 = \frac{b(a+b)\lambda}{a} 2\text{ m}$$

or
$$\lambda = \frac{a(x_{m+n}^2 - x_n^2)}{b(a+b)2m} \qquad ...(13.112)$$

Finding the values of distance of nth maximum and $(n + m)^{th}$ maximum with respect to point O, one can calculate the value of wavelength, λ, of light source from Eq. (13.112).

13.16 FRESNEL-KIRCHHOFF INTEGRAL FORMULA

An approximate solution of Helmholtz equation

$$(\nabla^2 + k^2)\, u(\vec{r}) = 0 \qquad ...(13.113)$$

$$u(\vec{x}) = -\frac{iA}{2\lambda}\int_{\partial V} \frac{e^{ik(r+S)}}{rS}(\hat{n}\cdot\hat{s} + \hat{n}\cdot\hat{r})\, dS \qquad ...(13.114)$$

is called **Fresnel-Kirchhoff diffraction formula.** Here, $\hat{n}$ is inward normal to surface (∂V) that bounds volume, $V \cdot k = \frac{2\pi}{\lambda}$ is wave vector incident spherical wave $\frac{e^{ikr}}{r}$ at point r on the aperture and diffracted spherical wave $\frac{e^{ikS}}{S}$ at point S of the diffraction pattern on screen. The derivation of Fresnel-integral formula is given in Section 11.5 of Chapter 11.

APPENDIX A

Calculation of path difference, $S_2A - S_1A = \frac{xa}{D}$

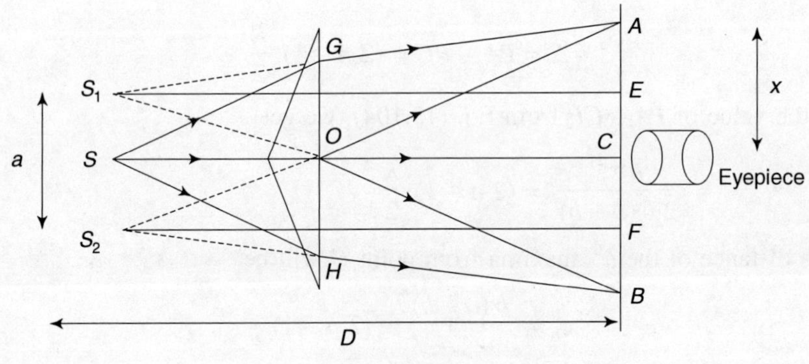

Fig. 13.16

In the right angle $\Delta S_2\,AF$

$$(S_2\,A)^2 = (S_2\,F)^2 + (AF)^2 = D^2 + \left(x + \frac{a}{2}\right)^2 \quad \text{...(A1)}$$

In the right angle $\Delta S_1\,AE$

$$(S_1\,A)^2 = (S_1\,E)^2 + (AE)^2 = D^2 + \left(x - \frac{a}{2}\right)^2 \quad \text{...(A2)}$$

or

$$S_2\,A = D\left[1 + \frac{\left(x + \frac{a}{2}\right)^2}{D^2}\right]^{1/2}$$

$$S_1\,A = D\left[1 + \frac{\left(x - \frac{a}{2}\right)^2}{D^2}\right]^{1/2}$$

or

$$S_2\,A = D\left[1 + \frac{1}{2}\frac{\left(x + \frac{a}{2}\right)^2}{D^2}\right]$$

$$S_1A = D\left[1 + \frac{1}{2}\frac{\left(x - \frac{a}{2}\right)^2}{D^2}\right]$$

or

$$S_2A = D\left[1 + \frac{1}{2}\frac{x^2 + \frac{a^2}{4} + xa}{D^2}\right]$$

$$S_1A = D\left[1 + \frac{1}{2}\frac{x^2 + \frac{a^2}{4} - xa}{D^2}\right]$$

or

$$S_2A = D + \frac{1}{2\,D}\left[x^2 + \frac{a^2}{4} + xa\right] \quad \text{...(A3)}$$

$$S_1A = D + \frac{1}{2\,D}\left[x^2 + \frac{a^2}{4} - xa\right] \quad \text{...(A4)}$$

Subtracting S_1A from S_2A, we get

Path difference

$$S_2A - S_1A = \frac{xa}{D} \quad \text{...(A5)}$$

Formulas at a Glance

S.No.	Subject	Formula	Explanation
1.	Position of bright band	$x = \frac{n\lambda D}{a}$, $n = 0, 1, 2, 3, \ldots$	a is separation of two monochromatic light beams D is the distance between light source and screen
2.	Position of dark band	$x = \frac{(2n+1)\lambda D}{2a}$, $n = 0, 1, 2, 3, \ldots$	
3.	Bright or dark band width	$\beta = \frac{\lambda D}{a}$	
4.	The distance between two slits	$a = \sqrt{a_1 a_1}$	a_1 and a_2 are the distance between the images of two, light sources
5.	Thickness of transparent slab	$t = \frac{x\lambda}{\beta(\mu - 1)}$	$\beta = \frac{\lambda D}{a}$, x is position of maxima, μ is refractive index of glass sheet
6.	Radii of zones	$\left[nb\lambda + n^2 \frac{\lambda^2}{4}\right]^{1/2}$	b is distance of pole from the zone
7.	Area of zone	$A = \pi b\lambda$	
8.	Average distance of half period zone from the pole	$d_n = b + (2n - 1)\frac{\lambda}{4}$	
9.	Ratio of area of zone and average distance	$\frac{A_n}{d_n} = \pi\lambda$	
10.	Amplitude at pole due to a zone $R_n = k \frac{A_n}{d_n}(1 + \cos\vartheta_n)$	$R_n = k\frac{A_n}{d_n}(1 + \cos\vartheta_n)$	ϑ_n is inclination angle of nth zone, $k = k\pi\lambda$, is constant of proportionality
11.	Amplitude at pole due to zones of zone plate	$R = \frac{1}{2}R_1$	R_1 is amplitude due to the first zone
12.	Intensity at pole due zones of zone plate	$I = k\frac{R_1^2}{4}$	K is constant of proportionality
13.	Radius of nth zone	$r_n = r_1\sqrt{n}$	r_1 is radius of the first zone

14.	Length of odd numbered $(2m - 1)^{th}$ foci	$f_{2m-1} = \frac{r_1^2}{(2m-1)\lambda}$	where, r_1 is the radius of the first Fresnel's first half period zone and m = 1, 2, 3, ... etc. For m = 1, first focal or primary or principal length, $f_1 = \frac{r_1^2}{\lambda}$
15.	If u and v are distance of light source and manage from zone plate, then focal length of odd numbered zones is given by	$\frac{1}{v} - \frac{1}{u} = \frac{1}{f_{2m-1}}$ where m = 1, 2, 3, ... For m = 1, first focal length of zone $f_1 = \frac{r_1^2}{\lambda}$ For m = 2, 3, 4, ..., the length of foci of second, third, ..., etc. zones $f_3 = \frac{r_1^2}{3\lambda}, f_5 = \frac{r_1^2}{5\lambda}$, ..., etc.	First focal lemgth of zone is called primary or principal focal length of zone. The length of foci of second, third, ... etc., zones are called secondary foci, Here r_1 is the radius of first Fresnel's first half period zone.

SOLVED EXAMPLES

13.1 Find the radius of half period zone having convex lens focal length is 60 cm, if wavelength of light source is 6000Å.

Solution: Formula required

First focal length, $f_1 = \frac{r_1^2}{\lambda}$

where, r_1 is radius of first zone.

Given: Focal length of first half period zone, f_1 = 60 cm

Wavelength of light source, λ = 6000 Å = 6000×10^{-8} cm

The radius of the first half period zone

$$r_1 = \sqrt{\lambda f_1} = \sqrt{6000 \times 10^{-8} \times 60}$$

$$= 6 \times 10^{-2} \text{ cm} = 0.06 \text{ cm}$$

Ans. The radius of the first half period zone is 0.06 cm

13.2 The radius of the first ring of a zone is 0.5 mm. If light of wavelength 5000 Å falls normally on it, where should the screen be placed to get the right spot.

Solution: Formula required

First focal length, $f_1 = \frac{r_1^2}{\lambda}$

wher, r_1 is radius of the first zone.

Given: The radius of the first half period zone, $r_1 = 0.5$ mm $= 5 \times 10^{-2}$ cm

Wavelength of light source, $\lambda = 6000$ Å $= 5000 \times 10^{-8}$ cm

Focal length of first half period zone,

$$f_1 = \frac{r_1^2}{\lambda} = \frac{(5 \times 10^{-2})^2}{5000 \times 10^{-8}}$$

$$= \frac{25}{5} \times 10^1 = 50 \text{ cm}$$

Ans. The radius of the first half period zone is 50 cm.

13.3 Light of wavelength 6000 Å falls on a zone plate where the first zone has a diameter of 3 mm. Find the position of brightest spot.

Solution: Formula required

Focal length of first zone

$$f_1 = \frac{r_1^2}{\lambda}, \text{ where, } r_1 \text{ is radius of first zone,}$$

Is the position of brightest spot.

Given, $\lambda = 6000 \times 10^{-8}$ cm, $r_1 = 1.5$ mm $= 0.15$ cm

$$f_1 = \frac{(0.15)^2}{6000 \times 10^{-8} \text{ cm}} = 375 \text{ cm}$$

Ans. Position of the brightest spot is 375 cm.

13.4 When point source of light of wavelength 6000 Å is incident on the zone plate, the brightest and the next brightest images are formed at 30 cm and 6 cm, respectively from the zone plate. Calculate (i) distance of source from the zone plate, (ii) radius of first zone, and (iii) principal focal length.

Solution: Formula required

Foci of zone plate, $f_m = \frac{r_n^2}{(2m-1)\,n\lambda}$

For first zone, $n = 1$

For strongest image, $m = 1$, so focal length of strongest image is $f_1 = \frac{r_1^2}{\lambda}$

For next strongest image, $m = 2$, so focal length of strongest image is $f_2 = \frac{r_1^2}{3\lambda}$

Let u be the distance of objet from zone plate, v_n is distance of mth image, the zone plate formula just like lens formula is

$$\frac{1}{v_m}-\frac{1}{u} = \frac{1}{f_m}$$

For $m = 1$, $\frac{1}{v_1}-\frac{1}{u}=\frac{1}{f_1}$ (i), for $m = 2$, $\frac{1}{v_2}-\frac{1}{u}=\frac{1}{f_2}$ (ii)

Given: $v_1 = 35$ cm, $v_2 = 7$ cm, $\lambda = 5000 \times 10^{-8}$ cm

So, $\frac{1}{35}-\frac{1}{u} = \frac{\lambda}{r_1^2}$ (iii), $\frac{1}{7}-\frac{1}{u} = \frac{3\lambda}{r_1^2}$ (iv)

From these Eqs. (iii) and (iv), we can have

$$\frac{1}{7}-\frac{1}{u} = \frac{3}{35}-\frac{3}{u} \quad \text{or} \quad \frac{2}{u} = \frac{3}{35}-\frac{1}{7} \quad \text{or} \quad u = -35 \text{ cm}$$

Putting the value of $u = -35$ cm, in Eq. (iii), we get

$$\frac{1}{33}+\frac{1}{35} = \frac{\lambda}{r_1^2} \quad \text{or} \quad r^2_1 = \frac{35\lambda}{2}$$

or

$$r_1 = \sqrt{\frac{35\lambda}{2}} = \sqrt{\frac{35}{2}\times 5000\times 10^{-8}}$$

$$= 2.958 \times 10^{-2} \text{ cm} = 0.02958 \text{ mm}$$

Putting the value $r_1 = 2.958 \times 10^{-2}$ cm

$$f_1 = \frac{r_1^2}{\lambda} = \frac{(2.958\times 10^{-2})^2}{5000\times 10^{-8}} = 17.5 \text{ cm}$$

Ans. (i) distance of source from the zone plate is 35 cm

(ii) radius of first zone 0.02958 mm

(iii) principal focal length is 17.5 cm

13.5 The diameter of the central zone of a zone plate is 3.5 mm. If a point source of light of wavelength 5890 Å is placed at a distance 7 m from it, calculate the position of first image.

Solution: Formula used

Formula of foci of zone plate is

$$f_n = \frac{r_n^2}{n\lambda}, \quad \text{for } n = 1 \quad f_1 = \frac{r_1^2}{\lambda} \text{ is focal length of first zone}$$

Formula for distance of nth image v from zone plate is

$$\frac{1}{v}-\frac{1}{u} = \frac{n\lambda}{r_n^2} = \frac{1}{f_n}$$

where, u is the distance of objet u from zone plate.

Formula for distance of first image v from zone plate is

$$\frac{1}{v}-\frac{1}{u} = \frac{n\lambda}{r_1^2}=\frac{1}{f_1}$$

where, u is the distance of objet u from zone plate.

Given:

Wavelength of light, λ = 5890 Å = 5890 × 10^{-8} cm

The radius of first zone, $r_1 = \frac{3.5}{2}$ = 1.75 mm = 0.175 cm

The distance of object, u = – 7 m

First focal length, $f_1 = \frac{r_1^2}{\lambda} = \frac{(0.175)^2}{5890 \times 10^{-8}}$ = 0.0519949 × 10^4 = 519.949 cm = 5.2 m

$$\frac{1}{v} = \frac{1}{f_1}+\frac{1}{u} \quad \text{or} \quad v = \frac{u\,f_1}{u+f_1}=\frac{7 \times 5.2}{7-5.2}=\frac{36.4}{1.8} = 20.22 \text{ m}$$

Ans. The distance of image from zone plate is 20.22m.

13.6 Calculate the radii of three transparent zones of a zone plate which has focal length 3, if a light beam of wavelength 6000 Å is incident.

Solution: Formula required

Radii of nth transparent zone, $r_n^2 = nb\lambda$

Given: λ = 6000 Å = 6000 × 10^{-8} cm, b = 300 cm

For n = 1, radius of first transparent zone, $r_1^2 = b\lambda$ or $r_1 = \sqrt{b\lambda}$

For n = 3, radius of second transparent zone, $r_2^2 = 3\,b\lambda$ or $r_2 = \sqrt{3\,b\,\lambda}$

For n = 5, radius of third transparent zone, $r_3^2 = 5\,b\lambda$ or $r_2 = \sqrt{5\,b\,\lambda}$

$$r_1 = \sqrt{b\lambda} = \sqrt{300 \times 6000 \times 10^{-8}}$$

$$= 10^{-1}\sqrt{1.8} \text{ cm} = 0.1341 \text{ cm}$$

$$r_2 = \sqrt{2\,b\,\lambda} = \sqrt{3 \times 300 \times 6000 \times 10^{-8}}$$

$$= 10^{-1}\sqrt{5.4} \text{ cm} = 0.2323 \text{ cm}$$

$$r_3 = \sqrt{3\,b\,\lambda} = \sqrt{5 \times 300 \times 6000 \times 10^{-8}}$$

$$= 10^{-1}\sqrt{9} \text{ cm} = 0.3 \text{ cm}$$

Ans. (i) Radius of first transparent zone 0.1341 cm.

(ii) Radius of second transparent zone 0.2323 cm.

(iii) Radius of third transparent zone 0.3 cm.

13.7 Calculate radius of the first zone in a zone plate of focal length 20 cm for light of wavelength 5000 Å.

Solution: Formula required

Focal length of nth zone, $f_n = \dfrac{r_n^2}{n\lambda}$

Radius of n^{th}zone will be $r_n = \sqrt{n\lambda f_n}$

Given: λ = 5000 Å = 5000 × 10^{-8} cm , focal length first zone, f_1 = 20 cm

For n = 1

Radius of first zone, $r_1 = \sqrt{\lambda f_1} = \sqrt{20 \times 5000 \times 10^{-8}} = 3.16 \times 10^{-2}$ cm

Ans. Radius of the first zone is 0.0316 cm.

13.8 A source of light of wavelength 6000 Å is placed 4 m from a zone plate, the diameter of the central zone is 2 mm. Find the position of primary and secondary images.

Solution: Formula required

Focal length of mth secondary zone is $f_{2m-1} = \dfrac{r_n^2}{(2m-1)n\lambda}$

We know that $r_n = r_1\sqrt{n}$

So, focal length of mth secondary zone, $f_{2m-1} = \dfrac{r_1^2}{(2m-1)\lambda}$

If u is the distance of source light from zone and v_n is the distance of image from the zone, the focal length of n^{th}zone is given by

$$\frac{1}{v_{2m-1}} - \frac{1}{u} = \frac{1}{f_{2m-1}}$$

Given: λ = 6000 Å = 6000 × 10^{-8} cm, u = – 4m = – 400 cm, r_1 = 1 mm = 0.1 cm

Position of primary image (v_1) is given by

$$\frac{1}{v_1} - \frac{1}{u} = \frac{1}{f_1}, \text{ where, for } m = 1, f_1 = \frac{r_1^2}{\lambda}$$

$$\frac{1}{v_1} - \frac{1}{u} = \frac{1}{f_1} - \frac{1}{u}$$

$$f_1 = \frac{(0.1)^2}{6000 \times 10^{-8}} = \frac{1000}{6} \text{ cm}$$

$$\frac{1}{v_1} - \frac{1}{400} = \frac{6}{1000} \quad \text{or} \quad \frac{1}{v_1} - \frac{1}{400} = \frac{6}{1000} - \frac{1}{400} = \frac{12-5}{2000}$$

or $$v_1 = \frac{2000}{7} = 285.714 \text{ cm}$$

Position of secondary image (v_3) is given by

$$\frac{1}{v_3} - \frac{1}{u} = \frac{1}{f_3}, \text{ where, for } m = 2\, f_3 = \frac{r_1^2}{3\lambda}$$

$$\frac{1}{v_2} - \frac{1}{u} = \frac{1}{f_3}$$

$$f_3 = \frac{(0.1)^2}{3 \times 6000 \times 10^{-8}} = \frac{1000}{18} \text{ cm}$$

$$\frac{1}{v_3} - \frac{1}{-400} = \frac{18}{1000} \text{ or } \frac{1}{v_2} - \frac{1}{400} = \frac{9}{500} - \frac{1}{400} = \frac{36-5}{2000}$$

or $$v_3 = \frac{2000}{31} = 65.516 \text{ cm}$$

Ans. (i) Position of primary image v_1 is 285.714 cm.

(ii) Position of secondary image v_3 is 65.516 cm.

13.9 Calculate the radii of three transparent zones of a zone plate which has focal length 2 m, if a light of beam of wavelength 5000 Å is incident.

Solution: Formula required

Radii of nth transparent zone $r_n^2 = nbf$

Given: $\lambda = 5000$ Å $= 5000 \times 10^{-8}$ cm, $f = 200$ cm

For $n = 1$, radius of first transparent zone, $r_1^2 = f\lambda$ or $r_1 = \sqrt{f\lambda}$

For $n = 3$, radius of second transparent zone, $r_2^2 = 3f\lambda$ or $r_2 = \sqrt{3\, f\lambda}$

For $n = 5$, radius of third transparent zone, $r_3^2 = 5f\lambda$ or $r_2 = \sqrt{5f\lambda}$

$$r_1 = \sqrt{b\lambda} = \sqrt{200 \times 5000 \times 10^{-8}} = 10^{-1} \text{ cm} = 0.1 \text{ cm}$$

$$r_2 = \sqrt{2b\lambda} = \sqrt{3 \times 200 \times 5000 \times 10^{-8}} = 10 \quad \sqrt{3} \text{ cm} = 0.1732 \text{ cm}$$

$$r_3 = \sqrt{3b\lambda} = \sqrt{5 \times 200 \times 5000 \times 10^{-8}} = 10^{-1} \sqrt{5} \text{ cm} = 0.2236 \text{ cm}$$

Ans. (i) Radius of first transparent zone 0.1 cm.

(ii) Radius of second transparent zone 0.1732 cm.

(iii) Radius of third transparent zone 0.2236 cm.

13.10 A point source of light of wavelength 6000 Å is incident at a zone plate. The brightest and next brightest images of the source are formed by the zone plate at a distance of 40 cm and 10 cm, respectively. Calculate (i) the distance of the point source from the zone, (ii) the radius of the first zone and (iii) the principal focal length of the zone plate.

Solution: Formula required

If u and v_{2m-1} are the distance of the source and image from the zone plate, then focal lengths of the foci are given by

$$\frac{1}{v_{2m-1}} - \frac{1}{u} = \frac{1}{f_{2m-1}}$$

where, focal length of mth secondary zone $f_{2m-1} = \frac{r_1^2}{(2m-1)\lambda}$, r_1 is radius of the first zone.

For $m = 1$ and 2 the formulas of brightest and next brightest focal lengths are

$$f_1 = \frac{r_1^2}{\lambda}, \quad \text{and} \quad f_3 = \frac{r_1^2}{3\lambda}$$

Now other formulas two focal lengths are

$$\frac{1}{v_1} - \frac{1}{u} = \frac{1}{f_1} \quad \text{and} \quad \frac{1}{v_3} - \frac{1}{u} = \frac{1}{f_3}$$

or

$$\frac{1}{v_1} - \frac{1}{u} = \frac{\lambda}{r_1^2} \quad \text{and} \quad \frac{1}{v_3} - \frac{1}{u} = \frac{3\lambda}{r_1^2}$$

Given: $\lambda = 6000$ Å $= 6000 \times 10^{-8}$ cm, $v_1 = 40$ cm, $v_3 = 10$ cm

Putting for $v_1 = 40$ cm, $v_3 = 10$ cm, we get

$$\frac{1}{40} - \frac{1}{u} = \frac{\lambda}{r_1^2} \quad \text{and} \quad \frac{1}{10} - \frac{1}{u} = \frac{3\lambda}{r_1^2}$$

Eliminating the value of $\frac{1}{u}$ from two equations

$$\frac{3\lambda}{r_1^2} - \frac{\lambda}{r_1^2} = \frac{1}{10} - \frac{1}{40}$$

or

$$r_1 = \sqrt{\frac{80}{3}\lambda} = \sqrt{\frac{8}{3} \times 6000 \times 10^{-8}}$$

$$= 10^{-2}\sqrt{16} = 0.04 \text{ cm}$$

$$f_1 = \frac{r_1^2}{\lambda} = \sqrt{\frac{80}{3}} = 26.67 \text{ cm}$$

$$\frac{1}{40} - \frac{1}{u} = \frac{\lambda}{r_1^2} = \frac{3}{80} \quad \text{or} \quad u = -80 \text{ cm}$$

Ans. (i) Radius of first transparent zone 0.04 cm,

(ii) Radius of primary focal length 26.67 cm.

(iii) Distance of source 80 cm.

QUESTIONS FOR EXAMINATION

13.1 What is the difference between Fraunhofer diffraction and Fresnel diffraction?

13.2 What is Fresnel biprism? Discuss Fresnel method to find fringes.

13.3 What are applications of Fresnel biprism method? Discuss any two applications.

13.4 How do you find the wavelength of light source using Fresnel biprism method.

13.5 How do you find the thickness of transparent glass sheet using Fresnel biprism method?

13.6 How do you find the width of bright fringe using Fresnel biprism method?

13.7 What are Fresnel's assumptions?

13.8 What is half period zone?

13.9 Find the area of half period zones. Show that the area of half period zones is almost equal.

13.10 Find the amplitude of the secondary wavelets at focal point due to the nth Fresnel's half period zone.

13.11 Discuss the phase difference among half period zones.

13.12 Find the resultant amplitude of secondary wavelets originating from all Fresnel's half period zones at focal point using Schuster's method.

13.13 How many types of zone plates are? Discuss them.

13.14 Find the intensity of secondary wavelets at focal point of zone plate?

13.15 What is the principle of zone plate?

13.16 Find out the intensity at focal point of multi-foci zone plate.

13.17 What do you mean by absence of even numbered foci?

13.18 Why do we call zone plate as half zone plate?

13.19 Discuss the Fresnel's rectilinear wave theory of light.

13.20 Derive an expression of focal length of zone plate and compare it the focal length of a thin lens.

13.21 Explain the construction of zone plate. Discuss its working as a lens.

13.22 What is Fresnel's integral formula?

13.23 Discuss the method to determine the wavelength of monochromatic light using diffraction at straight edge.

13.24 What are the similarities and dissimilarities between a zone plate and a lens?

13.25 Explain why higher order Fresnel's half period zones have greater area than that of lower order?

13.26 Prove that the ratio of area of the nth half period zone to its average distance from the pole of wavefront is constant for a particular wavefront.

13.27 Show that the resultant amplitude at a pole due to the entire wavefront is half the amplitude due to the first Fresnel's half period zone.

13.28 How does the radius of Fresnel's half period zone depend on the wavefront of light?

13.29 Differentiate between positive and negative half plates.

13.30 Prove that the radii of half period zones of a zone plate are proportional to the square root of natural numbers.

13.31 Prove that each half period of a zone plate has equal area.

13.32 Explain how the zone plate behaves like a lens?

13.33 Explain the presence of only odd numbered foci in a zone plate.

13.34 Explain the absence of even numbered foci in a zone plate.

13.35 Explain how the intensity of the focus of a zone plate decreases with increase in the order of focus.

OBJECTIVE QUESTIONS

13.1 A convex lens acts like a

(a) Converging lens only (b) Diverging lens only
(c) Converging as well as diverging lens (d) None of these

13.2 A zone plate acts like a

(a) Converging lens only (b) Diverging lens only
(c) Converging as well as diverging lens (d) None of these

13.3 If f_V, f_Y and f_R are first focal length of lens for violet, yellow and red light, then which expression is correct?

(a) $f_V > f_Y > f_R$ (b) $f_R > f_Y > f_V$ (c) $f_Y > f_R > fV$ (d) $f_R > f_Y > f_V$

13.4 If f_V, f_Y and f_R are first focal length of zone plate for violet, yellow and red light, then which expression is correct?

(a) $f_V > f_Y > f_R$ (b) $f_R > f_Y > f_V$ (c) $f_Y > f_R > f_V$ (d) $f_R > f_Y > f_V$

13.5 In case of convex lens the intensity of the focus is

(a) Minimum (b) Maximum (c) Moderate (d) None of these

13.6 For monochromatic light, a convex lens has

(a) Two foci (b) Multiple foci (c) Single focus (d) None of these

13.7 For monochromatic light, a zone plate has

(a) Two foci (b) Multiple foci (c) Single focus (d) None of these

13.8 Focusing action of light through the lens is due to

(a) Diffraction (b) Reflection
(c) Refraction (d) Both diffraction and refraction

13.9 Focusing action of light through the zone plate is due to

(a) Both diffraction and refraction (b) Reflection
(c) Refraction (d) Diffraction

13.10 The wavefront of light incident on slit in Fraunhofer diffraction is

(a) Spherical (b) Elliptical (c) Plane (d) Sinusoidal

13.11 The wavefront of light incident on biprism is

(a) Spherical (b) Elliptical (c) Plane (d) Sinusoidal

13.12 The secondary waves originating from the slitin case of Fraunhofer diffraction at every point

(a) Are in the same phase (b) Are in opposite phase

(c) Are not in the same phase (d) None of these

13.13 The secondary waves originating from two successive zones of a zone plate

(a) are in the same phase (b) are in opposite phase

(c) are in $\frac{\pi}{2}$ phase (d) none of these

13.14 The incident wavefront in case of Fresnel diffraction is

(a) Plane (b) Spherical (d) Cylindrical (d) Elliptical

13.15 The phase difference of wavefronts originating from two successive zones is

(a) 0 (b) $\frac{\pi}{2}$ (c) 2π (d) π

13.16 Amplitude at a point due to half period zones depends upon

(a) Area of the half period zone (b) Inclination factor

(c) Average distance of the half period zones from the pole

(d) All of above

13.17 The intensity on foci of zone plate depends upon

(a) Area of the half period zone

(b) Average distance of the half period zones from the pole

(c) Inclination factor (d) All of above

13.18 In a positive zone plate, the central zone is opaque.

(a) True (b) False

13.19 In a negative zone plate, the central zone is transparent.

(a) True (b) False

13.20 The shape of central zone in zone plate is

(a) Elliptical (b) Square (c) Circular (d) Rectangle

13.21 The intensity of foci decreases as we move towards the zone plate

(a) False (b) True

13.22 The radii of half period zones of a zone plate are proportional to

(a) Square root natural numbers (b) Square root odd natural numbers

(c) Square root of natural numbers (d) Square of natural numbers

13.23 The primary focus of a zone plate is situated at a point

(a) Farthest to the zone plate (b) Nearest to the zone plate

(d) Mid-way between zone plate and pole (d) None of these

13.24 Primary focal length of a zone plate does not depend on

(a) Wavelength of light (b) Radius of first half period zone

(c) Frequency of light (d) Distance of image from the zone

13.25 Primary focal length of a zone plate depends on

(a) Only wavelength of light

(b) Only radius of first half period zone

(c) Only area of first half period zone

(d) Both wavelength of light and radius of the first half period zone

13.26 If I_1, I_3 and I_5 are intensities of first three foci, then which of the following relation is correct?

(a) $I_3 > I_1 > I_5$ (b) $I_5 > I_1 > I_3$ (c) $I_1 > I_3 > I_5$ (d) $I_1 < I_3 < I_5$

13.27 If f_1, f_3 and f_5 are focal lengths of first three foci, then which of the following relation is correct?

(a) $f_3 > f_1 > f_5$ (b) $f_5 > f_1 > f_3$ (c) $f_1 < f_3 < f_5$ (d) $f_1 > f_3 > f_5$

13.28 Who gave the correct interpretation of rectilinear propagation of light?

(a) Fraunhoffer (b) Newton (c) Einstein (d) Fresnel

13.29 The phase difference of wavefronts originating from outer circumference of two successive zones is

(a) 0 (b) $\frac{\pi}{2}$ (c) π (d) 2π

13.30 The focal length of nth zone plate is equal to

(a) $\frac{r_n^2}{n\lambda}$ (b) $\frac{r_n^2}{\lambda}$ (c) $\frac{nr_n^2}{\lambda}$ (d) $\frac{n\lambda}{r_n^2}$

13.31 When light falls on the edge, then observed pattern on the screen is due to

(a) Refraction (b) Polarization (c) Reflection (d) Diffraction

Answers

13.1 (a)	13.2 (c)	13.3 (d)	13.4 (a)	13.5 (b)	13.6 (c)	13.7 (b)	13.8 (c)
13.9 (d)	13.10 (c)	13.11 (a)	13.12 (a)	13.13 (b)	13.14 (b)	13.15 (d)	13.16 (b)
13.17 (c)	13.18 (b)	13.19 (a)	13.20 (c)	13.21 (b)	13.22 (c)	13.23 (a)	13.24 (d)
13.25 (d)	13.26 (c)	13.27 (d)	13.28 (d)	13.29 (c)	13.30 (a)	13.31 (d)	

CHAPTER 14

Polarisation

14.1 INTRODUCTION

Light is an electromagnetic wave and exhibits interference and diffraction phenomena. But these phenomena do not indicate that light wave is transverse wave. Phenomenon of polarization shows that a light wave is a transverse wave. Polarization is a technique of restricting the vibrations of light wave to one direction or one plane perpendicular to the direction of propagation of it.

Transverse waves: According to quantum mechanics, transverse wave is that wave in which the direction of the oscillation of particle is perpendicular to the direction of propagation of the light wave. For example, light and sound waves transverse waves can also be produced in a string. The transverse waves can be polarized.

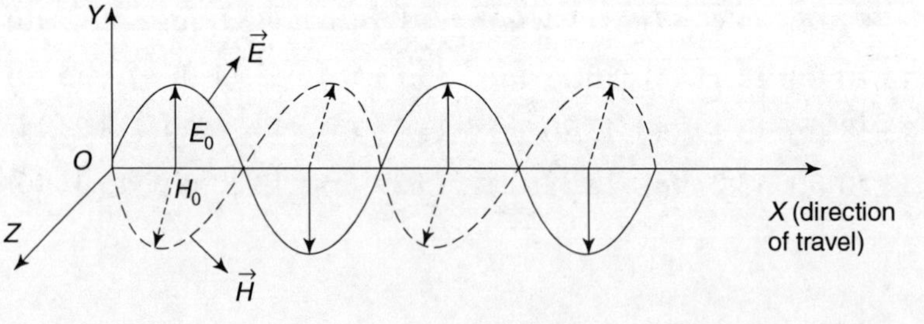

Fig. 14.1 $\vec{E}$ and $\vec{H}$ are perpendicular to each other

Longitudinal waves: Longitudinal wave is that wave in which the direction of oscillation of particles is parallel to the direction of propagation of the wave. For example, sound wave. The sound wave can be both transverse and longitudinal. Longitudinal waves cannot be polarized.

Transverse light waves: Light is an electromagnetic wave. An electromagnetic waves consists of a coupled oscillating electric field $\vec{E}$ and and magnetic field $\vec{H}$. Both $\vec{E}$ and $\vec{H}$ are always perpendicular to each other and also perpendicular to the direction of propagation of the wave as shown in Fig. 14.1.

That is, vibrations light electric vector $(\vec{E})$ are in the perpendicular direction of the propagation of the light wave. That is why it called a transverse light waves. The corresponding magnetic vectors also vibrate in the same plane in which the electric vectors vibrate.

It is to be noted further that

(i) $\vec{E}$ and $\vec{H}$ have same phase, i.e., when $\vec{E}$ is maximum (E_0) at the same time as $\vec{H}$ is also maximum (H_0).

(ii) The ratio of magnitudes of $\vec{E}$ and $\vec{H}$, at any time, $E_0/H_0 = c$, where, c is velocity of light.

Polarization: The "polarization" of electromagnetic waves is the process of making the vibrations of light in one perpendicular direction or one plane to its direction of propagation.

14.2 CLASSIFICATION OF POLARIZED LIGHT

The polarized light is of three types:

(i) Plane polarized

(ii) Circularly polarized

(iii) Elliptically polarized

Each of these is characterized by differing motion of the electric field vector with respect to the direction of propagation of the light wave.

Plane Polarized Light

Plane polarized light is also called linear polarized light. In plane polarized light, the vibrating electric vectors of light wave oscillate in a plane called the plane of vibration and the corresponding magnetic vectors also vibrate in the plane of polarization as shown in Fig. 14.2. From the figure it is to be noted that at any instant of time the vibration of electric vector $(\vec{E})$ in polarized light is one plane (like *PQRS*) perpendicular to the direction of propagation of the wave. The vibration of corresponding magnetic vector $(\vec{B})$ is at 90° with respect to electric vector and is in the same plane. The plane containing the vibrating electric vector $(\vec{E})$ and the propagation vector, $\vec{k}$ is called the plane of vibration. At the first instant the plane of vibration is *YZ* plane. The plane containing the magnetic vector $(\vec{B})$ and the propagation vector $(\vec{k})$ is called the plane of polarization. At the first instant the plane of polarization is *PQRS* plane. The vectors, $\vec{E}, \vec{B}$ and $\vec{k}$, are mutually perpendicular to each other at any instant of time.

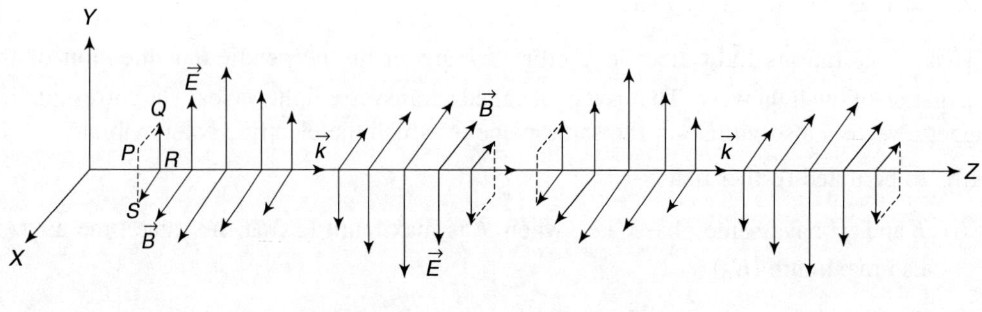

Fig. 14.2 Plane polarized light

Circularly Polarized Light

Circularly polarized light is that polarized light in which the electric vector is vibrating in single plane, where the tip of the light electric vectors move in circular path. Corresponding magnetic vectors also vibrate in the plane of polarization. The rotation can have two possible directions; if light electric vector rotates in a right hand sense with respect to the direction of wave travel, it is called right circular polarization. If light electric vector rotates in a left hand sense, it is called left circular polarization.

Elliptically Polarized Light

Elliptically polarized light is that polarized light in which the electric vector is vibrating in single plane, where the tip of the light electric vectors move in elliptically path. Corresponding magnetic vectors also vibrate in the plane of polarization. The rotation can have two possible directions; if light electric vector rotates in a right hand sense with respect to the direction of wave travel, it is called right elliptical polarization. If light electric vector rotates in a left hand sense, it is called left elliptical polarization.

14.3 PICTORIAL REPRESENTATION OF LIGHT

There are two ways of pictorial representation of unpolarized light. Light is electromagnetic wave and exhibits interference and diffraction phenomena. But these phenomena do not indicate that light waves are transverse waves. Polarization techniques show that light waves are transverse waves. Light is a transverse electromagnetic wave. Light ray consists of millions electromagnetic waves having vibrations perpendicular to the direction of its propagation. The light vectors have random orientations due to random emission of radiations from excited atoms or molecules of the source. But all these vibrations are confined to a single plane at right angle to the direction of propagation. Because of this reason, unpolarized light wave is represented by the star symbol as shown in Fig. 14.3.

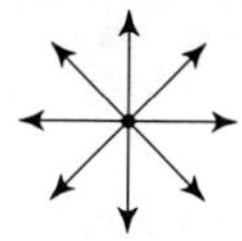

Fig. 14.3

Figure 14.4 shows a plane-polarized light wave in which the variation of $\vec{E}$ is in the *XY*-plane and that of $\vec{H}$ in the *XZ*-plane. The effect of the electric field usually dominates than the magnetic field during interaction of light with matter. For example, while taking the photograph, the electric component of the light affects photographic film. Similarly, only the electric component produces fluorescence. The plane of polarization of a light wave is therefore considered the plane of vibrations of electric field, $\vec{E}$. Therefore, whenever the word 'vibration' is used in the connection of a light wave, then normally it is the vibration of the associated electric field.

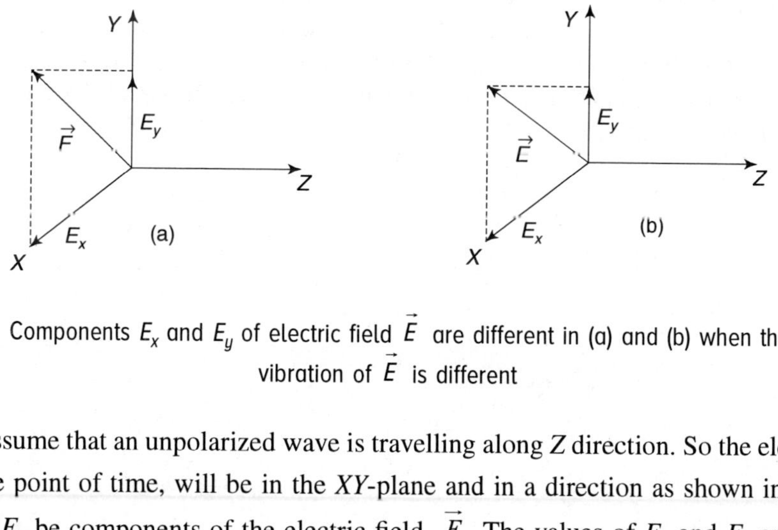

Fig. 14.4 Components E_x and E_y of electric field $\vec{E}$ are different in (a) and (b) when the angle of vibration of $\vec{E}$ is different

Let us assume that an unpolarized wave is travelling along *Z* direction. So the electric field, $\vec{E}$, at some point of time, will be in the *XY*-plane and in a direction as shown in Fig. 14.4. Let E_x and E_y be components of the electric field, $\vec{E}$. The values of E_x and E_y are changed, as the direction of $\vec{E}$ changes for every wave. $\vec{E}$ will have a large number of orientations, with equal probability for each component, E_x and E_y, and therefore the average value of E_x and E_y will be equal. The 'vibrations' of an unpolarized wave take place in every direction which is perpendicular to the direction of propagation. All such waves can be considered having components of electric field $\vec{E}$ only in two perpendicular directions, say, *X* and *Y* directions. Each component has same amplitude of vibration. According this criteria, the pictorial representation of an unpolarized light wave is shown in Fig. 14.5.

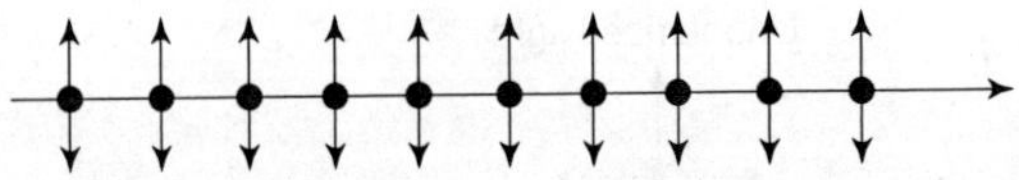

Fig. 14.5 Pictorial representation of unpolarized light

14.4 TRANSVERSE VIBRATIONS IN A STRING

The transmission of transverse wave of a rope through a slit is shown in Fig. 14.6. To show the plane of polarization, let us consider that one end of the rope fixed at B point after passing a rope through a slit. The width of the slit is more than the diameter of the rope. The other end point O of rope is forced to vibrate along YZ plane as shown in Fig. 14.6(a). It is observed that transverse wave formed by the rope does not pass through the slit, if the slit is kept perpendicular to the plane of vibrations, i.e., in ZX plane. When the slit is kept parallel to plane of vibrations, i.e., YZ plane then the transverse wave formed by the rope passes through the slit as shown in Fig. 14.6(b). The transverse wave formed by the rope is called plane polarized. Plane of polarized means that the vibrations of the transverse wave are restricted one direction.

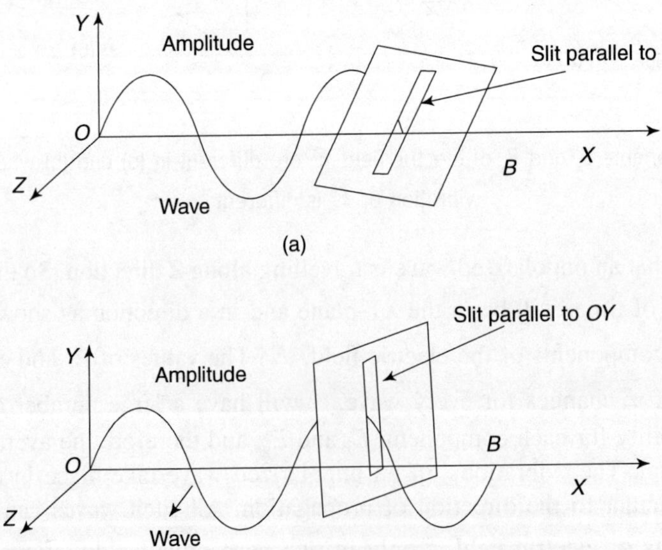

Fig. 14.6

14.5 PRODUCION OF PLANE POLARIZED LIGHT

14.5.1 Production of Plane Polarized Light by Using Polaroid

Polaroid is a material which allows to pass only a single component of light when they are kept in a particular direction. This direction is called the *optic axis.*

When a ray of ordinary light falls on the tourmaline plate N_1, cut with faces parallel to the vertical axis, only a part of the light is transmitted. The emerging light from tourmaline plate N_1 is plane polarized. If emerging light from N_1 is passed from the second tourmaline crystal N_2, kept with its vertical axis parallel to the vertical axes of N_1 as shown in Fig. 14.7(a), no change in the intensity is observed. If both the crystals, N_1 and N_2, are rotated with same angle, no change is observed in the intensity. If N_1 is kept fixed and N_2 is rotated clockwise as shown in Fig. 14.7(b), the intensity of light starts diminishing. It vanishes completely when second tourmaline crystal becomes perpendicular to N_1 as shown in Fig. 14.7(c). On rotation of N_2 clockwise further, the intensity of light increases and becomes maximum when vertical axis of N_2 becomes parallel to N_1, i.e., when the axes of both the crystals become parallel to each other. This experiment proves that light is transverse in nature. The tourmaline crystal acts just like slit as described in case of vibration of string in Section 14.4.

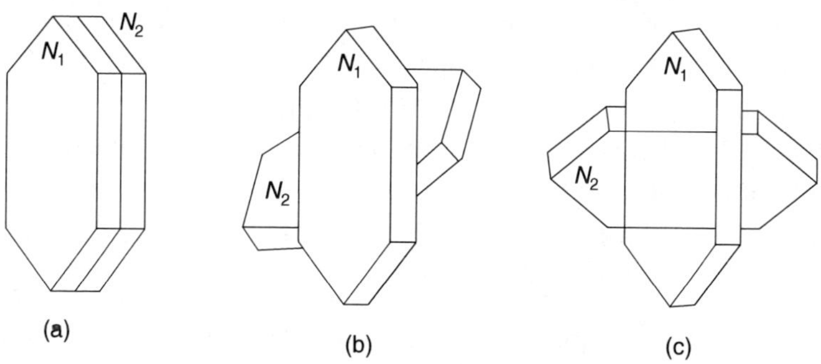

Fig. 14.7 Polarization of light using tourmaline crystal

14.5.2 Production of Plane Polarized Light by Reflection

When an unpolarized beam of light is incident on a glass surface at an angle, the beam is partially reflected and partially refracted as shown in Fig. 14.8. When the incident angle is of about 57°, the reflected light is completely plane-polarized. This can be checked by using apolaroid. When the reflected light is allowed to fall on a polaroid and then it is slowly rotated about the line of vision, the intensity of the polarized light emerging from the polaroid varies from some maximum value to zero. When the angle of incident is other than 57°, the reflected light is partially plane-polarized.

When light is incident on glass at 57° the angle between completely polarized reflected ray and the refracted ray is 90°.

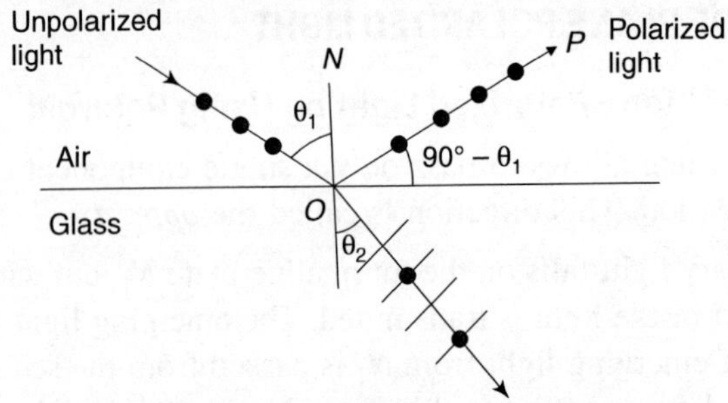

Fig. 14.8 Polarization by reflection

When an unpolarized light is incident on a glass surface at 57°, the vibrations which are perpendicular to the plane of the paper cannot be transmitted through glass. These vibrations are reflected in *OP* direction and are completely plane-polarized. When beam of light is incident on a glass surface at 57°, then the refracted ray is at 90° to the reflected ray as shown in Fig. 14.8. Thus, the vibrations of incident light which are in the plane of the paper and the direction of propagation of the refracted ray are parallel to each other. So, the vibrations of incident light which are in the plane of the paper can be transmitted through the glass. The refracted ray is partially plane-polarized, i.e., it contains more vibrations which are in the plane of the paper than which are perpendicular to the plane of the paper.

Let the angle of incident of unpolarized light is θ_1 [Fig. 14.8]. *OP* is the reflected ray and therefore $NOP = \theta_1$. Let the angle of refraction be θ_2 and n is the refractive index of glass, then according to Snell's law, we write

$$\frac{\sin\theta_1}{\sin\theta_2} = n \quad \text{...(14.1)}$$

From the ray diagram for polarization by reflection, we see that $\theta_2 = 90 - \theta_1$. Substituting for $\theta_2 = 90 - \theta_1$, in Eq. (14.1), we have

$$\frac{\sin\theta_1}{\sin(90 - \theta_1)} = n$$

or

$$\frac{\sin\theta_1}{\cos\theta_1} = n$$

or

$$\tan\theta_1 = n \quad \text{...(14.2)}$$

Equation (14.2) is known as Brewster's law θ_1, the angle of incident at which the reflected ray is completely plane-polarized, is called the angle of polarization or polarizing angle. For refractive index of glass $n = 1.54$, Eq. (14.2) gives $\theta_1 = 57°$.

14.6 CALCITE CRYSTAL

$CACO_3$ is known as calcite. Its another name is Iceland spar. Calcite crystals is transparent to visible and ultraviolet light. It breaks easily into simple small pieces. It crystallizes in rhombohedron form. Unit cell of calcitecrystal is shown in Fig. 14.9(a).

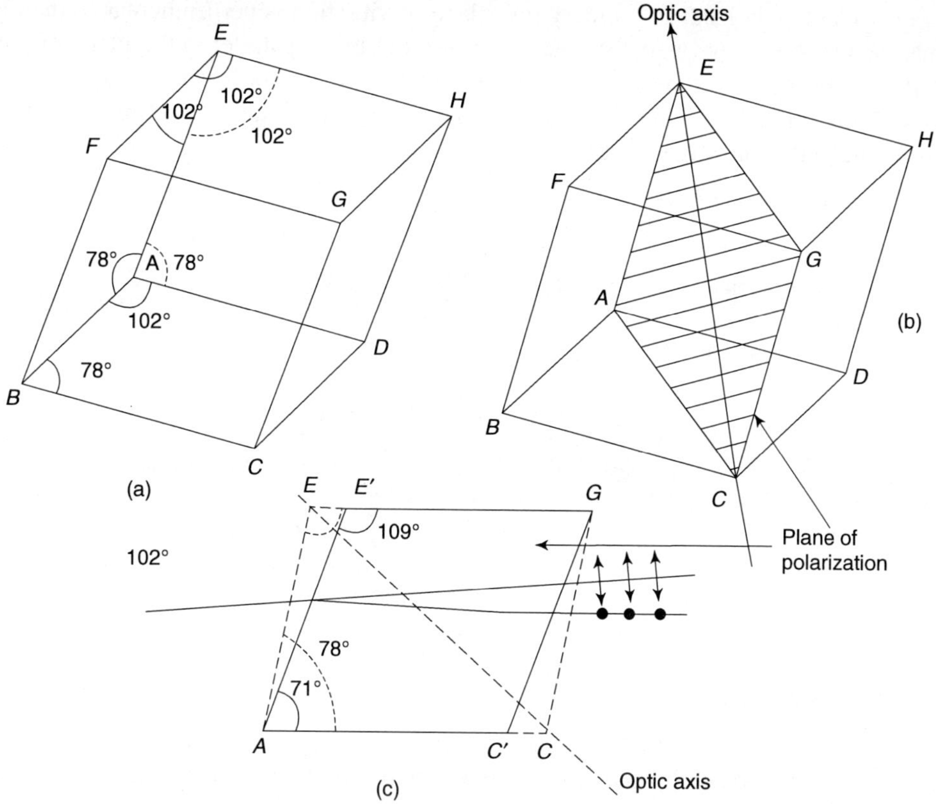

Fig. 14.9 (a) Rhombohedron unit cell of calcite crystal (b) Principal plane *EGCA* (c) Modified principal plane *E′GC′A*

Each face of the crystal is parallelogram with angles 78° and 102°. Two diagonally opposite solid angles formed by obtuse three angles of 102° each are called the blunt corners. The direction along the line joining two diagonally blunt corners is called optic axis of the crystal. The optic axis at the blunt corner makes equal angles with all the three planes. A crystal having only one optic axis is called a uniaxial crystal, e.g., calcite, quartz and tourmaline. On the other hand, a crystal having two optic axes is called biaxial crystal, e.g., mica, sulphur, topaz, aragonite. The two axes in biaxial crystal make certain angle with each other. In Fig. 14.9(b), plane *EACG* perpendicular to opposite faces *ABCD* and *EFGH* having optic axis is called the principal plane. A principal plane which is a section of crystal is called principal section. In Fig. 14.9(c), the principal plane *EACG* is shown. Two end of faces *EA* and *CG* of the crystal are cut down so as to reduce the angle from 78° to 71°.

14.7 DOUBLE REFRACTION

If an unpolarized ray is incident on a crystal of calcite (Iceland spar), it splits into two rays as shown in Fig. 14.10. One is known as the ordinary rayand the other extraordinary ray.

Each ray is plane-polarized. In ordinary ray, plane of vibration is perpendicular to the plane of paper. While in extraordinary ray, the plane of vibration is parallel to the plane of paper. The ordinary ray obeys the normal laws of refraction. The extraordinary ray does not obey the laws of refraction. If an object is viewed through a crystal of calcite, two images are seen. The phenomenon is known as double refraction.

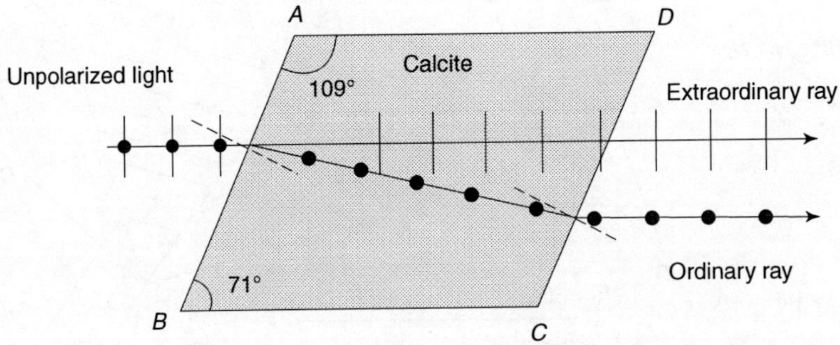

Fig. 14.10 Polarization by double refraction

14.8 NICOL PRISM

Nicol prism shown by Fig. 14.11 is a device for producing two separate plane polarized lights. Nicol prism is made by taking the crystal length and width in the ratio of about 3 : 1.

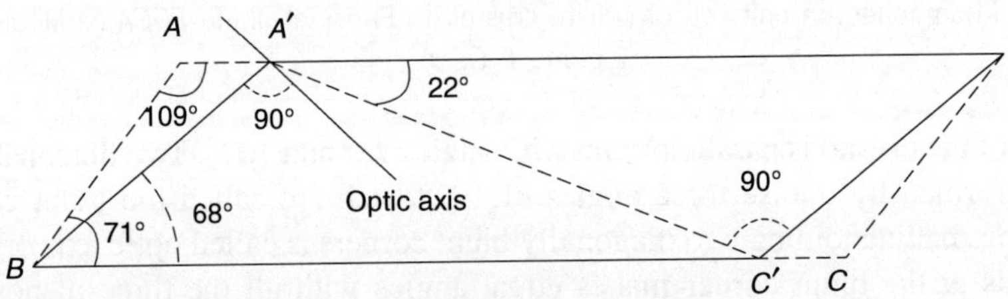

Fig. 14.11 Fabrication of Nicol prism

Two end faces, *AB* and *CD*, of the crystal are cut down so as to reduce the angle from 71° to 68° as shown in Fig. 14.11. The resulting crystal is then cut into two parts along the diagonal *A′ D*. These two pieces of calcite are cemented by Canada balsam. Canada balsam is hydrated calcium carbonate ($CaCO_3$). The refractive index of Canada balsam is less than calcite for ordinary ray, while it is larger than calcite for extraordinary ray. This property of Canada balsam makes it optically less dense than calcite for the ordinary ray, but optically more dense than calcite for extraordinary ray. Thus, the ordinary ray undergoes total internal reflection at the boundary with the balsam as shown in Fig. 14.12. The extraordinary ray passes straight through, as a beam of plane-polarized light.

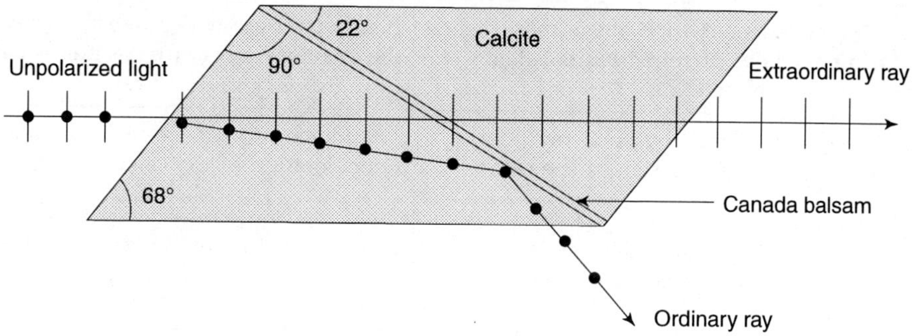

Fig. 14.12 Polarization by double refraction

The two images can be seen with the help of a polaroid sheet. When we rotate the polaroid, the two images disappear at certain points in each rotation.

Summary

Longitudinal waves are waves in which the vibrations are along the direction of travel of the wave.

Transverse waves are the waves in which the vibrations are perpendicular to the direction of travel of the wave.

Transverse waves can be polarized; longitudinal waves cannot.

The vibrations of an unpolarized light wave actually take place in every direction in a plane, which is perpendicular to the direction of propagation (A) transverse wave can be considered having vibrations only in two perpendicular directions to the direction of propagation. The amplitude of each vibration is the same.

Double refraction: The plane of polarization of the E-ray is at 90° to that of the O-ray.

Calcite, quartz, and tourmaline are uniaxial crystals, while topaz and aragonite are biaxial crystals.

Nicol prism is made of calcite crystal. Calcite is a doubly refracting crystal.

Due to difference in velocity of ordinary and extraordinary rays in calcite and quartz, these are used as phase retardation plate.

In calcite crystal velocity of ordinary ray is less than the velocity of extraordinary ray (i.e., $v_0 > v_e$). So calcite is known as a negative crystal.

In quartz $v_0 > v_e$, so quartz is known as positive crystal.

14.9 COMPARISON BETWEEN ORDINARY RAY AND EXTRAORDINARY RAY

S.No.	Ordinary ray	Extraordinary ray
1.	A refracted ray which obeys the ordinary laws of refraction is called ordinary ray or *O*-ray.	A refracted ray which doesnot obey the ordinary laws of refraction is called extraordinary ray or *E*-ray.
2.	The image formed by an *O*-ray does not rotate upon rotation of the calcite crystal.	The image formed by an *E*-ray rotates upon rotation of the calcite crystal.
3.	The speed of an ordinary ray is more than an *E*-ray in positive crystals like quartz.	The speed of an extraordinary ray is more than an *O*-ray in negative crystals like calcite.
4.	*O*-ray is always in the plane of incidence.	*E*-ray is not necessarily always in the plane of incidence.
5.	The speed of *O*-ray is same in all directions.	The speed of *E*-ray is not same in all directions.
6.	The plane of polarization of *O*-ray lies in the principal section, i.e., *O*-ray is polarized in the principal section.	The plane of polarization of *E*-ray is perpendicular to the principal section, i.e., *E*-ray is polarized perpendicular to the principal section.
7.	For an *O*-wave, the plane of the vibration is perpendicular to the principal section.	For an *E*-wave the plane of the vibration lies in the principal section.
8.	The ratio ($\sin i/\sin r$) for *O*-ray is same for different angle of incidence.	The ratio ($\sin i/\sin r$) for *E*-ray is different for different values of angle of incidence.
9.	The refractive index (μ_0) of the crystal is defined as $\mu_o = \dfrac{\text{Speed in vacuum}}{\text{Speed of } O\text{-ray}}$ for an *O*-ray.	The refractive index (μ_e) of the negative crystal is defined as $\mu_e = \dfrac{\text{Speed in vacuum}}{\text{Maximum speed of } E\text{-ray}}$ for an *E*-ray. The refractive index (μ_e) of the positive crystal is defined as $\mu_e = \dfrac{\text{Speed in vacuum}}{\text{Minimum speed } E\text{-ray}}$ for an *E*-ray.

14.10 LAW OF MALUS

In 1809, Etienne Louis Malus established that when a completely plane polarized light beam is incident on the analyser, the intensity of transmitted through the analyser varies as the square of cosine of the angle between the plane of transmission of the analyser and the plane of polarizer.

Let OD is the plane of analyser OP is the amplitude of the incident plane polarized light from a polarizer and θ is the angle between the plane of polarizer and plane of analyser as shown in Fig. 14.13. The amplitude of plane polarized light $OP = a$ can be resolved in two components. One component ($OD = a \cos \theta$) is along the plane of analyser and the other component ($OC = a \sin \theta$) is perpendicular to it. The component ($a \cos \theta$) is transmitted through the analyser.

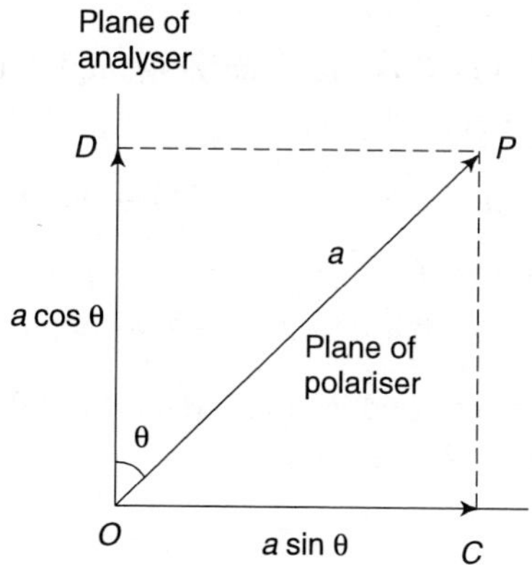

Fig. 14.13

Intensity of the transmitted light through the analyser

$$I_\theta \propto (\text{amplitude})^2$$

or

$$I_\theta = (a \cos \theta)^2 = a^2 \cos^2 \theta$$

or

$$I_\theta = I \cos^2 \theta, \qquad \text{...(14.3)}$$

where, $I = a^2$ is maximum intensity of light.

When $\theta = 0$, i.e., the two planes are parallel

$$I_\theta = I \quad \text{as} \quad \cos \theta = 1$$

When $\theta = \dfrac{\pi}{2}$, i.e. the two plane are perpendicular

$$I_\theta = 0 \quad \text{as} \quad \cos \frac{\pi}{2} = 0$$

Equation (14.3) is called the law of Malus. It states that the intensity of incident polarized light transmitted through the analyser varies as square of cosine of the angle between the plane of polarizer and the plane of analyser. This formula is found correct experimentally in case of two tourmaline and calcite crystals, Nicol prism, etc.

SOLVED EXAMPLES

Example 14.1 Calculate the Brewster angles for the following liquids (a) ethyl alcohol for which μ = 1.361, and (b) carbon tetrachloride for which μ = 1.461.

Solution: Formula used

$$\mu = \tan p,$$

where, μ is the refractive index and P is the polarizing angle or Brewster angle

(a) For ethyl alcohol, μ = 1.361,

$$\mu = \tan p \quad \text{or} \quad 1.361 = \tan p \quad \text{or} \quad p = 53^\circ\ 42'$$

(b) For carbon tetrachloride, μ = 1.461,

$$\mu = \tan p \quad \text{or} \quad 1.461 = \tan p \quad \text{or} \quad p = 55^\circ\ 36'$$

Example 14.2 A glass plate is to be used as a polarizer. Find the angle of polarization for it. Also, find the angle of refraction. Given μ for glass = 1.54.

Solution: Brewster law is

$$\mu = \tan p$$

Putting the value of μ, we get

$$1.54 = \tan p \quad \text{or} \quad p = \tan^{-1} 1.54 = 75^\circ$$

If r is the angle of refraction, then

$$p + r = \frac{\pi}{2} \quad \text{or} \quad r = \frac{\pi}{2} - p = 90^\circ - 57^\circ = 33^\circ$$

r may also found by using Snell's law $\dfrac{\sin p}{\sin r} = \mu$.

Example 14.3 Two polarizing sheets have their polarizing directions parallel to that the intensity of the transmitted light is a maximum. Through what angle must either sheet be turned so that the intensity becomes one-half the initial value?

Solution: Malus formula is

$$I_\theta = I \cos^2 \theta$$

Putting $I_\theta = \dfrac{I}{2}$, we get

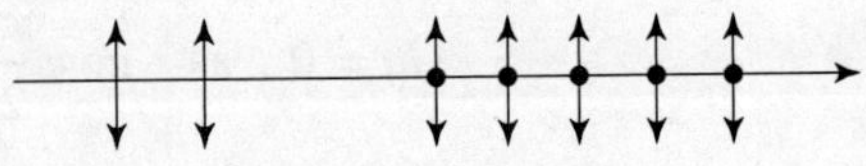

$$\frac{I}{2} = I \cos^2 \theta \quad \text{or} \quad \cos^2 \theta = \frac{1}{2}$$

or $$\cos \theta = \pm\left(\frac{1}{2}\right)^{1/2} \quad \text{or} \quad \theta = \cos^{-1}\left[\pm\left(\frac{1}{2}\right)^{1/2}\right] = \pm 45^\circ, \pm 135^\circ$$

14.11 QUARTER-WAVE PLATE

Quarter-wave plate is a calcite plate cut with optic axis parallel to the surface (Fig. 14.14). When a plane polarized light incident normally on this plate, the light splits up into ordinary (*O*-ray) and extraordinary (*E*-ray) plane polarized lights. The thickness of the crystal plate is such that it introduces a phase difference of $\pi/2$ radian or a path difference of $\lambda/4$ between *E*-ray and *O*-ray. That is why, this calcite plate is called a quarter wave plate. *E*-ray and *O*-ray travel normal to the surface along the same path but with different velocities. *E*-ray travels faster than *O*-ray. As a result a phase difference is introduced between them. Such a plate is known as retardation plate, because velocities of both the polarized light waves within the plate are less than respective velocities in air. The crystal, from which this plate is made, is called *negative crystal.*

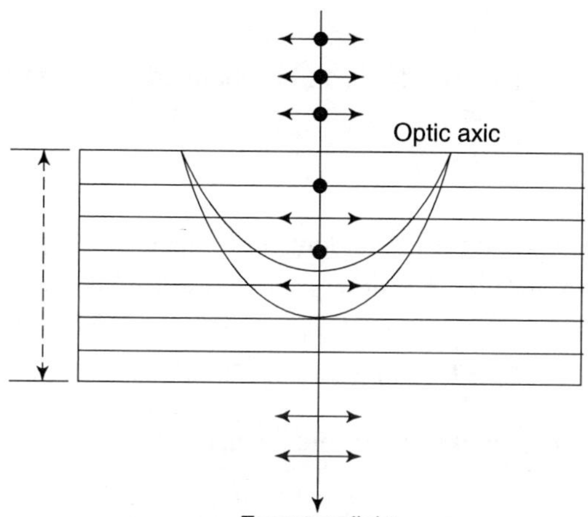

Fig. 14.14

Calculation of the Thickness Quarter Wave Plate

Let t be the thickness of quarter wave plate. If μ_o is the refractive index of crystal *O*-ray and μ_e is the refractive index of the crystal for *E*-ray, then path travelled by *O* and *E* waves in plate of thickness (t) is equivalent to $\mu_o t$ and $\mu_e t$, respectively. So, the path difference between *O* and *E* waves is

$$\Delta = \mu_o t - \mu_e t = (\mu_o - \mu_e)t$$

as $\mu_o > \mu_e$ in calcite. For a quarter wave plate this path differenceis $\lambda/4$, so

$$(\mu_o - \mu_e)t = \frac{\lambda}{4}$$

So the thickness of the quarter wave plate can be obtained by formula

$$t = \frac{\lambda}{4(\mu_o - \mu_e)}$$

14.12 HALF-WAVE PLATE

When the thickness of a calcite crystal plate, cut with its faces parallel to optic axis, introduces a phase difference of π or a path difference of λ/2 between *O* and *E* waves, then it is called a half-wave plate. For a half-wave plate, the path difference between *O* and *E* waves is

$$\Delta = \mu_o t - \mu_e t = (\mu_o - \mu_e)t$$

For a quarter wave plate this path difference is λ/4, so

$$(\mu_o - \mu_e)t = \frac{\lambda}{2}$$

So the thickness of the half-wave plate can be obtained by formula

$$t = \frac{\lambda}{2(\mu_o - \mu_e)} \quad ...(14.4)$$

Example 14.4 Plane polarized light is incident on a piece of quartz cut parallel to the axis. Find the least thickness for which the ordinary and extraordinary rays combine to form plane polarized light.

Given: $\mu_o = 1.5442$, $\mu_e = 1.5533$ and $\lambda = 5 \times 10^{-5}$ cm.

Solution: For half wave the thickness of mica sheet,

$$t = \frac{\lambda}{2(\mu_e - \mu_o)} \text{ in positive crystal}$$

Substituting, the given values, we have

$$t = \frac{5\times10^{-5}}{2(15533 - 1.5442)} = 2.75 \times 10^{-3} \text{ cm}$$

Ans. 2.75×10^{-3} cm

Example 14.5 Calculate the thickness of a mica sheet required for making a quarter wave, λ = 5460 Å. The indices of refraction for the the ordinary and extraordinary rays in mica are 1.586 and 1.592.

Solution: Formula used

For quarter wave the thickness of mica sheet,

$$t = \frac{\lambda}{4(\mu_e - \mu_0)} \text{ in positive crystal}$$

Given: $\lambda = 5460 \times 10^{-8}$ cm, $\mu_0 = 1.586$, $\mu_e = 1.592$

Substituting, the given values, we have

$$t = \frac{5460 \times 10^{-8}}{4\,(1.592 - 1.586)} = 2.275 \times 10^{-3} \text{ cm}$$

Ans. 2.275×10^{-3} cm

Example 14.6 Calculate the thickness of a quarter wave plate made of quartz to be used with sodium light, $\lambda = 6000$ Å, $\mu_0 = 1.544$, $\mu_e = 1.553$

Solution: Formula used

For quarter wave the thickness of mica sheet,

$$t = \frac{\lambda}{4\,(\mu_e - \mu_0)} \text{ in positive crystal}$$

Given: $\lambda = 6000 \times 10^{-8}$ cm, $\mu_0 = 1.544$, $\mu_e = 1.553$

Substituting, the given values, we have

$$t = \frac{6000 \times 10^{-8}}{4\,(1.553 - 1.544)} = 1.667 \times 10^{-3} \text{ cm}$$

Ans. 1.667×10^{-3} cm

Example 14.7 Determine the thickness of a crystal plate which can produce circular polarized light. For calcite, $\mu_e = 1.486$ and $\mu_0 = 1.658$ and $\lambda = 5893 \times 10^{-8}$ cm.

Solution: Circular polarized light is obtained by using quarter-wave plate. If t is the thickness of calcite plate, then

$$t\,(\mu_0 - \mu_e) = \frac{\lambda}{4}$$

The thickness of calcite sheet is

$$t = \frac{\lambda}{4\,(\mu_0 - \mu_e)}$$

Given: $\lambda = 5893 \times 10^{-8}$ cm, $\mu_e = 1.486$, $\mu_0 = 1.658$

Substituting, the given values, we have

$$t = \frac{5893 \times 10^{-8}}{4\,(1.658 - 1.486)} = 8.56 \times 10^{-5} \text{ cm}$$

Ans. 8.56×10^{-5} cm

Example 14.8 Calculate the minimum thickness of a plate if the ordinary and extraordinary rays coming out mix up to form plane polarized light.

Given: $\lambda = 6000 \times 10^{-8}$ cm, $\mu_e = 1.562$, $\mu_0 = 1.552$

Solution: In this case polarized light is obtained when crystal plate acts as half-wave plate. If t is the thickness of calcite plate, then

$$t\,(\mu_e - \mu_0) = \frac{\lambda}{2}$$

The thickness of calcite sheet,

$$t = \frac{\lambda}{2\,(\mu_e - \mu_0)}$$

Given: $\lambda = 6000 \times 10^{-8}$ cm, $\mu_e = 1.562$, $\mu_0 = 1.552$

Substituting, the given values, we have

$$t = \frac{6000 \times 10^{-8}}{2\,(1.562 - 1.552)} = 3 \times 10^{-3} \text{ cm}$$

Ans. 3×10^{-3} cm

Example 14.9 A half-wave plate is constructed for a wavelength of 6000 Å. For what wavelength does it works as a quarter-wave plate.

Solution: The thickness of the sheet in case it acts as a half-wave plate is

$$t = \frac{\lambda}{2\,(\mu_0 - \mu_e)} = \frac{6000 \times 10^{-8}}{2\,(\mu_0 - \mu_e)}$$

The thickness of the sheet in case it acts as a half-wave plate,

$$t = \frac{\lambda}{4\,(\mu_0 - \mu_e)}$$

Putting the value of t, we get

$$\frac{\lambda}{4\,(\mu_0 - \mu_e)} = \frac{6000 \times 10^{-8}}{2\,(\mu_0 - \mu_e)}$$

or

$$\lambda = 12000 \times 10^{-8} \text{ cm} = 12000 \text{ Å}$$

Example 14.10 Calculate the thickness of a doubly refractive plate capable of producing a path difference of $\lambda/4$ between ordinary and extraordinary rays with light of wavelength 5890 Å. The refractive indices for ordinary and extraordinary rays are 1.53 and 1.54, respectively.

Solution: The thickness of doubly refractive plate capable of producing a path difference of $\lambda/4$ is given by

$$t = \frac{\lambda}{4\,(\mu_e - \mu_0)}$$

Given: $\lambda = 5890 \times 10^{-8}$ cm, $\mu_e = 1.54$, $\mu_0 = 1.53$

$$t = \frac{5890 \times 10^{-8}}{4\,(1.54 - 1.53)} = 1.47 \times 10^{-3} \text{ cm}$$

Ans. 1.47×10^{-3} cm

Example 14.11 The values of μ_e and μ_0 for quartz are 1.5508 and 1.5418, respectively. Calculate the phase retardation for λ = 5000 Å when the plate thickness is 0.032 mm.

Solution: The phase relation is given by

$$\delta = \frac{2\pi}{\lambda}\Delta = \frac{2\pi}{\lambda}(\mu_e - \mu_0)\,t$$

Given: $\lambda = 5000 \times 10^{-8}$ cm, $\mu_e = 1.5508$, $\mu_0 = 1.5418$, $t = 0.032$ mm $= 32 \times 10^{-4}$ cm

Putting these values, we get

$$\delta = \frac{2 \times 3.14}{5000 \times 10^{-8}} (1.5508 - 1.5418) \times 32 \times 10^{-4}$$

$$= \frac{6.28}{5000 \times 10^{-8}} \times 0.009 \times 32 \times 10^{-4}$$

$$= 3.617 \text{ rad.}$$

Ans. 3.617 rad.

Example 14.12 Plane polarized light is incident on a plate of quartz cut with faces parallel to optic axis. Calculate the thickness for which the phase difference between the two rays is 60°, where, $\mu_0 = 1.5442$, $\mu_e = 1.5583$ and λ = 5000 Å.

Solution: When plane polarized light is incident on a plate of quartz cut with faces parallel to optic axis, the beam splits up into ordinary and extraordinary rays. The path difference between ordinary and extraordinary rays is given by

$$\Delta = (\mu_e - \mu_0)\,t,$$

where t is the thickness of quartz plate.

The phase difference (δ) corresponding to path difference (Δ) is given by

$$\delta = \frac{2\pi}{\lambda}\Delta = \frac{2\pi}{\lambda}(\mu_e - \mu_0)\,t$$

Given: $\lambda = 5000 \times 10^{-8}$ cm, $\mu_e = 1.5583$, $\mu_0 = 1.5442$, $\delta = 60° = 60 \times \frac{\pi}{180} = \frac{\pi}{3}$

Putting the given values, we get

$$\frac{\pi}{3} = \frac{2\pi}{5000 \times 10^{-8}} (1.5583 - 1.5442)\,t$$

or $$t = \frac{5000 \times 10^{-8}}{3 \times 2 \times 0.0141} = 0.00091 \text{ cm}$$

Ans. 0.00091 cm

14.13 PARALLEL AND CROSSED NICOLS

When the plane polarized light travels inside the crystal its plane of vibration is slowly rotated about the direction of propagation. In the emergent light, the vibrations are in some other plane. This rotation of plane of vibration can be demonstrated by the following experiment. In Fig. 14.15(a) two Nicols are set in a parallel position. When plane polarized light falls on the polarizer (first Nicol), the plane polarized light is divided into two parts: Ordinary ray (*O*-ray) and extraordinary ray (*E*-ray). *O*-ray has plane of vibration perpendicular to the plane of paper shown in Fig. 14.15(a) by solid circle (•). *E*-ray has plane of vibration parallel to the plane of paper shown in Fig. 14.15(a) by perpendicular line (|) with respect direction of propagation of the wave. *O*-ray is refracted out by the polarizer (first Nicol). *E*-ray travels straight and transmits through the analyser (second Nicol) without any deviation. In Fig. 14.15(b) the two Nicols are set in crossed position. The plane polarized light is divided into two parts by the analyser (first Nicol). Ordinary ray (*O*-ray) and extraordinary ray (*E*-ray). *O*-ray is refracted out by the polarizer (first Nicol) as shown in Fig. 14.15(b). *E*-ray travels straight through the

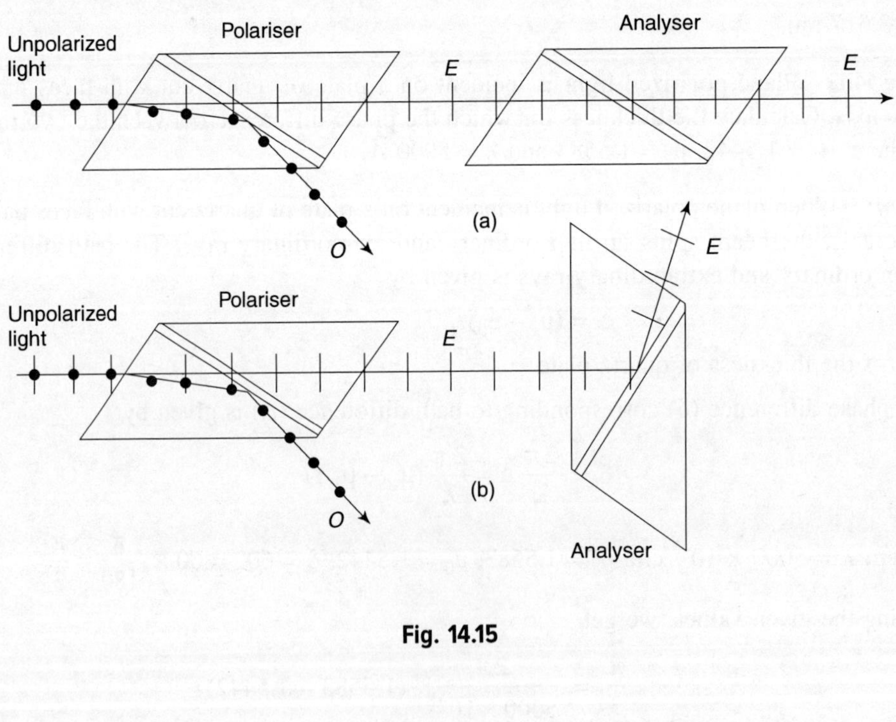

Fig. 14.15

analyser (first Nicol) and is refracted out by the polarizer (second Nicol) as shown in Fig. 4.15(b). When principal plane of second Nicol is perpendicular to the principal plane of the first Nicol, the *E*-ray is not allowed to pass through the second Nicol. The field of view will be completely dark, if viewed by telescope.

14.14 LAURENT'S HALF-SHADE POLARIMETER

Main arts of Laurent's half-shade polarimeter shown in Fig. 14.16 are two Nicol prisms, N_1 and N_2, Glass tube (T) filled with optical active liquid solution and half-shade device. The tube (T) is closed at the ends by cover slips and metal covers. Nicol prism N_1 acts as a polarizer and Nicol prisms N_2 act as a analyzer. This polarimeter is used to measure the angle of rotation on a graduated circular scale produced by optical active solution. The verniers are also provided to read the fraction of a degree. When polarimeter is calibrated to mesure the percentage of cane sugar in solution, it is called saccharin meter. Polarimeter is also used to find the specific rotation of sugar solution and its concentration.

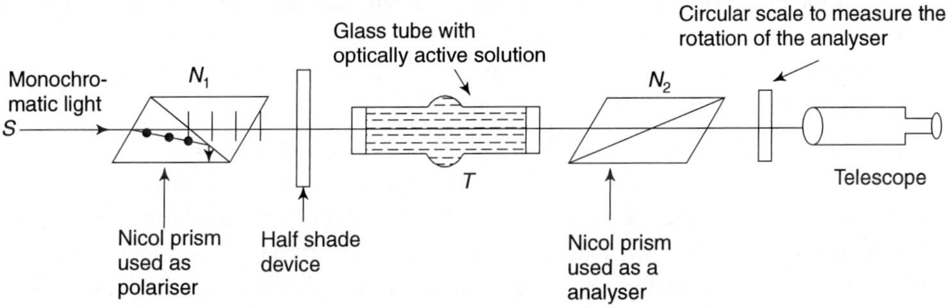

Fig. 14.16

Working of Laurent's Half-Shade polarimeter: When monochromatic light incident on polarizer N_1, it is polarized with its vibrations in the principal plane of N_1. This plane polarized light passes through a half-shade device and then through T containing the sugar solution, an optical active substance. A large diameter in the middle of hollow glass tube (T) is made to store air bubbles when tube (T) is filled with a liquid, so that no air bubble will be there in the path of light. The light emerging from tube T passes through analyser, N_2 and is viewed through the Galelian telescope by focusing it on the half shade. A circular scale is attached to telescope to measure the rotation produced by liquid filled in the tube.

Working of half-shade device: The Laurent's half-shade plate consists of a semicircular half-wave plate *ABC* of quartz (cut parallel to optic axis) and a semicircular glass plate *ADC* cemented along the diameter *AC* as shown in Fig. 14.17(a). The thickness of the semicircular quartz plate is such that it introduces a phase change of π (or path difference $\lambda/2$) between extraordinary and ordinary rays passing through it when polarised light from N_1 falls on it normally. Thus, half-shade device is half-wave plate. The thickness of the glass plate is such that it absorbs and transmits the same amount of light as the quartz plate.

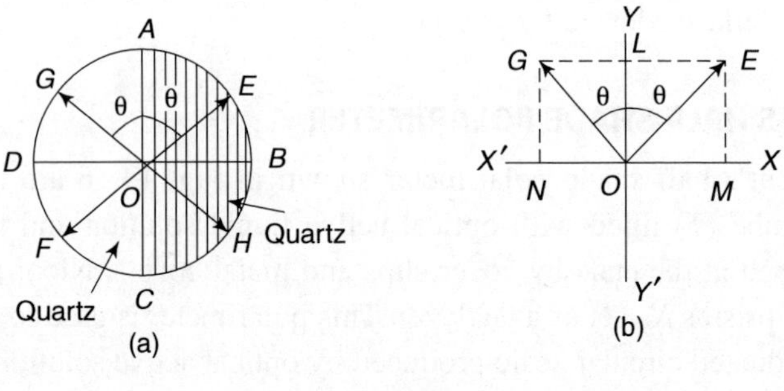

Fig. 14.17

Let the principal plane of N_1 be parallel to *EF*. Ten the plane of vibration of the plane polarized light incident normally on half-shade plate will be parallel to *EF*, which makes an angle θ with *AC* in Fig. 14.17(a) or with *YY′* in Fig. 14.17(b), an optic axis. The vibrations of light emerging from semicircular glass plate remain along the plane *EF* in Fig. 14.17(a) or along *OE* in Fig. 14.17(b). The quartz plate divides the polarized light into two components – one component *OM* called ordinary ray with vibrations in the direction of *X′X* [Fig. 14.17(b)], i.e., perpendicular to optic axis *YY′* and the other component *OL* called extraordinary ray with vibrations parallel to optic axis, i.e., along *Y′Y*. The two rays travel along the perpendicular direction to half-shade plate with different speeds. The ordinary ray moves faster than extraordinary ray. A phase difference of π is introduced between two emerging rays from quartz plate. Due to π phase difference, the direction ordinary ray or *OM* component is reversed, i.e., *OM* component now becomes *ON* [Fig. 14.17(b)]. Because of this, emerging component from quarts plate *OG* will be the resultant of *OL* component (extraordinary ray) and *ON* component *ON* (ordinary ray). The resultant wave *OG* makes an angle θ with *Y*-axis [Fig. 14.17(b]. Thus, the vibration of the beam emerging out of quartz will be along *GH* in Fig. 14.17(a) making an angle θ with *AC*. Thus, effect of quartz plate is to rotate the plane of polarization by an angle, 2θ.

There are two plane polarized beams emerging from half-shade device: one from glass plate with vibrations in the plane *OE* and the second from quartz plate with vibrations in the plane, *OG*. If the principal plane of the analyzer Nicol prism is parallel to *EOF*, then light emerging from semicircular glass plate will pass from analyser completely, while light emerging from semicircular quartz plate will pass only partially. Therefore, the glass half will appear brighter than the quartz half. On the other hand, if the principal plane of the analyser Nicol prism is parallel to *GOH*, the light emerging from semicircular quartz plate will pass through the analyser completely, while light emerging from the glass plate will pass only partially. Then, the quartz half will appear brighter than the glass half. If, the principal plane of analyser is parallel to *AOC* (*Y*-axis), it is inclined equally with two plane polarized beams: one from glass half and the other from quartz half, then equal amount of light will pass through the analyser. Therefore, the field of view will be seen from telescope equally bright. Thus, the half shade is used to divide the field of view equally in two halves. When the analyzing Nicol prism is rotated even slightly from the position where two parts of half-shade device are equally bright, a significant change in the intensity of two halves is seen.

Determination of specific rotation: In order to find the specific rotation (S) of a given substance, the following formula is used.

$$S = \frac{\theta}{l \times C}$$

where, θ is the angle of rotation in degrees, *l* is the length of the solution in centimeters and *C* is the concentration of solution in gm per c.c. The length of the solution can be measured by scale. We prepare the solution whose concentration © is known. The value of θ is obtained as follows:

First of all fill the glass tube with water and place it at position as shown in Fig. 10.15. Focus the telescope on the half-shade device and rotate the analyser till we obtain equally bright field of view. We note down two readings of with help of verniers on the circular scale. Now, fill the tube with the optically active solution of known concentration and place it at its position. We note that in this position, the field of view of two halves are not of equal intensity. Then, we rotate the analyser till we get whole field of view of equal intensity. We again note the new two readings using the same verniers. The difference of new and old readings gives the angle of rotation produced by solution. The procedure is repeated by taking solutions of different concentration. The graph between concentration and angle of rotation is a straight line.

Concentration of sugar solution: To find out the concentration of the given sugar solution, the experiment is performed as discussed above to find the specific rotation, θ. Length of the solution is measured with a scale. The specific rotation at the temperature of the solution can be obtained from the given table of constants. Then, the concentration of the given solution can be calculated from the formula,

$$C = \frac{\theta}{s \times l}$$

Definition Specific rotation: The formula for specific rotation,

$$S = \frac{\theta}{l \times C}$$

where, θ is the angle of rotation in degrees, *l* the of rotation in decimeter and *C* is concentration of solution in gm per c.c.

When $l = 1$ decimeter and $C = 1$ gm/c.c., then

$$S = \theta \text{ in degrees}$$

Now, the specific rotation (S) of a substance at a particular temperature for a given wavelength of light is defined as the rotation (in degrees) of the plane of vibration of polarized light by one decimeter length of solution having concentration 1 gm/c.c.

14.15 BIQUARTZ POLARIMETER

Biquartz polarimeter is used for finding the angle of rotation produced by an optically active substance with good accuracy. Biquartz poparimeter consists of two semicircular plates of quartz. The plate *CYD* is laevorotatory (left-handed quartz) and is denoted by *L*. The other plate *CYD* is dextrorotatory (right-handed) and is denoted by *R*. Both plates are cut perpendicular to the optic axis and joined together along the diameter *CD* as shown in Fig. 14.18. The thickness of each plate is taken 3.75 mm, because the plates of this thickness rotate the plane of polarization for yellow light by 90°.

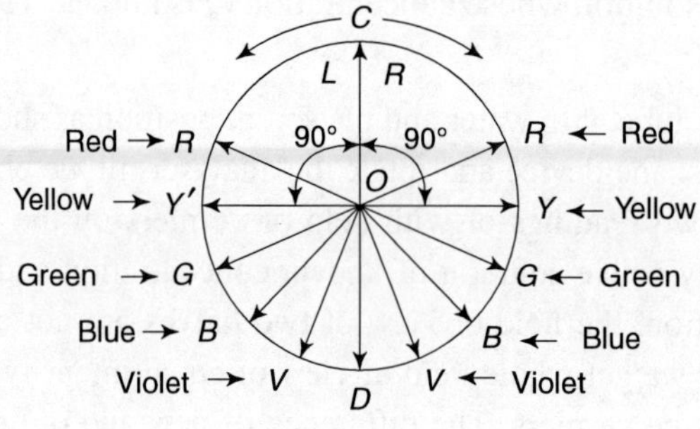

Fig. 14.18

When plane polarized white light is incident on biquartz normally, the waves of different colours rotate by different angles on passing through the biqartz polarimeter, because of different wavelengths. The phenomenon of splitting the vibrations of different colours in different angles is called rotator dispersion. Because of dispersion of white light into different colours, biquartz polarimeter is also called quartz biprism. The rotation of yellow colour is 90° from the axis of analyser, i.e., vibrations of yellow colour are along straight line *YOY′*.

If the principal plane of the analyser is parallel to *COD*, the yellow light will not be transmitted through the analyser, because the yellow light is polarized in perpendicular direction of the principal plane of the analyser. The colour of the two halves is same. Since all colours except the yellow colour upto some extent, the two halves have a grayish violet (GV) tint [Fig. 14.19 (a)], called the *tint of the passage* or *sensitive tint*. When the analyser is rotated in anticlockwise direction from position *COD*, right half looks blue while the left half looks red [Fig. 14.19(b)]. If the analyser is rotated in clockwise direction from position *COD* the colours are changed, i.e., the right half looks red and the left half looks blue [Fig. 14.19(c)]. The position of sensitive tint is used for finding the optical rotation with good accuracy.

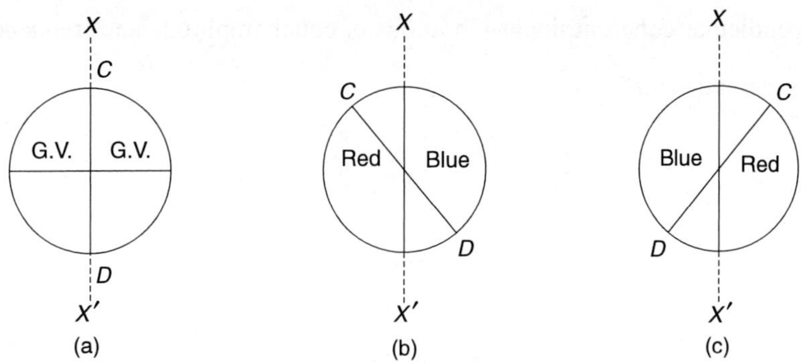

Fig. 14.19 Dispersion of colours through biquartz plate

14.16 PHOTOELASTICITY

Some materials, including glass, Bakelite and some other plastics, become doubly refracting when subjected to mechanical stress with help of polarized light. This phenomenon is called photoelasticity. When a plane polarized light is incident on two crossed Nicols, then the second Nicol (analyser) does not allow to pass the light and the field of view looks dark. When a piece of Bakelite or glass is placed between two crossed Nicol, it does not change the field of view, i.e., the field of view remains dark. When mechanical stress is applied to the material the stress, the emerging light is elliptically polarized and gets partly transmitted through the ananyser (second Nicol), thus illuminating the field of view. Different colours of light are affected differently because they have different wavelengths. Some very beautiful effects can be obtained. Models of components such as gear teeth, turbine blades or hooks can be made by plastic materials. After placing these models between two crossed Niclols and putting stress on these models, the interference patterns in them are observed to check their design. These patterns can be seen in car windscreens. The patterns stored in them are due to the stresses produced during their manufacture.

On putting a model of photoelastic materials between two crossed Nicols, if observed field of view is dark, then it indicates that the model is strain free. Photoelasticity is used in the measurement of strains in the buildings. A model of the building made of zylonite becomes doubly refracting when it is strained. It is also used for checking strains in glass before making optical instruments of it.

14.17 PRODUCTION OF CIRCULARLY POLARIZED LIGHT

14.17.1 Principle

When two mutually perpendicular coherent linear vibrations of equal amplitudes and frequency differing in phase by $\frac{\pi}{2}$ superimpose each other, then the resultant is circular motion. The two mutually perpendicular coherent linear vibrations of equal amplitude and frequency differing in phase by $\frac{\pi}{2}$ can be written as

$$x = \frac{a}{\sqrt{2}} \sin\left(\omega t + \frac{\pi}{2}\right) \qquad ...(14.5)$$

or

$$x = \frac{a}{\sqrt{2}} \cos(\omega t) \qquad ...(14.6)$$

$$y = \frac{a}{\sqrt{2}} \sin(\omega t) \qquad ...(14.7)$$

The sum of squares of x and y is

$$x^2 + y^2 = \frac{a^2}{2}\cos^2(\omega t) + \frac{a^2}{2}\sin^2(\omega t)$$

or

$$x^2 + y^2 = \frac{a^2}{2} \qquad ...(14.8)$$

This is equation of a circle of radius $\frac{a}{\sqrt{2}}$.

14.17.2 Production

The experimental setup for the production of circularly polarized light is shown in Fig. 14.20. The components of the experimental setup are two Nicol prisms, N_1 and N_2, one quarter-wave plate, two tubes T_1 and T_2 mounted on a stand. The plane polarized light from Nicol prism N_1 falls on another Nicol prism N_2 placed at a certain distance in the crossed position. Nicol prisms N_2 is placed in crossed with respect to Nicol prism N_1, therefore, no light emerges out from second Nicol prism. Now in this position, a quarter-wave plate mounted on the tube T_1 is held normal to the ray of light coming from Nicol N_1 between two Nicols.

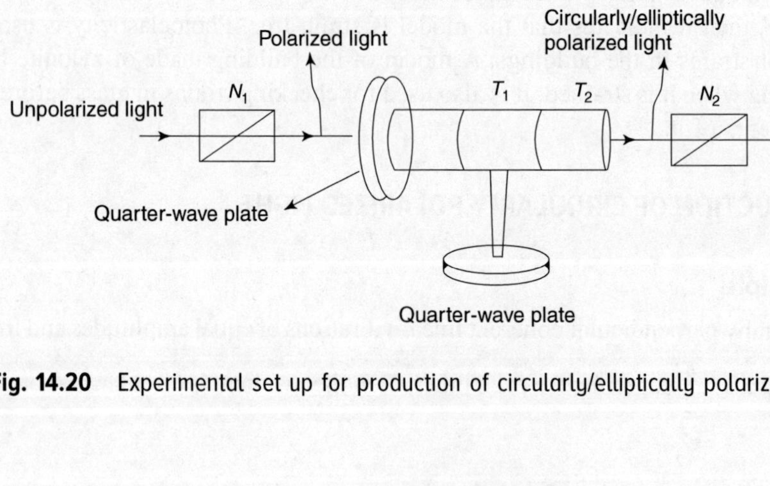

Fig. 14.20 Experimental set up for production of circularly/elliptically polarized light

Tube T_1, can be rotated, while the outer tube T_2 is kept fixed. The quarter-wave plate mounted on tube T_1 can be rotated about the horizontal axis at any desired angle. When we introduce a quarter-wave plate between N_1 and N_2, then we observe that some light is emerging out from Nicol prism N_2. Now, the quarter-wave plate mounted on tube T_1 is rotated till no light emerges out from Nicol prism N_2. In this position, the vibrations of plane polarized light incident normally on the quarter-wave plate are along the optic axis of the quarter-wave plate and perpendicular to Nicol prism N_2. Now, we rotate the quarter-wave plate mounted on tube T_1 by angle 45°. At this position, the vibrations of the plane polarized light incident normally on the quarter-wave plate make an angle 45° with the optic axis of the quarter-wave plate. As a result, the incident polarized light is split up into two rectangular components – the O-ray and the E-ray having equal amplitudes (because sin 45° = cos 45°) and equal frequency. Because, the quarter-wave plate introduces a phase difference of $\frac{\pi}{2}$ between the emergent O-wave and the E-wave, so the resultant beam emerging from the quarter-wave plate is circularly polarized as shown in Fig. 14.21.

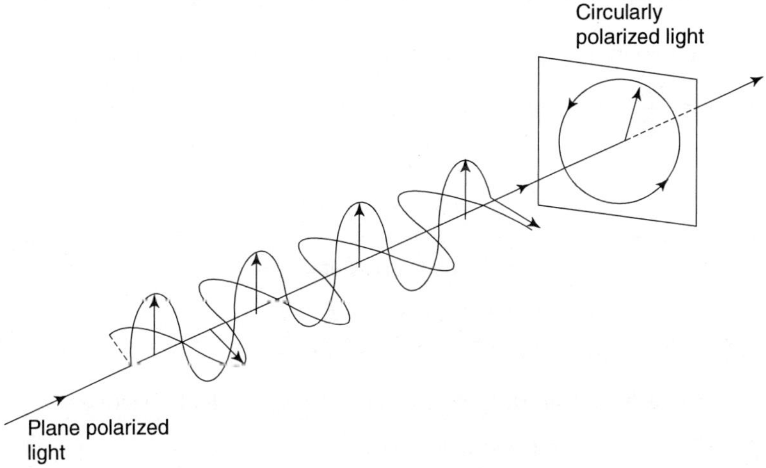

Fig. 14.21 Production of circularly polarized light by a quarter wave plate when the plane polarized light is incident on the quarter wave plate an angle of 45° with the optic axis

14.17.3 Analysis of Circularly Polarized Light

To analyze circularly polarized light, the light is incident on a Nicol prism. The intensity of the emergent light beam remains uniform when we rotate the Nicol prism. The emerging light from Nicol prism remains uniform in case of both unpolarized and circularly polarized light. Therefore, from the observation that light emerging from Nicol prism is uniform, we cannot conclude whether incident light is unpolarized or circularly polarized.

To decide whether incident light is unpolarized or circularly polarized, the beam is incident first on a quarter-wave plate and then on a Nicol prism. The circularly polarized after passing through a quarter-wave plate becomes plane polarized. Therefore, when the light emerging from the quarter-wave plate is incident on a rotating Nicol prism, the field of view becomes dark twice during one complete rotation of the Nicol prism. This observation shows that the incident original light is circularly polarized. On the other hand, when the emerging beam from the quarter-wave plate is incident on a rotating Nicol prism, the field of view does not become dark twice during one complete rotation of the Nicol prism, indicates that the incident original light is unpolarized. Therefore, we conclude that incident light on the quarter-wave plate is circularly polarized, if the emerging light from the quarter-wave plate becomes dark twice in one complete rotation after passing through the rotating Nicol.

14.18 PRODUCTION OF ELLIPTICALLY POLARIZED LIGHT

14.18.1 Principle

Two mutually perpendicular coherent linear vibrations of equal frequency differing in phase by $\frac{\pi}{2}$ and having different amplitude when superimpose to each other gives rise to elliptical motion of light. The two mutually perpendicular coherent linear vibrations of equal frequency differing in phase by $\frac{\pi}{2}$ having different amplitudes can be represented as

$$x = A \cos\theta \sin\left(\omega t + \frac{\pi}{2}\right) \quad \text{...(14.9)}$$

or

$$x = A \cos\theta \cos(\omega t) \quad \text{...(14.10)}$$

$$y = A \sin\theta \sin(\omega t), \quad \text{...(14.11)}$$

where, θ is the angle between vibrations of light and optic axis.

Let $a = A \cos\theta$ and $b = A \sin\theta$, then Eqs. (14.10) and (14.11) becomes

$$x = a \cos(\omega t) \quad \text{...(14.12)}$$

$$y = b \sin(\omega t) \quad \text{...(14.13)}$$

Sum of squares of $\frac{x}{a}$ and $\frac{y}{a}$ can be written as

$$\frac{x^2}{a^2} + \frac{y^2}{b^2} = \cos^2(\omega t) + \sin^2(\omega t)$$

or

$$\frac{x^2}{a^2} + \frac{y^2}{b^2} = 1 \quad \text{...(14.14)}$$

This is equation of an ellipse of radius with, $a = A \cos\theta$ and $b = A \sin\theta$ as major and minor axes, respectively. Angle θ should be other than 0°, 45°, 90°.

14.18.2 Production

The experimental setup for the production of elliptically polarized light is shown in Fig. 14.21. The components of the experimental setup are two Nicol prisms N_1 and N_2, one quarter-wave plate, two tubes T_1 and T_2 mounted on a stand. The plane polarized light from Nicol prism N_1 falls on another Nicol prism N_2 placed at a certain distance in the crossed position. Nicol prism N_2 is placed in crossed with respect to Nicol prism N_1, therefore, no light emerges out from second Nicol prism. Now in this position, a quarter-wave plate mounted on the tube T_1 is held perpendicular to the ray of light coming from Nicol N_1.

Tube T_1, can be rotated, while the outer tube T_2 is kept fixed. Therefore, the quarter-wave plate mounted on tube T_1 can be rotated about the horizontal axis at any desired angle. When we introduce a quarter-wave plate between two Nicol prisms, N_1 and N_2, then we observe that some light is emerging out from Nicol prism N_2. Now, the quarter-wave plate mounted on tube T_1 is rotated till no light emerges out from N_2. In this position, the vibrations of plane polarized light incident normally on the quarter-wave plate are along the optic axis of the quarter-wave plate and perpendicular to N_2. Now, we rotate the quarter-wave plate mounted on tube T_1 by angle 30°. At this position, the vibrations of the plane polarized light incident normally on the quarter-wave plate make an angle 30° with the optic axis of the quarter-wave plate. As a result, the incident polarized light is split up into two rectangular components – the *O*-ray and the *E*-ray having unequal amplitudes $A \sin 30°$ and $A \cos 30°$ (because sin 30 $\neq$ cos 30°), respectively and equal frequency. Because, the quarter-wave plate introduces a phase difference of $\frac{\pi}{2}$ between the emerging *O*-wave and the *E*-wave, so the resultant beam emerging fromthe quarter-wave plate is elliptically polarized light as shown in Fig. 14.22.

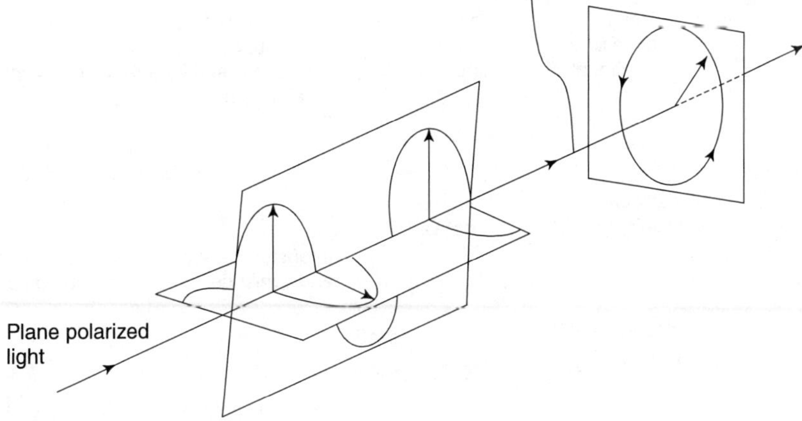

Fig. 14.22 Production of elliptically polarized light by a quarter-wave plate when the plane polarized light is incident on the quarter wave plate an angle excluding 0°, 45° and 90° with the optic axis

14.18.3 Analysis of Elliptically Polarized Light

To analyze elliptically polarized light, the light is incident on a Nicol prism. The intensity of the emergent light beam varies between minimum to maximum when we rotate the Nicol prism. Here minimum is not equal to zero. The emerging light from Nicol prism varies between minimum to maximum in case of partially polarized (a mixture of polarized and unpolariand light) elliptically polarized light. Therefore, from the observation that light emerging from Nicol prism varies between minimum to maximum, we cannot conclude whether incident light is partially polarized or elliptically polarized.

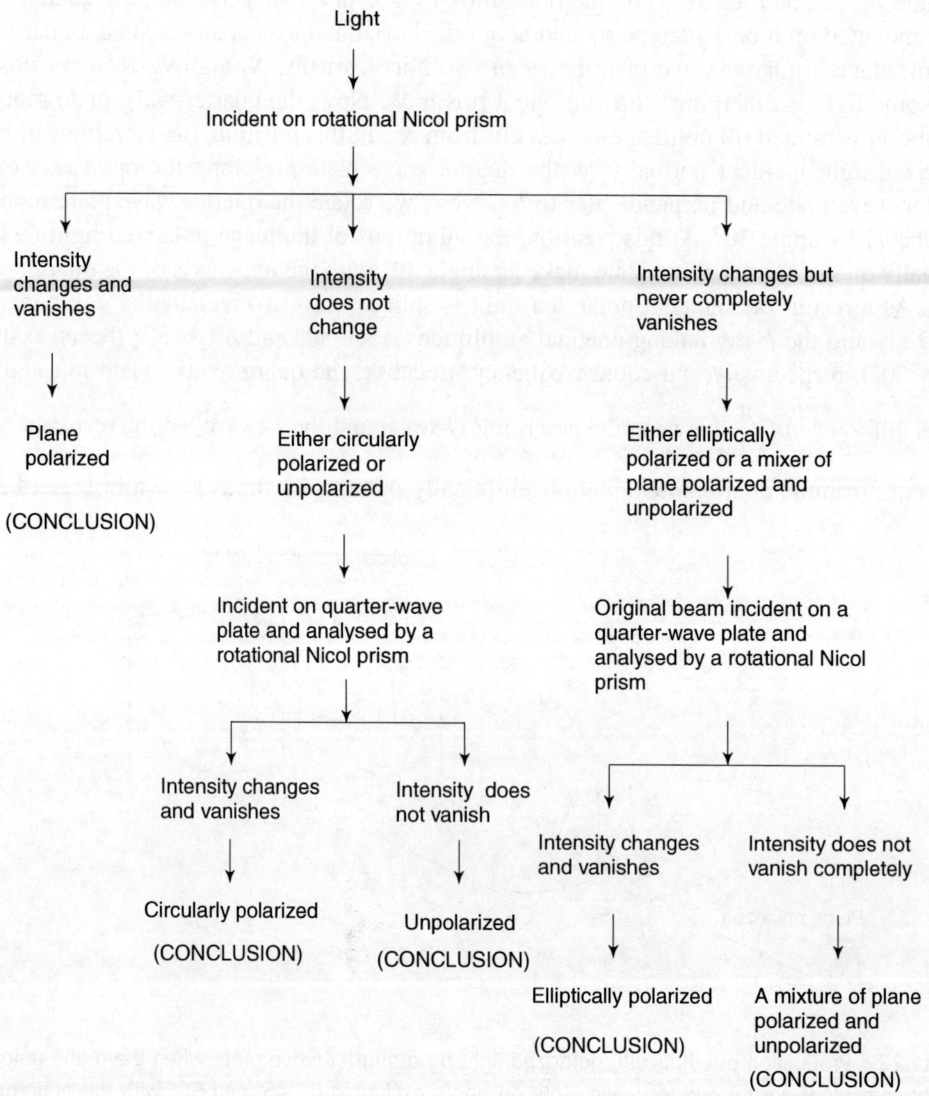

To decide whether incident light is partially polarized or elliptically polarized, the beam is incident first on a quarter-wave plate and then on a Nicol prism. The elliptically polarized after passing through a quarter-wave plate becomes plane polarized. Therefore, when the light emerging from the quarter-wave plate is incident on a rotating Nicol prism, the field of view becomes dark twice during one complete rotation of the Nicol prism. This observation shows that the incident original light is elliptically polarized. On the other hand, when the emerging beam from the quarter-wave plate is incident on a rotating Nicol prism, the field of view does not become dark twice during one complete rotation of the Nicol prism, indicates that the incident original light is partially polarized. Therefore, we conclude that incident light on the quarter-wave plate is elliptically polarized, if the emerging light from the quarter-wave plate becomes dark twice in one complete rotation after passing through the rotating Nicol.

Formulas at a Glance

S.No.	Subject	Explanation
1.	Live index of glass	$n = \dfrac{\sin\theta_1}{\sin\theta_2}$, where, θ_1 and θ_2 are angle of incident and refraction, respectively.
2.	Brewster's law	$n = \tan\theta_1$ In this formula, angle θ_1 is the angle of incident at which the reflected ray is completely plant-polarized Brewster's angle (θ_1) is called the angle of polarization or polarizing angle.
3.	Intensity of the transmitted light through the anlyser is given by formula	$I_\theta = I\cos^2\theta$. It is called law of Malus. Here, I = amplitude^2 is maximum intensity of light source.
4.	The thickness of the quarter-wave plate	$t = \dfrac{\lambda}{4(\mu_o - \mu_e)}$, where, μ_o is the refractive of the crystal for O-ray and μ_e is the refractive index of the crystal for E-ray. λ is wavelength of incident light.
5.	The thickness of the half-wave plate	$t = \dfrac{\lambda}{2(\mu_o - \mu_e)}$, where, μ_o is the refractive index of the crystal for O-ray and μ_e is the refractive index of the crystal for E-ray, l is wavelength of incident light.
6.	Specific rotation	$S = \dfrac{\theta}{l \times C}$, where, θ is the angle of rotation in degrees, l the of rotation in decimeter and C is concentration of solution in gm per c.c.

SOLVED EXAMPLES

14.1 Calculate the specific rotation of the plate if polarization is turned through 30, after traversing 25 cm length of 30% sugar solution.

Solution: Formula used

$$S = \frac{10\,\theta}{lC},$$

where, the length of solution is l cm.

$$S = \frac{10 \times 30}{25 \times 0.3} = 40°$$

14.2 A sugar solution of specific rotation 52 per decimetre per gm/c.c. causes a rotation of 12° in a column of 10 cm long. What is the concentration of the solution?

Solution: Formula used

$$S = \frac{10\,\theta}{lC},$$

where, the length of solution is l cm.

The concentration of solution,

$$C = \frac{10\,\theta}{Sl} = \frac{10 \times 12}{52 \times 10} = 0.23 \text{ gm/c.c.}$$

Ans. $C = 0.23$ gm/c.c.

14.3 Calculate the specific rotation if the plane of polarization is turned through 26.4°, traversing 20 cm length of 20% sugar solution.

Solution: Formula used

$$S = \frac{\theta}{l \times C}$$

Given: $\theta = 26.4°$, $l = 20 \text{ cm} = \frac{20}{10} = 2$ decimetre

$$C = 20\% = 0.2 \text{ gm/cm}^2$$

$$S = \frac{26.4}{2 \times 0.2} = 66°$$

Ans. Specific rotation, $S = 66°$.

14.4 A tube 20 cm long containing sugar solution rotates the plane of polarization through an angle of 13.2°. If the specific rotation is 66, find the amount of sugar present in one litre of the solution.

Solution: Formula used

$$S = \frac{10\,\theta}{lC},$$

where the length l is in cm.

The concentration of solution is

$$C = \frac{10\,\theta}{Sl} = \frac{10 \times 13.2}{66 \times 20} = 0.1 \text{ gm/c.c.}$$

The amount of sugar present in one liter of solution is 0.1 × 1000 = 100 gm.

Ans. 100 gm

14.5 A length of 25 cm of a solution, containing 50 gm of solute per litre causes a rotation of the plane of polarization of light by 5°. Find the rotation of plane of polarization by a length of 75 cm of a solution containing 100 gm of a solute per litre.

Solution: Formula used

$$S = \frac{\theta}{l\,C}$$

Given: $\theta = 5°$, $l = 25$ cm, $C = 50$ c.c.

$$\theta = ?,\ l = 75 \text{ cm},\ C = 100 \text{ c.c.}$$

$$S = \frac{5}{25 \times 50} = \frac{1}{250}$$

$$\theta = SlC = \frac{1}{250} \times 75 \times 100 = 30°$$

Ans. Rotation of plane is 30°.

14.6 A tube of sugar solution 20 cm long is placed between crossed Nicols and illuminated with light of wavelength 6×10^{-5} cm. If the capital of rotation produced is 13° and specific rotation is 65°, determine the strength of the solution.

Solution: Formula used

$$C = \frac{\theta}{lS}$$

Given: $\theta = 13°$, $l = 20$ cm = 2 decimetre, $S = 65°$

Find C = ?

$$C = \frac{13}{2 \times 65} = 0.1 \text{ gm/c.c.} = 100 \text{ gm/litre} = 10\%$$

Ans. 10%

14.7 80 gm of impure sugar when dissolved in one litre of water gives an optical rotation of 9.9 when placed in a tube of length 20 cm. If the specific rotation of sugar is 66 , find the percentage purity of the sample.

Solution: Formula used

$$C = \frac{\theta}{lS}$$

Given: $\theta = 9.9°$, $l = 20$ cm = 2 decimeter, $S = 66°$ (decimetre)$^{-1}$ (gm/c.c.)$^{-1}$

Find C = ?

$$C = \frac{9.9}{2\times 66} = 0.075 \text{ gm/c.c.} = 75 \text{ gm/litre}$$

Ans. 75 gm/litre

14.8 Calculate the specific rotation of sugar solution from the following data:

Length of the tube containing the solution 22 cm, volume of the solution equal to 88 c.c., amount of sugar in the solution is 6 gm and angle of rotation is 9°54′.

Solution: Formula used

$$S = \frac{\theta}{lC}$$

Given: $\theta = 9°\ 54' = 9 + \frac{54°}{60} = 9.9°$, l = 22 cm = 2.2 decimetre, $C = \frac{6}{88} = \frac{3}{44}$ gm/cc

Find S = ?

$$S = \frac{9.9\times 44}{2.2\times 3} = 66°$$

Ans. $S = 66°$

14.9 A certain length of 5% solution causes the optical rotation of 20°. How much length of 10% solution of the same substance will cause 35° rotation?

Solution: Formula used

$$S = \frac{}{lC}$$

Given: $\theta = 20°$, let length is l_1, C = 5% = 50 gm/litre

Find S = ?

$$S = \frac{20}{l_1 \times 50}$$

Given: $\theta = 35°$, C = 10% = 100 gm/litre

Find l_2 = ?

$$S = \frac{35}{l_2 \times 100} = \frac{20}{l_1 \times 50}$$

or $$l_2 = \frac{35\times 50}{20\times 100}\ l_2 = \frac{7}{8} l_1$$

Ans. $l_2 = \frac{7}{8} l_1$

14.10 A sugar solution in a tube of length 20 cm produces optical rotation of 13°. The solution is then diluted to one-third of its previous concentration. Find optical rotation produced by 30 cm long tube containing the dilute solution.

Solution: Formula used

$$S = \frac{\theta}{I \times C}$$

Given: $\theta = 13°$, $l = 20$ cm $= \frac{20}{10} = 2$ decimetre, let C_1 be the concentration of solution.

$$S = \frac{13}{2 \times C_1}$$

Given: $\theta = ?$, $l = 30$ cm $= \frac{30}{10} = 3$ decimetre, Now the concentration of solution is $\frac{C_1}{3}$.

$$S = \frac{\theta}{3 \times \frac{C_1}{3}} = \frac{\theta}{C_1} = \frac{13}{2 \times C_1} \quad \text{or} \quad \theta = \frac{13}{2} = 6.5°$$

Ans. 6.5°

14.11 20 cm length of a certain optically active solution causes right-handed rotation 40° and 30 cm of another solution causes left-handed rotation 24°. What will be the optical rotation produced by 30 cm length of the mixture of above solution in volume ratio 1 : 2 and the length of the tube is 30 cm.

Solution: Here, we assume that the tube is filled with 10 cm of first solution and 20 cm length of second solution because they are in volume ratio 1 : 2 and the length of the tube is 30 cm.

Optical rotation produced by first rotation $= 40 \times \frac{10}{20}$ (right-handed) $= -20°$

Optical rotation produced by second rotation $= 24 \times \frac{20}{30}$ (left-handed) $= +16°$

Total optical rotation $= -20° + 16° = -4°$

Ans. The resultant optical rotation is 4° right-handed.

QUESTIONS FOR EXAMINATION

14.1 What are transverse and longitudinal waves?

14.2 What is classification polarization of light?

14.3 Show transverse waves of a string.

14.4 What is polarization?

14.5 Discuss two methods for getting plane polarized light.

14.6 What is the difference between ordinary ray and extraordinary ray?

14.7 What is law of law of Malus?

14.8 What is the difference between quarter-wave plate and half-wave plate?

14.9 What is the difference between parallel and crossed Nicols?

14.10 What is Laurent's half-shade polarimeter? Discuss its use and working.

14.11 What is biquartz polarmeter?

14.12 What is photoelasticity?

14.13 What is Nicol prism? Discuss one use of it.

14.14 How Nicol prism is used to know circularly polarized light?

14.15 How Nicol prism is used to know illiptically polarized light?

14.16 What is circularly polarized light? Discuss the method for production of circularly polarized light?

14.17 What is elliptically polarized light? Discuss the production of elliptically polarized light.

14.18 Discuss the method for production of circularly polarized light. How do you know that it is circularly polarized light?

14.19 Discuss the production of elliptically polarized light. How do you know that it is elliptically polarized light?

OBJECTIVE QUESTIONS

14.1 Which is (are) light vector(*s*)

(a) Both Electric and magnetic vectors of light wave.

(b) Electric vector of light

(c) Magnetic vector of light

(d) Unit vector in the direction of propagation of the wave

14.2 Which of the phenomenon indicates that the light waves are transverse waves?

(a) Interference (b) Polarization (c) Reflection (d) Diffraction

14.3 Plane polarization of light is perpendicular to

(a) Magnetic vector of light wave (b) Electric vector of light wave

(c) Direction of propagation (d) The plane of vibration

14.4 What is plane of incident?

(a) The plane containing the incident ray and normal to the surface at point of incident

(b) The plane containing the light vector and normal to the surface at the point of incident

(c) The plane containing the refracted ray and incident ray

(d) The plane at which light incident

14.5 What is plane of reflection?

(a) The plane containing the light vector and normal to the surface at point of incident

(b) The plane containing the reflected ray and normal to the surface at point of incident

(c) The plane containing the reflected ray and incident ray

(d) The plane at which light reflected

14.6 What is plane of refraction?

(a) The plane containing the light vector and normal to the surface at the point of incident

(b) The plane containing the refracted ray and incident ray

(c) The plane containing the refracted ray and normal to the surface at point of incident

(d) The plane at which light refracted

14.7 Plane polarized light can be produced by

(a) Double refraction (b) Nicol rism

(c) Reflection from grass plate (d) All of them

14.8 Plane of vibration and the plane of polarization of light are perpendicular to each other.

(a) True (b) False

14.9 Which of the following waves cannot be polarized?

(a) Cosmetic wave (b) Sound waves (c) Light wave (d) X-rays

14.10 When unpolarized light falls on doubly refracting crystals, it splits into

(a) Both O and E rays are plane polarized

(b) Only O ray is plane polarized

(c) None of O and E rays are plane polarized

(d) Only E ray is plane polarized

14.11 If an unpolarized light is passed through a tourmaline crystal, then on rotation of the crystal the intensity of light

(a) Increases (b) Decreases

(c) Remains constant (d) First increases then decreases

14.12 If plane of analyser and plane of polarization are parallel, then intensity of emerging light from analyser is

(a) Zero (b) Maximum (c) Minimum (d) Half

14.13 If plane of analyser and plane of polarization are perpenicular, then intensity of emerging light from analyser is

(a) Zero (b) Maximum (c) Minimum (d) Half

14.14 In circularly polarized light

(a) Amplitude of vibration of the electric vector remains same, while its direction changes.

(b) Amplitude of vibration of the electric vector changes, while its direction remains same.

(c) Both amplitude and direction of electric vector vibrations change.

(d) Both amplitude and direction of electric vector vibrations remains same.

14.15 We pass circularly polarized light through Nicol prism. When Nicol prism is rotated then the intensity of it

(a) Decreases (b) Increases (c) Remains uniform (d) None of these

14.16 When circularly polarized light is passed through a quarter-wave plate,

(a) It remains circularly polarized (b) It becomes plane polarized

(c) It becomes elliptically polarized (d) It becomes unpolarized

14.17 We pass circular polarized light first from the quarter-wave plate and then from Nicol prism. When Nicol prism is rotated, the emergent light from Nicol prism

(a) Becomes bright once in one rotation (b) Becomes bright twice in one rotation

(c) Becomes dark twice in one rotation (d) Becomes dark once in one rotation

14.18 We pass elliptically polarized light through Nicol prism. When Nicol prism is rotated then the intensity of it

(a) becomes twice minimum to maximum and twice maximum to minimum during one rotation.

(b) becomes only minimum to maximum twice during one rotation

(c) becomes only maximum to minimum twice during one rotation

(d) remains uniform during one rotation

14.19 When elliptically polarized light is passed through a quarter-wave plate,

(a) It remains circularly polarized (b) It becomes unpolarized

(c) It becomes elliptically polarized (d) It becomes plane polarized

14.20 We pass elliptical polarized light first from the quarter-wave plate and then from Nicol prism. When Nicol prism is rotated, the emergent light from Nicol prism

(a) Becomes dark once in one rotation (b) Becomes dark twice in one rotation

(c) Becomes bright twice in one rotation (d) Becomes bright once in one rotation

Answers

14.1 (b) 14.2 (b) 14.3 (a)&(c) 14.4 (a) 14.5 (b) 14.6 (c) 14.7 (d)

14.8 (b) 14.9 (b) 14.10 (a) 14.11 (c) 14.12 (b) 14.13 (a) 14.14 (d) 14.15 (c)

14.6 (b) 14.17 (c) 14.18 (a) 14.19 (d) 14.20 (b)

CHAPTER 15

Laser

15.1 INTRODUCTION

Full form of LASER is Light Amplification by Simulated Emission of Radiation. The main features of lasers are coherence, directionality, monochromaticity and high intensity. Broadly, there are four types of lasers: (i) Gas laser (e.g., He-Ne gas laser), (ii) semiconducor junction or diode laser, (iii) solid state laser: Ruby (Al_2O_3 : Cr) laser and Neodymium Yttrium Aluminium Garnet (Nd: YAG) laser, and (iv) Liquid laser-dye laser. Diode laser 20%. Efficiency of Ruby, Nd:YAG laser and He-Ne lasers are about 1%, 6.5% and 0.1%, respectively. There are three major components of laser equipment: (i) external source for pumping the electrons to the excited state, (ii) active material for getting population inversion and laser action, and (iii) a resonator for stimulated emission of photons. Lasers have numerous applications in various fields like in medicine for painless and bloodless surgery, in industry for processing diamonds, in communication for increasing for long distance communication, in computers for storing large data on compact disk (CD) and in printers for getting hard copy of information, in military for invisible fencing at borders and blast the enemy target with good accuracy through missiles, etc. Nowadays, the power of available lasers is in the range 1 nw (= 10^{-9} W) to 10^5 PW (1 PW = 10^{15} W). The frequency range of these lasers is from 100 GHz (1 GHz = 10^9 Hz) to 100 PHz (1 PHz = 10^{15} Hz). The lasers with pulse duration as short as ~ 1 Fs (= 10^{-15} s) and pulse energies as high as 10 kJ are also available.

The power of 10 kJ laser is = $\dfrac{10 \text{ kJ}}{10^{\ 15}}$ W = 10^{19} W.

Metastable State

Let an electronbe moving in an atomic orbit of energy, E_0. When this jumps to higher orbit or higher level of energy (E_1) by absorbing energy, $\Delta E = E_0 - E_1$ from an external source, then it is said that electron or atom is in an excited state, E_1. All the excited atoms drop to ground state (E_0) within a finite time (Δt) called life time. Energy difference (ΔE) and lifetime (Δt) are related to each other by energy and time. Heisenberg's uncertainty principle as follows:

$$\Delta E \Delta t = \frac{\hbar}{2} \qquad ...(15.1)$$

This equation can also be written as

$$\Delta t = \frac{\hbar}{2\,\Delta E} \qquad ...(15.2)$$

If $\Delta E = 0$, i.e., $E_1 = E_0$, then $\Delta t = \infty$. It indicates that the atoms in the ground state have the infinite life time. Normally, the life time of an excited atom is 10^{-8} second, i.e., the excited atom drops to ground state in 10^{-8} second. The key point to get laser light is that many atoms have one or more energy levels whose lifetime is 10^{-3} second or more, instead of the usual 10^{-8} second. The one or more energy states in some materials, which have so much long life, are called metastable states. The metastable state lies in between the ground state and the excited state, but is more close to the excited state.

Transitions of Electrons

Let E_1 and E_2 be the lower and the higher energy levels of an atom. When an electromagnetic radiation of energy equal to the difference between the higher level and the lower level, i.e., $E_2 - E_1 = h\upsilon$, is incident on a material (i) electrons at the lower energy level are excited to the higher energy level by absorbing the incident photon energy, (ii) electrons at the higher energy level drop to the lower energy level by emitting photons. There are three types of transitions of electrons (sometimes called atoms) between two energy levels, E_1 and E_2 of an atom, when electromagnetic radiations are allowed to fall on a material: (i) induced or stimulated absorption of radiation, (ii) spontaneous emission, and (iii) induced or stimulated emission.

Absorption of Radiation

Absorption of radiation is the process by which the electrons in the ground state absorb energy from the photons to jump to the higher energy level. The absorption of radiation or light occurs only if the energy of incident photons exactly matches the energy difference of the two energy levels.

The orbits very close to the nucleus, in which the electrons are moving, are the lower energy levels or states, whereas the orbits far away from the nucleus in which the electrons moving are the higher energy levels or states. The electrons in the lower energy levels require some extra energy to jump to the higher energy level. The sources of these extra energy are heat, electric field or light.

Let us consider two energy levels, E_1 and $E_{2,}$ as shown in Fig. 15.1. E_1 is the lower energy or ground energy state of electrons, while E_2 is the higher energy or excited energy state of electrons. The electrons in the lower energy or ground energy state are called lower or ground energy electrons, whereas the electrons in the higher energy or excited energy state are called higher or excited energy electrons.

Let's consider that the atoms in the ground state at the lower level E_1 are excited to the higher energy level E_2 by absorbing a photon of energy, $E_2 - E_1 = h\upsilon$, from the incident electromagnetic radiation as shown in Fig. 15.1. This process is called induced or stimulated absorption of radiation or simply absorption. The electromagnetic radiations may light or heat radiations or electric discharge radiation.

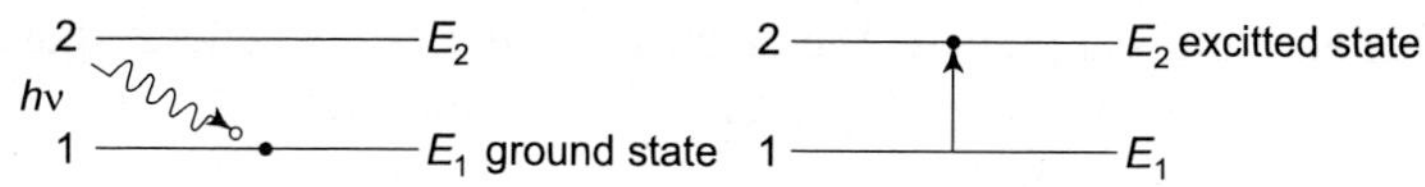

Fig. 15.1

This process is called *absorption.*

Spontaneous Emission

When the electrons in the excited state return to the ground energy state after completing their lifetime by emitting photons, this process is called the spontaneous emission. The lifetime of electrons in the excited state is 10^{-8}s. The time up to which the electrons stay in excited state is known as the lifetime of excited electrons. Therefore, the electrons stay in the excited state for a short duration. After short duration, they return to the ground state by emitting photons.

The spontaneous emission occurs naturally, as the electrons in the excited state return to the ground energy state after completing their lifetime by emitting photons. Therefore, we have no control over when the electrons in the excited state are going to lose their energy in the form of light.

The photons emitted in spontaneous process constitute the ordinary incoherent light. Ordinary light is incoherent, because the photon waves change their phase frequently and randomly. In other words, the emitted photons in spontaneous emission process do not move only in the direction of incident photons. Because of this reason, the light emitted from ordinary bulb is spherical.

When atoms from higher energy level drop to the lower energy level by emitting radiation without using any kind of external radiation, then this process is known as *spontaneous emission.*The radiation may be in the form of light or heat energy. For example, an atom is there in the excited state E_2 as shown in Fig. 15.2. This atom has lifetime of order of 10^{-8}s for staying in the upper state, E_2. After this duration it drops to lower state E_1 by emitting a photon of energy hv without any external force or energy. The light from the filament of an ordinary light bulb is generated in this way. In this process light or heat energy is emitted in different the directions.

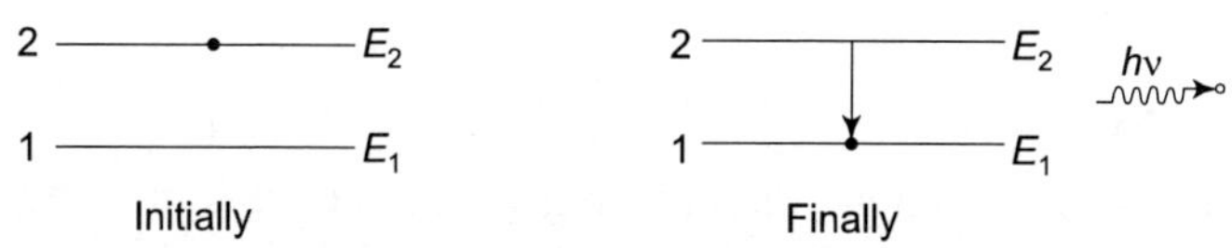

Fig. 15.2

Stimulated Emission

Stimulated emission is a process by which incident photon interacts with the excited electron and forces it to return to the ground state.

Unlike the spontaneous emission, the stimulated emission is not a natural process. It is an artificial process.

In the spontaneous emission process, the electrons in the excited energy state remain there until its lifetime is over. After competing their lifetime, they return to the ground state by emitting energy in the form of light.

However, in simulated emission, the electrons in the exited state need not wait for completion of their lifetime. External source is used to forcefully return the excited electrons to ground state before completion of their lifetime. This process is known as *stimulated emission.*

When incident photon interacts with the excited electrons, it forces the excited electron to return to the ground state. This excited electron releases energy in the form of light while falling to the ground state.

When atoms from excited state are forced to drop to the lower state by external radiation, then this process is called *induced* or *stimulated emission.* For example, seean atom is present in the excited state E_2 [Fig. 15.3]. When the light of energy $h\upsilon = E_2 - E_1$ is allowed to fall, the photon of light energy stimulates the atom in excited state to drop to ground state by emitting of a photon of energy, $h\upsilon = E_2 - E_1$. Thus, two photons are present in the output: one stimulated photon and second stimulating photon.

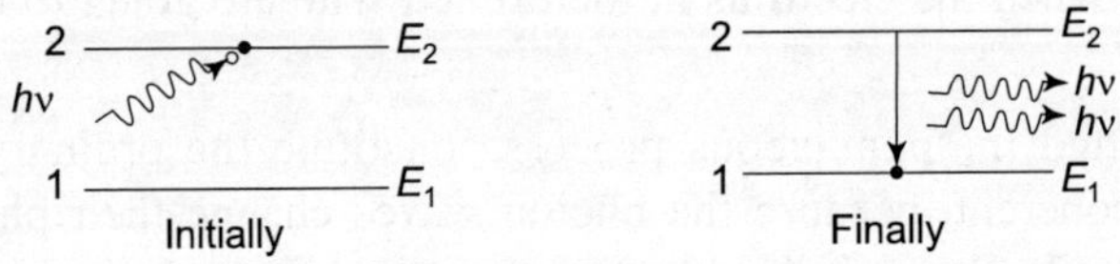

Fig. 15.3

The emitted photon in Fig. 15.3 is in every respect identical to the stimulating photon. That is, the waves associated with the photons have the same energy, phase, polarization and direction of travel, i.e., the waves are coherent. Now these two photons force two more electrons in the energy excited state E_2. As a result, two more photons are emitted making four photons. Similarly, a chain reaction occurs producing millions of photons of energy $h\upsilon = E_2 - E_1$. If all the waves of stimulated photons are coherent, they will produce a beam of light of powerful intensity. This process is called light amplification by stimulated emission of radiations (LASER) of light of powerful intensity.

Population Inversion

Population inversion is the technique of achieving greater population of higher energy state as compared to lower energy state. Population inversion technique is used mainly for light amplification. Population inversion is required for producing laser light.

Let E_1 and E_2 be the energy of lower and higher energy states, respectively. Let N_1 and N_2 be the concentration of electrons (number of electrons per unit volume) in the lower and higher energy states, respectively as shown in the Fig. 15.4.

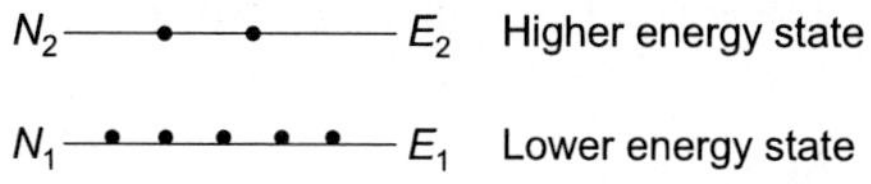

Fig. 15.4

Normally, the number of electrons per unit volume is more in lower energy state than the higher energy state, i.e., $N_1 > N_2$. At least equal population ($N_1 = N_2$) in two states can be achieved by supplying the energy. But no optical gain can be achieved with equal population in two states. Therefore, we need three or more states to achieve population inversion.

There are materials in which the electrons once excited remain in the excited state for longer time. Such materials are called *active materials*. These active materials are the mixture of different elements.

When active materials are formed, then their energy states are modified, and the energy levels of some elements acquire special properties. These active materials are used to produce laser light.

15.2 SPONTANEOUS AND SIMULATED EMISSION PROBABILITIES AND EINSTEIN COEFFICIENTS

Let us consider a radiation of frequency (υ) and energy density, E (υ). Let N_1 and N_2 be the number of atoms per unit volume in lower energy state (E_1) and higher energy state E_2, respectively. At any instant of time, all the three transitions processes: (i) induced absorption, (ii) spontaneous emission, and (iii) stimulated absorption can take place simultaneously.

Stimulated Absorption Probability

If the atom is initially in the lower energy state E_1, it can be excited to higher energy state E_2 after absorbing a photon of energy, $h\upsilon = E_2 - E_1$, where is the frequency of the photon absorbed.

The probability of absorption is dependent on the energy density of the radiation, because transition of atoms from lower state to higher state is simulated by absorbing radiation. The probability of transition of atoms from lower energy state (E_1) to higher energy state (E_2) after absorbing photon of light is proportional to the photons of frequency (υ) fall per second on unit area of material, i.e., rate of fall of radiation energy density, E (υ), i.e., rate of stimulated absorption transition, $\propto E$ (υ).

The simulated absorption rate is also directly proportional to the number of atoms per unit volume (N_1) in the energy state, E_1, i.e., rate of stimulated absorption transition, $\propto N_1$.

Combining these two conditions, we get rate of stimulated absorption transition, $\propto N_1 E$ (υ)

or rate of stimulated absorption transition = $B_{21} N_1 E$ (υ). ...(15.3)

The constant of proportionality (B_{12}) is called *Einstein's absorption coefficient.*

The Rate of Spontaneous Emission

The probability of spontaneous emission is independent of the energy density of the radiation. It is only proportional to the number of atoms (N_2) per unit volume present in the higher energy states, i.e., rate of spontaneous emission, $\propto N_2$

or rate of spontaneous emission = $A_{21} N_2$, ...(15.4)

where, proportionality constant (A_{21}) is called *Einstein's spontaneous emission transition coefficient.* Thus, the number of atoms drop from state E_2 to state E_1 per unit time is $A_{21} N_2$.

The Rate of Stimulated Emission

Since transition of atoms from higher state to lower state is simulated by stimulating radiation, therefore, the probability of stimulated emission is dependent on the energy density of the stimulating radiation. The probability of transition of atoms from higher energy state (E_2) to lower energy state (E_1) due to stimulating incidence light radiation is proportional to photons of frequency (υ) falling in one second on unit area of material, i.e., rate of energy density, E (υ), of induced radiation falling and hence, rate of stimulated emission, $\propto E$(υ).

The simulated emission rate is also dependent on the numberof atoms (N_2) per unit volume in the state E_2, therefore, it directly proportional to the rate number of atoms N_2 per unit volume in the energy state E_2, i.e. rate of stimulated emission, $\propto N_2$

Combining these two conditions, we get

rate of stimulated emission, $\propto N_2 E$ (υ)

or rate of stimulated emission = $B_{21} N_2 E$ (υ) ...(15.5)

where, the proportionality constant (B_{21}) is called *Einstein's stimulated emission transition coefficient.* Thus, the number of atoms that drop from state E_2 to state E_1 per second is $B_{12} N_2 E$ (υ).

15.3 THREE-LEVEL LASER

Let us consider an active material having three states E_1, E_2 and E_3 and N number of electrons. We assume that $E_1 < E_2 < E_3$. Let's us assume that initially the majority of electrons are in the ground energy state E_1 and only small number is in excited energy states E_2 and E_3 as shown in Fig. 15.5(a). When an external energy, which is equal to energy difference of E_3 and E_1, fall on electrons, by gaining sufficient energy some electrons jump to excited energy state, E_3. This process of supplying the energy is called pumping. Now the distribution of electrons becomes as shown in Fig. 15.5(b).

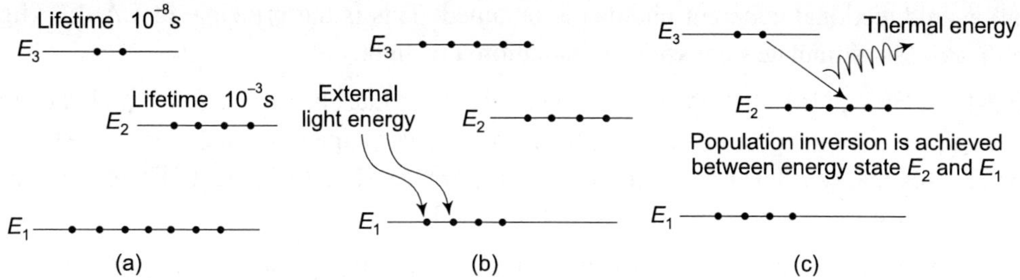

Fig. 15.5 (a) The distribution of atoms among excited state E_3, metastable state E_2 and ground state E_1 due to thermal agitation at equilibrium., (b) light from external source fall on electrons, (b) Population inversion is obtained when electrons drop from energy state E_3 and E_2 by spontaneous emission

The lifetime ~ 10^{-3}s of electrons in the energy state (E_2) is longer, than the lifetime (~ 10^{-8}s) of electrons in excited energy state, E_3. The state (E_2) which has longer lifetime is called *metastable state*. A material having metastable state(s) is called an*active material.*

The electrons in excited state (E_3) do not stay for long. After a short time, the electrons in the excited energy state (E_3) drop to energy state E_2 by radiationless spontaneous emission (e.g., thermal energy) instead of photons. Because of shorter lifetime only a small number of electrons stay in energy state E_3. As a result, a number of electrons accumulate in energy state, E_2. Thus, number of electrons in metastable energy state (E_2) becomes more than the electrons in the ground energy state (E_1) and excited state (E_3) as shown in Fig. 15.5(c). When population of atoms in metastable state becomes more than the ground state, the phenomenon is called *population inversion.* The method used for achieving population inversion is called *optical pumping.* For stimulated emission population inversion is a must. It can be obtained only in those materials in which there is metastable state having longer lifetime, ~ 10^{-3}s, in caparison to upper excited state. Such materials are called active materials.

Laser Action

After completing their lifetime in metastable energy state E_2, the electrons drop in energy state (E_2) by emitting photons by spontaneous emission as shown in Fig. 15.5(a). The emitted photon bounces back and forth between the two reflecting mirrors and interacts the electron in metastable energy state E_2, and forces one electron to drop down to energy state E_1 as shown in Fig. 15.5(b). As a result one more electron drops to ground state (E_1) and one photon is emitted as shown in Fig. 15.5(c) making the number of photons two: one stimulating photon and another stimulated photon. These two photons are of same wavelength and in the same phase. These two photons further force two more electrons in metastable energy state (E_2) to drop down to ground state (E_1) by emitting two more coherent photons making total number of coherent photons four. This chain reaction of stimulating the electrons in metastable state continues to produce millions of photons. Every stimulated electron releases a photon of energy, $h\upsilon = E_2 - E_1$. That is, the waves associated with the photons have the same energy, phase, polarization and direction of travel, i.e., the waves are coherent. Finally, an intense beam of unidirectional coherent photons is obtained. This is the principle of LASER (light amplification by stimulated emission by radiations) of light.

In 3-level laser, at least half the population of electrons must be excited to the higher energy state to achieve population inversion. Therefore, the laser medium must be very strongly pumped. This makes 3-level lasers inefficient to produce photons or light. Three-level laser is important historically, because it was first invented.

In 4-level laser, only a few electrons are excited to achieve population inversion. Therefore, a 4-level laser produces laser light more efficiently than 3-level laser.

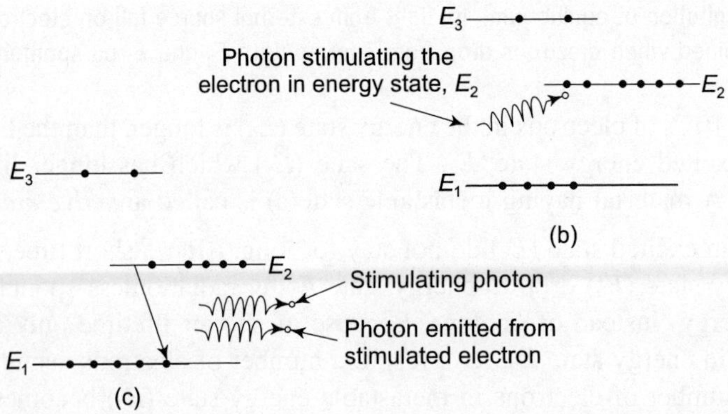

Fig. 15.6 Laser action (a) after competing its lifetime, one electron in metastable state E_2 drops to ground energy state E_1 by the emitting one photon by spontaneous emission. (b) This photon forces one electron in energy state E_2 to drop down to state E_1, (c) As a result, one electron drops to lower state E_1 and one photon of same wavelength and same phase is emitted by stimulating emission making number of photons two. These two photons have same wavelength and are in same phase. In the same wave, the chain reaction continues to produce millions of photons by stimulated emission. This is called laser action

15.4 FOUR-LEVEL LASER

Let us consider an active material having three states: E_1, E_2, E_3 and E_4 and N number of electrons. We assume that $E_1 < E_2 < E_3 < E_4$. Let us assume that initially the majority of electrons is in the ground energy state (E_1) and only a small number is in excited energy states, that is, E_2, E_3 and E_4 as shown in Fig. 15.7(a). When an external energy, which is equal to energy difference of E_4 and E_1, falls on electrons, by gaining sufficient energy some electrons jump to excited energy state, E_4. This process of supplying the energy is called pumping. Now the distribution of electrons becomes as shown in Fig. 15.7(b).

The lifetime, ~ 10^{-3}s, of electrons in the energy state E_3 is longer, than the lifetime (~ 10^{-8}s) of electrons in excited energy state E_4, The state E_3 which has longer lifetime is called metastable state. A material having metastable state(s) is called active material.

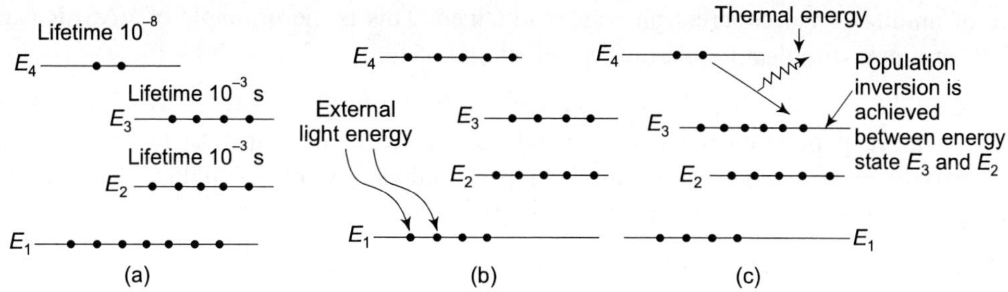

Fig. 15.7 (a) The distribution of atoms among excited state E_4, metastable state E_3, lower state E_2 and ground state E_1 due to thermal agitation at equilibrium. (b) Light from external source falls on electrons, (c) Population inversion is obtained when electrons drop from energy state E_4 and E_3 by spontaneous emission

The electrons in excited state (E_4) do not stay for long. After a short time, the electrons in the excited energy state (E_4) drop to energy state (E_3) by radiationless spontaneous emission (e.g., thermal energy) instead of photons. Because of shorter lifetime only a small number of electrons stay in energy state (E_4). As a result, a number of electrons accumulate in energy state, E_3. Thus, the number of electrons in metastable energy state (E_3) becomes more than the electrons in the ground energy state (E_1), lower energy state (E_2), and excited state (E_4) as shown in Fig. 15.7(c). When population of atoms in metastable state becomes more than the lower state, the phenomenon is called population inversion. The method used for achieving population inversion is called the optical pumping. For stimulated emission population inversion is a must. It can be obtained only in those materials in which there is metastable state having longer lifetime, ~ 10^{-3} s, in caparison to the upper excited state. Such materials are called active materials.

Laser Action

After completing their lifetime in metastable energy state (E_3), the electrons drop in energy state (E_2) by emitting photons by spontaneous emission as shown in Fig. 15.8(a). The emitted photon bounces back and forth between two reflecting mirrors and interacts with the electron in metastable energy state (E_3), and forces one electron to drop down to energy state E_2 as shown in Fig. 15.8(b). As a result one more electron drops to lower state E_2 and one photon is emitted as shown in Fig. 15.8(f) making the number of coherent photons two: one stimulating photon and another stimulated photon. These two photons bounce back and forth between two reflecting mirrors and force two electrons in metastable energy state (E_3) to drop down two more electrons to lower state (E_2) by emitting two more photons making total number of coherent photons four. This chain reaction of stimulating the electrons in metastable state continues to produce millions of photons. Every stimulated electron releases a photon of energy, $h\upsilon = E_3 - E_2$. That is, the waves associated with the photons have the same energy, phase, polarization and direction of travel, i.e., the waves are coherent. Finally, an intense beam of unidirectional coherent photons is obtained. This is the principle of LASER (light amplification by stimulated emission by radiations) of light.

In 3-level laser, at least half the population of electrons must be excited to the higher energy state to achieve population inversion. Therefore, the laser medium must be very strongly pumped. This makes 3-level lasers inefficient to produce photons or light. 3-level laser is important historically, because it was first invented.

In 4-level laser, only a few electrons are excited to achieve population inversion. Therefore, a 4-level laser produces laser light less efficiently than 3-level laser.

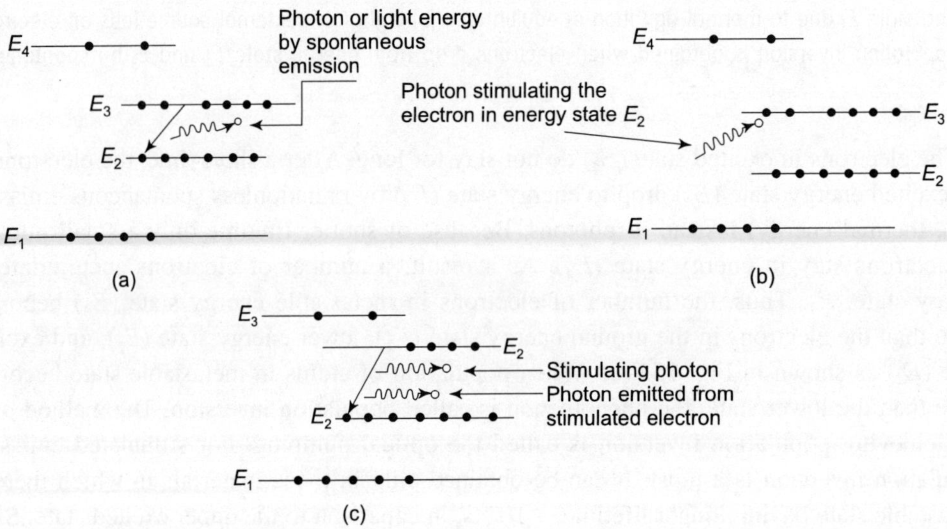

Fig. 15.8 Shows laser action (a) After competing its lifetime, one electron in metastable state E_3 drops to lower energy state E_2 by the emitting one photon by spontaneous emission. (b) This photon forces one electron in energy state E_3 to drop down to state E_2, (c) As a result one electron drops to lower state E_2 and one photon of same wavelength and same phase is emitted by stimulating emission making number of photons two. These two photons have same wavelength and are in same phase. In the same wave, the chain reaction continues to produce millions of photons by stimulated emission. This is called laser action

15.5 PRINCIPLE OF LASER PUMPING, POPULATION INVERSION, AND LASER ACTION

15.5.1 Pumping

When an external energy, which is equal to the energy difference of E_3 and E_1, falls on electrons, by gaining sufficient energy some electrons jump to excited energy state (E_3) as shown in Fig. 15.6. This process of supplying the energy is called pumping.

Consider an active material having energy states: E_1, E_2 and E_3, where, E_1 is ground state energy, E_2 is metastable energy state, and E_3 is the excited energy state. When photons having energy, $h\upsilon' = E_3 - E_1$ are incident on the materials, the atoms in ground energy state (E_1) absorbs incident photon energy and jumps over to the excited energy state, E_3.

15.5.2 Population Inversion

Population inversion is the technique of achieving greater population of higher energy state as compared to lower energy state. This technique is used mainly for light amplification. Population inversion is required for producing laser light.

The lifetime of electrons in energy state E_3 is 10^{-8} second. Therefore, the electrons stay for short duration in the excited state, E_3. The exited electrons present energy state E_3 quickly drop to metastable energy state (E_2) by spontaneous emission. Energy difference, $h\upsilon'' = E_3 - E_2$, disappears as heat radiations during spontaneous emission. The lifetime (10^{-3} second) of electrons in the energy state E_2 is large as compared to their lifetime (10^{-8} second) in energy state, E_3, so electrons the state E_2 stay for longer time than the electrons in the excited energy state, E_3. As a result, the number of electrons accumulated in state E_2 becomes more than the number of electrons in energy states E_3 and E_1. Clearly, number of atoms in metastable state (E_2) are more than ground state, E_1. This is known as *population inversion.*

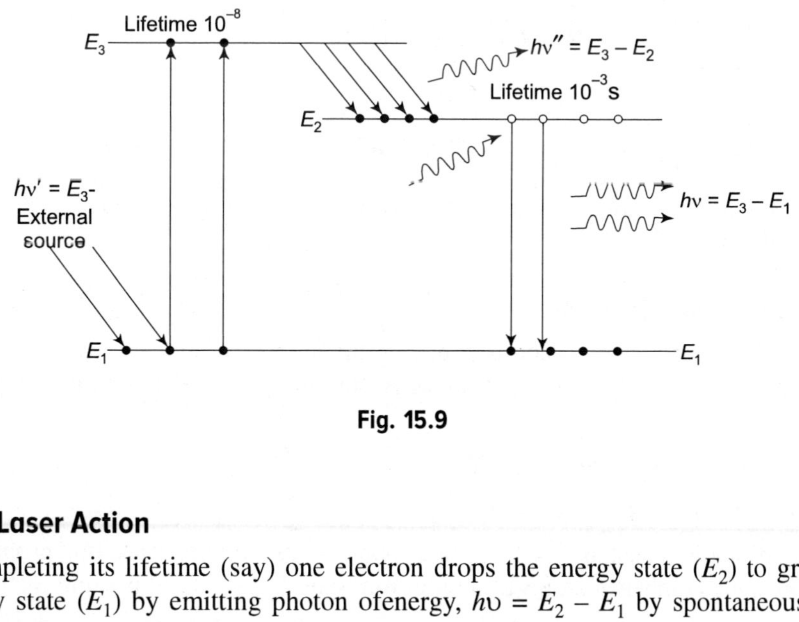

Fig. 15.9

15.5.3 Laser Action

After completing its lifetime (say) one electron drops the energy state (E_2) to ground state the energy state (E_1) by emitting photon ofenergy, $h\upsilon = E_2 - E_1$ by spontaneous emission as shown in Fig. 15.9. This photon bounces back and forth between two perfectly reflecting mirrors and collide with an atom to force an electron to drop down to ground state (E_1) by emitting one photon making the number of photons two: one stimulating photon and another stimulated photon as shown in Fig. 15.9. The emitted photon is in every respect identical to the stimulating photon. These two photons again bounce back and forth between two perfectly reflecting mirrors and in the process collide with two more atoms, and hence force two electrons to drop down to ground state (E_1) by stimulated emission. In the same way, a chain raction continues to produce millions of photons by stimulated emission. Every stimulated electron releases a photon of energy, $h\upsilon = E_2 - E_1$. That is, the waves associated with the photons have the same energy, phase, polarization and direction of travel, i.e., the waves are coherent. Finally, an intense beam of unidirectional coherent photons is obtained. This is the principle of LASER (light amplification by stimulated emission by radiations) of light.

15.6 EINSTEIN'S COEFFICIENTS AND ENERGY DENSITY OF EMITTED RADIATION

At any instant of time, all the three transition processes: (i) induced absorption, (ii) spontaneous emission, and (iii) stimulated absorption can take place simultaneously. Let us assume that N_1 and N_2 are the concentration of atoms in energy states E_1 and E_2, where, $E_1 < E_2$, respectively. E (υ) is the energy density of a radiation of frequency, υ.

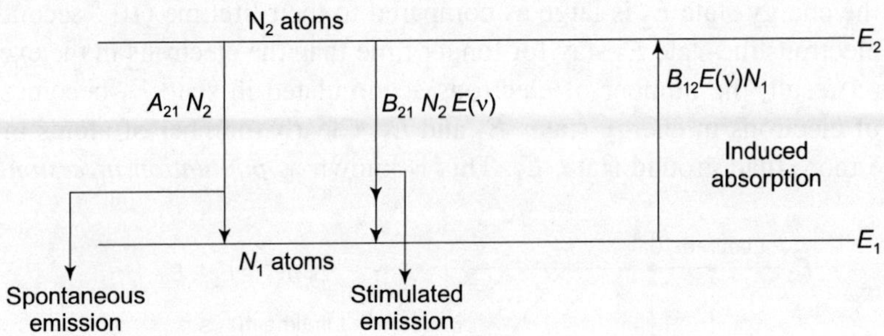

Fig. 15.10

Atoms in state E_2 may drop to state E_1 by two processes: (i) spontantaneous emission, and (ii) stimulated emission.

Rate of Spontaneous Emission

The rate of spontaneous emission depends on the number of atoms (N_2) per unit volume present in the state E_2 only, since spontaneous emission takes place without any external source, so,

the rate of spontaneous emission, $\propto N_2$

or rate of spontaneous emission = A_{21} N_2,

where, proportionality constant (A_{21}) is called spontaneous emission transition probability per unit time.Thus, the number of atoms dropping from state E_2 to state E_1 per unit time is A_{21} N_2.

Rate of Stimulated Emission

The stimulated emission takes place by external source (the incident radiation), so rate of stimulated emission will depend on of density of radiations, E (υ), falling in one second on the material and number of atoms (N_2) per unit volume in the state, E_2. Thus, rate of stimulated emission, $\propto N_2$

rate of stimulated emission, $\propto E(\upsilon)$

Combining these two conditions, we get

rate of stimulated emission, $\propto N_2 E(\upsilon)$

or rate of stimulated emission = $B_{21} N_2 E(\upsilon)$,

where, the proportionality constant (B_{21}) is stimulated emission transition probability per unit time per unit energy density. Thus, the number of atoms that drop from state E_2 to state E_1 per unit time is $B_{12} N_2 E(\upsilon)$

Rate of Absorption Transition

After absorbing energy $h\upsilon = E_2 - E_1$, the atoms in lower energy state E_1 get excited to higher energy state E_2. The rate of absorption transition takes place by external source, so rate of absorption transition will depend upon density of incident radiations $E(\upsilon)$ falling per second on the material and number of atoms (N_1) per unit volume in the energy state E_1. Thus,

rate of absorption transition, $\propto N_1$

rate of absorption transition, $\propto E(\upsilon)$

Combining these two conditions, we get

rate of absorption transition, $\propto N_2 E(\upsilon)$

or rate of absorption transition = $B_{21} N_1 E(\upsilon)$,

where, the proportionality constant (B_{12}) is absorption transition probability per unit time per unit energy density. Thus, the number of atoms that rise from state E_1 to state E_2 per unit time is $B_{12} N_1 E(\upsilon)$.

Total number of atoms that drop from state E_2 to state E_1 in one scecond is

$$[A_{21} + B_{21} E(\upsilon)] N_2.$$

The net rate of transition of atoms from state E_2 to state E_1 is

$$\frac{dN_2}{dt} = \text{rate of absorption transition - rate of (spontaneous + stimulated) emission}$$

$$\frac{dN_2}{dt} = B_{12} N_1 E(\upsilon) - [A_{21} + B_{21} E(\upsilon)] N_2 \qquad ...(15.6)$$

At equilibrium, $\frac{dN_2}{dt} = 0$, therefore we get

$$B_{12} N_1 E(\upsilon) = [A_{21} + B_{21} E(\upsilon)] N_2 \qquad ...(15.7)$$

i.e., number of absorption and number of emission per second between two states at equilibrium is equal.

Rewriting Eq. (15.7), we get

$$E(\upsilon)\,[B_{12} N_1 - B_{21} N_2] = N_2 A_{21}$$

or
$$E(\upsilon) = \frac{N_2 A_{21}}{B_{12} N_1 - B_{21} N_2}$$

or
$$E(\upsilon) = \frac{A_{21}}{B_{21}\left[\dfrac{N_1 B_{12}}{N_2 B_{21}} - 1\right]} \quad ...(15.8)$$

According to Maxwell-Boltzmann distribution method, the population N_1 and N_2 in energy states E_1 and E_2 at temperature T is written as

$$N_1 = N_0\, e^{-E_1/KT} \quad \text{and} \quad N_2 = N_0\, e^{-E_2/KT}$$

The ratio

$$\frac{N_1}{N_2} = e^{(E_2 - E_1)/KT}$$

As $E_2 - E_1 = h\upsilon$, energy of emitted photon, therefore

$$\frac{N_1}{N_2} = e^{h\upsilon/KT}$$

Putting the value $\dfrac{N_1}{N_2}$ in Eq. (15.8), we get

$$E(\upsilon) = \frac{A_{21}}{B_{21}} \frac{1}{\left[\left(\dfrac{B_{12}}{B_{21}}\right) e^{h\upsilon/KT} - 1\right]} \quad ...(15.9)$$

Here, $E(\upsilon)$ is energy density of photons of frequency υ in equilibrium with atoms in the two energy states E_1 and E_2 at temperature T. Now according to Planck's radiation law, heat radiation energy density is given by

$$E(\upsilon) = \frac{8\pi h\upsilon^2}{c^2} \frac{1}{[e^{h\upsilon/KT} - 1]} \quad ...(15.10)$$

Comparing Eqs. (15.9) and (15.10), we get

$$\frac{B_{12}}{B_{21}} = 1 \quad \text{or} \quad B_{12} = B_{21} \quad ...(15.11)$$

$$\frac{A_{21}}{B_{21}} = \frac{8\pi h\upsilon^3}{c^2} \quad \text{or} \quad \frac{A_{21}}{B_{21}} \propto \upsilon^3 \quad ...(15.12)$$

Equations (15.11) and (15.12) are known as Einstein's relations.

Equation (15.9) is similar to Planck's radiation law. In order to get the correct form of $E(\upsilon)$, Einstein predicted the existence of simulated emission, which was later confirmed by quantum theory. From Eq. (15.11), one infers that probability of induced absorption is equal to the probability of stimulated emission .

Putting $B_{12} = B_{21}$ from Eq. (15.11), Eq. (15.9) becomes

$$E(\upsilon) = \frac{A_{21}}{B_{21}} \frac{1}{[e^{h\upsilon/KT} - 1]} \quad ...(15.13)$$

or
$$\frac{B_{21}}{A_{21}} E(\upsilon) = \frac{1}{[e^{h\upsilon/KT} - 1]}$$

or
$$\frac{A_{21}}{E(\upsilon) B_{21}} = e^{h\upsilon/KT} - 1 \quad ...(15.14)$$

In Eq. (15.14), $\frac{A_{21}}{E(\upsilon) B_{21}}$ is the number of spontaneous emission and stimulated emission.

For normal optical source, $T = 10^3$ °K and $\lambda = 6000$ Å

Wavelength of ordinary source, $\lambda = 6000$ Å $= 6000 \times 10^{-8}$ cm $= 6 \times 10^{-7}$ m

Corresponding frequency, $\upsilon = \frac{c}{\lambda} = \frac{3 \times 10^2}{6000 \times 10^{-7}} = 0.5 \times 10^{15} = 5 \times 10^{14}$ c/s

$= 5 \times 10^{14}$ Hz

Photon energy, $h\upsilon = 6.6 \times 10^{-3} \times 5 \times 10^{14} = 3.3 \times 10^{-19}$ J

Thermal energy of source, $KT = 1.38 \times 10^{-23} \times 10^{-3} = 1.38 \times 10^{-20}$ J

{using the values of $h\upsilon$ and KT, $\frac{h\upsilon}{KT} = 2.39 \times 10 = 23.9$ and hence, $e^{h\upsilon/KT} = 2.39 \times 10^{10}$.

Now $\frac{A_{21}}{E(\upsilon) B_{21}} \approx 10^{10}$.

This ratio indicates that at optical frequencies, the emission from normal sources of light is dominated by spontaneous emission. Because of this reason, the emission from a normal light source is incoherent.

15.7 LASER PUMPING

Normally, the number of atoms in the ground state is more than the number of atoms in the metastable state. When the state of population inversion is achieved, then the number of atoms become more in metastable state than the number of atoms in ground state. The method used for raising the atoms from ground state to metastable state for achieving population inversion is known as pumping. Some of the methods used for laser pumping are:

(i) Optical pumping (ii) Chemical pumping (iii) Electrical pumping

(iv) Laser pumping (v) Solar pumping

(i) Optical pumping: It is most simple method for achieving population inversion. Most conventional sources of optical pumping arc lamp, flash discharge lamp (mainly xenon gas discharge lamp). This process is suitable for ruby laser and liquid (dye) lasers.

(ii) Chemical pumping: In chemical pumping population inversion is obtained by using suitable chemical reaction when no other sources of energy available.

(iii) Electrical pumping: In electrical pumping the population inversion is obtained by using electric discharge in the medium of gaseous form. This process is suitable for gas laser, e.g., Argon-ion, He-Ne and semiconductor lasers. In this process a high potential difference is applied between cathode and anode of a discharge tube. The electrons emitted from cathode move towards anode, on their way some of these electrons strike with atoms of active medium and excite them to higher energy state. This method is also called *direct conversion pumping method*, because electrical energy is used to get population inversion.

(iv) Laser pumping: In this method of pumping one laser beam is used for population inversion of another laser.

(v) Solar pumping: In this method, a sapphire sphere joined to one end of a ruby rod by means of parabolic mirror. When sunlight is focused on sapphire sphere, laser radiation emitted from the second end of the rod.

15.8 CHARACTERISTICS OF LASER BEAM

The ordinary light and laser light both are electromagnetic waves. The laser light is different from ordinary light, because the laser light has four unique characteristics:

(i) directionality, (ii) monochromatic, (iii) high intensity, and (iv) coherence

Laser beam is thousand times more intense.

15.8.1 Directionality

Lamp, sodium lamp, mercury lamp, sunlight, torch light, etc. are ordinary sources of light. The light from these sources is emitted by spontaneous emission. The light from these sources spreads in all possible directions. An aperture is placed in front of an ordinary light source for focusing itin a particular direction. On the other hand, the laser light travels only in one direction, i.e., laser light is highly unidirectional. This is called directionality of laser light. The width of laser light is extremely narrow. Hence, laser beam can travel to long distance without spreading.

For a typical laser, the spread of the beam is less than one milliradian. That is, when laser light travels by one metre distance, the spread of the beam is less than one millimetre. When laser light travels by 1 km distance, the spread of the beam is less than one cm. When an ordinary light travels a distance of 1 km, it spread to about 1 km in diameter.

15.8.2 Intensity

Intensity of light is defined as the amount of light energy passing from unit area in one second. The intensity of ordinary light source is very less as it spreads out more or less uniformly in all directions. The power from a 100 watt bulb entering the eye at a distance of 40 cm is less than $\frac{1}{1000}$ watt.

On the other hand, beam of laser light is very narrow, therefore, whole of its energy is concentrated in a small region. As a result, intensity of lasers beam is very high. If we look at 1 watt laser beam from 2 metres directly, it appears many thousands times more intense than 100 watt ordinary bulb.

Because of high intensity, laser beam can cut a huge block of steel by melting. Lasers are also used for recording and reproducing large information on a compact disc (CD).

Note: The light passing from 1 cm^2 area at a distance of 40 cm from the source (100-watt ordinary bulb) is

$$= \frac{100}{4\,\pi r^2} = \frac{100}{4\,\pi \times 40^2} = \frac{1}{64} = \frac{1}{200.96} \text{ watt/cm}^2$$

The area of eye lens is about $\left(\frac{1}{3}\right)^2$ cm^2, so the power entering the eye is

$$= \frac{1}{9 \times 200.96} = \frac{1}{1808.64} \text{W} < \frac{1}{1000} \text{W}$$

The intensity of a laser beam can be calculated by comparing the number of photons coming from a laser source and an ordinary source.

Photon energy of visible light, $h\upsilon = \frac{hc}{\lambda} = \frac{6.6 \times 10^{-34} \times 3 \times 10^2}{6 \times 10^{-7}} = \frac{19.8}{6} \times 10^{-19}\text{J}$

$$= 3.3 \times 10^{-19} \text{ J}$$

Power output of gas laser is ~ 10^{-3} watt

Power output of pulse solid state laser is ~ 10^9 watt

The number of photons emitted per second from a laser source is

$$\frac{P}{h\upsilon} = \frac{10^{-3}\, w - 10^9\, w}{10^{-19}\text{ J}} = 10^{16} - 10^{28}$$

Available laser 1 nW (10^{-9} W) → 10^5 PW (10^{20} W)

100 GHz (100×10^9 Hz) → 100 PHz (100×10^{15} Hz)

Available pulse laser 1 *fs* (10^{-15} s) → 10 kJ → $P = \frac{10^4}{10^{-15}} = 10^{19}$ W

15.8.3 Monochromaticity

The meaning of monochromatic light is that the light ray has single wavelength or frequency. The photons emitted from ordinary source have different energies, wavelengths, frequencies and colours. That is, ordinary light is a mixture of waves having different frequencies and wavelengths.

On the other hand, photons emitted from laser light have almost same energy, frequency or wavelength. In other words, the waves of laser light have almost single wavelength or colour. Every source of light consists of a bunch of waves having a little different wavelength. Hence, every source of light has some frequency spread, $\Delta\upsilon$. The frequency spread of ordinary light is order of several thousands of megahertz (~ 10^{10} Hz). The laser light, considered highly monochromatic, has frequency spread ~ 500 Hz.

The degree of monochromaticity is measured by the ratio $\Delta\upsilon/\upsilon_0$, where υ_0 is the frequency of spectral line. When the value of $(\Delta\upsilon/\upsilon_0)$ is lower, the degree of monochromaticity will be higher and vice versa. For zero monochromaticity, the frequency spread $(\Delta\upsilon)$ should be zero. The sources of light having zero monochromaticity are not available practically. The spread width for ordinary light and laser light for monochromatic light of frequency, $\upsilon_0 = 5.0 \times 10^{14}$ Hz and wavelength, $\lambda_0 = \frac{c}{\upsilon_0} = \frac{3\times10^8}{5.0\times10^{14}} = 6 \times 10^{-7}$ m = 6000 Å is shown in Fig. 15.11.

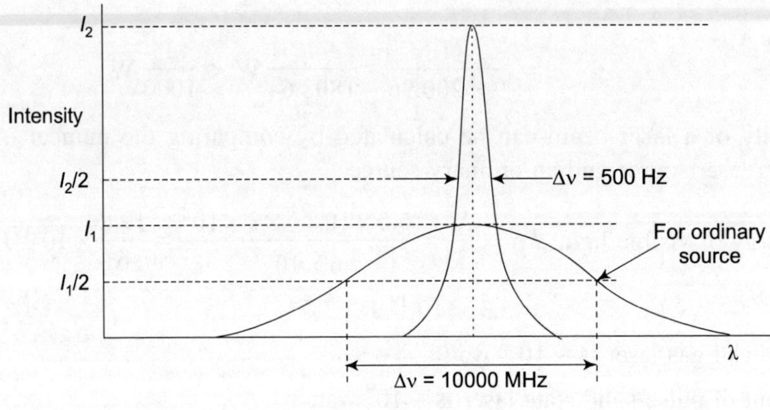

Fig. 15.11

Frequency spread of ordinary light is

$$\Delta\upsilon = 10^{10} \text{ Hz}$$

Degree of monochromaticity for ordinary light is

$$\frac{\Delta\upsilon}{\upsilon_0} = \frac{10^{10}}{5.0\times10^{14}} = 2 \times 10^{-5} = 0.00002$$

Frequency spread for laser beam is

$$\Delta\upsilon = 500 \text{ Hz} = 5 \times 10^2 \text{ Hz}$$

Degree of monochromaticity for laser light is

$$\frac{\Delta \upsilon}{\upsilon_0} = \frac{5 \times 10^2}{5.0 \times 10^{14}} = 10^{-12}$$

Hence laser beam is highly monochromatic.

15.8.4 Coherence

The laser light is obtained from electron transitions from metastable state to lower state forcibily. That is why the laser is obtained by electron transition in specific time. All the photons emitted in laser light have same energy, frequency or wavelength. That is, the laser lights have single wavelength and colour. The waves constituting the laser light are in same phase in space and time. The laser light is produced by stimulated emission. Thus, laser light is highly coherent.

On the other hand, in ordinary light from sources (ordinary bulb, sodium lamp and torch), the electron transition from excited energy state occurs naturally. The light is emitted from ordinary light source by spontaneous emission. Therefore, ordinary light is emitted randomly in all directions. The photons emitted from ordinary source have different energies, frequencies, wavelengths and colours. The photons of ordinary light have large phase difference.

Depending upon the distance and time upto which phonon waves maintain coherence, the coherence of laser light is of two types: (a) longitudinal or temporal coherence, and (b) transverse or spatial coherence.

(a) Longitudinal or temporal coherence: Let us consider two sinusoidal photon waves overlapping each other as shown in Fig. 15.12(a). Very close to the source, the two waves are almost in phase. As the distance of propagation increases, the phase difference of the two waves increases slowly and at a certain distance (L_l) the two waves become completely out of phase. Two waves become out of phase because there is difference in frequencies of two waves, even if it is too small. So the longitudinal coherence length (L_l) is defined as the distance from the source to the point at which the waves become completely out of phase. The phase difference of two waves of same frequency will always remain fixed at all points as shown in Fig. 15.13. Let λ_1 and λ_1 be the wavelengths of the two waves. The average value of wave lengths is given by root mean square formula as

$$\lambda = \frac{\sqrt{\lambda_1 \times \lambda_2}}{2},$$

longitudinal coherence length (L_l) is given by

$$L_l = \frac{\lambda^2}{\Delta \lambda} \qquad ...(15.15)$$

where, $\Delta\lambda = |\lambda_1 - \lambda_2|$. From Eq. (15.15) it is clear that if the difference between wavelengths of the two waves decreases, then longitudinal coherence increases. The derivation of Eq. (15.15) is given in Appendix B.

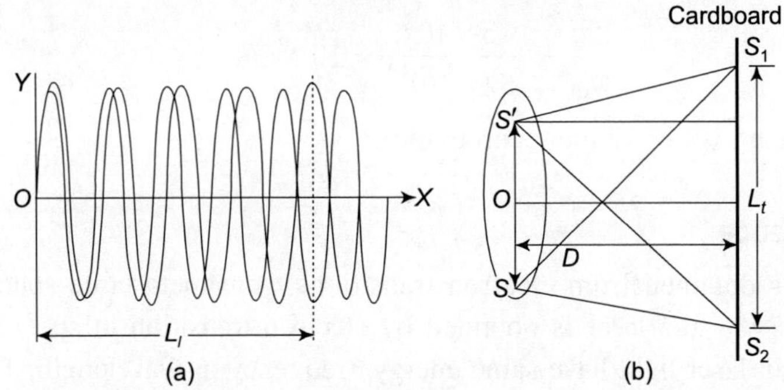

Fig. 15.12 (a) Longitudinal coherence or temporal coherence
(b) transverse coherence or spatial coherence

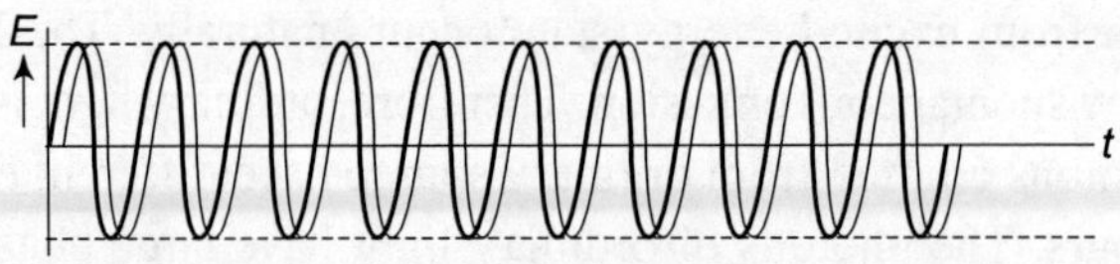

Fig. 15.13

The longitudinal coherence is also called temporal coherence. The temporal coherence time (τ) is defined as the time from the source to the point at which the waves become complexly out of phase. Let υ_1 and υ_2 be the frequencies of the two waves. The temporal coherence time is the inverse of the difference, $\Delta\upsilon = |\upsilon_1 - \upsilon_2|$, of the frequencies of the two waves as

$$\tau = \frac{1}{\Delta\upsilon} \quad \text{...(15.16)}$$

Longitudinal coherence length, $L_l = c\tau$, where, c is velocity of light.

(b) Transverse coherence or spatial coherence: Let us consider two waves emitted from two points, S' and S, of a single source as shown in Fig. 15.12(b). The points, S' and

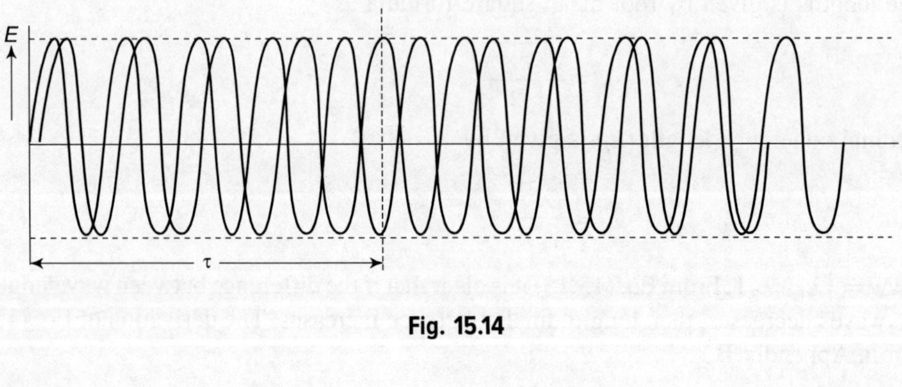

Fig. 15.14

S, are situated on a single plane perpendicular to the laser beam. Very close to the source at points S' and S the two waves are almost in phase, i.e., phase difference is zero or nearly zero. As two wave propagates towards cardboard the phase difference between two waves increases. After travelling a distance (D) equal to coherence length (L_l), two waves becomes out of phase laterally, i.e., the two waves at points S_1 and S_2 are out of phase. The transverse coherence length (L_t) is defined as the lateral distance between two points, S_1 and S_2, situated on the cardboard placed transverse to the laser beam at which the two waves are out of phase. If diameter of the beam, $S'S = a$ and distance between source and cardboard is D, then the transverse coherence length is given by

$$L_t = \frac{\lambda D}{a}, \quad ...(15.17)$$

where, λ is the wavelength of the laser source. The transverse coherence length given by Eq. (15.22) is also obtained with the help of Young's double slit experiment given below.

Let us place a cardboard having two pinholes S_1 and S_2 at distance D from the source of light as shown in Fig. 15.15. The wave emerging from two pinholes overlaps and will give interference fringes of good at the screen. Since the frequencies of photon waves differs from each other, of course, by very small amount. These waves having different frequencies after emerging from pinholes overlap and give rise to fringes on the screen. Because of this reason, let us now place another point source due to another wave S' near S as shown in Fig. 15.16. Now there will be interference pattern at the screen due to light waves from point sources S and S' through pinholes S_1 and S_2. If S' moves slowly away from S, then the contrast of interference fringes slowly becomes poorer. For distance, say $S'S = a$, $SS_1 - S'S_1 = \frac{\lambda}{2}$ and then the maxima of interference pattern produced by S falls on the minima of interference pattern produced by S' and vice versa. At this position of S', the interference fringes due to S will disappear and uniform illumination will be there on the screen.

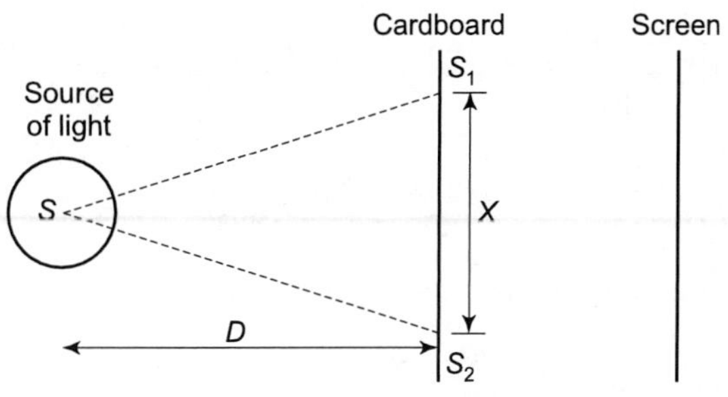

Fig. 15.15

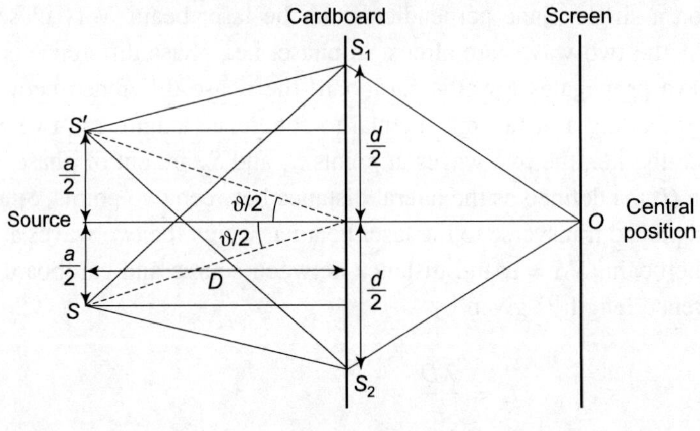

Fig. 15.16

From the geometry of Fig. 15.16, we have

$$SS_1 = \sqrt{D^2 + \left(\frac{d}{2} + \frac{a}{2}\right)^2} = D\left(1 + \frac{1}{4\,D^2}(d+a)^2\right)^{1/2}$$

$$= D\left(1 + \frac{1}{8\,D^2}(d+a)^2\right)$$

$$S'S_1 = \sqrt{D^2 + \left(\frac{d}{2} - \frac{a}{2}\right)^2} = D\left(1 + \frac{1}{4\,D^2}(d-a^2)\right)^{1/2}$$

$$= D\left(1 + \frac{1}{8\,D^2}(d-a)^2\right)$$

$$SS_1 - S'S_1 = D\left(1 + \frac{1}{8\,D^2}(d+a)^2\right) - D\left(1 + \frac{1}{8\,D^2}(d-a)^2\right)$$

$$= D\left(1 + \frac{1}{8D^2}(d^2 + a^2 + 2\,da)\right) - D\left(1 + \frac{1}{8D^2}(d^2 + a^2 - 2\,da\right)$$

$$= D\,(2\,da + 2\,da)\frac{1}{8\,D^2} = \frac{da}{2\,D}$$

Since two waves at points S_1 and S_2 are out of phase so

$$SS_1 - S'S_1 = \frac{da}{2\,D} = \frac{\lambda}{2} \qquad \text{...(15.18)}$$

Here, λ is the wavelength of the source. The distance between pinholes S_1 and S_2 (d) and distance between pin holes from Eq. (15.18) are

$$d = \frac{\lambda D}{a} \quad \text{and} \quad a = \frac{\lambda D}{d} \qquad \text{...(15.19)}$$

From extended sources the fringes will be observed when

$$d << \frac{\lambda D}{a} \quad \text{and} \quad a << \frac{\lambda D}{d} \qquad \text{...(15.20)}$$

From Fig. 15.10, one can see that

$$\tan\frac{\vartheta}{2} = \frac{\frac{a}{2}}{D} = \frac{a}{2D}$$

When ϑ is small,

$$\frac{\vartheta}{2} = \frac{a}{2D}$$

Thus, the angle by the extended source at the slit is

$$\vartheta = \frac{a}{D} \qquad \text{...(15.21)}$$

Putting the value of $\frac{a}{D}$ in Eq. (15.20), we get

$$d << \frac{\lambda}{\vartheta}$$

While performing experiment, pinhole S_1 is fixed and pinhole S_2 is moved. The area over which good contrast of interference fringes is observed is called the "coherent area" of the light wave. The distance between the pinholes (L_t) at which the fringes just disappears is called transverse lateral or spatial coherence length. Thus, the spatial coherence length

$$L_t = \frac{\lambda D}{a} = \frac{\lambda}{\vartheta} \qquad \text{...(15.22)}$$

The fringes disappear when the waves coming from two pinholes are out of phase.

Measurement of Coherence Length using Michelson Interferometer

Coherence length can be measured from Michelson interferometer [Fig. 10.1]. From the measured coherence length coherence time can be calculated.

In the beginning keep equal distance of both mirrors from partially reflecting glass plate. So path difference will be zero in the beginning. Now adjust both the mirrors to obtain circular fringes. Suppose two mirrors are placed at the distances of d_1 and d_2 from the partially reflecting glass plate. The difference in the distance of two mirrors from the partially reflecting glass plate will be $d = d_1 - d_2$. Then path difference between the interfering beams will be $2d$. If $2d < l$, then two interfering beams on overlapping produces fringes of good contrast. Move one of the mirrors slowly. When $2d > l$, then two interfering beams will not remain coherent. As the path difference increases continuously, contrast between dark and bright fringes falls off gradually. The path difference where fringes vanishes or disappear is an estimated *coherence length.*

Coherence length for sodium light is estimated to be 3 cm. So coherence time

$$\tau = \frac{l}{c} = \frac{3}{3\times 10^{10}} = 10^{-10} \text{ sec.}$$

For cadmium red light coherence length is 30 cm. The band width ($\Delta\upsilon$) is related to both coherence time τ of them. wave by the relation

$$\Delta\upsilon = \frac{1}{\tau}$$

Thus, the concept of temporal coherence is directly connected to monochromaticity. For mercury vapour lamp and He-Ne (helium-neon) laser, the longitudinal coherence lengths are 0.50 m and 300 m, respectively.

15.9 OPTICAL RESONATOR

Optical resonator or an optical cavity consists of two highly polished mirrors: one mirror has 99.99% reflectivity and the other has about 90% reflectivity. The second mirror used for laser output. These are placed parallel to each other at both the ends of quartz tube during laser action. When an active material is placed in between two perfectly parallel laser mirrors, the reflected laser light when passing through the active material induces more lasing action in the active material. Optical standing waves are formed in the active material due to reflections from both the mirrors. Let L be the length of the active material (distance between the two mirrors) and λ be the wavelength of the light. This is possible when the incident and reflected waves are out of phase. When two waves are out phase, then length (L) between two mirrors must be half integral multiple of λ, i.e.,

$$L = m\frac{\lambda}{2}, \quad \text{where } m = 1, 3, 5,... \qquad ...(15.23)$$

Velocity of the light within active material

$$\upsilon = \lambda\upsilon = \frac{2L\upsilon}{m}$$

The frequency of the oscillation modes will be

$$\upsilon = \frac{mv}{2\,L\upsilon} = \frac{mc}{2\,L\mu} \quad \text{as } c = \mu\upsilon \text{ within the active material} \qquad ...(15.24)$$

Here, μ is the refractive index of the active material. Frequencies of the oscillation of the modesare discrete. Oscillation mode frequencies are different for different values of m. These oscillation frequencies are, in fact, the longitudinal modes of the optical cavity. For an optical cavity, m is a very large number. From Eq. (15.24), it is clear that the output laser radiations consist of several closely spaced resonant lines lying within the bandwidth of the radiation.

The frequency difference (Δυ) between adjacent longitudinal modes is obtained as

$$\Delta\upsilon = (m+1)\frac{c}{2L\mu} - m\frac{c}{2L\mu} = \frac{c}{2L\mu} \qquad \text{...(15.25)}$$

The number of optical modes (N) within the bandwidth can be obtained by dividing the emission line width of the beam by Δυ, i.e.,

$$N = \frac{\text{Banwidth}}{\Delta\upsilon} \qquad \text{...(15.26)}$$

SOLVED EXAMPLES

Example 15.1 For a fully transparent active material of length 75 cm, calculate the value of m in Eq. (15.24) for a light of wavelength 6000Å.

Solution: Formula used

$$m = \frac{2L}{\lambda}$$

Given: $L = 75$ cm, $\lambda = 6000 \times 10^{-8}$ cm, $\mu = 1$

$$m = \frac{2 \times 75}{6000 \times 10^{-8}} = 2.5 \times 10^{6}; \text{ a large number}$$

Ans. Value of m is 25,00,000

Example 15.2 Determine the frequency difference between the cavity modes of a laser that is 50 cm long and contains gas of refractive index 1.0305. If the bandwidth of the laser beam is 1.6×10^9 Hz, find the number of optical modes in the output radiation.

Solution: Formula used

$$\Delta = \frac{c}{2L\mu},\ N = \frac{\text{Bandwidth}}{\Delta\upsilon}$$

Given: $\mu = 1.0305$, $L = 55$ cm, bandwidth $= 1.6 \times 10^9$ Hz, $\Delta\upsilon = ?$

$$\Delta = \frac{3 \times 10^{10}}{2 \times 55 \times 1.0305} = 2.647 \times 10^{8}\ \text{Hz}$$

The number of optical modes, $N = \dfrac{1.6 \times 10^9}{2.647 \times 10^8} = 6$

Ans. There are 6 optical modes in the output.

Example 15.3 The laser beam of wavelength 7000 Å has coherence time 5×10^{-5} s. Determine the temporal coherence length, spectral width and purity factor.

Solution: Formulae used

Longitudinal coherence length, $L_1 = c\tau$

Spectral width, $\Delta = \dfrac{\lambda^2}{L_l}$

The purity factor, $Q = \dfrac{\lambda}{\Delta\lambda}$

Given: $\lambda = 7000$ Å $= 7000 \times 10^{-8}$ cm $= 7000 \times 10^{-10}$ m, $\tau = 5 \times 10^{-5}$ s

The temporal coherence length, $L_1 = c\tau = 3 \times 10^8\ 5 \times 10^{-5}\ s = 15 \times 10^8 = 15{,}000$ m

Spectral width, $\Delta\lambda = \dfrac{\lambda^2}{L_l} = \dfrac{(7\times10^{-7})^2}{15000} = \dfrac{49}{15}\times10^{-17} = 3.28 \times 10^{-17}$ m

The purity factor, $Q = \dfrac{\lambda}{\Delta\lambda} = \dfrac{7\times10^{-7}}{3.28\times10^{-17}} = 2.13 \times 10^{-10}$

15.10 HE-NE LASER

He-Ne laser is a 4-level gas laser. There are four essential components of He-Ne laser.

- High voltage power supply for pumping source
- Quartz tube
- Resonator cavity
- A mixer of He and Ne gases as an active medium

For pumping the electrons from ground state to pumping state a DC power supply of 10 kV is required. A quartz tube is needed to keep the mixer of active material at pressure 1 mbar for Helium (He) gas and 0.1 mbar for neon (Ne). The resonator cavity consists of two highly polished mirrors. One mirror optically polished mirror of reflectivity 99.99% is placed at one end of the quartz tube. Another optically poished mirror of reflectivity about 90% reflectivity is place at another end of the quartz tube. This mirror also acts a window for getting the laser output. The mixture of helium and neon in the ratio in between 10 : 1 to 5 : 1 is used as the medium.

Principle

In He-Ne gas laser, neon atoms act as active centers, i.e., laser action occurs between the atomic levels of Ne only. Since the concentration of He atoms in the mixer is high so helium (He) atoms excited first by electron emitted by DC power supply. When sufficient number of He atoms are excited, then these He atoms collide with Ne gas atoms. After gaining the energy from He atoms, the Ne atoms get excited. When population inversion of Ne atoms is achieved, the stimulated emission occurs between Ne atomic levels only by emitting the red radiations. He atoms play only the role of excitation of Ne atoms. This method for producing He-Ne laser was first used by Ali Javan and co-workers at Bell Telephone laboratory in U.S.A. in 1951.

The energy level diagram of He-Ne laser is shown in Fig. 15.15. $5s$ and $4s$ bands of neon gas act as upper laser levels and $4p$ and $3p$ bands act as lower laser levels. One stimulated emission of wavelength 33900 Å occurs between $5s \rightarrow 4p$ in the infrared region. Second stimulated emission of wavelength 11520 Å occurs between $4s \rightarrow 3p$ also in the infrared region. The third stimulated emission occurs between $5s \rightarrow 3p$ by emitting radiation of red colour and of wavelength 6328 Å. The red laser is most popular and widely used.

Construction

It consists of a quartz tube of length 80 cm and diameter 1 cm. The tube is filled with the mixture of He and Ne in the ratio ranging from 10 : 1 to 5 : 1 at pressure 1 torr (= 1 mm Hg). However, the optimum ratio of He and Ne is 7 : 1. In the tube, pressure of helium gas is 5 to 10 times that of neon gas. At the ends of tube two mirrors are placed. One mirror is highly polished (99.99 reflectivity) and other mirror is partially polished (about 90% reflectivity), called output mirror. The mirrors act as cavity or resonator. For discharge of a gas in the tube, a high voltage of about 10 kV is applied by means of metal electrodes.

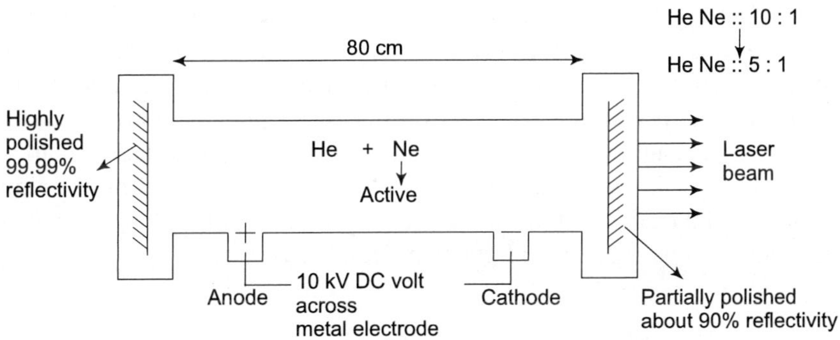

Fig. 15.17

Working

Basically, He-Ne laser is a 4-level laser. Ground state of He atom is $1s^2$. First excited states of He atoms are $2s$ energy states or $2\,^3S$ (triplet) and $2\,^1S$ (singlet) energy states. When high voltage DC voltage of about 10 kV is applied across the tube as shown in Fig. 15.17, the electrons move in the mixture from the cathode to anode. In the tube He atoms are in majority, so electrons strike first the Helium atoms. After gaining energy from electrons, He atoms get excited first to metastable states $2\,^3S$ (triplet) and $2\,^1S$ (singlet). The ground state neon gas atoms are $1s^2\,2s^2\,2p^6$. The exited states of neon gas atoms are $3s^1\,3p^1\,4s^1\,4p^1\,5s^1$. The excited helium atoms collide with neon atoms. The energy of $2\,^3S$ (triplet) state of He is identical to the energy state $5s$ of Ne gas atoms. Thus, resonant energy transfer takes place from He atoms to Ne atoms. After gaining energy from excited He atoms, neon gas atoms get excited to $5s$ metastable energy levels of Ne. While transferring energy to Ne atoms, helium atoms return to ground state. 5*s* metastable state of neon atoms have the longer lifetime. Therefore, a large number of electrons will stay for long time in the metastable state (5*s*) of Ne. As result, electrons get accumulated in metastable state (5*s*). So the number of electrons becomes more in metastable state than ground state of neon. This is called population inversion. The laser transition takes place only between the states of neon atoms, whereas helium atoms only help in the excitation of neon atom. When population inversion is achieved at level 5*s* of neon atoms, then after completing its lifetime (say) one electron drops from 5*s* energy state to 3*p* energy state by emitting a photon by spontaneous emission. The wavelength of photon wave by spontaneous emission from $5s \rightarrow 3p$ is 6328 Å in the red colour region.

This photon bounces back and forth between two highly polished mirrors and forces one electron in 5*s* metastable energy state to drop to 3*p* energy state. As a result, one photon of wavelength 6328 Å is emitted. The two photons: one stimulating photon and the second stimulated photon have same wavelength and same phase. These two photons again bounce back and forth between two highly polished mirrors and force two more electrons in 5*s* energy state of neon gas to drop down to 3*p* energy state making 4 photons of same wavelength and of same phase. Again these two photons force two more electrons in the 5*s* energy state to drop down to 3*p* energy state with emission of 4 more photons making 8 photons of same wavelength and of same phase. In the same way, a chain reaction continues to force the electrons in the 5*s* energy state with stimulated emission of photons of same energy producing millions of photons. Every stimulated electron releases a photon of wavelength 6328 Å. That is, the waves associated with the photons have the same energy, phase, polarization and direction of travel, i.e., the waves are coherent. Finally, an intense beam of unidirectional coherent photons is obtained. This is the principle of for producing LASER (light amplification by stimulated emission by radiations) light from He-Ne gas laser.

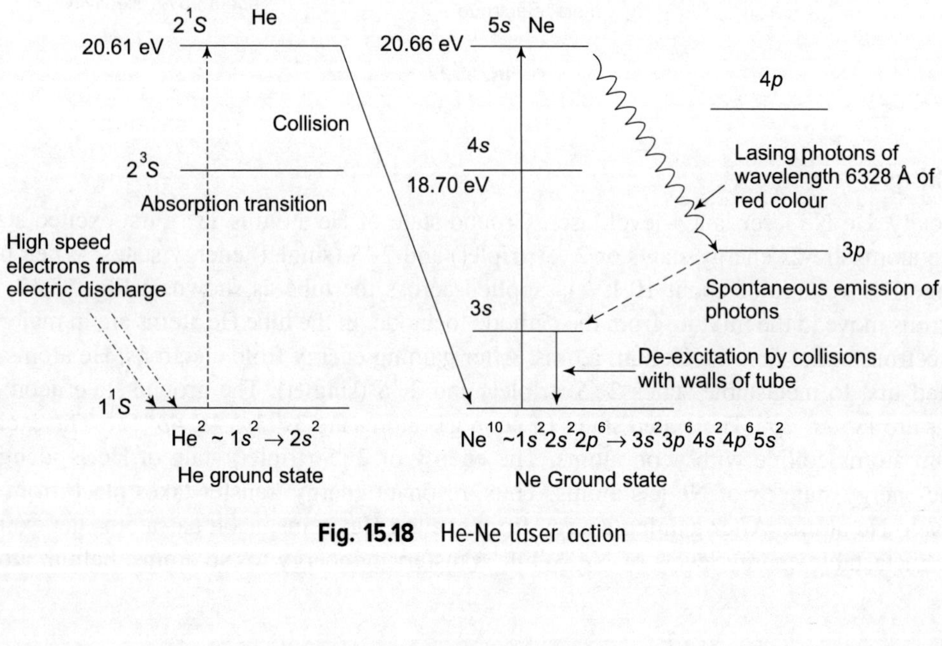

Fig. 15.18 He-Ne laser action

The laser beam emerges out from a semi-silvered/polished mirror. Power output of He-Ne gas laser is between 1 and 50 milliwatt. The electrons from $3p$ states drop down to $3s$ state by spontaneous emission of photons continuously. This is necessary for producing continuous laser beam. The wavelength of the incoherent radiations (photons) of the transition from $3p$ to $3s$ is 6000 Å. Finally, the electrons in the energy state $3s$ drop to ground state after colliding with walls of the quartz tube. As the lower lasing levels depopulate faster than the upper metastable states, it is easier to maintain the state of population inversion between the lasing levels throughout the laser operation. If discharge is maintained continuously, a continuous laser beam is produced by He-Ne gas laser. The continuous red laser beam from He-Ne laser action is most popular and widely used.

The He-Ne lasers have 0.1% efficiency. That is, 0.1% of the electric power supplied to the device is converted to laser light. For working of He-Ne gas laser in detail see Appendix A.

Advantages of gas laser over the solid state lasers.

A gas laser as number of advantages over the solid state lasers.

1. He-Ne gas laser produces a continuous beam rather than short pulses as produced in ruby laser.
2. He-Ne gas laser emits light in the visible region.
3. He-Ne gas laser is capable of operating continuously without cooling as the power of gas laser is very low.
4. He-Ne gas laser is highly stable.
5. He-Ne gas laser operates on low cost.
6. He-Ne gas laser operates at high temperature without damage.
7. Power output per unit band width of gas laser is large.

These advantages of a gas laser over solid state laser are because gas laser is free from crystal imperfection, thermal distortion and scattering present in solid state laser.

Disadvantages of He-Ne gas laser

1. He-Ne gas laser has low efficiency.
2. He-Ne gas laser has low gain.
3. He-Ne gas laser is limited to low power tasks.

Applications of He-Ne gas laser

1. He-Ne gas lasers are used in industrics.
2. He-gas lasers are used in scientific instruments.
3. He-Ne gas lasers are used in college laboratories.

15.11 SEMICONDUCTOR LASER

Semiconductor lasers play an important role in our everyday life. These lasers are economical, compact and consume less power. Semiconductor lasers are also known as laser diodes.

Semiconductor lasers are different from solid-state lasers. In solid-state lasers, light energy is used as the pump source whereas, in semiconductor lasers, electrical energy is used as the pump source.

In semiconductor lasers, a *p-n* junction of a semiconductor diode forms the active medium or laser medium. The optical gain is produced in the semiconductor material.

Forward biased diode or *p-n* junction in Ge and Si acts as a light emitting diode (LED). The emitted light from these diodes is incoherent due to spontaneous emission. It is observed that forward biased *p-n* diode of semiconductor gallium-arsenide emits coherent light due to stimulated emission. Thus, population inversion and simulated emission can be achieved in it or laser action will be possible in a semiconductor *p-n* junction. This device is called a semiconductor laser or diode laser or laser diode.

Principle

The forward biased *p-n* diode of semiconductor gallium-arsenide is shown in Fig. 15.19(a). When *p-n* diode is forward biased, the majority carrier holes (*p*) and electrons (*n*) starts moving toward depletion region. In this region electrons and holes combine and emit light. The energy band diagram of *p-n* diode is shown in Fig. 15.19(b).

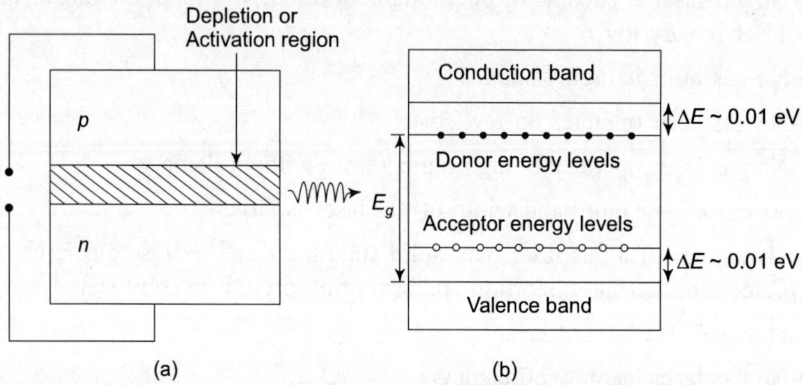

(a) Forward biased *p-n* diode (b) Band picture of *n-p* diode

Fig. 15.19

The energy bands of *n* type impurity (e.g., tellurium) are just below the conduction band, while energy bands of *p* type impurity (e.g., zinc) are just above the valence band as shown in Fig. 15.19(b). At about 10–25 °K, the electrons from valence band excited to *p* impurity band and electrons from *n* type of impurity band are excited to conduction band. Forbidden energy gap is written as

$$E_g \approx \frac{hc}{\lambda} \qquad ...(15.27)$$

When p-n junction is forward biased, the electrons from valence band drop to valence band and radiate phonon energy nearly equal to gap energy, E_g. The wavelength,

$$\lambda \approx \frac{hc}{E_g} \qquad ...(15.28)$$

associated with emitted light wave is in the visible range, i.e., 4000 Å $\leq \lambda \leq$ 7000 Å. Here, c is the velocity of light. Putting the value of $h = 6.623 \times 10^{-34}$ Js and $c = 3 \times 10^8$ m/s, λ_{min} = 4000 Å and λ_{min} = 7000 Å, 1 eV = 1.602×10^{-19} J. The energy range of energy gap of semiconductor laser material is

$$\frac{6.623\times10^{-34}\times3\times10^{8}}{7\times10^{-7}}\frac{1}{1.602\times10^{-19}} \leq E_g \leq \frac{6.623\times10^{-34}\times3\times10^{8}}{4\times10^{-7}}\times\frac{1}{1.602\times10^{-19}}$$

1.77 eV $\leq E_g \leq$ 3.10 eV

The material for laser semiconductor can be selected in this energy gap range.

Laser action in *p-n* junction

The p-n junction semiconductor diode is shown in Fig. 15.20.

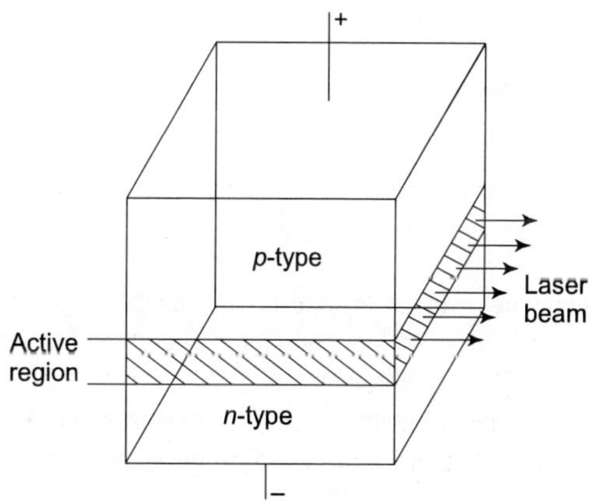

Fig. 15.20 Band picture of n-p diode

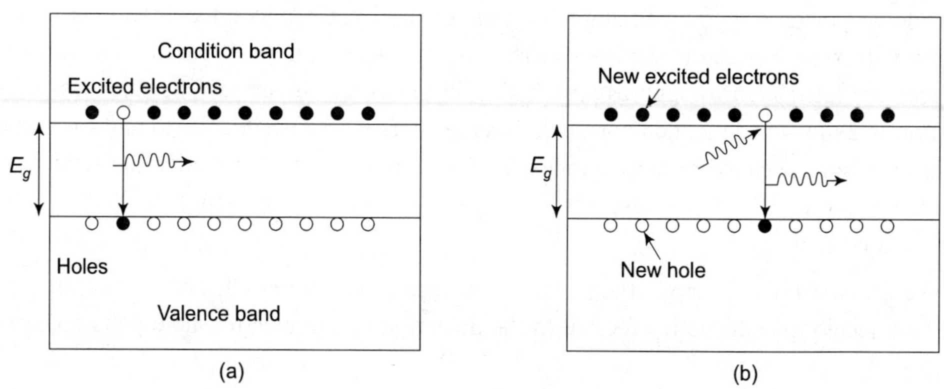

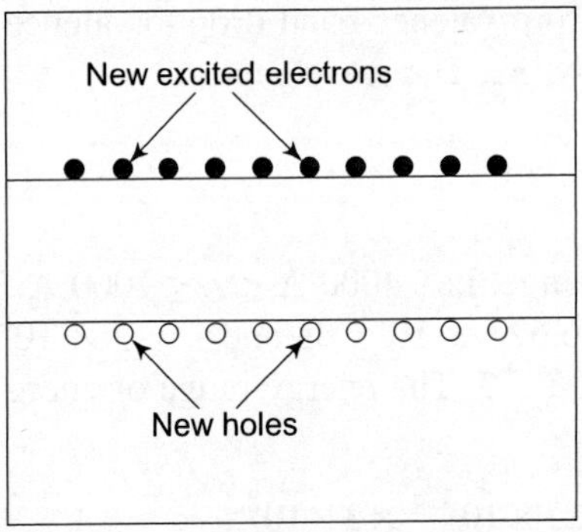

(c)

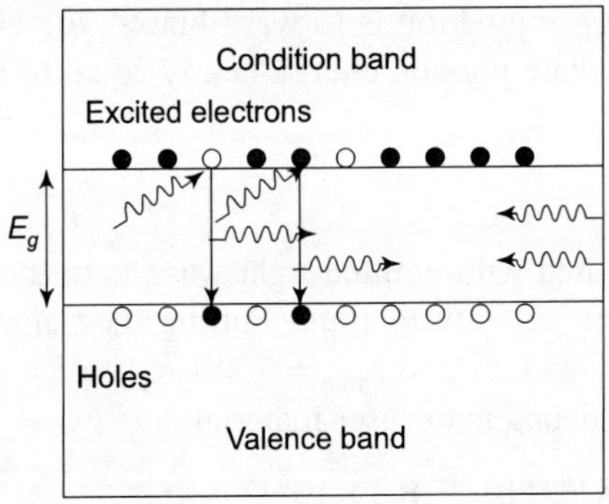

(d)

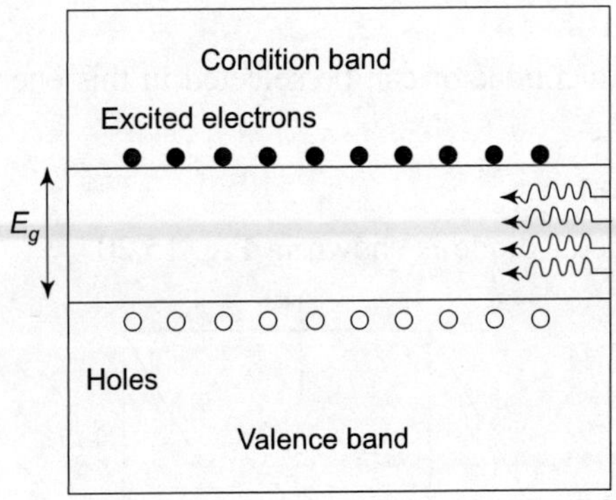

(e)

Fig. 15.21 Working of semiconductor laser. (a) one photon emitted by recombination of electron and hole, (b) stimulated emission one photon, (c) two photons: one emitted by recombination of electron and hole as mentioned in (a), another is stimulated photon, (d) two more stimulated photons making four photons (d) These four photons again stimulate 4 electrons to produce 4 more photons. In the same way, the chain reaction continues to produce millions of photons

First semiconductor laser of this type was made in 1962 using GaAs material. The gallium arsenide was made *n*-type by adding tellurium, while *p*-type semiconductor was achieved by doping with zinc. For achieving laser action, two end faces are made parallel and other two surfaces are left rough to suppress oscillation in undesired direction. The thickness of *p-n* junction is 2 μm. In semiconductor diode laser no mirrors are needed for to and fro motions for laser action, because refractive index is large enough to give considerable reflection at semiconductor interfaces. The laser output of GaAs oscillates at wavelength 8500 Å to 9000 Å near infrared region.

When forward bias is applied to *p-n* junction, electrons from *n*-side and holes from *p*-side will be injected into the active region. In the depletion region electron-hole pairs recombine and spontaneous emission of light takes place. As the forward biasing is increased gradually, electrons and holes recombine and the excess of energy is emitted as coherent light radiation.

In band picture model as shown in Fig. 15.21, the applied voltage excited the electrons from the valence band to the conduction band across the forbidden energy band resulting in the creation of holes in the valence band as shown in Fig. 15.21(a). Eventually, an electron and a photon recombine and emit a photon in spontaneous emission, a photon of wavelength given by Eq. (15.28) is emitted.

This photon bounces back and forth from two end parallel faces of diode and reflected back in the central active region, where it stimulates one electron in the conduction band producing one more photon as shown in Fig. 15.21(b).

These two photons have same wavelength and phase with one another, bounce back and forth from the end faces of the diode forces as shown Fig. 15.21(c) and reflected in the active central region. At the same time new excited electrons and new holes are generated by the biasing voltage.

These two photons force two more electrons to drop to valence band and recombine with holes. As a result two more photons of same wavelength are emitted as shown in Fig. 15.21(d) making four coherent photons of same wavelength. These four photons bounce back and forth from the end surfaces of the diode as shown in Fig. 15.21(e) and forces four more electrons to drop to valence band and recombine with holes emitting four more photons of same wavelength and in the same phase. In the same way a chain reaction continues to produce millions of photons. Every stimulated electron releases a photon of same energy. The waves associated with the photons have the same energy, phase, polarization and direction of travel, i.e., the waves are coherent. Finally, an intense beam of unidirectional coherent photons is obtained. Some of the laser beam escapes through a partially reflecting end surface of the semiconductor. This process is called laser action in a semiconductor laser. For laser action, two end faces of the diode semiconductor are made parallel so they act as perfect reflectors and other two surfaces are left rough to suppress oscillation in undesired direction.

Advantages of Semiconductor Laser

Diode lasers are economical, simple in construction, small in size, of light weight, good reliability, long service life and consume very less power (Continuous beam ~ 10 mW) as compared to 10 kW consumed by He-Ne lasers making them ideal for electronic devices, such as CD and DVD plyers, CD and DVD and other optical data storage devices. Since semiconductor lasers operate at low voltage, hence provide low power failure rate, safe operation and low cost of maintenance. It is easy in operation, has long life, is tunable and continuous beam. Besides, there are many types of semiconductor lasers, such as infrared and red lasers with long wavelength and green and blue lasers with shorter wavelength. The semiconductor lasers have 20% efficiency. That is, 20% of the electric power supplied to the device is converted to laser light as compared with 0.1% in the He-Ne laser. The signals of semiconductor laser take less than 10^{-10} s. That is, the light signals can be switched on and off within no time.

Disadvantages of Semiconductor Laser

- The temperature affects its output.
- Beam divergence is much greater as compared to other lasers.
- Its cooling requirement is considered to be its drawback in some cases.
- This laser is not suited due to low power production.
- They are not as suitable as He-Ne laser for high quality holography.

Applications of Semiconductor Laser

Diode lasers are economical, small in size, of light weight, good reliability, long service life and consume less power making them ideal for electronic devices, such as CD and DVD players, CD and DVD and other optical data storage devices. Laser printers, laser fax machines and supermarket barcode readers all use diode lasers. Using helium-neon laser increases the size of these devices by almost five times. Diode laser operates on low power, while gas and solid state lasers require a power supply of the order kilovolts.

Safety Risk

Normally, all lasers pose a significant safety risk due to high power and intensity. However, diode lasers are considered one of the safest types of lasers. Since diode lasers operate on low power, so they present minimal danger of electric shock. However, there is still harmful for human or animal eyes, if they are shined into the eyes directly.

15.12 RUBY LASER

A ruby laser is a 3-level solid state laser. It uses the synthetic ruby laser as is laser medium. Ruby laser is one of the few solid state lasers that produce visible light. It emits deep red light of wavelength 6943 Å.

Construction of Ruby Laser

A ruby laser consists of three important elements: Laser medium, the pump source, and the optical resonator. Ruby is a single crystal of Al_2O_3. It is a stable form of aluminum oxide. In the form of cylinder acts as a laser medium or active medium. The laser medium in the ruby laser is made of the sapphire (Al_2O_3) doped with small amount of chromium ions (Cr^{3+}).

In ruby laser we use the flash tube as the energy source or pump source for getting population inversion. When electrons in lower energy state gain sufficient energy from the flash tube, they jump to the higher energy state or excited state.

The cylindrical ruby rod is placed between two highly polished reflected mirrors. At one end of the rod, the mirror is highly polished (99.99% reflectivity), whereas at the other end the mirror is partially (90% reflectivity), that it allows a small portion of light through the produced laser light.

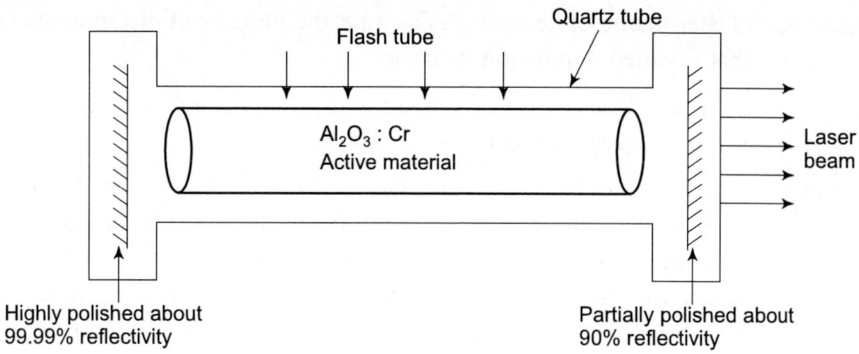

Fig. 15.22

Working of Ruby Laser

Ruby laser is a three-level solid-state laser. In ruby laser, optical pumping is used using flash tube. That light from flash tube raises the electrons from the ground level to excited level.

Let us consider the ruby laser medium consisting of energy levels E_1, E_2 and E_3 with N number of electrons. We assume that the energy level E_1 is the lower or ground energy level, the energy level E_2 is metastable energy level and the energy level E_3 is the third excited state.

In the beginning most of the electrons are in ground state (E_1) and a few of the electrons are in excited state (E_3).

The lifetime of excited state (E_3) is very small (10^{-8} s), the electrons will not stay in this for long. After a short time they fall into metastable state (E_2) by the emitted radiating energy in the form of phonons or heat energy. The lifetime of metastable state (E_2) is (10^{-3} s) which is much greater than the lifetime of excited state, E_3. Therefore, the electrons drop to metastable

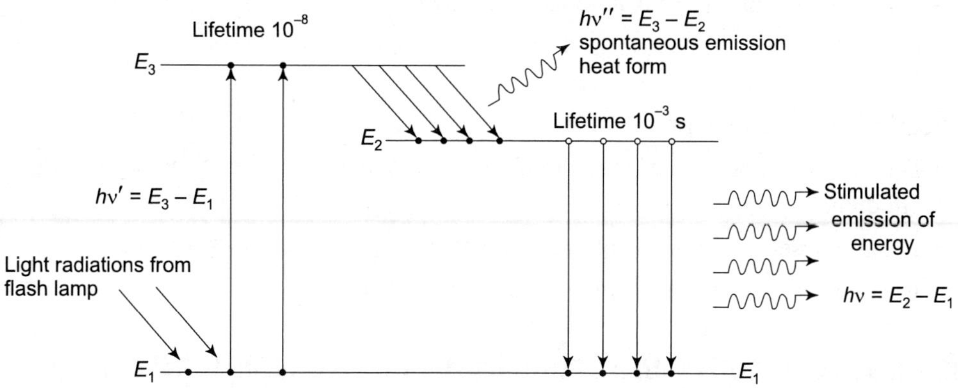

Fig. 15.23

energy state much faster than they leave it. As a result the number of electrons increases in metastable state. This is called population inversion.

After some time the electrons in metastable state drop to ground state by releasing the energy in the form of light by spontaneous emission.

The emitted photons bounce back and forth between two highly polished mirrors, they get amplified. These amplified photons interact with the electrons in the metastable state to force more electrons to drop into the ground state by emission of more photons. Further, these emitted photons interact with more electrons in the metastable state to force other electrons to drop into the lower state (E_2) by emission of more photons. Similarly, a chain reaction continues producing a million of photons. Every stimulated electron releases a photon of energy, $h\upsilon = E_2 - E_1$. That is, the waves associated with the photons have the same energy, phase, polarization and direction of travel, i.e., the waves are coherent. Finally, an intense beam of unidirectional coherent photons is obtained. This is how the LASER (light amplification by stimulated emission by radiations) light is produced by ruby laser.

The laser light comes out through the partially reflecting mirror. The semiconductor ruby lasers have 1% efficiency. That is, 1% of the electric power supplied to the device is converted to laser light as compared with 0.1% in the He-Ne laser.

Advantages of ruby laser

- Ruby lasers are economical.
- Output of ruby laser is not as less as that of He-Ne laser.
- Since ruby is in solid form so there is no chance of wasting material of active medium.

Disadvantages of ruby laser

- In ruby laser no significant emission occurs, until at least half of the ground state electrons have been excited to the metastable state.
- Efficiency of ruby laser is comparatively low.
- Optical cavity of ruby laser is short as compared to other lasers.

Applications of Ruby Laser

- Due to low output power they are class I lasers and so they may be used as toys for children.
- It can be used in schools, colleges, universities for science programmes.
- It can be used as decoration piece and artistic display.

15.13 ND : YAG (NEODYMIUM: YTTRIUM ALUMINIUM GARNET) LASER

Nd : YAG (Nd: Y_3 Al_5 O_{12}) laser is a 4-level solid-state laser. In Nd : YAG Laser, Neodymium-doped Yttrium Aluminum Garnet is used as a laser medium. This laser has many different applications in the medical and scientific field for doing laser surgery and to study laser spectroscopy.

Nd : YAG laser is a four-level laser, meaning that four levels are involved in laser action. This laser operates in both continuous and pulse modes.

Nd : YAG laser produces laser light in the infrared region of wavelengths 9400 Å, 10640 Å, 1320 Å and 1440 Å.

Construction of Nd : YAG laser

There are three important elements of Nd : YAG laser: energy source for pumping, active medium and optical resonator.

The source for pumping the electrons from ground state to excited state is laser diode or flush tube, because of low cost and high efficiency. The active medium of Nd : YAG laser is a synthetic crystal Yttrium Aluminum Garnet doped with neodymium (Nd). Optically coated two mirrors are used as resonator. One mirror is highly polished with reflectivity 99.99%. The second is also a highly polished mirror with about 90% reflectivity. The second mirror also acts as window for output of laser light.

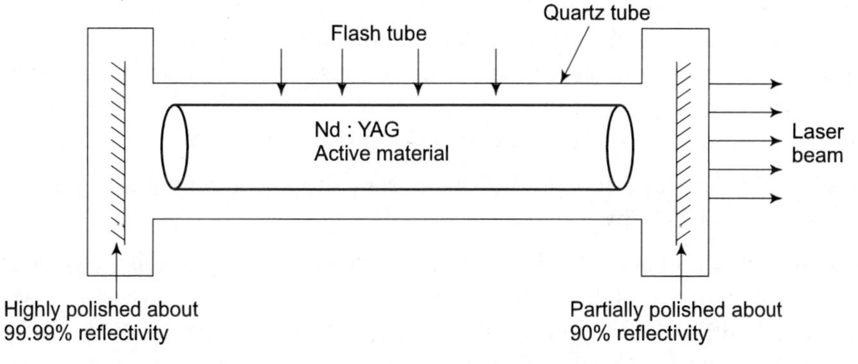

Fig. 15.24

Working of Nd : YAG laser

Nd : YAG laser is a four-level laser. This means that four levels are involved in the laser action. The diode or flash tube light source is used for exciting the electrons to the excited state to achieve population inversion.

Let us consider that Nd : YAG crystal active medium consisting of four levels: E_1, E_2, E_3 and E_4 with number of electrons. N_1, N_2, N_3 and N_4, respectively. We assume that energy E_4 level E_1 is the lower or ground energy level, the energy level E_2 is next higher energy level. The energy level E_3 is metastable state or excited state. The fourth level E_4 is the pumping or excited state. Let us assume that $N_1 > N_2 > N_3 > N_4$.

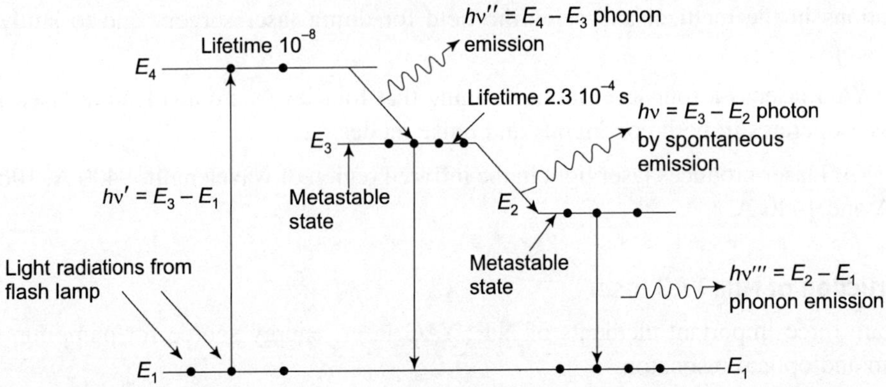

Fig. 15.25

When light energy using a diode or from flash tube falls on the Nd : YAG active material, the electrons from ground state (E_1) are raised to excited state, E_4.

The lifetime of excited state (E_4) is very small (230 10^{-6} s), therefore, the electrons do not stay in this state for long. After a short time, they fall into metastable state (E_3) by radiating energy in the form of phonons or heat. The lifetime of metastable state (E_3) is 10^{-3} s, therefore, the electrons stay in metastable state for long. Therefore, the electrons' drop to metastable energy state is much faster than they leave it. As a result the electrons get accumulated in metastable state. At one point of time, the number of electrons in metastable state becomes more than the ground state. This process is called population inversion.

After some time the electrons in metastable state drop to ground state by releasing the energy in the form of light by spontaneous emission.

The emitted photons bounce back and forth between two highly polished mirrors, they get amplified. These amplified photons interact with the electrons in the metastable state to force more electrons to drop into the lower state (E_2) by emission of more photons. Further, these emitted photons interact with other electrons in the metastable state to force more electrons to drop into the lower state (E_2) by emission of more photons. Similarly, a chain reaction continues producing a million of photons. Every stimulated electron releases a photon of energy, $h\upsilon = E_3 - E_2$. That is, the waves associated with the photons have the same energy, phase, polarization and direction of travel, i.e., the waves are coherent. Finally, an intense beam of unidirectional coherent photons is obtained. The amplified light comes out through the partially reflecting mirror is called the laser (light amplification by stimulated emission by radiations) light or Nd : YAG laser.

The lifetime of energy state (E_2) is very small just like energy state, E_4. Therefore, after a short time, the electrons in the energy state (E_2) fall to the ground state (E_1) by releasing energy in the form of phonons or heat.

The Nd : YAG lasers have 6.5% efficiency. That is, 6.5% of the electric power supplied to the device is converted to laser light as compared with 0.1% and about 1% in the He-Ne laser and ruby laser.

The spontaneous emission is a natural process, because the electrons drop down to lower state after completing their lifetime. But simulated process is not natural process, the electrons in the metastable state are fored to drop down to lower energy state. To achieve stimulated emission, external light energy is supplied to the active medium to achieve population inversion.

Advantages of Nd : YAG laser

- Low power consumption.
- Nd : YAG laser gives high gain.
- The efficiency of Nd : YAG laser is very high as compared to the ruby laser.
- It is easy to attain population inversion.
- It is easy to operate and maintain.
- Purchasing cost is relatively lower.
- Both continuous and pulsed output is possible from solid state laser.
- Construction of solid state laser is comparatively simple.

Disadvantages of Nd : YAG laser

- Low efficiency as compared to CO_2 laser.
- Low absorption of radiation very close to visible spectrum.
- Electron energy level structure of Nd^{3+} in YAG is complex.

Applications of Nd : YAG laser

Medicine:

Nd : YAG laser is used to remove skin cancers.

Industry:

Nd : YAG laser is used for etching or making a variety of plastics and metals.

Nd : YAG laser is used for cutting and welding steel.

15.14 APPLICATIONS OF LASER

Laser is an optical device that produces a beam of powerful intensity coherent monochromatic light by stimulated emission of radiation.

Laser light is different from an ordinary light in many respects. It is highly directional, and highly intense. In addition, it is highly coherent and monochromatic. Because of these unique properties, lasers have various applications.

Lasers are widely used for the following purposes:

- Medicine
- Science and technology
- Communications

- Industry
- Holography
- Military

Medicine

1. Lasers are used for bloodless surgery. Focused laser beam is like a small and sharp knife which not only performs bloodless surgery but also welds vessels being cut.
2. Lasers are used for painless microsurgery. Because the surgery is done very fast, there is not enough to respond to the incision and sense of pain.
3. Lasers are used to destroy kidney stones.
4. Lasers are used in cancer diagnosis and therapy.
5. Lasers are used to burn up brain tumours.
6. Lasers are used for retina operation.
7. Lasers are used in fibre-optic endoscope to detect ulcers in the intestines.
8. Lasers are used to remove the decayed portion of the teeth.
9. He-Ne laser is used in curing ulcer.
10. CO_2 laser is used in plastic surgery, including burn treatment.

Science and technology

1. Laser beam is used to measure distance accurately of the moon from the earth. LIDAR (Light detector and ranging) detects the laser light reflected from the reflectors places at moon by astronauts. Measuring the time for reaching and returning from the moon, the distance of moon can be measured, since velocity of light is known.
2. Lasers are used in computers to retrieve stored information from a Compact Disc (CD).
3. Lasers are used to store large amount of information or data in CD-ROM.
4. Lasers are used to measure the pollutant gases and other contaminants of the atmosphere.
5. Lasers helps in determining the rate of rotation of the earth accurately.
6. Lasers are used in computer printers to etch images on a photoconductor drum. The images are carried from drum to paper by means of electrostatic photocopying.
7. Lasers are used for detecting earthquakes and underwater nuclear blasts.
8. A gallium arsenide diode laser can be used to setup an invisible fence to protect an area.

Communication

Because a laser beam is coherent and unidirectional, so one laser beam carries as much information aa all the existing radio channels and travel long distance.

1. Laser light is used in optical fibre communications to send information over large distances with low loss.

2. Laser light is used in underwater communication networks.
3. Lasers are used in space communication, radars and satellites.
4. Band width and hence capacity and efficiency is large when laser light is used for combination.

Industry

1. Lasers are used to cut glass and quartz.
2. Lasers are used in electronic industries for trimming the components of integrated circuits (ICs).
3. Lasers are used for heat treatment in the automotive industry.
4. Laser light is used to collect the information about the prefixed prices of various products in shops and business establishments from the bar code printed on the product.
5. Ultraviolet lasers are used in the semiconductor industry for photolithography. Photolithography is the method used for manufacturing printed circuit board (PCB) and microprocessor by using ultraviolet light.
6. Lasers are used in cutting, welding and drilling with more accuracy and precision in even very hard materials because of their very high intensity like diamond processing.

Holography

The word holograph derived from Greek word "holo", means (complete or whole) and "graph" means (photo or picture).

Because of its brightness and coherence, lasers are suitable for holography, i.e., for three-dimensional imaging of a given object or scene without lens, just like ordinary light is used to make two-dimensional photo of an object. Two-dimensional photo of an object is called photograph, while three-dimensional picture of an object is called hologram. If the hologram is broken into (say) three pieces, then every broken piece again shows a complete picture of the object.

Military

1. Laser range finders are used to determine the distance of an object. The distance of an object is measured by time taken to reach an object and time taken to return from it, as speed of light is known.
2. The ring laser gyroscope is used for sensing and measuring very small angle of rotation of the moving objects.
3. Lasers can be used as secretive illuminators for reconnaissance during night with high precision.
4. Lasers are used to dispose of the energy of a warhead by damaging the missile.
5. Pulse laser is used in light radar called LIDAR (Light detection and ranging) to accurately detect an object at long distance.

APPENDIX A

He-Ne laser

Working in Detail

Ground state He atom is $1s^2$. First excited states of He atoms are $2s$ are enery states called $2\ ^3S$ (triplet) and $2\ ^1S$ (singlet) energy states. When high voltage DC voltage of about 10 kV is applied across the tube as shown in Fig. 15.17, the electrons move in the mixture from cathode to anode. In the tube, He atoms are in majority, so electrons strike first the Helium atoms. After gaining energy from electrons, He atoms get excited to metastable states $2\ ^3S$ (triplet) and $2\ ^1S$ (singlet). The ground state neon gas atoms are $1s^2\ 2s^2\ 2p^6$. The exited states of neon gas atoms are $3s^1\ 3p^1\ 4s^1\ 4p^1\ 5s^1$. The excited helium atoms collide with neon atoms. The energy of $2\ ^3S$ (triplet) and $2\ ^1S$ (singlet) states of He is identical to the energy states $4s$ and $5s$ states of Ne gas atoms, respectively. Thus, resonant energy transfer takes place from He atoms to Ne atoms. After gaining energy from the excited He atoms, neon gas atoms get excited to a $4s$ and $5s$ metastable energy levels of Ne. While transferring energy to Ne atoms, helium atoms return to ground state. The metastable states $4s$ an $5s$ of neon atoms have a longer lifetime. Therefore, a large number of electrons stay for long in the metastable states $4s$ and $5s$ of Ne. As result, electrons get accumulated in these metastable states. At one point of time the number of electrons in $4s$ and $5s$ energy states of neon atoms become more than the number of atoms in the ground state. This stage is called population inversion. When population inversion is achieved at levels $4s$ an $5s$ of neon atoms, after some time the laser transition takes place only between levels of neon atoms, whereas helium atoms only help in the excitation of neon atom. The spontaneous emission occurs from $5s \rightarrow 4p$ and $4s \rightarrow 3p$ by emitting radiations of wavelength 33910 Å and 11530 Å, respectively, in infrared region. The spontaneous emission also occurs from $5s \rightarrow 3p$ by emitting radiations of red colour and wavelength 6320 Å.

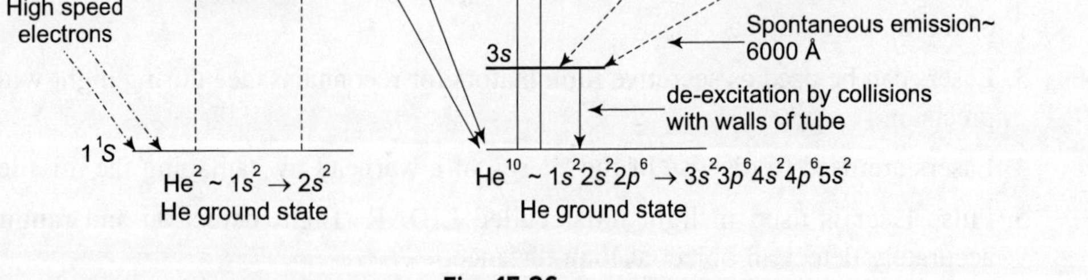

Fig. 15.26

Out of three laser beams, the red laser beam of wavelength 6328 Å is tapped and amplified by adjusting the distance between two mirrors. The photons of red laser bounce back and forth between two reflecting mirrors and trigger chain reaction to force electrons in 5*s* energy state of neon atoms with emission of photons of same energy. As a result, millions of photons are produced. Every stimulated electron releases a photon of wavelength 6328 Å. That is, the waves associated with the photons have the same energy, phase, polarization and direction of travel, i.e., the waves are coherent. Finally, an intense beam of unidirectional coherent photons is obtained. This is how the LASER (light amplification by stimulated emission by radiations) light is produced by He-Ne gas laser.

The laser beam emerges out from a semi-silvered/polished mirror. Power output of He-Ne gas laser is between 1 to 50 milliwatt. The electrons from 4*p* and 3*p* states drop down to 3*s* state by spontaneous emission continuously. This is necessary for producing a continuous laser beam. The wavelength of the incoherent radiations (photons) of the transition from 3*p* to 3*s* is 6000 Å. Finally, the neon gas atoms in the energy state 3*s* drop to ground state after colliding with walls of the quartz tube. As the lower lasing levels depopulate faster than the upper metastable states, it is easier to maintain the state of population inversion between the lasing levels throughout the time of laser operation. If discharge is maintained continuously, a continuous laser beam is produced by He-Ne gas laser. The continuous red laser beam from He-Ne laser action is most popular and widely used.

The He-Ne lasers have 0.1% efficiency. That is 0.1% of the electric power supplied to the device is converted to laser light.

Formulas at a Glance

S.No.	Subject	Explanation
1.	Longitudinal coherence length	$L_l = \frac{\lambda^2}{\Delta\lambda}$, where, $\Delta\lambda = \lvert\lambda_1 - \lambda_2\rvert$ and $\lambda = \frac{\sqrt{\lambda_1 \times \lambda_2}}{2}$
2.	Transverse coherence length	$L_t = \frac{\lambda D}{a} = \frac{\lambda}{\vartheta}$, where, λ is the wavelength of the laser source, a is diameter of the beam and D is the distance between source and slit, ϑ is the angle subtended by the source at the slit.
3.	Number of modes between cavity	$m = \frac{2L}{\lambda}$, where, L is length between two mirrors, and λ are wavelength of light waves within active material.

4.	The frequency difference ($\Delta \upsilon$) between adjacent longitudinal modes	$\Delta \upsilon = \frac{c}{2L\mu}$, where, L is distance two mirrors, μ is refractive index of active material.
5.	The number of optical modes (N) within the bandwidth	$N = \frac{\text{Banwidth}}{\Delta \upsilon}$
6.	Longitudinal coherence length	$L_l = c\tau$, Here, c is velocity of light in vacuum and $\tau = \frac{1}{\Delta \upsilon}$ coherence time
7.	Spectral width	$\Delta \lambda = \frac{\lambda^2}{L_l}$
8.	The purity factor	$Q = \frac{\lambda}{\Delta \lambda}$
9.	Velocity of the light within active material	$v = \lambda \upsilon = \frac{2L\upsilon}{m}$, where, L is length between two mirrors, λ and υ are wavelength and frequency, respectively, of light waves within active material.
10.	The frequency of the oscillation modes	$\upsilon = \frac{m\upsilon}{2L\upsilon} = \frac{mc}{2L\upsilon}$ as $c = \mu\ \upsilon$ within active material. Here, μ is the refractive index of the active material.
11.	Degree of non-monochromaticity	$\frac{\Delta \upsilon}{\upsilon_0}$
12.	Transition rate for active atom	$\frac{\text{Power}}{\text{Energy}} = \frac{P}{h\upsilon} = \frac{P\lambda}{hc}$
13.	Flux density of photon radiation	$Nh\upsilon$, where, N is number of photons per unit volume.
14.	Intensity	$I = \frac{P}{A}$ = power per unit area
15.	Momentum	$p = mc = \frac{2E}{c}$, Force, $F = \frac{P}{t}$, Power, $P = \frac{E}{t}$ as Energy, $E = \frac{1}{2}$ mc^2
16.	The ratio of Einstein's coefficients for spontaneous emission to stimulated emission	$\frac{A_{21}}{B_{21}} = \frac{8\pi h\upsilon^3}{c^3}$

APPENDIX B

(a) Longitudinal or temporal Coherence

Let λ_1 and λ_2 are the wave lengths of the two waves. The average value of these wavelengths is given by root mean square formula

$$\lambda = \sqrt{\lambda_1 \lambda_2} \quad ...(B1)$$

Let υ_1 and υ_2 be the frequencies of the two waves. The temporal coherence time is the inverse of the difference, $\Delta\upsilon = |\upsilon_1 - \upsilon_2|$, the frequencies of the two waves are

$$\therefore \quad \tau = \frac{1}{\Delta\upsilon} = \frac{1}{|\upsilon_1 - \upsilon_2|} \quad ...(B2)$$

Putting the values of υ_1 and υ_2 in Eq. (B2)

$$\tau = \frac{1}{\left|\frac{c}{\lambda_1} - \frac{c}{\lambda_2}\right|} = \frac{\lambda_1 \lambda_2}{c|\lambda_2 - \lambda_1|} \quad ...(B3)$$

$$L_t = c\tau = \frac{\lambda_1 \lambda_2}{|\lambda_2 - \lambda_1|} \quad ...(B4)$$

Longitudinal coherence length is $L_l = ct$, where, c is velocity of light.

Thus, the Longitudinal coherence length is L_l is given by

$$L_l = \frac{\lambda^2}{\Delta\lambda} \quad ...(B5)$$

where $\Delta\lambda = |\lambda_1 - \lambda_2|$. From Eq. (B5) it is clear that if the difference between wavelengths of the two waves decreases, then longitudinal coherence increases.

SOLVED NUMERICAL PROBLEMS

15.1 Calculate coherence length of a laser beam having band width of 1500 Hz.

Solution: Bandwidth, $\Delta\upsilon = 1500$ Hz

As $1 = c\tau$ and $\tau = \frac{1}{\Delta\upsilon}$

$$1 = \frac{c}{\Delta\upsilon} = \frac{3 \times 10^8}{1500} = 2 \times 10^5 \text{ m} = 2 \times 10^2 \text{ km} = 200 \text{ km}$$

Ans. Coherence length of a laser beam having bandwidth of 1500 Hz is 200 km.

15.2 If light from the sun subtends at an angle of 40′ at the slit, then calculate the transverse coherence length for wavelength 6000 Å.

Solution: $\lambda = 6000 \times 10^{-8}$ cm $= 6000 \times 10^{-10}$ m , $\vartheta = 40' = \frac{40}{60} \times \frac{\pi}{180}$ radians

Transverse coherence length is

$$l_t = \frac{\lambda}{\theta} = \frac{6 \times 10^{-7}}{\frac{40}{60} \times \frac{\pi}{180}} = \frac{6 \times 60 \times 180}{40 \times \pi} \times 10^{-7}$$

$$= \frac{1620}{\pi} \times 10^{-7} \approx 515 \times 10^{-7}$$

$$= 0.0000515 \text{ m}$$

Ans. Transverse coherence length is 0.0000515 m.

15.3 Calculate degree of non-monochromaticity for wavelength 6000 Å, if coherence time is 10^{-9} sec.

Solution: $\lambda = 6000 \times 10^{-8}$ cm $= 6000 \times 10^{-10}$ m, $= 6 \times 10^{-7}$ m, $\tau = 10^{-9}$ sec

Degree of non-monochromaticity is given as $\frac{\Delta\upsilon}{\upsilon_0}$

Now $$\Delta\upsilon = \frac{1}{\tau} = \frac{1}{10^{-9}} = 10^9 \text{ Hz}$$

and $$\upsilon_0 = \frac{c}{\lambda} = \frac{3 \times 10^8}{6 \times 10^{-7}} = 0.5 \times 10^{15} \text{ Hz}$$

Degree of non-monochromaticity is given as

$$\frac{\Delta\upsilon}{\upsilon_0} = \frac{10^9}{0.5 \times 10^{15}} = \frac{10}{5} \times \frac{10^9}{10^{14}} = 2 \times 10^{-5} = 0.00002$$

Ans. Degree of non-monochromaticity is 0.00002.

15.4 For ordinary source, calculate the degree of non-monochromaticity for wavelength 5800 Å, if coherence time of source is 10^{-9} sec.

Solution: **Given:** Coherence time, $\tau = 10^{-9}$ s , wavelength, $\lambda = 5500$ Å

$= 5500 \times 10^{-8}$ cm $= 5500 \times 10^{-10}$ m

Velocity of light, $c = 3 \times 10^8$ m, so frequency, $\upsilon = \frac{c}{\lambda} = \frac{3 \times 10^8}{55 \times 10^{-8}}$ Hz

Therefore, band width, $\Delta\upsilon = \frac{1}{\tau} = 10^9$ Hz

Degree of non-monochromaticity is

$$\frac{\Delta\upsilon}{\upsilon} = \frac{10^9}{\frac{3 \times 10^8}{55 \times 10^{-8}}} = \frac{10^9 \times 55 \times 10^{-8}}{3 \times 10^8}$$

$$= \frac{550}{3} \times 10^{-8} = 18.33 \times 10^{-7} = 1.83 \times 10^{-6} = 0.0000183$$

Ans. Degree of non-monochromaticity is 0.0000183.

15.5 Calculate the monochromaticity of sodium light of wavelength 5896 Å, if time of coherence is 10^{-9} s.

Solution: Wavelength of sodium light, λ = 5896 Å = 5896 × 10^{-8} cm = 5.896 × 10^{-7} m

$$\text{Corresponding frequency, } \upsilon = \frac{c}{\lambda} = \frac{3 \times 10^8}{5.896 \times 10^{-7}} = 0.5088 \times 10^{15}$$

$$= 5.088 \times 10^{14} \text{ c/s} = 5.088 \times 10^{14} \text{ Hz}$$

The time of coherence, $\tau = 10^{-9}$ s

The frequency spread is

$$\Delta\upsilon = \upsilon_2 - \upsilon_1 = \frac{1}{\tau} = 10^9 \text{ Hz}$$

Degree of monochromaticity of sodium light

$$\frac{\Delta\upsilon}{\upsilon} = \frac{10^9}{5.088 \times 10^{14}} = 0.1965 \times 10^{-5} \approx 0.000002$$

Ans. Degree of monochromaticity of sodium light is 0.000002.

15.6 Calculate the monochromaticity of ordinary light of wavelength 7000 Å, if time of coherence is 10^{-9} s.

Solution: Wavelength of ordinary light, λ = 7000 Å = 7000 × 10^{-8} cm = 7 × 10^{-7} m

$$\text{Corresponding frequency, } \upsilon = \frac{c}{\lambda} = \frac{3 \times 10^8}{7 \times 10^{-7}} = 0.428 \times 10^{15} = 4.28 \times 10^{14} \text{ c/s} = 4.28 \times 10^{14} \text{ Hz}$$

The time of coherence, $\tau = 10^{-9}$ s

The frequency spread,

$$\Delta\upsilon = \upsilon_2 - \upsilon_1 = \frac{1}{\tau} = 10^9 \text{ Hz}$$

Degree of monochromaticity of ordinary light

$$\frac{\Delta\upsilon}{\upsilon} = \frac{10^9}{4.28 \times 10^{14}} = 0.234 \times 10^{-5}\ 0.000002$$

Ans. Degree of monochromaticity of sodium light 0.000002.

15.7 What is transition rate for the neon atom in He-Ne laser, if the energy drop for the 6328 Å emission is 1.98 eV and the power output is 1.2 mW?

Solution: Formula used

$$\text{Transition rate for the neon atom} = \frac{\text{Power}}{\text{Energy}} = \frac{P}{h\upsilon} = \frac{P\lambda}{hc}$$

Given: Wavelength $\lambda = 6328 \times 10^{-10}$ m, $h\upsilon = 1.98$ eV $= 1.98 \times 1.602 \times 10^{-19}$ J, $P = 1.2 \times 10^{-3}$ W

$$\text{Transition rate for the neon atom} = \frac{1.2\times10^{-8}}{1.98\times1.602\times10^{-19}} = 3.78 \times 10^{10} \text{ atom per second}$$

Ans. Transition rate for the neon atom is 3.78×10^{10} atom per second.

15.8 If the flux density of radiations received on the earth's surface from the sunlight of wavelength 6000 Å is 0.15 W/cm^2, how many photons will arrive on each square metre per second of a solar cell panel placed on the earth?

Solution: Formula used

Flux density is $Nh\upsilon$, where N is number of photons per unit volume.

Given: $Nh\upsilon = 0.15 \times 10^4$ W/m^2, $\lambda = 6000 \times 10^{-10}$ m

$$Nh\upsilon = 0.15 \times 10^4$$

$$N = \frac{0.15\,\lambda}{hc} = \frac{0.15\times10^4\times6000\times10^{-10}}{6.63\times10^{-34}\times3\times10^8}$$

$$= 45.25 \times 10^{20} \text{ per square metre per second}$$

Ans. Number of photons is 45.25×10^{20} per square metre per second

15.9 Find the intensity of laser beam of 20 mW power and having a diameter of 1.2 mm.

Solution: Formula used

Intensity, $I = \dfrac{P}{A}$

Given: $d = 1.2 \times 10^{-3}$ m, $P = 20 \times 10^{-3}$ W

$$I = \frac{P}{A} = \frac{P}{\pi r^2} = \frac{4P}{\pi d^2} = \frac{4\times20\times10^{-3}}{3.14\times(1.2\times10^{-3})^2}$$

$$= 17.69 \times 10^3 \text{ W/m}^2 = 17.69 \text{ kW/m}^2$$

Ans. Intensity = 17.69 kW/m^2

15.10 In He-Ne laser the distance of resonator cavity is 0.4 m. What is the mode separation of longitudinal cavity in terms of frequency?

Solution: Formula used

$$\Delta\upsilon = \frac{c}{2L}$$

Given: $L = 0.4$ m

$$\Delta\upsilon = \frac{3\times10^8}{2\times0.4} = 3.75 \times 10^8 \text{ Hz}$$

Ans. Mode separation of longitudinal cavity is 3.75×10^8 Hz.

15.11 How many photons are required per second to generate laser beam of 2.4 mW power. The wavelength of laser light is 7000 Å.

Solution: Formula used

Flux density, $Nh\upsilon$, where N is number of photons per unit volume.

Power, $P = Nh\upsilon$

Given: $P = 2.4 \times 10^{-3}$ W, $\lambda = 7000 \times 10^{-10}$ m

$$N = \frac{P}{h\upsilon} = \frac{P\lambda}{hc} = \frac{2.4 \times 10^{-3} \times 7000 \times 10^{-10}}{6.63 \times 10^{-34} \times 3 \times 10^{8}} = 8.446 \times 10^{15}$$

Ans. Number of photons required per second = 8.446×10^{15}

15.12 The energy flux of a laser beam of 15 W/cm^2 falls on 1.5 cm^2 of the surface of an ideal reflecting mirror for 50 minutes. Calculate the momentum transferred and force exerted on the mirror.

Solution: Formula used

Energy, $E = \frac{1}{2} mc^2$, so momentum, $p = mc = \frac{2E}{c}$, force, $F = \frac{P}{t}$, power, $P = \frac{E}{t}$

Given: Laser energy flux, $\frac{E}{A} = 15$ W/cm^2, $t = 50 \times 60$ second, $A = 1.5 \times 10^{15}$ cm^2

$$E = P \times t \times A = 15 \times 50 \times 60 \times 1.5 \text{ J}$$

Momentum transferred, $p = \frac{2E}{c} = \frac{2 \times 15 \times 50 \times 60 \times 1.5}{3 \times 10^{8}} = 4.5 \times 10^{-4}$ kg m/s

Force exerted on the mirror, $F = \frac{P}{t} = \frac{2E}{ct} = \frac{2 \times 15 \times 50 \times 60 \times 1.5}{3 \times 10^{8} \times 50 \times 60} = 15 \times 10^{-8}$ newton

Ans. Momentum transferred, $p = 4.5 \times 10^{-4}$ kg m/s, force exerted on the mirror, $F = 15 \times 10^{-8}$ newton.

15.13 Imagine that we chop a continuous perfectly monochromatic laser beam of wavelength 6238 Å into 0.04 ns pulses using some sort of shutter. Compute the resultant line width, $\Delta\lambda$, bandwidth ($\Delta\nu$) and coherence length, L.

Solution: Formula used

Coherence length, $L = c\Delta t$, where, Δt is duration of pulse.

Bandwidth, $\Delta\upsilon = \frac{1}{\Delta t}$, line width, $\Delta\lambda = \frac{\lambda^2}{c}\Delta\upsilon$

Given: Duration of pulse, $\Delta t = 0.04 \times 10^{-9}$ s, $\lambda = 6238 \times 10^{-10}$ m

Coherence length, $L = c\Delta t = 3 \times 10^{8}\ 0.04 \times 10^{-9} = 12 \times 10^{-3}$ m

Bandwidth, $\Delta\upsilon = \frac{1}{\Delta t} = \frac{1}{0.04 \times 10^{-9}} = 2.5 \times 10^{10}$ Hz

Line width, $\Delta\lambda = \frac{\lambda^2}{c}\Delta\upsilon = \frac{(6238 \times 10^{-10})^2 \times 2.5 \times 10^{10}}{3 \times 10^{8}} = 0.0324 \times 10^{-9}$ m $= 0.0324$ nm

Ans. Line width, $\Delta\lambda = 0.0324$ nm.

15.14 In a He-Ne laser the two plane mirrors forming the resonant cavity are at a distance 0.8 m. What is the mode of separation of longitudinal cavity in terms of frequency?

Solution: Formula used

$$\Delta\upsilon = \frac{c}{2L}$$

Given: $L = 0.8$ m

$$\Delta\upsilon = \frac{3\times10^8}{2\times0.8} = 1.875 \times 10^8 \text{ Hz}$$

Ans. Mode separation of longitudinal cavity is 1.875×10^8 Hz.

QUESTIONS FOR EXAMINATION

15.1 What are the characteristics of laser? Explain all of them in brief.

15.2 Write short notes on the following:
(a) Absorption of light (b) Spontaneous emission
(c) Stimulated emission

15.3 Write short notes on the following:
(a) Pumping (b) Active medium (c) Resonator

15.4 Write short notes on
(a) Metastable energy state (b) Population inversion
(c) Laser action

15.5 Discuss various pumping sources.

15.6 Discuss longitudinal spatial and transverse coherence.

15.7 Discuss spatial and temporal coherence.

15.8 What are main components of laser equipment? Explain them in brief.

15.9 Define Einstein's coefficients of
(a) Absorption of light (b) Spontaneous emission
(c) Stimulated emission

15.10 Explain how does the 3-level system works?

15.11 Explain how the 4-level system works?

15.12 Explain construction and working of He-Ne laser. What advantages and disadvantages of He-Ne laser?

15.13 Discuss principle, construction and working of ruby laser. What are advantages and disadvantages of ruby laser?

15.14 Describe the working of YAG laser. What are its advantages and disadvantages of Nd : YAG laser?

15.15 Discuss construction and working of semiconductor laser with necessary diagrams. What are the advantages and disadvantages of semiconductor laser?

15.16 What is role of laser in

(a) Computer system (b) Medicine

(c) Industry (d) Military

15.17 What are the applications of the laser?

OBECTIVE QUESTIONS

15.1 Full form of laser is

(a) Light amplification by spontaneous emission of radiation

(b) Light absorption by stimulated emission of radiation

(c) Light amplification by stimulated emission of radiation

(d) Light absorption by spontaneous emission of radiation

15.2 The ratio of Einstein's coefficient of spontaneous emission and stimulated emission is proportional to

(a) υ^2 (b) υ (c) υ^4 (d) υ^3

15.3 Ruby laser is

(a) 2-level laser (b) 3-level laser (c) 4-level laser (d) 5 level laser

15.4 Nd : YAG laser is

(a) 2-level laser (b) 3-level laser (c) 4-level laser (d) 5 level laser

15.5 He-Ne laser is

(a) 2-level laser (b) 3-level laser (c) 4-level laser (d) 5 level laser

15.6 The expression for the ratio of spontaneous and stimulated Einstein's coefficient is.

(a) $\frac{8\,\pi h c^3}{\upsilon^3}$ (b) $\frac{8\,\pi h \upsilon^2}{c^3}$ (c) $\frac{8\,\pi h \upsilon}{c}$ (d) $\frac{8\,\pi h \upsilon^3}{c^2}$

15.7 In Ruby [Al_2O_3 : Cr] laser which is active atom?

(a) Al (b) Cr (d) O (d) both Al and Cr

15.8 In Nd : YAG (Yttrium Aluminum Garnet) laser which is active atom?

(a) Nd (b) Y (d) Al (d) both Nd and Y

15.9 In He-Ne laser which is active ion?

(a) He (b) both He and Ne (c) Ne

15.10 Which laser has highest efficiency?

(a) He-Ne (b) Ruby (c) Semiconductor (d) Nd : YAG

15.11 Which source is used for pumping in He-Ne laser?

(a) Electrical discharge (b) Flash lamp

(c) Thermal energy (d) Sodium lamp

15.12 Which source is used for pumping in ruby laser?

(a) Electrical discharge (b) Flash lamp

(c) Thermal energy (d) Sodium lamp

15.13 Which source is used for pumping in Nd: YAG laser?

(a) Electrical discharge (b) Thermal energy

(c) Flash lamp (d) Sodium lamp

15.14 Laser beam consists of

(a) Electrons (b) Ions (c) Photons (d) Plasmons

15.15 Characteristics of laser beam are

(a) Directionality, high intensity, coherence

(b) Directionality, monochromaticity, high intensity

(c) Directionality, high intensity and coherence

(d) Directionality, monochromaticity, high intensity, coherence

15.16 Main components laser equipment are

(a) Pumping (b) Active medium (c) Resonator (d) All of these

15.17 Which are solid state lasers?

(a) Ruby and He-Ne lasers (b) Ruby and Nd: YAG lasers

(c) Na: YAG and semiconductor lasers (d) Semiconductor and ruby lasers

15.18 Which of the atomic state has the longest lifetime?

(a) Metastable state (b) Excited state (c) Ground state (d) Intermediate state

15.19 Which of the following is correct?

(a) The ground state lies in between the excited state and the metastable state.

(b) The metastable state lies in between the ground state and the excited state.

(c) The excited state lies in between the ground state and the metastable state.

(d) None of them

15.20 The range of visible light is

(a) 0.1 μm to 0.9 μm (b) 0.2 μm to 0.8 μm

(c) 0.3 μm to 0.6 μm (d) 0.4 μm to 0.7 μm

15.21 The stimulated emission, stimulated absorption and spontaneous emission cannot take place simultaneously.

(a) True (b) False

15.22 During laser action, which of the following statement is correct?

(a) Number of atoms in the ground state and the metastable state are equal.

(b) Number of atoms in the ground state is more than in the metastable state.

(c) Number atoms in the metastable state is more than that in the ground state.

(d) None of these

15.23 During laser action, which of the following statement is correct?

(a) The stimulating photon is supplied directly by pumping source.

(b) The stimulating photon is emitted during spontaneous emission from metastable state.

(c) The stimulating photon is generated during back and forth movement of photon.

(d) The stimulating photon is emitted during population inversion.

Answers

15.1 (c)	15.2 (d)	15. 3 (b)	15.4 (c)	15.5 (c)	15.6 (d)	15.7 (b)	15.8 (a)
15.9 (c)	15.10 (c)	15.11 (a)	15.12 (b)	15.13 (c)	15.14 (c)	15.15 (d)	15.16 (d)
15.17 (b)	15.18 (c)	15.19 (b)	15.20 (d)	15.21 (b)	15.22 (c)	15.23 (b)	

CHAPTER 16

Holography

16.1 INTRODUCTION

Holograms are real and produced through an optical technique called holography. The word "holography" originates from the Greek word "holos" meaning "the whole". Dennis Gabor invented holography in 1948. He received Nobel Prize in 1971 for this invention. The waves may be of any kind—light, sound, X-ray, corpuscular waves, etc. Dennis Gabor, performed his first experiment with a mercury-arc lamp as the source of light, because he did not yet have a laser source. In 1963, E. Leith and J. Upatnieks prepared laser holograms for the first time.

In conventional photography, only distribution of intensity is recorded as a two-dimensional projection of an object on the plane of the photograph. Holography records both the intensity and the shape of the light wave to form 3-D images.

A hologram is an interference pattern. An interference pattern is created when two similar waves overlap with each other. The hologram must be recorded onto the detector (holographic emulsion) with an accuracy of 500 nanometers (one hundredth of the width of a hair). The 3-D image can then be seen by illuminating the hologram with a light wave, which is redirected by many tiny structures of the interference pattern due to a phenomenon called diffraction. To record interference patterns on the medium (emulsion), both light waves must be of the same wavelength. The recorded medium on developing is called hologram.

For a better understanding of the holographic process, it is necessary to understand interference and diffraction. Interference occurs when one or more wavefronts are superimposed. Diffraction occurs when two wavefronts scattered from the object are superimposed. In case of hologram formation, when a light from source (called reference waves) superimpose with the light of the same wavelength reflected from an object (called object waves). Two wavefronts of same wavelength are taken to maintain coherence in the waves, and hence for getting the interference pattern. The construction is the process of hologram formation, while the reconstruction is the process of viewing the image from hologram. The process of construction and reconstruction in holography purely in terms of interference and diffraction are explained below.

A plane wavefront picture is somewhat simplified but is accurate enough to give an understanding process of the holography. A diffraction grating is a structure with a repeating pattern. A simple example is a metal plate with slits cut at regular intervals. An incident light wave on a grating is split into several waves. The direction of these diffracted waves is determined by the grating spacing and the wavelength of the light. A hologram is short of grating with interference patterns of maxima and minima of light intensity.

A simple hologram can be made by the superimposition of two plane wavefronts: one reference wavefront and the second the object wavefront. The interference pattern is recorded on a holographic recording medium (emulsion). It is a straight-line fringe pattern whose intensity varies sinusoidally across the medium. This is called reconstruction of the hologram. The spacing of the fringe pattern is determined by the angle between the two waves, and by the wavelength of the light.

The recorded light pattern is a diffraction grating called hologram. When a hologram is illuminated by reference light wavefront, then the diffracted waves, emerging from all the points of the hologram, are called 'reconstructed' waves. The interference of these reconstructed waves produces the real image, that is, exactly like the original image.

16.2 PRINCIPLE OF HOLOGRAPHY

The principle of holography was first established by Dennis Gabor in 1948 when he was working on his research project for improving the resolving power of the electron microscope. In holography, three-dimensional images are produced on a photographic film which gives sensation of the exact and original 3-D object. In holography, the intensity and phase information are recorded on the photographic film. In holography to record phase information, phase information is converted into variation of intensity. The photographic image does not record phase information. In conventional photography only the variation of the intensity of light scattered by the object is recorded. The phase information can be recorded by the process of optical interference of the light waves. In the interference and diffraction pattern of light waves, both intensities and phases are recorded.

16.3 RECORDING AND RECONSTRUCTION METHOD

16.3.1 One-point Scatterer Holograms

With the help of Fig. 16.1, we can understand the basic principle of holography. The plane monochromatic wave of laser light is allowed to fall on a point object. The laser light is scattered by the point object. The scattered waves are known as *object waves* and the incident waves of laser light are known as *reference waves*. When these waves interfere with each other, they produce concentric circular interference fringes on a high resolution photographic plate placed at right angle to the incident laser light. When these two types of waves overlap, the bright and dark concentric rings, like Newton's rings, are formed on the photographic plate due to constructive and destructive interference. When photographic plate is developed, the light and dark partially absorbing circular fringes become visible in a reverse way. This developed photographic plate is called *Gabor's zone plate* and interference pattern on Gabor's zone plate is called a hologram. First of all, hologram plate was prepared by Dennis Gabor. This work of Gabor laid the foundation for holography. Gabor's zone plate is very similar to Fresnel's zone plate except (i) the light and dark fringes are continuous, and (ii) has a single focus.

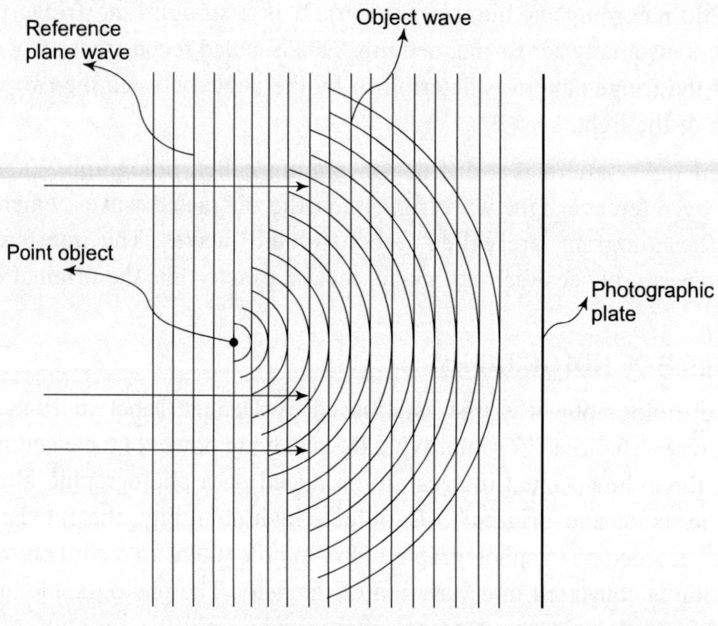

Fig. 16.1 The interference between directly incident laser light (reference wave) and corresponding scattered laser light (object wave) by a point object

The process of extraction of virtual and real images from Gabor's zone plate is called reconstruction. For reconstruction, the plate is illuminated by a reference wave, i.e., by the same original plane monochromatic wave laser light which was used to form the hologram as shown in Fig. 16.2. The light after passing through Gabor's zone plate the hologram produces a first-order interference maximum at S'. To the eye, this light appears to diverge from S. All the points on the hologram diffract the reference wavefronts and thus, virtual and real images are formed in the absence of the real scatterer.

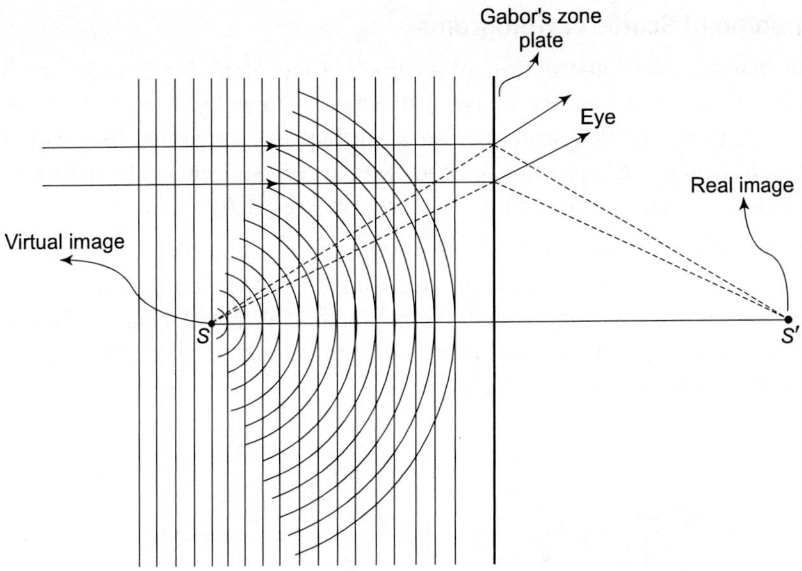

Fig. 16.2 Reconstruction of real and virtual image due to the illumination of Gabor's zone plate hologram by the same laser light. The virtual and real images have been formed at S and respectively as seen by the eye. The virtual and real images exactly look like point scatterer

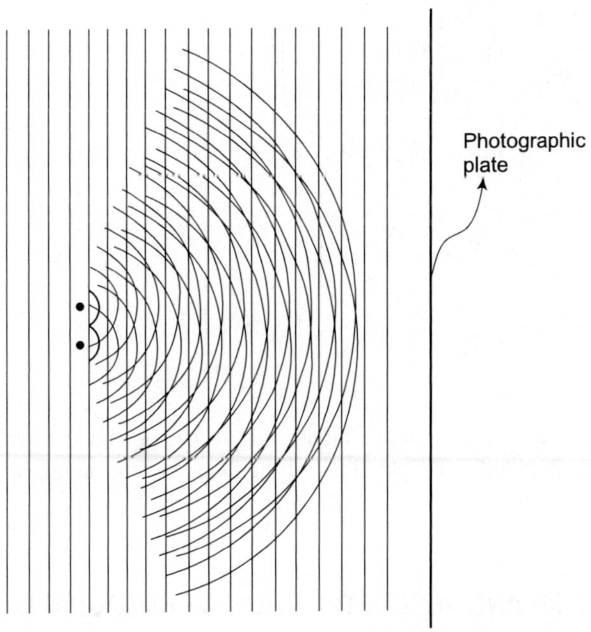

Fig. 16.3 The interference between directly incident laser light (reference wave) and corresponding scattered laser light (object wave) by two point objects.

16.3.2 Two-point Scatterer Holograms

Let us now discuss the construction of Gabor's zone plate for two-point objects and reconstruction of the corresponding image. It has been shown by drawing Fig. 16.3. Instead of one scatterer, there are two scatterers on a vertical line separated by a small distance. When laser light (reference light) falls on the two scatterers, each scatterer forms a separate diffraction pattern or Gabor's zone plate as shown in Fig. 16.3.

The intensity of each concentric fringe pattern on the zone plate is proportional to the scattered light intensity, if the photographic response of the photographic plate is linear. Reconstruction in this case is done using the same method as that of a single point. It produces two virtual and real images of both the scattering centres. Each concentric fringe formed with its proportionate intensity is shown in Fig. 16.4.

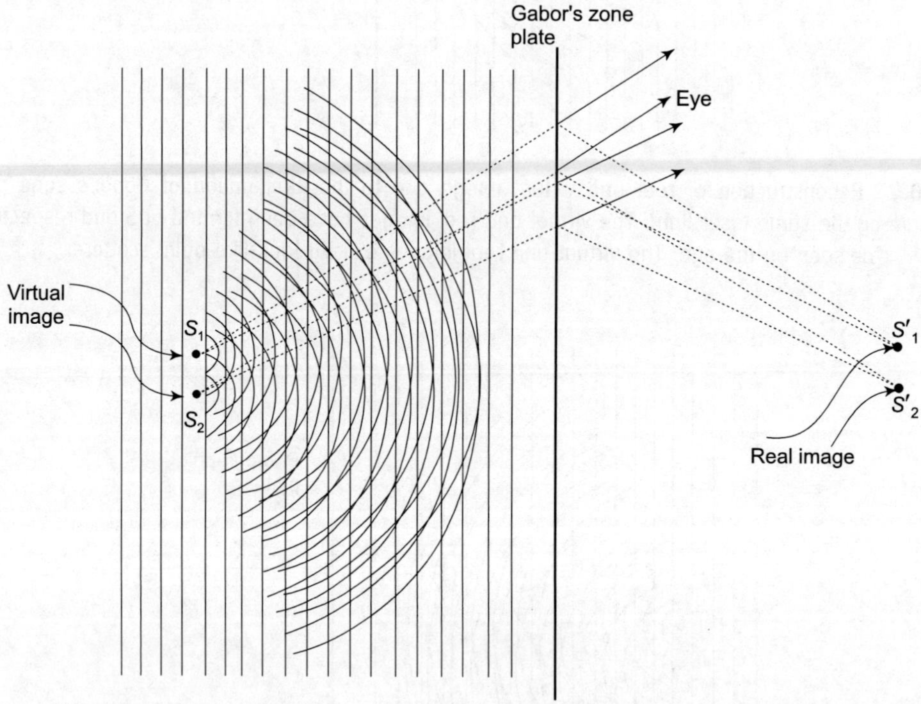

Fig. 16.4 Reconstruction of real and virtual images due to two-point scatterers. The virtual and real images have been formed at left and right sides of the Gabor's zone plate, respectively as seen by the eye

16.4 CONSTRUCTION AND RECONSTRUCTION OF HOLOGRAMS

The method of construction of Gabor's zone plate can now be extended to 3-D objects. 3-D objects consist of infinitely many points. Each point acts like a scatterer or scattering centre.

A 3-D object has continuous distribution of scattering centres. When a reference laser light falls on these scattering centres, the scattered light after overlapping forms an interference pattern on photographic plate of high resolution, placed in front of laser light. This interference pattern on the photographic plate is called Gabor's zone plate.

The reconstruction of this zone plate is shown in Fig. 16.5. When Gabor's zone plate is illuminated by a reference laser beam, all the continuous points on Gabor's zone plate diffract the reference wavefronts, producing virtual and real images on the left and right side of the hologram in the absence of a real 3-D object. After reconstruction, the distributed virtual image and real image should appear exactly like the original 3-D object as viewed from the right of the hologram.

The basic principle of the holography proposed by Dennis Gabor is presented in this section in a simple way. However, poor quality of the reconstructed image limits the application of these zone plates. This difficulty was removed in 1962 by the invention of off-axis holography by Leith and Upantnieks. On the basis of the method of recording the holograms, holography can be divided into two types: (i) in-line holography, and (ii) off-axis holography.

16.4.1 In-line Holography

The procedure followed above to construct a hologram is based on in-line holography as shown in Figs. 16.2 and 16.4. In Fig. 16.5, two light waves (one reference wave and the other object wave) incident perpendicular to Gabor's zone plate as shown in Fig. 16.5. The quality of reconstructed image using in-line holography is poor. The poor quality of the reconstructed image is due to the superposition of the conjugate image on the reconstructed image.

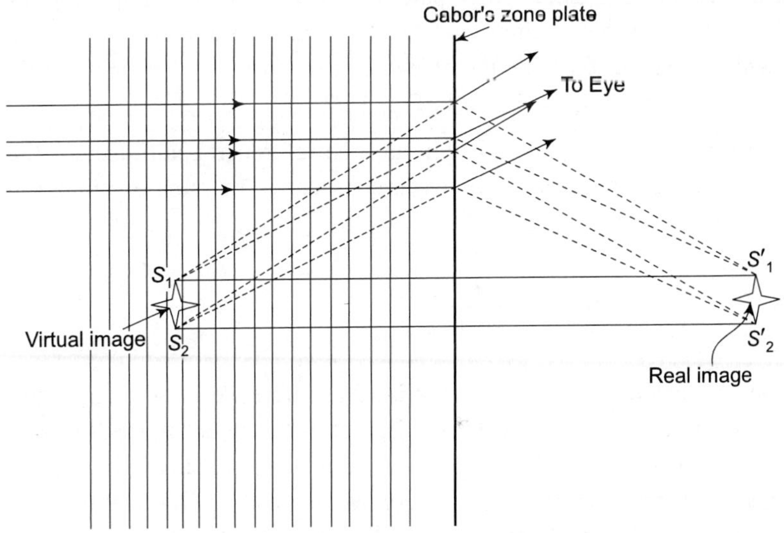

Fig. 16.5 Reconstruction of real and virtual image of 3-D object. The virtual and real images have been formed at the left and right sides of the Gabor's zone plate, respectively, as seen by the eye

16.4.2 Off-axis Holography

In 1962, Emmett Leith and Juris Upatnieks in the University of Michigan invented the off-axis holography using a laser source with increased coherence length. The problem of conjugate image of in-line holography was rectified in off-axis holography. A diagram for recording off-axis holography is shown in Fig. 16.6.

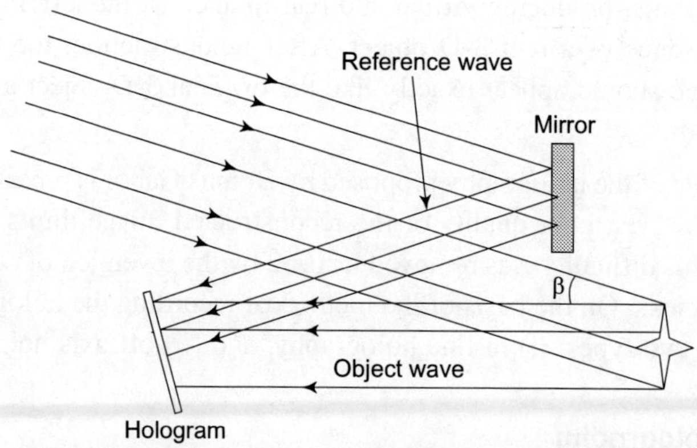

Fig. 16.6 Basics of off-axis holography

When a part of the monochromatic coherent reference laser light is reflected by the mirror and the other part reflected by the object incident on the high resolution photographic plate, then on interference a complicated fringe pattern is formed. When the photographic plate is developed, it reveals interference fringes called a hologram. After reconstruction, a 3-D image is produces, which looks exactly like the object. The reconstruction process of the hologram obtained using off-axis method is shown in Fig. 16.7.

The angle (β) between the scattered light and the reference beam decides the quality of the reconstructed image. The distance between consecutive maxima or minima is known as the pitch (p) and is given by

$$p = \frac{\lambda}{\sin \beta}$$

Here, we take an example of helium-neon laser as reference laser beam. The wavelength for a helium-neon laser is 632.8 nm. If β = 20°, then the pitch of He-Ne laser hologram, $p = 1.85\ \mu m = 1.85 \times 10^{-3}$ mm. The resolving power of the photographic plate is normally expressed in terms of lines per millimetre which can be resolved by it. When we use He-Ne laser as reference laser, the resolving power of the photographic plate comes out to be 540 lines/mm, which yields $p = 1.85 \times 10^{-3}$ mm. Hence, the He-Ne laser hologram is visible clearly, because the distance between the maxima and minima in the hologram produced by it is 1.85 μm = 1.85 × 10^{-3} mm.

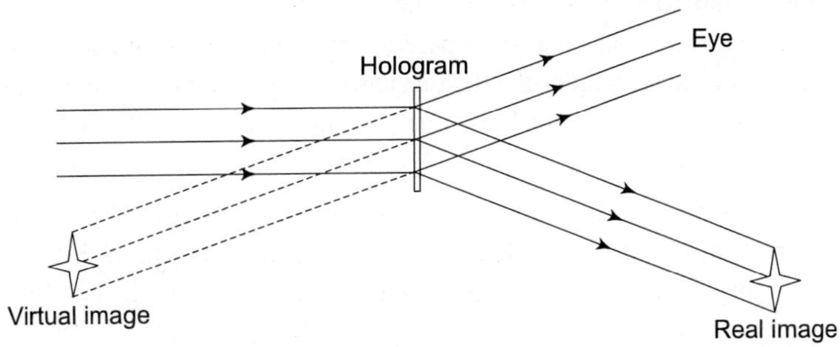

Fig. 16.7 Construction process of holography

The position of mirror and hologram as shown in Fig. 16.6 can be changed in a number of ways keeping the basic concept intact. One such change is shown in Fig. 16.8.

In this case, a narrow laser beam is expanded by a lens called the beam expander. Then it is incident on a semi-transparent mirror called a beam divider which divides the beam into two parts. One part of laser beam is reflected by the object and the other part by a mirror. When both reflected beams from mirror and hologram interfere with each other on the photographic plate, a complicated interference pattern is formed. Thus, a good quality hologram of the object is produced by an off-axis method.

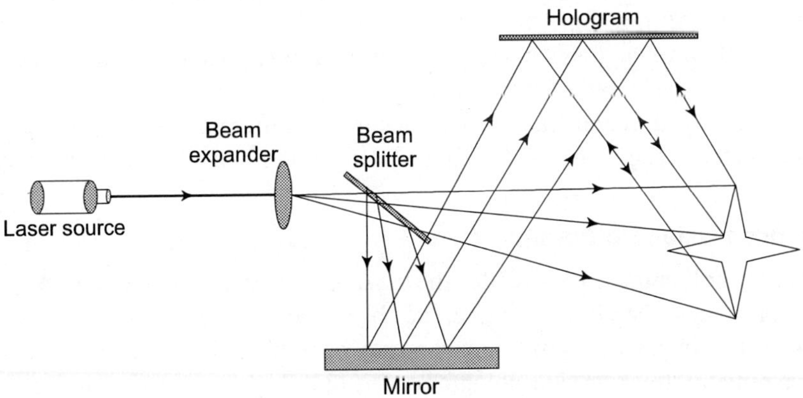

Fig. 16.8 Reconstruction of the setup as in Fig. 16.4 with the same basic concepts

16.5 CLASSIFICATION OF HOLOGRAMS

Classification of holograms is done on the basis of three important properties of a hologram: (i) an amplitude and a phase modulation, (ii) thin and (thick) volume, (iii) transmission and reflection. A given hologram will have one or other of each of these three properties.

16.5.1 Amplitude and Phase Modulated Hologram

An amplitude modulation hologram is one where the amplitude of light diffracted by the hologram is proportional to the intensity of the recorded light. A straightforward example of this is photographic emulsion on a transparent substrate. The emulsion is exposed to the interference pattern, and is subsequently developed giving a transmittance which varies with the intensity of the pattern – the more light falls on the plate at a given point, the darker the developed plate at that point.

A phase modulation hologram is made by changing either the thickness or the refractive index of the material in proportion to the intensity of the holographic interference pattern. This is a phase grating and it can be shown that when such a plate is illuminated by the original reference beam, it reconstructs the original object wavefront. The efficiency (i.e., the fraction of the illuminated object beam which is converted into the reconstructed object beam) is greater for phase than for amplitude modulated holograms.

16.5.2 Thin Hologram and Thick (Volume) Holograms

A thin hologram is one where the thickness of the recording medium is much less than the spacing of the interference fringes which make up the holographic recording. The thickness of a thin hologram can be down to 60 nm by using a topological insulator material Sb2Te3 thin film. Ultra-thin holograms have the potential to be integrated with everyday consumer electronic items like smartphones.

A thick or volume hologram is one where the thickness of the recording medium is greater than the spacing of the interference pattern. The recorded hologram is now a three-dimensional structure, and it can be shown that incident light is diffracted by the grating only at a particular angle, known as the Bragg angle. If the hologram is illuminated with a light source, with a broad spectrum of wavelengths, incident at the original reference beam angle, reconstruction occurs only at the wavelength of the original laser used for its construction. If the angle of illumination is changed, reconstruction will occur at a different wavelength and the colour of the reconstructed scene changes. A volume hologram effectively acts as a colour filter.

16.5.3 Transmission Holograms

The transmission holograms are also called the *transmission off-axis holograms*. A transmission hologram is one where the object and reference beams are incident on the recording medium from the same side. In practice, several mirrors may be used to direct the beams in the required directions. Normally, transmission holograms can only be reconstructed using a laser or a quasi-monochromatic source. But a particular type of transmission hologram, known as a rainbow hologram, can be viewed with white light.

Normally, the space between dark and bright fringes is 1.0 μm to 10.0 μm. This space is wider in transmission holograms than that of reflection holograms. The processing of transmission holograms is, therefore, simpler than in a reflection hologram. The transmission off-axis hologram can be viewed only with laser light, usually of the same laser light which is used for recording. The transmission of reference beams through the holograms makes images visible to the eye from the other side of the hologram. The virtual image can be clear and exact. If this hologram is broken into two or more small pieces, one can still see the entire scene through each piece. A real image of a transmission hologram can also be projected on a screen. The holograms used on different electronic cards are transmission holograms having a layer of aluminum on the back, which acts as a mirror.

16.5.4 Reflection Holograms

Reflection holograms are the most common holograms. In a reflection hologram, the object and reference beams are incident on the plate from opposite sides of the plate. The reconstructed object is then viewed from the same side of the plate at which the reconstructing beam is incident. Only volume holograms can be used to make reflection holograms, as only a very low intensity diffracted beam will be reflected by a thin hologram.

White light or ordinary light is needed to view them. The hologram is illuminated by white light at a specific angle and distance. The image is formed by light reflected by the hologram in which white light is located. The fringes forming the image of a reflection hologram are closely spaced. The fringe space is smaller compared to that of transmission holograms. Very closely spaced fringe patterns require the recording medium and processing method of very high standard. Another essential requirement is the excellent stability of the optical system. Reflection holograms produce usually dimmer images than that of off-axis transmission holograms. In the reflection holograms, the holographic image of the mirror reflects white light, but the holographic image of the diamond sparkles (sends flashes of light).

16.5.5 Volume Holograms

A reflection-type volume hologram can give a clear reconstructed image using a white light source. In theory, an image produced should be of approximately the same colour as the laser light used to make the hologram. In practice, the structure of the recording medium becomes compact due to chemical processing, so there is a colour shift towards a shorter wavelength. For example, a hologram recorded in a silver halide gelatin emulsion by red laser light, usually displays a green image. Deliberate alteration of the emulsion thickness before exposure, or permanent alteration after processing, has been used by artists to produce unusual colours and multi-colour effects.

16.6 COMPARISON BETWEEN TRANSMISSION AND REFLECTION HOLOGRAMS

In the following table, transmission and reflection holograms are compared.

Table 16.1

S.No.	Transmission hologram	Reflection hologram
1.	Transmission off-axis holograms are visible in laser light.	Reflection holograms are visible in white sun light.

Contd.

2.	The optical system involves a beam spreading lens and a beam splitter to obtain object waves and reference waves.	The optical system is simple. The laser beam is widened with a beam spreading lens and is incident on the object after passing through a transparent photo-sensitive plate.
3.	The range space between fringes is 1.0 μm to 10.0 μm in a transmission hologram.	The space between the fringes about half a wavelength or of the order of 3×10^{-1} μm in a reflection hologram..
4.	The recording medium and processing methods are simple because the distance between fringes is comparatively large.	The recording medium and processing methods are very complicated. Closely spaced fringes require high quality recording medium and processing.
5.	The image can be very clear since they are viewed by coherent laser light. White light cannot be used to view transmission holograms.	Reflection hologram images are usually much dimmer than off-axis transmission holograms. These are visible in white light.
6.	Transmission holograms is possible even when old developer is used.	Reflection holograms will be dark, if the developer is old.
7.	In this case, stability of the optical system is not essential against the motions.	In this case, stability of the optical system is essential against slightest motions.
8.	The reference waves and object waves incident on the photographic plate from the same side of the plate at a relatively small angle.	Here, the reference waves and object waves may come from the opposite sides of the photographic plate at angles close to 180°.

16.7 WHITE LIGHT HOLOGRAMS

The reflection and transmission holograms produce sharp images when they are viewed with highly monochromatic light only. These holograms have negligible thickness and are referred to as plane holograms. Reflection holograms can be viewed in white light. A brief description of white light holograms, which can be viewed with white light, is included for giving some idea of white light holograms.

For recording the reflection holograms, the reference waves and object waves fall from opposite sides of the photographic plate at angles close to 180°. Therefore, the resulting interference pattern is almost perpendicular to the plane of the film. For viewing the holograms in white light, one has to use a thicker (many times the fringe spacing) photographic emulsion for recording the surfaces of maxima within the photographic emulsion. The resulting interference pattern recorded is three dimensional, having twenty or more hyperboloid interference surfaces within the emulsion.These surfaces are semi-reflecting and semi-transparent. The distance between consecutive maxima surfaces is $\lambda/2$. The dark fringe and bright surfaces are reversed on development, and a variation of refractive index is created within it. This has been shown in Fig. 16.9.

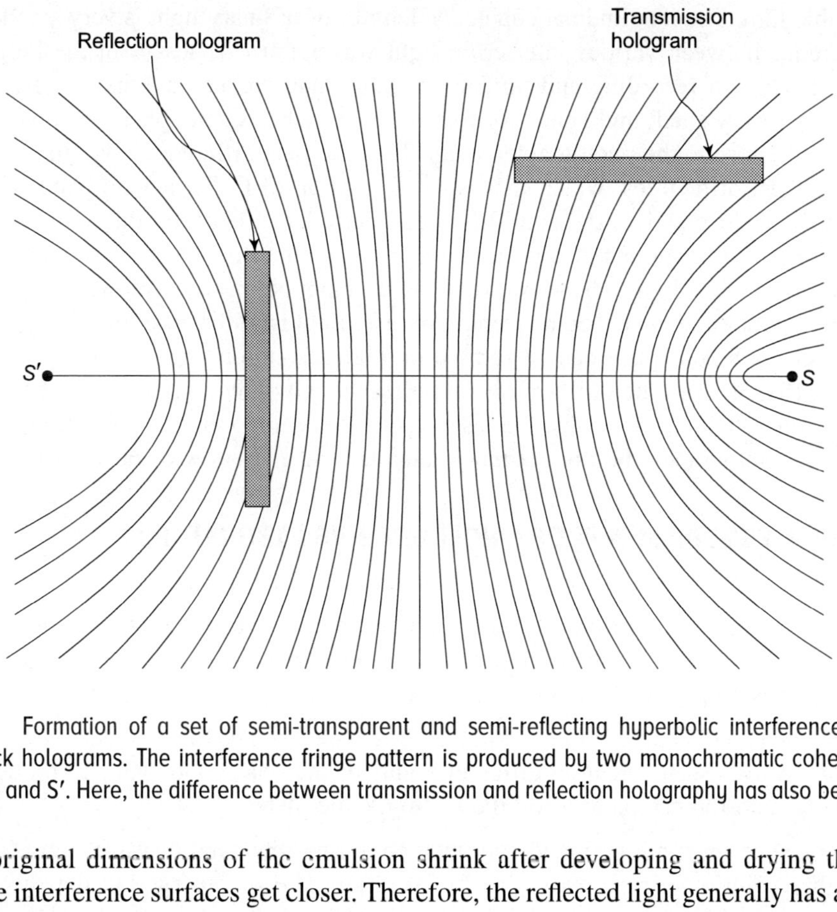

Fig. 16.9 Formation of a set of semi-transparent and semi-reflecting hyperbolic interference surfaces within thick holograms. The interference fringe pattern is produced by two monochromatic coherent point sources, S and S′. Here, the difference between transmission and reflection holography has also been shown

The original dimensions of the emulsion shrink after developing and drying the plate, hence the interference surfaces get closer. Therefore, the reflected light generally has a smaller wavelength.

When white beam falls on the hologram, some of the light is diffracted backward from the multiple layers. Due to the multiple reflections from these layers, a complicated interference pattern is formed which carries the whole 3-D visual information about the object. For each angle of incidence of white light spectrum satisfying the conditions of interference out its whole continuous spectrum, only a narrow range of wavelengths is reflected into the observer's eye. The wavelength to be reflected for each angle of incidence is selected by the hologram automatically. Thus, one can view the 3-D reflection holograms with white light, direct sunlight or light coming from a small light bulb.

Therefore, to view reflection holograms white light is required, instead the laser light used to record the hologram.

16.8 NECESSITY OF LASER SOURCE

In holography, high quality interference fringe patterns are produced by the interference of reference waves and object waves. The holograms are recorded on high resolution recording photographic film. The longitudinal coherence length for ordinary light is very small. So, the path difference between various interfering light waves must be less than the longitudinal coherence length for recording and producing stable interference patterns. For keeping the path difference very small and hence for formation of Gabor's zone plate, the object should be placed very close to the photographic plate. To make path difference very small, there are a lot of practical difficulties, which makes the formation of Gabor's zone plate impossible with white light. In practice, the path difference, introduced between different light waves reflected from different points of object, is not small, so that the interference pattern cannot be recorded with white light. Therefore, zone plate cannot be formed with white light. On the other hand, the longitudinal coherence length of laser light is of the order of a few kilometres. Therefore, Gabor's zone plate can be formed on high resolution photographic plates placed at a larger distance from the object. This is the reason why a hologram cannot be produced without a laser source. This is why, the laser light is very importantfor holography. Because of this reason, holography did not get much attention until lasers were developed.

16.9 BASIC REQUIREMENTS OF A HOLOGRAPHIC LABORATORY

The basic requirements of a holographic laboratory are given below:

(i) The maximum path difference between the object wave and the reference wave should be kept less than the coherence length. It is necessary for the reconstruction of stable interference patterns in Gabor's zone plate.

(ii) The spatial coherence is to be maintained for interference of reference waves and the object waves scattered from different points of the object and hence to produce high quality interference patterns on the photographic plates.

(iii) The source must not be broad and must emit a narrow band of wavelengths, because high resolution of reconstruction is essential. The reconstruction depends on the position of the reconstructing source and its the wavelength.

(iv) Ordinary photography and holography both suffer from aberrations. The aberration in reconstruction will be least if (a) the reconstruction light source is of the same wavelength as it was used during construction, and (b) is situated at the same relative position as it was situated during recording time.

(v) The photographic plate, the object, and any mirror used in producing the reference beam must be motionless with respect to one another during exposure, because the stability of all the recording components is highly essential for making fine holograms. This can be achieved by mounting the components tightly on supports attached to massive granite and marble plates. The plates must be placed on vibration absorbing base like inflated rubber tubes or sand bags. The vibrations of the fraction of wavelength of light may change the optical path lengths, hence can affect the recording of fringe patterns on the photographic plate. Even very small vibrations of interference fringes during the exposure time make the recording hologram blurred.

(vi) The high resolution of the photographic plate best output results, because, it can record the very closely spaced fringe patterns.

(vii) For holographic applications laser should be highly coherent light. In addition to this the laser should be powerful enough so that the exposure time is very short.

16.10 RECORDING A HOLOGRAM

To make a hologram, the following items are required:

- A suitable object or set of objects.
- A part of the laser beam which illuminates the object (the object beam) and another part which illuminates the recording medium directly (the reference beam).
- The arrangement is made so that the reference beam and the light which is scattered from the object are directed onto the recording medium to form an interference pattern.
- A recording medium which converts this interference pattern into an optical element, which modifies either the amplitude or the phase of an incident light beam according to the intensity of the interference pattern.
- A laser beam that produces coherent light with one wavelength.
- An arrangement which provides sufficient mechanical and thermal stability so that the interference pattern is stable during the time in which the interference pattern is recorded.

These requirements are inter-related, and are essential to record optical interference. When two light waves are superimposed the variation in intensity occurs. The maxima of intensity occur when two beams superimpose in same phase. The minima of intensity, which may be zero also, occur when two superimposing beams are out of phase. The interference pattern maps the relative phase between the two waves. Any change in the relative phases drifts the interference pattern. If the relative phase of the two waves changes by one cycle, then the pattern drifts by one whole fringe. One phase cycle corresponds to a change in the relative distances travelled by the two beams of one wavelength. Since the wavelength of light is of the order of 0.5 μm, it is observed that very small changes in the optical paths travelled by either of the beams in the holographic recording system move the interference pattern during holographic recording. Such changes can also be caused by relative movements of any of the optical components or the object itself, and by local changes in air temperature. For creating a clear well-defined recording of the interference pattern, it is essential that any such changes must be significantly less than the wavelength of light. The exposure time required to record the hologram depends on the laser power available, on the particular medium used and on the size and nature of the object(s) to be recorded, just as in conventional photography. Power of laser determines the stability requirements. Exposure times of several minutes are required, when quite powerful gas lasers and silver halide emulsions are used. All the elements within the optical system have to be stable to fractions of a μm over exposure period. It is possible to make holograms of much less stable objects by using a pulsed laser, which produces a large amount of energy in a very short time (μs or less). These systems have been used to produce holograms of live people.

Thus, the laser power, recording medium sensitivity, recording time and mechanical and thermal stability requirements are all interlinked. Generally, the smaller the object, the more compact the optical layout, so that the stability requirements are significantly less than when making holograms of large objects.

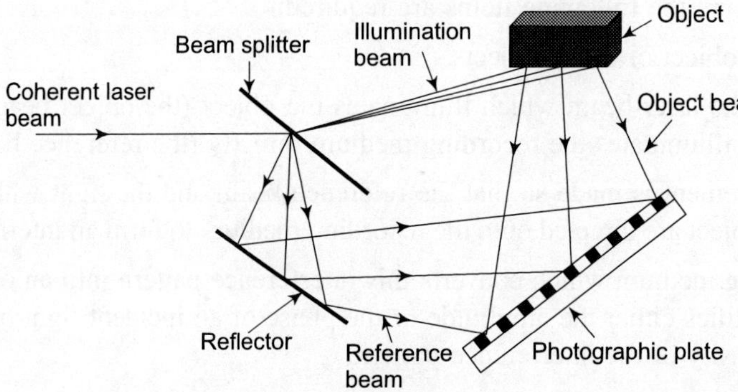

Fig. 16.10

Another very important laser requirement is its coherence. This can be envisaged by considering a laser producing a sine wave whose frequency drifts over time; the coherence length can then be considered to be the distance over which it maintains a single frequency. This is important because two waves of different frequencies cannot produce a stable interference pattern. The coherence length of the laser determines the depth of field recorded in the scene. A good holography laser will have a coherence length of several metres, sufficient for a deep hologram.

The objects for which a hologram is formed must, in general, have optically rough surfaces so that they scatter light over a wide range of angles. A plane or shiny surface reflects the light in only one direction from each point on its surface, so in general, most of the light will not be incident on the recording medium. A hologram of a shiny object can be made by locating it very close to the recording plate.

16.11 RECONSTRUCTING AND VIEWING THE HOLOGRAPHIC IMAGE

To view a hologram, the photographic plate containing the interference fringes must be kept under the same identical surroundings as that were during the recording time. When the hologram plate is illuminated by a laser beam identical to the reference beam which was used to record the hologram, an exact reconstruction of the original object wavefront is obtained. An imaging system (an eye or a camera) located in the reconstructed beam sees exactly the same thing as it would have done when viewing the original. When we look at the plate, as shown in Figs. 16.5, 16.7 and 16.11, the diffracted waves appear to diverge from the virtual image. When the observer moves his eyes to different positions, the image displays different parts of the object, as the rays of light entering the eye coming from different points of the fringe pattern on the hologram. Thus, the hologram enables us to see the object in a different way and from different angles. When a hologram is broken into many small pieces, even then each piece behaves as a complete hologram. One can see the complete object from each piece.

A hologram can also be obtained using a different laser beam from the original recording object beam, the reconstructed image will not match the original exactly. This is a major drawback in viewing a hologram by using different laser source other than the reference beam. White light consists of light of a wide range of wavelengths. Normally, if a hologram is illuminated by a whitelight source, each wavelength can be considered to generate its own holographic reconstruction, and these will vary in size, angle, and distance. These will be superimposed, and the summed image will wipe out any information about the original image. However, a holographic image can be obtained using white light in specific circumstances, e.g., with volume holograms and rainbow holograms. An extended source (e.g., a fluorescent lamp) will not reconstruct a hologram since its light is incident at each point at a wide range of angles, giving multiple reconstructions which will "wipe" one another out.

The real image is formed on the observer's side at the same distance from the photographic plate as that of the virtual image from it. This real image can be captured on a photographic plate placed at the position of image.

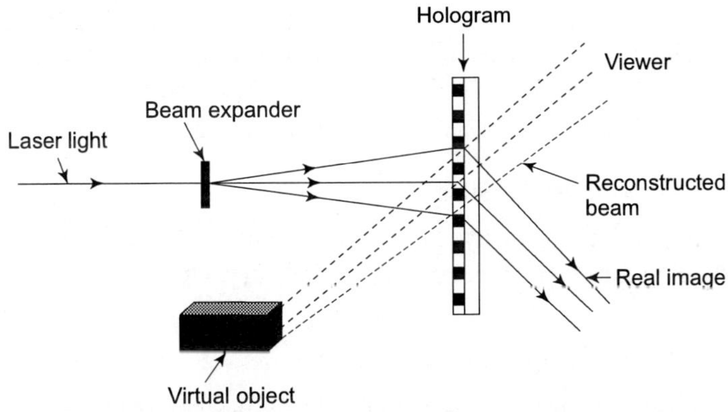

Fig. 16.11 Viewing the image

16.12 COMPARISON BETWEEN PHOTOGRAPHY AND HOLOGRAPHY

A hologram is different from a photograph in several ways. A few differences are cited in the following table.

Table 16.2

S.No.	Holography	Photography
1.	Each point in the holographic recording receives scattered light from every point in the object.	Each point in a photograph has light scattered from one corresponding point in the object.

Contd.

2.	The recorded scene in the hologram can be seen from a wide range of angles, because of the reason mentioned in point 1.	The photograph gives only a single view.
3.	The hologram has the same depth perception as are present in the original object.	The photograph is a 2-D representation of 2-D or 3-D objects.
4.	The developed hologram surface consists of a very fine, periodic pattern.	Each point of photograph has random intensity of light scattered from the original object.
5.	When a hologram is cut into pieces, each part is capable of reconstructing the entire object. However, the clarity and resolution decreases.	When a photograph is cut into pieces, each piece shows only that part.
6.	Holograms can only be seen with laser light.	A photograph can be seen with ordinary light.
7.	Holograms can only be constructed with laser source.	For photograph can be recorded from any ordinary light source.
8.	In holography no lens is required for focusing an image, therefore holography is called lensless photography.	In photography a lens is required for focusing the image.

16.13 APPLICATIONS OF HOLOGRAPHY

Holography has applications in many areas, for example, telecommunication, computers, scientific research, aviation, medicine, commerce, industry, etc.

16.13.1 Common Applications of Photography

A few common applications of photography have been discussed below.

(i) Telecommunications: Work of holographic telepresence technology plays an important role in telecommunications. This technology has brought digital participants and remote locations 3-D virtual space at our homes. The technology has reduced the necessity of travel for meetings and facilitates distance education. Nowadays, it is being used at large scale for distance education, meetings, teleconferencing, and office work from home.

(ii) Data Storage: Society generates incalculable amounts of data every day. Digital storage capacity increases every day. Our personal computers store hundreds of gigabytes of information in the form of family photos, videos and documents. The computer storage devices CD or DVD are prone to get corrupt and damaged. Now think about a storage disc gets corrupt. The losses are unimaginable.Moreover the data cannot be retrieved from corrupt or damaged CD or DVD. But, when a part of hologram becomes defective the remaining part will still keep all the data intact. If the hologram is broken into pieces, every part retains the complete data, which can be retrieved easily.

Though holograms create fascinating images, they do not just record and present only a visual object. Holograms are capable of recording pure data, may be mountains of it. Holograms have the potential to store very large amount of information. The current prototype systems can store 4.4 m individual pages of information on a DVD like disc. They also provide a lifelong security. The technology is simple to use and easy to store, retrieve and transport.

(iii) Photography: Holography is far superior to ordinary photography. Photography captures the intensity of light in a scene and records this onto a detector. This detector was traditionally photographic film, but is now more typically an electronic sensor in a digital camera.

Ordinary photography produces 2-D image of the object, while holographic photography produces its 3-D image . Holographic technology can be incorporated in digital cameras to detect the edge of the object, so that it can differentiate between object and the background. As a result, the holographic camera is able to focus accurately in dark conditions. Artists use holography to create pulsed holographic portraits and other art work.

(iv) Dynamic photography: In static holography, recording, developing and reconstructing occur sequentially, and a permanent hologram is produced. Here also exist holographic materials that do not need the developing process and can record a hologram in avery short time. This allows one to use holography to perform some simple operations in an all-optical way.Examples of applications of such real-time holograms include, optical cache memories, image processing (pattern recognition of time-varying images), and optical computing.

(v) Television: Holographic motion picture technology was implemented in the late 1970s. Holograms used to be the material of science laboratory, is now "coming to a theatre near you". However, the practical uses of holographic technology have eclipsed the film industry and become a common place feature in our everyday lives. We are only seeing the beginning of the usefulness of holograms and as the innovators and developers continue to improve the technology, holograms will become an even larger part of society soon.

(vi) Scientific research: Holographic technique is used in the characterization of materials. A hologram of an object can be made before and after it is subjected to stress. The change in the shape of the object due to that stress is recorded in the holograms. Particle physicists make holographic records of bubble chambers from which accurate measurements can be made. Electron holography is used to create images of materials at atomic dimensions. Nowadays, it is being used in the fields of nanostructured materials, high temperature superconductors, characterization of complex molecules, catalysts, polymers and different types of electronic materials.

(vii) Aviation: Holographic optical equipment are used for navigation by airplane pilots. A holographic image of the cockpit instruments appear to float in front of the wind shield. This helps the pilot to keep his eyes on the runway or the sky while reading the instruments. The same system is fixed on some models of automobiles.

(viii) Health science: Holography can revolutionise medicine, as a tool for visualising patient data while training students and surgeons. Many medical devices generate data in complex form using advanced imaging technology, such as magnetic resonance imaging (MRI) and ultrasound scans. Normally, that electronic information is used to display a flat image on a computer screen, but the technique of holography can be used to produce full coloured computer-generated 3-D holographic images.

Medical doctors use three-dimensional holographic CAT scans to make measurements without surgery. This technique is also used in medical education to show 3-D images of body parts.

A company in Scotland has been successful in using this kind of data to produce 3-D images for training and display. While viewing images from holograms, the advantage is that no special viewing devices, or glasses, are needed. Students and doctors can simply look, unhindered, at the three-dimensional images.

Storing several different images in the same hologram means that the viewer can move around the display, allowing them to examine different organs or body parts. To date, the company has produced 3-D holographic images of structures including the brain, liver, lungs, heart, skeleton, vascular system, nerves and muscles.

(ix) Commerce: Supermarket and department store scanners use a holographic lens system that directs laser light to read the bar codes of the product. Holograms are used in advertisements and packing of consumer products to attract buyers. They are also used on covers of magazines and publications.

(x) Industry: All types of industries use holograms to enhance the image of their brands in the market as genuine and authenticated. The holograms are almost impossible to counterfeit. They are used for attractive product packaging, security applications and so on. Holographic night vision goggles and fancy items have been made using this technology.

(xi) Security: You might have seen a security hologram on your debit or credit card. That small silver rectangle of a dove on your debit and credit cards is a transmission hologram backed by a white-light mirror. It displays a three-dimensional image which is visible as you move from side to side, and changes colour as you tilt your card up and down. It's very easy to manufacture en masse, but it is very difficult to forge.

Bank notes have also started the use of secure holograms. The reflective strip on the Indian 100 rupees note contains an image of Bharat produced with help of holography. It changes colours as you tilt the note, as well as a 3-D image Bharat can be seen "floating" above it.

For stopping tempering, holograms are fixed on the quality goods. Holograms are optical devices and difficult to make, which gives them an incredible advantage in the commercial security market.

(xii) Art: Artists use holography to create pulsed holographic portraits as well as other works of art.

Artists became involved with holography almost as soon as it became a practical optical process. In the UK, *Margaret Benyon*, is known as the "mother" of British creative holography.

She was awarded an arts fellowship in 1968 in the mechanical engineering department at Nottingham University. She managed to secure access to holographic recording equipment and laid the groundwork for her interest in the critical discussion around the process and its use by artists. She eventually submitted a doctoral thesis with the title "Art of Holography".

Around the world, there are pioneering artists using the three-dimensional recording holograms to construct multiple visually solid objects in the same volume; combine collections of still images or sculptures to produce animated 3-D works.

(xiii) Preservation of antiques: Holography has been used to make recordings of valuable antiques and/or fragile articles in the museum. Many museums have made holograms of valuable antiques in their collections, both for insurance purposes and to check for deterioration. Antiques are national treasures, their holograms can be made and sent to remote areas, enabling people there to see and appreciate their national heritage without long travelling. The original items can thus be preserved, without fear of theft.

(xiv) Multiple Scenes: One of the most dramatic features of a hologram is that it is capable of recording more than one independent scene in the same volume of space and then display them independently by changing the angle between the hologram and the reference beam. This phenomenon was discovered by a father and son team (William Henry and William Lawrence Bragg) through the study of X-ray diffraction by crystals. They were awarded the 1915 Nobel Prize in physics.

(xv) Military Mapping: Geographic intelligence is an essential part of military strategy and fully three-dimensional holographic images are being used to improve its vigilance with the help of 3-D pictures of dense forests at the borders. One American company has delivered over 13,000 3-D holographic maps of "battle-spaces" for the US army. This helps soldiers to view three-dimensional terrain, look around corners and perform mission training. The maps are also useful in disaster evacuation and rescue scenarios.

16.14 APPLICATIONS OF HOLOGRAPHIC INTERFEROMETRY

Holography interferometric method can be used for measuring the heat and mass transfer. Holographic interferometry has displaced the traditional interferometry completely, because it is much cheaper and very easy and convenient to handle. For traditional interferometry there is need of light source, machine or manufacture windows for test sections, mirrors and lenses of the optical components with special precision or accuracy. For holographic interferometry there is requirement of coherent laser source, a collimeter and photographic plate. No lens is required, so holographic interferometry is called lensless interferometry. In this case imperfections are automatically balanced by the holographic two-step procedure.

Holographic interferometry is a combination of interferometry and holography. Figure 16.12 shows a simple experimental setup for the examination of transparent media using holographic interferometry. The main difference between classical and holographic interferometry is that the object beam is compared with itself. Therefore, holographic and traditional interferometry differ from each other, because the first is a two-step method [see Fig. 16.13] and the second is a two-way method. In the holographic method the undistorted object wave, called the comparison wave, is stored on a photographic plate and can be reconstructed after its development (then it is called hologram) by laser light called the reference wave. The heat or mass transfer is then introduced, so that the momentarily object wave, which is called measuring wave, experiences an additional phase shift. Now the measuring (reference) and the comparison (object) wave interfere behind the hologram so describing the physical phenomenon which has been introduced and is to be investigated. (It must be pointed out, that macroscopic interference caused by the superposition of two object waves has to be clearly distinguished from the microscopic interference which takes place when the object and the reference wave are superimposed during the recording of the hologram.)

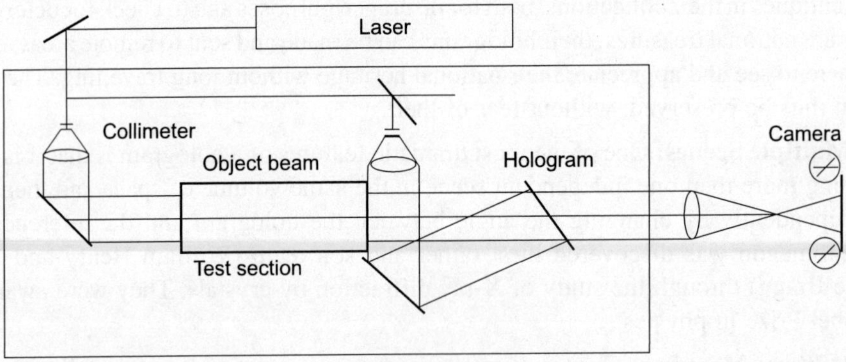

Fig. 16.12 Experimental set up for testing transparent medium using holographic interferometry

Double Exposure Technique

Several object waves, one after the other, can be recorded on one and the same hologram. When illuminating the hologram with the reference wave all are reconstructed simultaneously. This

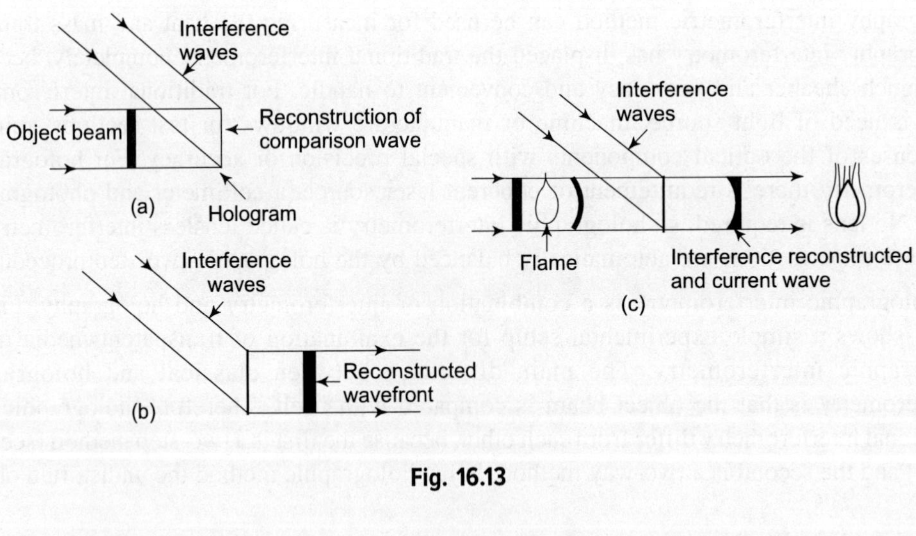

Fig. 16.13

principle is used for the double exposure technique, in the first exposure the comparison and in the second exposure the measuring wave is recorded. From the interference pattern one can determine the differences between the comparison and measuring conditions caused, for example, by heat transfer.

Disadvantages of this method are that short duration processes cannot be continuously observed and that the most favourable moment for the exposure with the measuring wave cannot be preselected, because the interference pattern only appears after the chemical process.

16.15 APPLICATIONS OF HOLOGRAPHIC MICROSCOPY

Holographic microscopy is also called *digital holographic microscopy* or *digital holography.* Digital holographic microscopy is suitable for high resolution investigations, because it is capable of quantitative phase contrast imaging. Though digital holography is based on the normal holographic principle, but the recording is done by digital image sensors like CCD (charge coupled device) or CMOS camera instead of photographic plate. The reconstruction process can be done numerically with a computer. For numerical reconstruction readymade pieces of software are available.

The property of holography to record information about the depth is used to study short duration microscopic events. These microscopic events can be recorded on the hologram. At any time later on, one can enjoy 3-D reconstructed images. In traditional microscopy, it is hard to record the three-dimensional shape of objects like living cells. In holographic microscopy, it is possible to record the 3-D shape of tiny objects such as cells with high resolution. This method is powerful and is very cheap. Holographic microscopy is based mostly on off-axis holography. Digital holography provides a new method for surface analysis and dynamic life cell imaging. Therefore, digital photography finds a wide range of applications in medical science such as cell structures, cell cycle analysis, morphology (it is a branch of biology which deals with form and structures of animals and plants) analysis of cells, nerve cell studies, red blood cell analysis, cell division and migration, tomography studies deep in living tissue and so on.

The principle of holography is not only limited to light waves, it can also be applied to sound waves, electron waves, etc. Acoustic holography using sound waves has become increasingly popular in the field of medical science. Electron holography has been developed for use in the field of scientific and industrial research. In recent years, atomic holography has been developed which produces much higher resolution holographic images. This is because the de Broglie wavelength of the atoms is much smaller than light waves. Holographic telepresence is being used at commercial scale in the world over.

Different types of holograms are available. All the 3-D images available commercially are not truly holographic. The visual effects produced by lenticular printing, the Pepper's Ghost illusion, tomography and volumetric displays give the sensation of holographic images.

1. CMOS amera: It operates on the principle of photoelectric effect. Image sensors convert photons into electrical charge. Unlike CCD (charge coupled device) sensors, CMOS (complementary metal-oxide semiconductor) sensors convert charge into voltage straight in the pixels.

2. Morphology: Morphology is a branch of biology which deals with the form and structures of animals and plants.

3. Tomography: Tomography is a method to produce 3-D images of solid object like human body using lasers. The method is used in radiology, archaeology, biology, atmospheric science, geophysics, oceanography, plasma physics, materials science, astrophysics, quantum information, and other areas of science. The word *tomography* is derived from Ancient Greek "*tomograph*", meaning "to write" or, in this context " to describe." A device used in tomography is called a tomograph, while the image produced is a tomogram.

In many cases, the production of these images is based on the mathematical procedure. Tomographic reconstruction, such as X-ray computed tomography technically produced from projections of multiple radiographs. Many different reconstruction algorithms exist. Most algorithms fall into one of two categories: filtered back projection (FBP) and iterative reconstruction (IR). These procedures give inexact results: they represent a compromise between accuracy and computation time required. FBP demands fewer computational resources, while IR generally produces fewer artifacts (errors in the reconstruction) at a higher computing cost.

4. Lenticular printing: Lenticular printing is a technology in which lenticular lenses (a technology that is also used for 3-D displays) are used to produce printed images with an illusion of depth, or the ability to change ormove when the image is viewed from different angles. Examples of lenticular printing include flip and animation effects such as winking eyes, and modern advertising graphics that change their message depending on the viewing angle.

5. Pepper's Ghost illusion: Pepper's Ghost is an illusionary trick. In this illusion, when a real or recorded image is reflected in a transparent screen at a 45° angle, viewers see a reflected virtual image that seems to have depth and appears like 3-D image. The illusion continues to be used in amusement parks, theatres, television and concerts.

6. Volumetric display: A volumetric display device is a graphic display device that forms a visual representation of an object in three dimensions, instead of the two-dimensional image, that simulate depth through anumber of different visual effects. One definition offered by pioneers in the field is that volumetric display form 3-D image via the emission, scattering, or relaying of illumination from well-defined regions in (x, y, z) space.

16.16 LIMITATIONS OF HOLOGRAPHY

Holography has many limitations. For example:

(i) The intensity ratio between the reference and object beams cannot be controlled. Thus, the quality of the hologram cannot be optimized.

(ii) The object is always illuminated with a single point source from a fixed direction, thereby casting shadows that cannot be controlled.

(iii) The location of the image is always behind the plate.

(iv) A laser is needed to observe the image formed by transmission holograms.

16.17 HOLOGRAPHIC RECORDING MEDIA

Holographic recording media are light sensitive materials. A list of these materials is given in Table 16.3 A light-sensitive media is required to record the original interference pattern. It modifies either the amplitude or the phase of an incident light beam in proportion to its intensity. The recording medium should be able to resolve completely all the fringes formed by interference between the object and reference beam. The range of these fringe spacings can be from tens of micrometers to less than one micrometer, i.e., range of these spatial frequencies are from a few hundred to several thousand cycles/mm. Ideally, the recording medium should have the same sensivity over this range. Photographic materials have a very low or even zero sensitivity at the frequencies, so cannot be used to make a hologram. For example, the resolution of Kodak's professional black and white film is less than 20 lines/mm, it is not possible to obtain any reconstructed beam using this film. The resolution limit given in the table indicates the maximal number of interference lines/mm of the gratings. The required exposure, expressed units of millijoules (mJ) of photon energy impacting the surface area, is for a long exposure time. Short exposure times (less than 1/1000 of a second, is for a pulsed laser) require much higher exposure energies, because exposure time is reciprocal to exposure energy.

Table 16.3 General properties of recording materials for holography

Material	Reusable	Processing	Type	Theoretical max. efficiency	Required exposure (mJ/cm^2)	Resolution limit (lines/mm)
Photographic emulsions	No	Wet	Amplitude	6%	1.5	5000
			Phase (bleached)	60%		
Dichromated gelatin	No	Wet	Phase	100%	100	10,000
Photoresists	No	Wet	Phase	30%	100	3,000
Photothermo-plastics	Yes	Charge and heat	Phase	33%	0.1	500–1,200
Photopolymers	No	Post-exposure	Phase	100%	10000	5,000
Photorefractives	Yes	None	Phase	100%	10	10,000

16.18 CONCLUSION

The general public is fascinated by holograms. A great deal of work and research is actually quietly taking place in labs and studios around the world. Its applications are changing lives, saving lives and stimulating creative and intellectual discussion. We are only seeing the beginning of the usefulness of holograms and as the innovators and developers continue to improve the technology, holograms will become an even larger part of society. There is hope that in near future the market for genuine, display holograms will be worth several billions.

The future is bright – the future might just be holographic.

Formula at a Glance

1.	$p = \dfrac{\lambda}{\sin\beta}$	The angle β between the scattered light and the reference beam. The distance between consecutive maxima or minima is known as the pitch p.

QUESTIONS FOR EXAMINATION

16.1 What is holography? Describe its basic principle.

16.2 What is Gabor's zone plate?

16.3 Describe the construction and reconstruction of images in holography.

16.4 Compare ordinary photography with holography.

16.5 Give a comparison between Gabor's zone plate and Fresnel's zone plate.

16.6 Distinguish between in-line holography and off-axis holography.

16.7 Describe the construction and reconstruction of images in off-axis holography.

16.8 What are reflection holograms?

16.9 What are transmission holograms?

16.10 Distinguish between reflection holograms and transmission holograms.

16.11 What are the basic requirements of a holographic laboratory?

16.12 Holograms cannot be produced without laser. Comment on the statement.

16.13 Explain how holography technology can be used in memory cards?

16.14 Mention a few examples of the application of holography technology in scientific research.

16.15 Write a few lines about holographic telepresence.

16.16 Explain how the use of holograms can help to prevent counterfeiting of currency.

16.17 Explain the basic principle of holographic interferometry.

16.18 Distinguish between digital holography and ordinary holography.

16.19 What are the advantages of digital holography over ordinary holography?

16.20 Write an essay about the future use of holography technique.

16.21 What is the most basic difference in the recording of reflection holograms and transmission holograms?

16.22 When a transmission hologram is broken into many pieces, each piece can reconstruct the image of the entire view of the object. Why?

16.23 When a reflection hologram is broken into many pieces, each piece cannot reconstruct the image of the entire view of the object. Why?

16.24 What type of hologram, even if it is thinner than a wavelength of light, can record and reproduce a complete three-dimensional image? Explain.

16.25 Can holograms be made using one laser for the reference beam and another laser for the object beam? Why?

16.26 Why can we use the sun to reconstruct the image of a reflection hologram but not of a transmission hologram?

16.27 Explain how a reflection hologram selects the colour from the light used to illuminate it.

16.28 You have made a reflection hologram using red laser light. Why is the image generally not red, but yellowish or greenish?

16.29 How is it possible to make a hologram with more than one scene that can be viewed one at a time through the full aperture of the hologram?

16.30 Why do we bleach the holograms after development?

16.31 Explain how a double-exposure hologram of an object that changes microscopically between exposures reveals the nature of the change.

16.32 When making holograms, we try to equalize the optical paths travelled by the reference and object beams. Why? Under what condition is this procedure not necessary?

16.33 Can a holographic image be obtained by illuminating the hologram with white light?

OBJECTIVE QUESTIONS

16.1 Which parameters of light are recorded in holography

(a) Intensities (b) Phases

(c) Speed (d) Both intensities and phases

16.2 Holography was discovered by

(a) Dennis Gabor (b) Huygens

(c) Newton (d) Einstein

16.3 Through holography technology, we can produce

(a) One-dimensional image (b) Two-dimensional image

(c) Three-dimensional image (d) Four-dimensional image

16.4 Holography produces

(a) Real images (b) Virtual images

(c) Both real and virtual images (d) None of these

16.5 A second holographic plate contains information in the patterns of

(a) Refection-refraction (b) Interference

(c) Diffraction (d) Polarization

16.6 In holography, interference pattern is produced due to superposition of

(a) Object wave (b) Reference wave

(c) Both object and reference waves (d) None of these

16.7 If a hologram breaks into pieces, then each piece contains

(a) Total image (b) No image

(c) Parts of image (d) Nothing

16.8 Holography is capturing pictorial details of 3-D objects by using the principle of

(a) Diffraction (b) Interference

(c) Refraction (d) Total internal reflection

16.9 In holography, three-dimensional image of an object may be reproduced

(a) By using convex lens (b) By using concave lens

(c) By using double convex lens (d) Without using lens

16.10 In holography, we record

(a) Only the phase of different parts of the object

(b) Only the intensity from different parts of the object

(c) Both the phase and intensity from different parts of the object

(d) None of these

Answers

16.1 (d) 16.2 (a) 16.3 (c) 16.4 (c) 16.5 (b) 16.6 (c) 16.7 (a) 16.8 (b)
16.9 (d) 16.10 (c)

Index

A

B

C

D

J

K

L

M

N